BIOGEOGRAPHY

BIOGEOGRAPHY

JAMES H. BROWN, Ph.D.

Department of Ecology and Evolutionary Biology,
The University of Arizona,
Tucson, Arizona

ARTHUR C. GIBSON, Ph.D.

Department of Biology,
University of California at Los Angeles,
Los Angeles, California

With 600 illustrations

The C. V. Mosby Company

ST. LOUIS • TORONTO • LONDON
1983

A TRADITION OF PUBLISHING EXCELLENCE

Editor: Diane Bowen
Assistant editor: Susan Dust Schapper
Manuscript editor: Mark Spann
Design: Kay M. Kramer
Production: Linda R. Stalnaker, Judy Bamert

Printed in the United States of America

The C.V. Mosby Company
11830 Westline Industrial Drive, St. Louis, Missouri 63141

Library of Congress Cataloging in Publication Data
Brown, James H.
Biogeography.

Bibliography: p.
Includes index.
1. Biogeography. I. Gibson, Arthur C. II. Title.
QH84.B76 1983 574.9 82-14121
ISBN 0-8016-0824-4

C/VH/VH 9 8 7 6 5 4 3 2 02/A/261

Preface

For years before we actually began to work on this book, we discussed the challenge of trying to write a basic textbook for college courses in biogeography. We realized it would be extremely difficult to produce a text that would do justice to the broad field of biogeography and satisfy the diverse needs of its teachers and students. Most large universities and many smaller institutions offer courses in biogeography, but these vary enormously in content and emphasis. Few scientists actually call themselves biogeographers. Consequently, biogeography courses tend to be taught by those whose real specialty is another discipline such as systematics, ecology, or paleontology. Although most courses are offered as part of a biology curriculum, sometimes they are offered by departments in related disciplines such as geography or the geosciences. Often zoogeography and phytogeography are taught as separate courses. Perhaps it is too much to expect that a single textbook could meet the needs of the diverse courses in biogeography or of the specialists who teach them.

Our book is intended to introduce advanced undergraduates and beginning graduate students to a very broad but exciting field. We have tried most of all to provide these students with a balanced, conceptual, and synthetic approach. We attempt to bring together specialized subdisciplines and information on both plants and animals in order to explain patterns of geographic distribution of organisms in terms of the historical geologic and contemporary ecological processes that have caused them. To do this we must deal with a time scale extending from near the beginning of life billions of years ago to the present, a spatial scale encompassing everything from local patches of habitat to the entire earth and the scala naturae, or variety of organisms, from the simplest microbes to the highest plants and animals. If biogeography is to be presented as a single, coherent field it is essential that patterns and processes on all of these scales be interrelated and synthesized. We hope that in this way biogeography will come alive, and the student will acquire a truly global perspective on the variation in the distribution of organisms and the different processes that have caused these patterns.

Like many diverse and active disciplines, biogeography is a field filled with competing ideas and clashing personalities. For any field this is a sign of vigor because much of the disagreement will stimulate research that will lead eventually to resolution of the controversial issues and to increased understanding. For authors of a textbook, however, this makes it difficult to achieve a balanced approach that will be acceptable to teachers and serve their students well. We have tried to convey a feeling for the dynamism of the field and to present the controversial issues and yet still retain a balanced coverage of topics. This is not a textbook of plate tectonics, techniques of phylogenetic reconstruction, vicariance biogeography, macroevolution, faunal and floral analysis, geographic ecology, or mathematical theory. All of these are important areas of current research in biogeography or allied fields, but to emphasize any one of them at the expense of the others and of more classical biogeographic subjects would be to present the beginning student with a biased view of what biogeography is all about.

The book is organized in four parts. These are preceded by an introductory chapter on the nature and history of the science of biogeography. The first unit describes the contemporary environmental setting that influences the distribution of living organisms. Individual chapters discuss the physical geography and climatology of the earth, the ecological factors that limit the geographic ranges, and the composition and distribution of ecological communities. The second part of the book discusses the historical events and evolutionary processes that have influenced the distributions of both extinct and living forms. A chapter on the theory of plate tectonics and the geologic history of the earth is followed by chapters on the processes of speciation and extinction, on the mechanisms and consequences of dispersal, on patterns of endemism and disjunction, and on the methods and difficulties of reconstructing the distributional histories of taxa. The third unit describes the distributions of contemporary animal and plant groups and considers the historical explanations that have been proposed to account for these patterns. Separate chapters on aquatic animals, terrestrial animals, flying animals, and plants are followed by a chapter on the role of the dramatic geologic and climatic changes that have occurred within the last 1.7 million years.

The last part of the book treats ecological biogeography. It emphasizes the development and testing of theories to explain general patterns of diversity and distribution that are relatively independent of the evolutionary histories of particular taxa. Two chapters are on islands, which have been a source of inspiration for biogeographers since the earliest beginnings of the field. A chapter on patterns of continental and marine species diversity is followed by a discussion of the ways that historical events and ecological processes interact to influence the present and past distribution of organisms.

The book is intended to be an introduction to the basic facts and concepts of biogeography, rather than an encyclopedia of data or a theoretical treatise. Although we have tried to use specific examples from a variety of organisms and geographic regions to illustrate our points, we have not done justice to the vast literature of relevant information. What we have offered, however, can be used as a starting point. Students can use the selected references at the end of each chapter in conjunction with the bibliography at the end of the book to pursue topics of interest. Instructors should have no difficulty embellishing or criticizing the themes that we present to develop their own distinctive emphasis.

Writing this book has been much more difficult than we anticipated when we began, but it has also been more rewarding. We have both learned a great deal, from the literature, from our colleagues, and from each other, as we have tried to distill the diverse data and ideas into an integrated conceptual framework. The extent to which we fail to convey this synthesis to readers reflects our own shortcomings. The degree to which we succeed depends largely on the contributions of others.

This book could not have been written without the help of many generous and dedicated people. We are particularly grateful to our colleagues and students at the University of Arizona, the University of California at Los Angeles, and elsewhere who have shared their interests and ideas and who have encouraged us to write the book. S. Carlquist, T.J. Case, J. Cracraft, L. Key, P.S. Martin, M.E. Mathias, P.L. Meserve, E.C. Olson, D.M. Porter, C. Robbins, H.J. Thompson, and R.F. Thorne read all or part of the manuscript and made many helpful suggestions that greatly improved the final version. For the shortcomings and errors that remain we alone are responsible. T.J. Case, D. Dunn, J.F. Eisenberg, R.D. Holt, R.B. Huey, D. Jablonski, J. Roughgarden,

T.W. Schoener, E.W. Stiles, E.E. Williams, and D.H. Wright provided unpublished information. A.K. Brown, W.A. Dunson, J. Faaborg, J.R. Hastings, V.C. LaMarche, and R.M. Turner allowed us to use their photographs. J. Smith drew nearly all of the maps, and S. Edwards and G. Ige drafted most of the other illustrations. Several people, especially B. Bonanno and R. McKinley, helped type the manuscript. Our editors at The C.V. Mosby Co. were increasingly supportive: S.E. Abrams got us started, and D. Bowen and S.D. Schapper advised us throughout the writing. To all of these people, and to any we may have forgotten to mention, we express our most sincere gratitude.

James H. Brown
Arthur C. Gibson

To

our families—

wives, parents, and children—

for their love, help, and encouragement

Contents

Chapter 1 The Science of Biogeography

There are more than 1.6 million known kinds of animals, plants, and microbes living on earth today, and certainly as many more undescribed forms have yet to be discovered. We can add to this list untold millions of species that lived sometime in the past but are now extinct, only a small fraction of which were recorded as fossils. Organisms are found in almost all conceivable environments, yet each extant and extinct species has or had a unique geographic distribution. Each species inhabits only a part of the earth's surface, occurs only in some habitats, and varies in abundance over its geographic range. These ranges change dynamically, usually starting small, experiencing increases and decreases in size, and finally decreasing to extinction. Contemporary forms will eventually become extinct as well, leaving future species in their places.

What is biogeography?

Biogeography is the study of distributions of organisms, both past and present. It is the science that attempts to describe and understand the innumerable patterns in the distribution of species and larger taxonomic groups. Few biological subjects are enjoyed by amateurs and professionals alike as is biogeography, because we possess an inherent interest in the organisms sharing this world and a desire to speculate about where they came from and why they occur where they do. Biogeography is in part a historical science. From the study of fossils, which provide valuable information on the history of life on earth, we can obtain the answers to some of the interesting questions. How did a species come to be confined to its present range? What are its closest relatives and where are they found? What is the history of the group, and where did their ancestors live? Why are the animals and plants of large, isolated regions, such as Australia, New Caledonia, and Madagascar, so distinctive? When did the distribution of a particular group expand or contract to assume the present boundaries, and how have geologic events, such as continental drift and Pleistocene glaciation, shaped this distribution? Why are some closely related species confined to the same region, and other pairs separated and found on opposite sides of the world?

Other questions that biogeographers ask are primarily ecological, because they concern the relationships between organisms and their environments. Why is a species confined to its present range? What enables it to live where it does, and what prevents it from expanding into other areas? What roles do climate, topography, and interactions with other organisms play in limiting the distribution? How do we account for the replacement of species that we observe as we go up a mountain or move from a rocky shore to the sandy beach nearby? Why are there so many more species in the tropics than in temperate or arctic latitudes? How are islands colonized, and why are there always fewer species on islands than in the same kinds of habitats on nearby continents?

The list of possible questions is nearly endless, but in essence we are asking: How do the number and kinds of species vary, from region to region, over the surface of the earth, and how can we account for this variation? This is the fundamental question of biogeography. It has always intrigued people who are curious about nature, but only within the last century or so have scientists called themselves biogeographers and focused their research on the study of the distribution of living things. They have not yet answered all the questions, but they have learned a great deal about where different kinds of organisms are found and why they occur where they do. Much progress has been made in the last two decades, stimulated in large part by exciting new developments in the related fields of ecology, systematics, paleontology, and geology.

Biogeography is a broad field. To be a complete biogeographer one must acquire and synthesize a tremendous amount of information, but not all aspects of the discipline are equally interesting to everyone, including biogeographers. Given biases in their training, their biogeography courses and writings tend to be uneven in coverage. A common specialization is taxonomic, e.g., phytogeographers study plants and zoogeographers study animals, and within these categories one finds specialists for groups at all taxonomic levels. Although viruses and bacteria play crucial roles in ecological communities and in human welfare, microbial geography is poorly known and rarely discussed. Some biogeographers specialize in historical biogeography and attempt to reconstruct the origin, dispersal, and extinction of taxa and biotas. This approach contrasts with ecological biogeography, which attempts to account for present distributions in terms of interactions between the organisms and their physical and biotic environments. Paleoecology bridges the gap between the two fields. Recently, some workers have emphasized different methods for understanding distributions: some approaches are primarily descriptive, designed to document the ranges of particular living or extinct organisms; whereas others are mainly conceptual, devoted to building and testing theoretical models to account for distribution patterns. All approaches to the subject are valid and valuable, and ridiculing or overemphasizing any division or specialization is counterproductive and unnecessary. Whereas no researcher or student can become an expert in all areas of biogeography, exposure to a broad spectrum of organisms, methods, and concepts leads to a deeper understanding of the science. As we hope to show, the various subdisciplines contribute to and complement each other, unifying the science.

Relationships to other sciences

Biogeography is a synthetic discipline, relying heavily on ecology, population biology, systematics, evolutionary biology, the geosciences, and natural history. Consequently, we do not want to draw sharp lines between biogeography and its related subjects, as some authors have attempted. For example, various authors have recommended that paleontology (the study of fossils and extinct organisms) and ecology should be divorced from biogeography; this would make biogeography a purely descriptive, mapmaking endeavor. Instead, biogeography is a branch of biology, and not surprisingly, a good knowledge of biology is an important starting point. This is why our treatment devotes considerable space to reviewing and developing ecological and evolutionary concepts that are used throughout the book (Chapters 3, 4, 6, and 7). In addition, we must be acquainted with the major groups of plants and animals and know something about their physiology, anatomy, development, and evolutionary history, so these topics are integrated in Chapters 10 to 13. For example, the distributions of frogs and salamanders begin to make sense once we know that these are amphibians, a group of vertebrates that are usually terrestrial as adults but aquatic as larvae. They thrive in moist places, but most are intolerant of salt water. This helps us to understand why amphibians are common in mesic habitats on continents but are poorly represented in deserts and on oceanic islands.

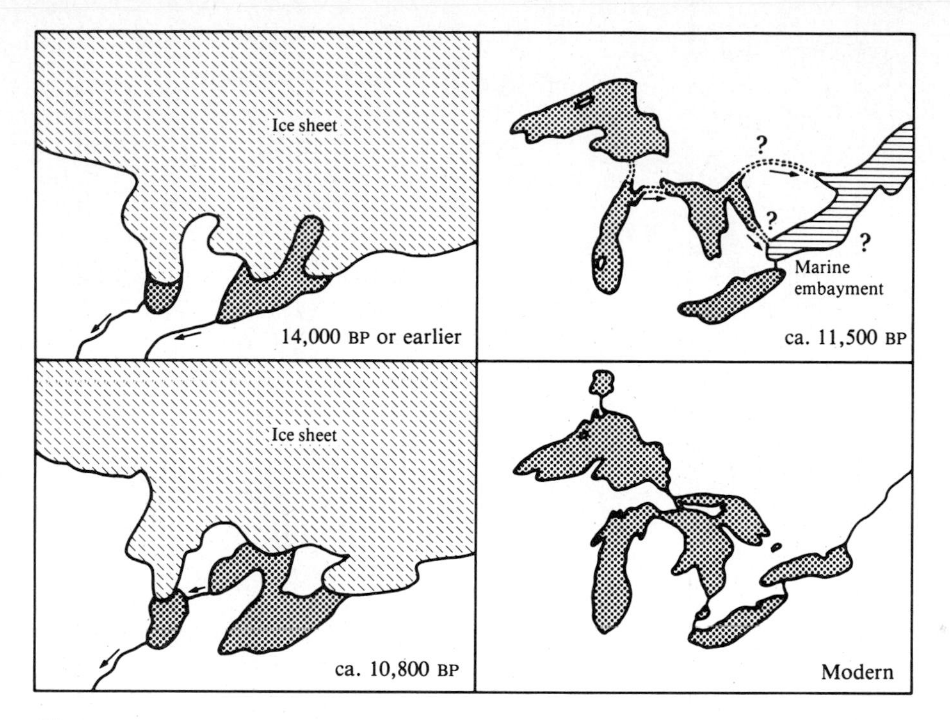

Figure 1.1
Changes in the geography of the Great Lakes region within the last 14,000 years, during and after the last episode of Pleistocene glaciation. Note the dramatic changes in the distribution of ice sheet, land, and water, and imagine the effects on the ranges of plants and animals. (Maps redrawn from Hutchinson, 1957.)

Naturally it is important to know some geography. Locations of continents, mountain ranges, deserts, lakes, major islands and island chains, and seas during the past as well as the present are indispensable information, as are past and present climatic regimes, ocean currents, and tides. Even looking back a mere 14,000 years, since the latest Pleistocene, we discover a vastly different topography in the Great Lakes region (Figure 1.1), where three of the present lakes were covered with glacial ice, and the other two, antecedents of Lake Erie and Lake Michigan, had markedly different shapes and were somewhat interconnected. Imagine how this could have influenced migration and speciation of organisms.

Contemporary climatic patterns are equally intriguing. For example, diverse and distinctive plant formations are found throughout the world in isolated regions where mediterranean-type climates prevail. Total annual precipitation is low, over two thirds occurring in the mild winter months, whereas summers are dry and often hot. Places sharing this climatic regime are widely disjunct in warm temperate latitudes: around the Mediterranean Sea, in coastal central Chile, in southwestern Australia and coastal southern and central Australia, in the Cape Region of South Africa, and in coastal and inland southwestern North America, especially California (Figure 1.2). These distinctive semiarid plant communities, named by local people as

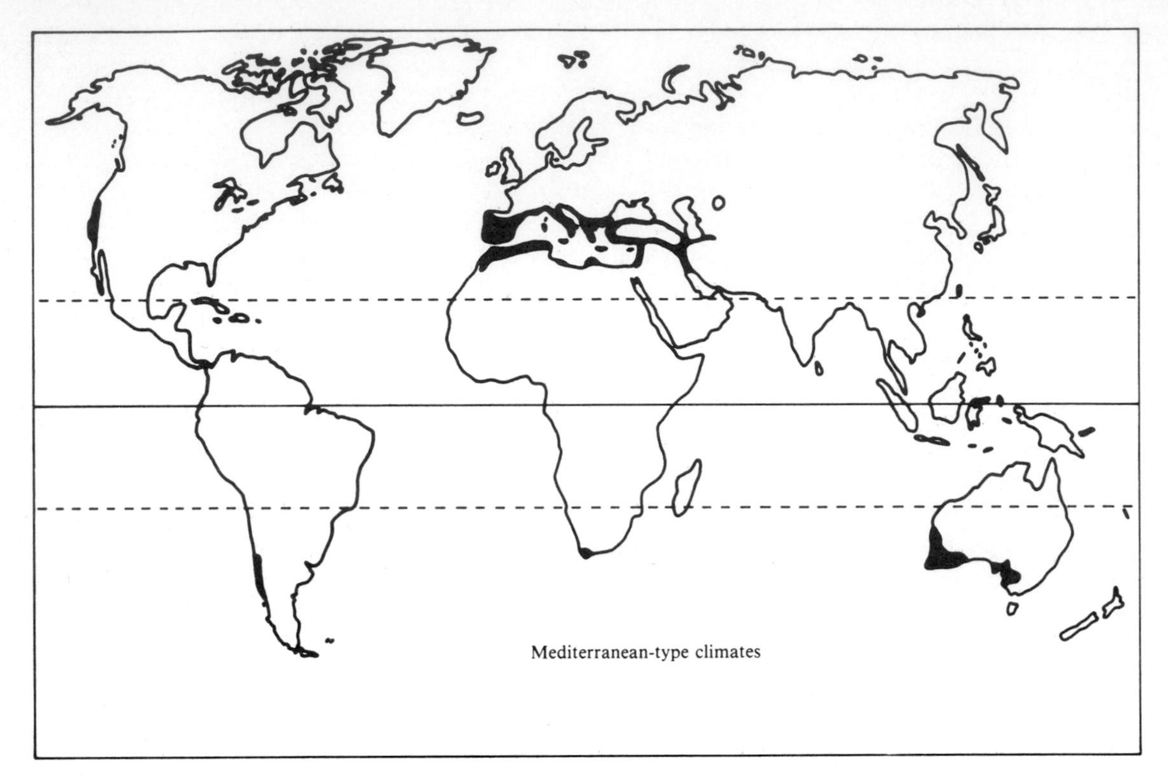

Figure 1.2
Worldwide distribution of regions experiencing mediterranean-type climates. These areas tend to occur as small patches on the western or southern coasts of continents between 30° and 40° latitude. Although these regions are isolated on different continents, they contain similar vegetation because distantly related plant species have evolved similar adaptations to the distinctive climate. (Map redrawn from Thrower and Bradbury, 1977.)

chaparral, matorral, macchia, maquis, fynbos, or sclerophyllous scrub, strongly resemble each other in vegetative structure and adaptations to periodic burning of the dense shrub cover. However, in each area the dominant vegetation is taxonomically unique, showing that the plants are derived from different ancestors and that they have converged in many traits in adapting to similar environments. Background knowledge of physical geography and climate presented in Chapter 2 is relevant to elucidating such patterns of evolutionary and ecological convergence.

Trying to understand distributions of organisms without knowing the past positions of continents and oceans means working with the same handicaps that plagued the pioneers of our science, who thought the geography of the earth has remained fixed over time. The modern biogeographer must appreciate not only the evolutionary history of different groups of organisms but also the geologic history of the earth on which they lived, because these processes have interacted to determine present distributions, especially over the last 250 million years. Reflect for a moment on the history of

North America. For most of its history North America has been in close proximity to if not connected with western Europe, and the final break of these two landmasses occurred only 60 million years ago when several northern seas and Iceland were born, resulting in the permanent separation of Europe and Greenland. Also about 60 million years ago western North America achieved a solid land connection with Siberia, which served as an important highway for biotic exchange until 8000 years ago. To the south lies South America, which was separated from North America during most of its late history until about 4 million years ago, when Panama began to emerge as an isthmus to connect Colombia with Central America. This is a different land connection than the one 80 million years earlier that was later displaced into the Caribbean region. Knowledge of such land connections is essential in providing biogeographic explanations for the distribution of some modern groups in terms of historical migrations. Similarities in some living fishes, insects, birds, earthworms, and certain extinct plants and reptiles among South America, Africa, and Australia make sense when we realize that at least until 135 million years ago, while these groups were evolving and expanding their ranges, all three continents were joined as part of a single giant southern landmass called Gondwanaland. For these reasons, considerable attention is given to geologic information, especially to the origin of current ideas and the latest information on plate tectonics (Chapter 5).

Philosophy and basic principles

Most people have a vague, misleading impression of what science is, how scientists work, and how major advances are made. Scientists try to understand the natural world by reducing its diversity and complexity to general patterns and basic laws. Philosophers and historians of science, viewing its progress with 20/20 hindsight, often suggest that it is possible to give a recipe for the most effective way to conduct an investigation. Unfortunately, most practicing scientists know that scientific inquiry is much more like working on a puzzle or being lost in the woods than baking cookies or following a roadmap. There are numerous mistakes and frustrations; luck, timing, and trial and error play crucial roles even in modest scientific advances. This is not meant to imply that intelligence, creativity, perseverance, and precision are not important. These attributes, plus sound technique, are as valuable to good scientific investigation as they are to solving a puzzle or finding the way when lost.

In essence, science is the investigation of the relationships between pattern and process. Pattern can be defined as nonrandom, repetitive organization. Occurrence of pattern in the natural world implies causation by a general process or processes. Science usually proceeds by the discovery of patterns, then the development of mechanistic explanations for them, and finally the rigorous testing of these theories until the ones that are necessary and sufficient to account for the patterns are accepted as scientific fact.

Traditional treatments of the philosophy of science usually devote considerable space to distinguishing between inductive methods, reasoning from specific observations to general principles, and deductive methods, reasoning from general constructs to specific cases. Several influential modern philosophers, especially Popper (1968a) have strongly advocated the so-called hypothetico-deductive method. Any good scientific theory has logical assumptions and consequences that must be verified before acceptance. The hypothetico-deductive method provides a powerful means of testing a theory by setting up alternative, falsifiable hypotheses. First, an author puts forth a new, tentative idea, stated in clear, simple language, that can be tested and thus falsified by means of experiments or observations. Only after a statement has withstood the severest tests should the statement be considered trustworthy or corroborated, but no theory is ever proven true.

New general theories are the ultimate source of most major scientific advances, and most of them are and have been arrived at by inductive methods. The theory of evolution by natural selection, the proposed double helical structure of DNA, and the equilibrium theory of insular biogeography all were derived largely by assembling factual data, recognizing a pattern, and then proposing a general explanation. Although it might be safe to generalize that theories usually arise by inductive methods and are tested by deductive ones, often the actual conduct of scientific research is much more complex. Empirical observations and conceptual generalizations are played back and forth against each other, theories are devised and modified, and understanding of the natural world is acquired slowly and irregularly. This is particularly true of biogeography. For this reason, some critics accuse the discipline of being metaphysical or pseudoscientific; they claim that some authors shield their ideas from contradictory evidence, dodging complaints and changing theories as they go.

Unlike most of contemporary biology, biogeography usually is not an experimental science. Questions about molecules, cells, and individual organisms typically are most precisely and conveniently answered by artificially manipulating the system. In such experimentation the investigator searches for patterns or tests specific hypotheses by changing the state of the system and comparing the behavior of the altered system with that of an unmanipulated control. It is impractical and often impossible to use these techniques to address many of the important questions in biogeography and the related disciplines of ecology and evolutionary biology. Historical evolution and historical biogeography are, as the names imply, history; they produce no exact predictions for the future.

Recently some biogeographers have used experimental techniques to manipulate small systems, particularly tiny islands, with spectacular success, e.g., Simberloff and Wilson (1969). However, most important questions have huge historical or geographic dimensions that make experimentation impossible. This methodological constraint does not diminish the rigor and value of biogeography, but it does pose major challenges. Other sciences, such as astronomy and geology, face the same problems. Copernicus, Galileo, Kepler, and Newton never moved a planet, but that did not prevent them from making tremendous contributions to our understanding of the motion of celestial bodies. Wallace and Darwin used patterns of animal and plant distributions observed in their world travels to develop important new ideas about evolution and biogeography. Islands had great influence on Wallace, Darwin, and numerous subsequent biogeographers, ecologists, and evolutionists because they represent natural experiments, replicated natural systems in which many factors are held relatively constant while others vary from island to island. Despite the difficulty of performing artificial experiments, it is possible to develop and rigorously evaluate biogeographic theories by the logical procedures used by other scientists: searching for patterns, formulating theories, and then testing the assumptions and predictions independently with new observations.

In dealing with historical aspects of their science, most biogeographers make one critical assumption that is virtually impossible to test: they accept the principle of uniformitarianism or actualism. This inferred concept holds that the physical processes now operating at the earth's surface have remained unchanged and are the result of the same fundamental laws that have acted throughout time. The principle of uniformitarianism is usually attributed to the British geologists Hutton (1795) and Lyell (1830), who realized that the earth was much older than had been previously supposed and that its surface was constantly changing as rocks were formed and weathered away and as mountains were uplifted and eroded down. In this same spirit, one of Darwin's great insights was the recognition that changes in domesticated plants and animals over historical time by selective breeding represent the same process as changes in organisms over evolutionary time through the process of natural selection.

As noted by Simpson (1970a), acceptance of uniformitarianism has never been universal, in part because authors have attached additional meanings to the concept. To many the term implies that the average intensities of processes have remained approximately constant over time and that changes are always gradual. Neither of these amendments is wholly acceptable. We have data to show how certain processes are now more or less intense than in the past; how forces are more active in one part of the globe than in others; where effects of forces have been sudden, not gradual; and where rates of change have not been constant over time. One must expect that intensity of forces varies from time to time and place to place; only the nature of the processes themselves is timeless.

To avoid unfortunate connotations associated with uniformitarianism, Simpson, after others, prefers to adopt the term *actualism*, conceptually similar to methodological uniformitarianism (Gould, 1965). Historical biogeographers in particular use this principle to account for present and past distributions, assuming that the processes of speciation, dispersal, and extinction operated in the past in the same manner that they do today. This premise is hard to falsify, of course, but fortunately most observations support the principle and have made it an accepted tool for understanding the past and predicting the future. The most serious problem in using the principle is that students must, of course, decide which timeless properties apply to a particular situation.

A brief history

Developments in the early nineteenth century. It is hard for us to appreciate that 200 years ago biologists had described and attempted to classify only 1% of all the plant and animal species we know today. Therefore biogeography really was founded and accelerated rapidly by world exploration and the accompanying discovery of new kinds of organisms, which gained great momentum in the 1800s. One of the early explorer-naturalists was Alexander von Humboldt, who usually is honored as the father of phytogeography. His treatises, beginning in 1805, were the first to conceptualize and quantify the primary role of climate in the distributions and forms of plants around the world. The close relationship between vegetation and climate was quickly expanded by Humboldt's contemporaries, such as A.P. de Candolle, J.F. Schouw, W.J. Hooker, and A. Grisebach, thus establishing the study of ecological phytogeography. Schouw (1823) published a remarkable textbook that not only included a classification of floristic regions of the world but also attempted to standardize descriptions of plant communities using Latin suffixes. From these humble beginnings, thousands of later scientists have attempted to quantify these botanical patterns.

Adolphe Brongniart is regarded as the father of paleobotany because he carried the theme of climate and vegetation into interpretations of the fossil record. From his studies, the use of plants as indicators of past climates became the basis of a new discipline, paleoclimatology. Brongniart (from 1827 to 1837) compared present and past floras of Europe; basing his discussions on the nature of fossil forms, he concluded that some of these localities once had tropical climates.

One of the great conceptual achievements of the early period was the law of the minimum, lucidly presented by Justus Liebig in 1840. Simply stated, the distribution of an organism is restricted by one particular basic requirement for life that is critically limiting, e.g., a foodstuff, a mineral, water, light, or temperature. This viewpoint is no longer widely held. We generally do not find that only one substance or factor is limiting; to the contrary, much evidence demonstrates that the interactions of these factors can be the mechanism for limiting distributions. Nevertheless, the investigation of limiting factors, both abiotic and biotic, stimulated interest in the fledgling disciplines of ecology and soil science in the late 1800s.

The study of animal distributions lagged behind phytogeography during its early history. Two factors contributed to this delay. First, because of the much greater number of animal species (about half of the 1.6 million described species are insects!), the task of describing and classifying animal life was several times greater than for plants, offering huge challenges for identifying general patterns. Second, the relationships between animal distributions and climate are mostly indirect; in fact, distributions of animals are more often closely associated with vegetation than with climate per se.

An early synthesis was produced by William Swainson (1835), but the first popular global classification of faunal regions that approaches our present-day classification of realms (Figure 1.3) was made for birds by W.L. Sclater (1858). As the pace of description and classification of animals was accelerated in the latter half of the

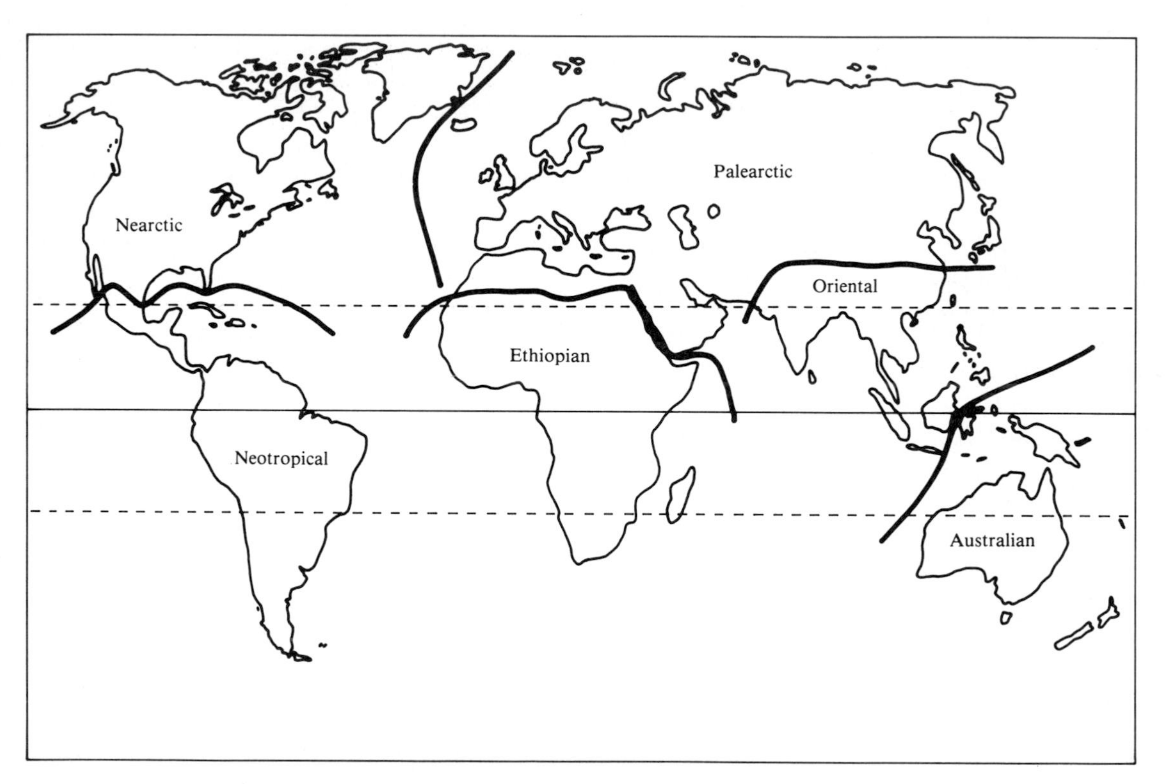

Figure 1.3
Major biogeographic regions reflect attempts of biogeographers to divide the landmasses into a classification reflecting the affinities of the terrestrial flora and fauna. The regions shown here are those described by A.R. Wallace in 1876 and are still widely accepted today. This classification is similar to that proposed by Sclater (1858) for birds.

nineteenth century, zoogeographic studies proliferated. Before the turn of the century, at least 20 such systems had been proposed.

Virtually all of the early biogeographers had noted three basic patterns that suggest that the distribution of living organisms has been importantly influenced by ancient historical events: (1) All taxa are endemic or restricted to particular areas; (2) certain kinds of organisms tend to occur together, providing a basis for dividing the earth into biotic regions or provinces; and (3) similar kinds of plants and animals sometimes occur in widely separated areas. Explanations for these patterns emerged as scientists began to realize that neither the earth itself nor its living inhabitants were immutable but rather that both had a long history of change dating back over thousands, perhaps even millions of years.

The Darwinian period. Arising out of a descriptive tradition were four English scientists (Figure 1.4) who revolutionized our view of the origins and distributions of species. The stage for this drama was set by Charles Lyell in the publication in 1830 of *Principles of Geology.* As discussed earlier, Lyell pictured the physical world as one gradually changing through eons of time, responding to timeless and predictable physical processes. This uniformitarianism replaced earlier catastrophic explanations for the history of the earth and its land forms. This explanation permitted new thinking about changes in living systems because, after all, the physical and living parts are historically inseparable.

With a copy of the first volume of *Principles* in hand, young Charles Darwin set sail in 1831 on a 5-year surveying voyage aboard *H.M.S. Beagle* as a scientist and gentleman companion for its captain, Robert Fitzroy. Darwin studied geology, native plants and animals, indigenous peoples, and domesticated animals during the journey in an attempt to understand the order of life. From his diary and vast collections of specimens, he later published a fascinating account (1839) of his adventures and observations during the voyage of the *Beagle*. Darwin was intrigued and perplexed by the patterns he observed: the fossils of extinct beasts in Argentina, the presence of seashells at high elevations in the Andes, and the occurrence of unique forms of life on islands. The patterns of variability in the Galápagos Archipelago, in which different species or races of tortoises and finches inhabit different islands, inspired him to the idea that geographic isolation facilitates inherited changes within and between populations. On his return to England, Darwin developed his theory of evolution, invoking natural selection as the primary mechanism by which new forms of life arose and are still arising today. The theory ranks as one of the most important scientific advances of all time and is woven into all aspects of biogeography.

The writing and eventual publication of Darwin's theory of evolution by natural selection is an interesting story that has been the subject of much review, too lengthy to repeat here. In a capsule, Darwin drafted a manuscript on the subject in the early 1840s but held the idea from print for 15 years while he continued to amass evidence to support his theory. He finally was forced to publicize his ideas when another brilliant scientist, Alfred Russel Wallace, sent him a manuscript that expounded the identical theory, developed independently but based on similar observations of the natural world. A paper by Darwin and one by Wallace were read together before the Linnean Society of London in 1858, and the following year Darwin's great book, *The Origin of Species,* was published and proved an immediate best seller.

Darwin made many other substantial contributions to biogeography, but Wallace is considered the father of zoogeography because he produced three massive works (1869, 1876, 1880) that synthesized the basic concepts and tenets of zoogeography using the theory of evolution through natural selection. Many of the concepts developed by Wallace were actually those of earlier workers, clearly restated, documented, and interpreted in an evolutionary context (see

Figure 1.4
Portraits of four English scientists who in the mid-nineteenth century revolutionized our understanding of the history of the earth and its organisms. **A,** C. Lyell. **B,** C. Darwin. **C,** A.R. Wallace. **D,** J.D. Hooker. (**A** and **C** by permission of the Council of The Linnean Society of London; **B** courtesy American Museum of Natural History; **D** courtesy Missouri Botanical Gardens Library.)

box below). As you read this book, you might refer to the following box periodically and note how many of those ideas are still being investigated by contemporary biogeographers. In addition, Wallace contributed immense knowledge on the biota of the East Indies and was the first person to analyze faunal realms based on the distributions of many groups of terrestrial animals, an analysis that supported Sclater's 1858 scheme. A distinctive original contri-

Biogeographic guidelines advocated by A.R. Wallace

These conclusions are summarized from Wallace's writings and have been verified many times by research in the twentieth century.

1. Distance by itself does not determine the degree of biogeographic affinities between two regions; distantly separated areas may share many similar taxa at the generic or familial level, whereas those very close may show marked differences, even anomalous patterns.
2. Climate has a strong effect on the taxonomic similarities of two regions, but the relationship is not always linear.
3. Prerequisites for determining biogeographic patterns are detailed knowledge of all distributions of organisms throughout the world, a true and natural classification of organisms, acceptance of the theory of evolution, detailed knowledge of extinct forms, and knowledge of the ocean floor and stratigraphy to reconstruct past geologic connections between landmasses.
4. The fossil record is positive evidence for past migrations of organisms.
5. The present biota of an area is very strongly influenced by the last series of geologic and climatic events; paleoclimatic studies are very important for analyzing extant distribution patterns.
6. Competition, predation, and other biotic factors play determining roles in the distribution, dispersal, and extinction of animals and plants.
7. Discontinuous ranges may come about by extinction in intermediate areas or the patchiness of habitats.
8. Speciation may occur by geographic isolation of populations that subsequently become adapted to local climate and habitat.
9. Disjunctions of genera show greater antiquity than those of a single species, and so forth for higher taxonomic categories.
10. Long-distance dispersal is not only possible but also the probable means for colonization of distant islands across ocean barriers; some groups have a greater capacity to cross barriers than others.
11. Good evidence of past land connections is the distribution of organisms not adapted for long-distance dispersal.
12. Animals isolated on landmasses may be allowed to survive and diversify without normal types of predation and competition.
13. When two large landmasses long separated are reunited, extinction may occur because many organisms will encounter new competitors.
14. The processes acting today may not be at the same intensity as those in the past.
15. Islands of the world may be classified into three major biogeographic categories: continental islands recently set off from the mainland, continental islands that were separated from the mainland in relatively old times, and distant oceanic islands of volcanic and coralline origin. Biotas of each type are intimately related to the island's origin.
16. Studies of island biotas are important because the relationships among distribution, speciation, and adaptation are easier to see and comprehend.
17. To analyze the biota of any particular area, one must determine the distributions of those organisms beyond that region as well as the distributions of their closest relatives.

bution was his observation of a sharp faunal gap between the islands of Bali and Lombok in the East Indies, where many species of Southeast Asia reach their distributional limit and are replaced by forms from Australasia (Wallace, 1860). This break has been called Wallace's line and subjected to subsequent discussion and modification (Mayr, 1944a; Carlquist, 1965).

A fourth great British contributor was Joseph Dalton Hooker, who was perhaps the world's most ambitious plant collector. At the age of 22 he became the assistant surgeon and botanist on an expedition to the Antarctic region led by Sir James Clark Ross, the discoverer of the magnetic north pole. The expedition visited many important southern lands, and Hooker's collections became his source for analyzing the affinities and probable origins of each of their floras, group by group. He developed the important concept of long-distance dispersal to account for occurrence on remote islands of those plants with easily dispersed seeds and fruits, and it was with this philosophy in mind that Hooker interpreted the vegetation of the Galápagos Archipelago, mainly based on specimens collected by Darwin.

Hooker's return to England in 1843 was significant in that this was when he formed a friendship with Charles Darwin, whom Hooker had admired from his writings on the *Beagle* voyage. Within a year of their meeting, Hooker read Darwin's manuscript on the theory of natural selection, the only person to see the manuscript until Asa Gray in 1857. Hooker shared with Darwin his ideas on geographic distribution of plants and was one of the few directly responsible for encouraging Darwin to work on and later publish *The Origin of Species*. In Darwin's original introduction to the book, Hooker is the only person singled out in acknowledgment, and Hooker is later acknowledged by Wallace in *Island Life* (1880) as the person to whom the book is dedicated.

Other contributions in the nineteenth century. Other scientists were attempting to look for important patterns in their data not previously considered. Chief among the early contributions were two generalized ecogeographic rules by C. Bergmann (1847) and J.A. Allen (1878). Bergmann's rule states that in warm-blooded vertebrates, races from cooler climates tend to be of larger body size and hence to have smaller surface-to-volume ratios than races of the same species living in warmer climates. The explanation for this pattern was that increasing volume per surface area in cold latitudes helps to conserve body heat and, conversely, small size and relatively large surface area facilitate the dissipation of heat in hot regions. Along this same line of reasoning, Allen's rule states that among species that maintain a high and constant body temperature (homeotherms), the limbs and other extremities are more compact and shorter in cold climates than in warmer ones. Thus birds and mammals of polar regions should tend to be stout with short limbs. Many authors have averred these general patterns but questioned the underlying causes of variation. Recent studies have revealed that body mass is closely correlated with basal metabolic rate, cost of transport, dominance in a community, success in mating, and size and types of food eaten. Because surface area and metabolic rate are closely coupled to body size, although not directly, natural selection for changes in size because of other advantages of being large or small would appear to favor changes in surface-to-volume ratios. Notwithstanding these problems, the early ecogeographic rules were pioneering efforts that stimulated development of the field of physiological ecology and led to important observations on allometry, how traits scale with body size.

Some evolutionary rules were described by paleontologists who were searching for patterns in the history of life and trying to interpret the fossil record. The theory of orthogenesis is an example of this. Orthogenesis states that evolution of a group continues in only one direction, this orientation being an intrinsic property of the organism and not controlled by natural selection. Of course, this theory was used to oppose Darwin's explanation for the role of selective factors in the environment in evolutionary change. A special type of orthogenesis was Cope's rule that evolution of a group involves a trend toward increased body size. Although there are many exceptions to this rule, it does seem that certain advantages in being large result in repeated increases in size in many animal lineages. Large body size, however, seems to make species susceptible to extinction, so that large forms (such as the dinosaurs and many now extinct groups of giant mammals) die out and are replaced by large representatives of new groups. Although subsequent research suggests that neither the patterns nor the explanations of these rules are as clear as their discoverers suggested, they did help to stimulate research on biogeographic patterns among both living and fossil organisms.

In the late 1800s plant researchers in Europe began to develop novel classifications in which plant taxa were grouped according to their external architectural designs or their tolerances of abiotic stresses, such as shortages of water or excesses of salts. Contributions from the Danish workers O. Drude (1887) and E. Warming (1895) quickly led, by the early twentieth century, to the now-famous classification of life forms by C. Raunkiaer (1934), who defined major types of plants based on the positions of the perennating tissues (Chapter 13). An ecological rather than a taxonomic approach was also adopted by the great German phytogeographer A.F.W. Schimper (1898, 1903), who summarized in elaborate detail the forms and habits of plants from around the world. These works later gave birth to two vital areas of biogeography and ecology: plant physiological ecology, which seeks to understand how various species are adapted to the habitats in which they are found, and plant sociology, a part of plant community ecology, which attempts to explain why certain combinations of plant species, but not others, cooccur in a given habitat.

Early botanists were aware that different types of vegetation occur at different elevations, but a zoologist, C. Hart Merriam (1894), provided one of the most valuable insights into these broad patterns. Merriam observed that changes in vegetation type and species going up a mountain in the southwestern United States are generally equivalent to latitudinal vegetational changes as one moves toward the poles but remains at low elevation (Figure 1.5). Belts of similar vegetation occurring both at low elevations in high latitudes and at high elevations in lower latitudes are termed life zones. Although Merriam tried unsuccessfully to generalize his concept of life zones to animals and to other regions of North America, he correctly concluded that elevational zonation of vegetation, like latitudinal zonation, is a response of species and communities to environmental gradients of temperature and rainfall.

In 1860 E.W. Hilgard demonstrated that certain climatic factors and plants are directly responsible for converting parent rock into different kinds of soils varying in pH, mineral composition, texture, and so forth. These interrelationships are discussed in more detail in Chapter 2. Shortly afterward, the Russian V.V. Dokuchaev recognized that each soil has a characteristic structure. These two contributions led us to understand that soils of a region are governed in large part by zonal climatic patterns that influence the breakdown of parent materials, the growth of plants, the decomposition of organic materials, and ultimately the kinds of plants and even animals that occur there, but the distribution of species may also be influenced by soil type on a local level.

First half of the twentieth century. From 1900 to the early 1960s several major trends in research had extraordinary impact on biogeography. Paleontology in particular deserves credit for new and fascinating descriptions of faunal changes on each continent. Numerous workers, but especially W.D. Matthew, E.H. Colbert, A.S. Romer, G.G. Simpson, E.C. Olson, and B. Kurtén, described the origin, dis-

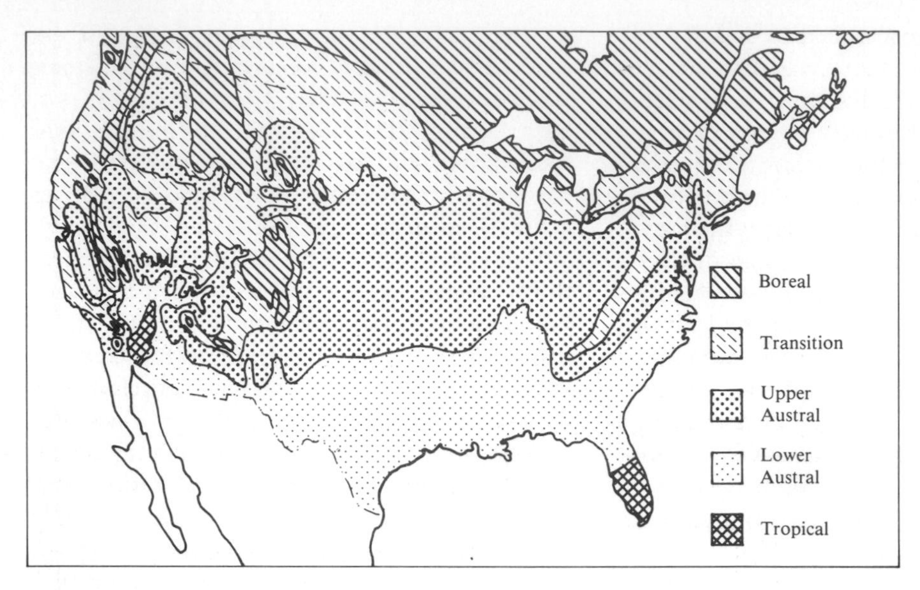

Figure 1.5
Redrawn copy of the original colored map used by C.H. Merriam in 1894 to delineate life zones for the United States and adjacent Canada. Merriam was one of the first to attempt to define precisely the relationship between climate and vegetation.

persal, radiation, and decline of many land vertebrates. Here we learned that new groups increase in number of species, radiate to fulfill new ecological roles, expand their geographic ranges, and become dominant over and contribute to the extinction of older forms. Thus our present-day continental faunas have extremely long and complex histories that can only be understood by elucidating the phylogenies of the groups and the history of their originations within and migrations between landmasses.

The explanations for how land organisms could have migrated from one landmass to another were legion. Investigators proposed an incredible number of short-lived land bridges or island archipelagos, now disappeared; former continents, now sunken; or once joined continents, now drifted apart. During this period tempers often flared as investigators hotly debated their explanations for how groups arrived on continents or islands, especially such distant and isolated spots as Australia and the Hawaiian Islands. Many such reconstructions are now rejected or considered unlikely, but the net result of this activity was to infuse much phylogeny, paleoclimatology, and geology into biogeographic syntheses.

A popular subject arising out of such inquiries was "center of origin." Where were the cradles of formation for various groups or biotas? This question was not new and had been discussed by earlier workers, such as the botanist A.P. de Candolle. One goal of each phylogenetic monograph was to propose a probable place of origin for the group and to describe its spread. The fossil record is spotty, so the information on which to base such reconstructions often was woefully incomplete. Some authors were bolder and more dogmatic than others on the ways to deduce centers of origin from limited data (see, for example, Matthew, 1915; Willis, 1922).

Early in this century researchers began to carefully investigate patterns of variation within single species. In the United States J. Grinnell and L.R. Dice and in Europe B. Rensch demonstrated close relationships between geographic and ecological properties of the environment and patterns of morphological variation within and between species. Subsequently, physiological and cytogenetic differences were related to distribution in nature through the pioneering studies of T. Dobzhansky on *Drosophila* and J. Clausen, D. Keck, and W. Hiesey on plants. Thus by the early 1940s workers were building on Darwin's synthesis to incorporate the causes and meanings of natural variation and the mechanisms responsible for the origin of new species. A long list of scientists contributed to our understanding of the modes of speciation. Out of this arose one unifying theme, the biological species concept, which states that a species is definable as a group of populations that is reproductively isolated from all others (Mayr 1942, 1963).

By the end of this period many authors attempted to synthesize the descriptive information on distribution. In addition to an impressive list of syntheses on restricted topics, too numerous to review here, major general syntheses included those mostly on vertebrates by P.J. Darlington (1957) and G.G. Simpson (1965); on marine zoogeography by S. Ekman (1953)· on vascular plants by R. Good (1947); and on island biogeography by S. Carlquist (1965). Moreover, there existed an impressive body of literature on ecological biogeography, summarized for animals by Hesse, Allee, and Schmidt (1951) and Niethammer (1958) and for plants by Dansereau (1957).

Biogeography since the early 1960s. Four major areas of research have revitalized biogeography in the last 20 years; acceptance of plate tectonics, development of new phylogenetic methods, exploration of new ways of conducting research in ecological biogeography, and investigation of the mechanisms limiting distributions.

Plate tectonics and continental drift, first introduced by Alfred Wegener in 1912, became widely accepted by biogeographers in the late 1960s and early 1970s when evidence for the process became irrefutable (Chapter 5). This revolutionized historical biogeography and required authors to rethink many distribution patterns. Changes in the relative sizes and positions of landmasses and oceans have resulted in important movements of biotas. Whatever one's interpretation of a distribution, the explanation must eventually be absolutely consistent with the geologic history of the earth's surface.

Biologists have made tremendous strides toward achieving phylogenetic classifications that trace the history and relationship of many groups, thus vastly improving our understanding of how biotas are and have been related. Guidelines for reconstructing phylogenies were already available for evolutionary systematics (Simpson, 1961) and on phylogenetic systematics or cladistics (Hennig, 1950), but the issues were really crystallized in 1966 when W. Hennig published his ideas in English. Phylogenetic research has been transformed from a discipline that discussed general similarities among taxa to one in which the degree of relationship of one species to another is carefully stated and quantified.

In the mid-1800s Asa Gray pioneered research on plant disjunctions, where two closely related species are spatially widely separated. Since then biogeographers have been fascinated by such disjunctions because they may reveal past land or water connections between the two regions. Interest in disjunctions has been renewed in the last 20 years, in part through the writings of L. Croizat (1958, 1960, 1964), as a way of evaluating past land connections. With new phylogenetic approaches in hand, the study of disjunct species, now called vicariants, has taken a central position, particularly in zoological research, and some of the older phylogenetic and biogeographic classifications are being tested and sometimes greatly revised.

Up to the 1960s, emphasis in biogeography

had been an evolutionary and historical one, emphasizing phylogeny of groups and their means of dispersing into and surviving in different areas and habitats. In the late 1950s G.E. Hutchinson began to focus attention on questions about the processes that determine the diversity of life and the number of species that coexist in local areas or habitats. Ecologists began to emphasize the importance of the biotic interactions of competition, predation, and mutualism in influencing the distribution of species and their coexistence to form ecological communities. Of all the work in ecological biogeography, perhaps the most influential was the attempt to develop a radically new theory to account for the distribution of species on islands by R.H. MacArthur and E.O. Wilson (1963, 1967). Their work changed the direction of ecological biogeography by focusing attention on a new set of questions. They asked abstract questions about patterns of distribution and diversity of species, and they suggested that these patterns reflected the operation of general processes that were to a large extent independent of the taxonomic affinities and phylogenetic histories of particular groups of organisms. What determines the capacity of an environment to hold species? Questions about species diversity and coexistence, in addition to dominating the fields of ecology and theoretical biogeography, have also spawned other abstract areas of inquiry, such as the extent to which dispersal, establishment, and radiation of one group versus another is a stochastic (i.e., random) or a deterministic (i.e., predictable because the underlying mechanisms are understood) process (Raup et al., 1973; Simberloff, 1974b; Stanley, 1979; Eldredge and Cracraft, 1980). Moreover, as mentioned earlier in this chapter, all these abstract questions stimulated experimental testing of biogeographic concepts (e.g., Simberloff and Wilson, 1969), as well as new mathematical ways of quantifying and analyzing observations (e.g., Pielou, 1977a, 1979).

Although some ecologists have emphasized the roles of biotic interactions in influencing the distribution of species and communities (MacArthur and Connell, 1966; MacArthur, 1972; Whittaker, 1975), there also has been an enormous research effort on the physiological ecology of plants and animals. These studies emphasize the importance of the abiotic environment in limiting the distributions of individual species and determining the diversity of species in different regions. Important advances in technique and instrumentation since the mid-1960s have permitted a flurry of activity, resulting in a tremendous amount of information that must be selectively integrated with biogeography.

Given this long list of conceptual achievements, in themselves the seeds of whole disciplines, one can easily comprehend how it has become impossible for one person to understand and follow completely all aspects of the field. Students of biogeography can be frustrated by their inability to comprehend all the subtleties of this awesome body of knowledge, or they can be challenged by the prospect of using biogeography as a focal point to synthesize many separate disciplines and to acquire a unique perspective on the history and distribution of life on earth.

Selected references

What is biogeography?

Briggs (1974); Cain (1944); C. Cox et al. (1980); Dansereau (1957); Darlington (1957, 1965); Daubenmire (1978); Ekman (1953); Good (1974); Illies (1974); Kellman (1975); Krebs (1978); MacArthur (1972); MacArthur and Wilson (1967); Müller (1974); Neill (1969); G. Nelson and Platnick (1981); G. Nelson and Rosen (1981); Newbigin (1968); Pears (1978); Pielou (1979); Polunin (1960); Seddon (1971); Simmons (1979); Szafer (1975); Tivy (1971); Udvardy (1969); Vermeij (1978); D. Watts (1971); Wulff (1943).

Relationships to other sciences

Ager (1963); Allee et al. (1949); Bannister (1976); M. Barbour et al. (1980); Butzer (1971); Daubenmire (1968); Elton (1966); Gates (1980); Grime (1979); J. Harper (1977); Horn (1974); G. Hutchinson (1957, 1967, 1978); Laporte (1968); Larcher (1980); Maarel and Werger (1978); Nairn (1961); Nobel (1974); E. Odum (1971); Raup and Stanley (1978); Rensch (1960); Ricklefs (1979); Solbrig et al. (1979); G. Stebbins (1950, 1974).

Philosophy and basic principles

Ackermann (1976); Dayton (1979); Felsenstein (1978); Gould (1965, 1977); G. Griffiths (1973); Hooykaas (1956); Hull (1970, 1974); R. Jeffrey (1975); Magee (1974); Medawar (1969); Nairn (1965); Platnick and Gaffney (1977, 1978a, 1978b); Platt (1969); Plumstead (1973); Popper (1968a, 1968b, 1972); Rosenkrantz (1976); Simberloff and Wilson (1970); G. Simpson (1970a); Strong (1980).

A brief history

Barbour et al. (1980); Barnett (1958); Beddall (1968); Carlquist (1965); Croizat (1958, 1960, 1964); Dansereau (1957); Darlington (1957, 1959a); C. Darwin (1859, 1962); F. Darwin (1887); Dice (1952); Dorf (1964); Egerton (1978a, 1978b); Eiseley (1958); Eldredge and Cracraft (1980); Good (1974); Hennig (1950, 1966, 1979); Hesse et al. (1951); Hull (1973); G. Hutchinson (1958); Kellogg (1957); Liebig (1840); MacArthur (1972); MacArthur and Connell (1966); MacArthur and Wilson (1963, 1967); Matthew (1915); Mayr (1942, 1944a, 1963); Merriam (1890, 1894); Moorehead (1969); G. Nelson (1978); Niethammer (1958); Pielou (1977a, 1979); Raunkiaer (1934); Raup et al. (1973); Rensch (1960); Rosen et al. (1979); Schimper (1898, 1903); Sclater (1858); Simberloff (1974a); Simberloff and Wilson (1970); G. Simpson (1953, 1961a, 1965, 1980a); Stanley (1979); Swainson (1835); Szafer (1975); A. Wallace (1869, 1876, 1880); R. Whittaker (1975); Wichler (1961); L. Wilson (1970); Wulff (1943).

UNIT ONE THE ECOLOGICAL SETTING

No species occurs everywhere on earth. The environments in which organisms can live are so diverse in nature that each species can tolerate only part of the entire range of conditions. Proximate ecological factors restrict each species to the limited area that we call its geographic range. The following three chapters discuss the ecological setting provided by various regions and the roles the abiotic and biotic environment play in limiting distributions.

Chapter 2 describes and explains large-scale variations in the physical environment. Simple physical principles account for global patterns in temperature, movements of winds and ocean currents, and patterns of precipitation. The terrestrial climate converts substrate materials and organic matter into soils, resulting in distinctive soil types in different geographic regions. Aquatic environments, which cover about three quarters of the earth, vary greatly in temperature, light, pressure, and the concentrations of dissolved gases, salts, nutrients, and other substances. Many of these aquatic patterns can be attributed to the relationship between the temperature, density, and salinity of water, which largely determine the vertical stratification and vertical and horizontal circulation of water masses. However, as we point out at the end of the chapter, these large-scale patterns of environmental variation do not adequately describe the conditions experienced by many organisms, which "read" the conditions on a very small scale and are able to take advantage of the tremendous microspatial variation in physical conditions to make a living in particular microenvironments.

The knotty problem of what sets the limits of the geographic range for a particular species is the subject of Chapter 3. In theory, the ecological niche of a species describes the environmental conditions that are necessary for its continued survival; but in practice it is impossible to measure precisely all parameters of the niche. Ranges expand and contract because spatial distributions are related to temporal fluctuations in abundance. The distribution of even a single species is often limited by different environmental factors in different parts of its geographic range, making it difficult to make satisfying generalizations for most species, except where the niche is narrow and defined primarily by a single resource. In particular cases the distributions of both plants and animals can be shown to be limited by physical factors, such as temperature, moisture, light, and salinity, and biotic interactions, such as competition, predation, or mutualism. Often several of these factors interact in complex ways so that it is difficult and unrealistic to understand their separate effects. In the long term, the capacity of a species to expand into adjacent areas is limited by the capacity of peripheral populations to adapt to new conditions, and this in turn can be limited by the continual influx of individuals and genes adapted to the more favorable environments near the center of the range. The study of peripheral populations and boundaries between two competing species helps us to determine the factors that limit species ranges.

As already intimated by our discussion of biotic interactions, species do not exist independently of each other. Rather, all species within a local region are integral parts of communities and ecosystems. Chapter 4 considers the organization and distributions of these associations. Although the interdependence of particular pairs of species and the organized flow of energy and materials have prompted some ecologists to view communities as superorganisms, there is little basis for considering these present-day communities as discrete biogeographic units. Most species appear to be distributed as if they were relatively independent of each other. Nevertheless, the way that individual species are organized into communities, as evidenced by such characteristics as body size and trophic status, importantly influences their geographic distributions. Despite the difficulty in justifying the recognition of discrete community types, climate, soils, and other environmental conditions profoundly influence the global distribution of the kinds of organisms that coexist in different regions. Terrestrial community types are usually recognized by characteristics of the dominant vegetation, whereas marine and freshwater ecosystems are classified primarily on the basis of features in the physical environment.

Chapter 2

Geographic Variation in the Physical Environment

Organisms can be found almost everywhere on earth: from the cold rocky peaks of high mountains to hot, windswept sand dunes of lowland deserts; from the dark, near-freezing depths of the ocean floor to the steaming waters of hot springs that may be 60° C. Yet no single kind of organism lives in all of these places. Each species has a restricted geographic range in which it encounters a limited range of environmental conditions. Polar bears and caribou are confined to the Arctic, whereas few palms and corals can be found outside the tropics. There are a few species, such as *Homo sapiens* and the peregrine falcon, that we call cosmopolitan because they are distributed over all continents and over a wide range of latitudes, elevations, climates, and habitats. But these are not only exceptional, they are also much more limited in distribution than they appear at first glance. In the cases mentioned, for example, they do not occur on the three fourths of the earth covered with water, or in many other places besides.

Although, as we point out in later chapters, we may need to invoke unique historical events or ecological interactions with other organisms to account for the limited geographic ranges of selected species, the most obvious patterns in the distribution of organisms occur in response to variation in the physical environment. In terrestrial habitats these patterns are determined by climate and by different types of soils. The distributions of aquatic organisms are limited largely by variation in temperature, salinity, light, and pressure.

Most geographic variation in the physical environment is regular and predictable. We all know that tropical lowlands are warm year-round, whereas the climate is colder and more seasonal at higher latitudes. Most streams and lakes contain fresh water, but the ocean is salty. Both the tops of the highest mountains and the depths of the deepest lakes and oceans are very cold. These and other climatic patterns are relatively easy to explain. To do so we need to know the orientation of the earth with respect to the sun and the size and location of major geographic features such as oceans, continents, and mountain ranges. We also need a rudimentary knowledge of the physical and chemical properties of air, water, and soil and of the principles of thermodynamics.

Climate

Solar energy and temperature regimes. Sunlight sustains life on earth; solar energy not only warms the earth's surface and makes it habitable but also is captured by green plants and converted into other forms of energy, which make possible the growth, maintenance, and reproduction of living things. According to the principles of thermodynamics, heat is transferred from objects of higher temperature to those of lower temperature by one of three mechanisms: (1) conduction, a direct molecular transfer, especially through solid matter; (2) convection, the mass movement of liquid or gaseous matter; or (3) radiation, the passage of waves through space or matter. Heat flows from the hot sun across the intervening space to the cooler earth as energy in the form of solar radiation. When incoming radiation strikes matter, such as water or soil, some of it is absorbed and the matter is heated. Some solar radiation is initially absorbed by the air, particularly if it contains suspended particles of water or dust, (e.g., clouds), but much radiation passes through the sparse matter of the atmosphere and is absorbed by the denser matter of the earth's surface. This surface is not heated uniformly. Soil, rocks, and plants absorb much of the radiation and may be heated intensely. Water also absorbs much solar radiation, but the heating effect is not confined to such a narrow surface layer as it is on land. Although the air is heated somewhat by incoming solar radiation, most of the heating occurs at the earth's surface. Here air is warmed by direct contact with warm land and water, by latent heat released by condensation of water, and by absorption of long-wave infrared radiation reflected from objects such as leaves.

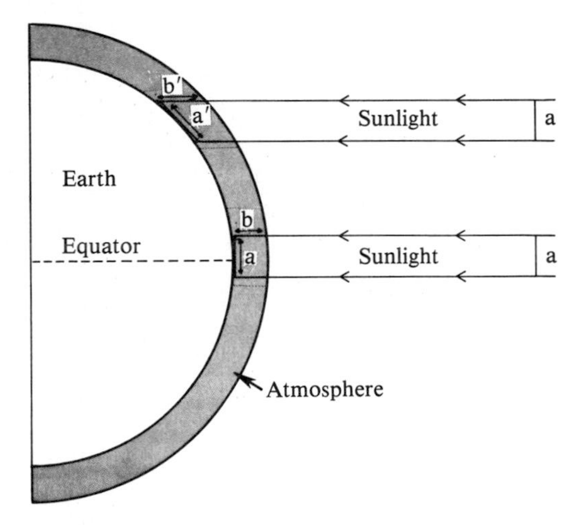

Figure 2.1
Heating of the earth's surface is most intense when the sun is directly overhead, as it is on the equator at the equinoxes. The high latitudes are cooler than the tropics because the same quantity of solar radiation must pass through a thicker layer of the filtering atmosphere (*b'* as opposed to *b*) and disperse over a greater surface area (*a'* as opposed to *a*).

The angle of incoming radiant energy to the earth's surface affects the quantity of heat absorbed. The most intense heating occurs when the surface is perpendicular to incident solar radiation, for two reasons: (1) the greatest quantity of energy is delivered to the smallest surface area; and (2) a minimal amount of radiation is absorbed during passage through the atmosphere because the distance traveled through air is minimized (Figure 2.1). This differential heating of surfaces at different angles to the sun explains why it is usually hotter at midday than at dawn or dusk, why the average temperatures in the tropics are higher than at the poles, and why south-facing hillsides are hotter than north-facing ones in the Northern Hemisphere (the reverse in the Southern Hemisphere). Because the earth is tilted 23.5° from vertical on its axis with respect to the sun, solar radiation falls perpendicularly on different parts of the earth during an annual cycle. This differential heating produces the seasons.

The seasons are also characterized by different lengths of day and night. Only at the equator are there exactly 12 hours of daylight and darkness every 24 hours throughout the year (Figure 2.2). At the spring and fall equinoxes (March 21 and September 22, respectively) the sun's rays fall perpendicularly on the equator, equatorial latitudes are heated most intensely, and every place on earth receives the same day length. At our summer solstice (June 22) sunlight falls directly on the Tropic of Cancer (23.5° N latitude), and the Northern Hemisphere is heated most intensely, experiences longer days than nights, and enjoys summer while the Southern Hemisphere has winter. On the other hand, at the winter solstice (December 22), the sun shines directly on the Tropic of Capricorn (23.5° S latitude), so the Southern Hemisphere enjoys its summer while the Northern Hemisphere has winter with cold

Equinox

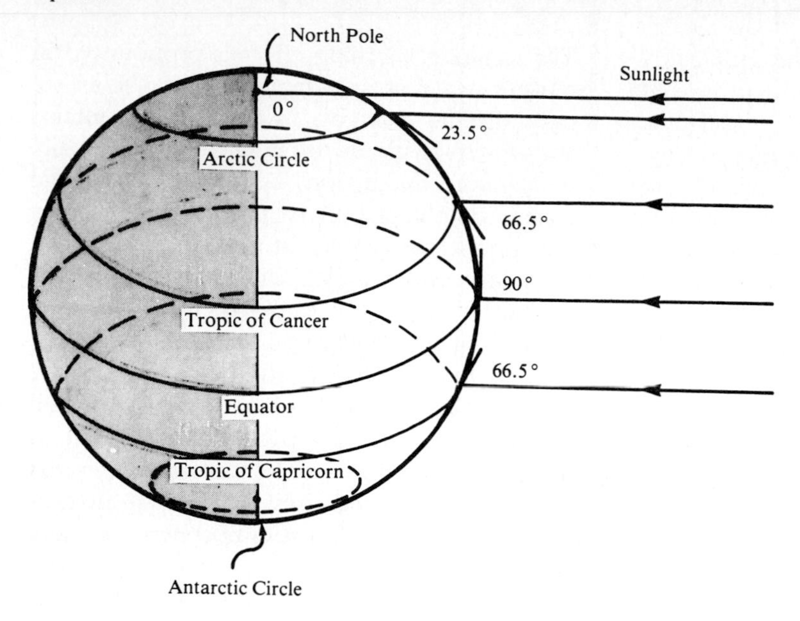

Summer solstice

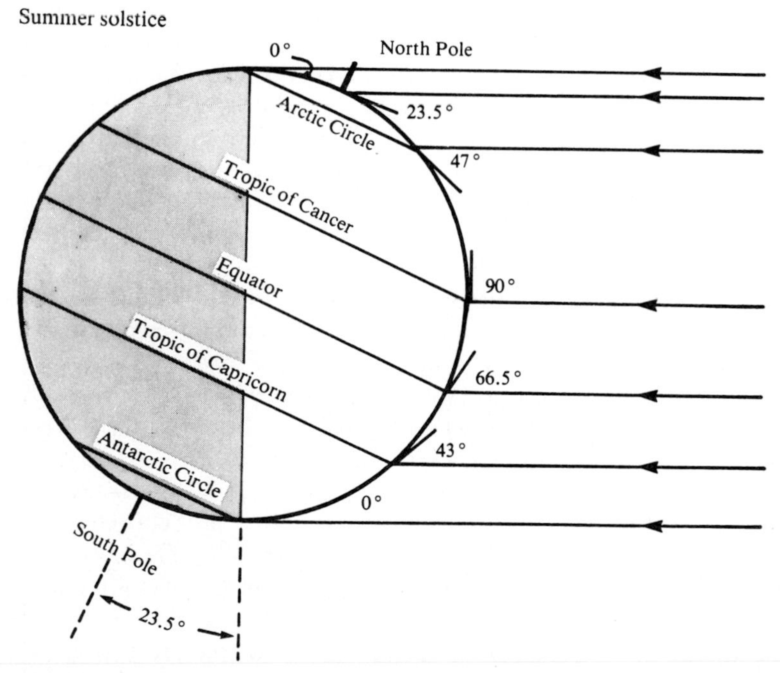

Figure 2.2
The earth's axis of rotation is inclined at an angle of 23.5°, which causes increasing seasonal variation in temperature and day length with increasing latitude. At the equinoxes the sun is directly overhead at the equator, and all parts of the earth receive 12 hours of light and 12 hours of darkness each day. At the summer solstice the sun is directly overhead at the Tropic of Cancer, and the Arctic Circle has 24 hours of continuous daylight, while all areas in the Southern Hemisphere experience less than 12 hours of light each day and the sun never rises below the Antarctic Circle. At the winter solstice (not shown) the situation is reversed; incoming solar radiation is perpendicular to the earth's surface at the Tropic of Capricorn, and all areas in the Northern Hemisphere experience less than 12 hours of light each day. (Greatly modified from Strahler, 1975.)

temperatures and long nights. The seasonality of climate increases with increasing latitude. At the Arctic and Antarctic Circles, 66½° latitude, there is one day each year of continuous daylight when the sun never sets and one day of continuous darkness, each marked at a solstice. Although every location on the earth theoretically experiences the same amount of daylight and darkness over an annual cycle, the sun is never directly overhead at high latitudes. Considerable solar radiation is absorbed during the long summer days, however, and temperatures in excess of 30° C are commonly recorded for July in Alaska.

The processes just described account for seasonal and latitudinal variations in temperature, but it remains to be explained why air gets colder as we ascend to higher altitudes. That Mt. Kilimanjaro in tropical East Africa is capped with permanent ice and snow seems in conflict with our intuitive expectation. Mountain peaks are nearer the sun, so why are they cooler than nearby lowlands? The answer lies in the thermal properties of air. Density and pressure of air decrease with increasing elevation. When air is blown across the earth's surface and forced upward over mountains, it expands in response to the reduced pressure. Expanding gases undergo what is called adiabatic cooling; this process also occurs in a refrigerator as freon gas expands after leaving the compressor. The rate of adiabatic cooling of dry air with increasing elevation is about 10° C/1000 m so long as no condensation of water vapor and cloud formation occurs. Higher elevations are also colder because the less dense air allows a higher rate of heat loss by radiation back through the atmosphere.

Water vapor and carbon dioxide in the atmosphere retard such radiant heat exchange and produce the so-called greenhouse effect. These gases have an effect similar to glass in a greenhouse: they allow shorter wavelengths of radiation to pass through, but the glass traps the longer wavelengths, which are emitted when solar energy strikes and warms surfaces. The resulting warming effect is pronounced in moist lowland areas, where water in the air retards cooling at night. In contrast, mountains and deserts typically experience extreme daily temperature fluctuations because there is little water vapor in the air to prevent heat loss by radiation to the cold night sky.

Winds and rainfall. Differential heating of the earth's surface also causes the winds that circulate heat and moisture. As we have already seen, the most intense heating is at the equator, especially during the equinoxes, when the sun is directly overhead. As this tropical air is heated, it expands, becomes less dense than the surrounding air, and rises. This rising air produces an area of reduced atmospheric pressure over the equator. Denser air from north and south of the equator flows into the area of reduced pressure, resulting in surface winds blowing toward the equator. Meanwhile, the equatorial air that has been heated and is rising cools adiabatically, becomes denser, is pushed away from the equator, and eventually descends again at about 30° N and S latitude (the horse latitudes). This vertical circulation of the atmosphere results in three convective cells in each hemisphere, with warm air ascending at the equator and at about 60° N and S latitude, and with cool air descending at the horse latitudes and the poles (Figure 2.3). These circulating air masses produce surface winds that typically blow toward the equator between 0° and 30° and poleward between 30° and 60°.

These zonal winds do not blow exactly in a north-south direction; they are deflected toward the east or west by the Coriolis effect. The Coriolis effect is often called Coriolis force, but it is not so much a force as a physical consequence of the law of conservation of angular momentum. Every point on the earth's surface makes one revolution every 24 hours. Because the circumference of the earth is about 40,000 km, a point at the equator moves from west to east at a rate of about 1700 km/h^{-1}. Points north or south of the equator move at a slower rate; remember that the lines of longitude con-

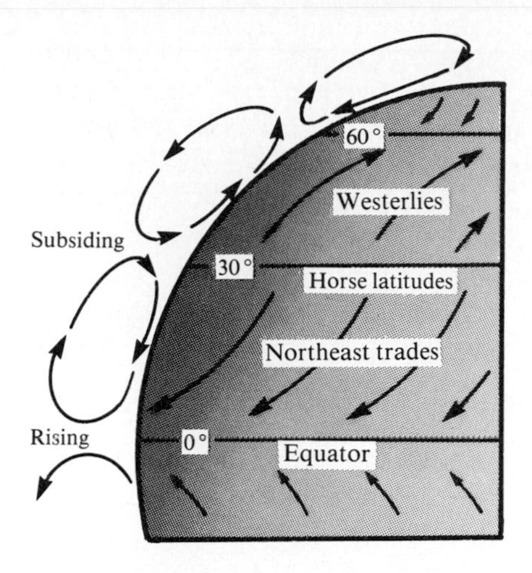

Figure 2.3
Relationship between the vertical circulation of the atmosphere and horizontal wind patterns at the earth's surface. There are three convective cells in each hemisphere. As the winds move across the surface in response to rising and subsiding air masses, they are deflected by the Coriolis effect, producing the easterly trade winds of the tropics and the westerlies of temperate latitudes. (Simplified from Trewartha et al., 1976.)

verge at the poles. Consider what happens at the equator if you throw a ball straight upward. Where does it come down? Right where it was thrown; the ball travels not only up and down but also eastward at a rate of 1700 km h^{-1}, the same rate as the earth moving beneath it. Now suppose the ball (or an air mass) moves northward away from the equator. It continues to travel eastward at 1700 km h^{-1}, but the earth underneath moves ever more slowly as the ball goes farther north, and consequently its path appears to deflect toward the right. This is the Coriolis effect. It describes the tendency of moving objects to veer to the right in the Northern Hemisphere and to the left in the Southern Hemisphere. The winds approaching the equator from the horse latitudes are deflected and therefore called northeast or southeast trade winds; winds blowing toward the poles between about 30° and 60° N and S latitude, called the westerlies, are deflected to the east (Figure 2.3). These winds naturally were very important in the days of sailing ships, when the trade winds, or "trades," got their name. Ships coming to the New World from Europe traveled south to tropical latitudes to intercept the northeast trades before heading westward, but they returned to Europe across the middle of the North Atlantic with the westerlies behind them.

Zonal winds, influenced by the Coriolis effect, initiate the major ocean currents. Trade winds push surface water westward at the equator, whereas the westerlies produce eastward moving currents at higher latitudes. Responding to the Coriolis effect, water masses are deflected toward the right or left, and the net result is that the ocean currents move in great circular gyres, clockwise in the Northern Hemisphere and counterclockwise in the Southern Hemisphere (Figure 2.4), and conforming to the shapes of the continents. Warm currents flow from the tropics along eastern continental margins; as these water masses reach high latitudes, they are cooled, producing cold currents down the western margins.

Now by superimposing the patterns of temperature, winds, and ocean currents we can begin to understand the global distribution of rainfall. We also need additional background in physics. When air warms it picks up water evaporated from the land and water, and as air cools it eventually reaches the dew point (when it is saturated with water vapor). Further cooling then results in condensation and formation of clouds, and when the particles of water or ice in clouds become too heavy to remain airborne, rain or snow falls. In the tropics the cooling of ascending warm air laden with water vapor produces heavy rainfall at low and middle elevations where rain forests occur. Rainy seasons in the tropics tend to occur when the sun is directly overhead and the most intense heat-

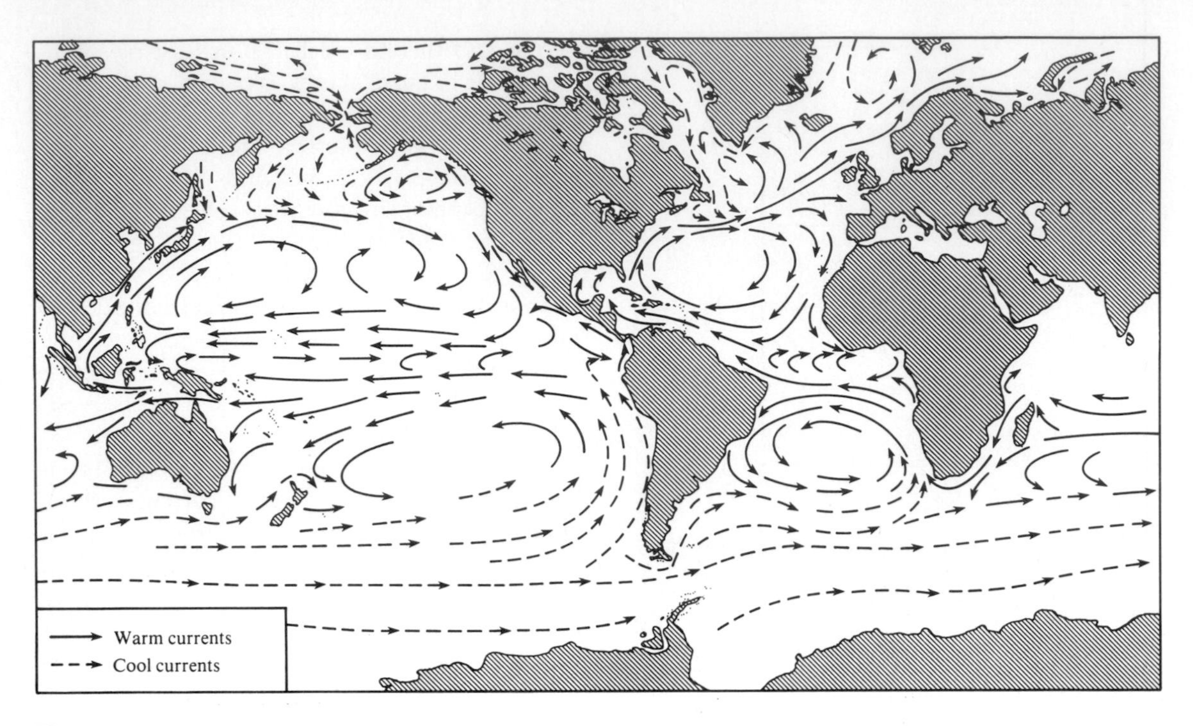

Figure 2.4
Main patterns of circulation of the surface currents of the oceans. In general, there are major circular gyres in each ocean, which move clockwise in the Northern Hemisphere and counterclockwise in the Southern Hemisphere. This results in warm currents along the eastern coasts of continents and cold currents along the western coasts.

ing occurs. The tropical grasslands of Kenya and Tanzania in East Africa, which lie virtually on the equator at higher elevations than rain forests, experience two rainy seasons each year, corresponding approximately to the equinoxes. On the other hand, the area around the Tropic of Cancer in central Mexico has only one principal rainy season, in the summer. Most tropical regions have at least one short dry season.

In the horse latitudes, where the cool air descends from the upper atmosphere, two belts of relatively dry climate encircle the globe. Descending air warms and so can therefore hold more moisture. This dries the land and has a desiccating effect on the organisms. In these belts the deserts have formed, and adjacent to the deserts are other semiarid climates. The seasonality of climate is very marked on the western sides of continents, which experience mediterranean-type climates. Parts of coastal California, Chile, the Mediterranean region, southwestern Australia, and southernmost Africa have dry, usually hot summers and a rainy mild winter (Figure 1.2). In summer the drying effect of descending air masses is reinforced by westerlies blowing inland from over cold ocean currents offshore. When land is warmer than the ocean, the air that passes over land takes up and holds more water vapor. In winter, when the land temporarily becomes cooler than the water, condensation occurs, foggy weather is common, and sometimes it rains. The effect of cold currents is even more pronounced in localized regions of western South America and

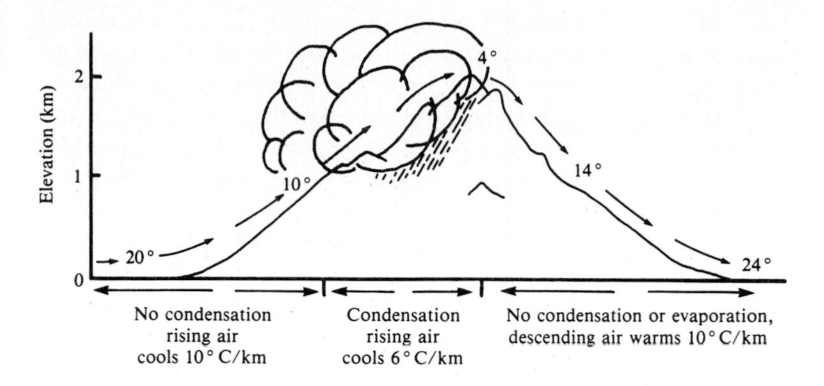

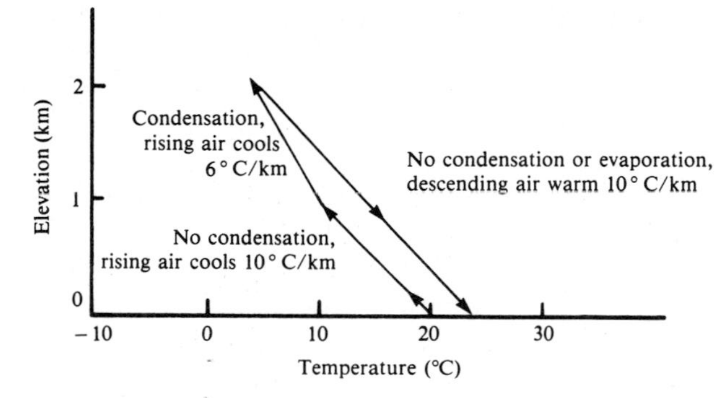

Figure 2.5
Schematic representation of the factors causing rain shadow deserts. Upper figure shows that warm moisture-laden air cools as it rises, condensing water vapor, and losing much of its moisture on the windward side of a mountain range so that the leeward side receives warm, dry winds. Lower figure shows how air on the windward side cools at the wet adiabatic lapse rate when condensation is occurring and warms almost entirely at the dry adiabatic lapse rate on the leeward side, resulting in warmer temperatures in the desert than at the same elevations on the windward side of the mountain range. (After Flohn, 1969.)

southwestern Africa, where it is responsible for the formation of coastal deserts (Amiran and Wilson, 1973).

Some deserts occurring between 30° and 40° N and S latitude are located on the western sides of continents but on the eastern sides of major mountain ranges. As westerly winds blow over the mountains, they are cooled until eventually the dew point is reached and clouds begin to form. Condensation releases heat, the latent heat of evaporation, which reduces the adiabatic lapse rate from 10° C/1000 m for dry air to 6° C/1000 m when water is either condensing or evaporating. As the air continues to rise and cool adiabatically, most of its moisture falls as precipitation on the western side. When the air passes over the crest and begins to descend, the remaining clouds quickly evaporate and the air warms at the adiabatic lapse rate for dry air. This rain shadow effect causes the warm, dry climate on the eastern sides of temperate mountains (Figure 2.5).

The same processes that we have just discussed on a global scale can often produce great climatic variation on a more local scale. The effect of mountains is particularly great, as we can illustrate with two examples. From Tucson, Arizona, it is only 25 km to the top of Mt. Lemmon at 2750 m in the Santa Catalina Mountains, where the climate and plants are far more similar to those in northern California and Oregon, 1500 km to the north, than to those in the desert just below (Table 2.1). Puerto Rico, which lies in the Caribbean Sea at 18° N latitude, is about 150 km long and 50 km wide and has a central mountainous backbone rising to about 1000 m. The lowlands on the northern and eastern sides are lush and tropical, but much more rain falls at higher elevations on the northeastern slopes, and this is where the best developed rain forests are found. So much moisture is lost as the northeast trade winds traverse the mountains that the southwestern corner of Puerto Rico is extremely dry; the cacti and shrubby vegetation that occur there remind a visitor of the deserts and tropical arid scrub or thorn forest of mainland Mexico (Figure 2.6).

Global patterns of temperature and precipitation frequently are summarized in climatic maps such as Figure 2.7. These are useful but can be misleading because they fail to show many features that influence the abundance and distribution of organisms. Typically such maps show only average conditions and provide no information on the extremes, seasonality, and predictability of the climate, features important for limiting species distributions. For example, a hurricane may pass over a Caribbean island only once in a century on the average, yet these rare, unpredictable storms wreak incredible devastation. Hurricanes probably are one of the primary causes of the extinction of terrestrial species on small Caribbean islands. Likewise, it is important to know when rainfall occurs. In many geographic areas, e.g., deserts, mediterranean-type climates, and certain tropical areas, rainy seasons may be followed by many months of drought, the type of drought that few organisms can survive.

Coastal regions or small islands may have climates markedly different from their climatic zones because of their proximity to a large body of water. Maritime locations have less extreme temperatures, warmer in winter and cooler in summer, because of the buffering effects of ocean water temperature and presence of abundant fog. Many maritime localities never receive frost. Fog changes the climate by decreasing light intensities and increasing moisture for plant life, and moisture from fog can be a principal source of water in certain coastal deserts.

Table 2.1

Weather data showing the influence of elevation on climate

Two of the sites are near one another in Arizona; the third site is in Oregon. Note that the climate of the high-elevation site in Arizona, Mt. Lemmon, is much more similar to that of Salem, Oregon 1700 km to the north than that of Tucson, only 25 km away but 1600 m lower in elevation. (Data from U.S. Weather Bureau.)

Site	Elevation (m)	Temperature (° C)				Mean annual precipitation (cm)
		Mean January	Mean July	Lowest	Highest	
Tucson, Arizona	745	10.8	30.7	− 9.4	46.1	27.3
Mt. Lemmon, Arizona	2373	2.3	17.8	−21.7	32.8	70.0
Salem, Oregon	60	3.2	19.2	−24.4	40.0	104.3

Figure 2.6
Comparison of the vegetation on opposite sides of a rain shadow, the central mountain range on the tropical island of Puerto Rico. Lush rain forests occur on the northeastern side, which receives the incoming trade winds, whereas arid shrub vegetation containing plants typical of desert regions are found in the rain shadow on the southwestern side. (Photographs by A. Gibson [*top*] and J. Faaborg [*right*].)

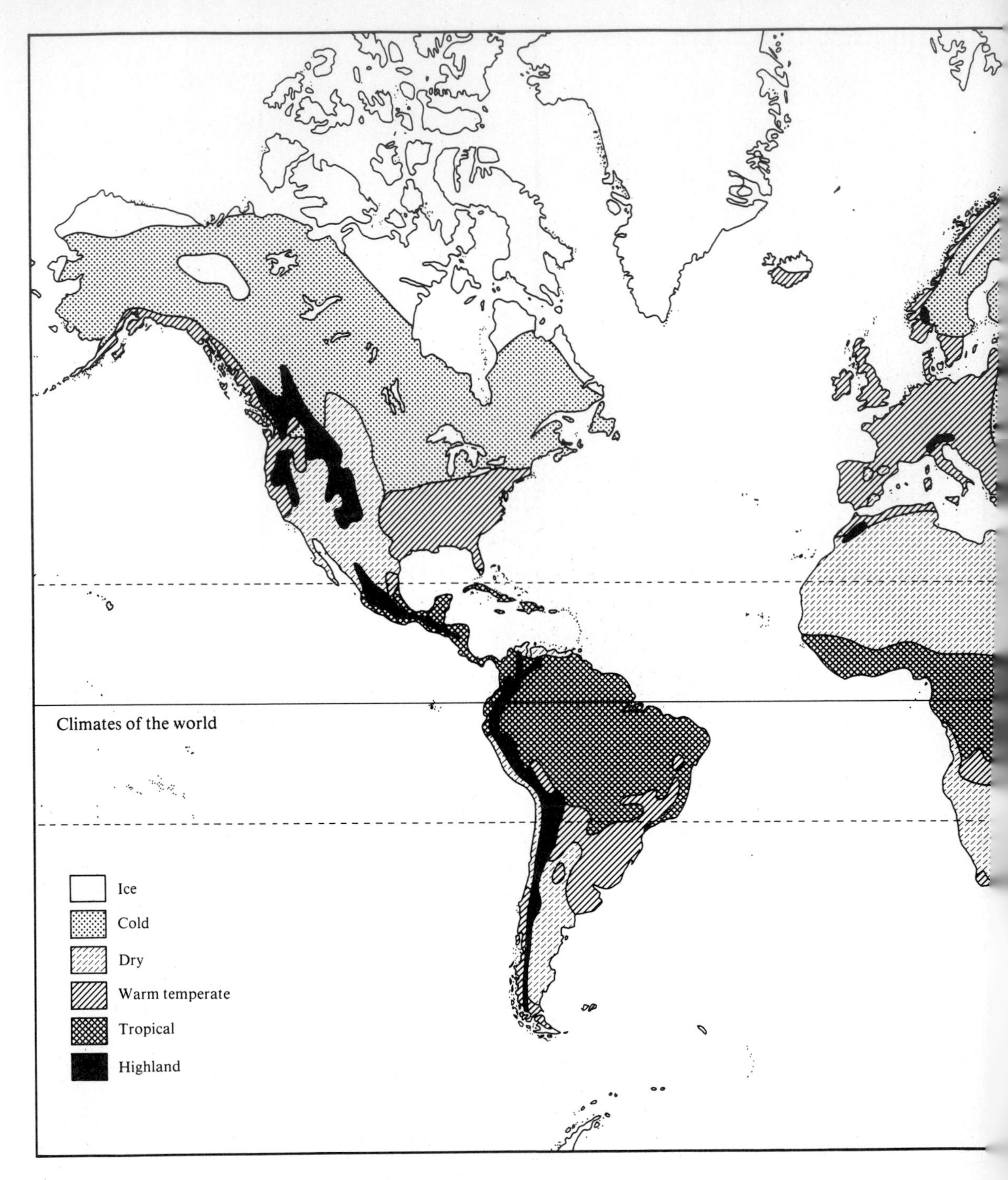

Figure 2.7
Major climatic regions of the world. Note that these occur in distinct patterns with respect to latitude and the positions of continents, oceans, and mountain ranges. (Simplified from Strahler, 1975.)

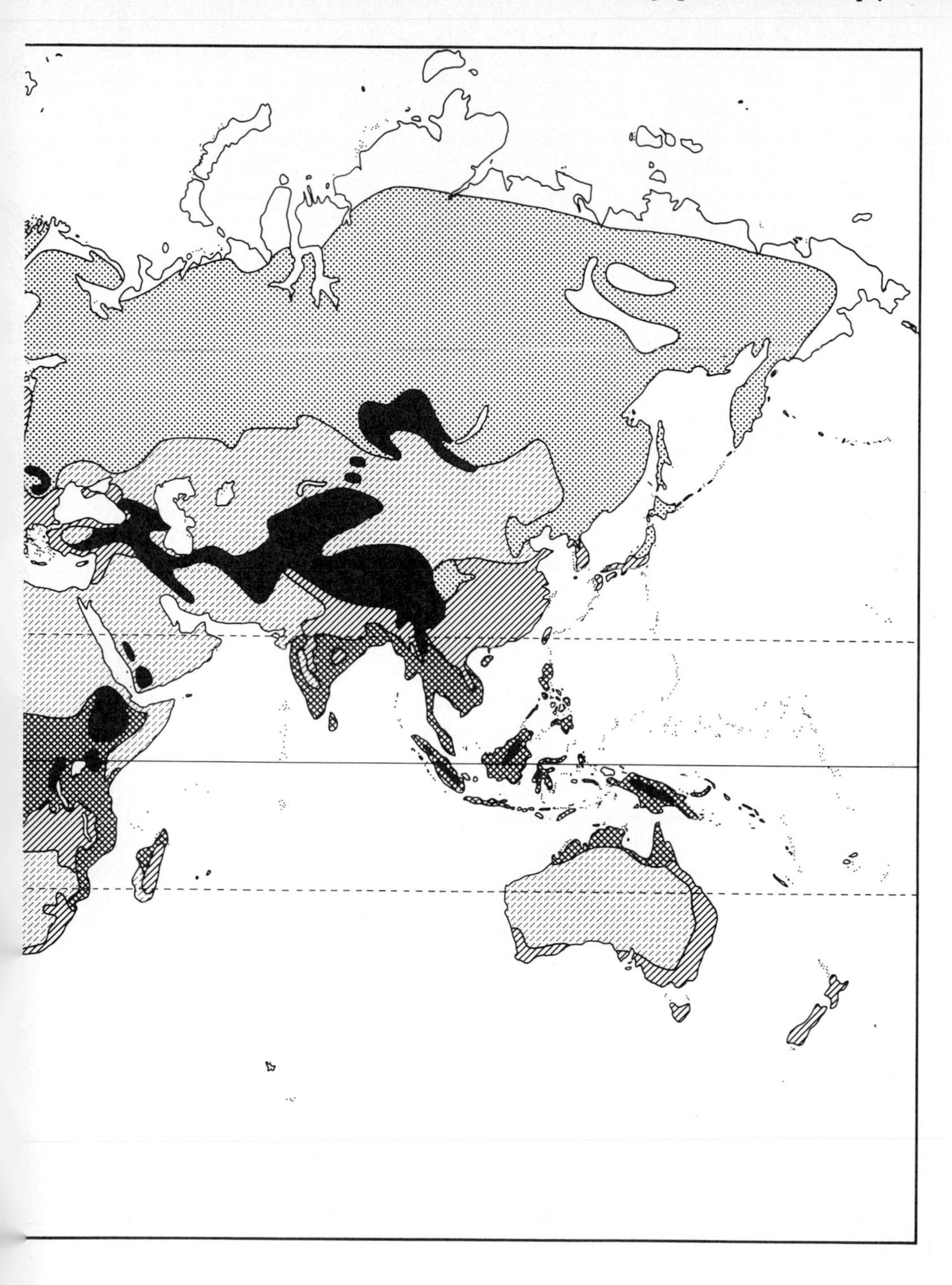

Soils

Primary succession. Except for the polar ice caps and the perpetually frozen peaks of the tallest mountains, almost all terrestrial environments on earth can and do support life. Areas of bare rock and other sterile substrates recently created by volcanic eruptions or other geologic events are gradually transformed into regions supporting living ecological communities by the process called primary succession. This process involves the formation of soil, the development of vegetation, and the assembly of a relatively stable complement of microbial, plant, and animal species.

The type of vegetation covering a region depends primarily on three ingredients: climate, type of soil, and stage in development of that vegetation and soil. For example, three distinctive vegetation types (temperate deciduous forest, pine barrens, and salt marsh) occur in northern New Jersey in close proximity to each other but on different soil types (Forman, 1979). Moreover, if a mature stand of deciduous forest is destroyed, as at the hands of humans or by natural fire, reestablishment of that plant community is not achieved directly. Instead, one group of plant species is gradually replaced by others until the mature combination may ultimately be achieved, a process termed secondary succession. Throughout this process both the microclimate and soil of the site are also changing, improving for some species and deteriorating for others. Thus discussions of soils cannot be divorced from either climatic or biotic factors.

Soil is formed by the weathering of rock plus the addition of organic material from dead and decaying organisms. In special situations, there are totally organic soils that lack rock entirely (histosols), such as peat. The process by which new soil is formed from bare rock is usually long and complicated. It involves breakdown of the parent material, colonization by simple plant and microbial forms, and gradual buildup and mixing of inorganic materials with decaying organic matter that enables larger plants to set down root systems. In arctic and desert regions, where temperature and moisture regimes are extreme, very shallow soils only a few centimeters in depth may have accumulated over thousands of years, e.g., in regions covered by the last Pleistocene ice sheets.

In other cases, especially soils formed from sand, lava, or alluvial materials, soil succession can be amazingly rapid. In 1883, the small tropical island of Krakatau in Indonesia experienced a tremendous volcanic eruption that killed the entire insular biota and left only bare volcanic rock and ash. Organisms rapidly recolonized from the large neighboring islands of Java and Sumatra, and by 1934, only 50 years after the eruption, 35 cm of soil had been formed and a lush tropical rain forest containing almost 300 plant species was rapidly developing (Docters van Leeuwen, 1936).

Formation of major soil types. Anything we write about soils has to be a gross oversimplification because both the classification and the distribution of soils are very complex, even controversial. Visit the vast flat plains of the United States or the Soviet Union and you will find one to several general soils distributed for as far as the eye can see, but in other geographic regions, such as those with numerous small and large mountain ranges, soil maps are mosaics that look like patchwork quilts (Figure 2.8). A country such as Great Britain has soils greatly modified by human manipulation over many centuries plus a series of unusual organic soils formed in cold, wet environments. However, if we want to begin appreciating ranges and qualities of soils, we can grasp this diversity by studying the four major processes that produce the primary or zonal soil types. These so-called pedogenic regimes are those of cold forested areas (podzolization), of the wet tropical forests (laterization), of calcareous bedrock with shrub or herbaceous vegetation (calcification), and of polar, waterlogged habitats (gleization).

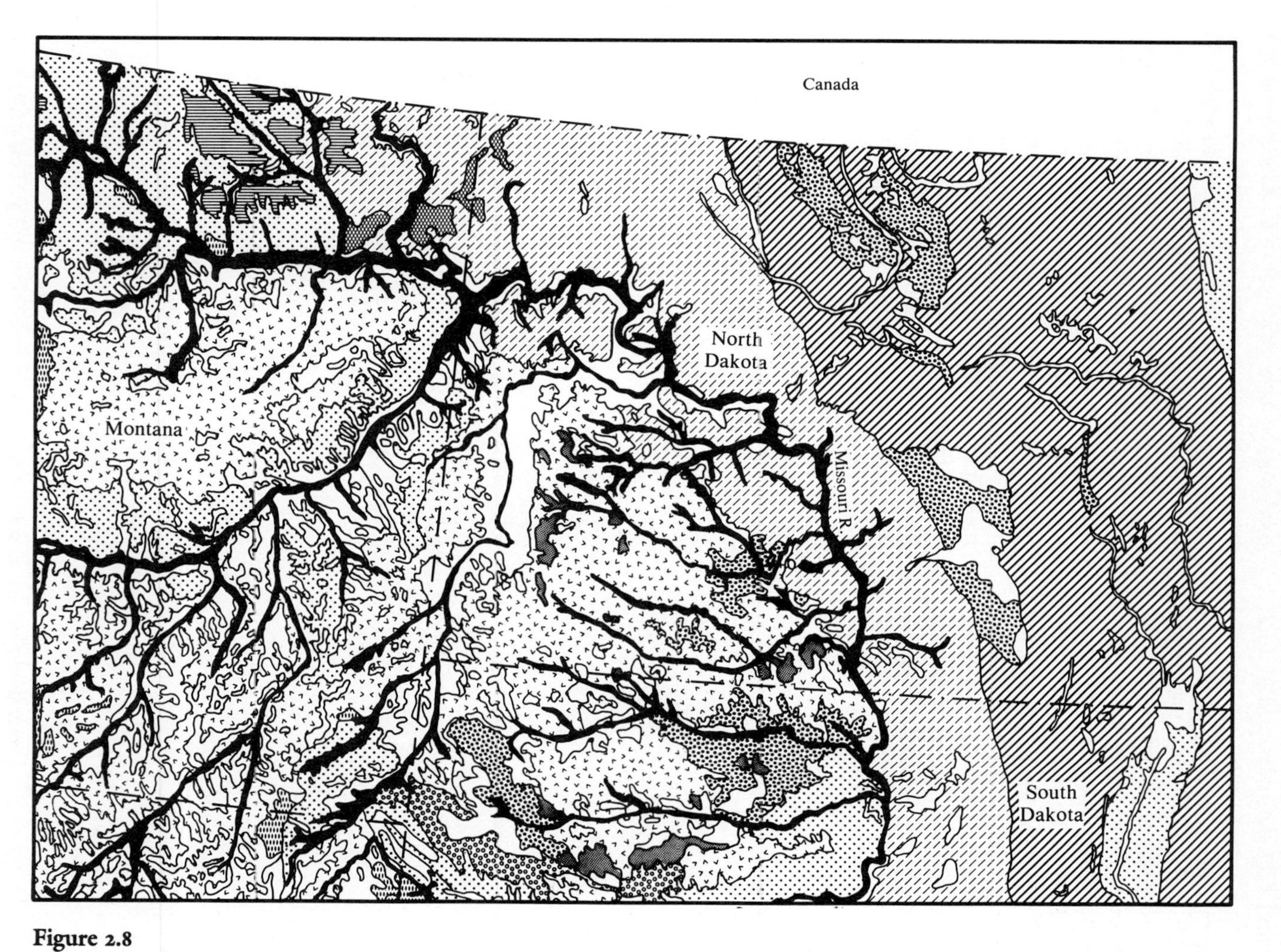

Figure 2.8
Soil map for the Great Plains and adjacent Rocky Mountains of North Dakota, Montana, and adjacent states. Note that large areas of the plains are characterized by homogeneous soils, whereas the mountain soils form a heterogeneous mosaic of diverse types. (Redrawn from Atlas of American Agriculture, 1936.)

Podzolization, producing podzolic soils, occurs in middle and high latitudes and at high elevations with cool temperature regimes and abundant precipitation. In cool climates plant growth may be substantial but microbial growth is slow, so organic matter, called humus, accumulates. During decay of humus, organic acids released into the soil are carried downward (leached) through the soil profile by percolating water. The hydrogen ions of these acids tend to replace important cations, such as calcium, potassium, magnesium, and sodium, which are removed by leaching from the soil (Figure 2.9, *A*). This leaves behind a silica-rich upper soil with oxidized iron and aluminum compounds but few cations. Consequently, where this soil is found in its extreme condition, plants that do best are those tolerant of acidic soils deficient in cations, such as conifers.

In warm to hot climates with heavy precipitation, as in the humid tropics, little or no humus can accumulate because decomposers and scavengers are extremely active. Iron and aluminum oxides precipitate as insoluble compounds in the absence of organic acids, and those form red clay or even bricklike layers (laterite). Excessive rainfall causes silica to be leached out of the soil (desilification) along

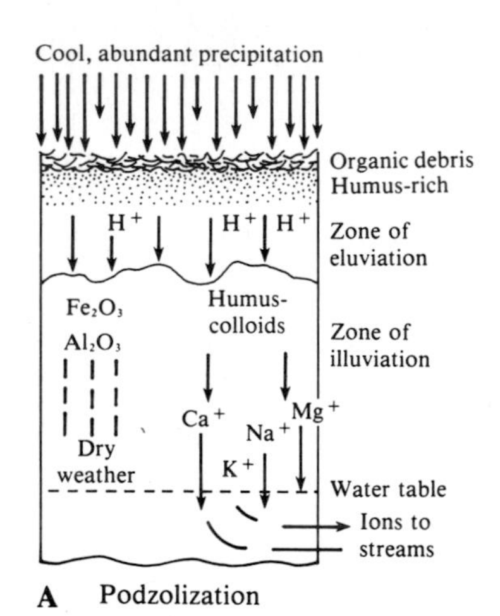

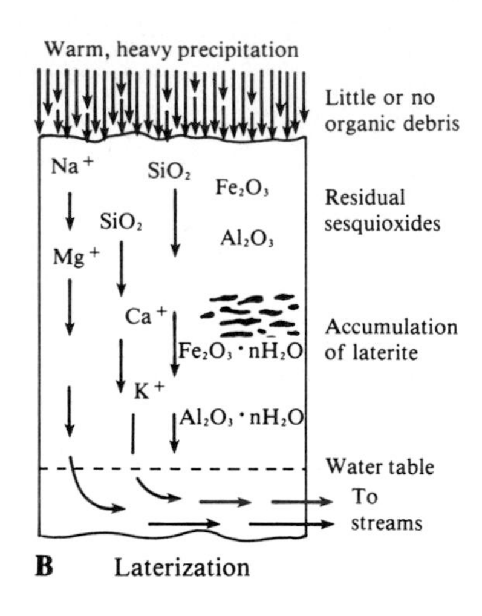

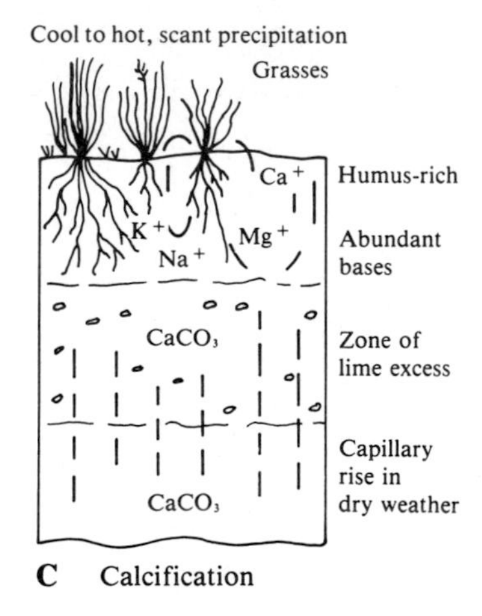

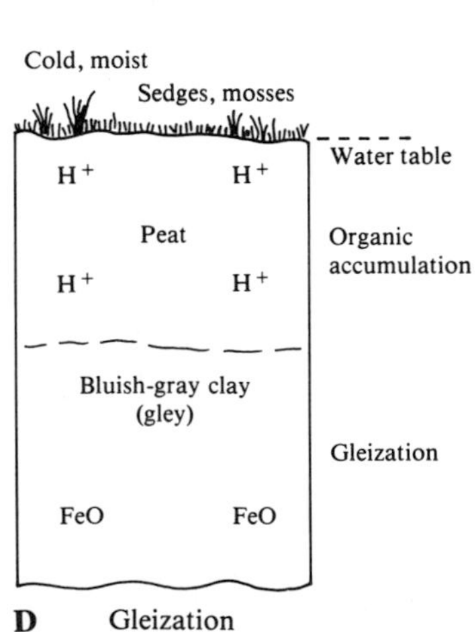

Figure 2.9
Schematic representation of the four major pedogenic regimes, showing how they give rise to soils of different composition and vertical profiles. **A,** Podzolization. **B,** Laterization. **C,** Calcification. **D,** Gleization. See text for further explanation. (After Strahler, 1975.)

with many important cations (Figure 2.9, *B*), leaving behind a firm and porous soil with very low fertility. In some areas, if the tropical forests are removed, the organic material and its bound nutrients are easily lost and the intense equatorial sun bakes the lateritic soils hard, making them useless for agriculture.

Wherever landmasses were once covered by shallow tropical seas one finds deep deposits of calcium carbonate. These regions are frequently found where precipitation is relatively low; i.e., evaporation and transpiration often exceed precipitation, although some regions with calcareous bedrock get abundant rainfall. The ions are generally not leached out of these soils, and where temperatures and precipitation are fairly high, calcareous soils support dense stands of grasses and forbs. These plants add large quantities of humus to the soil, making it highly fertile, and they send roots downward to great depths and thus return leached or new ions to the surface by incorporating these minerals into new plant structures (Figure 2.9, *C*). In semiarid and arid climates precipitation is so scanty that rainfall does not penetrate very far from the surface. Moreover, plant and microbial growth are greatly reduced. This results in the accumulation of cations, especially calcium, at the lowest level of water penetration, and if this pattern is reinforced year after year, a rocklike layer of calcium carbonate, called caliche, forms at the boundary. Caliche often restricts root growth to the upper few decimeters of the soil. Thus calcareous soils are usually either extremely fertile deep soils in areas where precipitation is high, or they are very shallow soils, poor in organic matter and especially in nitrogen, in arid regions.

In waterlogged soils of cold moist polar latitudes gleization occurs. Abundant organic matter builds up at the permanently wet (or frozen) surface, often as peat (Figure 2.9, *D*). This upper layer is rich in organic acids and therefore has a very low pH. Below this organic

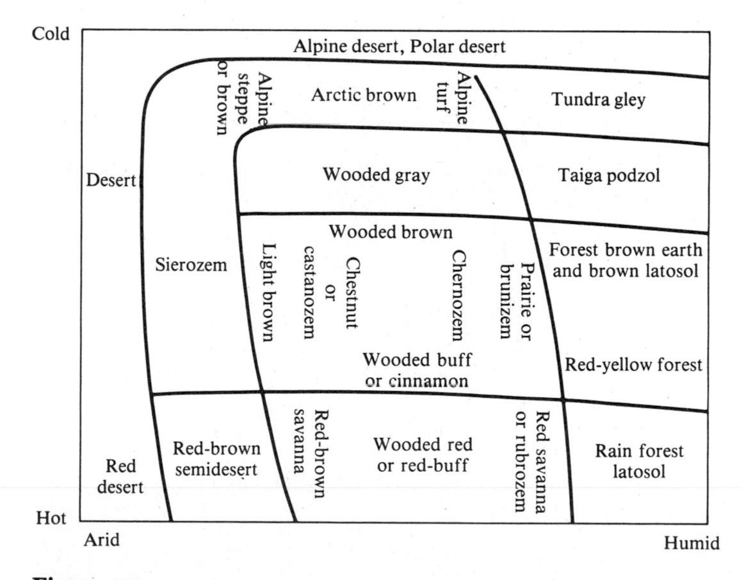

Figure 2.10
Relationship between the major zonal soil types and climate, showing that particular soil types are formed under the influence of certain conditions of temperature and precipitation. (Reprinted with permission of Macmillan Publishing Co., Inc., from Communities and Ecosystems, second edition, by R.H. Whittaker. Copyright © 1975 by Robert H. Whittaker.)

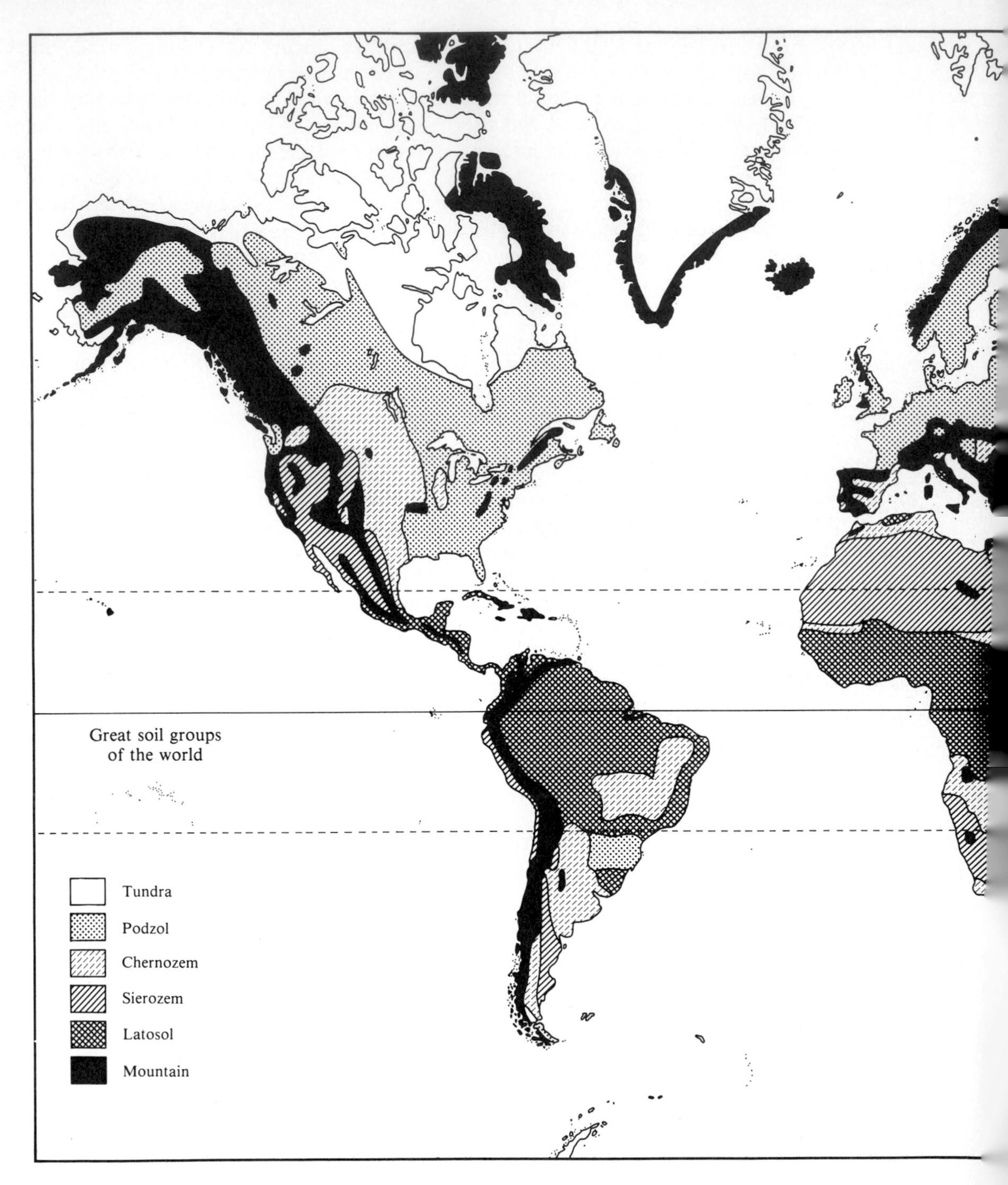

Figure 2.11
World distribution of major zonal soil types. Note that in general the pattern of soil types is closely correlated with climatic conditions. Refer to Figures 2.7 and 2.10. (Greatly simplified from Strahler, 1975.)

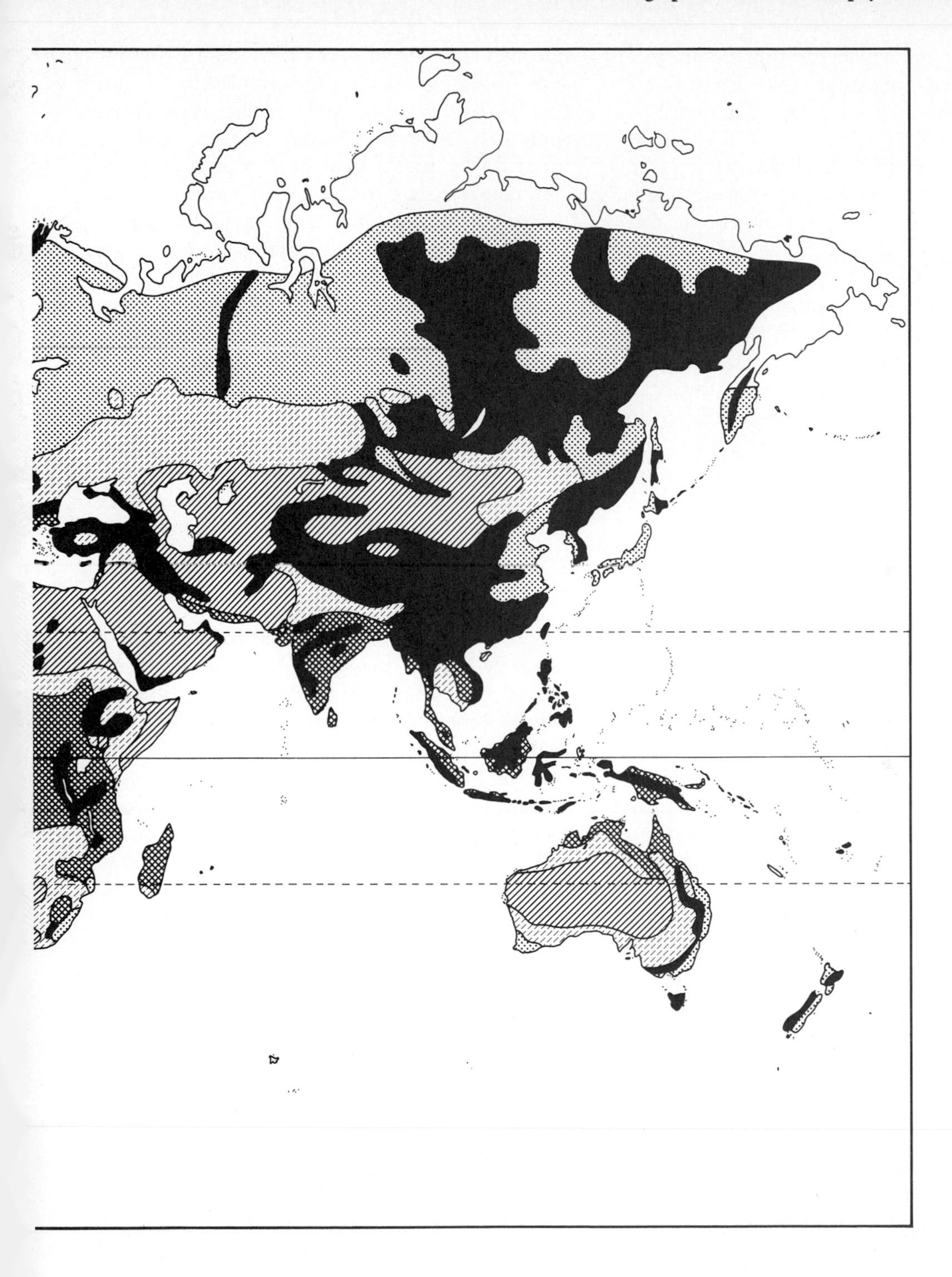

layer lies a clay, in which the iron occurs in a partially reduced condition, thus imparting a bluish gray color. No leaching occurs without runoff.

These are obviously idealized cases; given the gradual changes in climate from one latitude to another, we can correctly assume that pedogenic regimes change along temperature and moisture gradients in complex but predictable ways. A simplified summary of this relationship is illustrated in Figure 2.10, in which the major zonal soil types are outlined and related to climatic regimes. The world distribution of these zonal soil types (Figure 2.11) can be compared with the global climate map (Figure 2.7) to demonstrate the close relationships between soils and climate.

Unusual soil types requiring special plant adaptations. Of course, the chemistry of any soil depends not only on the leaching regime resulting from climate but also on what compounds were present in the original parent material. Soils made from special rock types, such as gypsum, serpentine, or limestone, are characterized by having excessive amounts of some compounds and little of others. For example, serpentine is particularly deficient in calcium, and gypsum has an excess of sulfate. Few plant species can deal with such special soils, and this can be a strong barrier to the occurrence of species in certain areas. A common class of soils that restricts plant growth is the halomorphic soil, rich in chlorides, sulfates, and sodium. Halomorphic soil may occur near the ocean or estuaries and in arid inland basins where excess salts accumulate from evaporation of temporary pools of salty water.

The pH of a soil can determine which ions are available for use by plants. Highly acidic or alkaline soils typically create distinctive conditions, e.g., by causing nitrogen and phosphorus to be bound in inaccessible compounds. Insectivorous plants are a spectacular group that is restricted to highly acidic, organic soils. These plants, which usually live in bogs, obtain their nitrogen and phosphorus by catching living insects, digesting them, and assimilating the nutrients. Also at low pH (acidic conditions) some metals become readily available and are taken up by plants in toxic amounts. A more common feature associated with acidic soils and low nutrient availability is evergreen vegetation (Beadle, 1966). Because nutrients are lost when leaves are dropped and more minerals must then be absorbed to produce new leaves, a possible plant strategy could be to use their limited nutrients more efficiently by retaining their leaves for longer periods. In mesic temperate climates where the predominant vegetation is deciduous forest, it is common to find evergreens growing on acidic and nutrient-poor soils. Examples are the pine barrens of the eastern United States and the sclerophyllous woodlands of Australia (Daubenmire, 1978; Beadle, 1981).

One of the basic properties of most plant species is that each is restricted to soils with a relatively narrow range of pH. A classic example of this is found in calciphilous species, which occur only on basic limestone soils. Recently Musick (1976) has provided some insight into the basis of calciphily. Under alkaline conditions ($pH > 8.5$) phosphorus is in a form mostly inaccessible to plants, so species that occur on such soils are probably adapted to take up as much as they can. However, when the pH shifts to slightly more acidic levels ($pH < 6.8$), phosphorus becomes available in a different ionic form, and the plant absorbs more than it can tolerate. Hydroponic experiments with the creosote bush *(Larrea tridentata)*, a desert calciphile, demonstrate that even these low levels of phosphorus taken up by seedlings in slightly acid conditions cause death from phosphorus toxicity in the leaves.

The structure, texture, and chemistry of a soil also influence how much water can be held and for how long, factors crucial to the survival of plants. Thus a person needing to understand how plant distributions relate to soil type must study this on a local level by trying to isolate the factor or factors that most influence the occurrence of each plant species.

Although we have concentrated here on the relationships between soil and the climate, vegetation, and parent material, readers must appreciate the fact that soils also can affect the distribution of animals, either indirectly by controlling which plant species are present or directly by effects of the chemical and physical environment on their life cycles. Fossorial rodents and many other groups of animals require particular types of soils for burrowing and making structurally sound chambers and tunnels; consequently, they avoid very rocky and heavy soils. Animals living on barren sand dunes have special adaptations to travel on and dig in this environment.

Physical environment of aquatic habitats

As anyone knows who has tried to keep tropical fish, appropriate temperature regimes are essential for their survival and reproduction. Salinity, light, inorganic nutrients, and pressure also play key roles in the distribution of aquatic animals, plants, and microbes. As with the terrestrial climate, the physical characteristics of water often exhibit predictable patterns along geographic gradients that can be understood from a basic background in physics.

Solar radiation and thermal stratification. When solar radiation strikes water, some is reflected but most penetrates the surface and is ultimately absorbed. Although water may appear transparent, it is much denser than air and the absorption of radiation is rapid. Even in exceptionally clear water, 99% of the incident solar radiation is absorbed in the upper 50 to 100 m, and this occurs more rapidly if many organisms or colloidal substances are suspended in the water column. Longer wavelengths of light are absorbed first, and the shorter wavelengths, which have more energy, penetrate farther, giving the depths their characteristic blue color.

This rapid absorption of sunlight by water has two important consequences. First, it means that photosynthesis can only occur in shallow surface waters where the light intensity is sufficiently high. Virtually all primary production that supports the rich life of oceans and lakes can be attributed to plants living in the upper 10 to 30 m. Along shores and in very shallow bodies of water some species, such as kelp, are rooted in the substrate and attain considerable size and structural complexity, but in the open waters that cover much of the globe the primary producers are tiny, often unicellular algae, called phytoplankton, which float in the water column. Second, the rapid absorption of sunlight by water also means that only surface water is heated directly. All heat that reaches deeper water must be transferred by convective movement or by currents. Consequently, extremely deep water may receive very little heat from the surface.

The density of pure water is greatest at 4° C and declines as its temperature rises above or falls below this point. One specific consequence of this unusual density property is that ice floats. This is significant for survival of many temperate and polar organisms because ice provides an insulating layer on the surface that prevents many bodies of water from freezing solid. The presence of salts in the water lowers the freezing point, and some organisms are therefore able to exist in unfrozen water below 0° C (de Vries, 1971).

A more general consequence of the relationship between water density and temperature is that water tends to acquire stable stratification. When solar radiation heats the water surface above 4° C, the warm water is lighter than deeper water and tends to remain on the surface, where it may be heated further and become even less dense. In tropical areas and also in cooler climates during the summer, the oceans and lakes usually have a thin layer of warm water. Unless these bodies of water are shallow, the deep water is very cold (often near 4° C). The rapid transition from cold depths to warm surface water is called a thermocline (Figure 2.12). The mixing of surface water by wave action determines the depth of the thermocline and maintains relatively constant temperatures in the water above the thermocline. In small temperate ponds and lakes that do not experience high winds and heavy waves, the thermocline is often shallow and swimmers can feel it by letting their feet dangle a short distance. In large lakes and in oceans, where there is mixing of surface waters, the thermocline is usually deeper and less abrupt.

Tropical lakes and oceans show pronounced permanent stratification of physical properties, with warm, well-oxygenated, and lighted surface water diminishing to frigid, nearly anaerobic, and dark (aphotic) deep water. Oxygen cannot be replenished at great depths where there are no green plants to produce it, and the stable temperature stratification prevents mixing and reoxygenation with surface water. Only a relatively few organisms can exist in these extreme conditions in deep water, and the lack of vertical circulation limits the availability of nutrients to the phytoplankton in the surface waters (photic zone). Consequently, deep tropical lakes are often relatively unproductive and depend on continued input from streams for the nutrients required to support life.

The situation is somewhat different in temperate and polar waters. The lakes, in particular, undergo dramatic seasonal changes; they develop warm surface temperatures and a pronounced thermocline in summer but freeze over in winter. Twice each year, in spring and fall, these lakes experience a time when the temper-

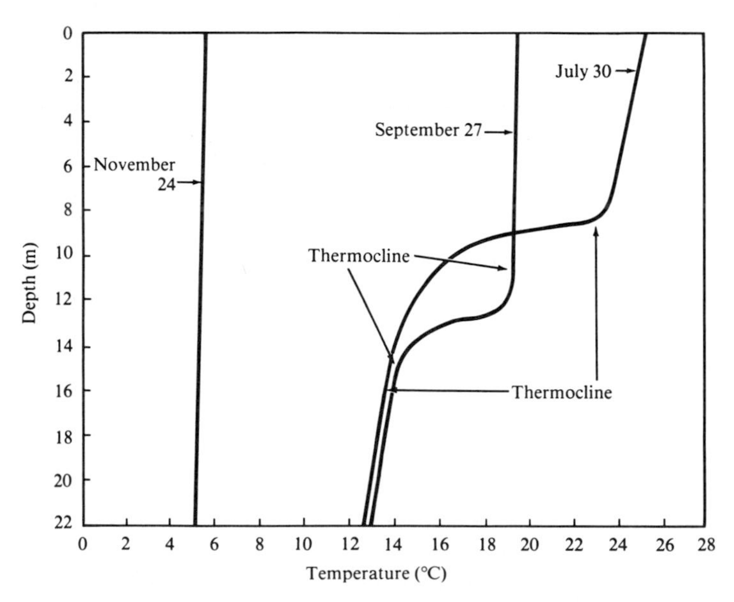

Figure 2.12
Vertical temperature profiles for Lake Mendola, Wisonsin, as it cools from midsummer to late fall. Note the pronounced vertical stratification and sharp thermocline in July. By late November the lake has cooled to virtually uniform temperature, permitting vertical circulation or fall overturn. (After Birge and Juday, 1911.)

ature-density stratification is eliminated, the entire water column attains equal temperature and equal density, and moderate winds may then generate waves that mix deep and shallow water, producing what is called overturn (Figure 2.12). Temperate lakes, such as the Great Lakes of North America, are often highly productive and support abundant plant and animal life, including valuable commercial fishes, because overturn occurs. Twice each year the mixing carries oxygen downward and inorganic nutrients upward. The latter stimulates growth of phytoplankton, which can be retarded by rapid depletion of phosphorus and other minerals during the summer when high temperatures allow algae to grow and reproduce at high rates. Overturn thus permits the minerals in feces and dead bodies of organisms, which sink to the bottom and are decomposed, to be recycled.

Circulation of oceans. The vertical and horizontal circulation of oceans is more complicated than that of lakes, in part because oceans are so vast, extending through many climatic zones, and in part because salinity affects the density of water. Salts are dissolved solids, carried into oceans and saline lakes by streams and concentrated by evaporation over millions of years. The presence of salts in water increases its density. Consequently, varying salinity has an important effect on circulation. Rivers and precipitation continually supply fresh water to the surface of oceans, and this light water tends to remain at the surface. If you have ever flown over the mouth of a large muddy river, such as the Mississippi, you may have noticed that fresh water remains relatively intact, flowing over the denser ocean water for many kilometers out to sea. In general, input of fresh water to the ocean from rivers and precipitation exceeds losses from evaporation in polar regions, but the reverse is true in the tropics. This creates a somewhat confusing situation, because warm tropical surface water tends to become concentrated by evaporation and increase in density, counteracting to some extent the density stratification owing to temperature. On the other hand, cold polar water, which would be expected to show little stratification, may be somewhat stabilized as low-density fresh water accumulates on the surface.

Vertical circulation occurs in oceans, but rates of water movement are so slow that a water mass may take hundreds or even thousands of years to travel from the surface to the bottom and back again. Areas of descending water masses tend to occur at the convergence of warm and cold currents in polar regions, where the colder, denser water sinks under warmer, lighter water. Areas of rising water, called upwellings, are found where ocean currents pass along the steep margins of continents. This happens, for example, along the western coast of North and South America, where there is little continental shelf and the land drops sharply right offshore. As the Pacific gyres sweep toward the equator along these shores, the Coriolis effect, and in tropical latitudes the easterly trade winds, tend to deflect the surface water offshore and water wells up from the deep to replace it. Because upwelling, like overturn in lakes, recycles nutrients to the surface, upwelling areas support highly productive ecosystems. Probably the greatest commercial fishery in the world is located off the upwelling coastal waters of Chile, Peru, and Ecuador.

Surface currents, such as the great gyres of oceanic water, (Figure 2-4), are relatively shallow and rapidly moving, so they tend to delineate discrete water masses, each of which has characteristic salinity and temperature profiles distinct from neighboring water masses. Some organisms with limited capacity for active dispersal may drift in currents for long distances without leaving a single water mass or ocean, whereas others, which may be able to move actively to overcome the currents, must also be able to tolerate the contrasting physical environments in different water masses.

Although oceanographers have recognized the existence of distinct water masses within oceans for many years, the extent of spatial heterogeneity in shallow ocean waters is just be-

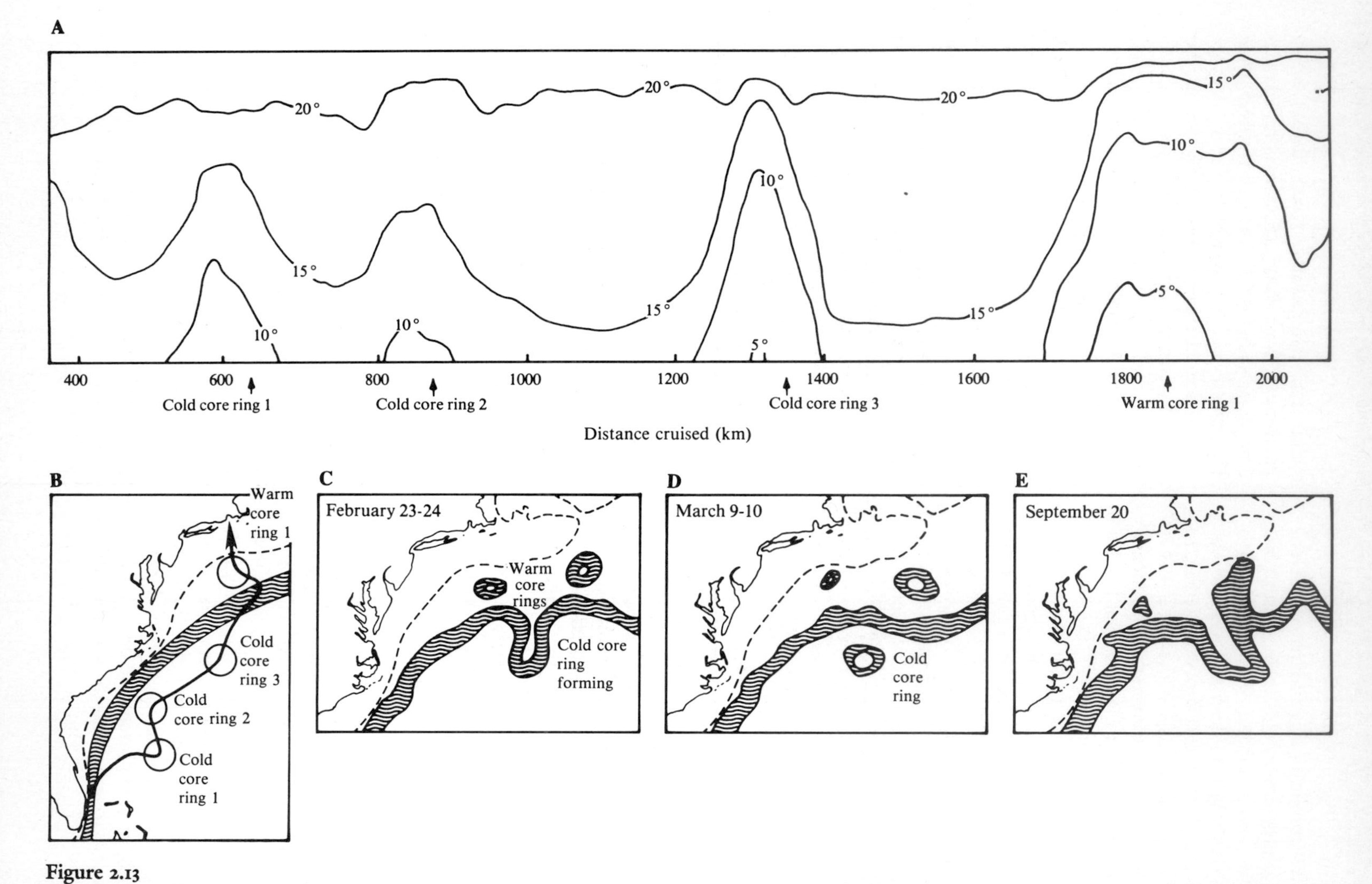

Figure 2.13
Pronounced spatial heterogeneity in the surface waters of the North Atlantic Ocean is caused by small water masses, called rings, that have split off from the Gulf Stream. **A,** Temperature-depth profile measured by an oceanographic cruise that passed through several rings. **B,** Path of the cruise. **C-E,** Changes in water surface temperatures, showing the formation and movement of rings, as mapped from infrared images taken by satellite during 1977. (Simplified from Wiebe, 1982.)

ginning to be appreciated. Recently investigators from the Woods Hole Oceanographic Institute have been studying the physical environment and the biota of Gulf Stream rings (Wiebe, 1976, 1982; Lai and Richardson, 1977). These are small masses of cold or warm water that have broken away from the southern or northern edges of the Gulf Stream, respectively, and are drifting through water of contrasting temperature in the North Atlantic (Figure 2.13). These rings not only have physical environments that are strikingly different from their surroundings, but they also contain a unique biota that can persist in these special conditions far from their normal distribution in the Gulf Stream. The possible roles of the floating warm- or cold-water eddies on trans-Atlantic dispersal, both now and in the past, are intriguing.

Pressure and salinity. Pressure and salinity vary greatly among aquatic habitats and have major effects on the distributions of organisms, because special physiological adaptations are necessary to tolerate the extremes. As every skindiver knows, water pressure increases rapidly with depth. It becomes a major problem in the ocean where the deepest areas are 2000 to 6000 m below the surface. Pressure increases at a rate of about one atmosphere (about 1.5 mega-Pascals) for every 10 m of depth. Even at the top of the abyss, pressures are 200 times greater than at the surface. Organisms adapted to withstand the pressure of the deep sea cannot survive in the surface waters and vice versa. Variation in salinity is relatively discontinuous. The vast majority of the earth's water is in the oceans and highly saline (greater than 34 parts per thousand), but in contrast, freshwater lakes and rivers contain very few dissolved salts. Consequently, most aquatic organisms are adapted either to fresh water, where the physiological problem is to obtain sufficient salts to maintain osmotic balance, or to salt water, where the osmotic problem, if any, is eliminating excess salt. Very few organisms have the physiological mechanisms that enable them to survive in both environments. Most organisms are restricted to either salt or fresh water, and habitats where the two meet and mix, such as estuaries, are inhabited by only a few widely tolerant (euryhaline) species.

Microclimates

It would be misleading to end here without a word of qualification. In this brief treatment we have tried to describe and explain the most important large-scale variations in global abiotic patterns that influence the distribution of organisms. At closer inspection these patterns may tell us surprisingly little about the actual conditions experienced by an organism living in a particular region. This point was best stated by the observer who pointed out that the climatic data taken by the U.S. Weather Bureau probably measures accurately only the climate experienced by the spiders living in the shelters that house the recording instruments. This is not so much of a problem in aquatic habitats because the physical properties of water (including its high specific heat) tend to prevent the occurrence of abrupt small-scale changes. But the existence of thermoclines and of rapid changes in salinity when rivers enter the ocean indicate that this generalization too has exceptions.

In terrestrial habitats, the climates of small places, called microclimates, may bear little relationship to the large-scale patterns. On the one hand, two organisms living only a few centimeters apart may live in radically different physical environments, humid or arid, hot or cold, windy or protected, and so forth. On the other hand, by selecting appropriate microenvironments an organism may be distributed over a wide range of latitudes and elevations and still experience virtually identical physical conditions. Examples of both situations abound. Lizards are conspicuous elements of most desert faunas because they are active during the day and able to tolerate the hot, dry conditions. These same deserts, however, may be inhabited by frogs and toads, which spend most of their lives buried in the cool, relatively moist soil, emerge to feed only on rainy or humid nights, and possess adaptations for breeding in ephemeral ponds that form after occasional heavy rains. Perhaps the best examples of organisms that live in similar physical environments over wide geographic ranges are internal parasites and microbial symbionts of birds and mammals. The same species may occur in tropical rain forest and arctic tundra but still live in virtually identical, homeostatically regulated environments within the bodies of their hosts.

The most distinctive microenvironments typically are small and widely dispersed sites. In a local area the capacity of organisms to exploit specific microclimates depends largely on mobility or vagility (regardless of whether they are actively or passively transported), body size, special physiological properties, and behavioral selectivity. We can readily imagine how mobile animals can behaviorally select a particular habitat, but we should keep in mind that some plants also can control their destinies somewhat by dispersing as seeds that germinate only when a narrow range of conditions are present and by facultative changes such as forming different types of leaves in different habitats (e.g., sun and shade leaves) that presumably optimize performance under different conditions.

Isolated localities and microclimates offer greater challenges for colonization. Plant species usually accomplish such jumps as seeds, fruits, or spores that may be adapted to be carried long distances and to tolerate extreme environments; however, their arrival is largely a result of chance. In contrast, many animals are able to use their sophisticated sensory and locomotor systems to seek out isolated microenvironments. To demonstrate this it is only necessary to create a small artificial pond and record the rapidity with which it is colonized by aquatic insects. Some microhabitats probably do not receive many immigrants and, in fact, the inhabitants arrived there when the climatic or edaphic condition was much more widespread. Examples of this would be strings of lakes that were once interconnected and moist montane plant communities that are now separated by dry intervening vegetation but were once lower and continuous. Still other microenvironments are so inaccessible that most residents are unique forms that have evolved in situ from neighboring communities, such as cave animals in limestone regions or plants in isolated pockets of serpentine soils.

Organisms can select microclimates that are appropriate for their life-styles, or they can modify the conditions to provide their preferred distinctive microclimate. Natural shelters used by plants or those built by animals often have an ameliorating, tempering effect on terrestrial microclimates by eliminating lethal temperatures. In deserts, woodrats and packrats (*Neotoma* spp.) construct large houses, piles of stems, stones, and other debris that may measure 1 m high and 2 or more m in diameter. As shown in Figure 2.14, these provide relatively stable temperatures and high humidities (J. Brown, 1968), and they are used for shelter, not only by their woodrat builders, but also by a large number of commensal invertebrates and vertebrates, including spiders, scorpions, insects, lizards, snakes, and mice. The animals that use subterranean burrows or caves have constant and very mild microclimates year-round because the sun's heat is not conducted more than a few decimeters through the soil.

The fact that many organisms are found only in particular microenvironments has important consequences for understanding geographic patterns. On the one hand, it means that some species may have much broader geographic ranges than we would have predicted from a cursory comparison of their physical tolerances and zonal climatic patterns. On the other hand, careful studies show that local distributions of many taxa may be patchy, because they are confined to unusual microenvironments. Thus comparisons between geographic ranges and macroclimatic parameters may provide little information or spurious correlations, and we should be particularly cautious about inferring causal relationships using only regional climatic data.

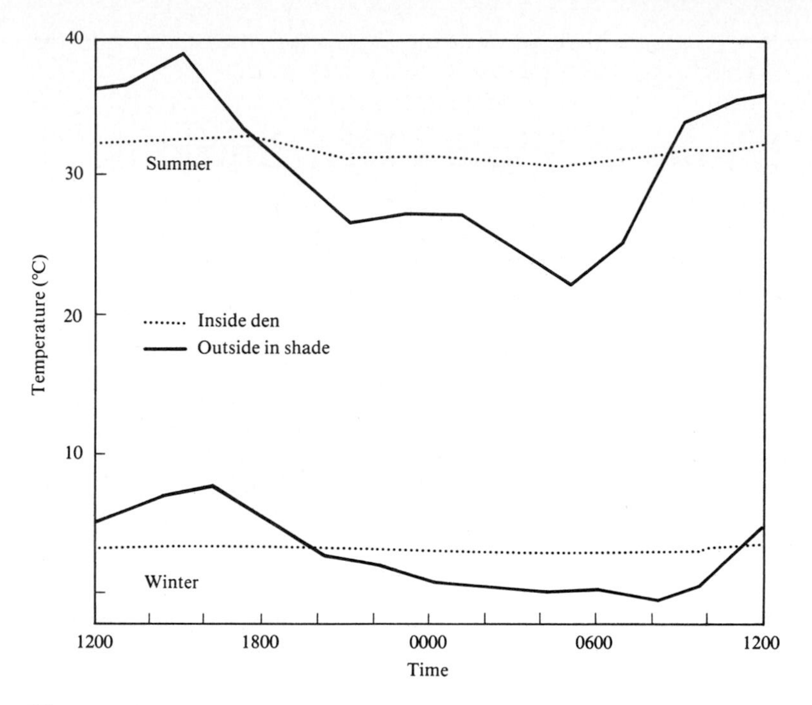

Figure 2.14
Temperatures inside and outside the den of a bushy-tailed woodrat *(Neotoma cinerea)* in the high desert of southeastern Utah during midsummer and midwinter. Note that the microclimate of the den, where the animal spends most of the time, varies much less in temperature than the macroclimate outside. This is particularly important to the rat during the daytime in summer, because outside temperatures frequently exceed 40° C, the upper lethal temperature. (After J.H. Brown, 1968.)

Selected references

Climate

Amiran and Wilson (1973); Barry and Chorley (1970); MacArthur and Connell (1966); Mather (1974); Navarra (1979); Sellers (1965); Strahler (1975); Strahler and Strahler (1978); Trewartha et al. (1976); U.S. Department of Agriculture (1941).

Soils

Antonovics et al. (1971); Beadle (1966); Brady (1974); Docters van Leeuwen (1936); Eyre (1968); Forman (1979); Grime and Hodgson (1969); Jenny (1979); Pears (1978); J. Proctor and Woodell (1975); Soil Conservation Service (1975); Tivy (1971); Walter (1979); R. Webster (1977).

Physical environment of aquatic habitats

Dykyjova and Kvet (1978); Hardy (1971); G. Hutchinson (1957); Kinne (1970-1972); Lauff (1967); Macan (1973); Menzies et al. (1973); Morris (1980); Perkins (1974); Pichard (1975); Popham (1961); Tait (1981); J. Valentine (1973); Vermeij (1978); Wetzel (1975); Weyl (1970); Wiebe (1982).

Microclimates

Bannister (1976); G. Bartholomew (1958); J. Brown (1968); Fritts (1976); Gates (1962, 1970); Geiger (1957); Gordon et al. (1981); Larcher (1980); Monteith (1973); Nobel (1974); Rosenberg (1974).

Chapter 3 The Limits of Species Distributions

The geographic distributions of organisms are not static, but dynamic. Over time the ranges of species shift, expand, and contract. The distributions of higher taxonomic groups such as genera, families, orders, and classes also change because their geographic ranges simply reflect the cumulative distributions of all the included species. These changes are the net result of two opposing processes: colonization, the expansion of populations into new areas, and extinction, the elimination of populations from all or part of their former range. These are ecological processes because they ultimately depend on how environmental conditions cause local populations to increase or decrease.

Biogeographers, ecologists, and evolutionists are concerned with patterns and processes that occur over a wide variety of temporal and spatial scales. For convenience the temporal scale is often divided into ecological and evolutionary time. By ecological time we usually mean periods from milliseconds to decades, in which populations may interact with their environment and respond to environmental fluctuations without undergoing evolutionary modification. By evolutionary time we normally mean periods of tens to millions of years, during which populations can evolve and become adapted to their environments by means of genetic changes. This dichotomy is artificial, because some evolutionary changes, such as the development of antibiotic resistance in bacteria, can be extremely rapid whereas certain ecological events, such as soil formation, may take a long time. Nevertheless, we have different perceptions about ecological processes, which we can study by direct observation and experimentation, and evolutionary mechanisms, which must be investigated indirectly by comparative studies of fossil and living organisms of varying degrees of relatedness. Ecologists are also more likely than evolutionists to investigate the small end of the spatial scale and to be concerned with the fate and characteristics of local populations rather than geographically widespread species and higher taxonomic groups. Because both ecological and evolutionary processes influence the distributions of organisms, biogeographers must be able to integrate phenomena that occur at all spatial and temporal scales. In the last analysis, the ultimate mechanisms controlling distributions are ecological: the influence of the environment on the birth, death, and movement of individuals.

Biogeography is an interesting and challenging science because the distribution of organisms is neither uniform nor random. Each species has a unique geographic range that reflects both its current ecological niche and its past history. Because related species share common ancestors from which they have usually inherited similar biological constraints and ecological requirements, higher taxonomic groups also tend to be confined to limited geographic areas that reflect their common heritage. But the fundamental problem is why each species inhabits only a small portion of the earth's surface. For many organisms this question can be answered at two levels. On the one hand, limited capacity for dispersal may prevent a population from colonizing distant areas where conditions are otherwise suitable for its establishment. We discuss such long-distance dispersal and colonization in Chapter 7. On the other hand, the local limits of a species range are determined by immediate ecological factors that prevent local populations from expanding into adjacent areas to which they have ready access. In some cases the explanations for local limits of species distributions are obvious. For example, island populations of terrestrial plants and animals may be entirely bounded by water, so that they could extend their range only by long-distance dispersal across water barriers or by evolving to become aquatic (which might theoretically occur but probably would take several million years). For the majority of species, however, the limiting factors that determine the boundaries of their geographic ranges are not so obvious. At some point in an apparently gradual gradient of temperature, moisture, salinity, vegetation density, or some other environmental factor or factors the species simply no longer occurs. Whether the limit of the range is abrupt or marked by a gradual decline in the abundance of individuals, its cause poses a challenging problem for the biogeographer and ecologist.

The ecological niche

For a species to maintain its population, its individuals must survive and reproduce. Certain combinations of environmental conditions are necessary for individuals of each species to tolerate the physical environment, obtain energy and nutrients, and avoid their enemies. The total requirements of a species for all resources and physical conditions determine where it can live and how abundant it can be at any place within its range. These requirements are termed the ecological niche. The term is usually applied to species, but we can also speak of the niche of a lower or higher taxonomic unit, such as a local population or a genus or family. The term *adaptive zone* is sometimes applied to the niche of a higher taxonomic category.

The concept of ecological niche was originally developed by Grinnell and Elton in the 1920s, but it has been refined and modified by subsequent workers (see Vandermeer, 1972; Whittaker and Levin, 1975; Hurlbert, 1981); especially by Hutchinson (1958). Hutchinson suggested that the niche could be viewed as a multidimensional hypervolume, i.e., an imaginary space of many dimensions, in which each dimension or axis represents the range of some environmental condition or resource that is required by the species (Figure 3.1). The niche of a plant, for example, might include the range of

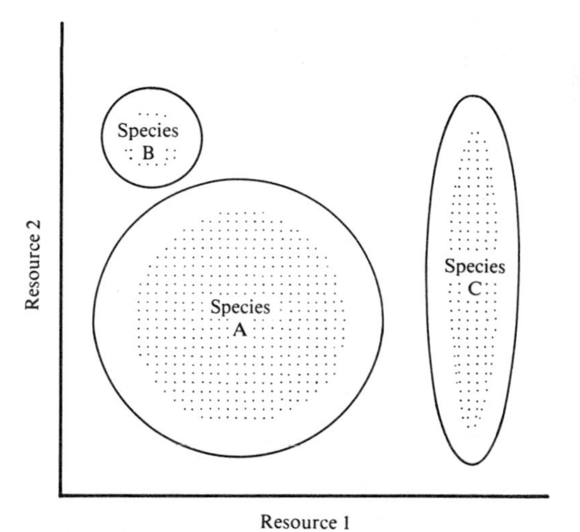

Figure 3.1
Diagrams of two dimensions of the niches of three hypothetical species. Circles indicate the extreme ranges of the two resources used by each species, and stippled areas show that the central parts of the niche spaces provide the most favorable environment. Species *A* is a generalist, using wide ranges of both resource types; species *B* is highly specialized in its use of both resources; and species *C* is a specialist in its use of resource 1, but a generalist with respect to resource 2. Although the real niches of species would have to be characterized by many more dimensions, there would still be much variation among species in the degree of specialization.

temperatures that it can tolerate, the soil water potentials at which its roots can survive and take up water, the intensity of sunlight required for photosynthesis, the concentrations of each of the essential nutrients between minimal and toxic levels, the densities of each of its herbivores and pollinators, and so on. Obviously no one has measured the entire niche for even a single species, but it is a useful concept because it indicates the large number of factors that must be considered to understand what determines distribution and abundance. The niche concept is valuable for biogeographers because it defines the kinds of places where each kind of organism can live.

A useful extension of the niche concept for biogeographers is the distinction between fundamental and realized niches (Hutchinson, 1958; Miller, 1967). The fundamental niche of a species includes the total range of physical environmental conditions that are suitable for its existence. The realized niche describes that part of the fundamental niche actually occupied by the species. Often the realized niche will be smaller in certain dimensions than the fundamental niche, because of the activities of other organisms, which are predators, competitors for resources, or mutualists that provide essential services (Figure 3.2). For example, we know that many cacti and other xerophytic plants can live in much wetter climates than the deserts to which they are naturally restricted. Many species from the deserts of the southwestern United States readily grow and reproduce in ornamental gardens in coastal southern California, many kilometers from the nearest natural populations of the species. These populations only persist, however, if competing plants are weeded out, if predators and diseases are controlled, and (sometimes) if the plants are artificially pollinated. If the gardens are abandoned, the desert species are soon replaced by native species characteristic of the wetter climate, indicating that these coastal regions lie within the fundamental niche of the xerophytic plants but not within the realized one. As we will see shortly, there are many cases of a species confined to only a small portion of its realized niche by one or a small number of identifiable species of competitors or predators. Often two closely related species have very similar fundamental niches, but they occupy different realized niches because competition has resulted in adaptations to use different resources. Sometimes one of these species is a specialist and is obligately restricted to an included niche, a narrow range of conditions (such as areas of high resource availability) where it is competitively

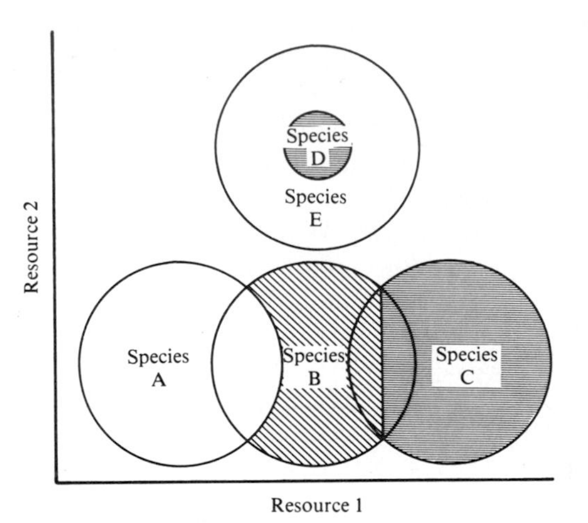

Figure 3.2
Diagrams of two dimensions of the niches of five hypothetical species, showing fundamental, realized, and included niches. Circles indicate the limits of the fundamental niches of the species, whereas shaded portions show the realized niches. For species *A* the fundamental and realized niches are identical, but the realized niches of species *B* and *C* include only part of their fundamental niches because they are excluded from the remaining niche space by other species. Species *D,* a specialist, occupies a niche included entirely within the fundamental niche of the more generalized species *E.*

superior, within the larger fundamental niche of its competitor (Figure 3.2).

There is much debate among ecologists as to whether there are unfilled niches. This may seem like an absurd elaboration of an already excessively abstract concept, but this idea has considerable importance to biogeography. Although we can recognize niches unequivocally only when they are occupied by organisms, it is clear that there must be unfilled niches in the sense that there are opportunities for organisms to live in places where they do not presently occur. If, heaven forbid, we tried to introduce many more mammals to Australia, New Zealand, and the Hawaiian Islands, undoubtedly some of them would be successful. If so, they would fill some sort of a vacant niche. Of course they might have a more devastating impact on the native biota and on human agricultural interests than rabbits in Australia and rats introduced in New Zealand and Hawaii have already had. Such ecological experiments (which we certainly do not recommend), like the already successful introductions documented by Elton (1958) and Udvardy (1969), clearly indicate that all environments are not absolutely filled with all the species they can hold. The distributions of contemporary species can be limited in part by their inability to disperse to regions where they otherwise could exist.

These vacant niches also could potentially be filled by the evolution of new species. Obviously this sometimes happens. The phenomenon of evolutionary convergence, discussed several places in this book but especially in Chapter 18, is evidence that distantly related organisms evolve to fill similar niches in similar environments in geographically isolated regions of the world. Niches, like the species that occupy them, are dynamic. We recognize the appearance of new niches unequivocally when existing species colonize or new species evolve to occupy them. Niches must also disappear as the environment changes, certain ways of life become untenable, and species become extinct.

The relationship between abundance and distribution

Some authors (notably Andrewartha and Birch, 1954; but see also Krebs, 1978) consider the abundance and distribution of organisms to be the central issue of ecology. In this view the basic units of ecological structure and function are populations. A population is arbitrarily defined; it is comprised of all the individuals of a single species that occur in a particular place. Thus we can speak of the local population of gray squirrels *(Sciurus carolinensis)* living in Central Park in New York City or a woodlot in eastern Michigan or of the entire species population that is distributed over much of eastern North America and has been introduced successfully to several urban areas in the western United States and to Great Britain and continental Europe. The abundance and distribution of entire species populations are closely interrelated because both depend on the dynamics of local, more or less panmictic (freely interbreeding) populations.

A local population increases when the combined rates of birth and immigration exceed the combined rates of death and emigration. We can express this mathematically as

$$r = b + i - d - e$$

where r is the per capita rate of population growth (if r is positive, the population increases; if r is negative, it decreases), b and d are the per capita rates of birth and death, respectively, and i and e are the respective per capita rates of immigration from and emigration to other populations.

Fluctuations in abundance and distribution. Over reasonably long periods of ecological time, the growth rate of most populations averages close to zero and neither abundance nor distribution changes greatly. This relative constancy in the long term may occur despite dramatic short-term fluctuations. A spectacular example is the local abundance of several species of voles and lemmings (mouselike rodents, subfamily Microtinae), which fluctuates on a 3- to 4-year cycle through several orders of magnitude at many local areas in northern North America, Europe, and Asia (for references see Finerty, 1980). Because voles have a limited capacity for dispersal, their large-scale geographic ranges do not vary significantly during these cyclic fluctuations in abundance. In contrast, some northern birds show modest year to year variations in abundance but great changes

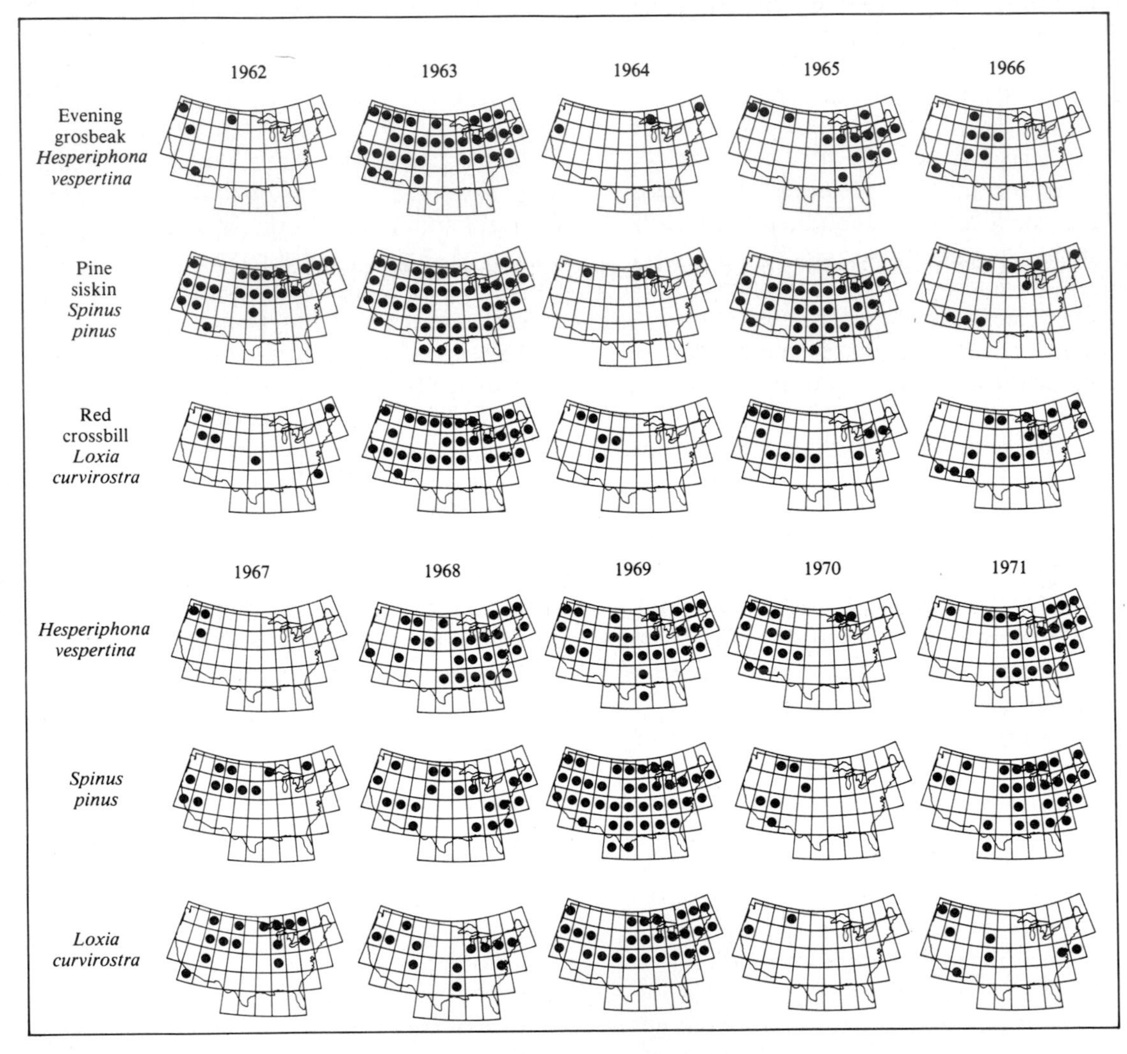

Figure 3.3
Year-to-year variation in the winter ranges of three species of seed-eating birds that breed in northern North America. Dots show large wintering populations in local regions. Note that in certain years all species tend to shift their winter ranges far to the south. Apparently these are times of low food availability on the normal wintering grounds. (From Synchronous eruptions of boreal seed-eating birds, by C.E. Bock and L.W. Lepthian, by permission of the University of Chicago Press. Copyright © 1976, The University of Chicago.)

in their winter ranges, which may extend hundreds of kilometers to the south in years of low food availability on their normal wintering grounds (Bock and Lepthian, 1976) (Figure 3.3). Similarly, some insects show great fluctuations in distribution, which in this case are correlated with huge variations in abundance. The migratory locusts of the Old World provide the most dramatic examples (Waloff, 1966; Albrecht, 1967; White, 1976). These grasshoppers persist in limited regions, called outbreak areas, where conditions are suitable for continued survival and reproduction. During periods when weather and food are particularly favorable, these populations increase fantastically, change their morphology and behavior, aggregate into huge swarms, and migrate outward from the outbreak areas to forage over an enormous region. The African migratory locust *(Locusta migratoria)* and the red locust *(Nomadacris septemfasciata)* have each plagued two or three times in the last century, invading most of southern Africa, an area of millions of square kilometers and more than 1000 times the size of the outbreak area from which they originated (Figure 3.4).

Historical changes in distributions. In the examples just described, abundance and distribution fluctuate dramatically over short periods of ecological time, but over longer periods the average density and geographic range remain relatively unchanged. This happens especially in species that are nomadic or migratory. However, the abundance and distribution of no species remains constant over long periods of evolutionary time. Some species temporarily increase in abundance and expand their geographic ranges, and eventually all species decrease and ultimately become extinct. Distributions also shift in response to long-term

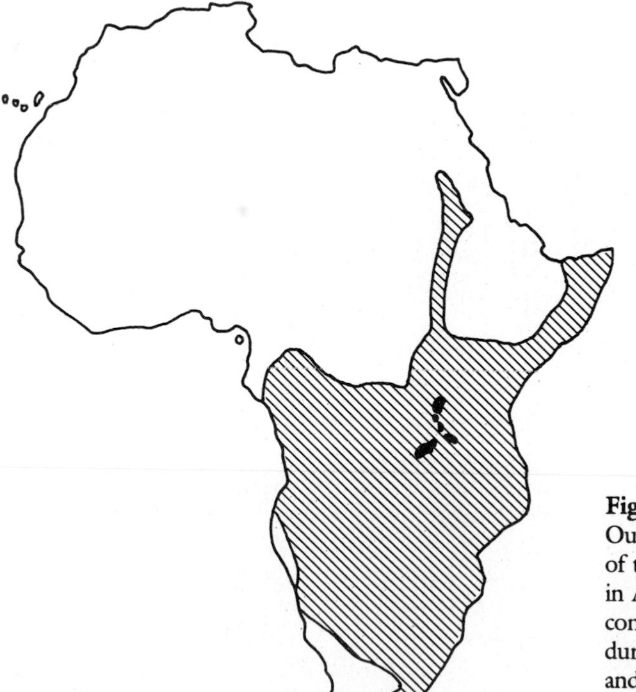

Figure 3.4
Outbreak *(black)* and invasion *(shaded)* areas of the red locust *(Nomadacris septemfasciata)* in Africa. Most of the time this species is confined to the small outbreak regions, but during plague years its populations expand and invade a vastly greater area. (After Albrecht, 1967.)

Figure 3.5
Photographs taken decades apart, showing vegetation change in southwestern North America. **A,** Vicinity of the small mining settlement of Rosemont, Arizona, between 1896 and 1897. **B,** The same area in 1982. Note that large live oak trees *(Quercus emoryi)* still occur in the valley bottom in the foreground, but that shrubby vegetation, especially junipers *(Juniperus)* and mesquite *(Prosopis)* have invaded both the valley and the formerly grassy hills in the background. Although the exact causes of these changes are hard to document, livestock grazing and resulting soil erosion have played an important part. (**A** Courtesy Special Collections, University of Arizona Library; **B** courtesy R.M. Turner; **C** and **D** by permission from The Changing Mile by J.R. Hastings and R.M. Turner. Copyright © 1965 by the University of Arizona Press, Tuscon.)

Figure 3.5, cont'd
C, Desert vegetation near Puerto Libertad, Sonora, Mexico, in 1932. **D,** The same area in 1965, showing almost no change. Note that the same individual shrubs and succulents are present, indicating the longevity of such species as ocotillo *(Fouquieria splendens)*, the tall, spindly plant in the left foreground, and cardón *(Pachycereus pringlei)*, the large columnar cactus in the center.

climatic and geologic changes. For many species these processes have been accelerated by human activities, so that major changes in distribution and abundance have occurred within recorded history. Much native vegetation has been completely eliminated by large-scale agriculture, mining, and other activities, but other less blatantly destructive events, such as grazing by domestic livestock, have also had major effects. Several authors (Shantz and Turner, 1958; Phillips, 1963; Hastings and Turner, 1965) have used photographs taken decades apart to document such vegetational changes (Figure 3.5). By preying on game animals, trying to eradicate pest species, destroying natural habitats, and polluting environments with toxic substances, humans have directly caused the extinction of thousands of animal species and drastically reduced the geographic ranges of many others. For example, the North American bison *(Bison bison)* once ranged over the entire central part of the continent, an area of at least 8 million square kilometers. The population in the 1700s was estimated at about 100 million but by the early 1890s hunting with rifles had reduced the vast herds to less than 1000 individuals in what is now Yellowstone National Park and a few private preserves. When finally protected by law from human predation, the bison population increased rapidly, and herds have been reintroduced into isolated parks and preserves in many parts of their former range (Rorabacker, 1970; Roe, 1970; Dary, 1974).

Other species have not been so fortunate. Many native fish species in the arid regions of the southwestern United States have become extinct and others have suffered drastic reductions in distribution as their marsh, spring, and stream habitats have been drained to obtain water for irrigation (Miller, 1961). Native brook trout *(Salvelinus fontinalis)* populations have disappeared from many parts of their original range in the Adirondack Mountains and elsewhere in the eastern United States, because acid rain, caused by atmospheric pollution, has acidified lakes to intolerable levels (Likens and Bormann, 1974).

On the other hand, human activities have led to increased abundance and distribution of some species. As we have modified the landscape to build our civilization and to supply it with food and water, we have created new habitats and food supplies for organisms. Some commensal animals such as the house mouse *(Mus musculus)*, house sparrow *(Passer domesticus)*, and housefly *(Musca domestica)* were adapted to live in close association with Europeans and accompanied them as they colonized new habitats and geographic areas around the world. Ships unintentionally transported a long list of species, including many serious pests, to new continents. The Mediterranean fruit fly *(Ceratitis capitata)*, which is capable of devastating fruit crops in warm regions, and tumbleweed *(Salsola iberica)*, perhaps the most common weed in the southwestern United States, were both introduced accidentally into North America from the Old World. The spread of some introduced species, such as that of the European starling *(Sturnus vulgaris)* in North America, has been well documented (Kessel, 1953) (Figure 3.6).

Perhaps more interesting examples are provided by native species that have benefitted from human activity and increased their geographic ranges. Over the last century several bird species, including the cardinal *(Richmondena cardinalis)*, tufted titmouse *(Parus bicolor)*, and mockingbird *(Mimus polyglottos)*, have expanded their geographic ranges northward from the southeastern United States to colonize Pennsylvania, New York, and in some cases even the New England states and southeastern Canada (e.g., Boyd and Nunneley, 1964). A well-documented history of the similar northward movement of the opossum *(Didelphis virginiana)*, a marsupial mammal, indicates that it apparently invaded southern Ontario in several waves, the first occurring about 1900 (Peterson and Downing, 1956). Two factors appear to have been important in causing these range expansions: (1) clearing of the extensive deciduous forests created abundant open and shrubby habitats around farms and suburbs,

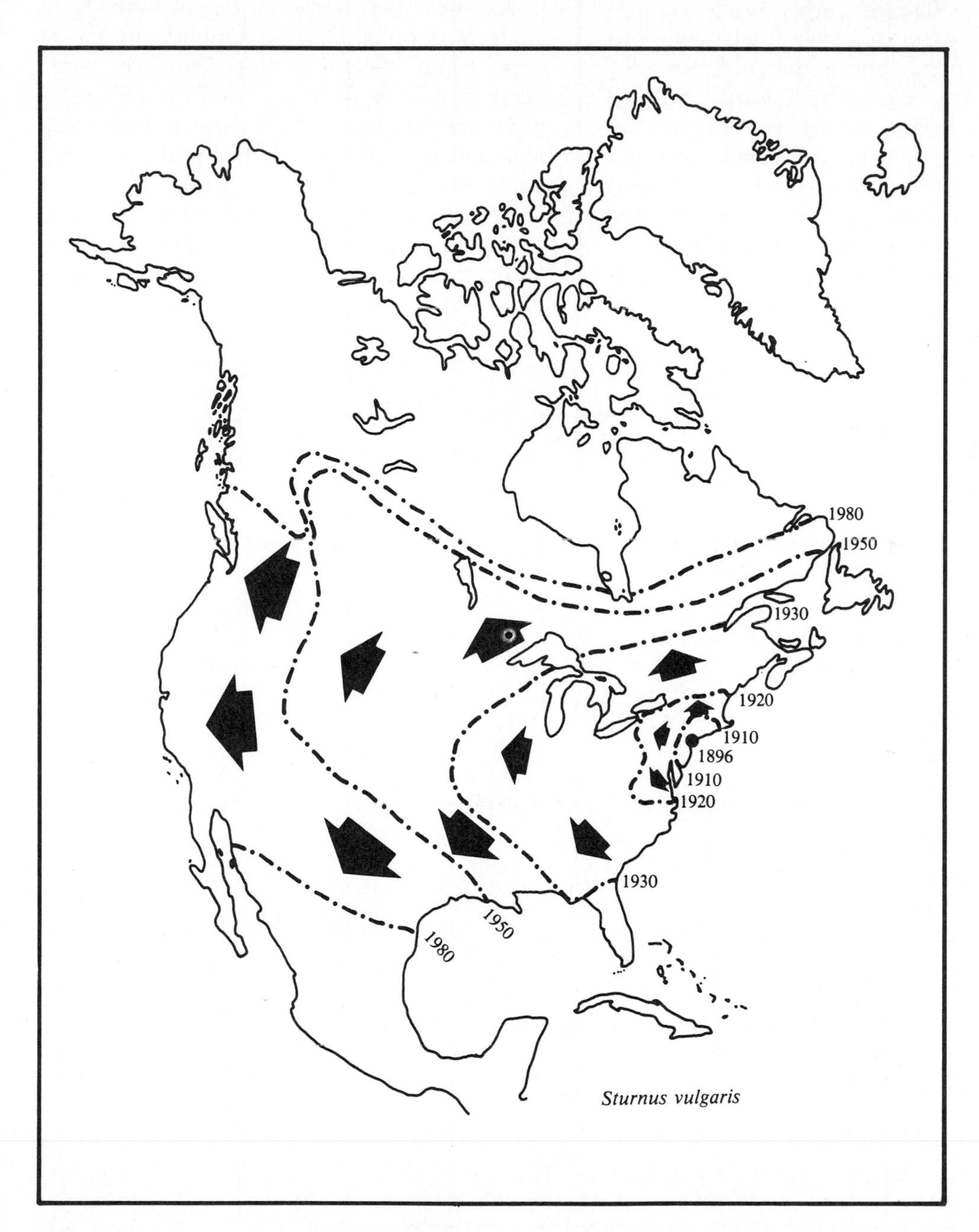

Figure 3.6
Rapid expansion of the geographic range of the European starling *(Sturnus vulgaris)* following its successful introduction to North America. (Map compiled from various sources.)

and (2) feeding stations, garbage cans, and garbage dumps provided reliable food supplies to support these nonmigratory species through the northern winters. However, Kalela (1949) suggested that climatic warming was primarily responsible for the similar northward expansion of several European bird species. Grassland bird and mammal species, such as the eastern meadowlark *(Sturnella magna)* and the prairie deermouse *(Peromyscus maniculatus bairdi)*, have expanded their ranges eastward to establish large populations on cleared agricultural lands in the northeastern United States (Hooper, 1942; Bent, 1965). The carnivorous coyote *(Canis latrans)* has also extended its range eastward, invading along a broad front from the southeastern United States to eastern Canada.

Dynamics of species boundaries. Since by definition the limit of the geographic range of a species occurs at the point in space where the

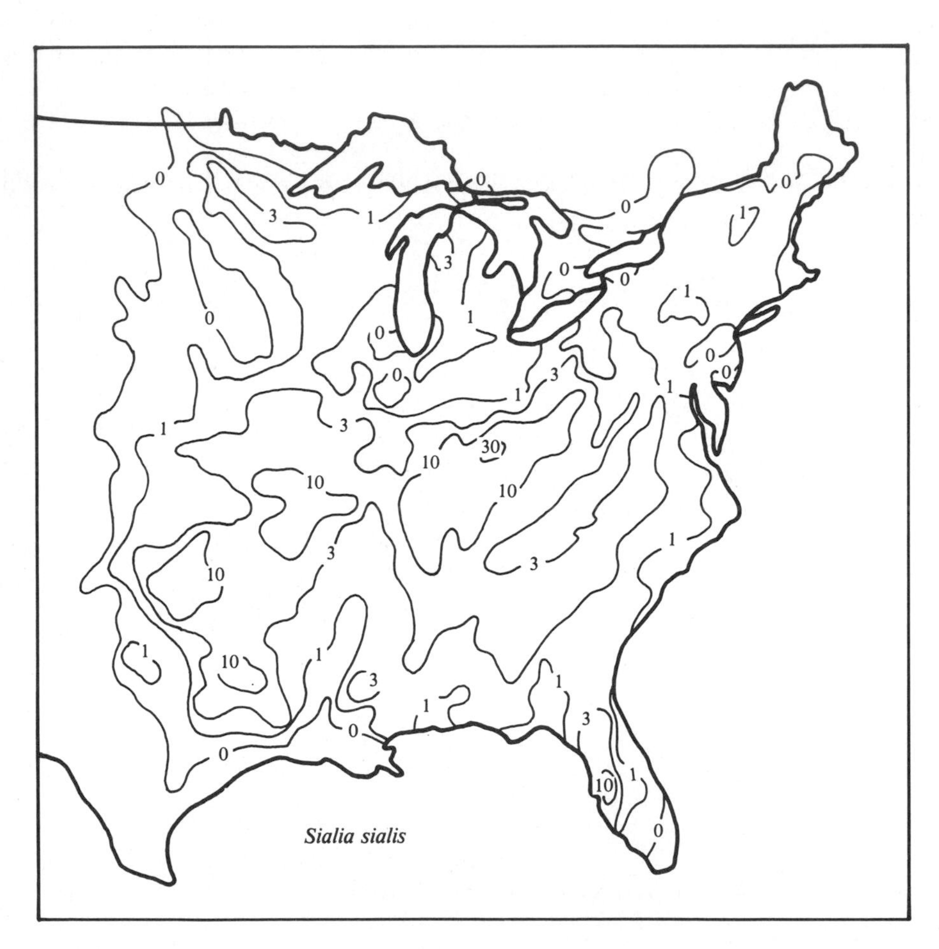

Figure 3.7
Variation in population density of the eastern bluebird *(Sialia sialis)* over its geographic range. Note that the most favorable regions are patchy, but they tend to occur toward the center of the range. (Redrawn from Bystrak, 1979.)

local population density declines to zero, changes in geographic distribution inevitably involve fluctuations in the sizes and distribution of peripheral populations. The great naturalist Joseph Grinnell (1922) suggested that many peripheral populations have death rates that exceed birth rates and are sustained by a continual influx of immigrants from central populations that produce a net excess of individuals. This may be true for the highly mobile birds that Grinnell studied. Recent maps of the local densities of bird populations over entire species ranges clearly show that abundance is usually greatest in central regions and declines toward the periphery of the range (Figure 3.7). This suggests that central areas tend to offer the most favorable environments and hence supports Grinnell's idea, although it does not necessarily imply centrifugal migration.

Even in less vagile organisms than birds, fluctuations in the geographic ranges and the densities of peripheral populations may reflect geographically widespread conditions and concomitant variations in abundance of the entire species population. On the other hand, environmental changes that enable some peripheral populations to increase and expand into new areas may be entirely local and have little or no effect on either the abundance of central populations or the distributions of other peripheral populations elsewhere along the margins of the species range. At the level of local populations, where ecologists often concentrate their studies, there is usually a direct, causal relationship between changes in the abundance of individuals and shifts in their spatial distributions. Biogeographers typically study distributions on a larger scale, where the relationship between population dynamics and geographic range may become obscured. Ecological processes determine the limits of geographic ranges by affecting the birth, survival, and dispersal of individuals in local populations; but different ecological factors may limit population increase and spatial distribution of a species in different parts of its geographic range.

Physical limiting factors

Conceptual and methodological problems. Many widespread species appear to be limited in at least part of their geographic range by physical factors, such as temperature regime, water availability, and soil and water chemistry. For example, many Northern Hemisphere plants and animals become increasingly restricted to low elevations and south-facing exposures as they approach the northern limits of their ranges, suggesting that their distributions are determined by ambient temperature (Figure 3.8). Such correlations provide only circumstantial evidence, however, and do not necessarily indicate direct causal relationships. The species in question might, for example, be limited not by their inability to tolerate low temperatures directly, but by competition from other species that are superior in cold climates.

It might seem an easy matter to investigate carefully the distributional limits of a particular species in order to identify the limiting factor and discover its mechanism of action on the organism. Many such studies have been undertaken, but few have been sufficiently rigorous and complete to produce definitive results. Most have raised more questions than they have answered, because the problem of what limits distribution is often complex. In 1840 Liebig suggested that biological processes are limited by that single factor that is in shortest supply relative to demand or for which the organism has the least tolerance. At one time ecologists accepted this idea so completely that they called it Liebig's law of the minimum and tried to identify the single factor that limited the growth of each population. Numerous recent studies have demonstrated that Liebig's concept is inadequate to account for many distributional limits and that it is often necessary to invoke interactions among several factors. Thus many temperate and arctic birds and mammals, for example, appear to be limited by their inability to tolerate cold temperature re-

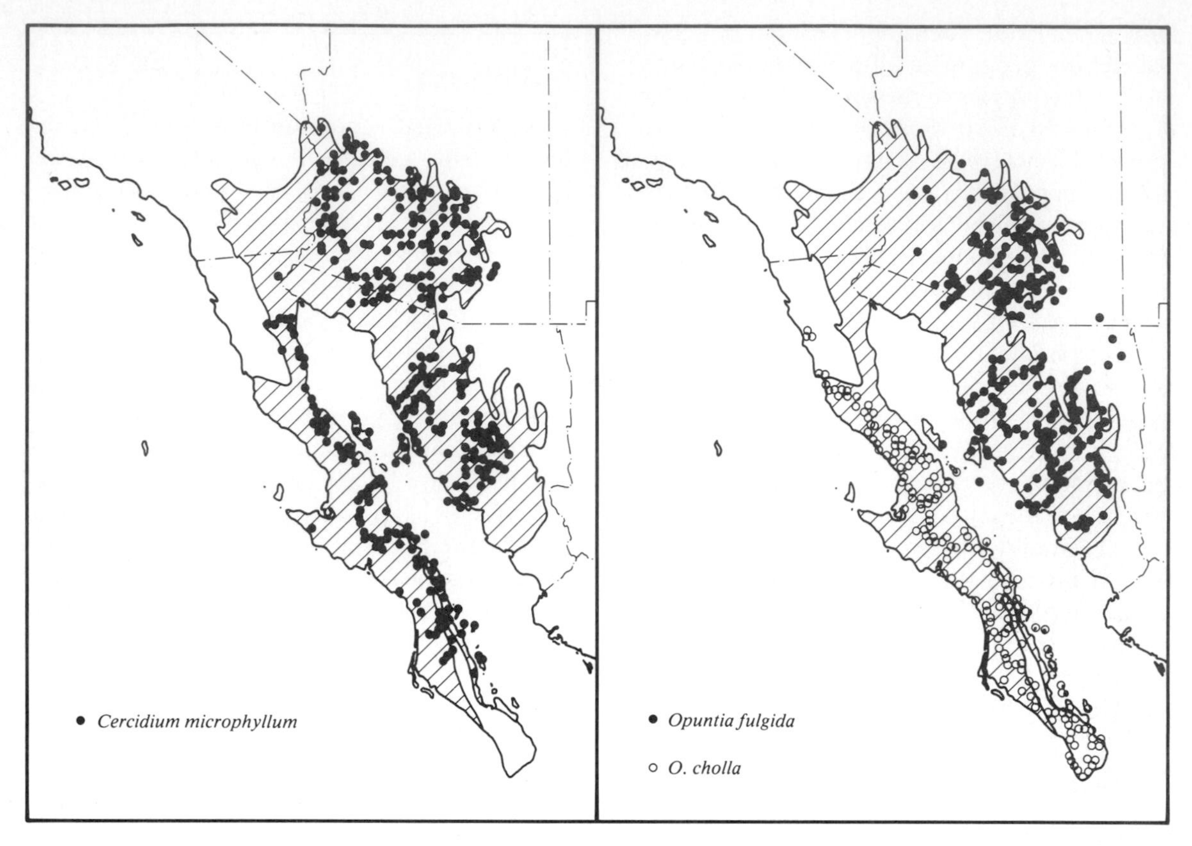

Figure 3.8
Distributions of endemic Sonoran Desert plants: the green-stemmed tree *(Cercidium microphyllum),* and the related cholla cacti *(Opuntia fulgida* and *O. cholla).* Note that the northern and eastern limits of the ranges correspond closely to the margin of the Sonoran Desert vegetation as a whole *(hatching),* which in turn is highly correlated with low temperatures owing to increasing latitude and elevation. (Redrawn from Hastings et al., 1972.)

gimes in the winter, not because they simply cannot survive at such low temperatures (Dawson and Carey, 1976), but because these thermal regimes increase the energy required for thermoregulation beyond the food supply available in the environment. In this case food and temperature interact to limit distribution, and increasing the food supply enables populations to inhabit colder climates. Most of those bird species, such as the cardinal, tufted titmouse, and mockingbird, which have expanded their ranges far northward in eastern North America (e.g., Boyd and Nunneley, 1964) are year-round residents. A new reliable winter food supply, provided by bird feeding stations, has undoubtedly contributed to the expansion of these species into colder environments.

Another problem in determining the cause of distributional limits is the difficulty in identifying the mechanisms by which environmental factors affect the growth and colonization of populations. Cold is not a single variable, and different aspects of low temperature regimes limit different populations in different ways. The adults of some plant species may be killed by critically low short-term temperatures, such as those experienced on a single, exceptionally cold winter night. Other species may be more susceptible to damage from prolonged freezing. Still other species may be limited by cold climates, not because they cannot withstand low winter temperatures, but because the summer growing season is too short for growth and successful reproduction.

The difficulties in investigating distributional limits can be overcome by careful observation and experimentation. Manipulative field experiments, such as introducing populations into areas where they do not presently occur and following their growth or decline, are particularly valuable for distinguishing correlation from causation and for identifying important interactions among limiting factors. Laboratory investigations by physiological ecologists are important in exploring the relationships between the physiological tolerance and performance of individual organisms at different phases of their life cycles and the physical and chemical dimensions of their niches. Although there are few cases in which these two kinds of studies have been combined to provide a relatively complete understanding of the factors limiting the distributions of species, there is abundant evidence that physical stresses play a major role in some populations.

Temperature and plant distributions. One of the best documented examples of cold temperatures limiting the upper elevational and latitudinal distribution of a species is provided by the saguaro cactus *(Carnegiea gigantea)*. This giant, multiarmed columnar cactus, which may reach 15 m in height and 200 years of age, is a conspicuous part of the landscape in much of the Sonoran Desert of southern Arizona and adjacent Sonora, Mexico. Although it lives where winter nighttime frosts are not infrequent, the saguaro is extremely sensitive to temperatures below −7° C. Individuals are killed by freezing of their tissues, especially destruction of the growing shoot tips. Young saguaros are more susceptible to frost damage than adults, but seedlings typically become established under the canopy of small desert trees that provide the young cacti with a protective microclimate for the first few decades of their lives (Nobel, 1980b). The canopy of these nurse trees shields the young saguaros from the cold night sky and prevents their freezing in much the same way that frost damage to tomato plants can be prevented by covering them at night with paper or plastic—the loss of heat by infrared radiation to the sky is retarded by the nurse trees and also by the dense spines and pubescence that cover the sensitive apical buds. Before they reach reproductive age, the saguaros grow above their nurse trees, often killing the trees in the process; but by then they are large enough not to be affected by overnight frosts. Nobel (1978, 1980a) has studied the thermal relations of the stems and shoot apices using computer simulations and direct field measurements. The results show that the large stem diameter of the saguaro enables it to maintain higher minimal temperatures of its apical buds and thus to have a more northern distribution than related species of columnar cacti.

Steenburgh and Lowe (1976, 1977) have studied populations of *Carnegiea gigantea* in Saguaro National Monument outside Tucson, Arizona, near the northeastern and upper elevational limit of the species range. Extensive mortality of both young and adult plants occurred as a result of exceptionally low temperatures in January of 1937, 1962, 1971, and 1978. The 1971 freeze killed about 10% of all individuals and severely injured about an additional 30%; many of the injured cacti died during the next few years as a direct result of this frost damage (Figure 3.9). These direct observations of episodic winter kill, together with close correspondence between the northern and eastern boundary of the species range and areas that ex-

Figure 3.9
Matched photographs of a stand of saguaro cacti *(Carnegiea gigantea)* near Redington, Arizona, near the upper elevational and northeastern margin of the species range. **A,** In 1961. **B,** In 1966, showing the loss of one individual (center foreground) and scars (white patches near the tips of the arms) as a result of the severe frost of 1962. **C,** In 1979, showing much additional mortality owing to severe frosts in 1971 and 1978. Several of the individuals still standing are dead or dying. (**A** and **B** courtesy J.R. Hastings; **C** courtesy R.M. Turner.)

perience below-freezing temperatures for more than 24 hours (Figure 3.10), suggest that low temperatures directly limit the distribution of saguaros in this region.

The distributions of many plant species appear to be limited by low temperature interacting with other environmental conditions such as water availability and soil chemistry. Hocker (1956) studied the distribution of loblolly pine *(Pinus taeda)* in the southeastern United States and concluded that the northern and western edges of the range were caused by low temperatures in concert with limited soil moisture. He suggested that this resulted from the inability of the roots to take up sufficient water when environmental temperatures were low to replace the quantities lost by evaporation. Shreve (1922) noted that the upper elevational limits of many desert plant species appear to be determined largely by temperature, because extensive mortality occurs in the populations at highest elevations during exceptionally cold winters. But many of these same species, such as the ocotillo *(Fouquieria splendens),* occur at much higher elevations and tolerate substantially colder winter temperatures on limestone soils than on granitic and other zonal soil types (Shreve, 1922; Whittaker and Niering, 1968).

Many investigators have attempted to determine the cause of timberline, the upper elevational limit of trees on mountains. The large-scale geographic position of timberline seems to be related to mean or maximum temperatures during the warm months of the growing season

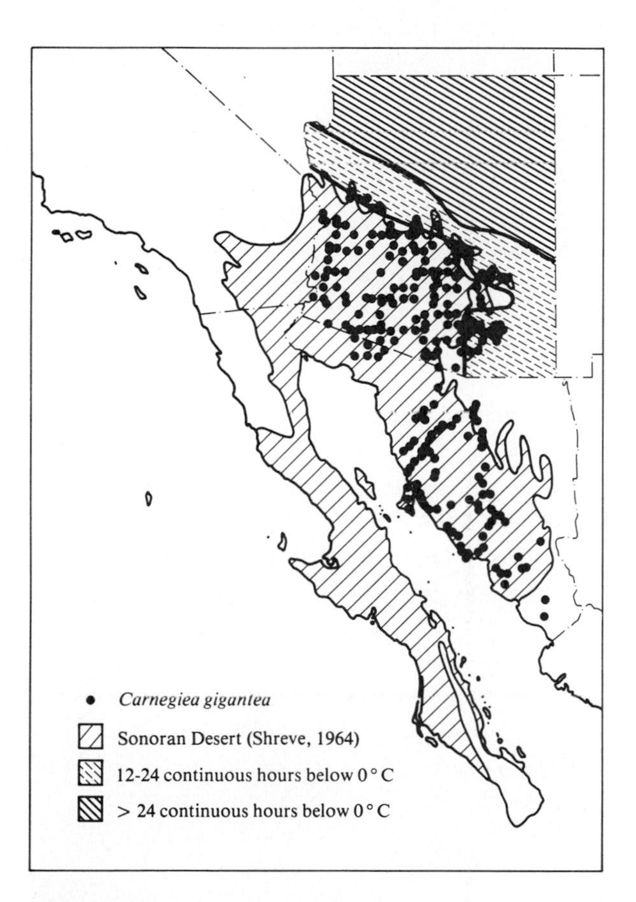

Figure 3.10
Map of the geographic range of the saguaro cactus *(Carnegiea gigantea)* showing the close correspondence between the northeastern limit in Arizona and climatic regimes in which the temperature remains below freezing for 24 hours. (Compiled from Hastings and Turner, 1965, and Hastings et al., 1972.)

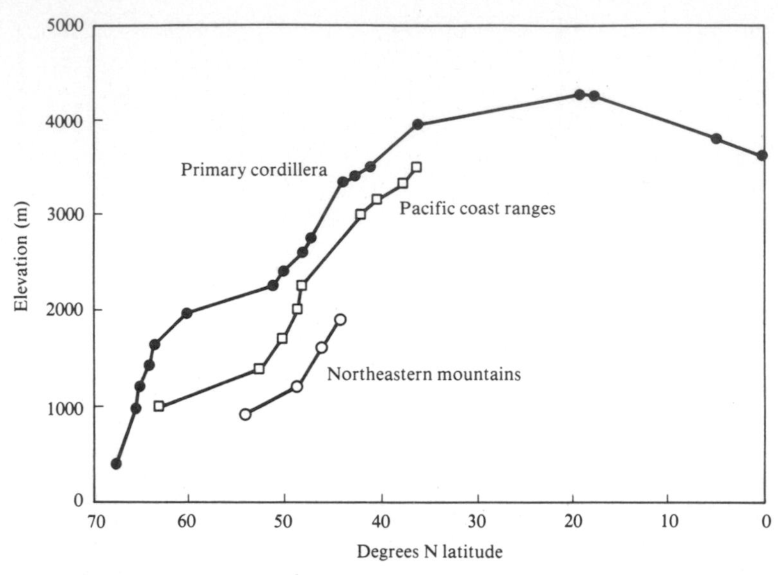

Figure 3.11
Relationship of timberline to elevation and latitude on three different mountain ranges in North America. The general pattern suggests a correlation between timberline and environmental temperature, and the differences between ranges led Daubenmire to conclude that high temperatures during the warm months were the most important factor determining the position of timberline. (Modified from Daubenmire, 1978.)

Figure 3.12
A bristlecone pine *(Pinus longaeva)* growing near timberline in the White Mountains of California. As is typical in individuals of this species growing in extreme environments, the tree has much dead wood and a highly contorted growth form. The dead logs in the foreground show that several other individuals once lived in the immediate vicinity. (Courtesy of Laboratory of Tree Ring Research, University of Arizona.)

(Figure 3.11), although it is influenced locally by other factors such as snow depth and wind. Daubenmire (1978) suggests timberline occurs at the point where trees are unable to accumulate sufficient energy during the growing season to reproduce. Warm temperatures are necessary for photosynthesis to produce sufficient energy above amounts required solely for maintenance that can be used for growth and reproduction. At timberline established trees often live for a long time, but growth is very slow and successful reproduction and seedling establishment are rare. Bristlecone pines *(Pinus longaeva)*, which often grow just below timberline on arid mountains in the southwestern United States, are the oldest known living things, some individuals being more than 4000 years old. They grow slowly and produce exceptionally hard, dense wood that is highly resistant to decay (Figure 3.12). The annual growth rings of living and dead bristlecones provide a valuable record of the climatic history of the Southwest, because the width of each ring is proportional to the suitability of the growing season in the year it was laid down (Fritts, 1976). In some places dead bristlecones demonstrate a "fossil timberline" above the timberline of living trees. Apparently this area was once favorable for tree growth, but climatic changes during the last few thousand years have killed the trees and caused a contraction in their elevational range (La Marche, 1973, 1978).

Effects of temperature on animals. Examples of temperature limiting distributions are often best documented for stationary organisms, such as terrestrial plants, because it is not hard to be impressed by the effects of limiting factors when we can readily observe large numbers of dead adults or the virtual absence of seedlings as a direct consequence of a period of physical stress. Nevertheless, environmental temperature regimes also influence the local and geographic ranges of many mobile terrestrial animals and aquatic plants and animals. The effects of physical limiting factors may be less apparent, however, in part because it is harder to observe death and reproductive failure directly and in part because these organisms can move to select more favorable microclimates, those that provide less stressful physical conditions. In highly mobile organisms, such as birds and mammals, physical factors may be more likely to interact with biotic constraints and to affect distributions by influencing the efficiency of foraging and reproduction rather than by causing wholesale mortality or absolute reproductive failure.

Sessile intertidal animals, such as barnacles and mussels, resemble plants in that adults are stationary and therefore must tolerate the full range of physical conditions in their local environments. Most intertidal organisms have evolved from exclusively aquatic marine ancestors. Despite adaptations to withstand periodic exposures at low tides (Vermeij, 1978), physical conditions become progressively more stressful higher in the intertidal zone as the frequency and duration of exposure to the terrestrial climate increases. The upper limits of many intertidal species appear to be determined largely by their inability to withstand extreme temperatures and desiccation while uncovered between tides. Because of the sessile nature of these animals, death of individuals can be observed directly and related to physical environmental conditions in much the same way as in plants. For example, Connell (1961) documented mortality of individuals of two species of barnacles in the rocky intertidal zone of Scotland and demonstrated that the upper limit of their distributions was determined primarily by their inability to tolerate physical stress during exposure between tides.

Even highly mobile animals such as fishes can be limited directly by physical factors such as environmental temperatures. Pupfish of the genus *Cyprinodon* are extremely eurythermal and euryhaline; i.e., they tolerate a wide range of temperature and salinity (Brown, 1971c; Brown and Feldmeth, 1971). Species of this genus occur in rigorous physical environments, including shallow streams and marshes in deserts and small pools in tidal flats, estuaries, and mangrove swamps, where temperature and salinity may fluctuate widely. Some populations also inhabit hot springs, although they generally cannot tolerate temperatures in excess of about 43° C (which is still amazingly high for a fish). Some pupfish occur in the cooler outlets of hot springs where temperatures of the source waters exceed lethal limits. The local distribution of one such population of *C. nevadensis* near Death Valley, California, is limited directly by high temperature (Brown, 1971c). Fish occur in all waters cooler than 42° C, including small pools only a few centimeters away from much hotter water (Figure 3.13). Occasionally individuals stray or are frightened into lethally hot water and die instantly. More infrequently, rapidly changing weather conditions cause lethally hot water to flow far downstream from the source, trapping thousands of fish inside pools where they are killed when the temperatures rise to over 43° C.

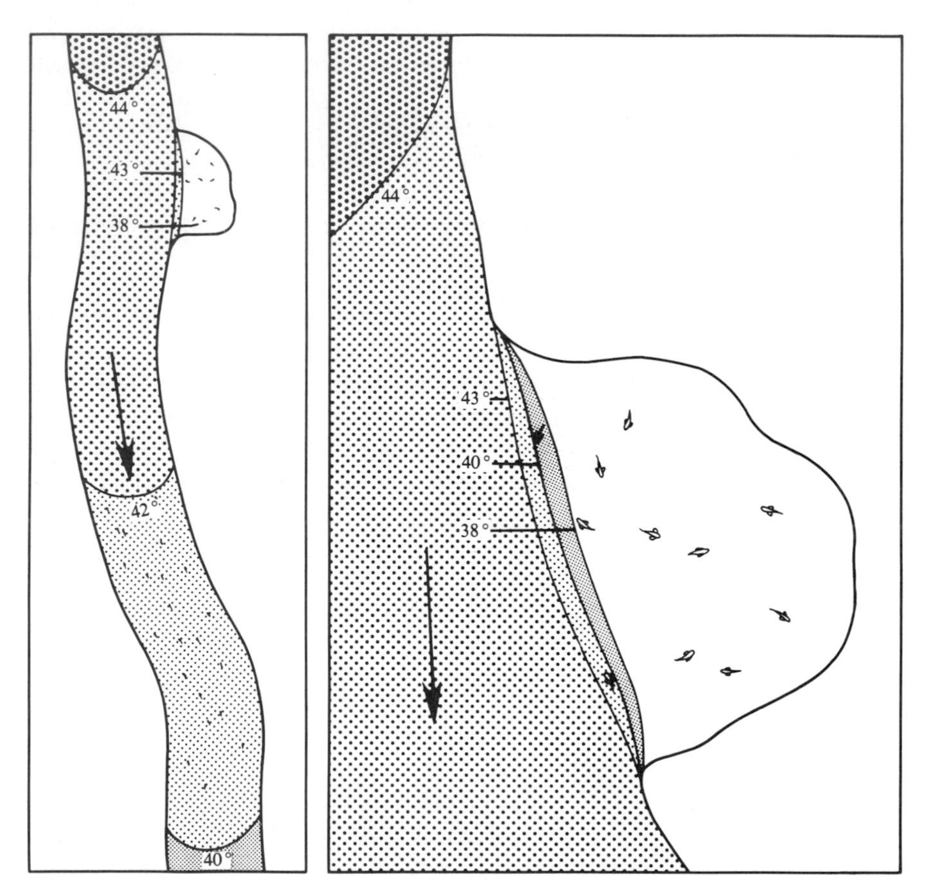

Figure 3.13
Effects of temperature on the local distribution of desert pupfish *(Cyprinodon nevadensis)* in the outflow of a hot spring near Death Valley, California. The fast-flowing water in the main channel is initially above the lethal temperature, but fish occur in the cooler side pools and in the main channel once it has cooled below the critical temperature of approximately 43° C. (From The desert pupfish by J. Brown. Copyright © 1971 by Scientific American, Inc. All rights reserved.)

Like many organisms, adult pupfish are tolerant of a wider range of physical conditions than are the early developmental stages. Eggs of *C. nevadensis* develop normally only in water between 20° and 36° C, although adults of this species can withstand temperatures between 0° and 42° C (Shrode, 1975). Eggs are also less tolerant than adults of extreme salinity. Consequently, adult pupfish are often found in places where successful reproduction is impossible so long as they can immigrate from nearby microhabitats that are suitable for egg development. On the other hand, species of *Cyprinodon* are conspicuously absent from some cold springs and other habitats where the adults can grow and survive but where there are no microclimates suitable for the earlier stages.

Other physical factors. Other physical and chemical factors, such as moisture, light, oxygen, pH, salinity, and soil and water elements, limit the distributions of animals. Most of these factors have been investigated less thoroughly than temperature because they are harder to measure, their effects on organisms are more difficult to assess because they are subtle and less rapid, and they interact with each other and with biotic factors in ways that are often difficult to interpret. A simple example of the last situation is provided by the interacting effects of temperature and oxygen on aquatic animals. As water increases in temperature, its capacity to hold oxygen and other dissolved gases decreases. The resulting combination of high temperature and low oxygen concentration is very stressful to many fishes and aquatic invertebrates, because high temperatures cause elevated metabolic rates and increased demand for oxygen.

Although various physical and chemical factors may have different physiological effects on individual organisms, they often limit geographic distributions in ways similar to those described for temperature—by affecting the survival, reproduction, and movements of individuals in peripheral populations. A few examples will suffice to illustrate distributional boundaries that can be attributed, at least in part, to such factors as moisture, salinity, and soil chemistry. Many terrestrial plants are limited by low soil moisture at the drier edges of their ranges, just as they are by low temperatures at the colder margins. In all vascular plants, photosynthetic rates decline as soil moisture decreases (Figure 3.14); plants can compensate for decreased water uptake by the roots by closing their stomates and reducing transpiration from the leaves, but rates of photosynthesis are reduced concomitantly.

Plants adapted to grow in full sunlight on dry soils show many specialized mechanisms for continuing to open their stomates with low levels of water in their leaves, whereas mesophytes close their stomates and their leaves suffer high, often fatal heat loads. On the other hand, xerophytes typically have relatively low rates of photosynthesis when abundant water is available, and they are intolerant of shade. A consequence of this tradeoff is that mesophytic plants are physiologically incapable of growing on dry soils, whereas xerophytic species can grow where there is little moisture but are competitively excluded from wetter soils by mesophytic species that have higher growth rates (see Odening et al., 1974). These physiological findings provide a mechanistic basis for the conclusions of Shreve (1922) and other early plant ecologists that limits of plant distributions in dry areas are determined largely by the inability to tolerate low soil moisture. Widespread diebacks in drought years are commonly observed in local populations at the margins of a species range (e.g., Sinclair, 1964; Westing, 1966). Similar kinds of tradeoffs between photosynthetic rate and ability to tolerate low nutrient levels, high salinity, extreme pH, or high concentrations of toxic minerals probably account, at least in part, for the failure of many otherwise widespread plant species to occur locally on soils with these characteristics, while species with special adaptations to these soil types are often restricted to them (Whittaker, 1975).

These kinds of physical and chemical factors

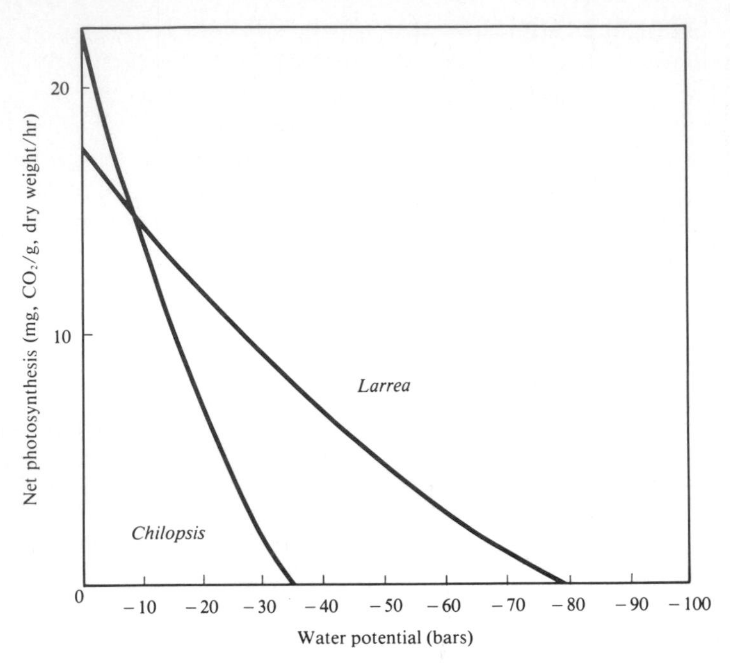

Figure 3.14
Relationship between photosynthetic rate and soil water potential for two desert plants, creosotebush *(Larrea tridentata)* and desert willow *(Chilopsis linearis)*. *Larrea tridentata* is a highly xerophytic species that occurs in some of the most arid parts of the North American deserts, whereas *C. linearis* is much more mesophytic and is found along watercourses where the deep soil is permanently moist. Note that photosynthetic rates of both plants decrease with decreasing soil moisture availability, but the xerophytic species can photosynthesize over a wide range of water potentials, whereas the mesophyte has higher photosynthetic rates when the soil is well hydrated. (From W.R. Odening, B.R. Strain, and W.C. Oechel. The effects of decreasing water potential on net CO_2 exchange of intact desert shrubs, *Ecology* **42**(3):594-598. Adapted by permission of Duke University Press. Copyright © 1974, Ecological Society of America.)

also limit animal distributions. Because of their osmoregulatory physiology, the vast majority of freshwater fishes and invertebrates are intolerant of salinities even approaching the concentration of seawater, whereas marine species cannot survive in fresh water. One consequence of this is that salt marshes and estuaries are inhabited by neither freshwater nor marine species but by specialized euryhaline species that can tolerate great, often daily, fluctuations in salinity caused by tides, floods, and storms. Only a few kinds of specialized organisms occur in the small number of lakes and springs that are even saltier than seawater. The Great Salt Lake in Utah, for example, is approximately seven times more concentrated than seawater and contains no fish and only two macroscopic invertebrates, the pelagic brine shrimp *(Artemia salina)* and the benthic brine fly *(Ephydra cinerea)*. Several fish species and numerous invertebrates inhabit the freshwater streams and marshes that empty into the lake. These animals have abundant opportunity to extend their ranges and colonize the lake, but they are prevented from doing so by their inability to tolerate the high salinity. This is dramatically illustrated by the wide-

Figure 3.15
Effect of a fire on vegetation near Elgin, Arizona. **A,** Several days after the fire in June 1975. **B,** Three months later, in September 1975. **C,** Four years later in November 1979. Note that the fire had a minimal lasting effect on this grassland habitat. Although most of the live oak trees had most of their leaves killed, only a few, such as the small individual in the center foreground, died or suffered major, lasting damage. The grass cover returned rapidly. (Photographs by R.M. Turner.)

spread mortality that occurs when occasional wind storms push salt water into the adjacent marshes.

Disturbance. A final class of physical factors that influences both local and geographic distributions of many organisms includes fires, hurricanes, volcanic eruptions, and other agents of sudden widespread disturbance and destruction. These natural disasters are capable of completely destroying habitats and their inhabitants. In regions where such disturbances occur with relative frequency, however, they are a natural part of the environment (Figure 3.15). Many species are dependent on periodic disturbance for their continued existence, and there is a regular pattern of colonization and replacement of species following a disturbance, which ecologists call secondary succession. For example, such successional processes occur in the drier forested regions of western North America and the chaparral shrubland of coastal California that are swept by periodic fires (Figure 3.16); in the forested islands and offshore coral reefs in the Caribbean region that are occasionally but inevitably decimated by hurricanes; and in intertidal habitats throughout the world that are subjected to heavy wave action and sand scouring during major storms. In many such situations, where periodic disturbance is a normal component of the environment, the major effects are on local distributions and abundances rather than on geographic ranges.

On the other hand, periodic natural disturbance may be sufficiently severe and frequent to prevent the expansion of some species into areas where they could otherwise survive. In

A

Figure 3.16
Photographs of forest fire succession at a site at Miller Creek in western Montana, showing the forest floor before the burn (**A**). (Photographs by Peter F. Stickney, U.S. Forest Service.)

Figure 3.16, cont'd
Just after the burn (**B**), and 3 years later (**C**), when a new community of herbaceous plants and saplings of woody perennials had become established.

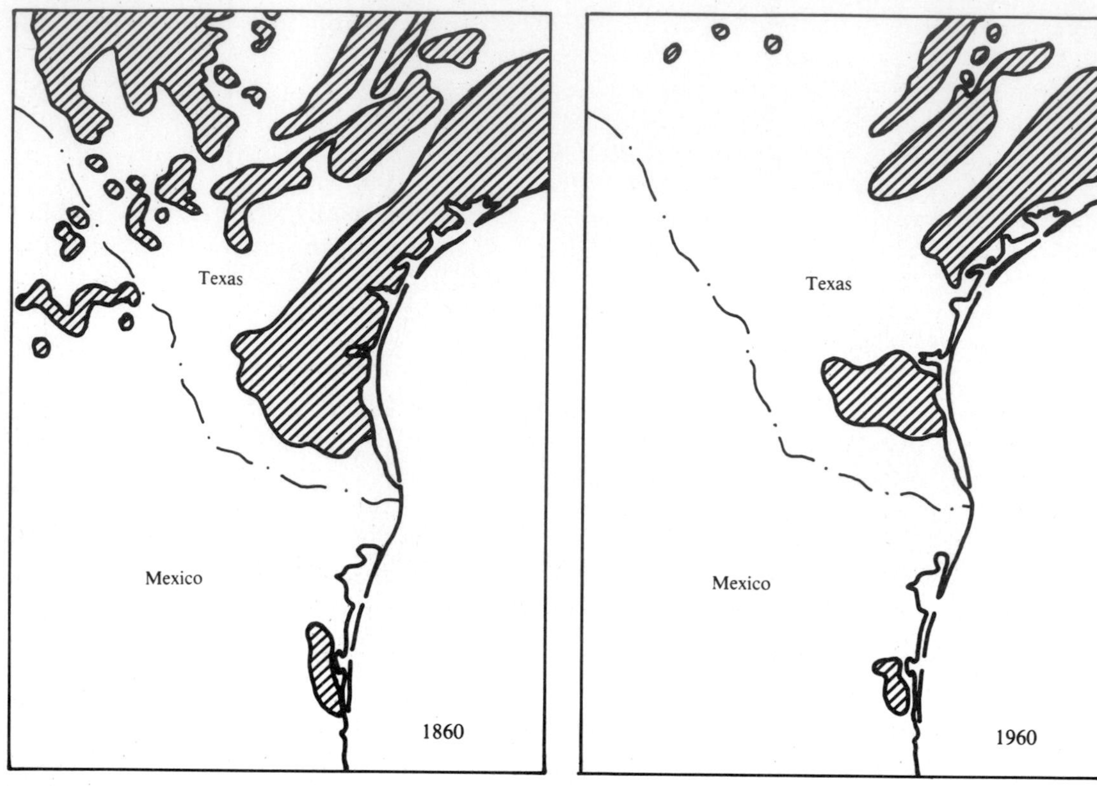

Figure 3.17
Contraction of grassland in Texas and northeastern Mexico within the last century. Map on the left shows the distribution of extensive areas of natural grassland habitat about 1860 as reconstructed from historical records. Map on the right shows the reduction in these areas, which can be attributed largely to invasion by trees and shrubs, especially mesquite *(Prosopis)*. Fire suppression probably has been a major cause of these changes. (Redrawn from Johnston, 1963.)

most grasslands, lightning-caused fires are a natural part of the environment (Figure 3.15). At forest-grassland boundaries, where soil moisture is sufficiently high for woody vegetation to become established, frequent fires may prevent trees and shrubs from extending their ranges. There is little doubt that artificial fire suppression within the last 200 years has contributed to the expansion of forest and shrubland at the expense of prairie and other grasslands in several regions of the United States (Beilmann and Brenner, 1951; Johnston, 1963) (Figure 3.17). However, many plant ecologists do not accept fire as a general explanation for all forest-grassland boundaries. They argue that climate and soils determine the transitions between vegetation types in most regions (e.g., Curtis, 1956, 1959).

Limitation by biotic interactions

In many cases geographic distributions are not limited directly by physical factors. The realized niches of many species are much narrower than their fundamental niches. Botanical gardens and zoos provide perhaps the most dramatic evidence that individuals can survive, grow, and even reproduce under physical conditions very different from those encountered anywhere in their natural geographic ranges. The fact that cultivated plants can often be grown only if they are protected from competing plants, animal herbivores, and microbial pathogens suggests that biotic interactions with other species may play a major role in limiting their distributions.

Competition. There are three major classes of interspecific interactions: competition, predation, and mutualism. All of these can influence the dynamics of populations and limit the geographic ranges of species. Competition is a mutually detrimental interaction between individuals. Organisms that share requirements for the same essential resources necessarily compete with each other and suffer reduced growth, survival, and reproduction if the resources are in sufficiently short supply. Plants may compete for light, water, nutrients, or pollinators, whereas animals most frequently compete for food but also sometimes for shelter, nesting sites, mates, or living space. Malthus (1798) and Darwin (1859) long ago pointed out that competition ultimately limits the capacity of every population to increase, and Darwin emphasized that many species distributions are limited by competition. He pointed out that competition normally will be most severe among individuals of the same species, because they have the most similar requirements, but often organisms share resources and compete with other closely and even distantly related species. These interactions may be purely exploitative, so that individuals simply use up resources and make them unavailable to others, or they may involve some form of interference, in which aggressive dominance or active inhibition is used to deny other individuals access to resources.

There is much circumstantial evidence that competition limits geographic ranges. There are many examples of ecologically similar, closely related species that occupy adjacent but nonoverlapping geographic ranges. Five species of large kangaroo rats *(Dipodomys)* are found in desert and arid grassland habitats in the southwestern United States and northern Mexico, but their geographic ranges are almost completely nonoverlapping (Figure 3.18). Two species, *D. ingens* and *D. elator,* have isolated or disjunct ranges, but *D. spectabilis* shares an extensive border with *D. deserti* in the west and with *D. nelsoni* in the south. Although such cases are highly suggestive that competition limits distribution by preventing coexistence, they are subject to alternative explanations and often there is no direct evidence of competitive interactions occurring on the boundaries.

Better evidence for the limiting effects of competition comes from natural experiments in which one species, simply by chance, is absent from regions that are apparently suitable for habitation. If a second species then expands its range to include habitats that are normally occupied by the first species, this implies that in other areas where the first species is present in these habitats the second is limited by competition. The forests and shrublands of the western United States are inhabited by about 20 species of chipmunks of the genus *Eutamias.* The distribution of these rodents is reminiscent of the pattern in large kangaroo rats, but more complex. Many species have distinct geographic ranges; those of other species appear to overlap on a geographic scale, but these species occupy different habitats and rarely coexist on a local level. On the mountains of the Southwest, two species typically occur in the woodlands and forests, but they are segregated by habitat and elevation. *Eutamias dorsalis* inhabits the open, xeric woodlands at lower elevations, and it is replaced in denser coniferous forests at higher elevations by a species of the *E. quadrivittatus* group. There are at least 18 isolated desert mountain ranges where appropriate habitats seem to be present but one of these species is absent, apparently because it either never colonized or else became extinct sometime in the

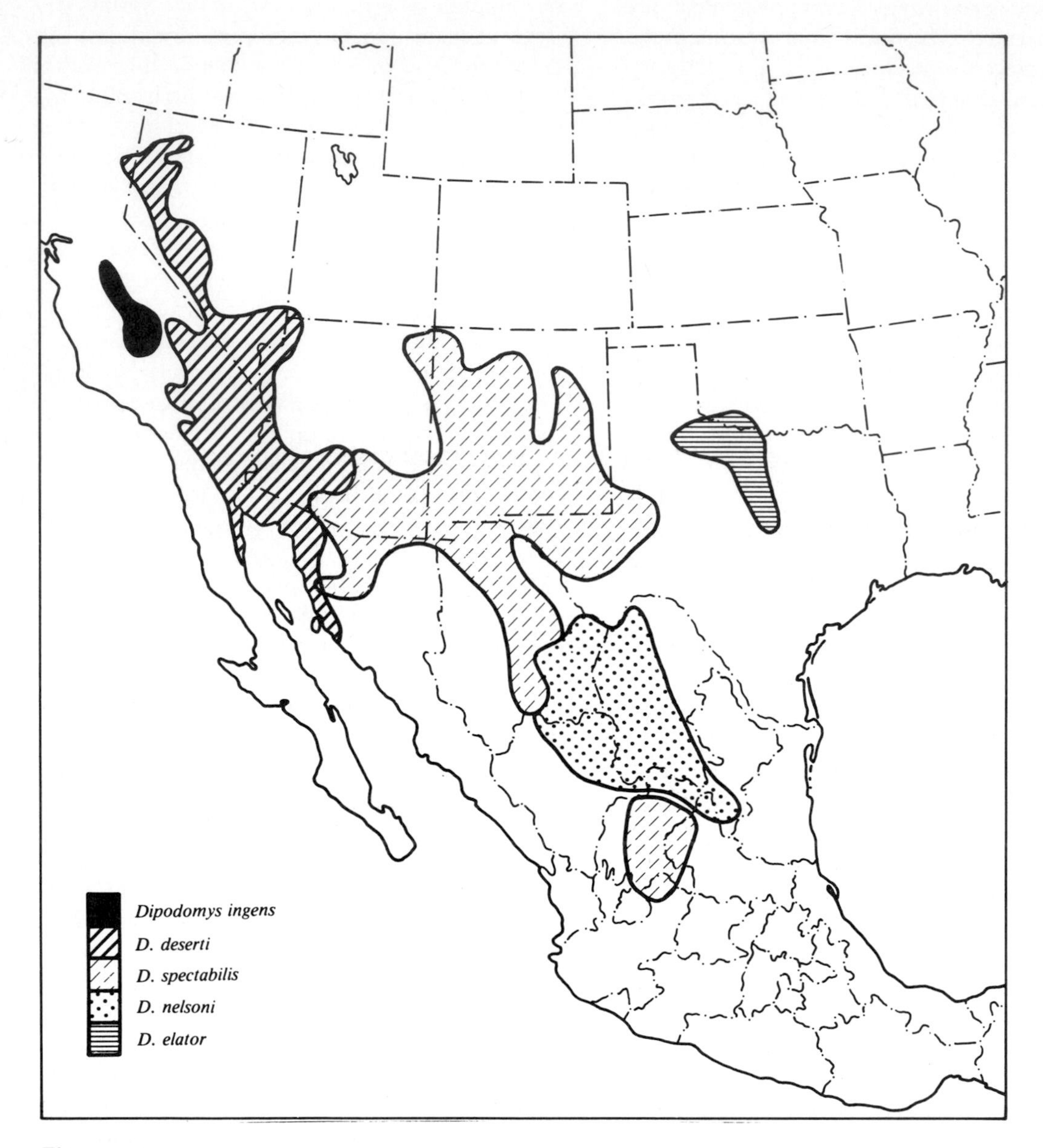

Figure 3.18
Geographic ranges of five species of large kangaroo rats *(Dipodomys)* in southwestern North America. These forms are similar in size and ecology, and the fact that their ranges frequently come into contact but almost never overlap significantly suggests that they competitively exclude each other. (Redrawn from Hall, 1981.)

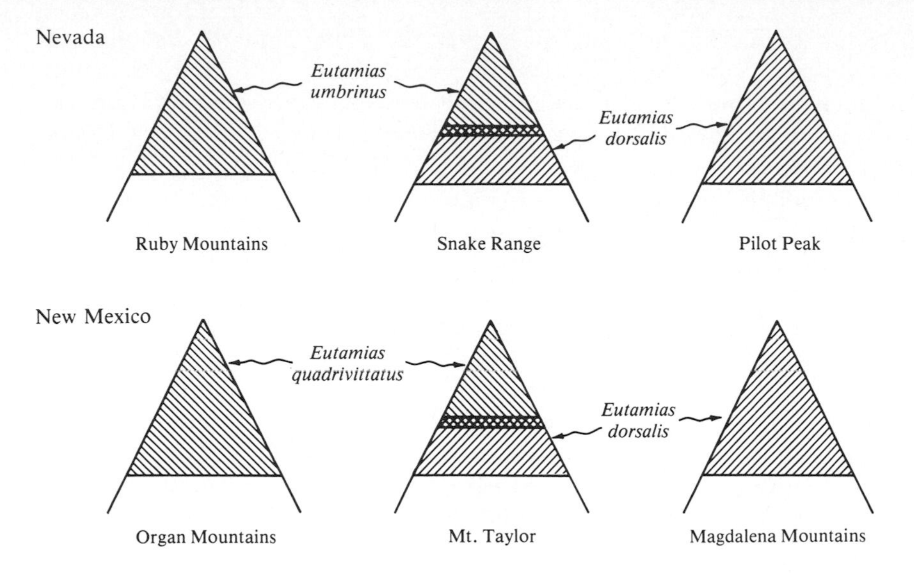

Figure 3.19
Diagrammatic representation of the elevational distribution of chipmunks *(Eutamias)* on mountains in the southwestern United States. On most mountains two species are present and their ranges overlap only slightly, but on several isolated ranges only one of the species occurs and then its range is expanded to include nearly all elevations and habitats usually occupied by both species. This is good circumstantial evidence for competitive exclusion, which in this case has been confirmed by direct ecological studies.

past. In every case, regardless of which species is absent, the remaining species has expanded its range to include all forested habitats from the edge of the desert to the timberline (Figure 3.19). Such examples of niche and range expansion are particularly convincing evidence of competition when, as in the present case, similar distributional shifts have occurred independently in several different places.

The mechanisms of competitive interaction among these chipmunk species have been investigated in some detail. Brown (1971a) placed feeding stations and observed behavioral interactions in the narrow zone where the ranges of *E. dorsalis* and *E. umbrinus* come into contact. Brown concluded that *E. dorsalis,* the more aggressive and terrestrial species, was able to exclude *E. umbrinus* from open woodlands, where chipmunks had to do most of their traveling on the ground, by aggressively defending patchy food resources. However, in denser forests where food is harder to defend because it is more abundant and the chipmunks can travel through the trees, *E. umbrinus* wins out because *E. dorsalis* wastes excessive time and energy on fruitless chases. Chappell (1978) studied a more complex situation where the ranges of several species come into contact in the Sierra Nevada of east-central California. He also found that the mutually exclusive distributions could be attributed primarily to the influence of habitat on the outcome of aggressive interactions. One species, *E. minimus,* however, is behaviorally subordinant to all other species; in sympatry with other species (coexisting populations) it is restricted to hot, dry, shrubby habitats where it alone can tolerate the stressful physical environment.

Such asymmetrical niche relationships, in which one species is limited by competition while the other is restricted by its inability to withstand physical stress, appear to be extremely common in many kinds of organisms. Earlier we mentioned that in gradients of physical stress such as increasing cold, aridity, or soil salinity, terrestrial plants that are adapted to tolerate the harsh conditions typically have slow growth rates so they are competitively excluded from more equitable environments by faster-growing species. In a classic study of barnacles in the rocky intertidal zone of Scotland, Connell (1961) used reciprocal removal experiments to investigate the vertical limits of distribution. *Chthamalus stellatus* is segregated above *Balanus balanoides* in a gradient of increasing exposure to the stressful terrestrial climate. Connell showed that the upper limit of *Balanus* was caused by its inability to withstand desiccation and extreme temperatures during exposure between tides, whereas the lower limit of *Chthamalus* was determined by competition with *Balanus* (Figure 3.20). However, after much subsequent work in other parts of the world, Connell (1975) concluded that lower limits of intertidal species were more often determined by predation than by competition.

Predation. Predation can be defined as any interaction between two species in which one benefits and the other suffers. According to this definition, relationships between herbivores and their food plants, parasites and their hosts, and Batesian mimics and their models would also be

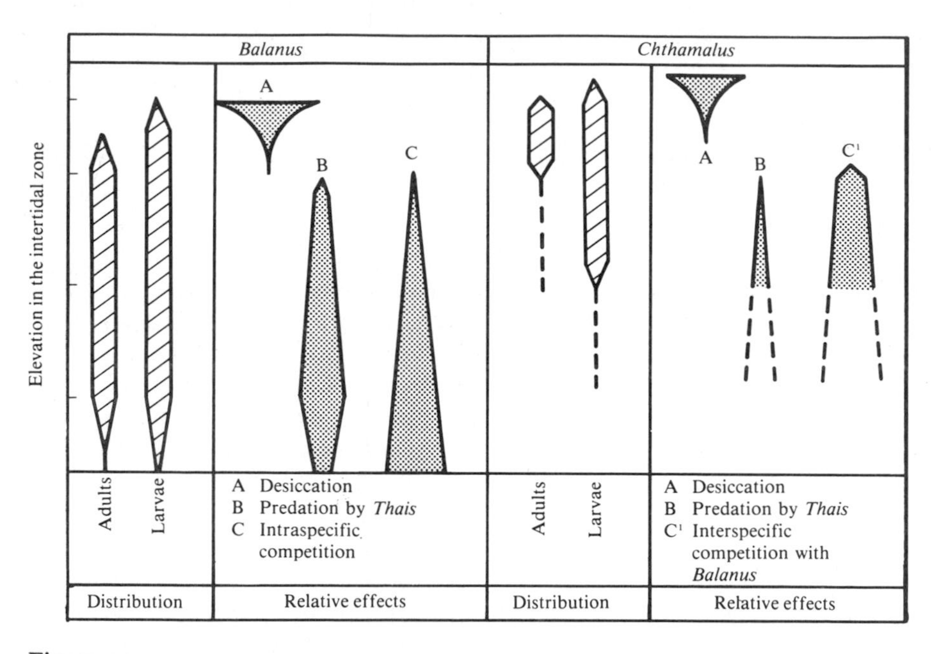

Figure 3.20
Diagrammatic representation of the effects of competition and other factors on the distribution of two barnacle species, *Chthamalus stellatus* and *Balanus balanoides,* on rocky shores in Scotland. J.H. Connell (1961) followed the fates of newly settled larvae and used manipulative experiments to demonstrate that competition from *Balanus* is largely responsible for the restriction of *Chthamalus* to the uppermost intertidal zone. The upper limits of both species are determined largely by abiotic factors (desiccation), and the lower limit of *Balanus* is probably caused largely by competition and predation from other organisms. (From J.H. Connell. The influence of interspecific competition and other factors on the distribution of the barnacle *Chthamalus stellatus, Ecology* **42**[4]:710-723.)

classified as predation. Predator-prey interactions can limit the distribution of either participant because, on the one hand, predators may depend on particular prey for food or other benefits necessary to support their own populations, whereas, on the other hand, predators may limit prey populations by killing or damaging individuals.

When predators are highly specific, it is obvious that their distributions must depend in part on the availability of appropriate prey. It is hardly surprising that the geographic ranges of many specific parasites and herbivores correspond almost precisely with those of their animal or plant hosts. Thus the distribution of the butterfly *Euphydryas editha* in coastal California is limited to the immediate vicinity of patches of serpentine soil to which its host plant, *Plantago hookeriana,* is restricted (Ehrlich, 1961, 1965). Of course, even highly specific predators may range less widely than their hosts, because in some areas their populations are limited by factors other than the availability of suitable prey.

It is much more difficult to document cases in which the distributions of prey populations are limited by their predators. Some of the best examples are artificial in that they involve introduction by humans of predators into regions where they did not originally occur. In some cases these introductions were made deliberately in an effort to control prey populations that were pests. Two conspicuously successful examples of such biological control involve drastic reductions in plant populations by introduced herbivores. The prickly pear cactus *(Opuntia stricta)* was introduced into Australia from North America in the mid-1800s to serve as an ornamental garden plant. By the early 1900s the cactus had escaped from cultivation and had become a serious pest in grazing lands. In 1926 Australian scientists introduced a moth *(Cactoblastis cactorum)* whose larva is a specific feeder on *Opuntia* in its native Argentina. By 1940 *Opuntia stricta* had been effectively checked as a pest species in eastern Australia, although small patches of cacti and local patches of the moth remain (Dodd, 1959). Unfortunately, other species of cacti have escaped and are becoming naturalized in native Australian habitats, species for which there is presently no known effective biological control.

Similarly, the Klamath weed *(Hypericum perforatum)* was introduced into the northwestern United States from Eurasia about 1900 and became an agricultural pest, but this was subsequently controlled by introduction of a specific leaf-eating beetle *(Chrysolina quadrigemina)*. In the southern part of its range Klamath weed persists only in small populations along roadsides and in shady areas, but the beetle populations do not do well in colder climates and the weed is more widely distributed in British Columbia (Huffaker and Kennett, 1959; Harris et al., 1969). In both cases, the prickly pear cactus in Australia and Klamath weed in North America, specific herbivores have drastically reduced plant populations, resulting in very limited local distributions, but they have not greatly altered the distributions of the weeds on a larger geographic scale.

Perhaps the best examples of complete elimination of prey populations from parts of their geographic ranges by voracious but nonspecific predators are provided by the artificial introduction of large predatory fishes into certain freshwater habitats. Many of the small native fishes in the southwestern United States have suffered great reductions in their geographic ranges and complete extinction of local populations as a result of introduction of large predatory game fish (especially largemouth black bass, *Micropterus salmoides*) into their habitats (Miller, 1961a). The native fishes are not adapted to large, specialized predators because they have evolved in isolated lakes, streams, and springs for thousands of years. Zaret and Paine (1973) document a similar example, the extinction of 7 to 11 species of native fishes from Lake Gatun, Panama, following the introduction of the predatory fish *Cichla orellaris*. The lake trout *(Salvelinus namaycush)* is widely distributed over northern North America but is absent from bodies of water that can be reached by lampreys of the genera *Petromyzon* and *Entosphenus,* which are voracious predators. Niagara Falls formerly prevented *Petromyzon* from entering the upper Great Lakes, which supported large populations of lake trout. Construction of the Welland Canal enabled the lamprey to colonize these lakes. The result has been a precipitous decline in populations of *S. namaycush* despite a major effort to control the lamprey and save a valuable commercial fishery.

In the previous examples, effects of predators on prey populations are particularly clear because we have been able to observe the responses to artificial introductions of the predators. Without this historical perspective, however, we would be hard put to infer the extent to which the prey are limited by their predators, because there is often very little evidence of predation when the populations are in equilibrium. Today, most patches of *Opuntia* in Australia are not infested with *Cactoblastis.* Similarly, it is difficult to observe black bass preying on pupfish in the southwestern United States, because the native fish have already been eliminated from most waters where bass are present. It is likely that many prey populations are limited, at least in part, by their predators, but it is difficult to obtain convincing evidence without performing manipulative experiments.

Much careful, experimental work has been done in the intertidal zone. For example, Lubchenco and Menge (1978; see also Lubchenco, 1980) used a variety of experiments to analyze factors influencing the distributions of sessile invertebrates and algae on rocky New England shores. The mussel *(Mytilus edulis)* is the dominant competitor for space in the lower intertidal zone, but it is normally restricted to sites of heavy wave action, the only areas where its predators, especially starfish (*Asterias* spp.) and a carnivorous snail *(Thais lapillus),* are ineffective. On protected sites *Mytilus* will monopolize space if these predators are excluded experimentally, but normally the surface is covered primarily by the shrubby red alga *Chondrus crispus.* Herbivorous periwinkle snails *(Littorina littorea)* play an important role in the distribution of this species. They feed selectively on fast-growing green algae, such as *Enteromorpha intestinalis* and *Ulva lactaca,* allowing the slower growing but less palatable *Chondrus* to dominate. The green algae are normally temporary seasonal or successional species that colonize and grow rapidly on bare areas after storms have removed both *Chondrus* and periwinkles. When the periwinkles are excluded by small cages, the green algae persist for over a year, and *Chondrus* is virtually eliminated. These experiments demonstrate not only the importance of predators but also the interactions between competition and predation in limiting local distributions.

Mutualism. The third class of interspecific interactions is mutualism, in which each species benefits the other. Examples of mutualistic associations are provided by plants and their animal pollinators, corals and photosynthetic zooxanthellae (algae), ants and aphids, and cleaner fishes and their hosts. Compared with competition and predation, mutualism has been little studied by ecologists and biogeographers. Much remains to be learned about the effects of mutually beneficial associations on the abundance and distribution of participating populations. Many mutualistic relationships are obligate and highly specific for at least one, and often for both, of the partners. In such cases, it is clear that the interaction may have a major influence on distributions. Some plant populations are dependent on the services of specific pollinators for sexual reproduction. For example, red clover *(Trifolium pratense)* did poorly after being introduced into New Zealand until its pollinator, the bumblebee (*Bombus* spp.), was also introduced (Cumber, 1953; Free, 1970). Janzen (1966) studied the association between ants of the genus *Pseudomyrmex* and trees and shrubs of the genus *Acacia* in the New World tropics. Although many species of *Acacia* have no ants associated with them, and some species

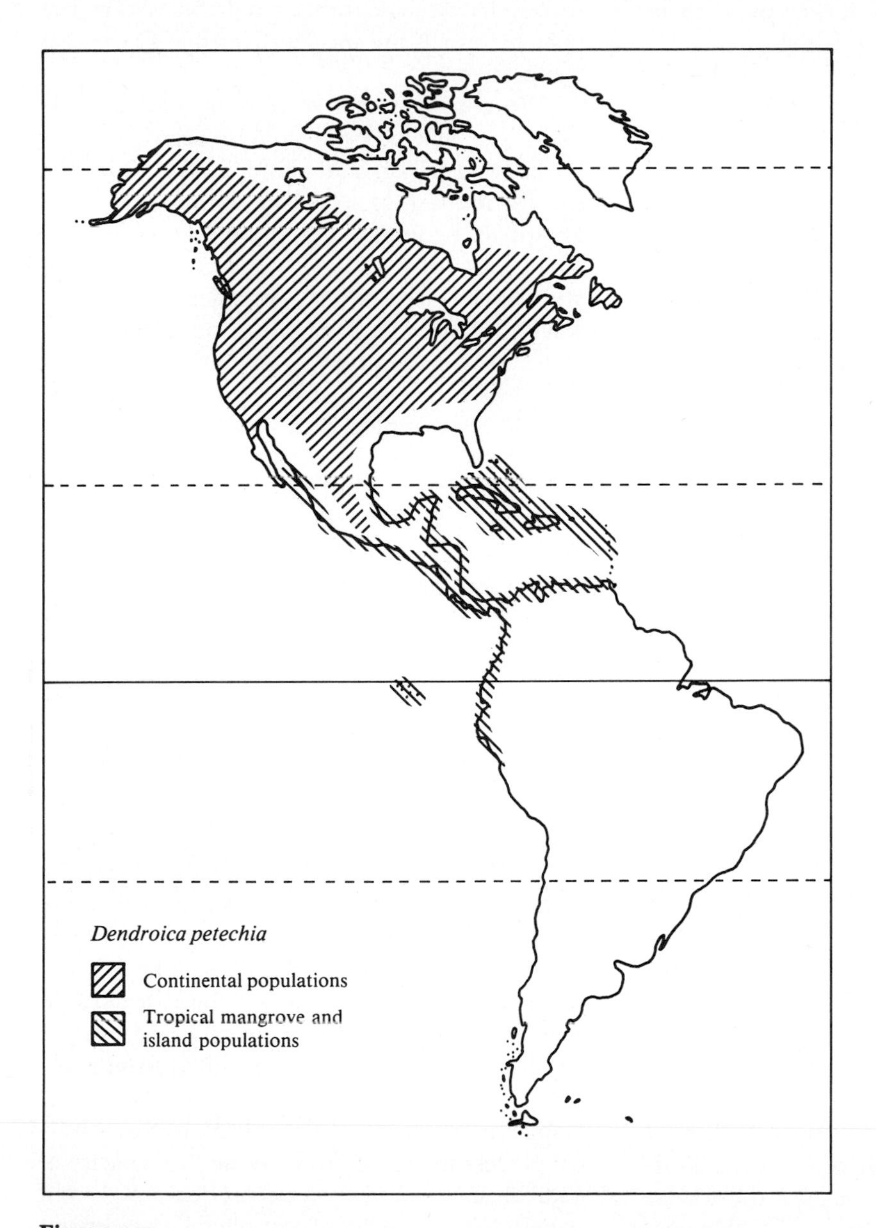

Figure 3.21
Geographic range of the yellow warbler *(Dendroica petechia)* showing that it is broadly distributed across temperate North America, but in the tropics it is restricted to coastal mangrove swamps and to islands. It has been suggested that the yellow warbler is excluded from most tropical habitats by diffuse competition from the large number of other insectivorous bird species.

of *Pseudomyrmex* may not be dependent on acacias, for numerous species the relationship is apparently obligate. The trees provide the ants with enlarged thorns in which they build their nests and with specialized foods rich in sugars, oils, and proteins. In return, the ants attack herbivorous insects and vertebrates and clear away surrounding vegetation, reducing competition from other plants. Such coevolved specializations apparently have made these mutualists so dependent on each other that they have virtually identical ranges.

Complex interactions. In addition to those cases in which it is possible to isolate the limiting effect of one species on the distribution of another, there are undoubtedly many situations in which ranges are structured by more diffuse biotic interactions. The limits may be the result of different, interacting effects of several species. MacArthur (1972) noted that the southern limits of the ranges of many North American bird species apparently could be attributed neither to climate (because the climate becomes more equable at lower latitudes) nor to habitat (although it might be important for some species) nor to any particular species of competitor or predator. MacArthur suggested that for many species the limiting factor must be the diffuse competition from an increasing number of tropical species. He noted, for example, that of 202 land bird species that breed in Texas, only 29 also breed in Panama, but Panama has a total of 564 breeding land bird species. One of the few species that breeds in both the United States and Panama is the yellow warbler *(Dendroica petechia)*, a small, insectivorous foliage gleaner. In the United States the yellow warbler is abundant in a wide variety of shrubby and forested habitats, but in Panama it is restricted to mangrove swamps and small offshore islands (Figure 3.21). Because the forests of Panama contain many species of highly specialized foliage-gleaning birds whereas mangroves and islands have few species, MacArthur attributed the distribution of the yellow warbler to diffuse competition, the combined negative effects from many bird species. However, it is difficult to rule out a significant influence of predation, because the number of species of snakes, hawks, and other predators shows the same pattern as the species diversity of potential competitors.

We conclude this section by emphasizing that it may not be productive to search for single limiting factors and simple explanations for the distributions of species. Not only may a single species be limited by different factors in different parts of its range, but even in one local area several factors may interact in complex ways to prevent expansion of populations. In some cases, particularly in gradients of increasing physical stress, climate and other physical factors may combine with biotic interactions to limit distributions. In other cases, especially in gradients of decreasing physical harshness, diverse biotic interactions act in concert to prevent range expansion. It will often be difficult and misleading to try to assess the relative contributions of single factors if there are complex interactions between various niche dimensions.

Adaptation and gene flow

Although the geographic distributions of most species and higher taxa appear relatively stable over short periods of ecological time, they undergo major changes over evolutionary time. Some of these changes are primarily the result of long-distance dispersal or of geologic events that have altered the geography and climate of the earth. These will be discussed in later chapters. However, many historical changes in species ranges must have resulted from gradual expansions or contractions of peripheral populations. Since natural selection is a universal process that tends to increase the capacity of individuals to survive and reproduce, we might expect that, in general, peripheral populations would gradually adapt to their local environments, increase in densities, and expand their ranges to colonize adjacent areas (see Baker and Stebbins, 1965). Lewontin and Birch (1966) describe an apparent example in the Australian fruit fly, *Dacus tryoni*. During the last century this species has expanded its range several hundred kilometers to the south along the eastern edge of the continent. The flies are limited by low temperature, and the expansion has been accompanied by adaptation of the peripheral populations to increasingly cold climates. But clearly not all species are increasing their ranges in this fashion. A historical perspective suggests that over the last 100 million years or so the distributions of some species and higher taxa have indeed increased, but these have been almost equally matched by contractions in the ranges of other organisms.

Genetic and ecological processes may interact to limit the capacity of peripheral populations to adapt to local conditions and thereby extend their ranges. The exchange of genes between local populations, called gene flow, may prevent populations from acquiring and maintaining the combinations of genes necessary for continual adaptation. Such gene flow is caused by the migration of individuals or gametes between populations, introducing characters that have been selected in the area where they originated. Sufficiently high rates of gene flow obviously can swamp a local population with genes from outside, effectively preventing adaptation to local conditions.

The critical question is whether gene flow is high enough to overwhelm natural selection, prevent continual adaptation of peripheral populations, and thereby preclude expansion into new areas of even more extreme environments. The answer to this question for many organisms is not clear and has been the subject of much debate among population geneticists and evolutionists. Certainly, many species show adaptive genetic changes to variation in their environments over their geographic ranges. Patterns of genetic differentiation in response to environmental gradients are termed clines and have been analyzed carefully by Endler (1977). Even though environmental variation is gradual, the responses of animals and plants need not be. Discontinuous genetic variation, called step clines, frequently occurs in small, relatively sedentary organisms, especially when either genetic or dispersal barriers inhibit gene flow between adjacent populations. Endler suggests that the coincidence of genetic and dispersal barriers can eventually lead to the differentiation of populations along a cline into distinct species.

North American pocket gophers (*Thomomys* spp.) provide perhaps the best example of such differentiation in progress. These fossorial rodents spend almost their entire lives underground in the extensive burrow systems they excavate. They are confined to local areas where the soil is sufficiently deep and friable for tunneling. There is tremendous geographic variation in characteristics such as coat color and body size, as evidenced by the fact that there are described 229 subspecies of *T. umbrinus* and 58 of *T. talpoides* (Hall, 1981). This morphological variation is accompanied by genetic differences between populations in allozymes (proteins coded by specific genes) and in the number and arrangement of chromosomes (Patton et al., 1979). Low rates of dispersal across areas of unsuitable soils and genetic barriers to interbreeding apparently limit gene flow between nearby populations and facilitate adaptation to local conditions. A comparable situation occurs in the ecologically similar Old World burrowing rodents of the genus *Spalax*, suggesting that differences in chromosome numbers between populations may have been important in facilitating colonization of the interior deserts of Israel by populations of *S. ehrenbergi* by reducing gene flow with populations adapted to mesic, coastal areas (Wharmann et al., 1969; Nevo and Bar-El, 1976) (Figure 3.22).

Chromosomal rearrangements also have been implicated in influencing the distributions of vascular plants. Populations at the geographic margins of the species range as well as those inhabiting extreme soil types or climates often are characterized by major chromosomal

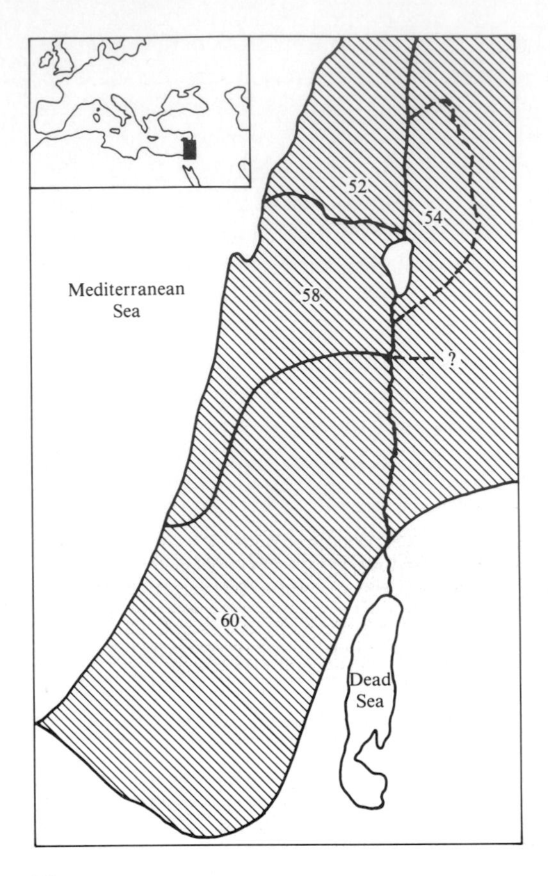

Figure 3.22
Geographic distribution of chromosomal races of the burrowing rodent *Spalax ehrenbergi* in the Middle East. Populations with different numbers of chromosomes replace each other abruptly, with little overlap and hybridization. In general, populations with higher chromosome numbers are found in more arid environments. This suggests that reduced gene flow between the races may be important in permitting populations to adapt to arid conditions and thus to expand from the mesic coastal regions into desert habitats. (After Nevo and Bar-El, 1976.)

changes, especially polyploidy (Stebbins, 1971b). Although these rearrangements of genetic material may sometimes confer specific adaptations, perhaps their most important effect is to reduce the frequency of crossing with other populations and thereby reduce or completely block gene flow, permitting adaptation to the local environment and facilitating colonization of new areas.

Ultimately of course, the capacity of populations to adapt to new environments and to colonize different habitats is limited. All organisms are constrained by their biological characteristics so that they can live in only a subset of the range of habitats and geographic areas available on earth. Although some species are more widely distributed and more tolerant of varying conditions than others, there are no superorganisms that occur everywhere. Adaptation inevitably involves tradeoffs and compromises. In order to tolerate the physical conditions and deal with the biotic interactions in some environments, a species must sacrifice its ability to do well in other habitats (e.g. Clausen et al., 1948). In interbreeding populations, gene flow prevents different individuals from adapting to widely different environments. When gene flow between populations is interrupted, as it is during the process of speciation (see Chapter 6), then over evolutionary time populations can and usually do diverge in response to natural selection, adapt to widely different conditions, and expand their ranges to occupy new habitats and geographic areas.

Selected references

The ecological niche

Cody (1974); Cody and Diamond (1975); Grinnell (1917); Hurlbert (1981); G. Hutchinson (1958, 1978); Krebs (1978); MacArthur (1958, 1972); R.S. Miller (1967); Pianka (1975, 1978); Whittaker and Levin (1975); J. Valentine (1969).

The relationship between abundance and distribution

Albrecht (1967); Andrewartha and Birch (1954); Bock and Lepthian (1976); Boyd and Nunneley (1964); Bystrak (1979); Crosby (1972); Grinnell (1922); Hastings and Turner (1965); Hooper (1942); Kalela (1949); Kessel (1953); Krebs (1978); Likens and Bormann (1974); MacArthur (1960, 1972); R.R. Miller (1961b); Naiman and Soltz (1981); Peterson and Downing (1956); Phillips (1963); Shantz and Turner (1958); Waloff (1966).

Physical limiting factors

Antonovics et al. (1971); G. Bartholomew (1958); Beilmann and Brenner (1951); J. Brown (1971c); Connell (1961); Crisp (1964); Curtis (1959); Dansereau (1957); Daubenmire (1954, 1978); Dayton (1971); Ekman (1953); Griggs (1946); Hocker (1956); Howarth (1980); Hutchins (1947); M. Johnston (1963); Krebs (1978); La Marche (1973, 1978); Levitt (1980); Liebig (1840); Nobel (1978, 1980a, 1980b); Odening et al. (1974); Parker (1963, 1969); Shreve (1915, 1922, 1951); Shrode (1975); Steenbergh and Lowe (1976, 1977); Strain and Billings (1974); Vermeij (1978); Westing (1966); Whittaker (1975); Whittaker and Niering (1968).

Limitation by biotic interactions

B. Bartholomew (1970); J. Brown (1971a); Chappell (1978); Cody and Diamond (1975); Connell (1961); Crowell (1973); Darwin (1859); Dodd (1959); Gause (1934); L. Gilbert (1975); Gilbert and Singer (1975); P. Grant (1972b); Hardin (1960); P. Harris et al. (1969); Huffaker and Kennett (1959); Janzen (1966); Krebs (1978); Lindsey (1964); Lubchenco (1980); Lubchenco and Gaines (1981); Lubchenco and Menge (1978); Malthus (1798); Means (1975); R.S. Miller (1964, 1967); Orians and Willson (1978); Park (1948, 1954, 1962); Pianka (1978); Price (1980); Rosenzweig and MacArthur (1963); Vaughan and Hansen (1964); Yeaton et al. (1981); Zaret and Paine (1973).

Adaptation and gene flow

Antonovics et al. (1971); H. Baker and Stebbins (1965); Clausen et al. (1948); Endler (1977); J. Emlen (1978); Ford (1975); Johnson and Packer (1965); Levin and Kerster (1964); Lewontin and Birch (1966); Patton et al. (1979); Schaal and Levin (1978); G. Stebbins (1971b); Wahrmann et al. (1969); B. Wallace (1960); T. White (1973, 1978).

Chapter 4 Community Ecology

No living thing is so independent that its abundance and distribution are unaffected by other species. The physical and chemical composition of the earth's surface has been drastically altered by the activity of organisms that, among other things, have created the oxygen in the atmosphere and contributed to the development of soil. Organisms vary greatly, however, in the extent to which they are dependent on others. Some autotrophic organisms, such as certain kinds of algae and lichens, not only make their own food from sunlight, carbon dioxide, water, and minerals but also inhabit extremely rigorous physical environments, such as hot springs and bare rocks, where they may encounter and interact directly with few if any other species. Heterotrophic organisms, including all animals, fungi, and some nonphotosynthetic vascular plants, cannot make their own food and are dependent on autotrophs for usable energy. However, even many photosynthetic plants are directly dependent on specific kinds of other organisms, such as nitrogen-fixing bacteria and mycorrhizal fungi, which make available essential mineral nutrients, and insects and vertebrates, which pollinate flowers and disperse fruits and seeds.

Historical and biogeographic perspectives

Species occur together in complex associations called ecological communities. In the early twentieth century, as the new science of ecology was becoming firmly established, some of its foremost practitioners debated the nature of relationships among species that determine the organization of communities. On the one hand, F.E. Clements (1916) suggested that a community could be regarded as a type of superorganism with its own life and structure as well as spatial and temporal limits. According to this view, individual organisms and species could be analogized to the cells and tissues of an organism and the process of secondary succession could be likened to the growth and development of an individual.

In contrast to Clements' concept of the community as a discrete and highly integrated unit, H.L. Gleason (1917, 1926) viewed a community as merely the coexistence of relatively independent individuals and species in the same place at the same time. Focusing primarily on plants, Gleason pointed out that the occurrence of species in an area depends primarily on their individual capacities to immigrate and to grow in the local environment.

As is often the case with such debates, both sides made some important points. Communities do have certain properties, analogous to those of individuals, that can be measured and studied in their context. These properties include photosynthesis (primary production) and metabolism (respiration) as well as the more complex processes associated with the transfer and use of energy and nutrients and with the changes in species composition and habitat during succession (e.g., Odum, 1969, 1971). We also know that some species are interdependent because they require others as mutualists, prey, or hosts (Chapter 3). On the other hand, plant ecologists such as J.T. Curtis (1959) and R.H.

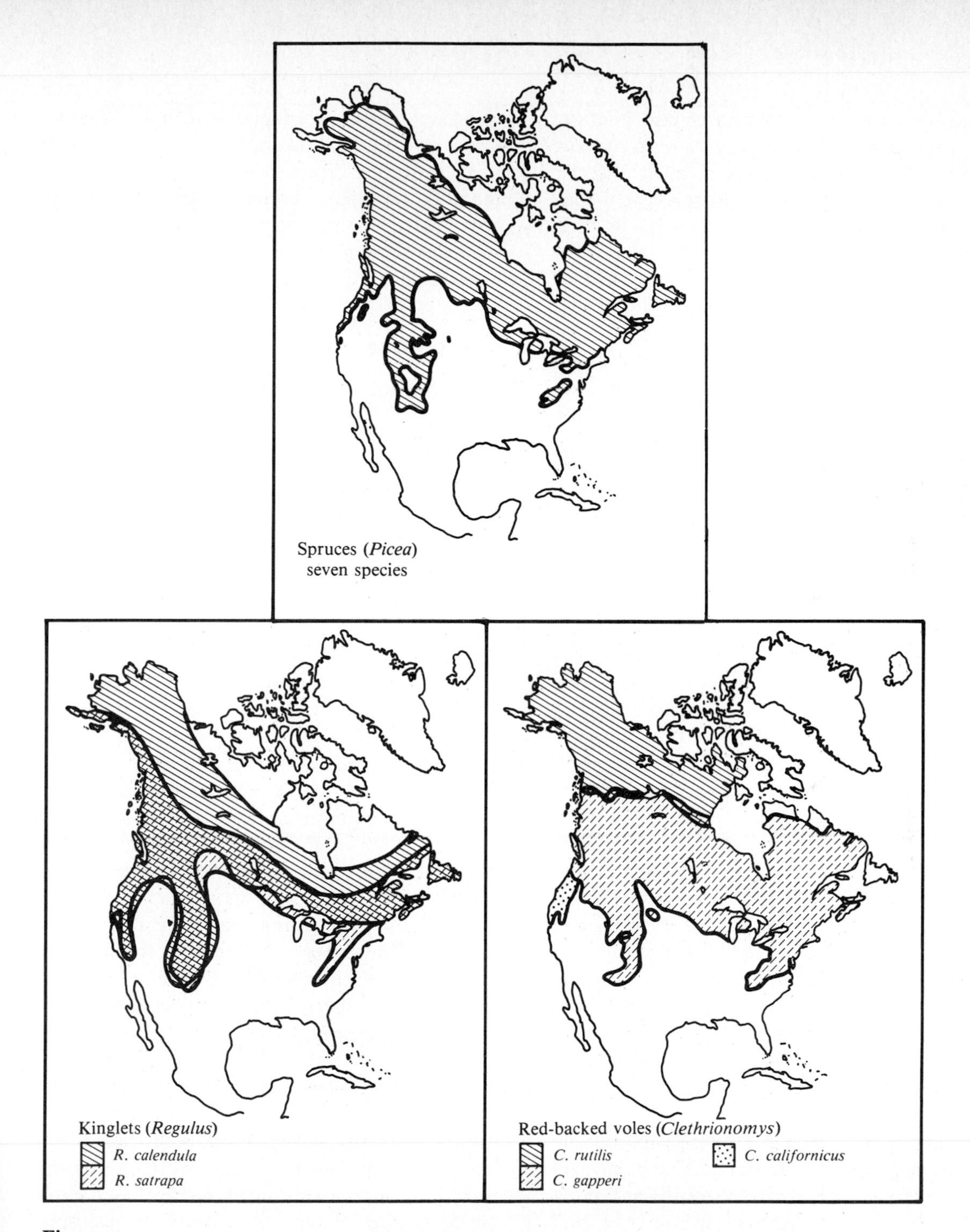

Figure 4.1
Three groups of organisms inhabiting coniferous forests exhibit similar geographic ranges in North America. These seven species of trees (not shown individually), two species of birds, and three species of mammals are typical inhabitants of the boreal forests that spread across the northern part of the continent and extend southward at high elevations in the mountains.

Whittaker (1967, 1975; Whittaker and Niering, 1965) have amassed much data showing that the abundance and distribution of most vascular plant species vary in time and space as if they were independent of other members in their communities (McIntosh, 1967).

The question of particular interest to biogeographers is this: To what extent are species distributed together as entire communities as opposed to being limited individually by direct competitive, predator-prey, or mutualistic interactions with a few other species, or to being distributed essentially independently of each other? If communities are integrated sets of species that are adapted to each other and tolerant of similar physical environments, then we might expect communities to be distributed as discrete units.

Even the casual observer will notice that certain kinds of plants tend to occur together in particular climates to create distinctive vegetation types. Ecologists and biogeographers refer to these as life zones or biomes and recognize that specific kinds of animals and microorganisms as well are associated with these vegetation formations. For example, a broad band of coniferous forest, sometimes referred to facetiously as the spruce-moose biome, extends around the world at high latitudes in the Northern Hemisphere. Not only similar vegetation, but also many of the same species and genera of plants, animals, and microbes are distributed over the Old World from Scandinavia to Siberia and across the New World from Alaska to Nova Scotia and New Brunswick. These organisms are generally adapted to living with each other and in similar physical environments. Their ranges extend southward together when mountain ranges provide appropriate habitats at higher elevations. In North America the coniferous forests extend southward along the Appalachian Mountains in the east and along the Cascades, Sierra Nevada, and Rocky Mountains in the west. Several kinds of organisms are restricted to coniferous forest habitats and have similar geographic distributions (Figure 4.1). We should be cautious, however, about inferring from such patterns that communities represent discrete, highly integrated ecological and biogeographic units, because there are alternative explanations, including historical ones (see Unit 2).

Communities and ecosystems

Definitions. Communities and ecosystems, the highest levels of ecological organization, are rather arbitrarily defined. A community consists of those species that live together in the same place. The member species can be defined either taxonomically or on the basis of more functional ecological criteria such as life form or diet. The place where member species occur can be designated by arbitrary boundaries or on the basis of more natural topographical features. Thus we can speak of the fish community in some ecologist's study area of a 2-hectare coral reef on the north shore of Jamaica; of the grazing community (which would include representatives of several invertebrate phyla as well as some species of fish) on the entire reef, which might extend uninterrupted over many square kilometers; or of the entire community of all coral reef–inhabiting organisms in the Caribbean Sea.

An ecosystem includes not only all the species inhabiting an area but also all the features in the physical environment. Because ecosystem ecologists have been interested primarily in the exchange of energy, gases, water, and minerals among the biotic and abiotic components of a particular ecosystem, they often try to study naturally confined areas where the input and output of energy and material are restricted and hence easier to control or monitor (e.g., Odum, 1957; Teal, 1957, 1962). Small, relatively self-contained ecosystems are often called microcosms because they represent miniature systems in which most of the ecological processes characteristic of larger ecosystems operate on a reduced scale. A sealed terrarium is a good example of an artificial microcosm, and a small pond is an example of a natural one.

Few systems are as isolated and independent as they appear. The largest and only really complete ecosystem is the biosphere, which encompasses the earth. The interdependence of all ecosystems is vividly demonstrated by the widespread impact of human activity. Acid rain falls on virgin forests and is carried into pristine lakes far from the source of pollution. Deforestation, especially the cutting of tropical forests, and the burning of fossil fuels are changing the composition of the atmosphere and perhaps altering the climate throughout the world.

Many ecologists are actively investigating the structure and function of communities and ecosystems. As is usually the case in a rapidly developing field, many of the important questions remain unanswered and some tentative conclusions are highly controversial. Because many of the unresolved issues are relevant to biogeography, this can be both challenging and frustrating for those trying to account for plant and animal distributions. On the one hand, biogeographers have the opportunity to use their own methods and data to make important advances in community ecology. Those with a broad biogeographic perspective are in a position to make unique contributions to the unraveling of interacting ecological processes that influence the number and kinds of species that live together in different parts of the earth. On the other hand, until these ecological processes are understood, the biogeographer will be unable to integrate them with the effects of historical events to provide a solid conceptual basis for interpreting and predicting distribution patterns.

Community organization: energetic considerations. In Chapter 3 we emphasized that each species has a unique ecological niche that reflects the biotic and abiotic environmental conditions necessary for its survival. How are these individual niches organized to produce complex but stable associations consisting of many species? This is the ultimate question of community ecology, and one for which we can advance only tentative answers. Two characteristics, body size and trophic status, particularly influence the roles species play in communities.

If you wanted to take a single measurement that would provide the greatest information on the biology of an organism, you would be best advised to take its body weight. The primary reason is that the larger an organism, the more energy it requires for maintenance, growth, and reproduction. The rate of energy uptake and expenditure, the metabolic rate (m), varies with body mass (M) according to the relationship $m = cM^{0.75}$. Although the value of the constant (c) varies somewhat among taxonomic groups, the rate of increase of metabolic rate with increasing size, given by the exponent, $^{0.75}$, is very general (Hemmingsen, 1960) (Figure 4.2). The ecological consequences of this simple formula are profound. Because the exponent is positive, total energy requirements increase with increasing body size. It takes about 14,000 times more energy to maintain a 5000 kg elephant than a 15 g mouse. The same applies to all organisms; a mature redwood tree uses much more energy than a strawberry plant. On the other hand, because the exponent is less than 1, the metabolic rate per unit mass of a small animal is greater than that of a larger organism. It requires about 25 times more energy to maintain a gram of mouse than a gram of elephant. Apparently for this reason almost all rate processes are accelerated in small organisms, which are more active and have higher reproduction rates and shorter life expectancies (presumably because they wear out sooner) than large ones.

Body size also has important ecological consequences because it influences the scale on which organisms use the environment. Because small organisms require fewer resources per individual than large ones, they can become more specialized and still maintain sufficiently high

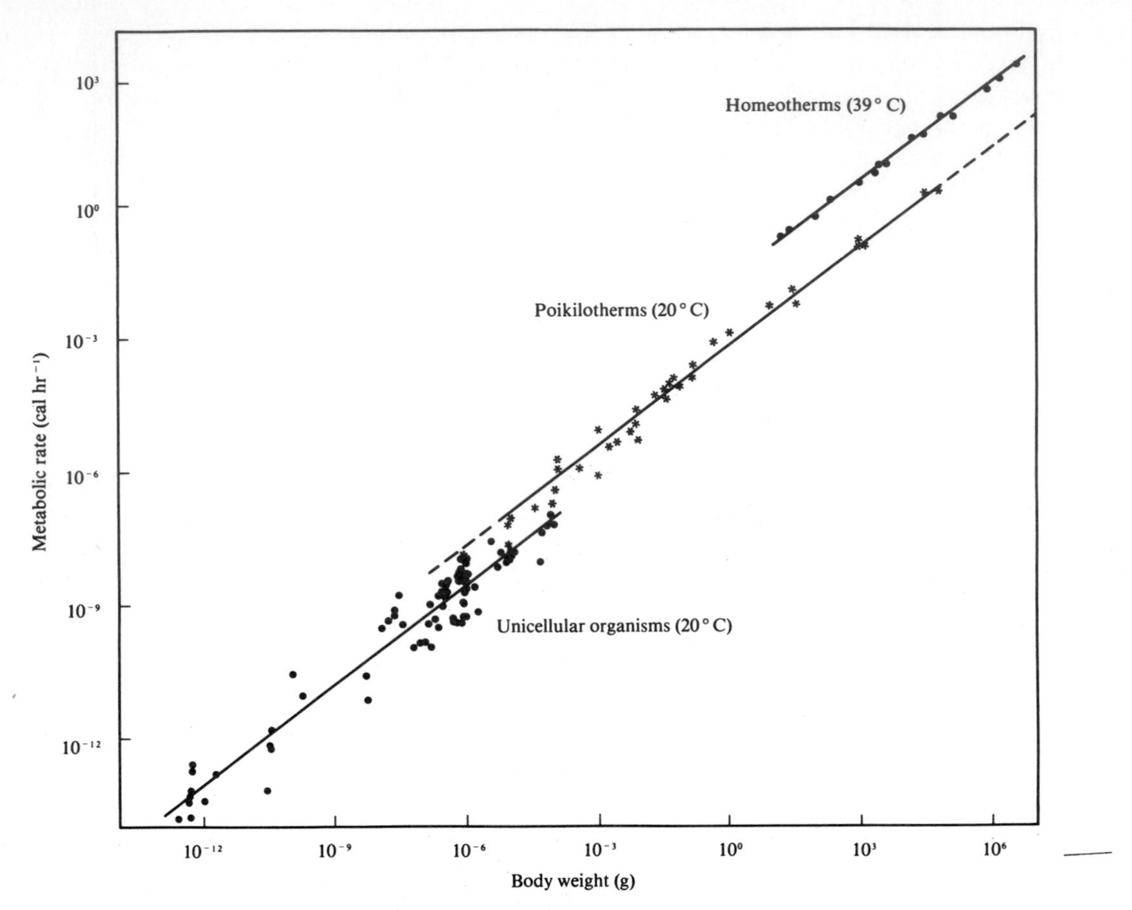

Figure 4.2
Relationship between metabolic rate (m) and body mass (M) for a wide variety of organisms, from unicellular forms, to poikilothermic (cold-blooded) animals, to homeothermic birds and mammals. Note that the axes are in a logarithmic scale, so the relationship is described by a power function of the form $m = cM^{0.75}$, where the constant (c) varies slightly for the three different groups, but the exponent (slope) of $^{0.75}$ is remarkably constant. (After Hemmingsen, 1960.)

population densities to avoid extinction. Small organisms are better able than large ones to respond to what Hutchinson (1959) termed the mosaic nature of the environment, the spatial heterogeneity or patchiness of the environment that is pronounced on a small scale. Consequently, small organisms have been able to divide the environment more finely so that any geographic area contains a greater number of small-bodied species than of large ones (Van Valen, 1973a; May, 1978) (Figure 4.3). Consider the tremendous diversity (about 800,000 known species) and specializations of insects as compared with terrestrial vertebrates (about 20,000 species). A biogeographic correlate of this pattern is that large organisms are constrained to have wide geographic ranges (Figure 4.4). The reason for this should be obvious. Because each individual requires more space (McNab, 1963; Schoener, 1968b), the carrying

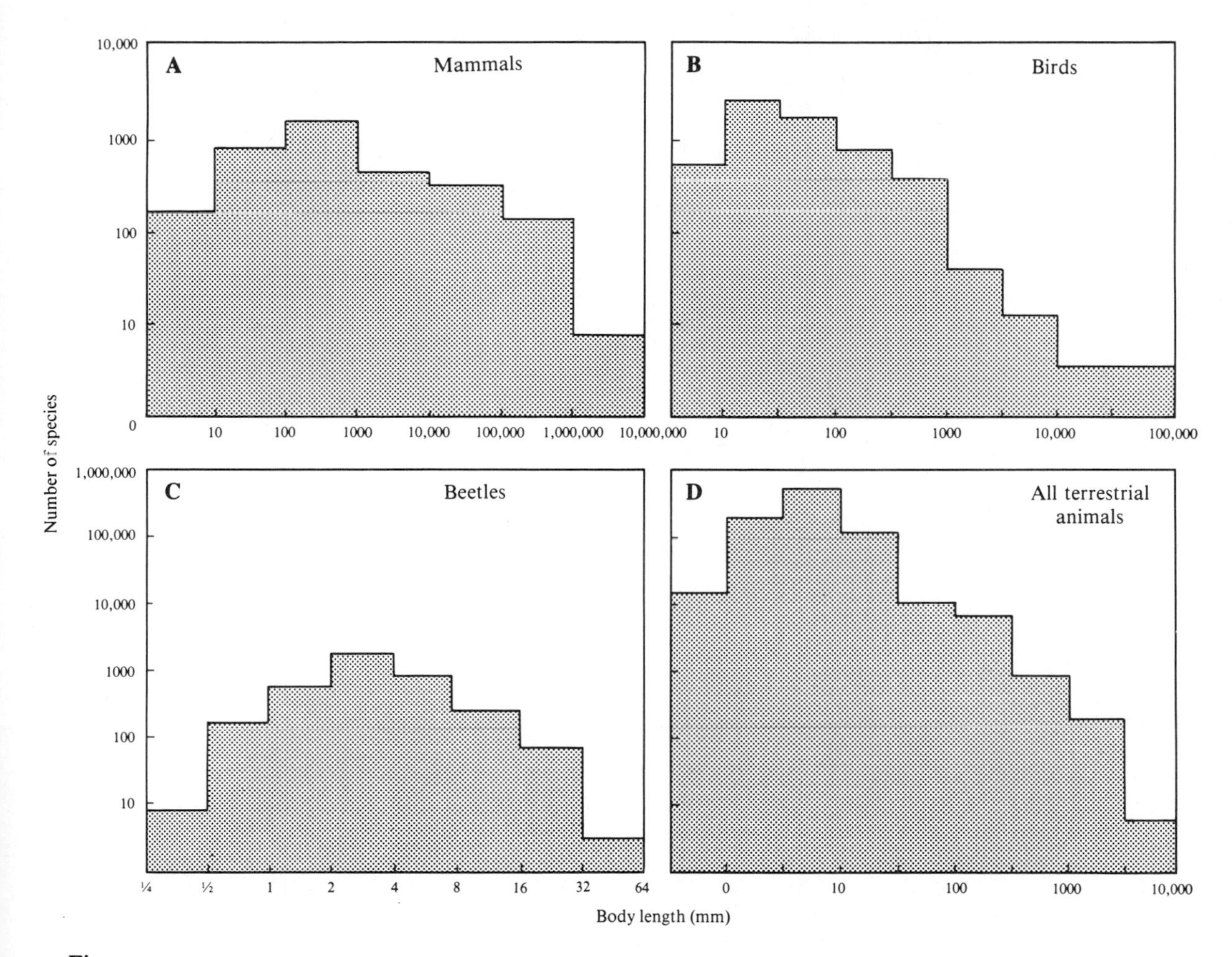

Figure 4.3
The frequency distribution of body size among species for several different kinds of organisms. **A**, Terrestrial mammals of the world. **B**, Land birds of the world. **C**, Beetles of Great Britain. **D**, All terrestrial animals of the world (approximately). Note that the axes are in a logarithmic scale, so small species are much more numerous than large ones. This pattern is very general and accounts for the obvious fact that insects are much more numerous (in numbers of species as well as individuals) than vertebrates in terrestrial environments. (After May, 1978.)

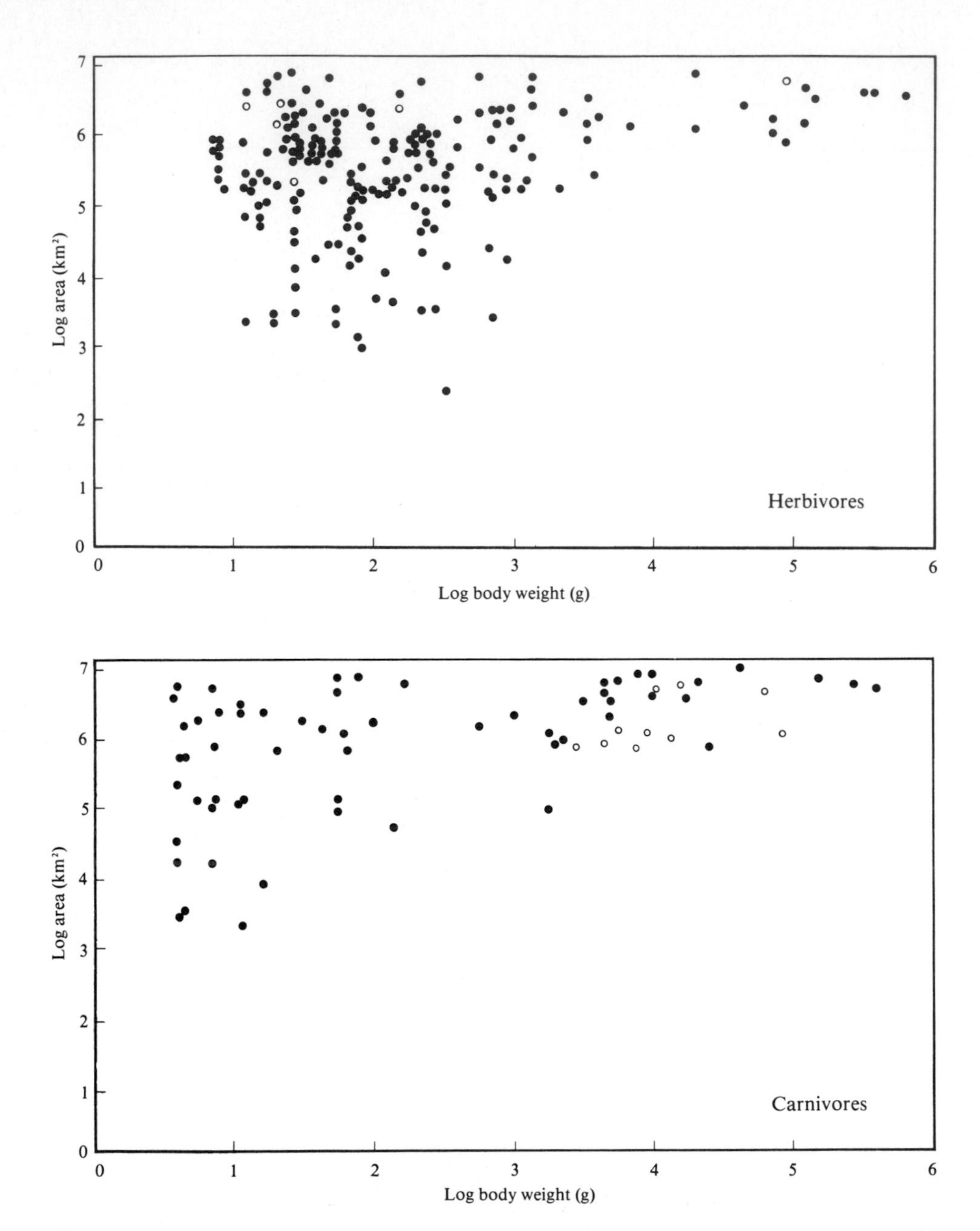

Figure 4.4
Relationship between area of geographic range and body size in North American terrestrial mammals. Note that although there is much variation, the areas of the smallest ranges increase with increasing body size and are smaller for herbivores than for carnivores. Unshaded circles indicate species that have larger ranges than indicated, because they range into Central America. Only the North American area of the distribution has been measured. (From Brown, 1981.)

capacity of the environment for these animals is low and only large areas can support sufficient numbers of individuals to maintain the species over evolutionary time. In contrast, small areas can support large populations of small organisms that consequently have low probability of extinction. Small species may or may not have very restricted ranges (Figure 4.4).

The trophic status of organisms, or how they acquire energy, also influences the role they play in community structure. Essentially all of the energy used by living things ultimately

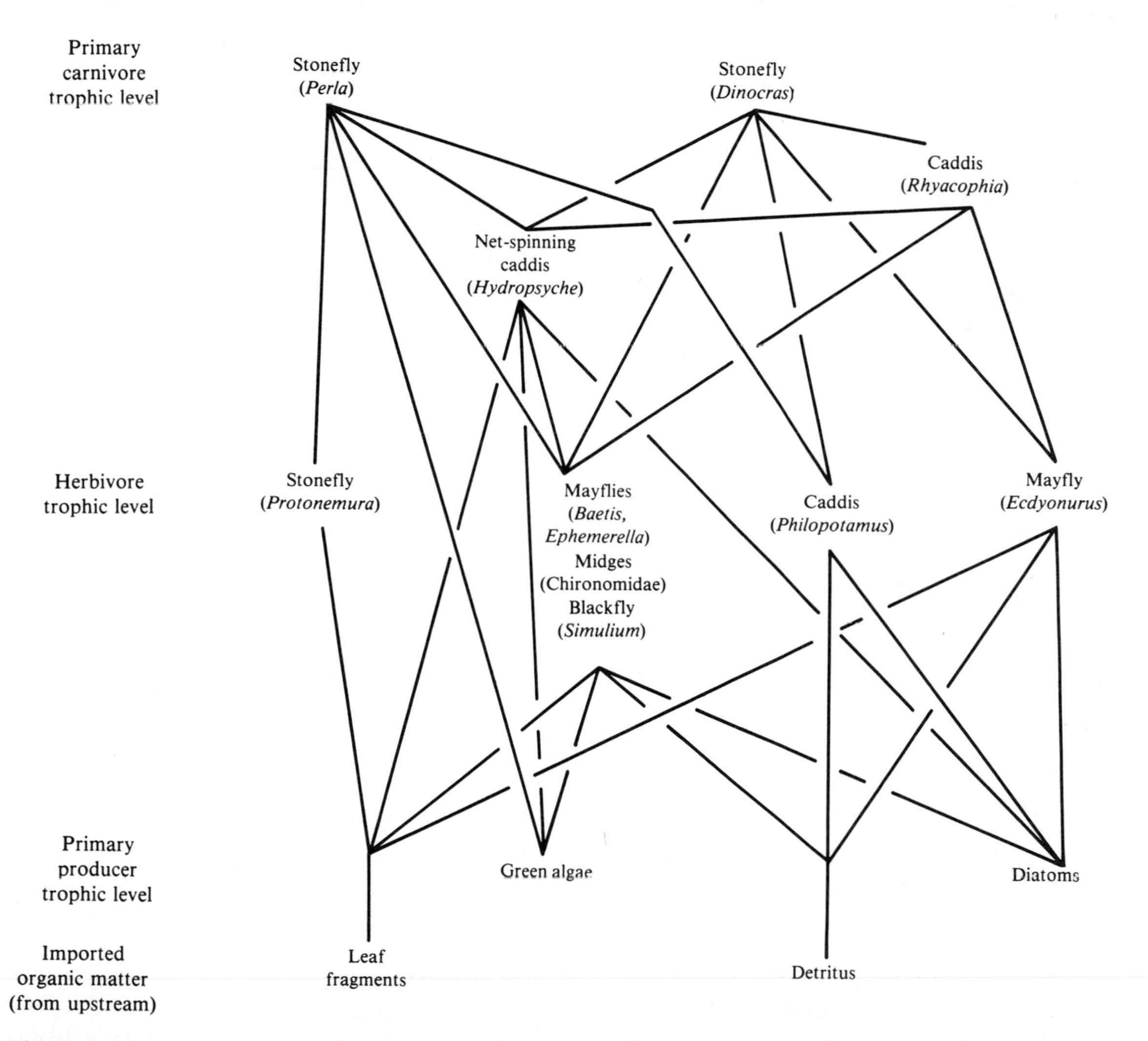

Figure 4.5
A part of the food web for an aquatic community inhabiting a small stream in Wales. The diagram shows the interconnections of food chains to form food webs. There are three trophic levels, but some organisms such as *Hydropsyche*, which feeds on both plant and animal material, may occupy intermediate positions. The many individual species of leaves, green algae, and diatoms that make up the primary producer trophic level are not shown individually. (Redrawn from Jones, 1949.)

comes from the sun. Autotrophic green plants use solar radiation, carbon dioxide, water, and minerals to synthesize organic compounds, and oxygen is produced as a by-product. The organic compounds are not only used by the plants for making structures and fueling basic metabolism but are also the sole source of energy for the heterotrophic organisms in the community. Solar energy, trapped in organic molecules, is transferred from one species to another and gradually used up as herbivores eat plants, carnivores eat herbivores, and so on. These heterotrophic species oxidize the organic compounds to obtain usable energy and in the process consume oxygen and release carbon dioxide. The unidirectional paths of energy flow between species through communities are termed food chains. The different links in a food chain are called trophic levels (Figure 4.5). The first level contains green plants or primary producers, the second herbivores or primary consumers, the third carnivores or secondary consumers, and so on. At the ends of food chains are the decomposers or detritivores, mostly bacteria and fungi, which break down the last organic matter and release the inorganic minerals into the soil or water where they can be recycled and again taken up and used by plants. Although each tiny packet of energy would follow a linear path along a food chain, the actual trophic relationships among species are complex. Because most consumers feed on several species and are in turn consumed by several kinds of predators and decomposers, the aggregate paths of energy flow through a community form interconnected branching patterns called food webs (Figure 4.5).

As energy is distributed in this fashion through a community, it obeys the laws of thermodynamics, which have important implications for community organization. Simply stated, the first law of thermodynamics says that energy is neither created nor destroyed, but it can be converted from one form to another, as, for example, from sunlight to the high-energy bonds of organic molecules to a muscular contraction. The second law of thermodynamics states that as energy is converted into different forms, its capacity to perform useful work diminishes and the disorder (entropy) of the system increases. Organisms use the energy stored in organic molecules to perform the work of moving, growing, and reproducing. As the organic compounds are oxidized to provide energy for these activities, much of the energy is dissipated as heat. In fact, most organisms are very inefficient. Most animals are able to incorporate only 0.1% to 10% of the energy they ingest into the organic molecules of their own bodies, and the remaining 90% to over 99% is lost as heat.

One ecological consequence of these thermodynamic properties of organisms is that substantially smaller quantities of energy are available to successively higher trophic levels. This can be shown diagrammatically by portraying the community as an ecological pyramid (Figure 4.6). In any community the green plants have the highest rates of energy uptake, and they usually comprise the largest number of individuals, the greatest number of species, and the greatest biomass (the total quantity of organic material). Successively higher trophic levels tend to have only 0.1% to 10% the rate of energy uptake of their prey and usually have proportionately lower biomass and fewer individuals and species. Availability of energy must decrease substantially with each successive trophic level in accordance with the laws of thermodynamics, but there are exceptions to the other patterns. The distribution of biomass, individuals, and species among trophic levels depends on how energy is acquired and used by the species in the community. In some communities, such as those inhabiting caves (Poulson and White, 1969) and the dark depths of ocean trenches, there are no autotrophic organisms and all energy is imported in organic form, usually as dead material called detritus. In the deciduous forests of eastern North America and western Europe there may be fewer individuals and species of primary producers (autotrophs) than of primary consumers, because most of the

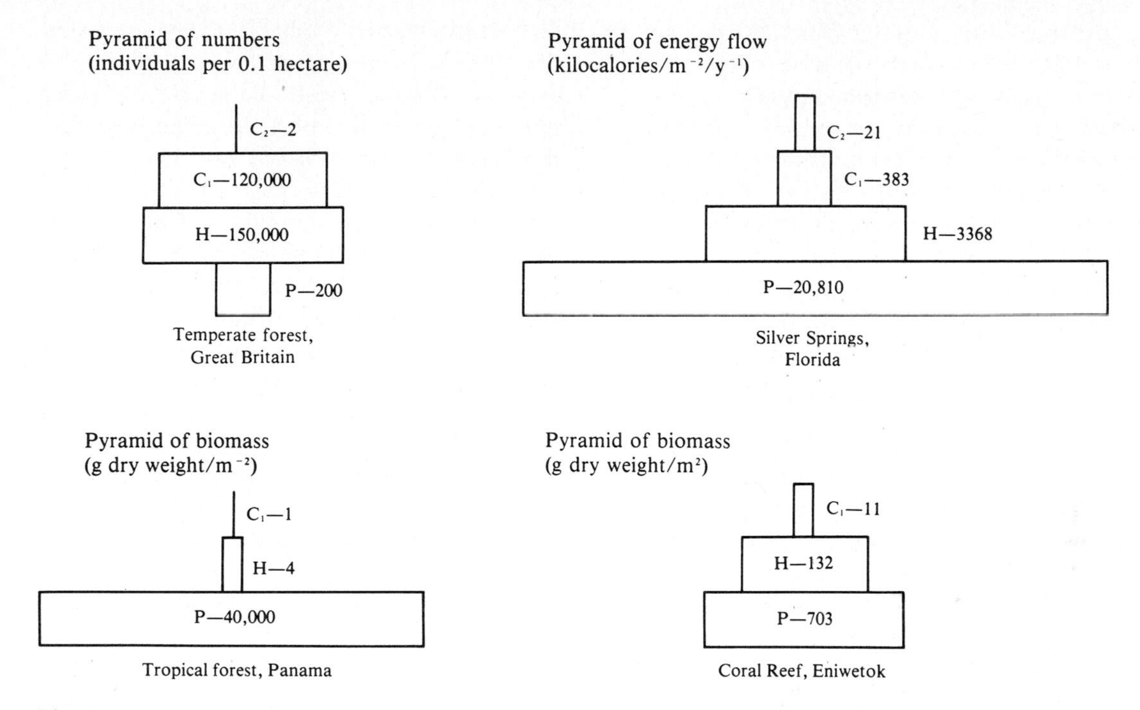

Figure 4.6
Ecological pyramids of energy flow, biomass, and number of individuals for several diverse communities. Note that three of the pyramids are regular, with each successive trophic level reduced compared to the one below, but the pyramid of individuals for the temperate forest has fewer primary producers (mostly trees) than herbivores (mostly insects). Trophic levels are indicated as follows: *P*, Primary producers; *H*, herbivores; C_1, primary carnivores; C_2, secondary carnivores. (After Odum, 1971.)

energy used by plants is monopolized by a relatively small number of large trees whereas the energy used by herbivores is allocated among many individuals and species of small insects (Elton, 1966; Varley, 1970).

Because the carrying capacity, measured in units of usable energy, of any area is lower for successively higher trophic levels, the organisms occupying these levels exhibit predictable characteristics that affect their ecological roles and their geographic distributions. Not only are there fewer species of carnivores than of herbivores and plants but also the carnivores tend to be large and generalized. Carnivores usually have to be large enough to overpower their prey. They also tend to feed on several prey species and to have broad habitat requirements and wide geographic distributions (Figure 4.4) (Brown, 1981; see also Van Valen, 1973a). For example, the cougar or mountain lion *(Felis concolor)* is one of the top carnivores in the New World. It weighs 50 to 100 kg and takes a variety of prey, mammals ranging in size from rabbits to elk. The cougar also has the widest geographic distribution of any American mammal. It ranges from Alaska to the southern tip of South America and (at least originally) from the Atlantic to Pacific coasts on both continents. The cougar inhabits tropical rain forests, coniferous and deciduous hardwood forests, shrublands, and deserts. Other top carnivores, such as the wolf *(Canis lupus)* and jaguar *(Felis onca)*, also have broad geographic and habitat distributions, whereas herbivores of comparable size tend to have more restricted ranges.

Parasites are an exception to these patterns, but they still illustrate the fundamental importance of energetic relationships in community organization. Parasites are usually much smaller and more highly specialized than their hosts. Consequently, individual parasites need not consume large numbers of host individuals to meet their energy requirements, and parasites can be more numerous than their hosts. One or more parasites usually inhabit a host for an extended period, often for the entire lifetime of either parasite or host. Parasites are often highly specialized to infect only one or a few, often taxonomically related, host species. Although these adaptations have allowed parasites to maintain geographic ranges and numbers of individuals roughly comparable to their hosts in the next lower trophic level, many species have had to solve the problem of finding and infecting sparsely distributed hosts. The elaborate, highly specialized life cycles of many parasites

Table 4.1
Net primary production and biomass of major kinds of continental and marine habitats
Modified from Whittaker and Likens (1973).

Habitat	Area (10^6 km^2)	Net primary production per unit area (g/m^2/yr)	Total net primary production (10^9 MT/yr)	Mean biomass per unit area (kg/m^2)
Continental				
Tropical rain forest	17.0	2000.0	34.00	44.00
Tropical seasonal forest	7.5	1500.0	11.30	36.00
Temperate evergreen forest	5.0	1300.0	6.40	36.00
Temperate deciduous forest	7.0	1200.0	8.40	30.00
Boreal forest	12.0	800.0	9.50	20.00
Savanna	15.0	700.0	10.40	4.00
Cultivated land	14.0	644.0	9.10	1.10
Woodland and shrubland	8.0	600.0	4.90	6.80
Temperate grassland	9.0	500.0	4.40	1.60
Tundra and alpine meadow	8.0	144.0	1.10	0.67
Desert shrub	18.0	71.0	1.30	0.67
Rock, ice, and sand	24.0	3.3	0.09	0.02
Swamp and marsh	2.0	2500.0	4.90	15.00
Lake and stream	2.5	500.0	1.30	0.02
TOTAL CONTINENTAL	149.0	720.0	107.09	12.30
Marine				
Algal beds and reefs	0.6	2000.0	1.10	2.00
Estuaries	1.4	1800.0	2.40	1.00
Upwelling zones	0.4	500.0	0.22	0.02
Continental shelf	26.6	360.0	9.60	0.01
Open ocean	332.0	127.0	42.00	0.003
TOTAL MARINE	361.0	153.0	55.32	0.01
WORLD TOTAL	510.0	320.0	162.41	3.62

appear to be largely adaptations to get into appropriate hosts. Price (1980) gives an interesting account of some special features of parasite ecology and evolution.

These kinds of energetic considerations lead to the prediction that the capacities of habitats or geographic areas to support many individuals and diverse species of organisms ultimately depend on total productivity. Everything else being equal, the higher the fixation rate of sunlight into organic material, the more usable energy should be available to be subdivided among individuals, species, and trophic levels. Productivity varies greatly among different habitats, depending on such factors as climate, soil type, water availability, and the influence of human activity (Jordan, 1971; Lieth, 1973; Whittaker and Likens, 1973) (Table 4.1). In general, the predicted relationship between productivity and diversity is observed. Widespread, highly productive habitats such as tropical rain forests and coral reefs are renowned for their great diversity of specialized species. In contrast, small, isolated areas, such as small islands, and widespread, unproductive habitats, such as deserts and tundra, contain fewer and for the most part more generalized species. We will examine such patterns of species diversity in more detail in Chapters 16 and 17.

Distribution of communities in space and time

Spatial patterns. We are left, then, with an apparent paradox. On the one hand, we might expect species to be distributed together as discrete communities, because they are adapted to coexist with each other and to tolerate the same physical conditions. On the other hand, we expect many species not to be closely associated, because close relatives will be displaced in response to competition to depend on different resources, and more distantly related species often will be of different body size and trophic status and consequently will use the environment on different spatial and temporal scales. How do we resolve this dilemma?

It is obvious that abrupt changes in the environment, such as those found at the shore of a lake or the edge of a forest, are usually accompanied by rapid transitions between two associations of species. If the environmental discontinuity is rapid and severe, many of the species living on each side will be limited almost simultaneously when they encounter inhospitable conditions at the border. There may, however, be some interesting and unexpected results. If the two habitat types are rather similar, the edge or ecotone may actually contain more species than either pure habitat. Species from either side of the boundary may be able to mix and occur together in the narrow area where the environments meet. If the transition zone is fairly productive and not too narrow it may even support its own community of organisms uniquely adapted to live there. This is especially likely if the boundary is not simply an ecotone, with conditions intermediate between those on either side, but an environment in which conditions are unique or fluctuate back and forth between those found on either side. The most dramatic example of an entire community interposed in the narrow transition zone between two much more extensive habitat types is provided by the intertidal zone. Special adaptations are required to withstand prolonged periodic inundation in seawater followed by exposure to a desiccating terrestrial environment. On the narrow strip of shore there are entire communities of such plants and animals, which show their own small scale patterns of association and segregation within the vertical gradient of tidal exposure (e.g., Connell, 1961, 1975; Menge and Sutherland, 1976; Lubchenco and Menge, 1978; Lubchenco, 1980; Souza et al., 1981).

How are species distributed along gradual environmental gradients, such as the variation in exposure within the intertidal zone, or in climate within the elevational gradient on terres-

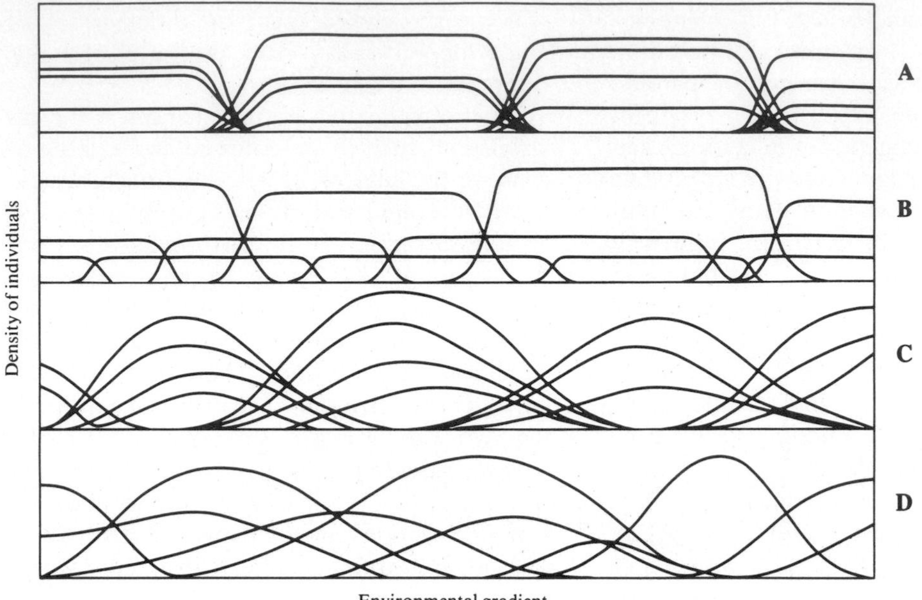

Figure 4.7
Four hypothetical patterns of distributions of species along an ecological or geographic gradient. **A,** Species distributed as discrete communities that replace each other abruptly. **B,** Species not segregated into communities, but some sets replace each other abruptly. **C,** Species distributed as discrete communities, which gradually replace each other. **D,** Species behave as if they are independent of each other, neither associating in discrete communities nor replacing each other abruptly. (From Whittaker, Communities and ecosystems, second edition. Reprinted with permission of Macmillan Publishing Co., Inc. Copyright © 1975 by R.H. Whittaker.)

trial mountains? Some of the most thorough studies have been those of Whittaker and his colleagues on the distribution of tree species on mountainsides in the United States. The data can be used to evaluate four alternative hypotheses (Whittaker, 1975) (Figure 4.7):

1. Species are distributed as discrete communities with sharp boundaries between them. This could be caused by competitive exclusion between dominant species and by other species evolving to coexist with the dominants and with each other.
2. Individual species abruptly exclude one another along sharp boundaries, but most species are not closely associated with others to form discrete communities.
3. Species form discrete communities that gradually replace each other. This could happen if species evolved to coexist with one another, but competitive exclusion did not cause rapid replacement of species along the gradient.
4. Individual species gradually appear and disappear, seemingly independent of the presence or absence of other species. Species neither competitively exclude each other nor associate together to form discrete communities.

Whittaker and his associates (e.g., Whittaker, 1956, 1960; Whittaker and Niering, 1965) gathered data on the elevational distributions of tree species in several mountain ranges in order to test these hypotheses. They sampled several

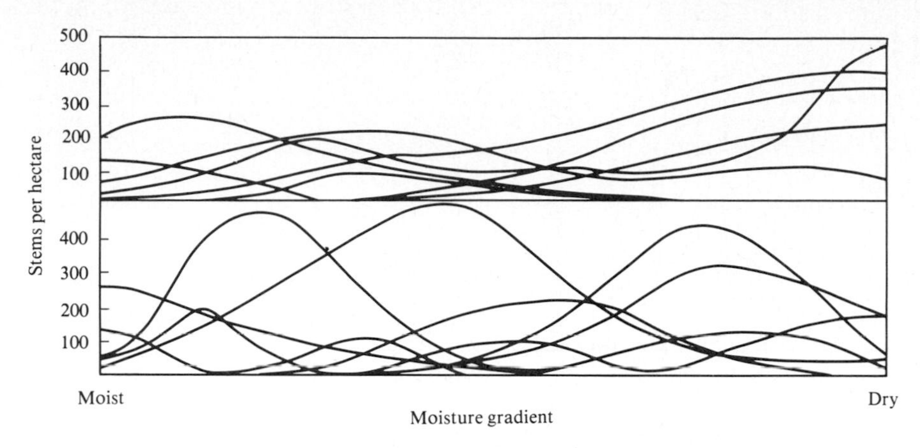

Figure 4.8
The actual distribution of tree species in two elevational gradients. *Top,* In the Siskiyou Mountains, Oregon, 760 to 1070 m elevation. *Bottom,* In the Santa Catalina Mountains, Arizona, 1830 to 2140 m elevation. Note that the species replace each other gradually and seemingly independently of each other. The species have narrower elevational ranges in the steeper moisture gradient in Arizona. (After Whittaker, 1967.)

sites at each elevation on a mountain. Sites were chosen so as to keep soil type, slope, and exposure as constant as possible but to allow natural variation in temperature and rainfall. Results of two such studies, shown in Figure 4.8, clearly support the hypothesis that species are distributed as if they were independent of each other, with no evidence of either abrupt replacements that might be attributed to competitive exclusion or association of species to form discrete communities.

It should be pointed out, however, that Whittaker's methods of collecting and analyzing data may have contributed to his obtaining the patterns shown in Figure 4.8. By censusing relatively large plots and then averaging the results to obtain his species abundance curves, Whittaker would tend to miss abrupt replacements of species owing to competitive exclusion and mediated by the distribution of microsites in a spatially heterogeneous environment. Yeaton (e.g., Yeaton, 1981; Yeaton et al., 1980) has made careful analyses of the elevational distributions of pines *(Pinus)* in western North America. He finds many instances of abrupt replacements by species of similar growth form, apparently as a result of interspecific competition (Figure 4.9). For example, digger pine *(P. sabiniana)* reaches its upper elevational limits and is abruptly replaced by ponderosa pine *(P. ponderosa)* at approximately 840 m on southeastern-facing slopes of the Sierra Nevada near Yosemite National Park. Digger pine attains its greatest density and growth rate in the mesic environments at higher elevations just before being supplanted by ponderosa pine, suggesting that it is competitive exclusion, not inability to tolerate the physical environment, that determines the upper limit of its range.

In general, however, Whittaker's results quite accurately reflect the typical distribution of species along gradients. Abrupt replacements by competing species, as documented above for pines and in Chapter 3 for chipmunks, undoubtedly occur in many cases in which ecologically similar, closely related species come into contact. Careful studies of highly coevolved

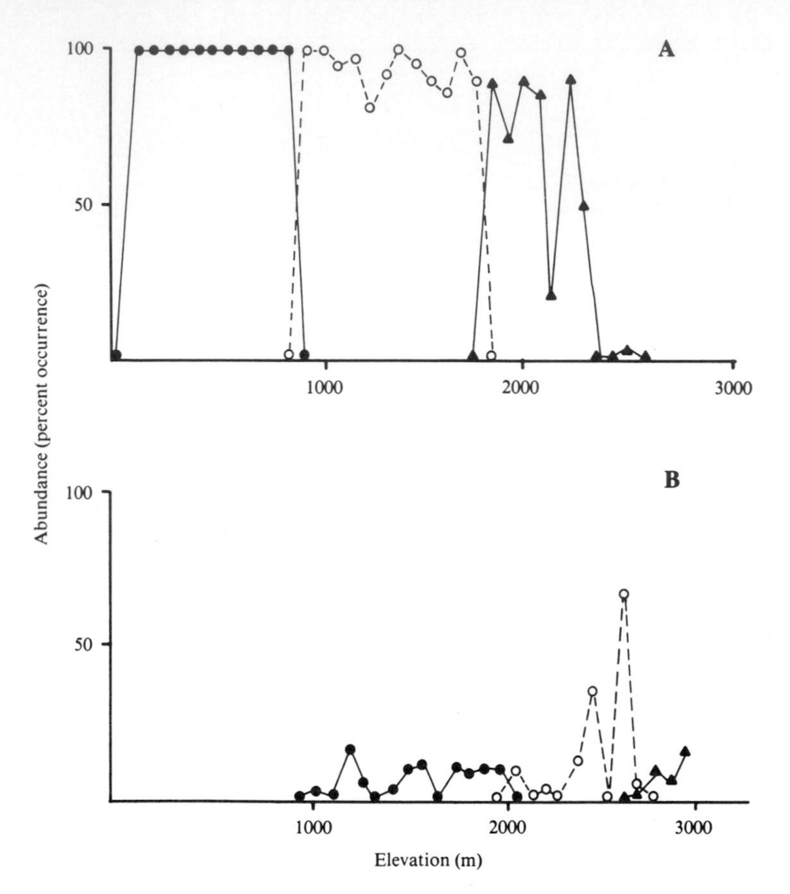

Figure 4.9
Elevational distributions of pine trees on the western slope of the Sierra Nevada in California. **A,** Three species of three-needled pines. **B,** Three species of five-needled pines. The species with the same number of needles are morphologically and ecologically similar and they overlap little in elevation on sites with similar slope, exposure, and soil type. (Redrawn from Yeaton, 1981.)

species in different trophic levels, such as parasites and their hosts or plants and their pollinators, reveal many instances in which such pairs of species have, as we might expect, virtually identical ranges. Nevertheless, the vast majority of species in any reasonably diverse community are probably distributed relatively independently of each other, because most species do not interact strongly and directly with each other. This is almost a necessary consequence of the large numbers of species present in most communities. The number of possible pairwise interactions between species varies with the number of species (S) as $(S^2 - S)/2$. If a community contained only 50 species, each species could interact with each of the 49 others, resulting in a total of 1225 possible direct pairwise interactions. Clearly each species cannot be finely adapted to all of the other species with which it coexists. An organism may be involved in a few strong competitive, predator-prey, or mutualistic interactions that have important influences on its abundance and distribution, but it must be relatively independent of

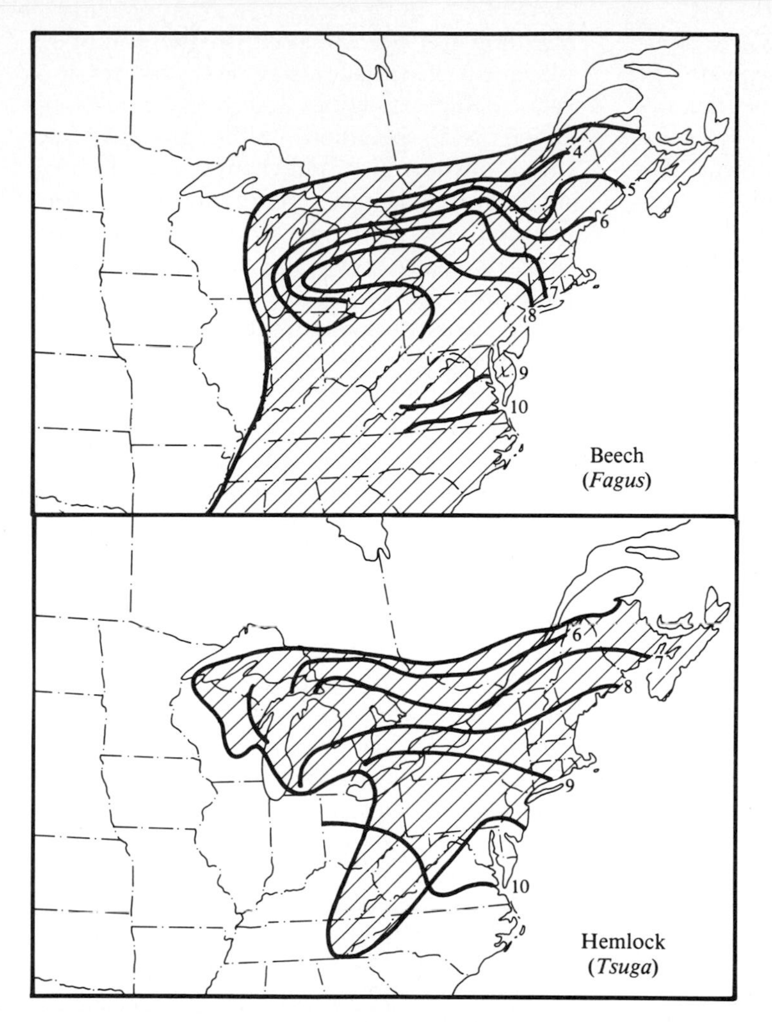

Figure 4.10
Reconstruction from fossil pollen records of the recolonization of North America by two species of trees, beech *(Fagus)* and hemlock *(Tsuga)*, since the last Pleistocene glaciation. Numbered lines indicate the fronts of each species at 1000-year intervals BP, showing the progressive northern migration of each species. Note that the migration of these trees was quite different, although the northern borders of their present ranges *(shaded areas)* are virtually identical. (After Bernabo and Webb, 1977.)

the direct effects of most of the other species in the community.

Temporal patterns. Individual species and entire ecological communities replace each other in time as well as in space. These changes have usually been of more interest to ecologists, studying relatively short time scales, and to paleontologists, studying very long ones, than to biogeographers. Ecologists, especially those studying sessile organisms, such as terrestrial plants or the attached animals and plants of the intertidal zone, have devoted much effort to the study of secondary succession (Clements, 1916; Beckwith, 1954; Monk, 1968; Odum, 1969;

Drury and Nisbet, 1973; Horn, 1974). Secondary succession is the predictable sequence of colonization and replacement of species that occurs following the clearing of space, a sequence eventually leading to the reestablishment of a final or climax community. Succession is a normal process in any ecosystem in which disturbances such as fires or storms repeatedly eliminate entire communities from patches of local habitat. The pattern of temporal replacement of species in succession is quite comparable to the spatial replacement of species along environmental gradients. Although a few species may have strong positive or negative associations with each other, in general species invade, increase, and die out seemingly independently of one another until succession has run its course and the climax community has been reestablished.

Of perhaps more interest to biogeographers are temporal changes in communities that take place over larger geographic areas and longer periods of time. Davis (1969, 1976; see also Bernabo and Webb, 1977) used the analysis of fossil pollen to document the reestablishment of eastern deciduous forest communities following the climatic and geologic changes that accompanied Pleistocene glaciation. Since the retreat of the continental glaciers and the associated colder climates and coniferous forests (over the last 10,000 years), individual species of deciduous trees have expanded their distributions northward and recolonized northeastern North America at different rates (Figure 4.10). Apparently, rates of invasion depended in large part on mechanisms of seed dispersal and other aspects of the life history. Clearly, however, the present communities of local regions were established by the gradual colonization and elimination of individual species, not by the abrupt replacement of entire, discrete communities.

The taxonomic specialization of most paleontologists and the fragmentary nature of the fossil record have posed challenges for those attempting to reconstruct the changes in communities that occurred over long spans of evolutionary time (Gray et al., 1981.) Much attention has been focused on the abrupt replacement of virtually entire biotas as a result of catastrophic extinctions followed by recolonization and speciation. Although the exact causes of these historical changes are still hotly debated, they almost certainly must have been triggered by drastic environmental perturbations. In many ways these rapid historical replacements are comparable to the abrupt spatial replacements of communities across major environmental discontinuities. Although these catastrophic events were spectacular, historical changes in communities more frequently were gradual. Species colonized, speciated, and became extinct relatively independently of each other, much like the pattern of spatial replacement of contemporary species in gradual gradients. As evidence of this we can point to the coexistence in modern communities of both living fossils, forms that have survived virtually unchanged for hundreds of millions of years, and recently evolved species that have been formed by speciation and adaptive radiation within the last few million years.

Terrestrial biomes

The fact that communities do not represent discrete associations of species in either time or space obviously complicates any attempt to classify the communities of species that inhabit particular physical environments or geographic regions. Where physical and geographic changes are abrupt it is relatively easy to recognize distinct community types, but when environmental variation is gradual we are faced with the problem of dividing an essentially continuous variation in species composition, life form, or other community traits into a discrete number of arbitrary categories. The human mind seems to require or at least to depend heavily on such classifications (e.g., the stages of mitosis, the geologic periods, the biogeographic provinces) even when we are aware that they represent artificial divisions of continuous processes or variables.

In the last few decades mathematicians and

biologists have developed sophisticated multivariate statistical techniques for quantifying the degree of similarity (or difference) between two samples based on a large number of variables. Pielou (1975, 1979) and others have applied these methods to ecological and biogeographic data. If comparable measurements are available for a large number of communities, these techniques can be used to group the associations into hierarchical clusters reflecting their similarities in species composition, life form, or other attributes of interest. As we will see in later chapters, these statistical methods are extremely useful for detecting quantitative patterns of floral and faunal resemblance that may suggest the effect of historical geologic or contemporary ecological mechanisms. Unfortunately, this approach has not been used systematically to classify community types on a worldwide basis, probably because the task is so awesome and the data required are simply not available for many regions.

There has, however, been no lack of attempts to produce simpler, less objective groupings of community types on a worldwide basis. Beginning with the pioneering classifications of Schouw (1823) and the early phytogeographers, continuing with Merriam's (1894)

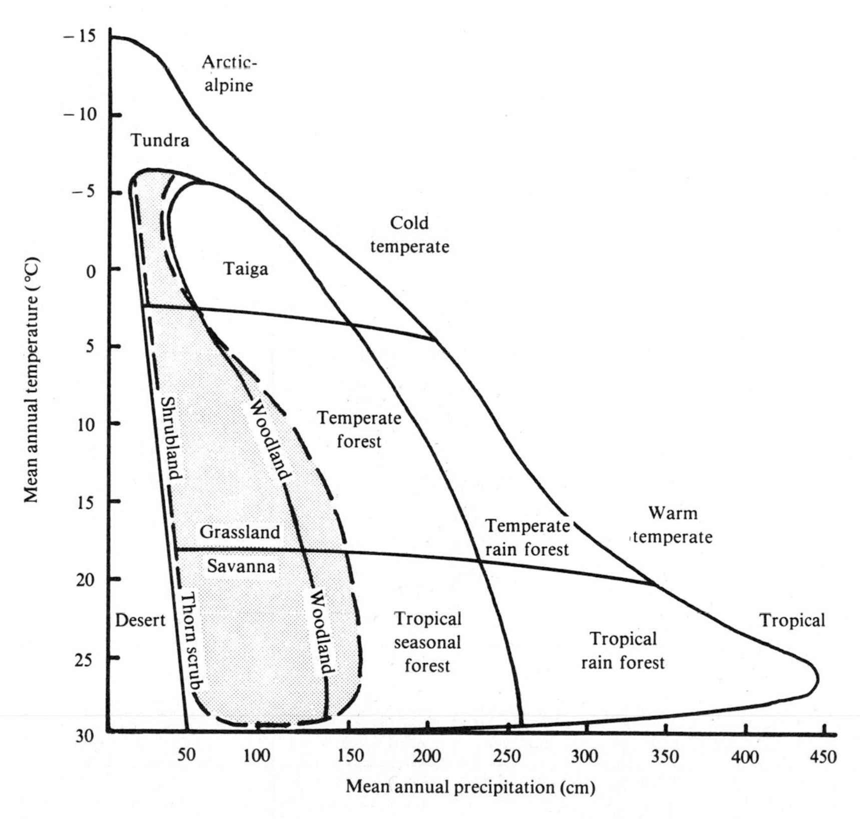

Figure 4.11
Simple diagram quantifying some aspects of the relationships between climate and vegetation types. Some authors have proposed much more elaborate classifications, such as that shown in Figure 13.5. (From Whittaker, Communities and ecosystems, second edition. Reprinted with permission of Macmillan Publishing Co., Inc. Copyright © 1975 by R.H. Whittaker.)

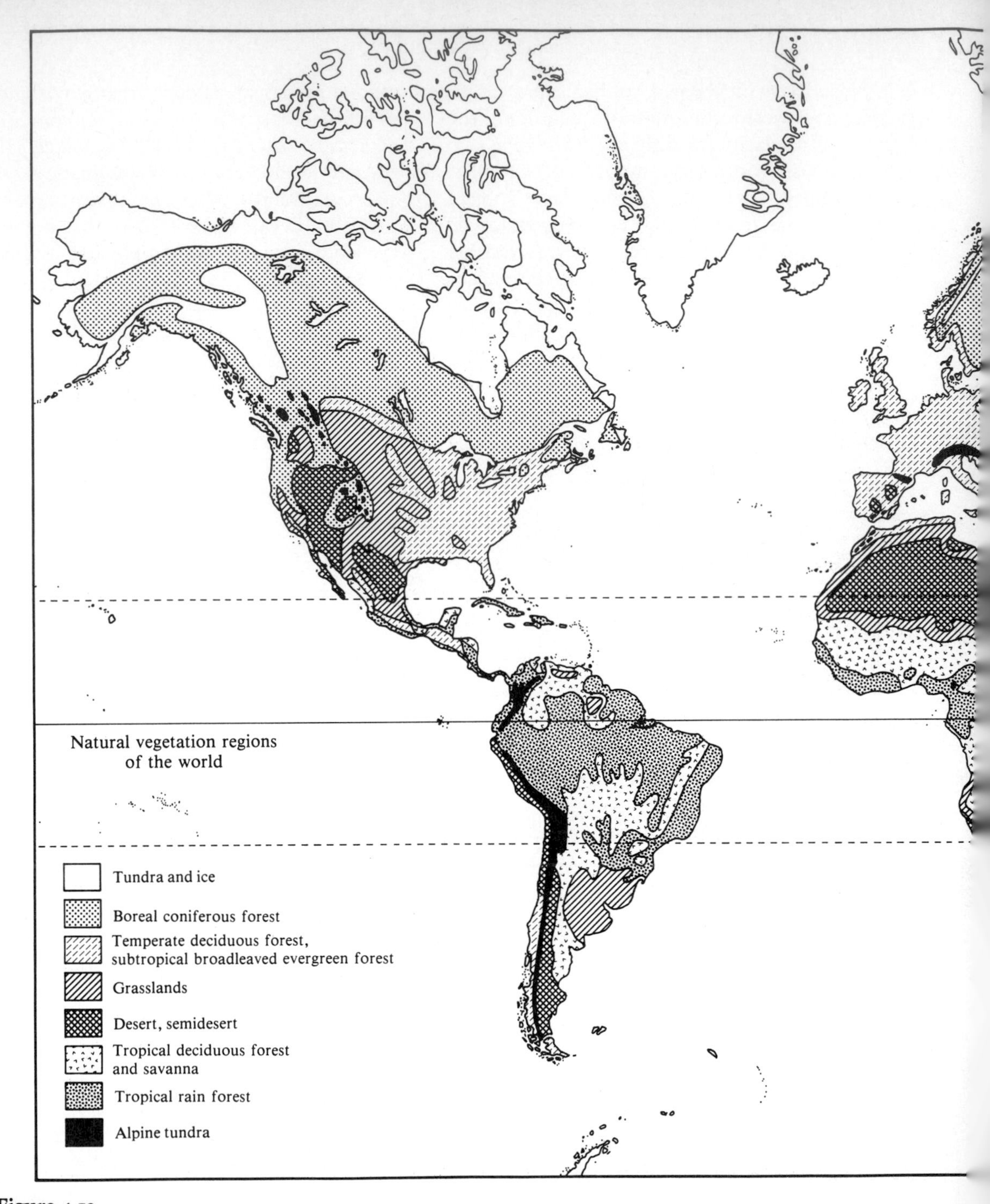

Figure 4.12
The world distribution of the major terrestrial biomes. Note that the locations of these vegetation types correspond closely to the distribution of climatic regimes and soil types (Figures 2.7 and 2.11). Several different vegetation types are grouped together so the general zonal pattern of biomes is observed, e.g., tropical and paratropical rain forest and tropical deciduous forest and savanna. (Modified from several sources and highly simplified.)

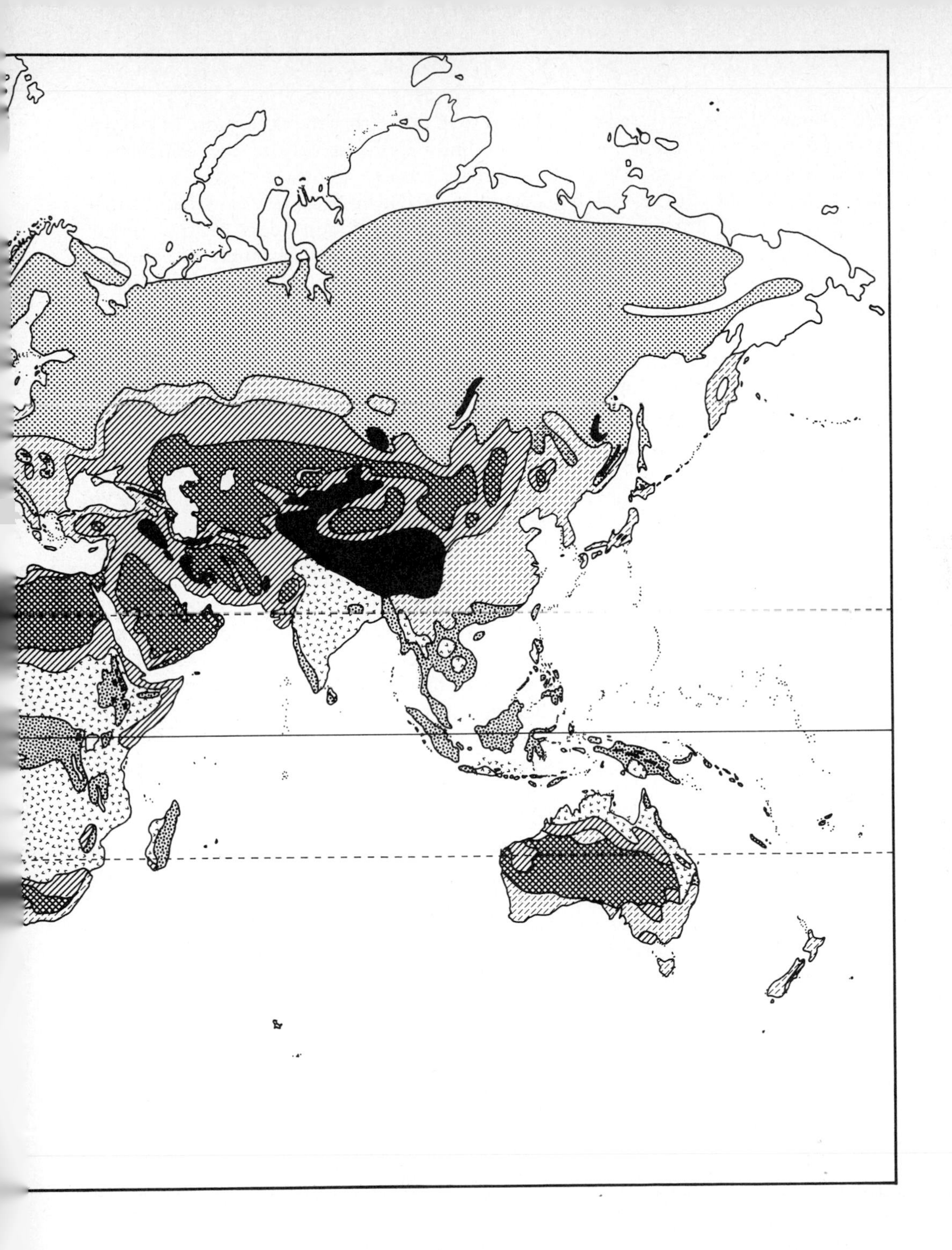

formulation of life zones (Figure 1.5), and extending to the contemporary concept of biomes, ecologists and biogeographers have almost without exception classified terrestrial communities on the basis of the structure or physiognomy of the vegetation. Implicit in all these classifications is the recognition that the life forms of individual plants and the resultant three-dimensional architecture of the vegetation reflect the predominant influence of climate on the kinds of plants occurring in a region. Some authors, such as Holdridge (1947; Chapter 13) and Dansereau (1957), have attempted to depict these relationships more quantitatively, showing the fairly tight relationships between ranges of climatic variables (such as temperature and precipitation) and specific vegetation types (Figure 4.11; see also Figure 13.5). Similar climatic regimes tend to support structurally and functionally similar vegetation in disjunct areas throughout the world. Often the similarities result from convergence; i.e., quite unrelated plant species in geographically isolated regions have evolved under the influences of similar climates to possess similar forms and to play similar ecological roles.

There are almost as many different classifications of vegetation types as there are textbooks in ecology and phytogeography. In general, most workers recognize six major forms of vegetation on land: forest, a tree-dominated assemblage with a fairly continuous canopy; woodland, a tree-dominated assemblage in which the specimens are widely spaced, often with grassy areas or low undergrowth between them; shrubland, a fairly continuous layer of shrubs, up to several meters high; grassland, where grasses and forbs predominate; scrub, mostly a shrubby assemblage in which the specimens are discrete or widely spaced; and desert, a very sparse plant cover in which most of the ground is bare. These terms are combined with a descriptor of the climate so the reader can envision the general nature of each biome.

We recognize 12 common terrestrial biomes, whose geographic distributions are mapped in Figure 4.12. Note that the occurrence of biomes corresponds approximately to the distribution of climatic zones (Figure 2.7) and soil types (Figure 2.11). This reflects not only the fact that vegetation is highly dependent on the climate and the underlying soil separately but also on the influence of climate on soil formation. We briefly consider the features of the major biomes.

Tropical rain forest. Tropical rain forests (Figure 4.13) are found at low elevations in tropical latitudes (chiefly 10° N to 10° S) where rainfall is abundant (over 180 cm annually). Although most tropical rain forests occur in regions that receive some precipitation most of the year, nearly all experience at least one pronounced dry season lasting 2 to 3 months. Temperature is uniform year-round (21° to 30° C range of daily temperatures), without any freezing, and humidity is high. The dominant plants are large evergreen trees that form a closed canopy at 30 to 50 m. The trees are often similar, having buttressed bases and smooth, straight trunks, but the height and shape of their crowns are highly variable. Their evergreen leaves also tend to be convergent in form. There may be several levels of trees below the uppermost canopy, and many palms occur. In the upper layers grow numerous vines (lianas) and epiphytes (orchids, ferns, and, in the New World, bromeliads), and very little light reaches the forest floor, which is usually surprisingly open and devoid of vegetation. Annual plants are conspicuously absent. Lowland tropical rain forests are the most diverse and productive of the major terrestrial biomes. Hundreds of tree species can be present in a few hectares, and here one also finds an incredibly diverse fauna, especially of arboreal insects and other invertebrates. Decomposition of dead organic matter occurs so rapidly that little litter accumulates on the forest floor.

At higher elevations and latitudes true lowland tropical rain forest is replaced by vegetation with fewer evergreen species, and some trees and shrubs present in the mountains tend

Figure 4.13
Tropical rain forest. **A,** Tropical cloud forest, elevation 1400 m, Cerro Pirre, Daríen Province, Panama. **B,** Close-up of plants in lowland tropical forest near Laguna de Arenal, Costa Rica. (Courtesy Thomas B. Croat, Missouri Botanical Gardens.)

Figure 4.14
Tropical deciduous forest, southern Puebla, Mexico. (Photograph by Arthur C. Gibson.)

to have temperate relatives. This is in keeping with the fact that these rain forests of the subtropics and the montane tropics have cooler temperatures, and the subtropical climate has more pronounced dry spells.

Tropical deciduous forest. Tropical deciduous forests (Figure 4.14) usually occur in hot lowlands outside the equatorial zone, where rainfall is more seasonal than in regions of tropical rain forest. Compared with tropical rain forest, canopies are lower and more open, and there is often more understory vegetation because more light reaches the ground. Many of the trees and understory plants are leafless during the long dry season, but much flowering and fruit maturation occur at this time.

This vegetation is often called rain-green forest because the foliage is produced during the first heavy rains following the dry spell. The most luxuriant form is the monsoon forest, the layman's "jungle." The monsoon forest is especially well developed in southern Asia, with many large leaves and much dense undergrowth rich in bamboos. These areas are drenched with torrential rainstorms, and some are among the rainiest habitats in the world. Tropical deciduous forests form where rainfall is scantier than in the rain forests; consequently, the deciduous forests are shorter and more open, and have relatively few epiphytes.

Thorn forest. Tropical and subtropical thorn forests (Figure 4.15) are low arborescent vegetation types that grow in hot, somewhat dry to semiarid lowlands. The dominant plants are small, spiny or thorny shrubs and trees. Members of the genus *Acacia* and other legumes (Fabaceae) are typically common in these biomes on all continents. Succulents, such as cacti (Cactaceae) in the New World and convergent forms of the genus *Euphorbia* (Euphorbiaceae) in Africa, are often abundant. Most plants lack leaves during the prolonged dry season, but the trees leaf out and a dense herbaceous understory develops during the wet season.

Figure 4.15
Tropical thorn forest near Todos Santos in southern Baja California, Mexico. The vegetation here is 4 to 5 m high, quite open, and composed of a variety of small shrubby trees and cacti. (Photograph by J.R. Hastings.)

Thorn forests are often found on drier sites adjacent to tropical deciduous forests, and as the climate becomes even drier along a gradient, thorn forests give way to thorn scrub. A minimum of 30 cm of annual rainfall usually is necessary to establish a thorn scrub, and generally the region is essentially without rainfall for 6 months.

Tropical savanna. Tropical savannas (Figure 4.16) are tall grasslands with widely scattered trees or shrubs. Savannas usually occur at low to intermediate elevations where seasonal drought and fire combine to favor perennial grasses and limit tree growth. The most extensive savannas are found in central and eastern Africa, where they support the most abundant and diverse community of large grazing mammals in the world, as well as large carnivorous mammals that prey on herbivores.

Desert. Deserts and semideserts (Figure 4.17) occur around the world at low to intermediate elevations, especially in belts of dry climate from 30° to 40° N and S latitude (the horse latitudes) between the humid tropics and the mesic temperate biomes. Rainfall is not only scanty (often less than 25 cm per year) and seasonal but also highly unpredictable. In true deserts rainfall is less than 10 cm per year, and some extremely arid regions may not experience any rainfall for several years in a row and therefore have no perennial vegetation. In less arid areas, the dominant vegetation consists of widely scattered low shrubs, sometimes interspersed with succulents. Ephemeral forbs and grasses may grow rapidly to carpet the normally bare ground during brief periods following heavy rains. Where shrubs predominate, the vegetation is usually called desert scrub. Desert vegetation has expanded into many new regions in the last few centuries, following the destruc-

Figure 4.16
Tropical savanna, between Betioky and Tongobory, Madagascar. (Courtesy Thomas B. Croat, Missouri Botanical Gardens.)

Figure 4.17
Desert vegetation near Puertecitos in northern Baja California, Mexico. Note that the vegetation, which consists almost entirely of small shrubs, is extremely sparse. A few small trees of palo verde *(Cercidium microphyllum)* grow in the sandy bottom of the dry watercourse in the center, where the plants in general are larger and denser than on the surrounding rocky hillsides. (Photograph by J.R. Hastings.)

Figure 4.18
Sclerophyllous scrub. **A,** Chapparal, Santa Monica Mountains, California. **B,** Fynbos, Pakhuispas, Cape Province, South Africa. (**A,** Photograph by Arthur C. Gibson; **B,** courtesy Martin L. Cody, U.C.L.A.)

tion of native vegetation and the erosion of soil owing to activities of humans and domesticated livestock.

Sclerophyllous woodland. Sclerophyllous woodlands and chaparral (Figure 4.18) occur in mild temperate climates where they receive moderate winter precipitation but experience long, usually hot, dry summers. The dominant plants have sclerophyllous (hard, tough, and evergreen) leaves. Sclerophyllous woodlands can be tall communities that receive over 100 cm of annual rainfall, as in the eucalypt woodlands in southwestern Australia, whereas less rainfall is characteristic of oak woodlands (with evergreen *Quercus* species) of the Holarctic region, especially in western North America. Those sclerophyllous communities that receive less than 60 cm per year tend to be shrublands. The scrublands, called chaparral, matorral, maquis, fynbos, or macchia, are characteristic of mediterranean-type climates (Chapter 2). This is a dense and almost impenetrable mass of vegetation only a few meters in height. Fires frequently sweep through these habitats, burning off the above-ground biomass and apparently playing a major role in preventing the establishment of trees. Shrubs resprout from their root crowns to reestablish the vegetation.

Subtropical evergreen forest. Subtropical broad-leaved evergreen forests (Figure 4.19), some of which have been called oak-laurel forests, montane forests, or cloud forests, are common in subtropical mountains at intermediate elevations and cover extensive areas of China and Japan, the southeastern United States, and disjunct areas in the Southern Hemisphere. These areas may receive as much as 150 cm of annual rainfall evenly distributed throughout

Figure 4.19
Subtropical broad-leaved evergreen forest, La Selva, Costa Rica. (Courtesy Mildred E. Mathias, U.C.L.A.)

the year, but subtropical evergreen forests cannot occur where mean annual temperature is much below 13° C. Most dominant species are dicotyledons with entire-margined, sclerophyllous evergreen leaves, such as laurels (Lauraceae), oaks (*Quercus*, Fagaceae), and magnolias (Magnoliaceae). Stratification is usually not present, and understory plants, especially mosses, can be exceedingly common where fog persists. A number of temperate broad-leaved deciduous trees occur in these forests, and wherever the climate becomes colder, broad-leaved evergreens are gradually replaced by deciduous trees or conifers. The factor most likely limiting these evergreen dicotyledons is a cold month mean temperature below 1° C (Wolfe, 1979).

Temperate deciduous forest. Temperate deciduous forests (Figure 4.20) grow throughout temperate latitudes almost wherever there is sufficient water to support the growth of large trees. These are also called summer-green deciduous forests because they have a definite annual rhythm, dormant and leafless in the cold and snowy winter and leafing out in the spring. Temperate broadleaf deciduous forests are extremely variable in structure and composition in eastern North America, western Europe, and parts of eastern Asia. In the southwestern United States these species reappear in the deserts, foothills, and mountains along permanent watercourses (riparian deciduous woodland). The height and density of the canopy and the importance and composition of understory plants vary greatly depending on climate, soil type, and the frequency of fires. Temperate forest trees of climax communities grow slowly, and most forests have been significantly affected by logging over the last few centuries.

Figure 4.20
Temperate deciduous forest, west bank of the Little Meramec, Franklin County, Missouri. (Courtesy Carol Sutherland.)

Figure 4.21
Boreal coniferous forest, near Mineoka-yama, Honshu, Japan. (Courtesy Joe Goodin, Texas Tech University.)

In many parts of the Northern Hemisphere, temperate deciduous forests are located next to other arborescent communities, especially temperate evergreen forests, sclerophyllous woodlands, and coniferous forests, on both a latitudinal or an elevational scale. Consequently many phytogeographers recognize a long list of hybrid associations between these communities, such as mixed evergreen-deciduous forest.

Boreal forest. Boreal forest or taiga (Figure 4.21) extends in a broad band across northern North America, Europe, and Asia in regions of subtemperate climate. This biome also extends well southward into the temperate latitudes at higher elevations. Much of highland Mexico is covered by boreal forest. The vegetation is typically dominated by a few species of narrow, needle-leaved evergreen tree conifers such as spruce *(Picea)*, fir *(Abies)*, pine *(Pinus)*, and Douglas fir *(Pseudotsuga)*, species able to grow in continually cool climates and soils low in available cations. Often the canopy is not dense, and a well-developed understory of shrubs, mosses, and lichens may be present in the most mesic sites.

Temperate rain forest. Temperate rain forest (Figure 4.22) is an uncommon but interesting biome found in the Northern and Southern Hemispheres where precipitation exceeds 150 cm per year and falls in at least 10 months. Cool temperatures predominate year-round, but these regions are always above freezing and have much fog and high humidity, permitting the growth of large evergreen trees. Cool temperatures account for the absence of any true tropical plants, such as palms, and the relatively low number of tree species. Moreover, cool temperate rain forests do not have many kinds of lianas; the epiphytes are mostly mosses, lichens, epiphyllous fungi, and some ferns.

Figure 4.22
Temperate rain forest Olympic Penninsula, Washington. (Courtesy Mildred E. Mathias, U.C.L.A.)

Figure 4.23
Temperate grassland at Missouri Botanical Gardens Arboretum, Gray Summit, Missouri. (Courtesy Carol Sutherland.)

The best example of a temperate rain forest in North America is the spectacular forest on the Olympic Peninsula in Washington, which receives nearly 200 cm of annual precipitation (rain plus fog). This is a coniferous forest, composed of huge spruces and firs with narrow leaves. In contrast, the temperate evergreen forests of the southern part of the Southern Hemisphere are dominated by large conifers with large evergreen leaves, such as *Agathis* (Australasia) and *Podocarpus;* large tree ferns (mostly *Dicksonia*); and the various evergreen species of the Southern Hemisphere beech *(Nothofagus)*.

Temperate grassland. Temperate grasslands (Figure 4.23) occur in areas that experience a moderately dry and cold continental climate. The vegetation is confined to a single stratum, which varies in height and density depending largely on water availability, decreasing from prairie (veldt of South Africa, puszta of Hungary, or pampas of Argentina and Uruguay) to short-grass plains or steppe in colder latitudes and desert grassland adjacent to warm arid regions. Although perennial grasses usually predominate, a large number of forbs are often present. Fires are frequent in grasslands, and in some regions fires play an important role in preventing invasions of trees and shrubs. Because they are often found on deep, fertile soils in the centers of continents, many temperate grasslands, such as the prairies of North America, have been largely converted to agricultural uses.

Tundra. Tundra (Figure 4.24) and other low scrubland or matlike vegetations are characteristic of high latitudes and elevations. Even more than taiga, the tundra is characterized by plants adapted to low temperatures and short growing seasons. Many classifications treat these as a series of semidesert or desert biomes, because precipitation is scant and cold temperatures limit the water available for plant

Figure 4.24
Tundra, Alaska. (Courtesy Stanwyn G. Shetler, Smithsonian Institution.)

growth. Arctic, antarctic, and alpine tundra are low vegetation types, usually only a few centimeters or decimeters in height, and are dense and complex. The dominant plants tend to be dwarf perennial shrubs, sedges, grasses, mosses, and lichens. Many tundra regions receive less precipitation than some deserts, but evaporation is usually also scant in these cold climates so that soils become saturated with water.

Above the timberline on mountaintops in the equatorial zone the vegetation, called tropic alpine scrubland, is taller than arctic tundra. The dominant groups are bizarre, erect rosette perennials with thick stems and tussock grasses. These vegetation types are found at elevations above 3300 m in the Andes (paramo) in South America, the upper slopes of the highest mountains in east Africa, and mountaintops in New Guinea.

• • •

In concluding this discussion of biomes we should sound a note of caution. The biomes described above and their distributions on the map in Figure 4.12 indicate only the general kind of climax vegetation that may be found in a region, primarily based on climate. However, someone visiting many of the areas on the map would have difficulty finding good stands of the vegetation typical of these biomes and might find some unexpected vegetation types. In some cases, this might be the result of secondary succession occurring in response to natural disturbances such as fire or violent storms. More often, it would be caused by human destruction of the original vegetation and great modification of the landscape. Originally tall-grass prairie, the most productive temperate grassland, covered much of Illinois, Iowa, and adjacent regions of the central United States. Now the area is mostly converted to cornfields and it is extremely difficult to find relatively undisturbed stands of native prairie. A few remnants, totalling less than 1% of the original area, remain mostly in a small number of preserves. Yet the saddest tale of all is the rapid destruction of virgin lowland tropical rain forest, whose boundaries are ever shrinking, giving way to secondary successional communities of reduced biotic diversity and diminished economic value.

In some places, local variations in topography or soil type strongly control the types of vegetation to be found in regions where one would not predict them from the general map. Throughout temperate grasslands, sclerophyllous scrubland, tropical thorn scrub, and deserts, for example, there are galleries of riparian forest vegetation along permanent streams. Such diversity of vegetation types contributes greatly to the biotic richness of a region, because the distributions of many other plants and numerous animal species are strongly influenced by the dominant vegetation. Bird species and distributions, and the diversity of their communities, for example, are highly dependent on vegetation structure (MacArthur and MacArthur, 1961) and the riparian deciduous woodlands in the deserts of the southwestern United States are local areas of exceptionally high bird species diversity (Carothers and Johnson, 1974).

Kinds of aquatic communities

Marine and freshwater ecologists and biogeographers do not classify aquatic communities into categories analogous to terrestrial biomes. For one thing, the relatively simple arrangement of sessile plants growing on surfaces is not comparable to the three-dimensional diversity of aquatic communities. Terrestrial habitats are essentially two-dimensional, in the sense that organisms do not remain permanently suspended in the air above the soil surface. To the extent that a third dimension is present, it has been formed by the vertical growth of sessile plants. The three-dimensional organization of aquatic communities is very different. On the one hand, a well-developed architecture of attached, vertically growing organisms is absent from most aquatic habitats, although there are obvious exceptions, such as kelp forests and coral reefs. On the other hand, many aquatic organisms spend much or all of their lives suspended in the third dimension, either drifting passively or swimming actively in the water column.

The physical factors that vary in time and space to affect the abundance and distribution of aquatic organisms are quite different from those that determine both the terrestrial climate and the organization of terrestrial communities. Because of the high specific heat of water, temperature varies less on a daily, seasonal, and latitudinal basis in aquatic environments than in terrestrial ones. On the other hand, variation in pressure, salinity, and light are relatively more important in aquatic systems. Tidal cycles, which fluctuate bimonthly with the phase of the moon, are more important to many marine shore communities than daily or seasonal cycles.

Oceanographers, limnologists, and aquatic ecologists have developed classifications for marine and freshwater ecosystems. Like the division of terrestrial communities into biomes, these are arbitrary groupings that break up a continuous spectrum of biological associations into a number of convenient categories; but they reflect the ways that physical environmental conditions determine the kinds of organisms that live together in a community. Salinity, depth, movement, and relationship to substrate are the physical characteristics that most influence the abundance and distribution of aquatic organisms and are most often used in classifying aquatic communities.

The first major division is into marine and freshwater communities. On biological as well as geographic grounds, the earth's bodies of water can be divided into the oceans, which form a huge interconnected water mass covering 70% of the earth's surface, and the tiny, highly fragmented lakes, ponds, rivers, and streams, which together cover only a small fraction of the remaining surface. The sea and these bodies of fresh water differ greatly in salinity. The salt concentration of the oceans varies slightly around 35 parts per thousand, whereas even the hardest fresh waters have salinities of less than 0.5 parts per thousand. As stressed in Chapters 3 and 10, this difference in salinity has dramatic effects on distributions. Only a tiny fraction of aquatic organisms can live in both salt and fresh water, so this effectively divides the aquatic communities into two nonoverlapping groups. Because of this strong dichotomy, marine and freshwater ecosystems have largely been studied independently by different groups of ecologists, oceanographers and limnologists, respectively, who have developed different classifications of communities. These are best considered separately.

Marine communities. Compared to terrestrial and freshwater environments, the sea is large and essentially continuous. Organisms live everywhere in the ocean, but the abundance and kinds of life vary greatly depending on the local physical environment. Perhaps the most important features are light and substrate. As mentioned earlier (Chapter 2), the sea can be divided into two vertical zones, the photic and aphotic zones, based on the penetration of sunlight (Figure 4.25). Because sunlight is gradually absorbed with increasing depth, the demarcation of these zones is somewhat arbitrary. The depth of the photic zone increases from coastal waters, where light rarely penetrates more than 30 m because of organisms and inanimate particles suspended in the water column, to the open ocean, where it may extend to a depth of 100 m or more. The significance of this zonation of light, of course, is that photosynthesis can occur only in the photic zone. Essentially all of the organic energy that sustains marine life is produced in the shallow surface waters. Most organisms in the aphotic zone obtain their energy by consuming organic material that is produced in the photic zone and imported to deep water as feces and dead bodies sink to the bottom. Recently, however, oceanographers have discovered an entire flourishing community of organisms, including unique kinds of worms, mussels, and crabs, in local areas on the otherwise barren slopes of the

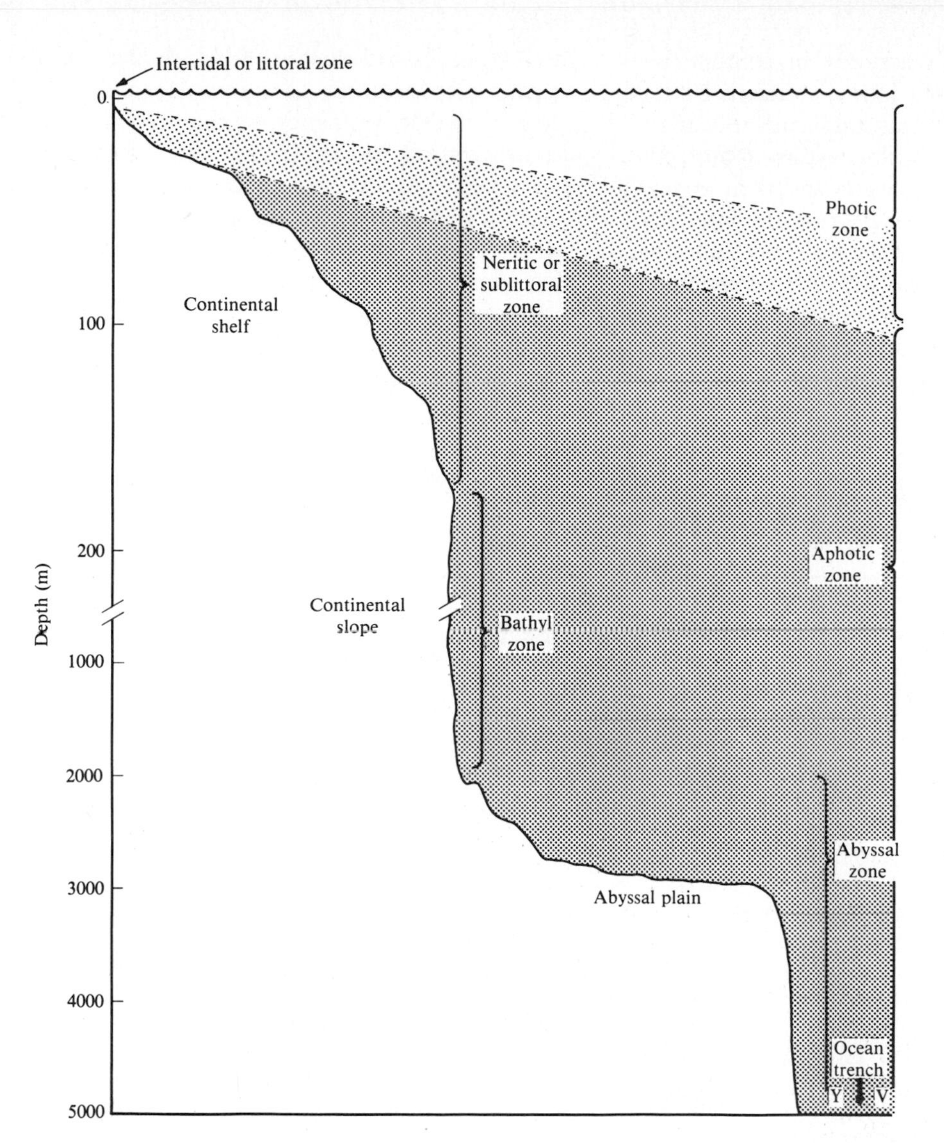

Figure 4.25
Classification of marine communities into major zones. Note that the zones are based primarily on water depth and relationships between organisms and substrates.

Galápagos rift in the eastern tropical Pacific where submarine hot springs emit hydrogen sulfide. Chemosynthetic bacteria obtain energy by oxidizing the hydrogen sulfide, and these unusual autotrophs serve as the base of the food chain for this unique community (Jannasch and Wirsen, 1980; Karl et al., 1980).

Marine communities are also classified into another set of zones on the basis of bathymetry, i.e., the depth and configuration of the bottom (Figure 4.25). The most shallow zone is the intertidal or littoral zone, which occurs on the shore where sea meets land. Although it is inhabited almost exclusively by marine organisms, the intertidal zone is actually an ecotone between land and ocean. It is dominated by the tides that periodically advance and retreat on semidiurnal and bimonthly cycles, alternately covering the shore with water and then leaving it exposed (Figure 4.26). Beyond the intertidal zone is the neritic or sublittoral zone, encompassing waters of a few meters to about 200 m deep that cover the continental shelves. At the edges of the continental plates (see Chapter 5) are regions of highly varied relief, called the bathyal zones, with the marine equivalent of mountainsides and canyons in which the waters rapidly drop away to the great ocean depths. Most of the ocean comprises the abyssal zone, extensive areas in which the water ranges in depth from 2000 to more than 6000 m. Deep ocean waters provide some of the most constant physical environments; they are continually dark, cold (4° C), and virtually unchanging in chemical composition.

The organisms that inhabit the sea are often classified as either benthic or pelagic, depending on whether they are closely associated with the

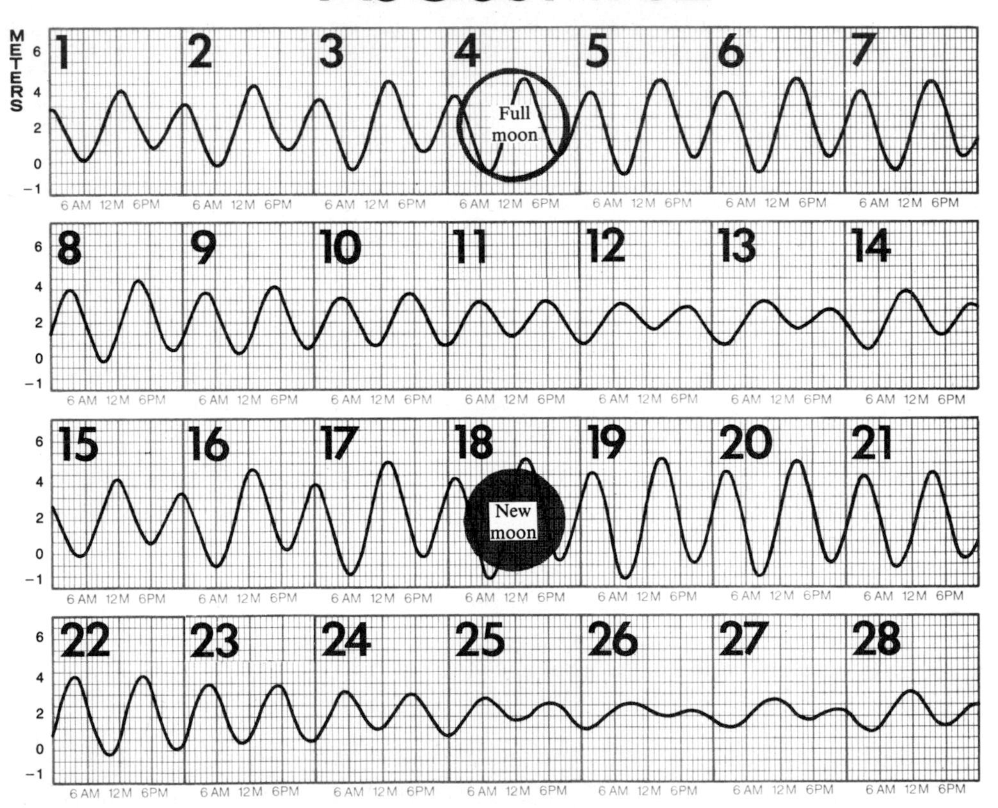

Figure 4.26
Tide calendar for the northern Gulf of California for August 1982. Note that there are two high and two low tides each day and two periods of extreme tides each month, corresponding to the time of new and full moons. Because of its orientation and configuration, the northern Gulf of California is characterized by tides of extreme amplitude that range over 7 m of vertical shore.

substrate or distributed up in the water column (Figure 4.25). Communities of benthos vary greatly in composition depending on the nature of the substrate. On hard substrates attached benthic organisms often form a three-dimensional architecture that varies in complexity from low crusts and turfs of algae and sessile invertebrates to tall "forests" of kelp and coral. On soft sandy or muddy substrates there is often a comparable three-dimensional complexity, but it is formed of burrowing invertebrates that live beneath the surface. Pelagic organisms are usually divided into two groups, plankton and nekton. The former consists of those primarily microscopic organisms that float in the water column. The plankton typically includes many simple plants or phytoplankton, such as diatoms, and tiny animals or zooplankton, such as small crustaceans and the larvae of many invertebrates and fish. The nekton comprises the actively swimming animals, including fish, whales, and some large invertebrates, which usually occupy higher trophic levels than planktonic organisms.

Freshwater communities. Freshwater communities are distributed as small, isolated lakes, ponds, and marshes, sometimes connected by long, branching streams and rivers. These environments are usually divided into two categories: lotic or running-water habitats, such as springs, streams, and rivers, and lentic or standing-water habitats, such as lakes, ponds, or marshes. Lotic habitats are often divided into rapids and pools. In the former the velocity of

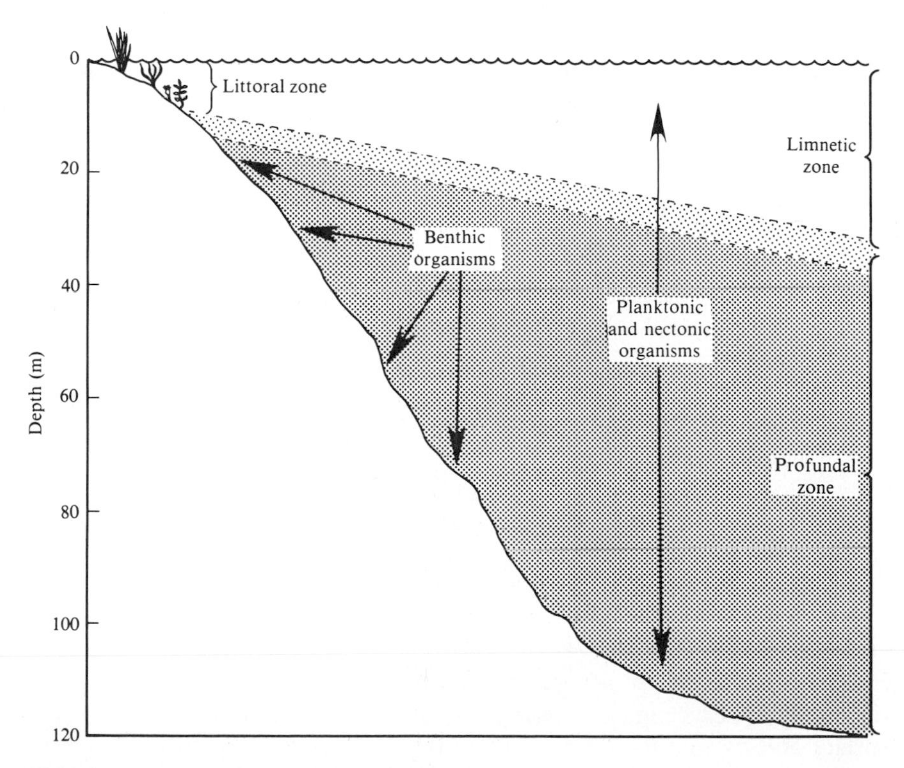

Figure 4.27
Division of freshwater lentic habitats into major zones. This classification is similar to that used for marine environments because it is based on variation in light and water characteristics with depth and on relationships between organisms and substrates. Nevertheless, somewhat different terms are used.

water is sufficient to keep the water well oxygenated and the substrate clear of silt. Rapids zones of streams are usually inhabited by organisms that live on the surfaces of the rocky substrates or swim strongly in the current. Pool zones are characterized by deep, slowly moving water and silty, often poorly oxygenated bottoms. Swimming animals are common, and many of the benthic species burrow in the substrate. Although some of the organic material produced in streams is manufactured in place by benthic plants or phytoplankton, often much of it is washed in from the surrounding watershed.

Lentic habitats are often divided into zones reminiscent of the oceans, although somewhat different terms and meanings are applied (Figure 4.27). The littoral zone consists of shallow waters where light penetrates to the bottom and rooted aquatic vegetation may be present. The offshore waters are divided into a surface limnetic zone, where light penetrates sufficiently for photosynthesis to occur, and a deep profundal zone beyond the depth of effective light penetration. Lakes can be highly productive, supporting extensive food webs based on both attached vegetation in the littoral zone and phytoplankton in the limnetic zone. Productivity is usually limited largely by the availability of inorganic nutrients, which wash in from surrounding watersheds and, in temperate and subarctic lakes, are recycled from the organic material on the bottom when the vertical temperature and density stratification of the water disappear seasonally and overturn occurs (see Chapter 2). Temperate lakes are often classified as either eutrophic or oligotrophic. Eutrophic lakes are shallow lakes that are highly productive because light penetrates almost to the bottom and vertical circulation of the water column occurs each spring and fall, recycling limiting nutrients to the surface and oxygen to the depths. Oligotrophic lakes are usually so deep that little or no vertical circulation occurs, and limited nutrients and sunlight restrict primary productivity.

The preceding classification omits a few communities. Some of these are rare and atypical but interesting because they represent unusual physical environments that pose special problems for the few organisms that are able to live there. Examples of such harsh environments are hypersaline lakes (undrained lakes, such as the Great Salt Lake in North America and the Dead Sea in the Middle East, which are much more saline than seawater), caves (which admit no light and have communities supported entirely by imported organic matter), and hot springs (among the most physically rigorous of all environments). A group of small but important communities occurs in special areas where fresh water, the sea, and the land meet. The estuaries, salt marshes, and (in tropical regions) mangrove swamps that occur in these sites are highly productive ecosystems. They usually contain only a few species that can tolerate the physical rigor of successive exposure to fresh water, seawater, and the terrestrial climate. On the other hand, the circulation provided by the tides and the input of nutrients from the rivers, and ultimately the land, often permits these habitats to support great biomass and densities of individuals.

Selected references

Historical and biogeographic perspectives

Clements (1916, 1949); J.T. Curtis (1959); Dice (1952); Egerton (1978a, 1978b); Gleason (1917, 1926); McIntosh (1967); Merriam (1894); Schouw (1823); Tansley (1920); Warming (1909); Whittaker (1967, 1975).

Communities and ecosystems

Barbour et al. (1980); Bormann et al. (1974); J. Brown et al. (1979); Daubenmire (1968); Elton (1966); Forman (1979); Golley (1960); Hemmingsen (1960); G. Hutchinson (1959); Jordan (1971); Krebs (1978); Küchler (1964); Lieth (1972, 1973); Likens and Bormann (1974); McNab (1963); May (1978); Mueller-Dombois (1981); H. Odum (1957); E. Odum (1971); Pianka (1978); Pomeroy (1970); Poulson and White (1969); Price (1980); Ricklefs (1979); Ryther (1969); T. Schoener (1968b); Teal (1957, 1962); Varley (1970); Whittaker (1975); Whittaker and Likens (1973).

The distribution of communities in space and time

Beals (1969); Beckwith (1954); Bernabo and Webb (1977); Braun-Blanquet (1965); Clements (1916, 1949); Connell (1961, 1975); Connell and Slatyer (1977); J.T. Curtis (1959);

Davis (1969, 1976); Dayton (1971); Drury and Nisbet (1973); Gray et al. (1981); Horn (1974); Kozlowski and Ahlgren (1974); Lubchenco (1980); Lubchenco and Menge (1978); McIntosh (1967); Menge and Sutherland (1976); Monk (1968); E. Odum (1969, 1971); Shreve (1915, 1922); Souza et al. (1981); Sprugel and Bormann (1981); Whittaker (1956, 1960, 1967, 1975); Whittaker and Niering (1965, 1968); Yeaton (1981); Yoda (1967).

Terrestrial biomes

Barbour et al. (1980); Beadle (1981); Braun (1950); Brenan (1978); Burbridge (1960); Chapman (1976); Dansereau (1957); Daubenmire (1978); Eyre (1968, 1971); Hare (1954); Holdridge (1947); Holdridge et al. (1971); Kellman (1975); Major (1977); Meggars et al. (1973); Merriam (1894); Neill (1969); Newbigin (1968); E. Odum (1971); Pears (1978); Richards (1969); Ricklefs (1979); Schouw (1823); Simmons (1979); Tivy (1971); Walter (1971, 1973, 1979); T. Whitmore (1975); Whittaker (1975, 1978a); Wielgolaski and Roswall (1972).

Kinds of aquatic communities

Birge and Juday (1911); Briggs (1974); A. Brown (1977); Dykyjova and Kvet (1978); Ekman (1953); Frey (1963); Hardy (1971); Hedgpeth (1957); G. Hutchinson (1958, 1967); Hynes (1970); Kinne (1970-1972); Lauff (1967); Macan (1970, 1973); Macan and Worthington (1952); E. Odum (1971); Perkins (1974); Popham (1961); Steele (1974); Tail (1981); Wetzel (1975); Weyl (1970).

UNIT TWO THE HISTORICAL SETTING

THE ENVIRONMENTAL setting that we have just described has not been constant. We assume that the fundamental nature of the ecological processes limiting distributions has not changed over time, but certainly the surface of the earth and the distributions of its organisms have changed markedly during the several billion years that living things have been in existence. Present distributions and especially the evolutionary relationships of now widely separated relatives can only be understood by reconstructing these historical events. This is difficult, because our evidence of the past is fragmentary and often indirect, but much progress has been made within the last two decades toward this goal. The next five chapters consider the historical setting in which organisms have become distributed across the globe, how organisms have responded to historical events, and by what methods we can reconstruct this history.

Chapter 5 describes our current understanding of the geologic and geographic history of the earth. Within the last few decades we have had to abandon the traditional notion that land and water masses have always been in their present locations. Evidence is overwhelming to describe earth's history as a dynamic process in which thin but solid surface plates of oceans and continents float on a molten layer. These plates move over the surface of the earth as new material is added along ocean ridges, and at the older margins the plates may be consumed and melted to become part of the earth's mantle again or transformed into mountain ranges. Well-documented physical evidence can be used to demonstrate not only that such plates exist but also that the history of their movements can be reconstructed in a fairly accurate fashion. This evidence supports and completes a model for continental drift first proposed by Alfred Wegener in 1912. About 200 million years ago most if not all major landmasses were joined as a single giant continent called Pangea, which broke up shortly after into a northern supercontinent, Laurasia, and a southern unit, Gondwanaland; these in turn became subdivided by the formation of new oceans over the last 125 million years. Not only did these major geomorphological events occur within a time frame that could have profoundly influenced the distributions of major taxa of plants and animals but they also subsequently brought landmasses such as North and South America together in secondary contact, resulting in the mixing of previously distinct biotas that had developed in isolation.

Chapter 6 discusses the processes of speciation and extinction that determine the historical fates of all populations in space and time. Since the ancient origin of life from nonliving matter, new species have formed from existing ones. The history of diversification is a branching process in which populations of a species become genetically and ecologically or geographically separated from related populations so that they ultimately constitute independent evolutionary lineages. The fate of these distinct populations is not predetermined. Some have colonized favorable environments, flourished, differentiated, and sometimes given rise to diverse forms by a process called adaptive radiation. Other lineages have survived almost unchanged for millions, even tens of millions of years. Still others, in fact the majority of all forms that once lived, have become completely extinct. Although extinction and origination have occurred throughout the history of life, there have been a few well-documented episodes of catastrophic extinctions, which eliminated most of the organisms then in existence. Following these episodes there have been rapid phases of speciation and radiation to fill the empty niches, thus returning to a level of organic diversity similar to that before the mass extinction.

The subject of dispersal is discussed in Chapter 7. We know from observations on natural colonization and artificial introductions that some species can exist in regions far from their present ranges, but they do not occur in these areas because colonists cannot readily cross the unfavorable intervening environments. However, over long periods of time some successful long-distance dispersal may occur. It is difficult for biogeographers to deal methodologically, if not conceptually, with these events because they tend to be infrequent and unique. Nonetheless, certain taxa possess characteristics that facilitate movement across seemingly insurmountable barriers and establishment of populations in new regions. These groups are particularly well represented in the biotas of oceanic islands and other highly isolated habitats. Organisms may have had opportunities to disperse between regions at times when the barriers were temporarily diminished or removed by geologic, climatic, or habitat changes. Such connections between regions that permitted relatively complete movement and interchange of biota are termed corridors, whereas areas of interchange that still posed significant barriers for some kinds of organisms, resulting in differential dispersal of taxa, are called filters. The existence of a historical corridor can often be established quite unequivocally, especially when it is corroborated by geologic or paleoclimatic evidence. On the other hand, long-distance dispersal across a severe filter can be so idiosyncratic that it is almost impossible to eliminate alternative hypotheses for observed distributions.

The most apparent cumulative biogeographic effect of historical events is endemism. The vast majority of taxa are confined to restricted regions that rarely correspond exactly to common features in the abiotic and biotic environment; instead they apparently reflect a history of local origin and limited spread. Chapter 8 examines the patterns and causes of endemism. Related taxa tend to occur together in the same or nearby regions. This provincialism was first discussed by such famous biogeographers as De Candolle, Sclater, and Wallace in the nineteenth century. They used the taxonomic affinities of biotas to subdivide the world into major biogeographic regions and smaller provinces within these areas. Although these regions do not correspond precisely to the boundaries of continents and oceans, they do describe a pattern of taxonomic relationships within and between biotas that we now perceive to be amazingly general for large numbers of unrelated taxa, both animals and plants. Although modern quantitative methods can in principle be used to define biogeographic provinces more precisely than in the past, practical difficulties have precluded application of these techniques except in a few limited cases, namely for only one taxon at a time. Disjunction, the occurrence of related organisms in widely separated regions, is the perplexing opposite of provincialism. Disjunctions are interesting because we do not expect to find closest relatives living on opposite sides of the globe, and these splits in ranges must be explained by one of three alternative hypotheses: long-distance dispersal, reduction of an originally wide range by extinction of forms in intervening areas, or movements of land masses or water masses carrying organisms to distant locations as a preformed biota.

Chapter 9 discusses the problem of trying to reconstruct the biogeographic history of particular taxa and regions using distribution records of living and extinct forms. The recent development of cladistic methods for inferring phylogenetic relationships among taxa has done much to stimulate interest in historical biogeography. The cladistic approach, pioneered by Willi Hennig, provides a logical basis for inferring evolutionary relationships on the basis of shared derived characteristics, called synapomorphies. Once such a phylogenetic hypothesis, called a cladogram, is available for a taxon, similar methods can be used to infer the biogeographic history of the group from shared distributional patterns with other taxa occupying the same areas. This approach, first crudely used by the botanist Asa Gray in the nineteenth century and recently refined by workers such as Croizat, Nelson, Platnick, and others, is called vicariance biogeography. Each reconstructed distributional history can be regarded as an hypothesis. Such hypotheses and their underlying assumptions, e.g., the cladograms, can be tested for agreement with evidence not used in the original synthesis. Biogeographic hypotheses can be tested with information of the geologic history of the regions, reconstructions of additional taxa inhabiting the same disjunct areas, and fossils of the groups involved. Although the conceptual development of the vicariance approach has received much attention, there have as yet been relatively few attempts to apply the method to many specific cases and to test rigorously all assumptions and predictions. Nonetheless, such methods have permitted historical biogeographers to define patterns on objective grounds, to develop methods of establishing ages and modes of origin for disjunct taxa, and to evaluate hypotheses about the existence of past corridors and filters for which we do not yet have unequivocal evidence from the earth sciences.

Chapter 5

Past Changes in the Physical Geography of the Earth

The earth's surface has changed continually during the history of life. Continents have moved, seas have expanded and contracted, mountain ranges have risen and been eroded away, islands have appeared and disappeared, and glaciers have advanced and retreated. In addition, the fossil record reveals that the climate has experienced profound changes. The width of the tropics has varied, influencing global patterns of vegetation and associated animal life. Positions of the equator and poles have not changed, but because continents have moved, regions that are tropical or polar today were not always so in the past. A sound knowledge of past physical changes is essential to understand the influence of historical events on present distribution patterns.

The geologic time scale

Anyone studying historical biogeography needs to be familiar with the time scale used to date the history of the earth (Table 5.1). Early geologists recognized that each layer in a stratigraphic column contains a unique assemblage of fossils, characteristic of a single time span, and therefore these assemblages could be used to correlate the ages of rock units in one locality with those in distant localities. For such reconstructions, the most reliable fossils were wideranging species whose life styles were independent of the sedimentary environment, including forms freely dispersed in marine habitats by currents and that fell randomly into any sedimentary situation. These are called index or guide fossils. Examples are the chitinous colonial graptolites (phylum Hemichordata), floating animals of the Ordovician and Silurian; the swimming ammonoid cephalopods (phylum Mollusca) of the Mesozoic, whose buoyant calcareous shells also probably drifted with currents after death; and the planktonic, calcareous foraminifers and siliceous radiolarians (phylum Protozoa) of the Cenozoic. On the continents, the most widespread fossils were in coal beds of the Paleozoic, called the Carboniferous, characterized by numerous distinctive vascular plants.

Correlations of fossils around the world only gave relative estimates of ages of various rocks and fossils, however; scientists had no specific knowledge on actual dates of rocks. For example, Alfred Russel Wallace (1880) accepted an estimate of 400 million years for the absolute age of the earth. The discovery in the twentieth century of radioactive materials finally led to more exact dating procedures. Radioactive elements are unstable and decay through a series of intermediate unstable products to stable atoms, and during the disintegration process atomic particles are released. The rate of decay, which is independent of environmental factors, can be quantified, and this rate is expressed as a half-life, the amount of time needed for half the radioactive material to decay to the stable element. By calculating the ratio of radioactive substance to the stable end-product, one can determine, within limits, the age of a sample. Thus by using isotopes of uranium and thorium, whose stable end-product is lead, the esti-

Table 5.1
Phanerozoic time scale; subdivisions of the Cretaceous are emphasized

Eras	Periods	System	Epochs or Series	Stages (N. America)	Age before present (× 10^6 yr)
Cenozoic	Quaternary		Recent		0.011
			Pleistocene		0.011-1.7
	Tertiary	Neogene	Pliocene		1.8-5.0
			Miocene		5.1-23
		Paleocene	Oligocene		23-38
			Eocene		38-53.6
			Paleocene		53.6-65
Mesozoic	Cretaceous			Maestrictian	65-71
				Senonian	71-88
				Campanian	
				Santonian	
				Coniacian	
				Turonian	88-90
				Cenomanian	90-94
				Albian	94-106
				Aptian	106-112
				Barremian	112-118
				Hauterivian	118-124
				Valanginian	124-130
				Berriasian	130-135
	Jurassic				135-192
	Triassic				192-223
Paleozoic	Permian				223-280
	Carboniferous				
	Pennsylvanian				280-321
	Mississippian				321-345
	Devonian				345-405
	Silurian				405-432
	Ordovician				432-495
	Cambrian				495-570

mate of earth's age has been pushed back to 4.5 billion years ago. The oldest known fossils are dated at 3.4 billion years BP. The potassium-argon method is another valuable technique for dating Phanerozoic rocks. Radioactive potassium ($^{40}K_{19}$) decays to stable calcium ($^{40}Ca_{20}$) and the inert gas argon ($^{40}Ar_{18}$); the half-life of potassium 40 is 1.31 billion years. A major problem with this technique is that argon gas will escape from rock heated above 300° C, e.g., during metamorphism, and is therefore not wholly reliable. Measurement of the decay of rubidium 87 (^{87}Rb) to strontium 86 (^{86}Sr) is used mainly for rocks older than 100 million years.

Paleomagnetism provides another method for dating old rocks because iron particles are frozen into position during solidification, oriented toward the then-existing and closest magnetic pole. These patterns are like fingerprints and have been used to estimate ages of undated sedimentary rocks.

For very recent material, radiocarbon dating is extensively used. Carbon 14 decays to carbon 12 at a fairly rapid rate (half-life 5730 ± 30 years). After 50,000 years so little radiocarbon remains that detection is very difficult. Consequently, the radiocarbon dating method is presently only useful in late Quaternary research.

Because many trees put down an annual ring each growing season, analysis of tree rings in temperate latitudes has been a reliable method for dating fossils formed within the last 10,000 years wherever sampling procedures have been rigidly employed, and results compare closely to carbon 14 values. In fact, tree rings are used to calibrate carbon dating. Paleoclimatic reconstructions are possible if one assumes that a fossil growth ring wider than one today represents moister conditions and narrower ones represent drier conditions (Fritts, 1976).

Problems arise when different methods give vastly different ages, e.g., stratigraphic, potassium-argon, and paleomagnetic techniques. The disagreement is usually resolved by accepting the two dates that agree most closely, especially if there is evidence of some disturbance causing argon leakage or magnetic domain realignment. However, many fossils have never been dated by more than one method, in part because radioactive methods are expensive.

Dates in the Phanerozoic time scale (PTS) (Table 5.1) for the epochs of the Mesozoic are not universally accepted. For example, the scale accepted in 1971 divided the Cretaceous into 12 equal epochs of 6 million years each, but recent radioactive dating has revealed that some epochs were longer and others quite short (Baldwin et al., 1974). Reexamination of Triassic deposits may eventually lead to a drastic shortening of that period and to increased time spans for the Jurassic and Permian. Accurate dates in the Mesozoic are crucial because the early evolution and dispersal of major lineages of land vertebrates and seed plants, as well as extinctions of certain marine groups and radiations of others, occurred in the Mesozoic (Table 9.1). Likewise, the length of the Pliocene has been shortened considerably and is now treated as having ended about 1.7 million years BP (Haq et al., 1977).

Emergence of ideas on continental drift

No contribution to biogeography has had more impact than the development of the concept of continental drift from a speculative theory in the early 1900s to a well-established fact by the 1960s. Today we call this subject plate tectonics. Simply defined, the theory of continental drift states that continents and portions of continents are or have been separate crustal entities that have been rafted across the surface of the globe on the weak upper mantle. Thus the lithosphere is not composed of fixed ocean basins and continents, as was once supposed, but instead is a changing landscape in which once distant lands are now in juxtaposition and others once attached are now distantly removed. The evidence in favor of past continental movements is conclusive, and within the last few decades the theory has given rise to a respected science. However, acceptance of continental movement in a horizontal plane was strongly resisted for many years.

Scholars have searched diligently to determine who first proposed continental drift. Various authors have attributed the germ of the idea to early writers, for example, to Sir Francis Bacon (1620), although Bacon never really discussed the subject. Credit certainly belongs to Antonio Snider-Pelligrini (1858), who first recognized the geometric fit of coastlines on opposite sides of the Atlantic Ocean and postulated that the continents split apart there. Yet at the beginning of the twentieth century geologists had little understanding of past relationships of continents. Most believed, after James D. Dana, in the permanence of the extant continental-oceanic pattern, the so-called permanence theory. Two versions of that idea existed. Either the earth cooled and by contraction

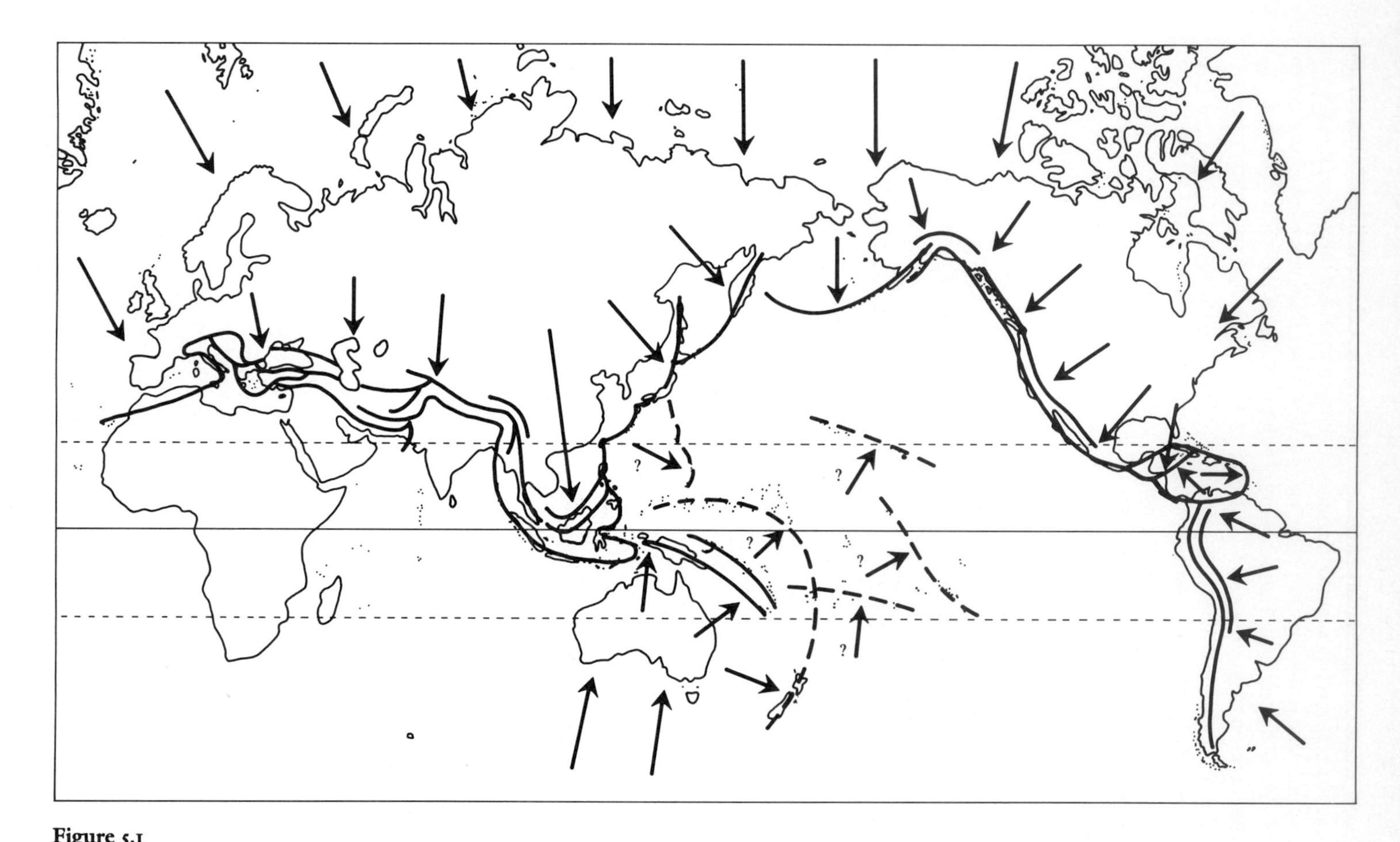

Figure 5.1
An early model by Taylor (1910) showing crustal movement. This scenario suggests a general drift of the major landmasses toward the equator.

formed the oceans, or continents were once much larger and therefore closer, but portions became submerged to form oceans.

In 1908, F.B. Taylor, an American geologist, presented a model (privately published in 1910), in which the continents move, distorting crustal materials into mountain ranges and island chains (Figure 5.1). Taylor postulated that moving continents form mountains at their forward margins and leave oceans behind them. Although Taylor did not correctly perceive the directions of continental movements, his ideas were innovative, and later research showed that crustal upheaval and continental movement are indeed intimately related.

Wegener's theory. Alfred L. Wegener (Figure 5.2), a German meteorologist, championed the theory of continental drift, which he con-

Figure 5.2
Alfred Wegener (1880–1930), who developed the ideas leading to the modern science of plate tectonics and its confirmation of continental drift. (From Wegener, 1966, The origin of continents and oceans. Translation of 1929 edition by J. Birnam. Dover Publications, Inc.)

ceived and presented with scanty evidence but which anticipated much of our current knowledge of continental movements, global vulcanism, wandering poles, and paleoclimates. Wegener developed his ideas on continental displacement in 1910, independently of Taylor, while observing on a world map the congruence of opposing coastlines across the Atlantic. In January 1912, Wegener unveiled his working hypothesis with supporting evidence in two oral reports, published later that year (1912a, 1912b). These observations were expanded into his classic book, *Die Entstehung der Kontinente und Ozeane* (1915). Wegener's theory on horizontal movements not only discussed all continents but also synthesized evidence from many disciplines: geology, geophysics, paleoclimatology, paleontology, and biogeography. The strongest attribute of the theory was that it integrated many types of phenomena for the first time.

Like many revolutionary ideas, Wegener's conclusions were accepted at first by only a few geologists, especially those in the Southern Hemisphere. The second edition (1920) received some attention when it was criticized by several prominent geologists, but wide knowledge of the theory came when the third edition (1922) was translated into five languages, including English (1924). A fourth edition, the one generally used now, contained more information, but throughout the various editions, the substance of Wegener's ideas was not altered. The following are some of the pertinent conclusions.

1. Continental rocks, called sial, are fundamentally different, less dense, thicker, and less highly magnetized than those of the ocean floor (basaltic, called sima). The lighter sialic blocks, the continents, float on a layer of viscous, fluid mantle.

2. Major landmasses of the earth were once united as a single supercontinent, Pangea (1920 term). Pangea broke into smaller continental plates that moved apart as they floated on the upper mantle. Breakup of Pangea began in the Mesozoic, but North America was still connected with Europe in the north until the late Tertiary or even the Quaternary (Figure 5.3).

3. Breakup of Pangea began as a rift valley, which gradually widened into an ocean, apparently by adding materials to the continental margins. The midoceanic ridges mark where opposite continents were once joined, and the ocean trenches formed as the blocks moved. Distributions of major earthquake centers and regions of active vulcanism and mountain building are related to the movements of these crustal plates.

4. The continental blocks have essentially retained their initial outlines, except in regions of mountain building, so the manner in which the continents were joined is seen by the matchup of their present margins. When this is done, similarities in the stratigraphy, fossils, and reconstructed paleoclimates of now distant landmasses are evidence that those blocks were once united. These patterns are inconsistent with any explanation that assumes fixed positions of continents and ocean basins.

5. Estimated rates of movement for certain continents range between 36 and 0.3 m yr^{-1}, the fastest being Greenland, which separated from Europe only 50,000 to 100,000 years ago.

6. Radioactive heating in the mantle may be a primary cause for block movement, but other forces are probably involved. Whatever the causal processes, they are gradual and not catastrophic.

First opposition to continental drift. Strong criticism of Wegener's theory arose in the mid-1920s. Some scientists resisted the new idea, because it conflicted with their preconceived ideas of fixed continents and a solid earth and was proposed by a man who was not part of the geologic establishment. The theory aroused the skeptical interest of scientists in many fields, but each discipline recognized major factual errors in Wegener's presentations, and these inconsistencies had to be resolved. Even du Toit, Wegener's strongest proponent, who published two books (1927, 1937) in favor

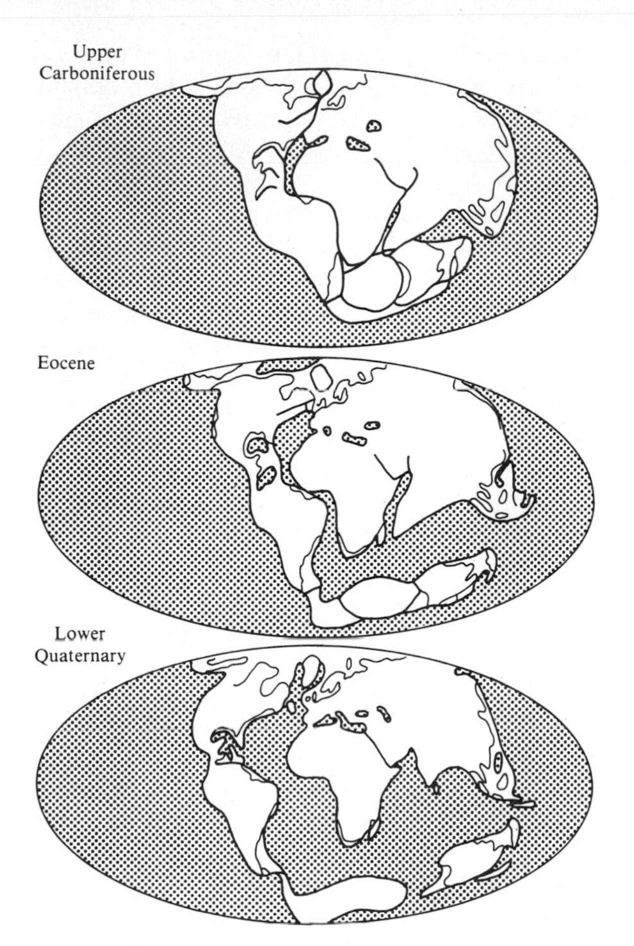

Figure 5.3
Wegener's (1929) model of continental drift, showing how he envisioned the continents, initially united in one giant landmass, to have moved apart during the Mesozoic and early Tertiary. In Wegener's time the geologic epochs and periods were thought to have been more recent than has been indicated by modern dating methods. Nevertheless, comparison with Figures 5.10 and 5.11 shows that Wegener's view was extremely similar to current reconstructions of continental movement. (Modified from Wegener, A. 1966. *The origin of continents and oceans.* Translation of 1929 edition by J. Birnam. Dover Publications, Inc., N.Y.)

of the theory, had to concede unquestionable errors by Wegener.

A major blow to Wegener's early synthesis was the reaction of participants at an important symposium held in 1926 by the American Association of Petroleum Geologists (van Waterschoot van der Gracht, 1928). Influential speakers challenged the ill-fitted or equivocal continental matchups, the stratigraphy, the lack of plausible mechanisms, and so forth. This symposium marked a declining interest in Wegener's theory, especially in North America.

The prevailing attitude in biogeography and paleontology was that first expressed by du Toit himself: "Geological evidence *almost entirely* must decide the probability of this hypothesis"

(1927, p. 118). Paleontologists in particular were disenchanted by putative biogeographic and fossil evidence marshalled to support the continental drift model. For example, in 1943 G.G. Simpson published an analysis of past and present mammalian distributions to show how these data fit the various explanations of past intercontinental connections. After correcting prevalent errors in the literature concerning these distributions, he pointed out that most of the known Cenozoic patterns could be explained without invoking continental drift. It was difficult, using paleontological data, to discriminate between opposing theories such as stable continents with periodic flooding, existence of ancient land bridges, and continental drift.

At the time of Simpson's article, lines were clearly drawn between the few biogeographers who supported and the many who denied continental drift (du Toit, 1944; Longwell, 1944), and new geologic evidence was needed to break the stalemate.

Resurgence of interest in continental drift

After World War II a second generation of scientists, often those not directly involved in the initial debates, made important discoveries about ocean basins and rock magnetism that encouraged reexamination of the ideas and evidence advanced by Wegener and du Toit.

Advancements in marine geology. When Wegener proposed his ideas, very little was known about the structure of ocean floors. Geologists suspected that the floor was composed of basalt (sima, consisting primarily of silicon and magnesium), based on loose samples obtained in dredgings, but no one had actually made core samples of the deep basins. Sialic continental rocks (composed largely of silicon and aluminum) were well known. Echo soundings from several transoceanic expeditions had portrayed ocean bottoms as smooth structures (abyssal plains) 4 to 6 km beneath the ocean surface. A midoceanic ridge was known only in the Atlantic Ocean. Lastly, trenches, deep cuts in the ocean floor, had been found on the ocean side of island arcs and were known to display unusual gravitational properties.

Oceanographic research was just beginning to accelerate before the outbreak of World War II, when charting ocean topography became a practical goal. During the war, H.H. Hess, a marine geologist, discovered flat-topped submarine volcanoes 3000 to 4000 m high when he used an echo sounder as he sailed aboard a United States troop transport. Peaked submarine volcanoes, called seamounts, had been previously identified. The new structures, which Hess later named guyots in honor of a Princeton geologist, were thought to be volcanic islands formed above the ocean surface, later truncated by wave action, and finally sunk to 1 or 2 km beneath the waves (Figure 5.4). These guyots are common in the northern and western Pacific Ocean, and as we shall see shortly, they figured prominently in models of continental drift.

Following the war, marine exploration blossomed through funding by the Allied navies. Initial discoveries were made using new deep-sediment piston corers and explosive charges. With these techniques geologists learned that under recent sediments all ocean floors are composed of basalt, and the basement is young, at a maximum dating back only to the Jurassic (150 million years BP). Thus oceans are considerably younger than the continents, whose ancient foundations, called cratons or Precambrian shields, are older than 1 billion years!

By the mid-1950s a team of scientists recognized that the submarine mountain ranges bisecting oceans are really segments of a global system 65,000 km long (Figure 5.5), having a central rift valley closely associated with a belt of shallow earthquakes. New instruments could measure remarkably high temperatures, suggesting the release of molten mantle material at

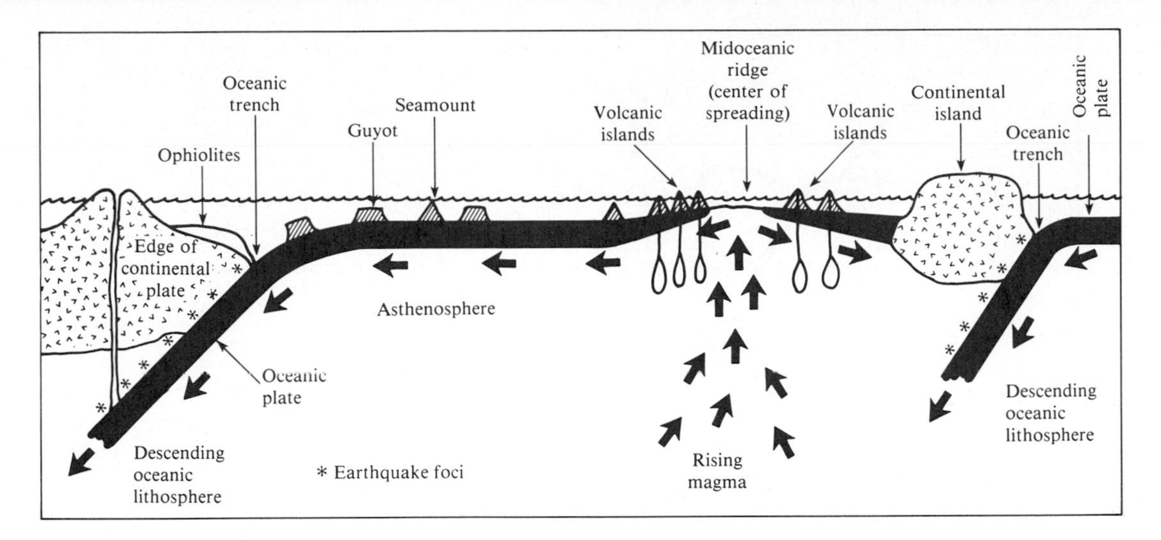

Figure 5.4
Highly simplified model of seafloor spreading that depicts how oceanic plates are pushed apart at the center by the upwelling of magma from the mantle, which causes the plate to slide away from the ridges over the viscous asthenosphere. Magma may also produce volcanic islands near the spreading ridge, but as a point on the plate is displaced from the ridge it also descends to 4 to 6 km beneath sea level and islands become submerged. These submerged volcanic structures (seamounts or guyots) eventually disappear into an oceanic trench where the oceanic plate meets another plate. In the case illustrated, the heavier oceanic plate descends beneath the lighter continental plate, which causes the metamorphosis of the surface material on the oceanic plate into ophiolites and deposition on the continent, the consumption of the volcanic islands, and eventually the remelting of the plate itself. The asterisks indicate the earthquake epicenter (the Benioff Zone) resulting from the contact of two plates.

these rifts. Scientists began to interpret the midoceanic ridges and rifts as zones where the oceans expand, establishing the concept of seafloor spreading.

Bottoms of ocean trenches are so deep that until the 1950s most knowledge of them was obtained by taking soundings. Trenches are V shaped troughs about 10 km deep (Figure 5.4). Through the use of seismic refraction techniques, marine geologists learned that the crust is extremely thin in the trenches, and heat flow is half that found in the abyssal plain, implying that heat is consumed in trenches. Gravity measurements in trenches are lower than any other place on earth. Geologists therefore postulated that the crust is being pulled downward into these trenches and its material reincorporated into the mantle.

Around the Pacific Ocean is a belt of great vulcanism and earthquake activity known as the Ring of Fire (Figure 5.5). It was thought that subduction (downward movement) of crustal materials in trenches was the direct cause of these violent geologic events. Hugo Benioff (1954) provided the first convincing evidence. By plotting the positions and depths of earthquake epicenters he demonstrated that epicenters closest to trenches are shallow and those farther away are progressively deeper. The epicenters therefore are aligned along a zone of about 45° dipping downward behind the trench, indicating that earthquakes are caused

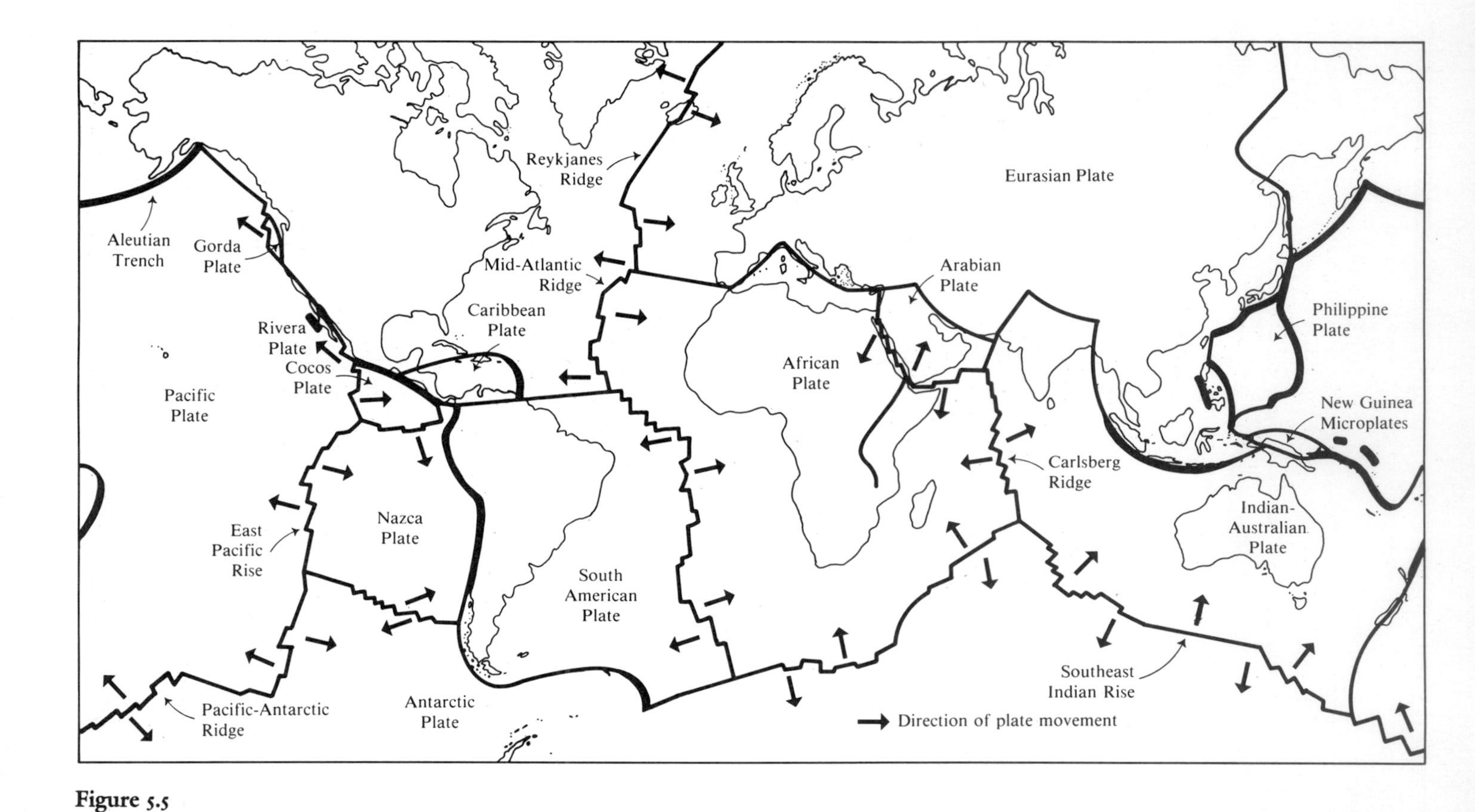

Figure 5.5
Global map showing the location of the major oceanic and continental plates discussed in the text as well as the oceanic trenches *(thick lines)* where plates are consumed and the oceanic ridges or rises along which seafloor spreading occurs. Arrows indicate the direction of spreading and hence the relative movement of each plate. (After various sources.)

as the cold, rigid crustal slab descends into the mantle. These zones are now termed Benioff zones.

Paleomagnetism. Paleomagnetism is the study of rock crystal orientation at the time of mineral formation, i.e., when molten rock solidifies. Rocks containing iron and titanium oxides become magnetized as they solidify and cool, and this is reflected in the crystalline structure, which remains "frozen" in the rock, oriented as a fossil compass in the direction of the then-prevailing magnetic field. This high-temperature magnetization, referred to as remanent magnetism, is very stable unless the rock is reheated to extremely high temperatures (the Curie point). Hence, by measuring the direction of magnetism in cooled lavas in their original and permanent positions, it is possible to determine the relationship of any landmass to the magnetic poles at the time the rock was formed and, by triangulation and computer techniques, to reconstruct the positions of landmasses relative to each other.

In the early 1950s the British physicist P.M.S. Blackett invented a new supersensitive magnetometer (magnetic detector) that could be used to determine continental orientation throughout geologic history. First, the magnetometer was used to show that the British Isles had rotated 34° clockwise since the Triassic (Clegg et al., 1954). A major breakthrough came when British scientists (Creer et al., 1954, 1957; Runcorn, 1956) analyzed geologic columns of Europe and North America and provided strong evidence that the two continents had once been joined but later drifted apart. Subsequent studies around the world reaffirmed the necessity of continental movements to explain such paleomagnetic patterns (Irving, 1956, 1959; Runcorn, 1962).

General acceptance of the drift model

Based on the above information, the groundwork had been laid for a new model of continental drift. Herman Hess proposed the first modern and thorough synthesis using the growth, movement, and destruction of crustal plates to account for major tectonic events of the world (plate tectonics). This synthesis was presented at Princeton University and informally published in 1960 and finally published for general readership in 1962. By that time, R.S. Dietz (1961) had published similar but less detailed accounts of global continental movements.

The current model. Midoceanic ridges occur where magma (molten rock) is upwelling from the mantle to the surface (Figure 5.4). It is believed that the parent rock of mantle is partially melted and the basaltic portion is then brought to the surface. The addition of basaltic magma at the center of the ridge causes older rocks on either side to spread, literally to be pushed apart, a phenomenon called seafloor spreading. Older rocks are therefore found progressively displaced from the center.

Each plate of ocean crust and some mantle is about 100 km thick and consists of a thin basaltic floor covered by recent marine sediments and a lower ocean floor composed of serpentine rock, peridotite, from which the basaltic magma has been removed. This plate is rigid and rests on a region in the upper mantle, a region called the asthenosphere, which is partially melted and can therefore slide over this viscous layer as the plates grow at the ridges.

Volcanoes are formed near the ridges by the buildup of magma (Figure 5.4). These may become emergent as oceanic islands. Once formed, such an island is eventually carried away from the ridge down a slope to the abyssal plain, thereby decreasing the elevation of the island relative to sea level, and eventually drawing it beneath the surface, making it into a submarine seamount. In classic models, wave action wears an island flat to form a guyot; however, some evidence suggests that flat-topped guyots may actually be formed that way without having been emergent.

At its oldest edge an ocean plate contacts

another geologic structure, for example, a sialic continental block. The denser of the two plates is generally subduced beneath the lighter one and consumed in the resulting trench, causing earthquakes. This is what occurs along the subduction zone on the Pacific coast of South America, a region of severe and frequent earthquakes. Behind a trench form mountains or island arcs. When ocean sediments are piled up on continents, they become metamorphosed rocks of glaucophane greenschists and blueschists (Ernst. 1975) called ophiolites.

Geophysicists liken the upwelling of magma at the ridges and subduction and cooling of basalt in the trenches to a large conveyor belt, a convection system. Materials gained in the trenches replace those used in seafloor spreading. Unfortunately, there is as yet no completely satisfactory and widely accepted version of this mechanism to explain plate movements.

Testing predictions of the model

Matching shapes of continental plates. One of Wegener's strongest arguments in favor of continental drift was the geometric matchup of landmasses when reassembled as Pangea. From the start, opponents criticized the apparent liberty used by early authors in achieving a fit. A good modern reconstruction was achieved by S.W. Carey (1955, 1958b), an Australian geologist, by using plasticene shapes of landmasses sliding over a globe. Nonetheless, the fit was not widely accepted until three workers (Bullard et al., 1965) combined computer mapping techniques and statistical analysis of the results to test continental fits. Their analysis showed that Wegener was correct if one uses the submarine contours of the continental shelf to delineate the margins of the continental plates. Subsequently, greater attention was given to the fits of various regions, and the Smith and Hallam model (1970) is now generally regarded as the best overall scheme.

Magnetic stripes on ocean floors. At the beginning of this century in central France, Bernard Bruhnes (1906) first discovered reversely magnetized lavas, i.e., lavas magnetized in a direction opposite that in presently formed ones. Since then, many investigators have found reversals, occurring every 10^4 to 10^6 years. On the ocean floor these alternating patterns of normally and reversely magnetized basalt appear as magnetic stripes that retain their spacing and shapes for long distances (Cox, 1973).

Marine geologists F.J. Vine and D.H. Matthews (1963) were the first to perceive the significance of magnetic stripes for drift theory. They were able to demonstrate several important properties of ocean floors: (1) basaltic rocks at the ridges have normal field (present-day) magnetic properties, (2) the width of alternating magnetic stripes on the opposite sides of a ridge are often roughly symmetrical and generally parallel to the long axis of the ridge, and (3) the banding pattern for one ocean matches closely that of the others, and they correspond approximately to reversal timetables from terrestrial lava flows (Figure 5.6). This is further evidence that oceans formed by seafloor spreading, each successive magnetic stripe in each ocean recording the direction of the prevailing magnetic pole at the time the basalt was formed. Differences in stripe width from one ocean to another from the same period indicate that seafloor spreading was more rapid in some places than in others. Spreading does not proceed uniformly for a given segment of ocean over long periods of time or in different latitudes in the same ocean at any given time.

Other corroborative evidence. If the model is correct, the ocean floor and its associated chains of volcanic islands, seamounts, and guyots should be youngest at the source ridges and oldest in the subduction trenches. Beginning with Tarling (1962) and Wilson (1963b), various workers have shown that these predictions are correct.

Wegener and du Toit argued that when landmasses are fit together, opposing rock strata should and do match for the period during connection. Early opponents often disre-

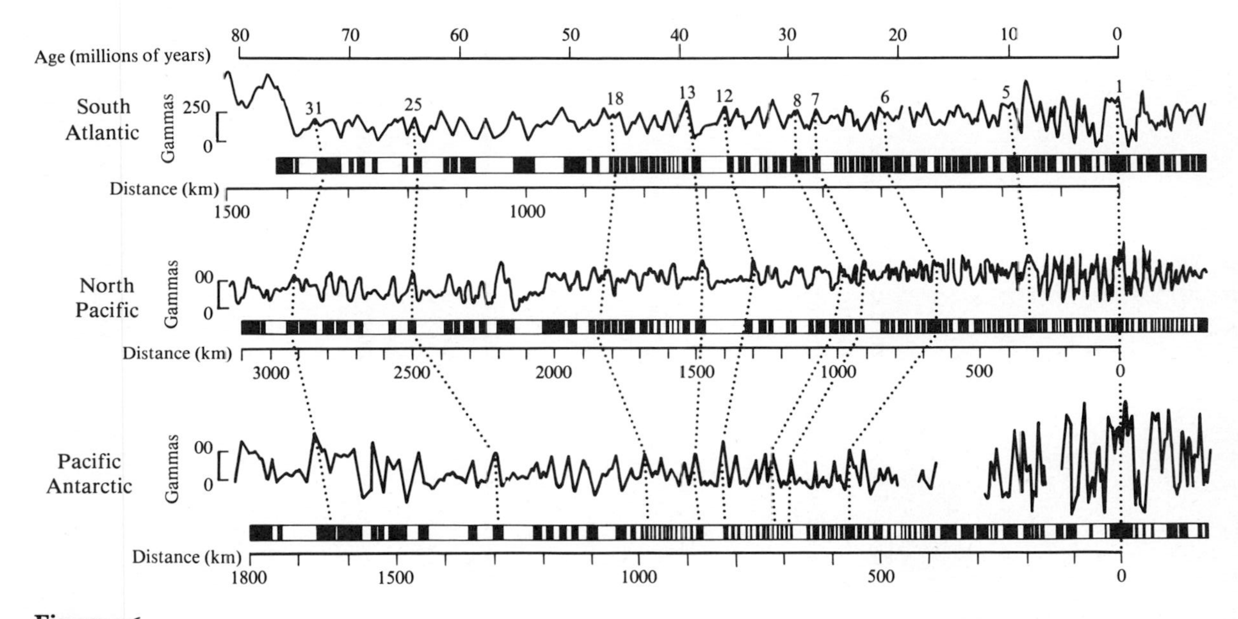

Figure 5.6
Comparison of the magnetic stripes on the floors of several oceans. The banding patterns are roughly comparable, showing that the basaltic rocks were formed in similar sequences of seafloor spreading relative to the periodic reversals in the earth's magnetic field. The stripes are not identical, however, indicating that crust formation proceeded at somewhat different rates in different regions. (After Heirtzler et al., 1968.)

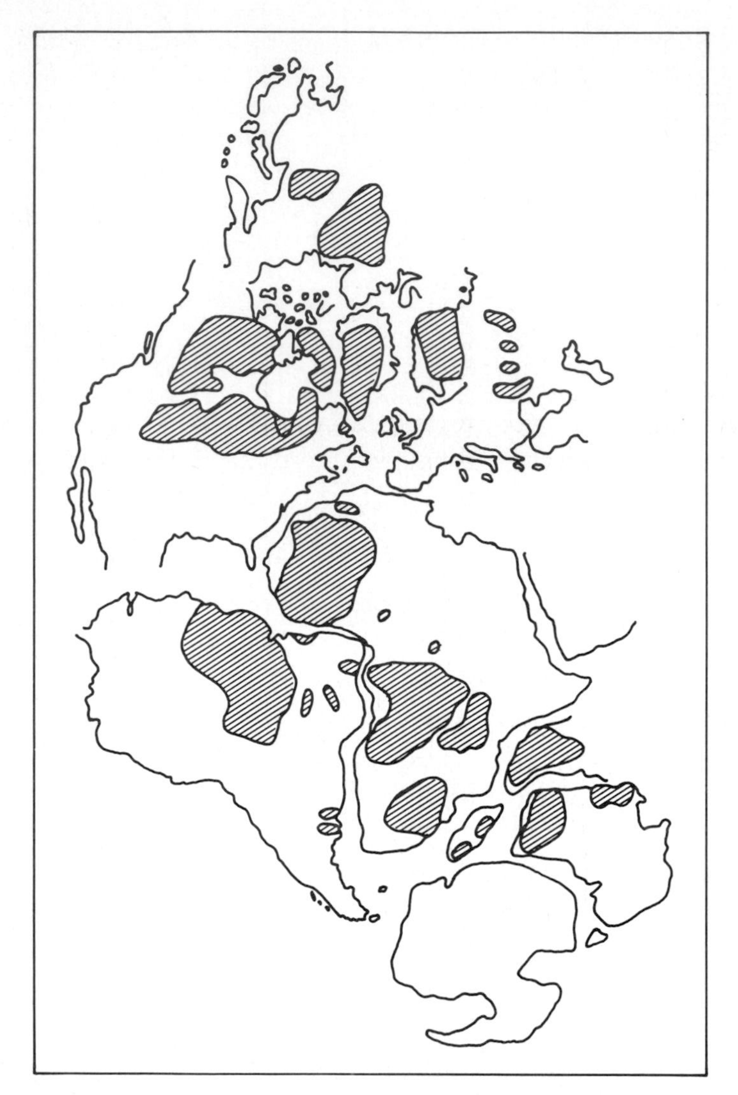

Figure 5.7
Distribution of the present locations of Precambrian shields, showing how they match up if the southern continents are reassembled in the configuration of Gondwanaland before its breakup. (After P.M. Hurley and J.R. Rand. "Pre-Drift Continental Nuclei." Science **164**:1229-1242, June 13, 1969. Copyright © 1969 by the American Association for the Advancement of Science.)

garded such stratigraphic evidence as unreliable and perhaps owing only to coincidence, but after computer fits of continental plates were presented, stratigraphic patterns could be reexamined to test predictions of the model. One striking pattern arising from the reconstructions was the amazingly close match of Precambrian shields in the southern continents, which were once united as a single continent called Gondwanaland (Hurley, 1968; Hurley and Rand, 1969). For example, fragments of the Congo and West African shields are found in coastal Brazil (Figure 5.7), and continuity of the Samfrau Geosyncline from southern South America to Australia is also considered fairly good confirmation of fit for the southern continents. Other stratigraphic comparisons show close similarities in sedimentary rocks of opposing geologic columns across an ocean, e.g., the Mesozoic rocks of Brazil and Gabon (Allard and Hurst, 1969). Such stratigraphic examples and matches of faults abound; but they can be used to reconstruct historical relationships only where the formations have not been subsequently disturbed, and often they do not provide sufficiently precise data to discriminate between alternative hypothesized patterns of continental fit.

All continental fragments in the southernmost Southern Hemisphere, including the Falkland Islands and India, have the unusual coincidence of Late Paleozoic glacial deposits (tillites) covered with Permian rocks bearing the so-called Glossopteris flora (Schopf, 1970a, 1970b). The glossopterids were arborescent gymnosperms with broad leaves (Figure 5.8)

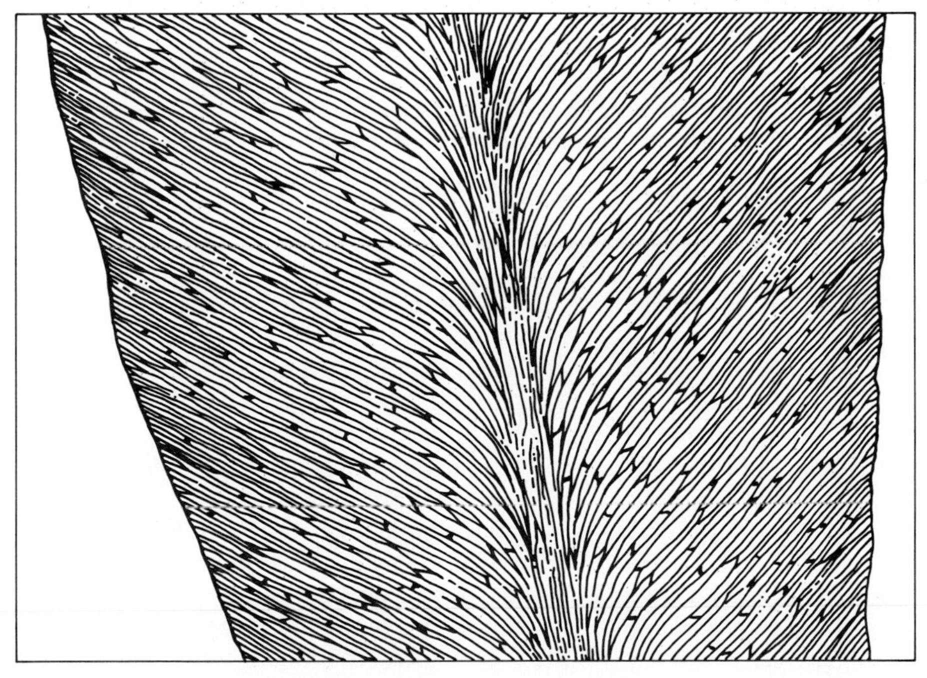

Figure 5.8
Drawing showing the distinctive structure of part of a leaf of *Glossopteris* from a Permian fossil found in Antarctica. Great numbers of these deciduous leaves, and those of the related glossopterid genus, *Gangamopteris,* are found in the Mesozoic sedimentary rocks of the southern continents that were once part of Gondwanaland. (After Schopf, 1970a.)

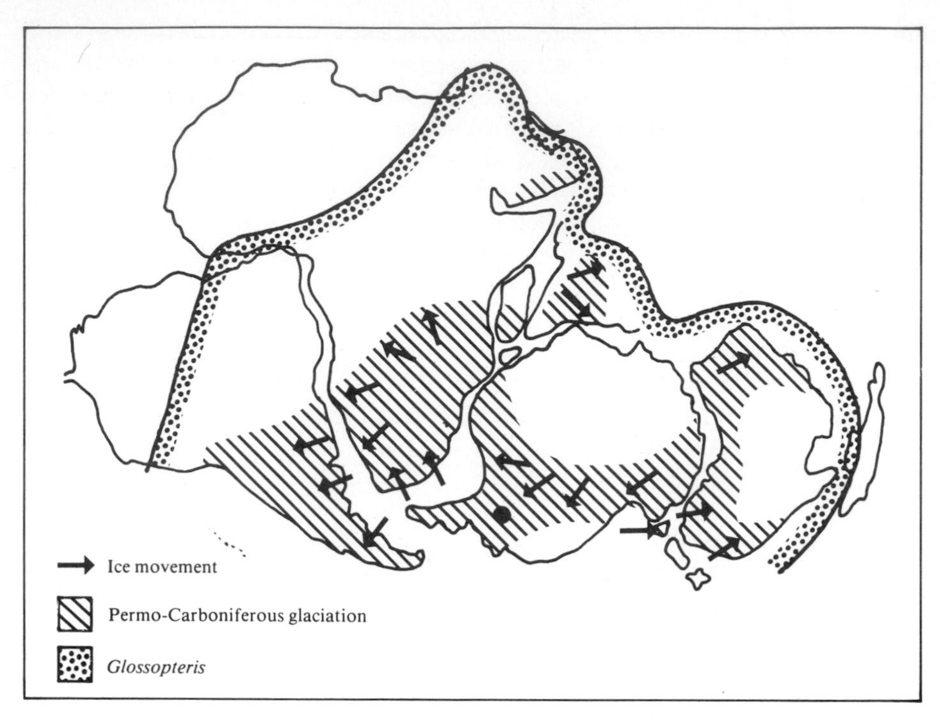

Figure 5.9
Northern distributional limits of the ancient *Glossopteris* flora, showing how well the fossil localities correspond to the original union of the southern continents to form Gondwanaland. The direction of the early Mesozoic glacial movements also correspond when Gondwanaland is reassembled. (From The evolving continents, by B.F. Windley. Copyright © 1977 by John Wiley & Sons., Ltd. Reprinted by permission of John Wiley & Sons, Ltd.)

and large seeds and are believed to have been closely related to Cordaitales, a fossil group that eventually gave rise to modern conifers (Schopf, 1976). This flora consisted of a small taxonomic assemblage of plants, seemingly adapted to a temperate climate because the plants had deciduous leaves and conspicuous growth rings in their woods. When occurrence of the Glossopteris flora is plotted on a map of Pangea (Figure 5.9), the points circumscribe a discrete region in Wegener's Gondwanaland. These observations convinced many workers that tropical India once occupied a temperate latitude and that Antarctica was once much warmer.

Most geologists in Great Britain who worked with global tectonics were converted to drift theory by 1964, convinced especially by the soundness of the model and the discoveries by Vine and Matthews on magnetic anomalies. Acceptance in North America lagged behind for several years. Meanwhile, widely circulated, popular articles appearing in *Scientific American, Science,* and *Nature* did much to make the entire scientific community aware of the rebirth of Wegenerism. Young scientists became aware of the latest evidence and helped to create a wide wave of acceptance following the mid-1960s. The biological community was, in general, rather slow to embrace drift theory after their experience in initial controversies.

Lystrosaurus: a bittersweet tale. Paleontological evidence was required for many biologists to embrace drift theory. Distribution of

Table 5.2
Tetrapods known from the Lystrosaurus zone of the Lower Triassic of Gondwanaland landmasses and parts of China

	Africa	Antarctica	India	China
Lystrosaurus	+	+	+	+
Thrinaxodon	+	+		
Chasmatosaurus	+		+	+
Procolophon	+	+		
Lydekkerina	+			
Uranocentrodon	+			
Small labyrinthodonts		+		
Labyrinthodonts			+	
Large labyrinthodonts	+			

the glossopterids was, after all, an old argument, but little was known about their means of dispersal. If such plants could have dispersed across long distances by wind or water, the floristic similarities would not necessarily indicate direct land connections.

During the austral summer of 1969-1970, a team of paleontologists discovered a fossil fauna in Antarctica that persuaded many to accept the drift model. From mudstone and volcanic sandstone in the Fremouw Formation of the Lower Triassic, D.H. Elliot and E.H. Colbert unearthed bones of the first Antarctic tetrapods (Elliot et al., 1970). Additional finds provided convincing evidence that many of the bones belonged to *Lystrosaurus,* a dicynodont theraspid reptile also found in rocks of similar age in the Karroo of southern Africa and the Panchet Formation of southern India. *Lystrosaurus* was as large as a sheep and apparently herbivorous and aquatic, resembling in some ways a reptilian hippopotamus, and the animal fossilized well in the freshwater sediments.

Found with *Lystrosaurus* were a variety of other tetrapods, including reptiles and amphibians (Table 5.2). As shown in Table 5.2, the Antarctic assemblage of the Lower Triassic shares similarities with those of South Africa and India. Because these forms are not found in North America and Europe, which are characterized instead by *Chirotherium* and different genera of amphibians, the unmistakable conclusion was that Antarctica, India, and Africa were joined at that time. This was consistent with the data of Permian glaciation and the Glossopteris flora mentioned earlier. Moreover, the strata just above *Lystrosaurus* in South Africa, India, and South America contains a different faunal assemblage, the Cynognathus zone, which also required a land connection across those southern continents. *Cynognathus* has not been discovered in Antarctica or Australia.

The fly in the ointment was, and still is, the occurrence of *Lystrosaurus* and *Chasmatosaurus* and the overlying *Cynognathus* and *Kannemeyeria* in Lower Triassic rocks of Sinkiang and Shansi, China (Table 5.2). The linear disjunction between fossil localities of southern Gondwanaland and China is now 15,000 km. Because it is unlikely that these animals crossed the ocean, two plausible explanations are available if continental drift is assumed. First, China could have been a secondary range extension, receiving migrating populations from northeastern Africa. However, no fossils have been found between these two locations. The second possibility is that parts of eastern or southeastern Asia were once closer to or part of a southern landmass. Neither explanation can yet be ruled out.

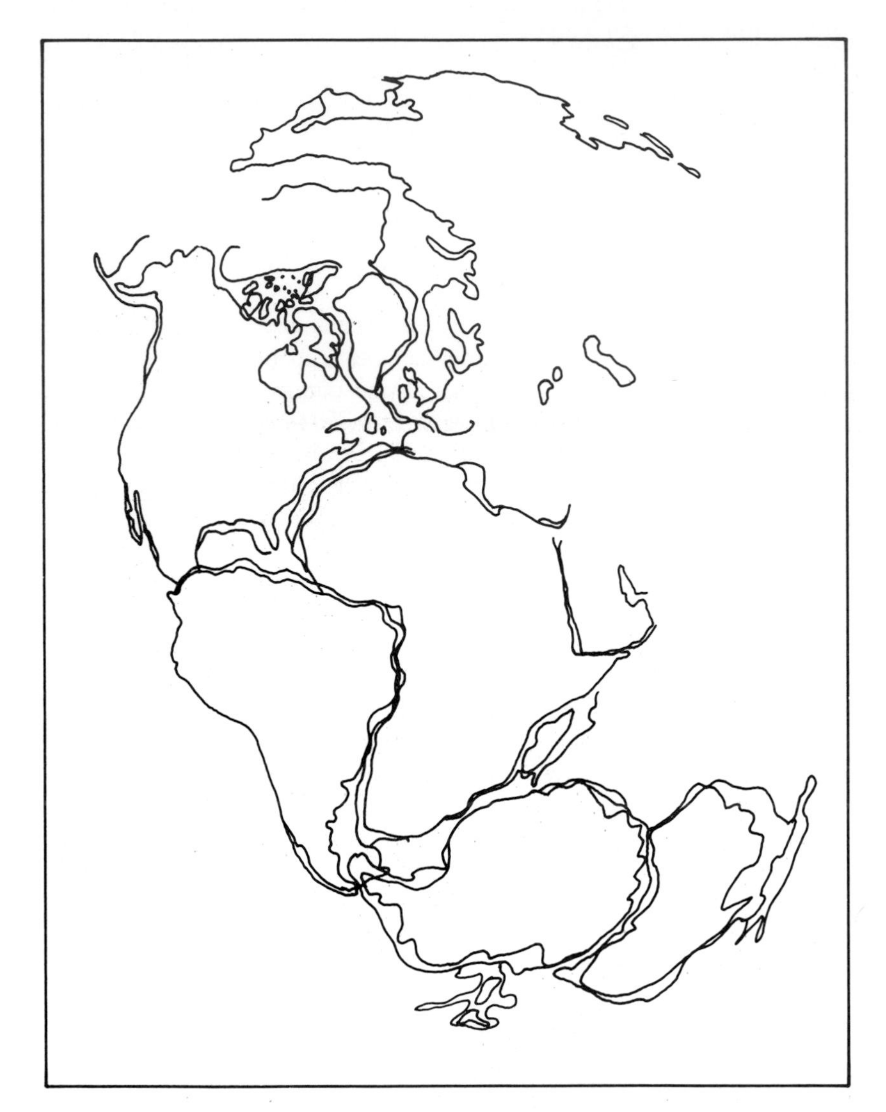

Figure 5.10
Reconstruction showing the composition and orientation of Pangea at the beginning of the Mesozoic before the beginning of fragmentation of the single ancient landmass. Details in the Gulf of Mexico region are probably incorrect. (After Smith and Briden, 1977.)

The moral to this story is that even in the early 1970s available biological evidence favoring the drift theory caused some difficulties in interpretation.

Continents of the Paleozoic and early Mesozoic

Supporters of the continental drift theory originally believed that prior to the Mesozoic all landmasses had been continuously united as a supercontinent, Pangea (Figure 5.10), which occupied one third of the earth's surface. Now geophysicists tell us that Pangea was a more temporary structure that probably existed as a single unit only during late Paleozoic and early Mesozoic time. The northern half, called Laurasia, not only had a very complex early history but also in ancient times had a substantially different history from the southern half, Gondwanaland. Paleomagnetical data show where the landmasses were at any given time, and suture zones show where past oceans disappeared and landmasses were welded together (Burke et al., 1977; Dewey, 1977).

Gondwanaland. Gondwanaland (named for a region in eastern India with fossil glossopterids) appears to have been a single, fairly stable supercontinent from the oldest comprehensible Precambrian records until the middle Mesozoic. This landmass included the foundations of present-day South America, Africa, Madagascar, Arabia, India, Australia, Tasmania, New Guinea, New Zealand, New Caledonia, and Antarctica. There were, of course, local geologic disturbances within and on the margins of Gondwanaland. For example, a small region in southeastern Australia apparently has rotated about 90° anticlockwise since the Middle Silurian (Embleton et al., 1974). Gondwanaland has also drifted great distances across the globe. Examining the history of Queensland, Australia, now at 12° S latitude, one discovers that it was located near the north pole in the Proterozoic (1 billion years ago), at the equator in the Silurian, and at 40° S latitude in the Mesozoic (Embleton, 1973). By the Permian, Gondwanaland was connected with Europe and North America.

Laurasia. We use the term *Laurasia* here with the realization that its history was not as simple as biogeographers describe it. For example, from Precambrian to Middle Ordovician time an early Atlantic Ocean separated these landmasses. Greenland was then part of North America. This ocean then closed, causing the continents to collide in the Permian and fuse, probably along the 3000 m (submarine) contour. About 100 million years later the present Atlantic Ocean began to open. Even when North America and Europe were very close, they were not always connected with solid land; in fact, western Europe, Greenland, and eastern North America were covered from time to time with shallow seas (Anderton et al., 1979). Thus the picture showing how much actual land connected the Old and New World at any point in time is very confused.

There is convincing evidence that most of eastern Asia is composed of land fragments that were welded together in the Paleozoic and Mesozoic (McElhinny, 1973a, 1973b; Dickinson, 1973). At last count, these fragments were the following: the Siberian platform, which sutured with Europe finally in the Triassic, forming the Urals; the Tarim and Tibet blocks, including the fossil localities of *Lystrosaurus;* the Sunda Peninsula block, including Indochina, Malaysia, Sumatra, and Borneo (Kalimantan); the Sino-Japanese block or blocks; and the Kolyma block of northeastern Siberia. Most reconstructions of plate movements either are purposely vague about the history of eastern Asia or avoid noting the problems by using present-day contours. Eastern Asia was mostly assembled during the Mesozoic. The Indian subcontinent (including Sri Lanka), originally part of Gondwanaland, was added in the Lower Eocene (53 million years BP, forming the Himalayas at the suture.

Most biogeographers have also avoided complex plate movements in two other areas.

First, there are many unresolved problems in interpreting the outlines of southern Europe through time, especially following the breakup of Pangea. Second, many maps tend to show western North America as we find it today; but in fact, the North American Cordillera, the mountainous backbone in the western part, was mostly absent in the Paleozoic and has apparently been added mostly in the Mesozoic and early Cenozoic, from contact of plates and presumably from the addition of some materials in the eastern Pacific Ocean (Coney et al., 1980). These details may or may not have important biogeographic meaning—we have not had time to evaluate them—so our interest is drawn mainly to the generalized reconstructions (called cartoons) for the Upper Triassic to the lowermost Cretaceous, when 99% of all landmasses were somehow interconnected. Great evolution of most land families and many orders, especially those of vertebrates, seed plants, and insects, has occurred since then; we believe now that the diversification of many taxa was in fact a direct outcome of this breakup of Pangea. For marine biogeographers, the formation and splintering of Pangea produced deep water between formerly continuous shallow water communities and vastly changed the amount of shallow water habitats for sublittoral organisms.

Breakup of Gondwanaland

Wegener had correctly identified the major pieces of Gondwanaland and how they moved apart (his dates were very wrong!), but even today a fully accurate reconstruction of the breakup has not been achieved because only limited information is available on seafloor spreading in the southern oceans. Nonetheless, Figure 5.11, *A,* gives a reasonable description of Gondwanaland just prior to its breakup in the Upper Jurassic (Norton and Sclater, 1979). Several aspects are noteworthy. First, Africa and South America were broadly attached at their centers only. This land connection was equatorial in the Jurassic, and these same regions are equatorial today. Second, the point in Antarctica that is currently at the south pole was at 50° S latitude in the Jurassic. And third, around Antarctica the landmasses were closely assembled, but some were still separated by water gaps.

Opening of the South Atlantic. Most accounts agree that active spreading between Africa and South America began in the Lower Cretaceous (127 million years BP). At least until 115 million years BP the continents remained more or less united at the equator (Figure 5.11, *B*), acting as a pivot while the southern portions spread. Marine sediments appeared in west Africa and Brazil at this time (Allard and Hurst, 1969). Spreading was slow at first, so that the gap really started to widen in the Upper Cretaceous and was essentially completed by the Eocene (Figure 5.11, *B* to *D*). The spreading rate was 1.2 to 2.0 cm per year, meaning that the distance between the two continental shelves increased up to 40 km every million years.

By the end of the Cretaceous the distance between Brazil and west Africa was over 800 km of ocean, but a system of oceanic islands, part of the Walvis Ridge, occurred between them. These volcanic islands, now submerged, formed at the midoceanic ridge, marked by the present location of Tristan da Cuhna. Those islands nearest the ridge would have been emergent while islands closest to the continents became submerged. Thus some important stepping-stones for migration and sites for speciation probably existed until Tertiary times. Some other submarine ridges have been suggested as emergent structures of the past, but certainly no solid land bridge connected Africa with the New World after Pangea started to break apart.

The Canary Islands opposite northwestern Africa have a complex origin (Schmincke, 1976). Geologists suspect that the eastern islands were probably connected to Africa because they have continental rocks, whereas the western islands have oceanic origins.

Within the continent of Africa, a spreading

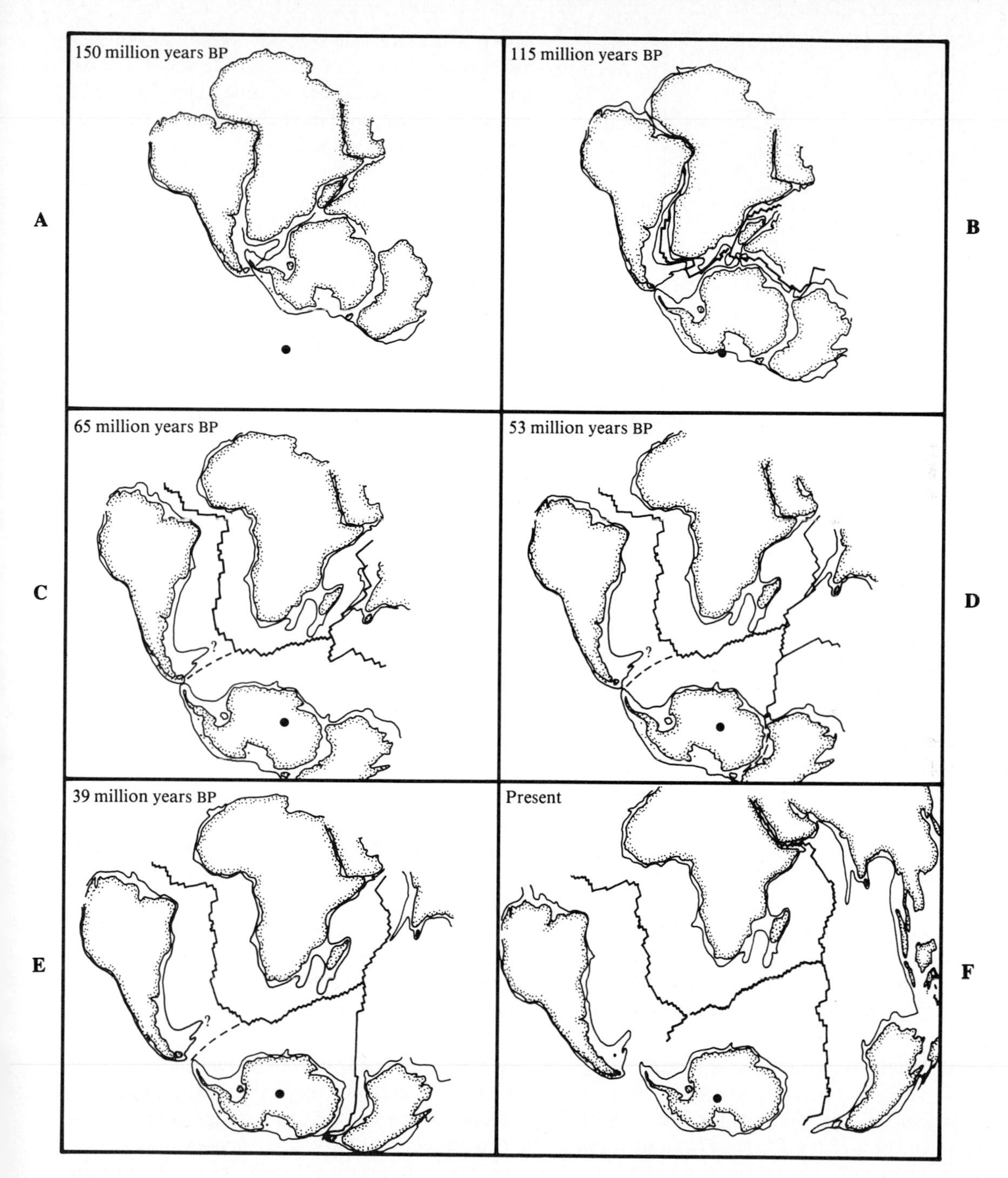

Figure 5.11
Reconstructions of successive stages in the breakup of Gondwanaland, showing the separation of the southern continents and the formation of the South Atlantic and Indian oceans. (After Norton and Sclater, 1979.)

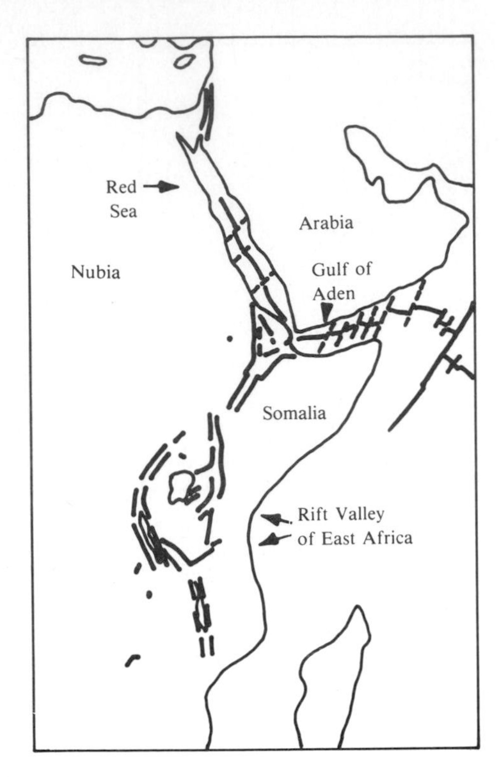

Figure 5.12
Map of eastern Africa showing the position of the crustal plates and the relationships between the Rift Valley, the Red Sea, and the Gulf of Aden. (After Mohr, 1970.)

event is now occurring. This is the well-publicized and studied rift valley of eastern Africa, which extends northward through the Afar Triangle and accounts for the opening of the Red Sea (Figure 5.12), which began about 20 million years BP. If we could return 50 million years from now, East Africa might be separated from mainland Africa by a substantial seaway, and the Sinai Peninsula would be separated from Egypt.

Evolution of the Indian Ocean. Figure 5.11, *A* and *B*, shows how Madagascar and India broke away from Gondwanaland as a unit. By 80 million years BP the deep Mozambique Channel had formed, and most authors suggest a separation of Madagascar at or before 100 million years BP, even though the rocks to date this event have not been found. Madagascar has remained more or less at the same latitude for the past 150 million years.

Between 80 and 90 million years BP, the Indian subcontinent broke away from Madagascar and began its northward journey (Figure 5.11, *C* and *D*). For a period of 20 million years this landmass was displaced rapidly toward Eurasia, averaging 180 km per million years. When India collided with Eurasia, the force of the impact caused great deformation and started the creation of the Himalayas. The date of the initial collision was 55 to 53 million years BP, which coincided with the union of the Indian and Australian plates along the Ninetyeast Transform Fault and the separation of Australia from Antarctica. Sri Lanka is part of the Indian Plate, to which it has a long history of direct land connections, most recently in the Pleistocene.

Several islands in the Indian Ocean are formed of continental rocks, implying that they were once attached to Gondwanaland. The Kerguelen Plateau, including Kerguelen Island, was very close to Antarctica prior to 80 million years BP but became isolated by the southward movement of Antarctica beginning 53 million years BP. At 65 million years BP the Mascarene Plateau, including the Mascarenes and Seychelles, were very close to India, if not part of it, and resting aside the submarine Central Indian Ridge. The Mascarene Plate was subsequently displaced several hundred kilometers to the southwest.

Movements of South America and Australasia. As far as we know, the southern tip of South America and Antarctica were close during the Cretaceous, experiencing relatively little movement. Tectonic maps depict this as a series of islands, not a solid land bridge, but the last word has certainly not been written on the subject. The gap between Tierra del Fuego and the Antarctic islands widened as Antarctica moved southward by seafloor spreading in the Eocene.

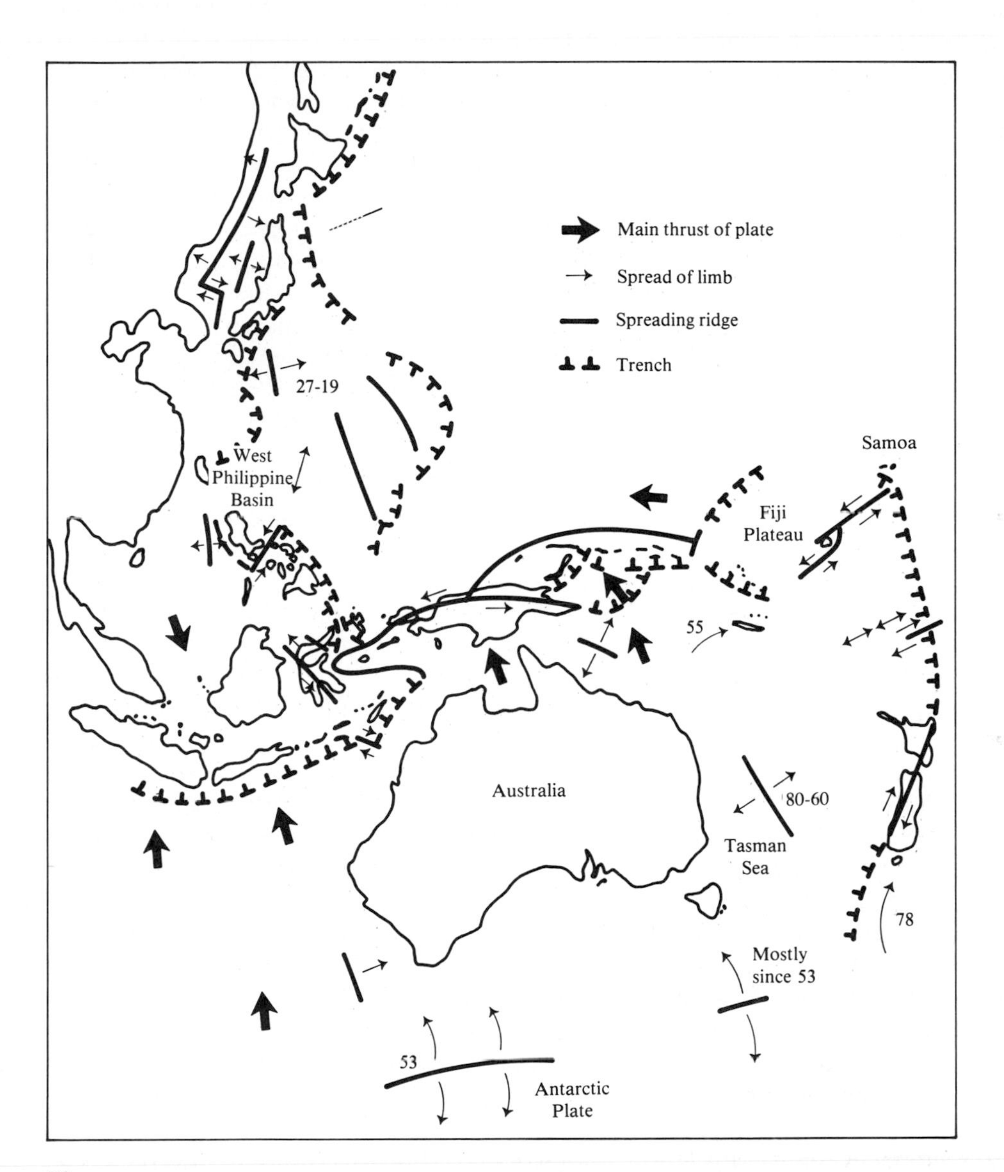

Figure 5.13
Major tectonic events of the Australasian region. The movements of the crustal plates and the locations of the ocean trenches are shown. Numbers refer to millions of years BP during which spreading occurred.

Final disruption of the connection occurred in the middle Tertiary, when the South Georgia island group was displaced eastward to their present position in the Scotia Arc adjacent to the Scotia Trench (Dalziel et al., 1975; DeWit, 1977).

The Falkland Islands in the South Atlantic are part of the margin of the South American continental shelf. A land bridge apparently connected the Falklands with Argentina during the Pleistocene.

New Zealand was attached to Antarctica but broke away and drifted northward around 80 million years BP (Figure 5.13). New Zealand was not connected directly to eastern Australia either before or after this event, only indirectly through Antarctica. In fact, from 80 to 60 million years BP, the distance between eastern Australia and New Zealand increased by the expansion of the Tasman Sea basin. During some periods, New Zealand was a single large island, e.g., in the Pleistocene, although at other times it was subdivided into two or more units, islands different in size and shape than the two present islands (Fleming, 1975).

Australia and the large islands of New Guinea and Tasmania must be treated together as a single geologic structure. A land connection between southeastern Australia (southern Tasmania) and Antarctica may have been lost in the Mesozoic (150 million years BP), but at least southwestern Australia was solidly fused with Antarctica at the start of the Tertiary. Spreading there began around 53 million years BP, sending Australia northward and Antarctica southward (Figure 5.13), but some land connection probably persisted during the initial stages of rifting.

Tasmania and Victoria, Australia, are now separated by the very shallow Bass Strait, but during periods of low sea level a land bridge has connected them, most recently in the late Pleistocene. In the north, New Guinea and Queensland are separated by the very shallow Torres Strait, which also has been a land bridge during periods of low sea level. It is not surprising, therefore, that many organisms are the same or very closely related on opposing sides of each strait.

In the early Tertiary, Australia's only solid land connection was with Antarctica and, through this, to other parts of Gondwanaland—Asia was very far away. On the northern margin of the Australian Plate was open ocean with Timor (which was probably submerged at the time) and New Guinea at the leading edge. Compared with today, the emergent portion of New Guinea was much smaller then; the island increased in size with the uplift of the northern side beginning in the Miocene, when the Australian Plate collided with several plates to the north (Curtis, 1973).

Also in the Miocene, the oceanic northwestern margin of the Australian Plate struck the continental shelf of southeast Asia. Deep ocean trenches formed along the plate boundary and behind them were produced large islands. This boundary is easily detected as a line drawn south of the Andaman Islands and Indonesia and through the Sunda Strait and Bunda Sea toward New Guinea (Figure 5.14). These zones are now sites of violent earthquakes and some vulcanism, for example, the eruption of Krakatau (Chapter 3). Most of these islands were small or not emergent before the plates made contact. Thus by the Pliocene Australia had achieved approximately its present position, closer to southeast Asia than ever before but connected with the mainland only indirectly by a series of large and small islands separated by much open ocean.

An introductory textbook on biogeography is no place to describe in detail all the hypothesized and controversial tectonic events surrounding the Australian Plate. Perhaps the biggest controversy affecting us is the origin of New Caledonia, surrounded now by open ocean and approximately equidistant from New Guinea, Australia, and New Zealand. One model proposes that New Caledonia was once

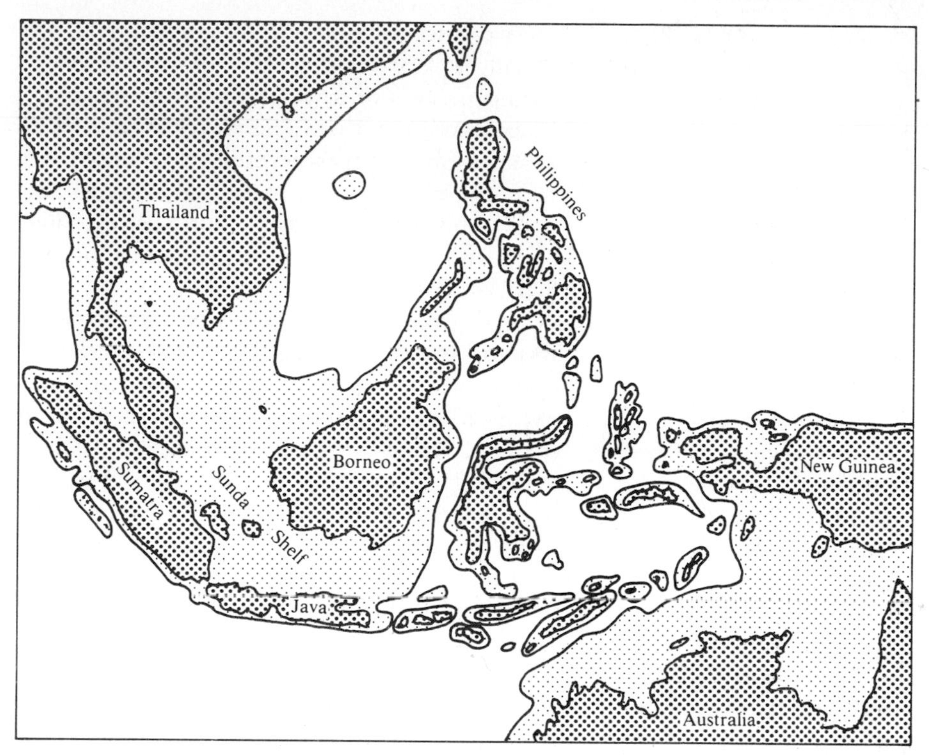

Figure 5.14
Map of southeast Asia and the western Pacific Basin, showing the position of the 200 m isobathyl (water depth) contour. This depth indicates the location of the continental shelves, and these are of great biogeographic importance because much of them were exposed when sea levels were lowered during the Pleistocene.

part of eastern New Guinea, and in the Upper Eocene the island was displaced to the southeast along a subduction zone to its present isolated position. Another model favors a geologic relationship of New Caledonia with New Zealand to the south, suggesting that submarine rises roughly extending between New Caledonia and New Zealand may have been emergent, as a land bridge or island arc, an idea that has many geologic problems and uncertainties.

During all this activity in the tropics, Antarctica was quietly moving toward the south pole. The great polar ice cap of Antarctica formed in the Miocene, according to a model presented by Kvasov and Verbitsky (1981).

Important tectonic events in Laurasia

For our purposes there are relatively few tectonic events that are necessary for interpreting broad biogeographic patterns.

Opening of the North Atlantic. The breakup of Pangea was actually initiated by the opening of the North Atlantic in the Upper

Triassic–Lower Jurassic (200 to 180 million years BP). Most of the eastern coast of North America was opposite Africa (Figure 5.14), western Europe being adjacent to Greenland, Newfoundland, and the New England region of North America. Active seafloor spreading was clearly occurring by 170 million years BP, separating North America from both Africa and Europe at the same time except in the far north. By the Upper Jurassic spreading began to form the Labrador Sea between Greenland and Canada, but much of the spreading here occurred later, between 80 and 45 million years BP. Likewise, the opening of the northernmost Atlantic really developed in the earliest Tertiary with the development of the Reykjanes Ridge, which is located beneath Iceland and is its origin. Land connections between North America and Europe and Africa consequently were severed almost immediately in the low latitudes, but a land connection of some sort persisted at poleward latitudes until the Lower Eocene. This northern connection has been called the De Geer Route, and it must have been an important corridor for the migration of animals and plants between the Old and New World (McKenna, 1972a, 1972b). There were probably very broad continuous or nearly continuous land connections through the British Isles into Greenland and Canada in the Lower Cretaceous (Hallam and Sellwood, 1976), and the British Isles have many times been connected by land to the European mainland, most recently in the very late Pleistocene.

As North America drifted away from Europe the eastern coast, including land from New Jersey to Yucatán, Mexico, became submerged to form a great sedimentary basin. The Atlantic Gulf Coast was under water from the Upper Jurassic until the Eocene, so that the forms living there today are colonists since that time.

The emergence of the Atlantic Gulf Coast was probably synchronous with the collision of Alaska with Siberia of the Eurasian Plate, which occurred in the Eocene. This established a land connection known as the Bering Land Bridge, actually a very shallow strait that became land during periods of low sea level. Consequently, about the time that a land connection with western Europe was severed, a new one between western North America and eastern Asia was established, permitting continued exchange of organisms between North America and the Old World.

The geologic history of Alaska and most of western North America is extremely complex and too complicated for a brief summary (Coney et al., 1980). Suffice it to say that much of the land found west of Salt Lake City, Utah, including the North American Cordillera, has been added, created de novo, or greatly deformed during the last 100 million years.

Formation of the Mediterranean Sea. After being freed from North America and South America, Africa began to swing anticlockwise toward Eurasia and closed the formerly extensive Tethys Sea. A bridge was formed between Asia and Africa through Arabia following the collision in the middle Tertiary, which not only created the Zagros Mountains of Iran but also confined the Mediterranean Sea. As Africa approached southern Europe, a number of deformations were initiated, including the anticlockwise rotation of Italy, which produced its characteristic diagonal orientation (McElhinny, 1973a).

Plate tectonic events in and around the Mediterranean Sea have been the subject of several recent symposia (Biju-Duval and Montadert, 1977; Bureau de Recherches Géologiques et Minières, 1980a, 1980b), in which the many divergent opinions were expressed. Not only are the movements of microplates extremely difficult to interpret because much deformation has occurred, but also at least one parcel, a portion of southern Spain, is believed now to have been a chunk of northern Africa at one time. Moreover, during the Cenozoic the Mediterranean has apparently been completely drained and then reestablished at least once, a phenomenon known as the Messinian crisis (Hsü, 1972).

Continental islands of eastern Asia. Most accounts of the continental islands of eastern Asia lack information on when these islands were connected to the mainland and when and how they were connected with each other. Geologists do not generally publish such diagrams because these reconstructions are very unreliable and highly speculative, even when based on detailed studies of rock formations. Imagine for a moment an island group like the Philippines, which has a long and complicated geologic history (Gervasio, 1973). There are stable regions, perhaps continuously emergent regions, that date back to the Lower Permian, including the island of Palawan, parts of Mindoro and Zamboanga, and regions under the Sulu Sea. The Palawan–Sulu Sea area may have been part of Borneo at one time, but there are other equally likely explanations. Some areas of the Philippines were emergent in the Permian and through the Triassic, but in the Jurassic and Cretaceous several large areas in the island group became submerged basins. At the close of the Cretaceous and in the early Paleocene some new island arcs were produced. In the Upper Paleocene and Eocene numerous areas became submerged and received sediments, and into the Oligocene much of the island group was covered by the sea, except in southern Palawan. In the Miocene there occurred much vulcanism in the Philippines and major uplifts of the sedimentary basins, a pattern that has been altered by subsequent faulting. When coastal shelves were exposed in the Pleistocene,

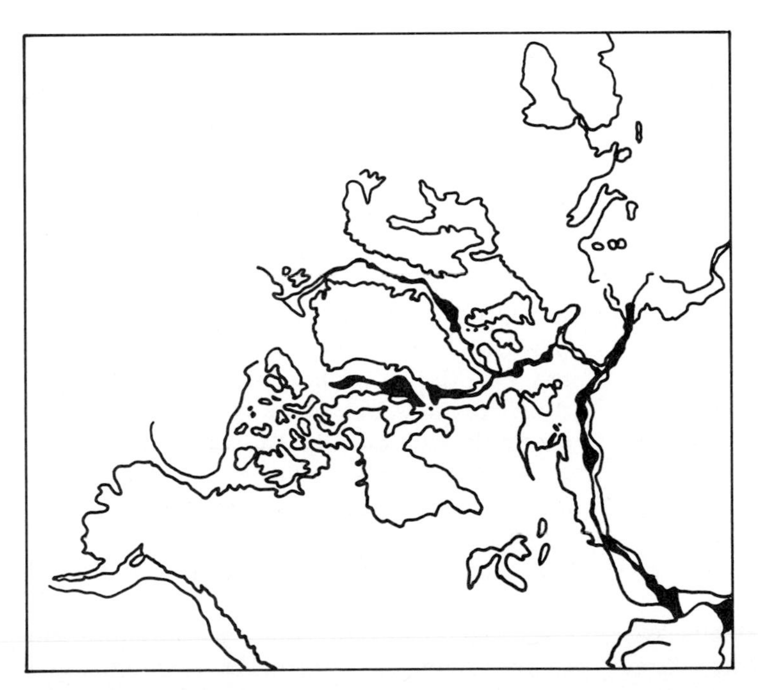

Figure 5.15
Illustration of the fit of North America, Greenland, Europe, and Northern Africa, showing the presumed positions of these landmasses prior to the commencement of seafloor spreading. The continents are juxtaposed, matching the fit of the continental shelves. The shaded areas show regions where the continental shelves either overlap or fail to meet, thus indicating the imprecision of the matching. (After Sullivan, 1974.)

the Philippines were probably one or several large islands. Now, along with this changing landscape in the Philippines, envision complex events occurring in the surrounding island groups and connections with eastern Asia, and you can soon appreciate the vast number of scenarios that can be invented.

There are two practical ways of estimating whether an island was ever connected to the Asian mainland: (1) by the occurrence of many different mainland forms that cannot jump water gaps and (2) by showing that islands would have been connected with each other and with the mainland if the sea level was lowered 100 m or so, as it was in the Pleistocene. When this is done, we learn that all the large, close islands could have been attached to the mainland (Figure 5.14). These criteria were the same ones used by Wallace (1869) to explain why distributions of certain land animals and freshwater fishes of the Oriental region frequently terminate at Bali and Borneo and do not reach Lombok and Celebes, respectively. Wallace's line is supposed to mark the boundary of the continental islands.

Dynamic changes in the Pacific Ocean

Early articles on continental drift discussed mainly the continental plates, but soon considerable attention was focused on the oceanic plates in the Pacific Basin that have relatively little emergent land. It may surprise you to learn that our largest ocean, the Pacific, is now getting smaller; it once covered two thirds of the globe when Pangea existed.

Hot spots and triple junctions. In the basic model of seafloor spreading, volcanic islands are formed either at the midoceanic ridges in files perpendicular to the ridge or behind the trenches as parallel island arcs. However, in the Pacific Ocean there is no single central ridge, and many of the islands are not associated with ridges or trenches. The Hawaiian Islands are a perfect example of this: a narrow, linear chain of islands located far from any ridge or trench. The oldest emergent island is Midway and the youngest is Hawaii, which is presently forming. J. Tuzo Wilson (1963a) proposed a unique solution to explain the pattern, what he called a hot spot. This is a fixed weak spot in the mantle at which magma is being released. As the oceanic plate passes over this hot spot, volcanoes are produced at the surface, causing the formation of an island (Figure 5.16). The spot is very narrow, so the islands form in a narrow chain, the chain being linear if the plate has moved strictly in one direction. The youngest islands in a chain are emergent and the progressively older ones gradually become submerged.

The nice aspect of this model was that it explained the occurrence of most islands in the eastern Pacific (Figure 5.17). The Hawaiian Islands, many of which are still emergent, are really only the most recent part of a long series extending to the Aleutian Trench, including the submerged Emperor Seamount Chain (Morgan, 1972a; Dalrymple et al., 1973). The seamount closest to the trench is naturally the oldest (about 80 million years BP). Morgan (1972a, 1972b) noted that three very similar chains of emergent and then submerged volcanoes occur in the Pacific basin (Figure 5.17): the Tuamoto Archipelago (including Easter Island) and the Line Island Chain, the Austral Seamount Chain and the Marshall–Ellice Island Chain, and a chain from the Cobb Seamount opposite Washington through the Gulf of Alaska.

Another interesting quality of three of these four chains is that the recent islands are all roughly parallel, then there is a sharp bend (elbow), and all the older sections are parallel again. This is the pattern you would expect if a rigid plate was passed over separate points and at some point the whole plate was shifted in a different direction. The Hawaiian elbow is dated at 44 million years BP. Consequently, we can use the distance and orientation of the chains to infer the movement of the plate through time.

At this time, there is some disagreement whether hot spots account for all the roughly

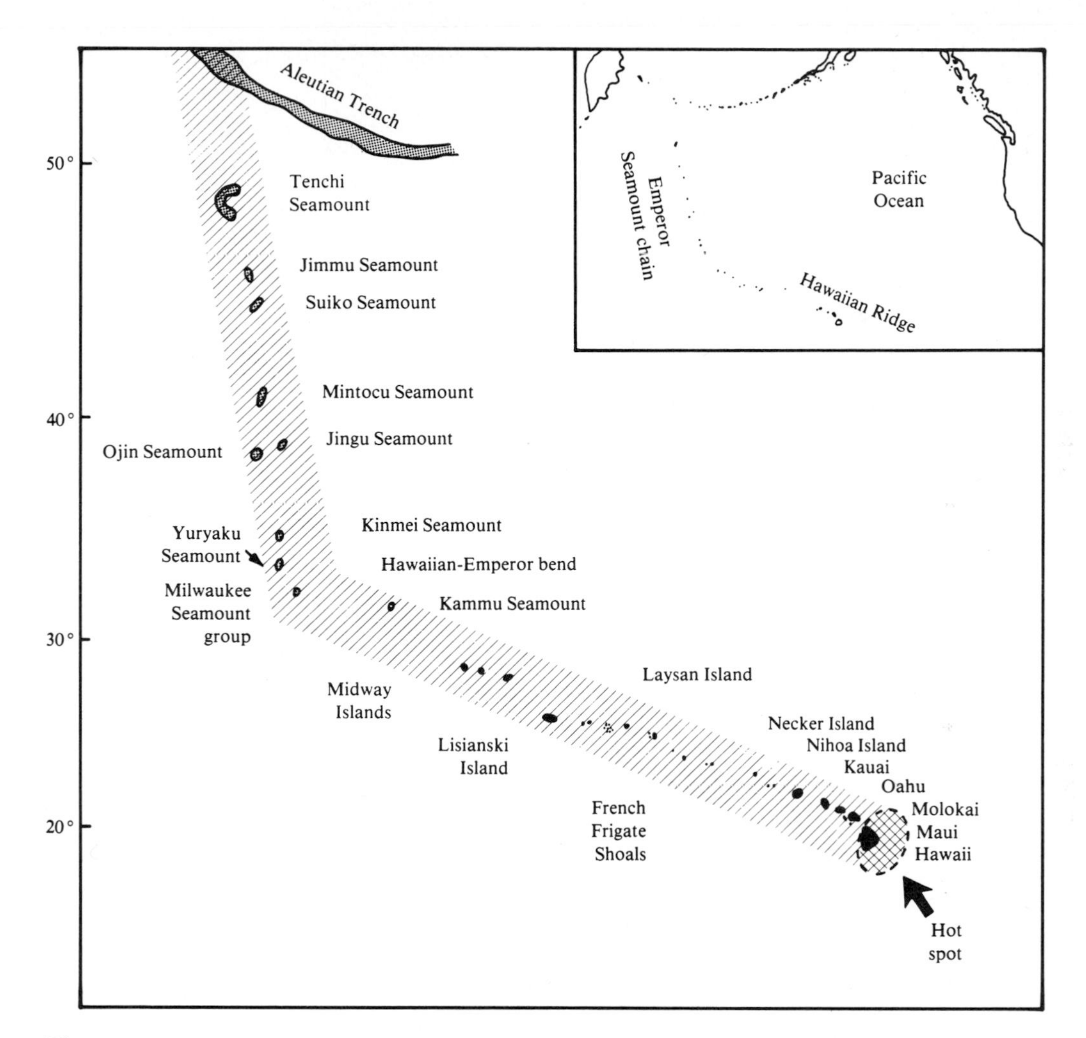

Figure 5.16
A hot spot has produced the volcanic Hawaiian Islands and the Emperor Seamount Chain in the central Pacific. The hot spot, presently located near the Hawaiian Deep, is presumably responsible for the intense volcanic activity on the island of Hawaii. Several authors agree that the narrow chain of islands has been created as their locations on the oceanic plate successively drifted over the hot spot. The older islands to the north were formed first; they have disappeared beneath the surface to become seamounts, and the oldest have disappeared into the Aleutian Trench. The sharp bend between the Emperor and Hawaiian chains occurred when the drifting oceanic plate changed directions about 44 million years BP. (After Jackson et al., 1972. Courtesy the Geological Society of America.)

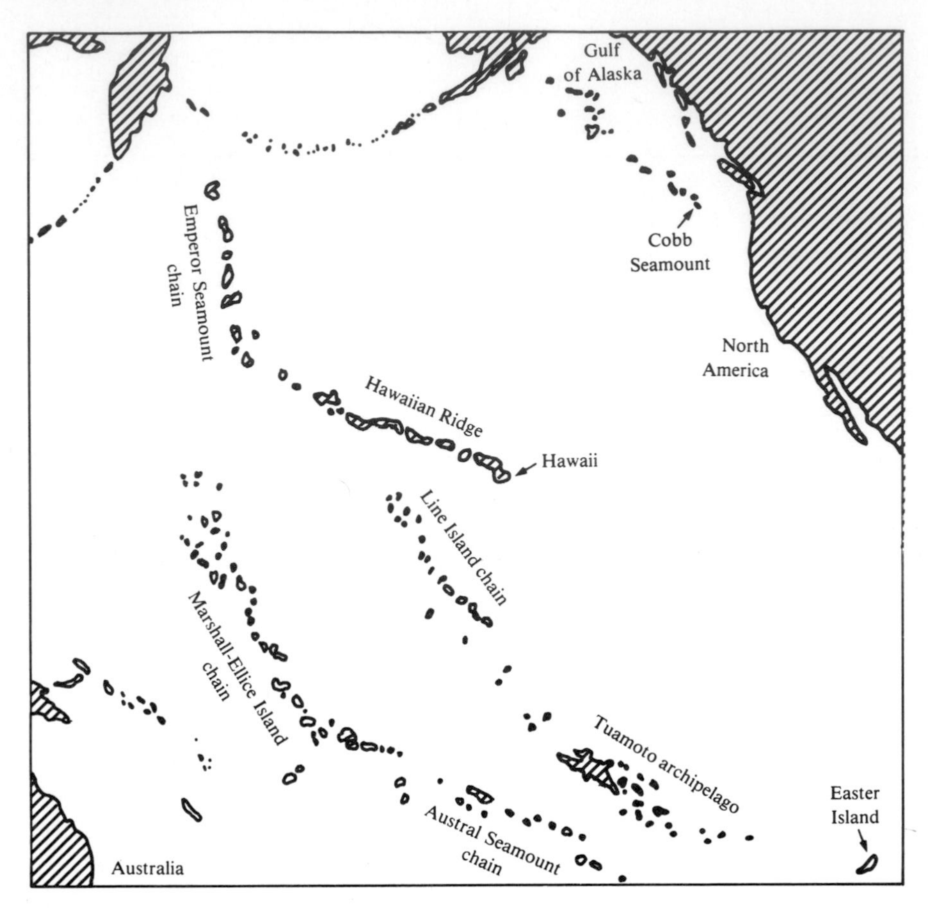

Figure 5.17
Map of the Pacific Ocean showing the four chains of emergent and submerged islands that are presumed to have formed over hot spots. However, as discussed in the text, there is reason to question this interpretation because the progression in island age from southeast to northwest is not always as regular as in the Hawaiian-Emperor Chain. (After Dalrymple et al., 1973.)

parallel island chains in the Pacific basin. Jarrard and Clague (1977) noted that the hot spot model does not adequately explain age patterns of islands in certain chains. For example, in the Austral-Cook chain the oldest islands are distributed at both ends, and active vulcanism occurs on both ends of the Samoan chain. Hence, some of these island chains could have been formed along fracture zones rather than from a fixed point. Nonetheless, the hot spot model is applicable for some chains, and all of these are still considered oceanic islands, mostly of Cenozoic age.

Volcanic islands may also be formed where three plates rest against each other, so that three trenches meet. Each plate has its own rate of disappearance into the trench; consequently, the junction where the three meet can shift through time (Figure 5.18). This is called a triple junction (McKenzie and Morgan, 1969), an example of which is the present location of the Galápagos Islands at the intersection of the east

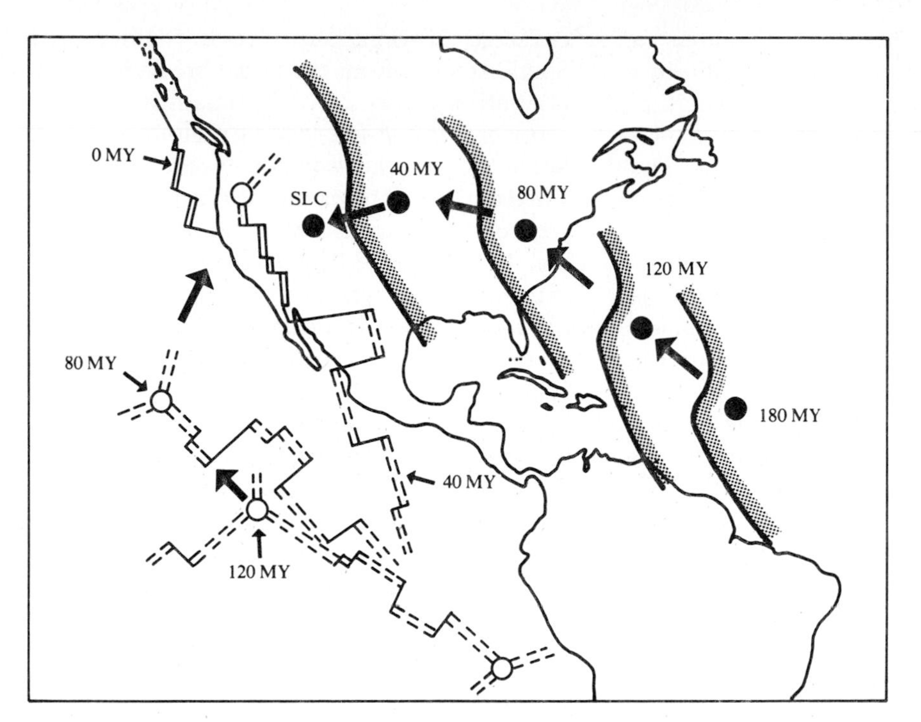

Figure 5.18
Reconstructed sequence of movement of ocean plates in the eastern Pacific Ocean. The movement of a triple junction, where three plates meet, is followed as is the shift in the edge of the continental plate and the position of present day Salt Lake City (SLC), Utah, and of the Hawaiian hot spot (open circle). (After Coney, 1978.)

Pacific rise and the Galápagos Rift Zone, the point where the large Pacific Plate meets the Nazca Plate west of South America and the Cocos Plate opposite Central America.

Inferred history of the Pacific Basin. Using the data on hot spots and triple junctions, many workers have attempted to reconstruct the early history of the Pacific Basin. Unfortunately, the oldest rocks, those entering the Marianas Trench, are 150 million years BP; without older materials the early events will be difficult to reconstruct.

There are six major oceanic plates used in most Pacific reconstructions. The Phoenix Plate was located in the south Pacific and disappeared by the enlargement of the Pacific Plate, and the Kula Plate has disappeared under Alaska and Siberia. A triple junction existed at the intersection of the Kula, Pacific, and Farallon plates (Figure 5.18), which migrated to come in contact with the North American Plate; at 155 million years BP this junction was just east of the Hawaiian hot spot. Most of the Farallon Plate has been subduced under western North America, being responsible for much of the mountain building there, but small pieces of this plate remain in the vicinity of the Pacific Northwest (called the Gorda or Juan de Fuca Plate) and opposite central mainland Mexico (called the Rivera Plate). Thus the East Pacific Rise moved eastward so much that the ridge became superimposed on the trench around 30 million years BP. The former trench was uplifted to form the coastal ranges of southern California and became fused to the northward-moving Pacific Plate, creating the San Andreas Fault, famous for earthquake activity in California. The Pacific Plate also became fused with the section of mainland Mexico to become Baja California, and its northward movement began the formation of the Gulf of California. This gulf has been spreading mainly over the last 4 million years. The convergence of the Cocos Plate has produced the present mountainous backbone of Central America, and the meeting of the Nazca Plate and South America induced the formation of the Andean Cordillera, mostly since the Oligocene.

As these plates were swept out of the central Pacific and subduced under the continents, they must have carried on board some small islands or chunks of land that were accreted to the continents (Coney et al., 1980), along the Andes, from Alaska to California, and in eastern Asia. Whether some of the large fragments of eastern Asia and northwestern North America came from the breakup of an ancient continent, Pacifica (Nur and Ben-Avraham, 1977; Melville, 1981; Nelson and Platnick, 1981), is a question that needs to be resolved by paleomagnetic and paleobiogeographic data. Certainly the biological examples used now to promulgate Pacifica are mostly highly specialized Tertiary clades of angiosperms, groups with modes of long-distance dispersal and certainly not in existence when Pacifica might have existed. The idea of Pacifica will have to be tested primarily using invertebrate taxa.

History of Central America and the West Indies

No matter what brand of historical biogeography you practice, the history of Mesoamerica and the Caribbean region is a central concern. The reasons for this should be obvious. First, an investigator needs to know how much geographic opportunity there was for the exchange of organisms between North and South America from the initial breakup of Pangea until the Isthmus of Panama fused the two continents in the Pliocene (4 million years BP). Secondly, an investigator needs to know whether the rich biota on the Caribbean Islands arrived there exclusively by long-distance dispersal, as suggested by Darlington (1938) and many others, or whether some species are present because the islands either are the remnants of an ancient landmass or had an ancient land connection through the Gulf of Mexico.

Before going too far, one must forget about the present profile of eastern North America. As we described earlier, this was submerged as the trailing eastern margin of the continent, beginning in the Jurassic. Also, in the Jurassic the Gulf of Mexico was probably not in existence, and in the Middle Jurassic a transform fault (break between adjacent segments of a plate) developed between Mexico and the United States, extending from the Gulf of Mexico to near Los Angeles, California. As a consequence, the western margin of Mexico shifted southeastward about 800 km. This means that southern Central America remained in close proximity to South America and North America north of Mexico was drifting away from Europe (Figure 5.19, *A*). Thus until the Cretaceous great biotic exchange could have occurred between North and South America.

Events during the Cretaceous of Central America are still rather sketchy. A land bridge between Mexico proper and South America was apparently severed early in the Cretaceous, probably to the south of and adjacent to Yucatán. Between the Isthmus of Tehuantepec in southern Mexico and Colombia there existed a series of volcanic islands, manufactured in the first half of the Cretaceous. Beginning in the Turonian (90 million years BP), this series of islands was pushed eastward by seafloor spreading of a new Caribbean Plate (Figure 5.19, *B* and *C*). What was proto–Central America thus became the Greater Antilles. From this magmatic arc the northernmost element became the

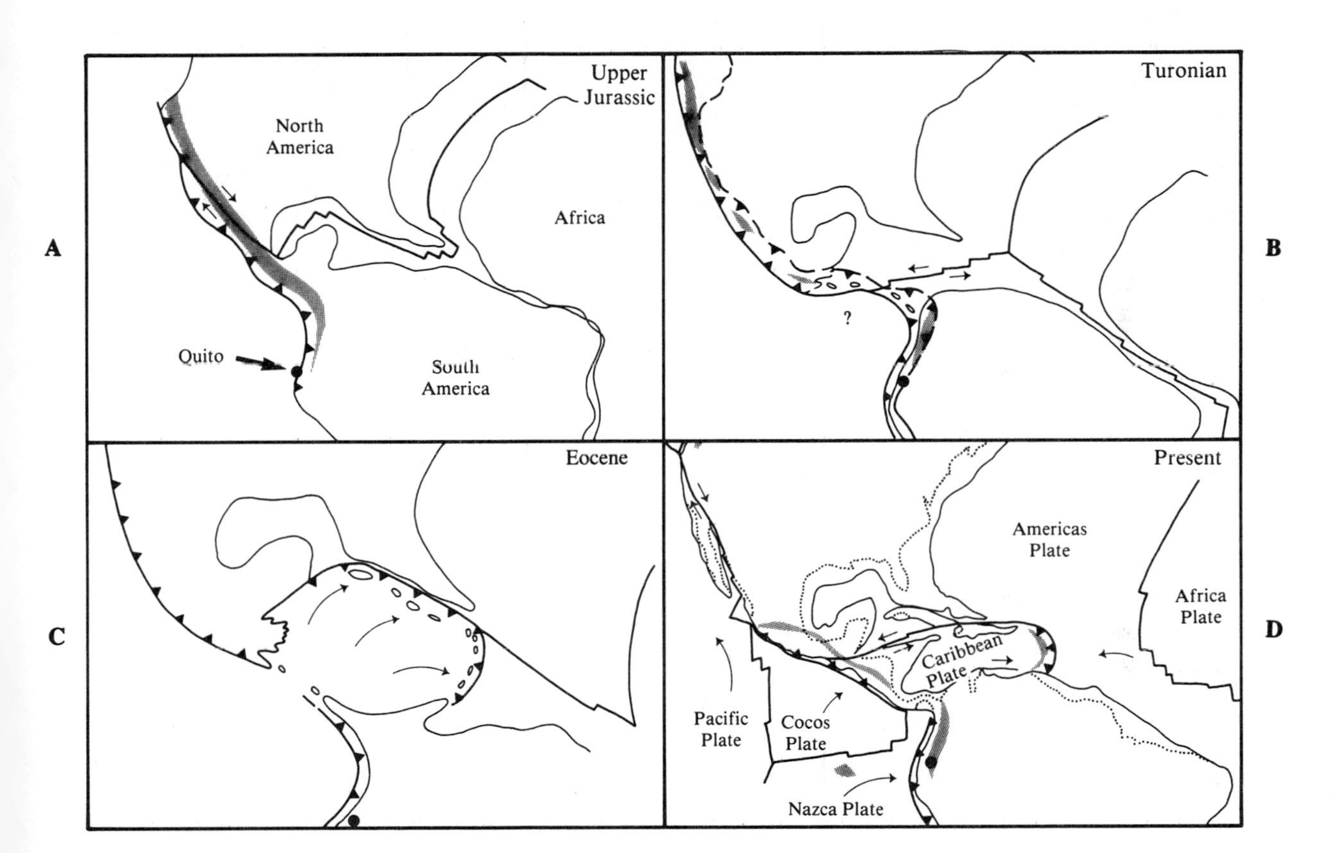

Figure 5.19
Sequential maps showing the reconstructed configuration of Central America and the Caribbean Sea relative to the nearby landmasses and oceans at various times in the past. The events described by these maps are explained in some detail in the text. (After Coney, 1982.)

backbone of Cuba, the central elements became other parts of the Greater Antilles, and the southernmost element became the Caribbean Mountains of Venezuela and the ABC islands off the coast of Venezuela (Aruba, Bonaire, and Curaçao). Some small fragments may also have been displaced to the region of the Lesser Antilles. By Eocene times (Figure 5.19, *D*), the core of the Greater Antilles had achieved their present positions. After this, new islands were created across Central America, becoming the backbone for present-day Central America (Figure 5.19, *C* and *D*). Island hopping could have occurred along this chain, but no solid land connection existed probably until the Pliocene, when all islands were fused and the isthmus of Panama and northwestern Colombia (the Bolivar Trough) became emergent. This final rise of Central America in the Neogene was produced by the converging plates of the Pacific Ocean. In the last 40 million years, the original Lesser Antilles were consumed in a trench, the trench has migrated eastward, and a new volcanic island arc has been created adjacent to the trench (within the last 3 million years).

This geologic scenario (Coney, 1982), an enlargement of numerous similar earlier ones, suggests that from the latest Jurassic to late Tertiary, organisms did not have a continuous land bridge between North and South America. However, in the Cretaceous and Miocene at least there were significant stepping-stones between them. It is considered unlikely that land ever connected North America and islands of the Greater Antilles after their initial separation. Florida and Cuba are very close (150 km), and when seas were lowered in the Pleistocene, the water gap must have been fairly short; small islands around Cuba, e.g., the Isle of Pines, would have been joined to Cuba. A submarine ridge from Yucatán to Jamaica exists, but this was probably never a land bridge. Moreover, Jamaica was totally submerged in the Miocene, meaning that its rich biota has become established by oversea dispersal and differentiation in isolation only in the last 15 million years (Robinson and Lewis, 1971).

Epeiric seas and other intracontinental phenomena

Epeiric seas, also called epicontinental seas, are formed when the ocean levels rise and spill onto continental areas. We happen to be living today in one of the driest periods in the history of the earth. However, anyone who has hunted for fossils in the interiors of continents, e.g., the United States or Eurasia, knows that vast seas covered these areas many times, leaving thick deposits of calcium carbonate from animals with exoskeletons.

Epeiric seas usually act as barriers, subdividing a landmass into smaller emergent regions. An example of this is illustrated in the Mesozoic and Cenozoic history of Australia (Figure 5.20). This has important biogeographic consequences for allopatric speciation and extinction (Chapters 6 and 16). Some of the more interesting examples since the Paleozoic are the marine transgression in North America from the Gulf of Mexico as far west as Arizona, a north-south seaway separating Europe and Asia, and a subdivision of mainland Africa (Cooke, 1972). India is one of the few large regions that has had relatively little change in its coastal outline.

On a smaller scale, we can find important changes in river systems. In South America, the Amazon drained westward in the early Cenozoic until the Pacific slope was uplifted to form the Andes. Subsequently, the drainage was reversed to the east. In Africa, where much of the continent is elevated and relatively flat, the regions of lowland tropical rain forest have been innundated by the sea, even very recently. River systems such as that of the Mississippi in the United States have changed courses many times, and in the western states, the Great Basin, which includes much of Utah, Nevada, and eastern California, now has no drainage to the sea, only numerous alkali sinks and dry lake and river beds. However, within the last million years, during wetter conditions, there were con-

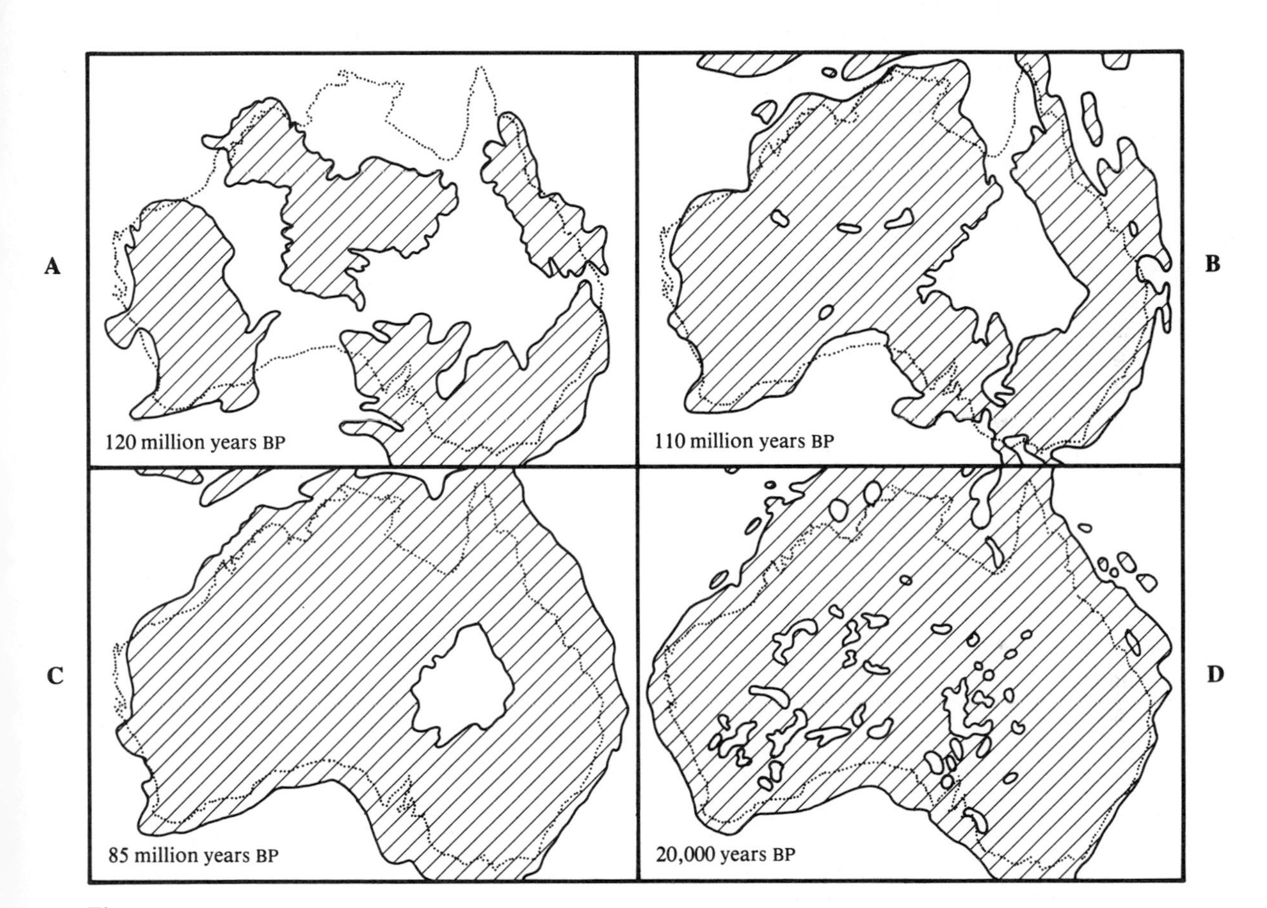

Figure 5.20
Four sequential maps showing the extent to which Australia has been covered by epeiric seas at various times in the past. Shaded areas indicate exposed portions of the landmass. Note that during much of the Cretaceous a large portion of the continent was submerged, whereas during the Pleistocene much of the continental shelf was exposed and Australia had an extensive land connection with New Guinea. (After Laseron, 1969.)

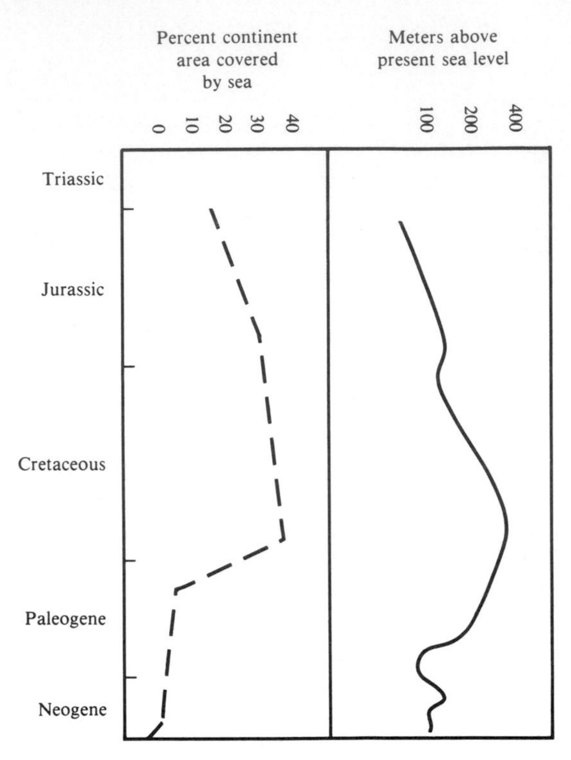

Figure 5.21
Major fluctuations in sea level during the Mesozoic and Cenozoic. The graphs show generalized worldwide eustatic changes in sea level and the approximate percentage of the present land area of North America covered by sea. (After Hallam, 1981.)

nections to the Colorado River in the south and the Columbia River (via the Snake River) in the north.

Several times in the text we have mentioned changes in sea level. Figure 5.21 shows the estimated historical changes in sea level relative to that of today. Fluctuations had greatest amplitude in the Pleistocene, when in short time intervals exposed land became submerged and emergent, as a result of the melting and formation of glaciers, respectively.

When geophysicists make reconstructions of drifting continents, they plot the positions of plates using the margins of the continental shelves, usually at the 2000 or 3000 m level. For land animals, this boundary is meaningless; instead, it is necessary to know the linear distances from shore to shore (Howden, 1974), which are not usually known. Authors try to simplify this important issue by finding a region of shallow water and adding or subtracting the presumed depth of water to determine if adjacent land or water masses were ever connected. In practice, however, such simple methods rarely represent what actually occurred.

Paleoclimates and paleocirculations

Biogeographers often attempt to reconstruct the climatic history of the earth using the biological properties of fossilized organisms. For example, the presence of coral in a fossil locality indicates that the location was shallow and tropical. On land, pollen is often used to infer the climatic regime of the organisms living at that place. Reconstructions for land and sea habitats must assume actualism (Chapter 1), that the niches (requirements and tolerances) of fossil organisms are close to those of their living relatives. For aquatic organisms, reconstruction can be made by certain physical techniques, e.g., quantifying the ratios of oxygen isotopes deposited in calcite shells (Chapter 14). Existing physical methods are less reliable for determining climatic regimes past the Quaternary, and workers are forced to rely on inferences based on fossil organisms and the principle of actualism.

As discussed in Chapter 2, global circulation patterns of wind and ocean currents are controlled not only by the equatorial to polar gradients but also by the relative proportions and distributions of land and water. Using knowledge of plate tectonics, and reconstructing the size and location of continents and oceans during each period, some earth scientists have attempted to deduce paleocirculation patterns and detailed models for climate (e.g., Webster and Streten, 1978). The potential for error in these reconstructions is very great, and most often authors have not considered all factors. Paleocirculation models are presently impossible to test on a broad scale, because different kinds of independent data are unavailable or inadequate.

Selected references

The geologic time scale

Damon et al. (1978); Stearn et al. (1979).

Emergence of ideas on continental drift

Du Toit (1937); Georgi (1962); Hallam (1973c, 1975a); Holmes (1931, 1944); Marvin (1973); Meyerhoff (1968); Schwarzbach (1980); G. Simpson (1976a); Sullivan (1974); F. Taylor (1910); van Waterschoot van der Gracht (1928); Wegener (1966).

Resurgence of interest in continental drift

Benioff (1954); Bullard (1975); Carey (1958a); Clegg et al. (1954); A. Cox (1973); Creer et al. (1954, 1957); R. Fisher and Revelle (1955); Heezen et al. (1959); Hess (1946, 1955); Irving (1956, 1959); Menard (1960); Runcorn (1962).

General acceptance of the drift model

Allard and Hurst (1969); ARCYANA (1975); Axelrod (1963); Ballard et al. (1975); Bird and Isacks (1972); Blackett et al. (1965); Craddock et al. (1965); Crawford (1974); Dietz (1961); Dietz and Holden (1970a, 1970b); J. Griffiths and Burnett (1973); Haile et al. (1977); Hallam (1972); Heirtzler and Bryan (1975); Hess (1962); Hurley and Rand (1969); Isacks et al. (1968); Keast (1972a); Klein (1972); Le Pichon (1968); McElhinny (1973a); McElhinny et al. (1974); McKenzie (1972); McKenzie and Richter (1976); Morgan (1968); Ridd (1971, 1972); J. Schopf (1970b); G. Simpson (1970b); A.G. Smith and Hallam (1970); Tarling (1972); Tarling and Tarling (1975); Vine (1966); Vine and Matthews (1963); Vine and Wilson (1965); J.T. Wilson (1963a, 1963b, 1963c, 1965); Windley (1965); Uyeda (1978).

Continents of the Paleozoic and early Mesozoic

Bird (1980); Boucot and Gray (1979); Burke et al. (1977); Dewey (1977); Dickinson (1973); Embleton and Schmidt (1977); Embleton and Valencio (1977); G. Johnson (1973); J. Schopf (1975); A.G. Smith and Briden (1977); Tarling and Runcorn (1973).

Breakup of Gondwanaland

Bergh and Norton (1976); J. Cole and Lewis (1981); Crook (1981); Crook and Belbin (1978); J.W. Curtis (1973); Dalziel et al. (1975); DeWit (1977); J. Griffiths and Austin (1977); P. Gunn (1975); Hackman (1980); Hallam (1981); Kvasov and Verbitsky (1981); McElhinny (1976); Norton and Sclater (1979); Schmincke (1976); Scrutton and Dingle (1976); A.G. Smith and Briden (1977); A.G. Smith and Hallam (1970); Zeller (1966).

Important tectonic events in Laurasia

J. Allen et al. (1977); Anderton et al. (1979); Biju-Duval and Montadert (1977); Boulin (1981); Bureau de Recherches Géologiques et Minières (1980a, 1980b); Coney (1978, 1980); Cooper et al. (1976); Dickinson (1973); Gastil et al. (1981); Girdler and Styles (1978); Hallam and Sellwood (1976); Hopkins (1967); Kristoffersen and Talwani (1977); Le Pichon et al. (1977); McElhinny (1973b); McKenna (1972a, 1972b); Mohr (1970); G. Murray (1961); J. Phillips and Forsyth (1972); Pitman and Talwani (1972); Stearn et al. (1979); J.T. Wilson (1966).

Dynamic changes in the Pacific Ocean

Atwater (1970); Burke and Wilson (1976); Clague and Jarrard (1973); Coleman (1973); Coney (1980); Coney et al. (1980); Hey (1977); Hilde et al. (1977); E. Jackson et al. (1972); McKenzie and Morgan (1969); Morgan (1971, 1972a, 1972b); Nur and Ben-Avraham (1977); Stone (1977); Stone and Packer (1977); Sutton et al. (1976); A. Watts et al. (1977); J.T. Wilson (1963a); Winterer (1976); Uyeda et al. (1979).

History of Central America and the West Indies

Coney (1982); Ladd (1976); Malfait and Dinkelman (1972); Perfit and Heezen (1978); Rosen (1975).

Epeiric seas and other intracontinental phenomena

Anderton et al. (1979); H. Cooke (1972); Gray and Boucot (1979); Hallam (1973b, 1981); Howden (1974, 1981); Hughes (1973); Laseron (1969); Middlemiss, et al. (1971); Por (1977); G. Smith (1981); Stearn et al. (1979); J. Valentine (1973).

Paleoclimates and paleocirculations

Berggren and Hollister (1977); Butzer (1971, 1977); Fell (1967); Gray et al. (1981); Hambrey and Harland (1981); Hopkins (1959, 1967); P.S. Martin (1964); Street (1981); P. Webster and Streten (1978).

Chapter 6 Speciation and Extinction

Species have histories that are somewhat analogous to the lives of individuals. New species originate by the multiplication of old ones (speciation); they survive for varying periods of time, during which they may or may not leave descendents; and they die (extinction). There are, however, important differences between the histories of individuals and of species. Whereas the birth and death of most individuals are discrete, easily recognizable events, it is often difficult to determine when a species is fully formed. Rather than becoming extinct abruptly when its last individual dies, some of a species population may disappear in time by evolving into a new kind of organism that will soon be classifiable as a different species.

Related species possess a particular combination of traits, at both the structural and the molecular levels, that are inherited from a common ancestor. If the group has been diverse and successful, one can usually infer that the traits that made the group distinct were adaptive—that is, they represented an innovative solution to a problem that limited survival and reproduction in other populations. It is not necessarily true, however, that all new traits of an evolutionary line are adaptive and have arisen as a direct result of natural selection. Although a particular lineage may have flourished for a period of time and radiated to produce a diverse group of species, this success is usually ephemeral on a geologic time scale. Eventually the rate of extinction exceeds the rate of speciation, and the group diminishes and finally becomes extinct. These repeated episodes of speciation, radiation, and extinction have occurred many times during the history of the earth to produce the variety of living and extinct organisms.

The importance of speciation to biogeography is that it is an evolutionary branching process. Through speciation, evolutionary lineages split and are freed to adapt to different environments and to colonize new regions. Where and when these evolutionary events took place naturally did much to determine present and past geographic distributions.

Speciation

What is a species? Not all taxonomists and evolutionary biologists agree on how the term *species* should be defined, although the specialists on most taxonomic groups share an idea of how different populations must be in order to be classified as separate species. The classical and conventional definition of a species is the morphological species concept. This concept recognizes that each species usually is morphologically distinguishable from its closest relatives. However, the criteria for determining how much a population must differ and in which particular traits vary from group to group.

In sexually reproducing animals and plants, the biological species concept frequently is employed to define a species as a group of organisms that is reproductively isolated from any other group. When a population is isolated reproductively, it constitutes a separate evolutionary lineage that is prevented by geographic or biological barriers from interbreeding with other populations. In the absence of gene flow from related populations in other habitats, the new population is free to follow its own course in response to those processes determining evolutionary change.

As straightforward as it may sound, the criterion of being reproductively isolated is a difficult one to employ. This criterion cannot be applied at all to fossil organisms; the best one can do is to determine whether the morphological gaps between specimens are as large or larger than for living species that are reproductively isolated.

The biological species concept is unwieldy for extant groups as well. Many species have never been mated with their closest relatives to determine the status of reproductive isolation. Such mating tests may not be definitive. If mating tests are conducted and no hybrids can be produced between the two taxa, the conclusion is obvious, but if the crosses yield some fertile progeny, for argument say 10% or 20% of the time, are they still separate species? What happens in the case of plants in which two morphologically differentiated taxa live in completely different habitats, even flower at different times, but crosses between them are totally fertile? They are certainly separate evolutionary lineages and should be recognized as such. The absence of postfertilization reproductive isolation between close relatives of plants signifies two things: (1) that plants are simply constructed, with open growth, and can tolerate more modification in early structure than can animals, which exhibit very precise early developmental sequences with more rigorous genetic control, and (2) that there has been no selective pressure to develop postfertilization barriers to reproduction, perhaps because prefertilization isolating mechanisms, such as differences in flowering time, are effective in preventing interbreeding. Imagine closest relatives living on different continents or on an island and a mainland. If gene flow between the two is zero, the course of evolution of each can proceed unimpeded without establishing biological barriers to gene exchange.

Of course, it is impossible to apply the criterion of the biological species to asexual organisms. In a completely asexual taxon, each individual would represent a genetically independent evolutionary line and hence a new species. Consequently, such organisms are classified as distinct species if they differ from others in the traditional morphological measures, arbitrary as they may be.

It is also difficult to apply the biological species concept to extinct organisms. Paleontologists often try to classify fossils so they will be comparable to extant groups, but they must make inferences from the limited remains that are preserved. In many contemporary groups, such as the rodents and certain taxa of flowering plants, many species are so similar in basic structure that they cannot be distinguished on the basis of skeletal characteristics, leaf morphology, or other traits that are likely to be preserved as fossils. Paleontologists also have the problem of dividing continuous evolutionary lineages into discrete taxonomic categories. Fossil organisms have limited distributions not only in space but also in time. Approaches to resolving these problems are discussed in several recent treatments of evolution and the fossil record (e.g., Stanley, 1979; Eldredge and Cracraft, 1980). In general, new taxa, including species, are recognized when organisms possessing their distinctive combination of traits first appear in the fossil record. Naturally, much emphasis is placed on the first appearance of derived character states in which the new taxon differs from its immediate ancestors. In practice, differences in evolutionary rates and gaps in the fossil record are frequently capitalized on to divide evolutionary lineages into species and higher taxonomic categories.

Mechanisms of genetic differentiation. Speciation is a branching process in which different kinds of organisms originate from a single common ancestor. The magnitude of this process is staggering when we consider that all species of green land plants ultimately share a common ancestor in a simple green alga living

500 million years ago, all vertebrates are traceable to some ancient chordate, and all the hundreds of thousands of species of insects have evolved in the 400 million years since their common ancestor first invaded land. How can such differentiation occur?

Populations can diverge in their genetic characteristics primarily by two evolutionary processes: genetic drift and natural selection. Genetic drift is by far the weaker of these, because it involves changes in the genetic constitution of a population caused solely by chance events. Given sufficient periods of time, which usually means many generations, the frequencies of alleles (alternative genes at the same locus) in a population will tend to change as different individuals survive, mate, and produce offspring at random. Genetic drift, the genetic consequence of these chance events, has relatively little effect in large populations, but the random sampling of genes from each generation can have important influences on the evolution of small populations. How small is a matter of considerable discussion in evolutionary biology, but if new species start as small populations, genetic drift could be very important in the initial shifting of characteristics from parental populations.

Natural selection, on the other hand, can be a potent force for evolutionary change in both large and small populations. Natural selection is the inherited change in a population that occurs because certain individuals leave more offspring than others. Over many generations, genes for traits that cause their bearers to survive and leave more offspring tend to increase in frequency at the expense of genes that confer less fitness. Because survival and reproduction depend largely on interactions between organisms and their environment, a population can diverge in response to natural selection if there is sufficient variation in the environment for different genetic lines to adapt to different parts of the environment. This fine-tuning of phenotypes to environmental heterogeneity is an extremely common observation among all types of organisms, even at the subpopulation level.

There is often a geographic component to speciation, because both genetic drift and natural selection are facilitated by geographic isolation. Genetic drift can be an important force in small, isolated populations, such as those that inhabit small outlying patches at the periphery of the species range or that have recently been founded by long-distance colonization. Genetic drift that occurs in a population started by only a few organisms has been termed the founder principle. Mayr (1942) suggested that this process accounts for apparently random differences among the insular bird populations of islands (Figure 6.1), which were probably derived from a few successful colonists. Recent experimental evidence suggests that the founder principle can play a major role in speciation, because the initially random genetic sampling also has effects on the subsequent genetic differentiation of small colonizing populations (Templeton, 1980a).

Geographic separation of populations also facilitates genetic differentiation by natural selection. Different environmental regimes tend to select for different traits, and spatially isolated populations are likely to occur in different environments. The effects of natural selection caused by spatial environmental heterogeneity are suggested especially by data on small mammals. In the deer mouse *(Peromyscus maniculatus)*, coat color varies greatly over the range of the species (Figure 6.2). Each color of the fur closely matches the color of the soil or other substrate on which the animal is active. Apparently this color matching is the result of selection by predators. Dice (1947) performed a classic experiment showing that owls selectively killed mice that contrasted with their background.

Patterns of geographic variation can take many forms. In the *Peromyscus* example, coat color may appear to be random on a large map, but the background matching pattern is apparent when it is analyzed in relation to the patchy distribution of soil color. The term *cline* is used

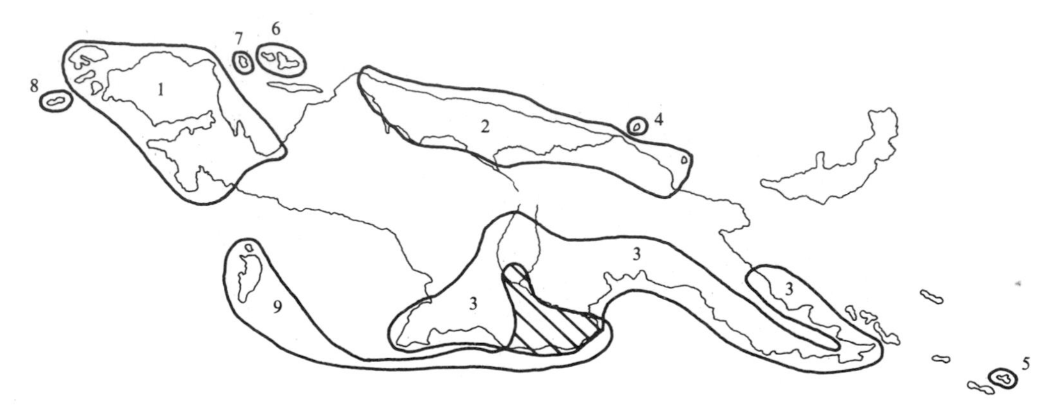

Figure 6.1
Geographic variation in the kingfisher, *Tanysiptera galeta,* on New Guinea and adjacent islands. Numbers indicate the ranges of geographically isolated populations recognized as distinct subspecies. Variation among these forms, many of which inhabit tiny islands, appears essentially random, leading E. Mayr to suggest that it arose largely as a result of genetic drift, or what he called the founder principle. Subspecies 3 and 9 apparently were formerly isolated but have come into contact recently, perhaps as a result of changes in sea level. (After Mayr, 1942.)

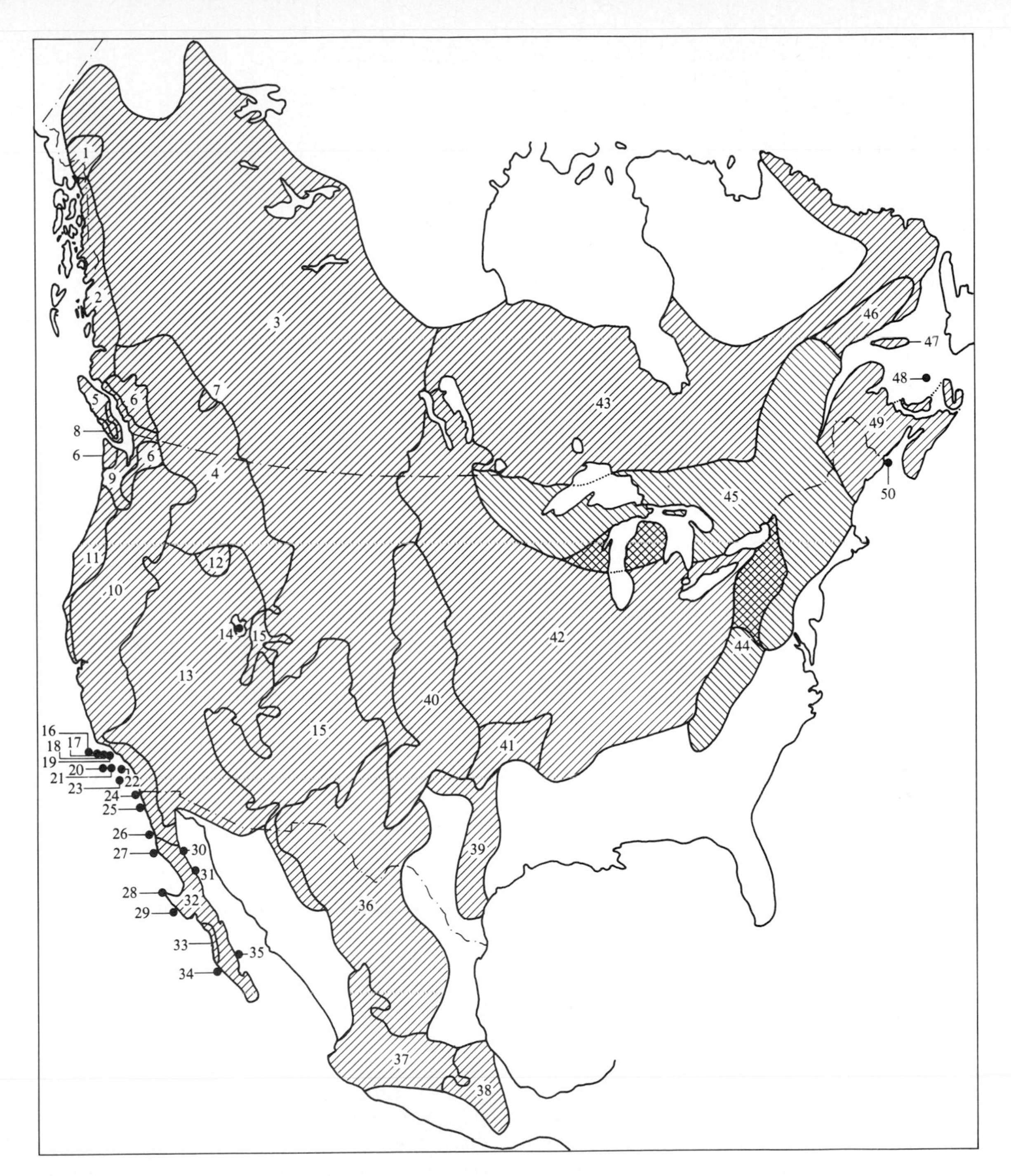

Figure 6.2
Geographic variation in the deer mouse, *Peromyscus maniculatus,* indicated by 50 formally recognized subspecies. This rodent, the most widely distributed small mammal in North America, varies greatly over its range, especially in coat color, which is similar to background color and provides camouflage, and in tail and foot length, which are related to habitat structure and climbing ability. (After Hall, 1981.)

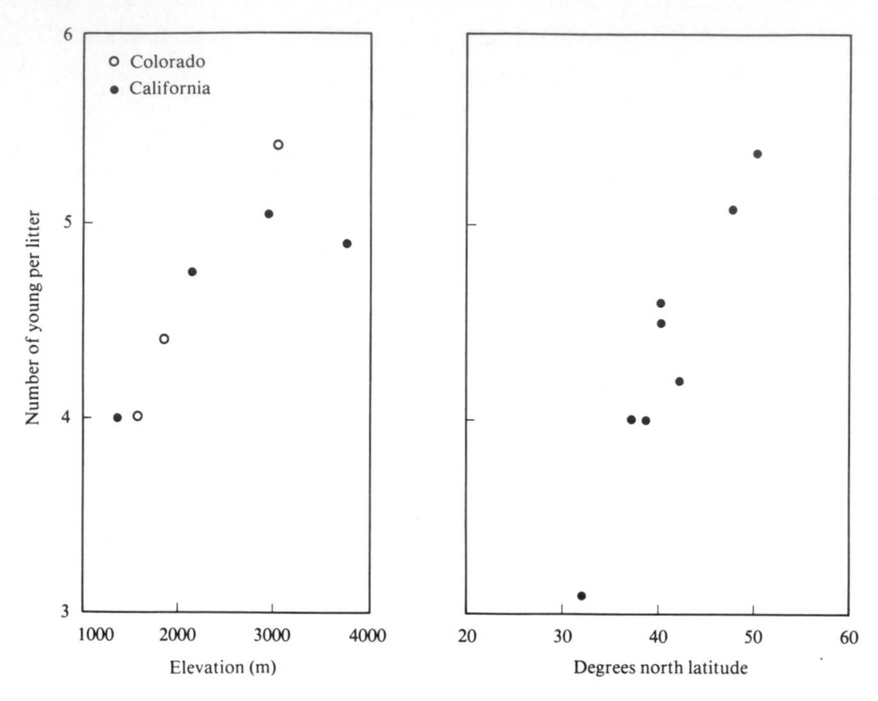

Figure 6.3
Latitudinal and elevational variation in litter size in the deer mouse, *Peromyscus maniculatus,* in North America. Note that litter size increases consistently with both increasing latitude and increasing elevation at the same latitude. Apparently, large litters are adaptive in cold environments with short growing seasons, perhaps because in warmer climates the animals can have more litters per year with fewer young per litter. The lowest latitude point is for the sibling species *Peromyscus polionotus* from Georgia.

to describe the gradual change in one or more features along a single environmental gradient. Many birds and mammals exhibit clinal variation in clutch or litter sizes, respectively, with latitude and elevation (e.g., for *Peromyscus* see Dunmire, 1960; Lord, 1960; Smith and McGinnis, 1968; Spencer and Steinhoff, 1968) (Figure 6.3). Such variation presumably reflects adaptations in life history traits to environments that differ in temperature, seasonality, productivity, and other factors. Similarly, clines in physiological characteristics of plants and insects show that populations at progressively higher latitudes survive and reproduce best under cooler temperature regimes.

Allopatric speciation. The simplest way to model speciation is to assume that genetic differentiation cannot lead to new species unless the populations are geographically isolated, so that gene flow between them is virtually cut off. This is termed allopatric speciation, meaning the divergence occurs in different places. The classic model of allopatric speciation by geographic subdivision (geographic speciation) (Figure 6.4) was championed by Mayr (1942, 1963). If the environment is heterogeneous, a geographically widespread ancestral population will develop regional genetic differences in response to either natural selection or genetic drift. Because of barriers that limit dispersal,

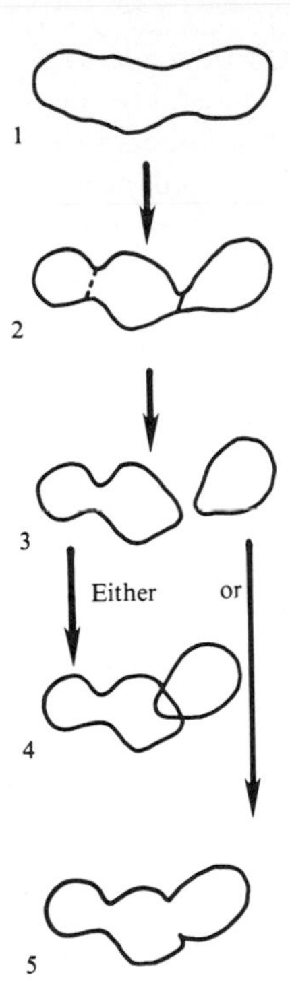

Figure 6.4
Speciation by geographic isolation. A once widespread species *(stage 1)* becomes subdivided into two isolated populations by the formation of a geographic barrier *(stages 2* and *3)*. After a period of isolation the barrier disappears and one of two outcomes are possible: if sufficient differentiation has occurred in allopatry, the two populations remain distinct and speciation has occurred *(stage 4)*; alternatively, if the populations have not differentiated sufficiently in isolation, they hybridize and eventually merge to form a single widespread species again *(stage 5)*. (After Mayr, 1942.)

free gene flow from one end of the range to the other rarely occurs. Thus populations of particular regions tend to become somewhat differentiated from each other, but some gene flow maintains the genetic integrity of the species. At some point one or more of these populations may become completely geographically isolated, e.g., by the intrusion of a new barrier, such as the advance of a continental glacier or the rising of an epeiric (epicontinental) sea. Without any dispersal and gene flow between them, the isolated populations continue to differentiate. Of course, the divergence will proceed more rapidly when substantially different environments subject the isolates to different selective regimes.

There are several possible outcomes after the isolates have become differentiated. The barrier may be abolished so that the populations come back together, i.e., they become sympatric. There are then three possible outcomes: (1) the sympatric populations can no longer produce fertile offspring and therefore reproductive isolation is complete and speciation has occurred; (2) the two populations can produce many fertile, fit hybrids, so that as a result of free interbreeding the populations merge, meaning that speciation has not occurred; or (3) the two populations can successfully hybridize. If, however, the genetic differences are so great that the resulting interpopulation hybrids are less fit than offspring produced from matings within the populations, then natural selection will favor those individuals that choose mates from within their own population, which will tend to lead to reproductive isolation and the completion of the speciation process. A final scenario is one in which the isolated populations never do come in contact. In this case reproductive isolation may or may not develop. Purists argue that speciation is only demonstrated if the two formerly isolated populations come into contact in nature and are shown to be reproductively isolated.

Another type of allopatric speciation is that described earlier as the founder principle, in which one or very few individuals initiate a geographically isolated population. Because this usually occurs by colonization of an insular habitat and the founder population size is very small, the genetic changes producing new species can be very different from those in geographic speciation (Carson, 1971; Bush, 1975;

Templeton, 1980a, 1981). Allopatric speciation by disjunct founders (the founders principle) may occur following colonization of isolated habitats, such as oceanic islands. A mainland version of allopatric speciation by the founder principle may be catastrophic speciation (Lewis, 1962), in which a catastrophe eliminates all but a few individuals that subsequently increase to establish a population with a different set of characteristics.

Today the term *vicariance* is widely used to describe the biogeographic consequences of allopatric speciation (Chapter 7). The term generally refers to disjunct populations that have never come back into contact after geographic isolation. Normally the term *vicariance* is reserved for describing those cases in which a once continuous distribution becomes fragmented into two or more disjunct populations. It is not used for cases in which new populations are founded by long-distance dispersal. A vicariant pattern is simply the biogeographic consequence of geographic isolation owing to the intrusion of barriers. It does not depend on the genetic details of the process of geographic speciation; in fact, one of the implied assumptions of many vicariance models is that differentiation of the two entities occurs only after the barrier is formed.

There is no question that speciation can and often does occur through geographic isolation. This must be the common mode of speciation in many groups. The tortoises and finches that Darwin observed in the Galápagos Islands provide examples of populations in different stages of the process. There are distinctively different populations of tortoises on each of the large islands. These have diverged from a single ancestor that long ago colonized the archipelago and subsequently dispersed to found populations on the different islands. However, there are no cases in which two or more species occur in sympatry on the same island. Even in the unlikely event that individuals colonized from another island and were sufficiently differentiated so that widespread hybridization did not occur, the largest islands are probably too small and unproductive to support two sympatric species of giant tortoises. On the other hand, in Darwin's finches the final stages of speciation are represented. Here again, all species are believed to be derived from a single common ancestor, whose descendents colonized the archipelago. In the case of the finches, however, after diverging in isolation on different islands or in different habitats, some populations have successfully reinvaded already inhabited areas so that several species (as many as 10 on certain islands) now coexist in sympatry. Not only have these species evolved specific mating behaviors that prevent interspecific hybridization but also they have diverged to occupy different ecological niches (Figure 6.5).

Sympatric speciation. Although many earlier workers maintained that speciation in most organisms occurs primarily or solely as a result of geographic isolation, numerous investigators have argued recently that much speciation occurs sympatrically, i.e., within spatially contiguous populations. Two classes of mechanisms have been proposed. If strong selection causes the population to adapt to two or more envi-

Figure 6.5
Speciation and adaptive radiation in Darwin's finches (Fringillidae) on the Galápagos islands. Figure shows the males of the 13 species. Ground finches: *1, Geospiza conirostris; 2, G. magnirostris; 3, G. fortis; 4, G. scandens; 5, G. difficilis;* and *6, G. fuliginosa.* Tree finches: *7, Camarhynchus pallidus* (woodpecker finch); *8, C. crassirostris; 9, C. pauper; 10, C. psittacula; 11, C. parvulus;* and *12, C. heliobates.* Warbler finch, *Certhidia olivacea.* (The Cocos Island finch, *Pinaroloxias inornata,* is not shown.) Although some of the smallest islands have only one or a few species, many of these species occur together on the largest islands. (From Biological science: molecules to man, Biological Sciences Curriculum Study, Boston, 1963, Houghton Mifflin Co.)

Figure 6.5
For legend see opposite page.

ronmental regimes, it can progressively pull an ancestral population apart and, through the disruption of gene flow by the evolution of isolating mechanisms, eventually result in speciation. Endler (1977; see also Slatkin, 1973; Rosenzweig, 1978) suggested that such disruptive selection, acting along an environmental gradient, could gradually sharpen clinal variation until a single ancestral population fragmented into two or more species. Such speciation is often termed parapatric or stasipatric because the differentiating populations are geographically distinct but remain in contact with each other.

Bush (1975), Price (1980), and others have argued that sympatric speciation by disruptive selection is common in certain groups of phytophagous (herbivorous) insects and animal parasites that are highly specialized for specific host species. In these organisms successful colonization of a new kind of host must be a rare event, but when it occurs, the colonists are immediately subjected to selection for the ability to survive and reproduce in a drastically different environment. Usually this selection is intensified by counter evolution of the host to escape from or to reject the parasite. Selection to meet the challenge of a new host can potentially lead to rapid differentiation and speciation of totally sympatric populations, in which the organisms are spatially close enough to mate but they do not, often because mating occurs on the host.

Sympatric speciation can also occur through chromosomal changes. Chance rearrangements of the genetic material sometimes result in changes in the number of chromosomes or in the sequence of genes on the chromosomes (Figure 6.6). Chromosomal number can change either by the addition or subtraction of one or more chromosomes (aneuploidy) or by the acquisition of additional sets of chromosomes, e.g., doubling or tripling the number, called polyploidy. In other cases, chromosomal number can remain the same, but some of the genetic material can be either rearranged within a chromosome (inversion) or transferred to another chromosome (translocation). In diploid organisms precise pairing during meiosis of the genes and chromosomes inherited from each parent usually is necessary to ensure transmission of a complete set of genes to each gamete and hence the production of viable offspring. Consequently, mutant individuals with brand new arrangements often have impaired fertility when they mate with an individual with the original chromosomal arrangement but unimpaired fertility when they mate with another individual with the new arrangement. If fertility is reduced in this manner, it is difficult for a population with a new arrangement to become established. However, if one does become established, especially in a small, isolated population, it is genetically isolated from its parental population and can diverge rapidly as a new species.

Sympatric speciation by way of polyploidy appears to have occurred frequently in some groups of organisms, especially in plants (e.g., Stebbins, 1971b; de Wet, 1979; Lewis, 1979). There are many documented ways of achieving polyploidy in plants, but only one process is considered very common (de Wet, 1979). Diploid (2N) organisms have two sets of chromosomes and produce haploid (N) gametes by meiosis. Occasionally, a female gamete may be formed without undergoing meiosis and remains diploid (2N). This unreduced gamete can then fuse with a haploid pollen grain (N) to produce a triploid (3N) plant that will produce 3N gametes because of complications during meiosis. In the next generation, a triploid female gamete can then fuse with a haploid pollen grain to yield a tetraploid (4N) zygote, which can survive and produce fertile offspring from diploid (2N) gametes, either by self-fertilization or by crossing with other rare tetraploids in the population. The resulting polyploid population is immediately genetically isolated from the diploid population.

This mechanism can result in the formation of polyploids by the accumulation of additional sets of chromosomes, either from within the

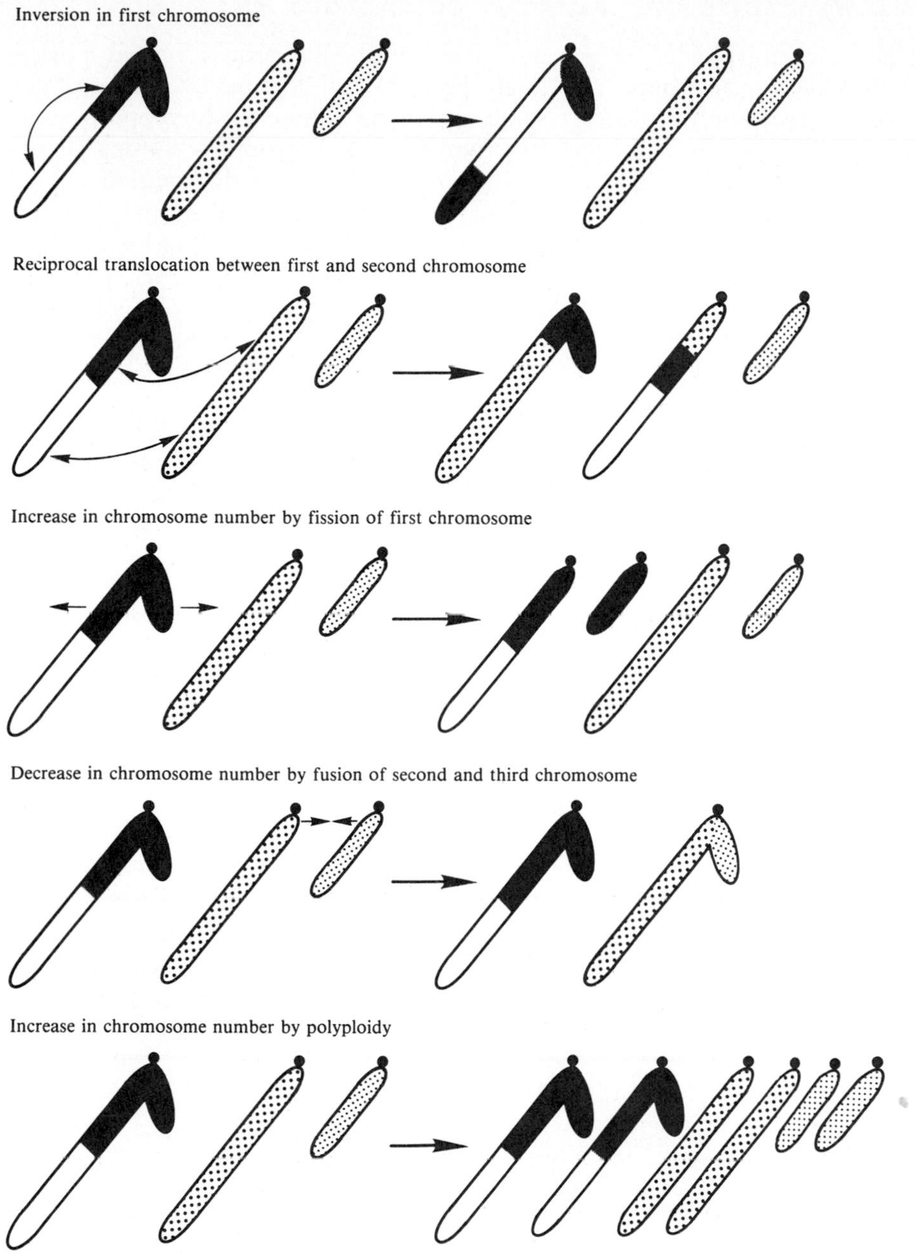

Inversion in first chromosome

Reciprocal translocation between first and second chromosome

Increase in chromosome number by fission of first chromosome

Decrease in chromosome number by fusion of second and third chromosome

Increase in chromosome number by polyploidy

Figure 6.6
Kinds of chromosomal rearrangements and changes in chromosome number that may, in some circumstances, substantially reduce fertility of hybrids between parents and offspring, resulting in rapid sympatric speciation. In this simplified diagram only one member of each diploid chromosome pair is shown, and the different shading patterns are used only to identify particular chromosome segments. The last diagram illustrates autopolyploidy, in which additional sets of chromosomes are derived from the same species. In allopolyploidy the additional sets of chromosomes are derived from another species (through interspecific hybridization) so they would not be perfectly homologous.

same population, called autopolyploidy, or from different but usually closely related populations or species, called allopolyploidy. Allopolyploidy is thought by many to be more common. Because the chromosomes from different species may not pair and segregate properly, interspecific hybridization often results in abnormalities in the meiotic process that can facilitate the process described above. Allopolyploids may not only arise more frequently than autopolyploids but also may be more likely to become established. In addition to possessing a larger genome than the parental species, allopolyploids tend to be intermediate in their characteristics, enabling them to be superior competitors in certain habitats.

Historical patterns. Technically speaking, speciation has not occurred if one species is gradually transformed completely into another one over time. This process, called phyletic gradualism, is often depicted by a series of very similar creatures whose body proportions or structural features are gradually shifting (Figure 6.7). Each major stage in this continuous evolutionary sequence is called a chronospecies and is given a Latin name for descriptive purposes, but many authors do not call this speciation, because there is no branching of evolutionary lines—only one species is present at any one time. Many examples have been cited in the literature, but relatively few have been studied carefully enough to demonstrate that the chronospecies are an unbroken series and not in fact products of successive speciation events that were followed by rapid extinction of all but one lineage.

Recent investigations of the process of speciation have given rise to new, often conflicting

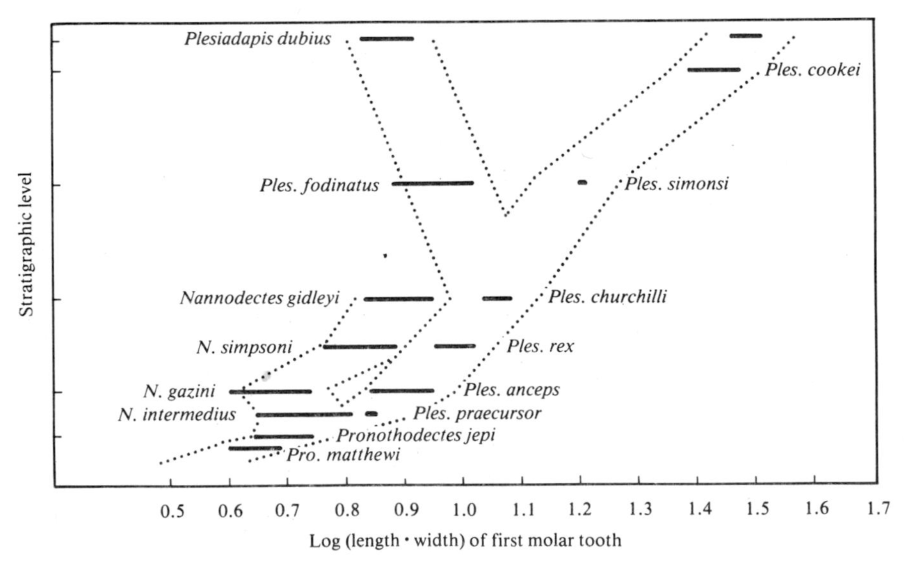

Figure 6.7
Apparently gradual evolution in a group of fossil North American mammals, the primitive primate family Plesiadapidae, showing how one species over time changes form until it is classified as a different species (chronospecies) and how speciation results in the branching of phyletic lineages. This reconstruction is based on changes in tooth size among fossils deposited in a stratigraphic sequence representing about 10 million years in the Paleocene in Wyoming. (After Gingerich, 1976a.)

views of the temporal pattern of the evolution of phyletic lines. Traditional evolutionists have assumed that evolution usually proceeds gradually by the successive incorporation of relatively small phenotypic changes as populations respond to environmental change and diverge in allopatric speciation. However, many proponents of phyletic gradualism (e.g., Simpson, 1953; Mayr, 1963) have emphasized that evolutionary rates can vary widely, even within the same evolutionary line. Mayr and others have also suggested that major, rapid changes can occur during speciation as a consequence of both the founder principle and the selection for traits that promote reproductive isolation and ecological divergence in newly sympatric species. The unique importance of speciation has been emphasized even more by those who advocate a punctuated equilibrium model of evolution (Eldredge and Gould, 1972; Stanley, 1979). They suggest that the evolution of a phyletic lineage usually consists of periods of rapid evolution accompanying speciation, followed by relatively little change until the next speciation event. The punctuated equilibrium model has been developed and supported primarily by paleontologists, especially those working on aquatic invertebrates, who note that new species often appear abruptly in the fossil record and then persist for millions of years with virtually no morphological change. Perhaps the best evidence of rapid change associated with speciation events comes from P.G. Williamson's (1981) recent work on the molluscan fauna of the Lake Turkana Basin in eastern Africa. In a stratigraphic sequence representing several million years of fossil deposition, Williamson found that all kinds of molluscs showed a history of long periods of virtually no change in shell morphology punctuated by rapid shifts (Figure 6.8). These changes took place in a time interval of less than 50,000 years, so rapidly with respect to the fossil record that often no shells with intermediate characteristics were found. These shifts frequently were accompanied by the splitting of phyletic lineages and they often occurred in conjunction with rapid changes in lake level.

The generality of the punctuated equilibrium mode of evolution has not been accepted by many investigators, especially those who study fossil and recent vertebrates. Gingerich (1976a, 1979) has studied changes in early Cenozoic North American mammals in stratigraphic sequences in Wyoming. His reconstructions provide little evidence for either long periods of stasis or short periods of rapid evolution (Figure 6.7). Instead, lineages appear to change relatively gradually and continually, even during periods in which speciation occurs. Vertebrate evolutionists working with extant forms are impressed with the extent of geographic variation within species, which is often relatively gradual and continuous (Figures 6.2 and 6.3), suggesting that species often adapt to small-scale spatial and temporal variation in their environment.

As is often true with conflicting ideas, both may contain an element of truth. Either the gradualist or the punctuated equilibrium model may apply to certain cases, and the majority of lineages may evolve in some intermediate fashion. Gradual evolution may be particularly important in groups such as terrestrial vertebrates, in which limited dispersal and geographic barriers disrupt free gene flow and promote continual adaptation of relatively small populations to local conditions. In contrast, in marine forms, especially those with widely dispersed planktonic larvae, virtually complete gene flow throughout the range of a species may prevent local differentiation (e.g., see Scheltema, 1971, 1977), and major changes may occur only when small founder populations give rise to new species. It will be interesting to see how these differences are resolved in the coming years. Whatever the answers, they will have important biogeographic consequences. Rates of evolution during and between episodes of speciation must have important influences on how

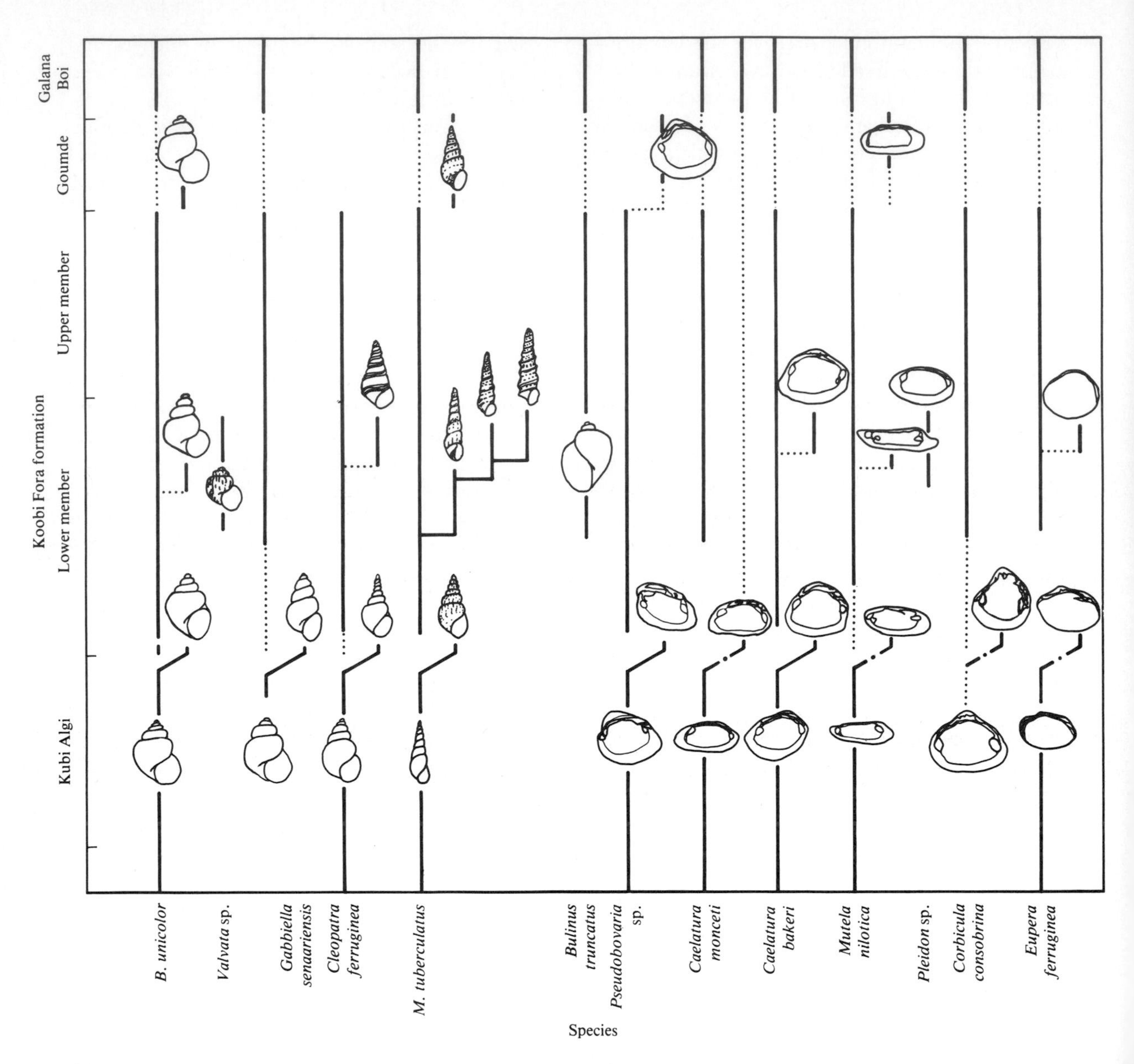

Figure 6.8
Apparently punctuated evolution in the fossil molluscs of Lake Turkana Basin in eastern Africa, showing long periods of stasis occasionally interrupted by rapid evolution that is often associated with a speciation event. This reconstruction is based on changes in shell morphology in several different lineages of fossils recovered from a stratigraphic sequence representing several million years in the late Cenozoic. Note that the rapid changes, which could have occurred over as much as 50,000 years, often occurred simultaneously in several lineages. Those periods of speciation often correspond to major changes in lake level. (After P.G. Williamson, 1981. Reprinted by permission from *Nature* **293**[5832]:437-443. Copyright © 1981, Macmillan Journals, Ltd.)

phyletic lines adapt to new environments and shift their geographic ranges over evolutionary time.

Adaptive radiation

Ecological differentiation. Once new species are formed, what happens to them? After a speciation event, the resulting species are usually quite similar to each other. They can increase in abundance and expand their geographic ranges only by evolving different ways of dealing with the physical and biotic factors that limited their ancestral populations. As pointed out earlier (Chapter 3), this may be facilitated by the effect of speciation in disrupting gene flow from other populations adapted to different environments. Nevertheless, adaptive innovations are required for a newly evolved species to increase its range and expand into that of a vigorous, related species. Gause (1934) showed in laboratory experiments with protozoans, such as closely related species of *Paramecium,* that two species with identical requirements for resources cannot persist in the same environment: one species eventually outcompetes and causes the extinction of the other. Ecologists have termed this the principle of competitive exclusion (see Hardin, 1960; Miller, 1967; Hutchinson, 1978).

A biogeographic corollary of the principle of competitive exclusion is that species that are extremely similar in their niches will tend to have nonoverlapping geographic distributions, whereas species that coexist in the same area and habitat will exhibit substantial differences in resource use. Perhaps the most striking example of this is provided by species that have recently been produced by chromosomal changes, such as in aneuploidy. These forms are genetically isolated from each other but are nevertheless virtually identical ecologically. As demonstrated by numerous examples in both plants and animals (Figure 6.9; see also Figure 3.22), sibling species usually exhibit almost perfectly nonoverlapping geographic ranges, although their distributions may be extensively contiguous. Even species that have apparently originated by allopatric speciation and are no longer extremely similar in all attributes may competitively exclude each other from local habitats or extensive geographic areas. There are about 20 species of chipmunks *(Eutamias)* in western North America, but these species exhibit almost perfectly nonoverlapping spatial distributions. Although there is some overlap in ranges on a geographic scale, there is usually only one chipmunk species per habitat. There is abundant evidence that this pattern is maintained by competitive exclusion (e.g., Brown, 1971a; Chappell, 1978).

On the other hand, numerous field studies show that when closely related species coexist in nature, they differ substantially in their use of limiting resources (e.g., MacArthur, 1958, 1972; Cody and Diamond, 1975). Often these niche differences are reflected in pronounced morphological, physiological, or behavioral differences among closely related sympatric species. One of the best examples is provided by the Galápagos Island finches. As the finches have reinvaded inhabited islands after speciation, they have diverged morphologically, behaviorally, and ecologically from coexisting species. This process, called character displacement (Brown and Wilson, 1956), has resulted in species being more different where they coexist than where they live in allopatry (Figure 6.10). In the finches, character displacement is most apparent in the size of the bill, which enables coexisting species not only to specialize on different kinds of foods (Lack, 1947; Abbott et al., 1977) but also to select mates from their own populations and thus prevent interspecific hybridization (Downhower, 1976).

Adaptive radiation is the diversification of species originating from a common ancestor to fill a wide variety of ecological niches. It occurs when a single species gives rise, through repeated episodes of speciation, to numerous kinds of descendents that remain sympatric within a small geographic area. The coexisting

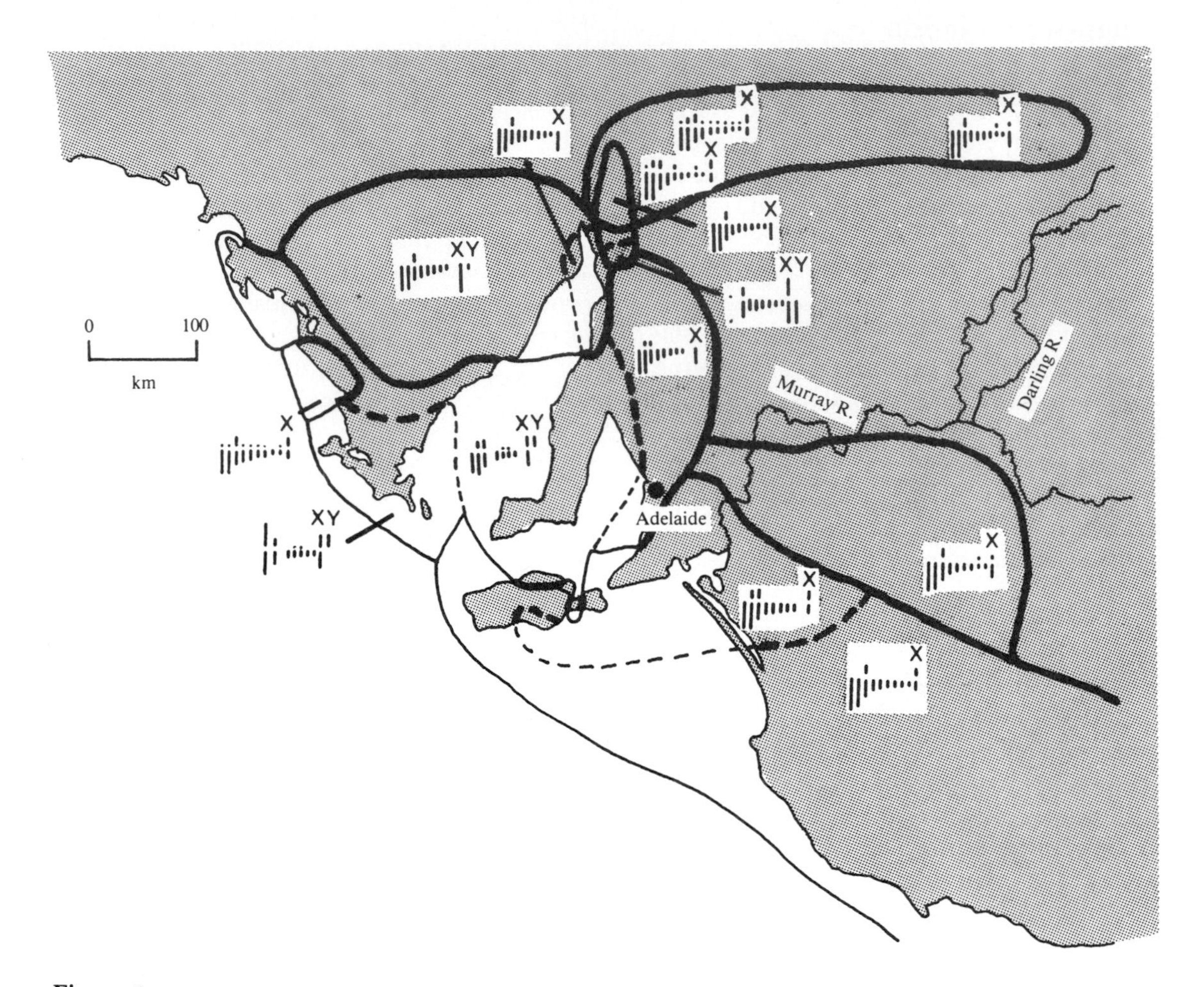

Figure 6.9
Contiguously allopatric ranges of different chromosomal forms of the grasshopper, *Vandiemenella*, in southern Australia. These populations are genetically isolated because the different chromosomal arrangements (illustrated) effectively prevent hybridization, but they have virtually identical niches and apparently cannot coexist. (From Modes of speciation by Michael J.D. White. W.H. Freeman & Co. Copyright © 1978.)

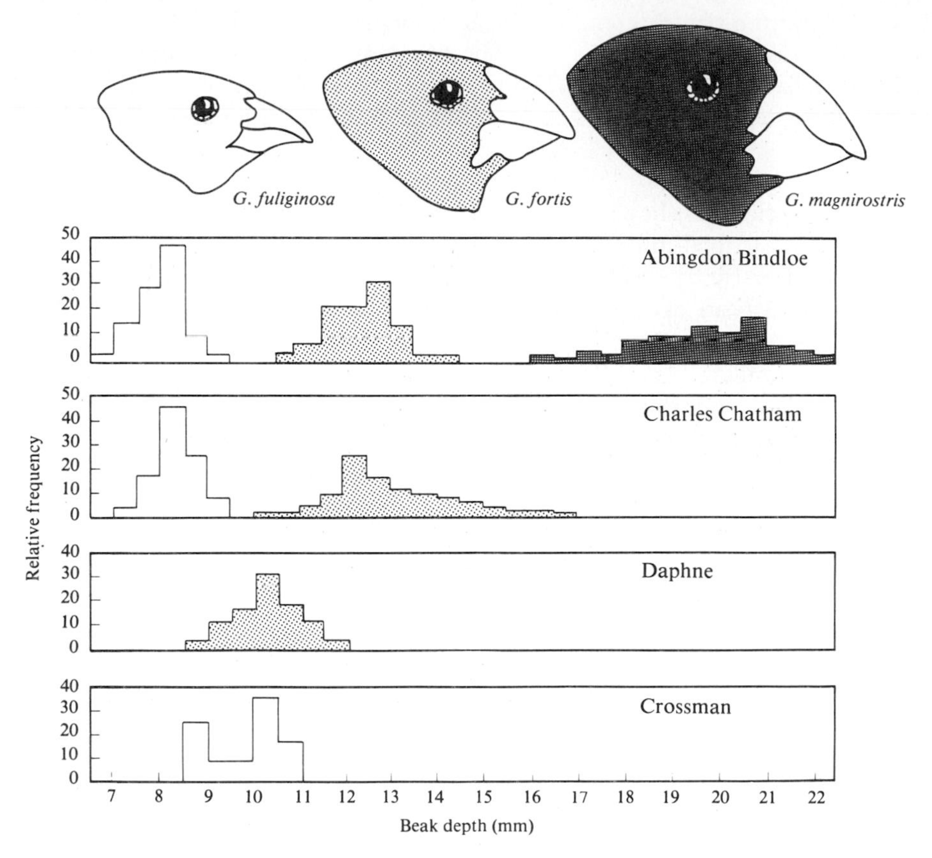

Figure 6.10
Character displacement in ground finches *(Geospiza)* on the Galápagos Islands. Note that when *G. fuliginosa* and *G. fortis* occur alone on the islands of Crossman and Daphne, respectively, they have similar-sized beaks, intermediate between the sizes they exhibit when they occur sympatrically on other islands. (After Lack, 1947.)

species tend to diverge in their use of ecological resources in order to reduce interspecific competition. Such character displacement in response to competition will not occur if the ancestral species merely becomes subdivided by physical barriers, i.e., vicariance, but differentiation will still tend to occur as the allopatric species adapt to different environments.

Examples. Today, looking at the variety of living things, we can find numerous examples of successful phyletic lines that have radiated to produce diversity at many levels. We can, for example, consider the adaptive radiation of Hawaiian honeycreepers (Drepaniidae), a group of small perching birds, which probably colonized the archipelago within the last few million years (Amadon, 1950; Raikow, 1976), or we can examine the radiation of terrestrial vertebrates, which began when the crossopterygian fishes started to come out of water onto land in the Devonian period, 350 million years ago (Romer, 1968). In all cases, however, the basic ecological and evolutionary processes are very similar. A species colonizes a new area, acquires a special new adaptation, or, most frequently, does both more or less simultaneously. This event opens a large number of ecological opportunities that are realized as the original form repeatedly speciates, diversifies, and specializes to fill the various ecological niches that are available to it.

One of the most dramatic examples of adaptive radiation is provided by the cichlid fishes of the lakes of east Africa and is described in a fascinating book by Fryer and Iles (1972; see also Greenwood, 1974). The great African lakes, such as Victoria, Tanganyika, and Malawi, were formed beginning perhaps 3 to 5 million years ago when this region was uplifted and fractured, creating rift valleys and mountains that interrupted rivers and created large lake basins. The rivers contained several kinds of fishes that became isolated in the newly formed lakes. One group, representatives of the teleost family Cichlidae, was well suited for lentic (still water) habitats. This group speciated and radiated to produce a diverse fauna. From a handful of ancestral species these radiations have produced numerous endemic species in each lake (Table 6.1). The fish have diversified morphologically (Figure 6.11) and specialized behaviorally and ecologically to fill many different niches. There are herbivores and carnivores; species with mouths and teeth adapted for catching tiny zooplankton, crushing snails, and eating other fishes whole; even forms specialized to feed just on the fins, scales, or eyes of other fishes (Figure 6.12).

Comparable adaptive radiations have occurred in other groups, although sometimes these have taken place on different temporal or spatial scales. An ancient radiation occurred in the cryptically colored marine pediculate fishes (anglerfishes, Lophiiformes), producing the shallow water frogfishes *(Antennarius)*, the open ocean sargassumfish *(Histrio histrio)*, the stingray like batfishes (Ogcocephalidae), and a variety of deep water bioluminescent forms (suborder Ceratoidea). In plants, a family that has radiated over the past 20 million years in North

Table 6.1
Species diversity and extent of endemism in fishes
Note in particular the great number of endemic cichlid species produced by speciation and adaptive radiation within the largest lakes. (From Fryer and Iles, 1972.)

Lake	Surface area (km^2)	Maximum depth (m)	Cichlids		Noncichlids	
			Total	Endemic	Total	Endemic
Victoria	68,635	93	170+	All but 6	39	17
Malawi	29,604	704	200+	All but 4	44	28
Tanganyika	34,000	1,470	126	All	67	47
Albert	6,800	58	10	4	38	4
Rudolf	8,547	73	5	2	32	4
Edward/George	2,594	117	28+	19+	19	2
Kivu	2,370	485	9	8	9	0
Rukwa	3,302	6+	2	1	20	1
Bangweulu	2,072	10	9	1	59	3
Mweru	4,413	37	12	4	76	?
Chilwa	673	1.5	5	0	8	0
Nabugabo	28.5	4.6	10	5	14	0

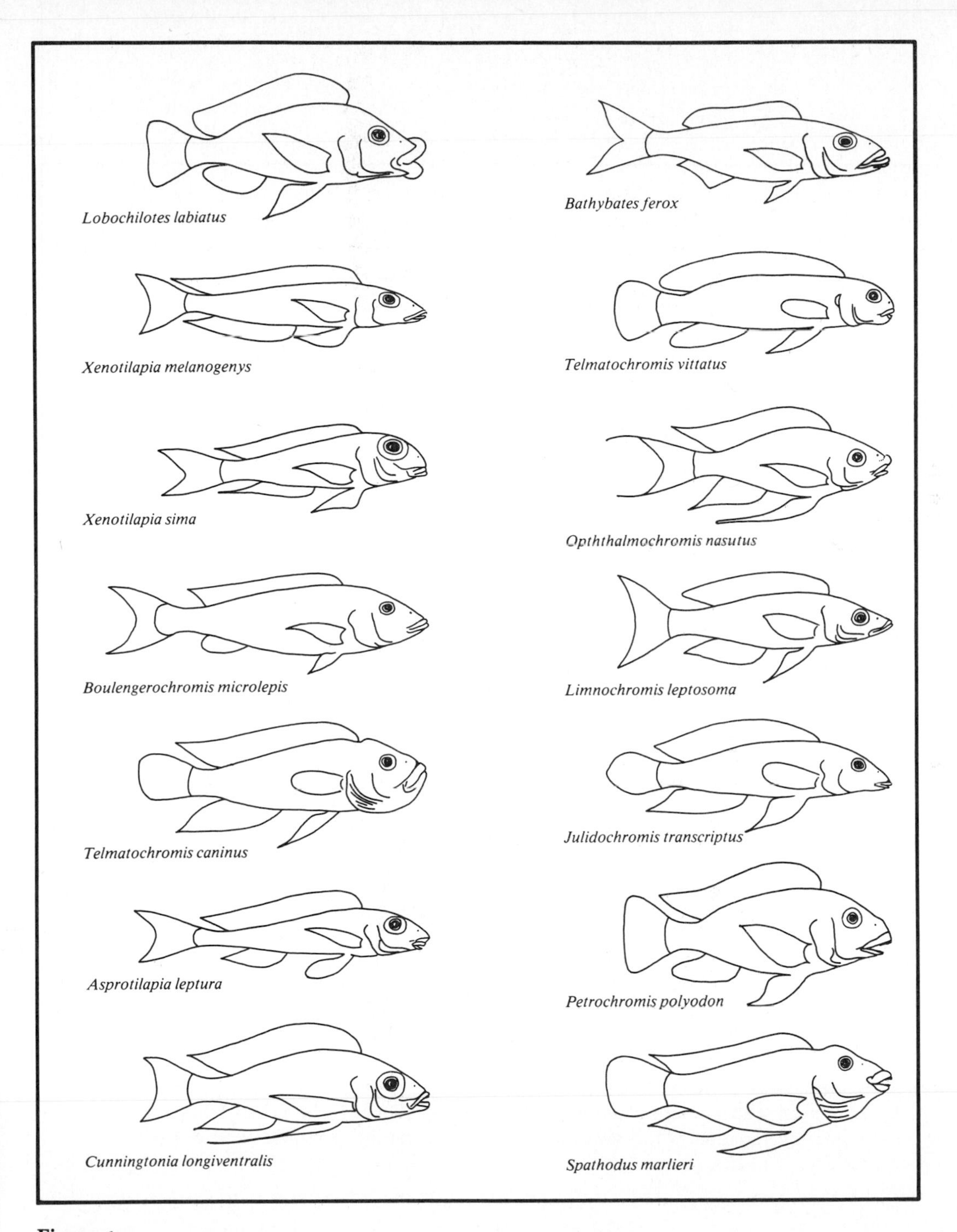

Figure 6.11
Adaptive radiation in body form in some of the cichlid fishes in Lake Tanganyika. (After Fryer and Iles, 1972.)

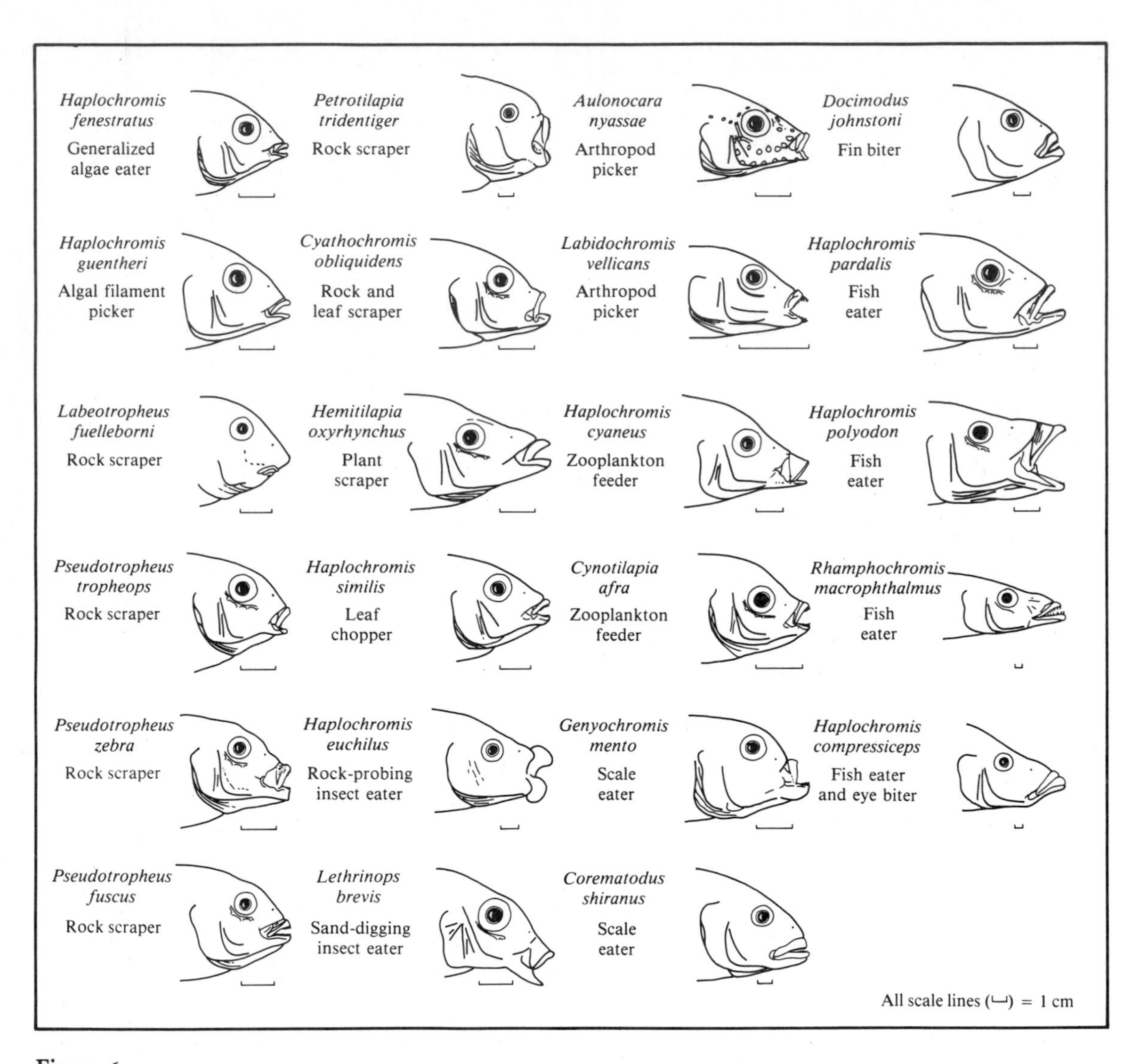

Figure 6.12
Adaptive radiation in head shape, mouth parts, and feeding habits in some of the cichlid fishes in Lake Malawi. Note the amazing variation, which sometimes reflects tremendous specialization in diet (e.g., scale eaters). (After Fryer and Iles, 1972.)

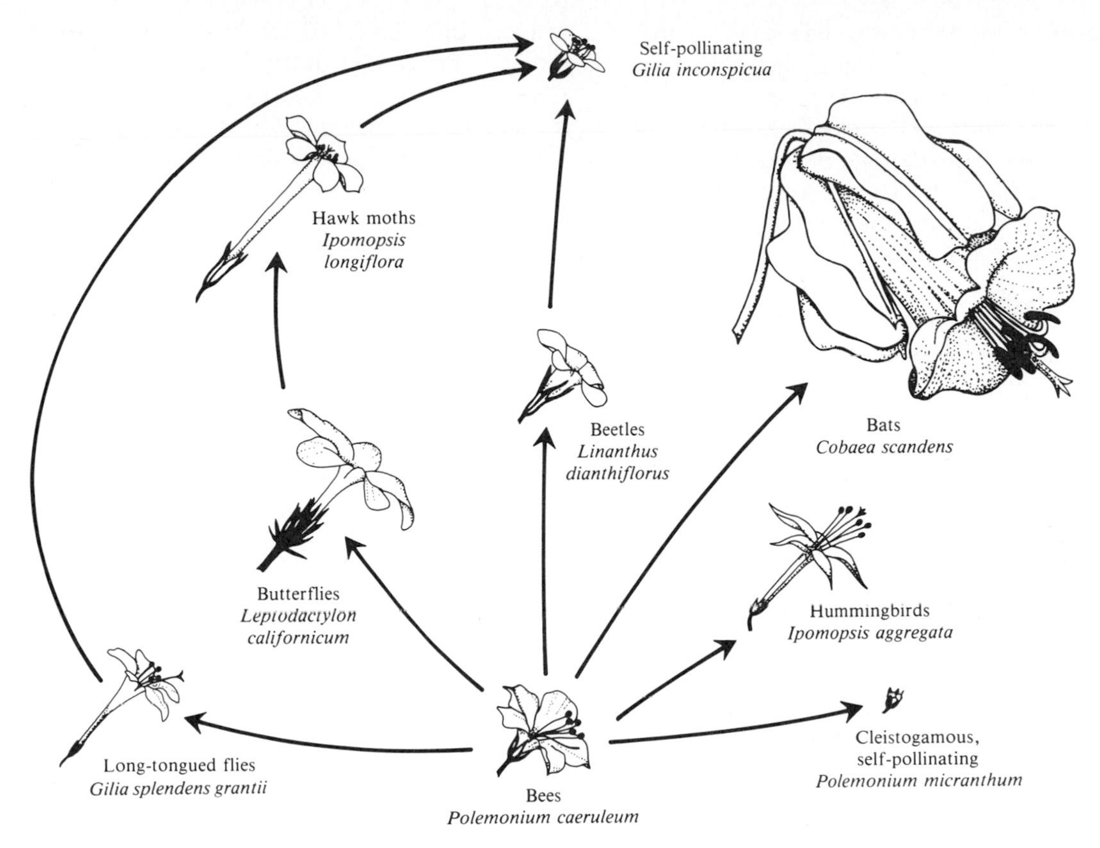

Figure 6.13
Adaptive radiation in floral characteristics related to pollination in the phlox family, Polemoniaceae. It is inferred that a generalized bee-pollinated ancestor similar to *Polemonium caeruleum* gave rise to flowers of different form, color, and attractants, specialized to use different kinds of animals as pollinators or (in some cases) to be self-pollinated. Although members of this family have radiated to exploit different pollinators, plants in some other families have convergently evolved flowers similar to some of these to use the same pollinators. (From The Process of Evolution by P. R. Erlich and R. W. Holm. Copyright © 1963 by McGraw-Hill Book Co. Used with the permission of McGraw-Hill Book Co.)

America is the phlox family (Polemoniaceae). Within a single tribe, flower form and color have become amazingly variable as different species have adapted to be pollinated by hawkmoths, bees, butterflies, beeflies, or hummingbirds, and some have specialized for self-pollination (Figure 6.13). A comparable example is the radiation in form and pollinator attractants in columnar cacti in Mexico (Cactaceae, tribe Pachycereeae) (Gibson and Horak, 1978).

Whereas many genera, tribes, and families exhibit a fairly broad range of approaches to life, many others are extremely monotonous, differing mainly in minute structural details. Taxa that have changed only slightly over evolutionary time apparently are very good at what they do but lack the ability to shift adaptive strategies. Perhaps this is because their genome is inflexible and does not permit much evolutionary experimentation, whereas other genomes promote opportunism. This leads to enormous problems in generalizing about the biogeography of higher taxa. For some, the acquisition of a suite of features was obviously important for their speciation and radiation, whereas in others the origins of species and genera are more attributable to chance events.

Extinction

Ecological processes. Although all living organisms represent a continuous evolutionary lineage extending billions of years back to the origin of life, the ultimate fate of all species is extinction. This can be appreciated by a brief glance at the fossil record. The earth was teeming with life 100 million years ago. Both terrestrial and aquatic habitats were occupied by diverse biotas that formed complex ecological communities. However, the taxonomic groups that were dominant then have been eliminated or drastically reduced by extinction, and they have been replaced by the adaptive radiation of new lines. Dinosaurs and other reptilian groups have been replaced by birds and mammals, ferns and gymnosperms have been largely supplanted by angiosperms, and teleost fishes have replaced cephalopod molluscs as the dominant group of large, actively swimming marine animals. Although extinctions apparently occurred at a fairly constant rate throughout most of the history of life, the fossil record also catalogs occasional episodes of widespread disaster, when much of the earth's biota was wiped out, apparently by rapid and drastic environmental change (Raup and Sepkoski, 1982). For example, Raup (1979) estimates that 88% to 96% of all species of marine organisms became extinct during a brief period at the end of the Permian about 225 million years ago. Not surprisingly, extinctions have had a major influence not only on the kinds of organisms in existence at any period, but also on the geographic distributions of these extinct or contemporary forms.

Several authors have likened the evolutionary history of life to a continual race with no winners, only losers—those species that became extinct. This view is probably best expressed in Van Valen's (1973b) Red Queen hypothesis. It is based on an analogy with the Red Queen in Lewis Carroll's *Through the Looking Glass* who said, "It takes all the running you can do to keep in the same place." The idea is that a species must continually evolve in order to keep pace with an environment that is perpetually changing, because all the other species are also evolving, altering the availability of resources and the patterns of biotic interactions. Those species that cannot keep pace with this change became extinct, but others do well temporarily and speciate to produce new forms. Van Valen further points out that the probability of a species becoming extinct is independent of its evolutionary age, but not of its taxonomic and ecological status: certain taxonomic and ecological groups have consistently higher rates of extinction than others. For example, Brown (1971b) showed that larger and carnivorous mammals had higher rates of extinction on isolated habitat islands than smaller or herbivorous species. This appears to be a general pattern (see Van Valen, 1973a).

All populations experience fluctuations in size as a result of variations in environmental resources and the activities of their enemies. When populations become very small, purely chance factors, such as random variations in the sex ratio, can also affect abundance and contribute to extinction. In general, the smaller a population becomes and the longer it remains at low numbers, the more vulnerable it is to extinction. MacArthur and Wilson (1967; see also Richter-Dyn and Goel, 1972; Leigh, 1981) developed a model showing how the probability of extinction would be expected to increase with decreasing per capita birth rate *(b),* increasing death rate *(d),* and decreasing equilibrium population size or carrying capacity *(K)* (Figure 6.14). Birth and death rates are important, because a population with a high birth rate, especially when coupled with a low death rate, can rapidly recover from a temporary reduction

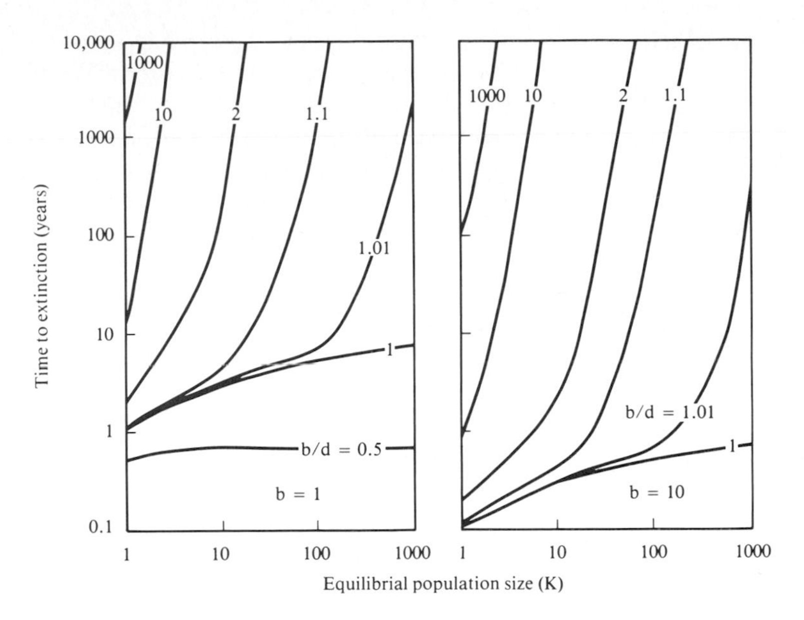

Figure 6.14
MacArthur and Wilson's model of how extinction probability depends on equilibrial population size or carrying capacity *(K),* per capita birth rate *(b),* and per capita death rate *(d)*. Note that the expected time to extinction increases with increasing carrying capacity, birth rate, and ratio of birth rate to death rate. (After MacArthur and Wilson, The theory of island biogeography. Copyright © 1967 by Princeton University Press. Reprinted by permission of Princeton University Press.)

in numbers. The carrying capacity is probably the most important factor. MacArthur and Wilson's calculations suggest that when this equilibrium population size is low, say less than 100, extinction is likely to be fairly rapid; whereas when populations are large, in the tens of thousands, the probability of extinction is infinitesimal. One shortcoming of this model is that it does not take into account the probability of environmental change, including stochastic variation in the carrying capacity. Nevertheless, it is valuable because it suggests that very large populations are likely to become extinct only when a major stress, such as a change in environment, drastically reduces the population size.

Information on ecological factors that have contributed to the extinction of species is difficult to obtain and interpret, because extinctions, with the exception of those caused by humans, are rarely observed. One particularly well-documented example, however, is provided by the work of Smith (1974, 1980) on pikas *(Ochotona princeps)*. Pikas are small relatives of rabbits that live in rock slides and boulder fields in the mountainous regions of western North America. Smith carefully monitored a population in the Sierra Nevada of California that had colonized the rock piles left by a mining operation. These mine tailings appear as islands in a sea of sagebrush habitat. Smith was able to document extinctions in two ways. First, he found rock piles where pikas had been present (as evidenced by piles of droppings and stored dried vegetation that last for many years) but no longer occurred. Second, he carefully censused some 77 rock piles in 1972 and then returned and made another census in 1977. The second method revealed 11 islands where the pikas had been present but became extinct and about an equal number (8) where pikas had been absent but became established in the 5-year period. From data on birth and death rates and on the carrying capacities of mine tailing of varying sizes, Smith found that the population size (less than three individuals) predicted to have a high probability of extinction by MacArthur and Wilson's model (Table 6.2) was similar to the sizes of small, isolated populations that actually became extinct. Of course, this example is concerned only with the extinction of small, isolated populations over a short time scale, but it illustrates the importance of population size and growth rate.

Recent extinctions. Over the last 200 years humans have undoubtedly caused the extinction of thousands of species. We have probably been unaware of many species of microbes and small animals and plants that have disappeared, but the demise of some larger, more spectacular organisms is well documented. In many cases population size was rapidly and drastically reduced by hunting, by habitat destruction, or by the introduction of new predators, parasites, and diseases. A couple of examples will illustrate this.

The passenger pigeon *(Ectopistes migratorius)* was incredibly abundant in eastern North America when the first European colonists arrived. Estimates of the total population size are in the billions. Because the pigeons traveled in dense flocks numbering in the millions, feeding on beechnuts, acorns, and other abundant seeds and fruits and nesting in huge aggregations, they were very vulnerable to humans, who hunted them for food. In the 1870s, 2000 to 3000 birds were often taken in one net in a single day, and 100 barrels of pigeons per day were shipped to New York City for weeks on end. By 1890 the birds had virtually disappeared. In 1914 the last passenger pigeon died in the Cincinnati Zoo (Pearson, 1936).

Equally dramatic was the demise of the American chestnut tree *(Castanea dentata)*. Along with beech, maple, oak, and hickory the chestnut was one of the most abundant trees in the deciduous forests of eastern North America. In 1904 a pathogenic fungus *(Endothia parasitica)* was accidentally introduced, apparently from Asia, where it attacks a related species of chestnut that is somewhat resistant. The disease spread rapidly (Metcalfe and Collins, 1911), and within 40 years mature chestnuts were virtually

Table 6.2
Estimated time to extinction for certain pika populations
The populations represented here occupy habitat islands of varying size calculated by Smith (1974) based on the model of MacArthur and Wilson (1967) (see Figure 6.14). Per capita birth and death rates were determined from the age structure of this population. The model predicts that only very small populations ($K < 3$ individuals) should have measurable rates of extinction owing to ordinary random fluctuations. This prediction was confirmed by Smith's (1980) subsequent measurement of population turnover on these same islands during a 5-year period.

Carrying capacity (number of individuals) K	Birth rate (per capita per year) b	Death rate (per capita per year) d	Time to extinction (years) E
1	0.35	0.00	2.9
2	0.35	1.63	6.9
3	0.35	2.44	46.2
4	0.35	2.84	405.1
5	0.35	3.15	3,751.5

eliminated from their entire range. In some areas, scattered small trees that have sprouted from surviving root stock can still be found. Unfortunately, these are usually attacked by the disease and killed before they can reach reproductive size, so it is doubtful that the chestnut can much longer avoid absolute extinction.

One of the best documented cases of widespread extinction is the loss of bird species from Barro Colorado Island in Panama. Prior to the early 1900s Barro Colorado Island did not exist; it was simply a hill in a tract of tropical lowland forest. During the construction of the Panama Canal the Chagres River was dammed and the rising waters of Lake Gatun covered adjacent lowland areas, creating Barro Colorado, an island of about 16 km^2. Despite its relatively large size, Barro Colorado no longer contains many of its original bird species. Because the Smithsonian Institution has long operated a biological research station on the island, its biota is well known and some of the extinctions are well documented. Using these records, Willis (1974) calculated conservatively that at least 17 species had been lost from a total land bird fauna in excess of 150 species. More recently, by comparing the avifauna of the island with a mainland site of similar size and habitat, Karr (1982) concluded that at least 50 species have become extinct on the island since its isolation. Although habitat changes may have contributed to the demise of some forms, most of the species that have become extinct normally occur at low densities, and it appears that when the small populations present on the island died out, they were not replaced by colonists from across the narrow water gap.

At present there is much concern about rare and endangered species, because continued extinctions, as a result of human activity, seriously threaten to reduce the diversity of living things inhabiting the earth. Some species are very vulnerable, because destruction of their habitats has reduced population sizes to dangerously low levels. It is questionable, for example, whether or not the black-footed ferret *(Mustela nigripes)* can survive, because this large weasel needs large areas of prairie to support sufficient populations of its prey, prairie dogs *(Cynomys)*. The California condor *(Gymnogyps californianus)* is almost certainly doomed. This large, carrion-feeding bird has an extremely low reproductive rate. Mertz (1971) has shown that survival of the condor population is dependent on adults having a long lifespan, and the species has little capacity to recover rapidly from a catastrophic reduction in numbers. The condor has probably been declining gradually since the end of the Pleistocene, 10,000 years ago, but within the last century its death rate has been increased by habitat destruction and by the poisoning of carcasses to kill mammalian predators of livestock.

On the other hand, the capacity of some species to recover from low numbers is amazing. Two species of marine mammals, the elephant seal *(Mirouga angustirostris)* and the sea otter *(Enhydra lutris)*, which inhabit the Pacific Coast of North America, were hunted almost to extinction for their fat and fur, respectively. Tiny populations managed to escape detection, however, and once protected, these increased rapidly to produce large, healthy populations. Another Pacific marine mammal, the gray whale *(Eschrichtius robustus)*, has been increasing reassuringly following the protection of its breeding grounds off Baja California and the cessation of hunting (Rice and Wolman, 1971).

Extinctions in the fossil record. The fossil record provides abundant evidence of extinctions but often provides only tantalizing clues to their causes. In many cases the fossil record is complete enough to show that many diverse species became extinct virtually simultaneously over relatively short periods of geologic time. In such instances we can infer that some drastic, widespread environmental change occurred, but the exact nature of the perturbation and its effect on the organisms concerned is often difficult to deduce clearly. Long-standing, vigorous debates about the causes of episodic extinctions mark the paleontological literature. One of the

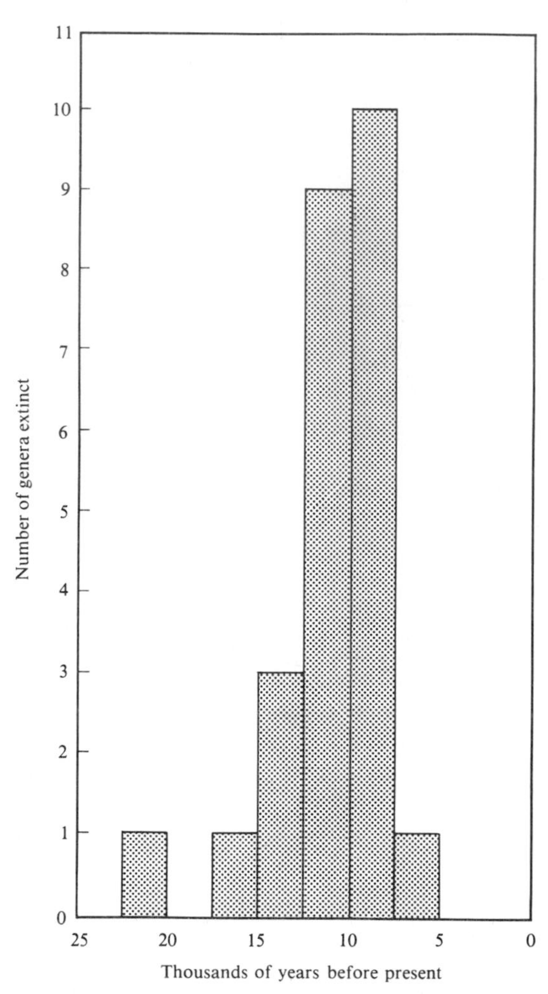

Figure 6.15
Extinctions of North America mammal genera during the last part of the Pleistocene. Note the large peak 7500 to 12,500 years ago, indicating that many species, especially of large mammals, survived most of the Pleistocene, only to become extinct during the last interglacial and glacial period. (From Webb, 1969.)

most recent episodes of extinction was the elimination of the so-called Pleistocene megafauna of North America between 15,000 and 8000 years ago (Figure 6.15). Martin (1967, 1973) has argued that the extinction of many genera and species of large mammals, such as ground sloths, mammoths, horses, and camels, was caused primarily by human hunters, who had recently colonized the continent by crossing the Bering Sea from Asia. Stone weapons embedded in fossils indicated that people definitely hunted these animals, but whether or not this predation was sufficient to account for the wholesale extinctions is a question that is hotly debated. Grayson (1977) pointed out that there were also wholesale extinctions of North American birds at the end of the Pleistocene, and because these were not hunted intensively, he used this information to argue against the hypothesis that humans played a major role. Recent evidence indicates that Australia once also supported a Pleistocene megafauna of now extinct giant marsupials. Further research should determine when these mammals became extinct relative to the time of colonization by aborigines and thus either support or cast doubt on Martin's ideas.

The causes of other episodes of wholesale extinctions that occurred even longer ago are also controversial. Several explanations have been proposed for two of the most dramatic extinctions: the elimination of the dinosaurs and many other groups of terrestrial and marine organisms at the end of the Cretaceous, about 65 million years ago, and wholesale extinctions of many lines, particularly of marine organisms, that occurred at the boundary between the Permian and the Triassic, about 225 million years ago (see Raup, 1979; Raup and Sepkoski, 1982). Large rapid changes in the climate of the earth are the causes most often invoked for such catastrophic extinctions, but other explanations, including continental drift (e.g., Schopf, 1974) and collision of the earth with meteors (Alvarez et al., 1980) have their proponents.

Species selection

Differential survival and reproduction of species. A brief look at the fossil record and at the diversity of contemporary forms reveals that all groups of organisms have not been equally successful. Some lineages have radiated to leave many derived taxa, others have survived virtually unchanged for millions of years, and still others have disappeared quickly, leaving no descendents. The historical pattern of speciation and extinction has had a major influence on the taxonomic diversity and geographic distribution of living things. Episodic events such as explosive adaptive radiations and wholesale catastrophic extinctions have been particularly important. Paleontologists and biologists have long recognized that certain lineages appear to possess particular traits that result in high speciation rates and low extinction rates, thereby leading to adaptive radiation and relative evolutionary success compared to those groups that die out or barely maintain themselves.

Recently the differential survival and splitting of species over geologic time has come to be termed species selection (Stanley, 1979) by analogy to the differential survival and reproduction of individuals (individual selection) that has traditionally been thought to be the primary mechanism of evolutionary change by natural selection. Species selection and individual selection should be viewed not so much as totally different biological processes but as the consequences of generally similar ecological and genetic processes operating on different levels of biological organization. The same conditions (e.g., rapid environmental changes) that cause large differences in the birth and death rates of individuals of different genotypes and thus result in rapid evolution by individual selection are also likely to lead to differential multiplication and survival of species and thus result in rapid evolution by species selection as well. Rapid evolution by individual selection is likely to be a primary cause of high speciation rates that will contribute to rapid phyletic evolution by species selection.

To some extent the fate of evolutionary lineages is a matter of chance opportunities. It depends on a species with particular traits being in a favorable place at an opportune time. Many adaptive radiations began when a population either colonized a virgin area or evolved some new trait that substantially increased its fitness in a particular environment. In either case, the species was suddenly faced with new ecological opportunities that stimulated further rapid evolution and speciation. For example, the eutherian or placental mammals acquired several traits, including higher metabolic rate, improved temperature regulation, and new mechanisms of nourishing the developing young, which represented major advances over their ancestors and over other kinds of primitive mammals including monotremes, marsupials, and other now extinct groups. During the late Mesozoic and early Cenozoic the placental mammals radiated at the expense of their more primitive relatives, except in Australia and, during its early history, South America. On these isolated continents, surrounded by ocean barriers, marsupials and other kinds of primitive mammals persisted. South America subsequently was colonized by many groups of placental mammals from North America, and many of the original South American taxa became extinct (Simpson, 1950, 1980b). Australia remained isolated, however, and marsupials radiated and flourished. A few egg-laying monotremes have survived up to the present. Two groups of placental land mammals, rodents and bats, managed to colonize Australia from Asia across water barriers and have subsequently undergone their own adaptive radiations. The rodents have diversified to produce many endemic genera and 77 species that include desert hopping mice *(Notomys)* and stick nest builders *(Leporillus),* ecological counterparts of North

American kangaroo rats *(Dipodomys)* and packrats *(Neotoma)*, respectively (Keast, 1970, 1972c).

However, the course of adaptive radiation is not dictated by chance alone. The fate of evolutionary lineages also depends on predictable relationships between the characteristics of the organisms and features of their environments. The phenomenon of evolutionary convergence, discussed in more detail in Chapter 18, indicates that the physical and biotic characteristics of particular habitats limit the numbers and kinds of organisms that can live there. There appears to be a finite number of ways to make a living, and distantly related organisms have independently evolved similar forms and functions to fill similar ecological niches in similar habitats around the world. A splendid example is provided by the convergence between the marsupials and rodents of Australia and the placental mammals found in similar habitats on other continents.

There is some suggestion that the pattern of adaptive radiation may differ, depending on whether or not it occurs at the expense of other groups of organisms that are already present and competing effectively for the resources required by the new group. Simpson (1952a) has noted that when a species colonized a completely new area or acquired a major adaptive innovation, radiation was often explosive. The lineage diversified rapidly, forming several major lineages that each subsequently underwent its own radiation (Figure 6.16). An example of this pattern is provided by the evolution of the echinoderms, which underwent their major radiation in the Cambrian, about 525 million years

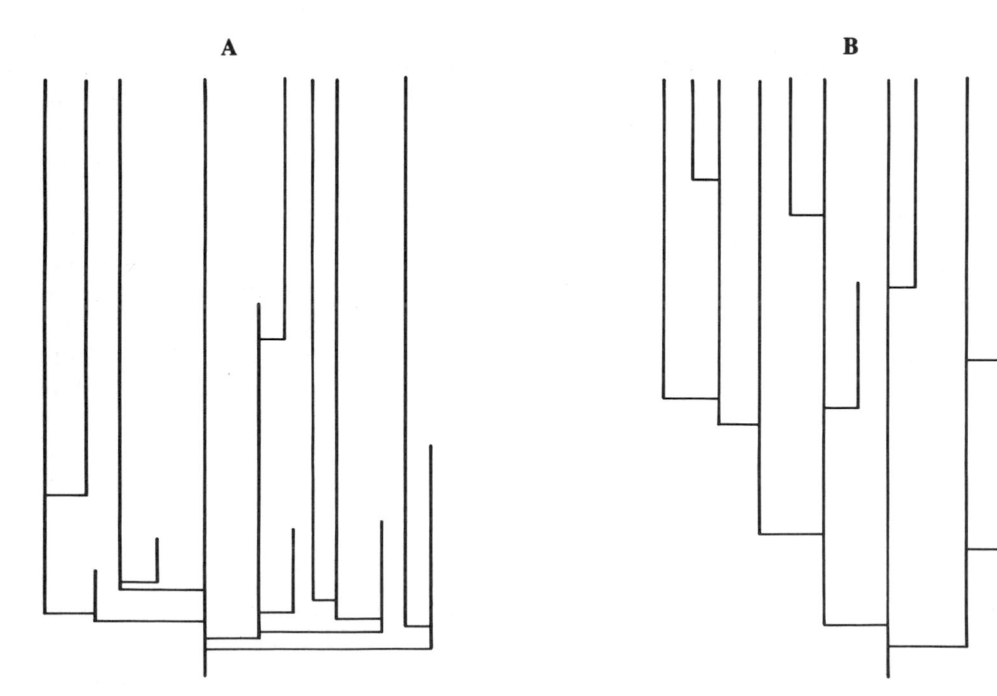

Figure 6.16
Two generalized patterns of adaptive radiation. **A,** Explosive radiation of a taxon into a previously unoccupied niche (adaptive zone) or uninhabited geographic region. **B,** More gradual radiation of a taxon in an environment relatively saturated with species so that increases of diversity occur at the expense of extinctions in other groups. (After Jablonski, in preparation.)

ago. Other groups evolved later in environments already filled with efficient competitors and predators. Jablonski (in preparation) suggests that often these lineages increased and branched more slowly, only gradually producing diverse taxa adapted for widely different ecological niches (Figure 6.16). An example is provided by the neogastropods, a group of specialized, predatory marine snails that originated in the Cretaceous, 130 million years ago, which has gradually diversified to become a dominant invertebrate group.

Because divergence and adaptation depend on genetic change, it is hardly surprising that genetic processes influence the rates of speciation and adaptive radiation. The diversification of certain groups of higher plants, such as the largest plant family, the sunflower family (Asteraceae), appears to have been facilitated by high incidences of polyploidy and by population structures and breeding systems that have promoted speciation by this mechanism of chromosomal change. A probable advantage of polyploidy is that it creates duplicate sets of genes that may ultimately provide the raw material for evolutionary innovations. Bush et al. (1977) have suggested that chromosomal mechanisms have also played a major role in the adaptive radiation of mammals. They point to the high incidence of chromosomal rearrangements in groups such as the horses, primates, rodents, and artiodactyls (even-toed grazing mammals such as antelopes, sheep, and cattle) that have undergone major, rapid radiations. They hypothesize that the social and breeding structures of these groups facilitated rapid speciation by subdividing the species into small, inbred local populations in which new chromosomal arrangements could be fixed by means of genetic drift.

Patterns of speciation and extinction in the fossil record. An interesting relationship between speciation rate, geographic distribution, and mode of larval distribution has been described for marine invertebrates (Jackson, 1974; Hansen, 1980; Jablonski and Lutz, 1980; Jablonski, 1982). Within certain groups of molluscs some species have planktotrophic larvae, i.e., juvenile stages that drift passively in the ocean, whereas others brood their offspring or have other specializations so their young are not dispersed in the plankton. The mode of larval dispersal can be determined for fossil as well as for recent forms, because the original larval shell is preserved as part of the adult shell that is of course readily fossilized. Molluscan species with planktotrophic larvae tend to have larger geographic ranges (Figure 6.17) and lower speciation rates than related species in which the young rarely disperse far from their parents. Limited dispersal of offspring presumably reduces gene flow, enables populations to adapt to local conditions, and thus facilitates genetic differentiation and allopatric speciation. On the other hand, because small, specialized populations are also particularly vulnerable to extinction, species with low rates of larval dispersal tend to persist in the fossil record for shorter periods of time than those with planktotrophic larvae (Figure 6.17).

As implied by the previous example, patterns of extinction, which depend in part on characteristics of the organisms themselves, can also influence the evolutionary histories of lineages by species selection. Over a period of more than 400 million years one group of marine, bottom-dwelling invertebrates, the clams (Mollusca, order Pelecypoda), have been replacing a phylum, the brachiopods (Brachiopoda) (Figure 6.18). Because these two groups have superficially similar bivalved morphology, feeding habits, and habitat requirements, it has long been thought that the clams were supplanting the brachiopods by competitive exclusion (Elliot, 1951). In fact, species of the two groups may compete significantly when they occur together, but careful examination of the fossil record indicates that the increase of clams and the dramatic decline of brachiopods was owing in part to the catastrophic effects of extinctions that eliminated many more taxa of brachiopods than of clams (Gould and Calloway, 1980). Fur-

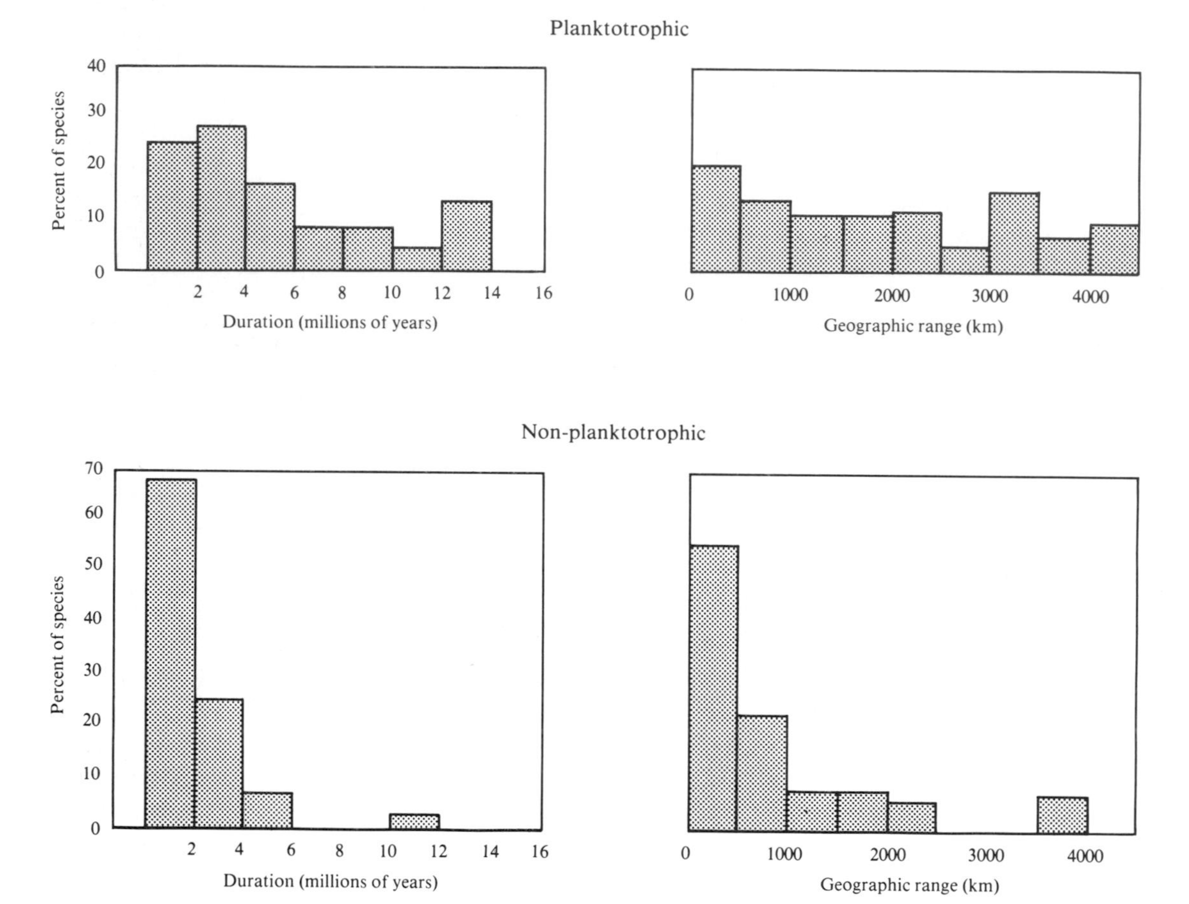

Figure 6.17
Effect of mode of larval dispersal on the geographic range and survival time of species of fossil molluscs in the Late Cretaceous on the east coast of North America. Note that species with widely dispersed planktotrophic larvae had wider geographic ranges and persisted for longer periods than those with less vagile, nonplanktotrophic larvae. (After Jablonski, 1982.)

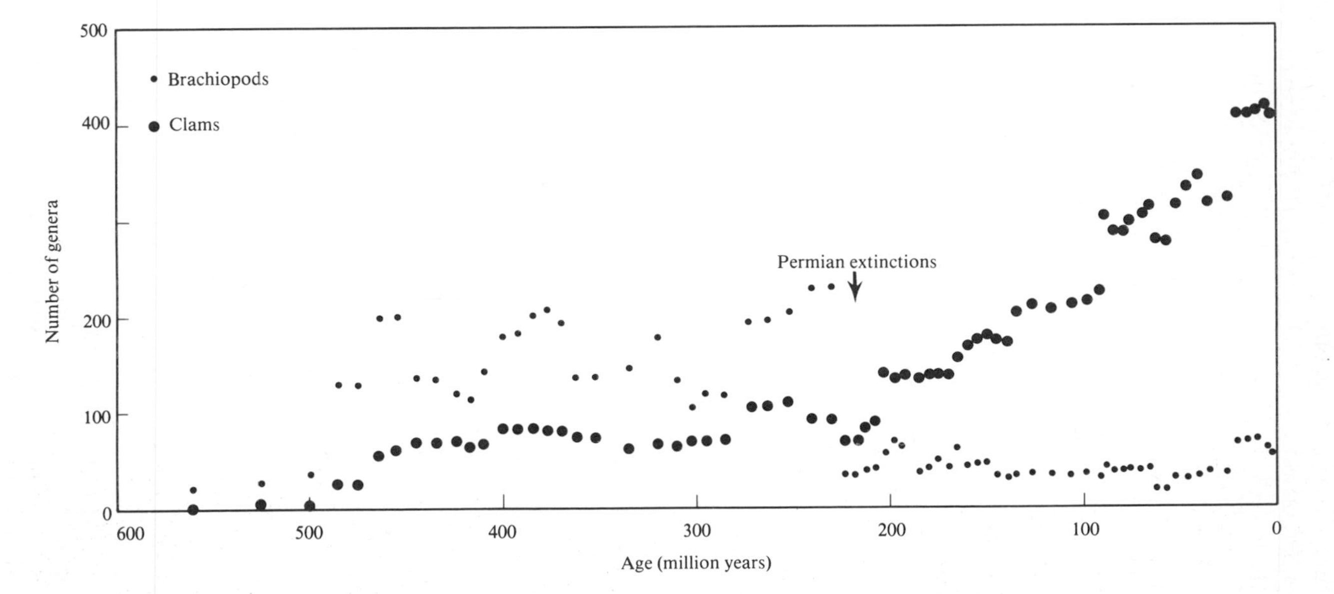

Figure 6.18
Replacement of the brachiopods (Brachiopoda) by the clams (Mollusca, order Pelecypoda) during evolutionary history. Because of superficial similarities in shell morphology and ecology, some authors have attributed the decline of the brachiopods and the diversification of the clams to competition. Some competition may have occurred between the two groups, but the single most important event in their history was the wave of extinction at the beginning of the Permian, which decimated the brachiopods but only temporarily inhibited the adaptive radiation of the clams. (Data from Gould and Calloway, 1980.)

thermore, Valentine and Jablonski (1982) have suggested that the brachiopods are particularly susceptible to extinction because they do not have planktotrophic larvae but instead brood their young or produce larvae that spend only a brief time in the plankton. Marine invertebrates with nonplanktotrophic larvae tend to have restricted geographic distributions and high extinction rates. Brachiopods have largely been eliminated from shallow water habitats where they are exposed to environmental fluctuations, but some forms persist in deep waters where their mode of reproduction is not disadvantageous.

This is not to imply that many cases of replacement of one group by another cannot be attributed in large part to competition. Simpson (1953) provided circumstantial evidence for the importance of competition in the evolutionary history of the mammals. In one example, the multituberculates (see also Landry, 1965; Van Valen and Sloan, 1966), an early group of

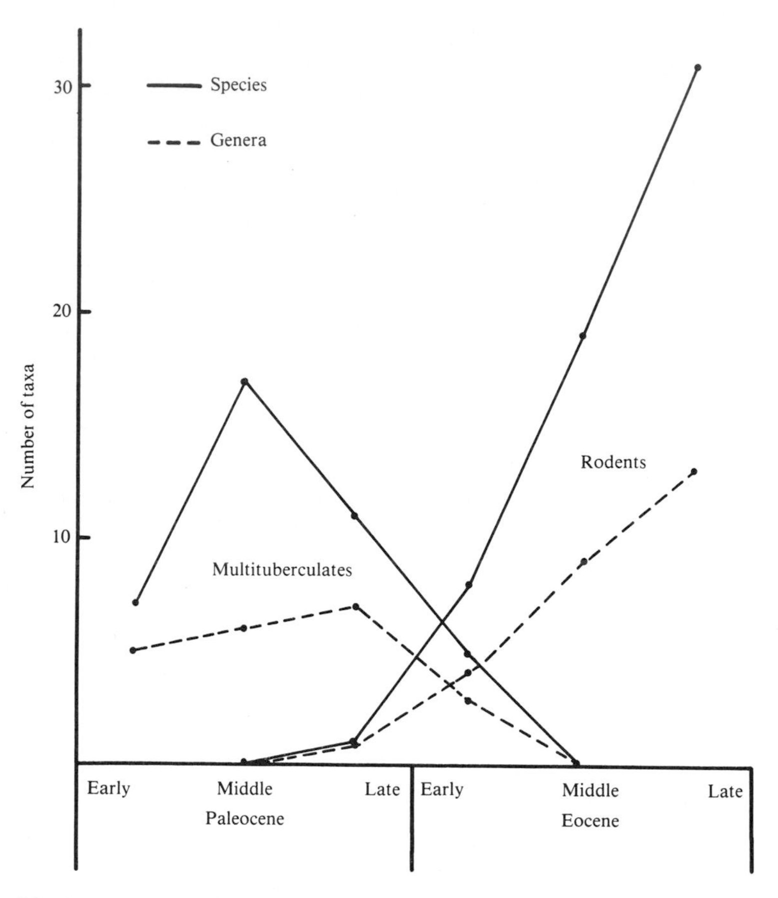

Figure 6.19
Replacement of one order of mammals, the multituberculates, by another, the rodents, during the early Cenozoic. There has been a continuing debate about the extent to which these reciprocal changes in diversity can be attributed to competition, but some kinds of multituberculates had remarkably similar morphology (as inferred from teeth and bones) and perhaps also similar ecology to some rodent groups. (Data from Simpson, 1953.)

small to medium-sized gnawing mammals, radiated early in the Cenozoic and attained maximum diversity in the Paleocene about 60 million years ago. Within about 20 million years they had virtually disappeared and were replaced by the rodents, which include the rats, mice, and squirrels, the dominant group of small, nonflying mammals today (Figure 6.19). Simpson (1950, 1980b) also documented the rapid replacement of many native South American mammals by similar but more advanced forms that invaded from North America in the Pliocene, about 5 million years ago. As will be discussed in more detail in Unit Three, during most of the Cenozoic South America was isolated by water barriers from the other continents. A distinctive mammalian fauna, including many endemic groups of large herbivores and carnivores, had evolved in isolation. These forms were largely replaced when species that had evolved in the more diverse, and presumably more competitive, communities of the Northern Hemisphere were able to invade.

A few patterns of extinction appear to be very general. Ecological considerations would suggest that large body size, upper trophic levels (i.e., top carnivores), specialized diet, unusual habitat requirements, and restricted geographic ranges should tend to diminish carrying capacities and increase rates of extinction. This prediction seems to be supported by available data. Earlier (Chapter 1) we mentioned Cope's rule, the tendency for evolution of increased body size within evolutionary lineages. There have long been niches available for very large herbivorous vertebrates, and these have been filled by such groups as the dinosaurs, the titanotheres and uintatheres (extinct primitive mammals), and the diverse Pleistocene megafauna. All of these groups have been ephemeral, however, because large animals have high extinction rates. In general, species with low carrying capacities often tend to be specialists, and excessive specialization usually leads to extinction. Occasionally, however, an evolutionary change that at first appearance represents a highly specialized condition enables the organism to exploit new opportunities and to radiate to fill a new adaptive zone. The first bird was a very specialized reptile indeed, but its highly modified scales were the feathers that enabled it to be endothermic and to fly and thereby opened up a whole new way of life so that a new group of vertebrates was able to radiate and colonize the world.

Selected references

Speciation

Atchley and Woodruff (1981); Berigozzi (1982); Bush (1975); Bush et al. (1977); Carson (1971, 1975); Cuellar (1977); Eldredge and Cracraft (1980); Eldredge and Gould (1972, 1976); Endler (1977); Gingerich (1976a, 1979); Gould and Eldredge (1977); V. Grant (1963, 1971); Hopper (1979); Keast (1961); Lande (1982); H. Lewis (1953, 1962, 1966); W. Lewis (1979); Mayr (1942, 1963, 1970); G. Nelson and Platnick (1980, 1981); Platnick and Nelson (1978); P. Raven (1964, 1976); Rosenzweig (1978); Scheltema (1971, 1977); G. Simpson (1937, 1951, 1953); Soumalainen (1962); Stanley (1979); Stanley et al. (1981); G. Stebbins (1950, 1971b); Templeton (1980a, 1980b, 1981); M. White (1968, 1973, 1978); Wiley (1981); P. Williamson (1981).

Adaptive radiation

Abbott et al. (1977); Amadon (1950); Carson et al. (1970); Dressler (1981); Eisenberg (1981); Fryer and Iles (1972); P. Grant (1972b); V. Grant and Grant (1965); Greenwood (1974); Keast (1977b); Lack (1947, 1969); Maglio (1973); Mayr (1963); Norris (1966); Olson (1971); Raikow (1976); Rosenzweig (1973); G. Simpson (1953); G. Stebbins (1971a, 1974); J.R. Turner (1980); Vaurie (1980); E. Williams (1983).

Extinction

Alvarez et al. (1980); Boucot (1975b); J. Brown (1971b); Fischer and Arthur (1977); Foin et al. (1975); Gould and Calloway (1980); Gray et al. (1981); Grayson (1977, 1979); Karr (1982); Leigh (1981); MacArthur and Wilson (1967); P.S. Martin (1966, 1967, 1973); P.S. Martin and Wright (1967); Mertz (1971); R.R. Miller (1961a); Raup (1979); Raup and Sepkoski (1982); Richter-Dyn and Goel (1972); Schopf (1974); G. Simpson (1950, 1980b); A.T. Smith (1974, 1980); Van Valen (1970, 1973b); S. Webb (1969, 1977, 1978); E. Willis (1974).

Species selection

Bush et al. (1977); Gould and Calloway (1980); Hansen (1980); Jablonski (1982); J. Jackson (1974); Newell (1967); G. Simpson (1950, 1952a, 1953); Stanley (1975, 1979).

Chapter 7 Dispersal

What is dispersal?

All organisms have some capacity to move from their birthplaces to new sites. Movement of offspring away from their parents is a normal part of the life cycle of virtually all plants and animals. Often dispersal is confined to a particular stage of the life history. Higher plants and some aquatic animals are sessile as adults, but in their earlier developmental stages they are usually capable of traveling small but significant distances from the source (Figure 7.1). Mobile animals may shift their locations at any time during their lives, but many settle down and confine their activities to a limited home range for long periods of time. Dispersal should not be confused with dispersion, an ecological term referring to the spatial distribution of individual organisms within a local population.

Dispersal as an ecological process. Dispersal is basically an ecological process. Natural selection favors individuals that move some distance from their natal site. A more distant location is always likely to be more favorable than the exact birthplace, in part because intraspecific competition between parent and offspring and among siblings is reduced, and in part because the environment, and hence the quality of the natal site, is always changing. Plants and animals have evolved a wide variety of mechanisms of dispersal in response both to the spatial variation in the environment, which determines the probability of survival and reproduction as a function of distance traveled, and to evolutionary constraints on the organism, which influence the kinds and distances of movements that are possible. From an ecological point of view dispersal must be considered simply as an adaptive part of life history of every species. (For an extensive discussion and many examples see Krebs, 1978.)

Dispersal as a historical biogeographic event. The role of dispersal in biogeography is very different. Although dispersal is continually occurring in all organisms, most of it does not result in any significant change in their geographic distributions. As pointed out in Chapter 3, the geographic ranges of most species are limited by environmental factors and remain relatively constant over ecological time. Biogeographers are concerned primarily with the exceptions: those rare instances in which species shift their ranges by moving over large distances. This occurs so infrequently that we seldom see it happening, and even less often are we able to study it. Usually dispersal must be viewed as a historical process, and the nature and timing of past long-distance movements must be inferred from indirect evidence, such as the distributions of living and fossil forms. This is a monumental task. The distribution of every taxon reflects a history of local origin, dispersal, and local extinction extending back to the very origin of life. Patterns of endemism, provincialism, and disjunction of geographic ranges (Chapter 8) indicate that the dispersal of many groups has been so limited that their histories are reflected in the distributions of living and fossil representatives. However, reconstructing this history requires that we deal with a process

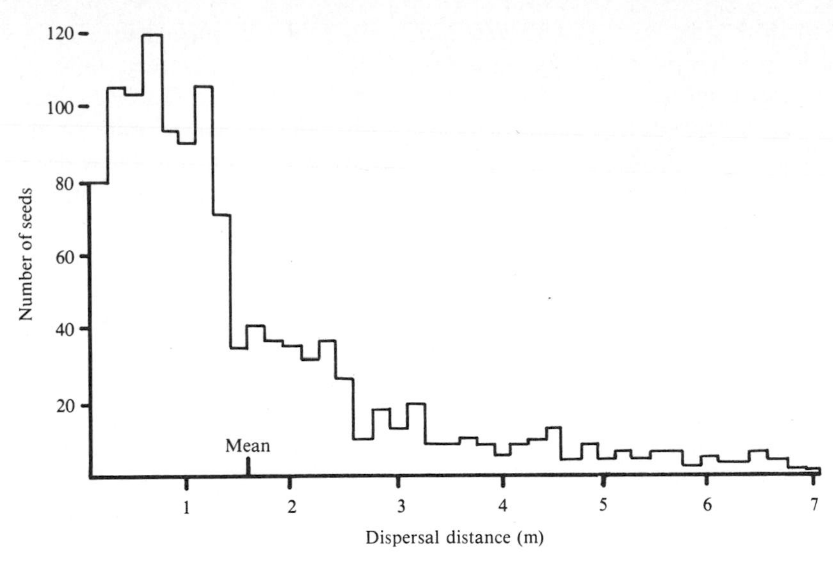

Figure 7.1
Frequency distribution of seeds (achenes) of the composite *Liatris cylindracea* (Asteraceae) showing the distance traveled from the parent plant. This is a typical pattern for dispersal of seeds and other passively transported disseminules; most end up very near the source, but the distribution is highly leptokurtic and a few are carried long distances. (After Schaal and Levin, 1978.)

that operates so infrequently that its effects may often appear highly random and idiosyncratic.

The problem of dealing with rare but important events is not unique to biogeography. In many ways the role of beneficial mutations in evolution is analogous to the role of successful dispersal in biogeography. Beneficial mutations provide the raw material for evolutionary change, but they occur essentially randomly and so infrequently that they are difficult to observe and study. Every inherited feature of an organism has its history in a series of unique genetic changes. Trying to reconstruct the history of mutations that resulted, for example, in a reptilian scale evolving into a feather, is conceptually similar in many ways to attempting to reconstruct the biogeographic events that led to the present disjunct distribution of large predaceous birds (e.g., hawks, kestrels, and owls) in Africa, Australasia, New Zealand, and South America. We draw this analogy to make two points. First, long-distance dispersal events may be infrequent and somewhat stochastic, but that does not mean they have been unimportant. On the contrary, these movements are among the most important events that have shaped present distributions. Second, we cannot afford to ignore the role of dispersal in biogeography just because it is difficult to study. One of the challenges in understanding the evolution of life is to develop ways to evaluate the influence of rare but important events such as beneficial mutations and successful long-distance dispersal.

Biogeographers often distinguish two kinds of dispersal events (e.g., Platnick and Nelson, 1978). In some cases a species may successfully cross an existing barrier, such as an ocean or mountain range, and establish a population on the other side. In other instances species may simply expand their geographic range by moving outward at one or more of the boundaries and inhabiting a larger area. In both cases an initially restricted species spreads from its place of origin to new regions. Both can eventually result in disjunct ranges. The term ***vicariance event*** is used to describe the situation in which a once widespread taxon becomes restricted to geographically isolated regions by extinction of intervening populations.

Although major shifts in geographic ranges are not common, it is possible to cite observations that unequivocally document both kinds of dispersal. In Chapter 3 we reported several examples in which species had gradually expanded their geographic ranges by colonizing new regions at the boundaries. Most of those cases that have been documented by written records involve shifts in response to habitat modification by humans, perhaps sometimes accompanied by adaptations of the species to the new niches. Presumably the same processes, habitat change and adaptation to new conditions, were responsible for historical expansions that cannot be attributed to human influence, such as the invasion of South America by many groups of North American mammals during the Pliocene (Chapter 18) or the northward movement of deciduous forest trees in eastern North America following the retreat of the last Pleistocene ice-sheet (Figure 4.10).

Long-distance or waif dispersal. Similarly, there is abundant evidence of long-distance dispersal of some taxa across barriers. Anyone who has built a small pond in the backyard cannot help but be impressed by the rate at which populations of aquatic insects, snails, other invertebrates, vascular plants, and algae become established. The same process of colonization also occurs on a much larger geographic scale. When a volcanic eruption blasted the Indonesian island of Krakatau in 1883, it covered both what was left of Krakatau and the neighboring island of Verlaten with a deep blanket of ash, obliterating all life. Biological surveys, primarily of birds and plants (Figure 7.2), document the rapidity with which new populations of organisms became established on these islands (Docters van Leeuwen, 1936; Dammermann, 1948; see also Brown et al., 1919). By 1933, only 50 years after the eruption, Krakatau was once again covered with a dense tropical rain forest, and 271 plant species and 31 kinds of birds were recorded on the island. Where did these organisms and the numerous invertebrates come from and how did they get there? In the case of Krakatau and Verlaten, the answer to at least the first question is relatively clear: they dispersed across the water gap from the large islands of Java and Sumatra, which lie 40 and 80 km, respectively, from Krakatau.

The case of Krakatau is unusual only in that the creation of completely virgin habitat was so dramatic, that the sources of colonizing organisms can be identified easily, and that their immigration could be carefully documented. The Galápagos and Hawaiian archipelagos lie far out in the Pacific Ocean, 800 km west of Ecuador and 4000 km west of Mexico, respectively. These are completely oceanic islands that are actually the tops of volcanoes that arose from the ocean floor. Although they have never been much nearer to continents than they are today, they have acquired diverse biotas from plants and animals that have dispersed across the ocean. Both the Galápagos and the Hawaiian islands have land snails and bats as well as numerous species of trees, insects, and birds. In addition, giant tortoises and native rats inhabit the Galápagos Islands.

Oceanic islands illustrate the potential of some organisms to disperse long distances over barriers of inhospitable habitat and to establish new populations on arrival in suitable areas. Some authors deny the importance of long-distance dispersal to champion alternative explanations. Nevertheless, the evidence is undeniable that many groups of organisms have reached distant islands by traveling through the air or floating on the sea (Ridley, 1930). Those species that have special adaptations for long-distance travel or that send out more than one immigrant to found a population have been especially successful at colonizing islands. We also

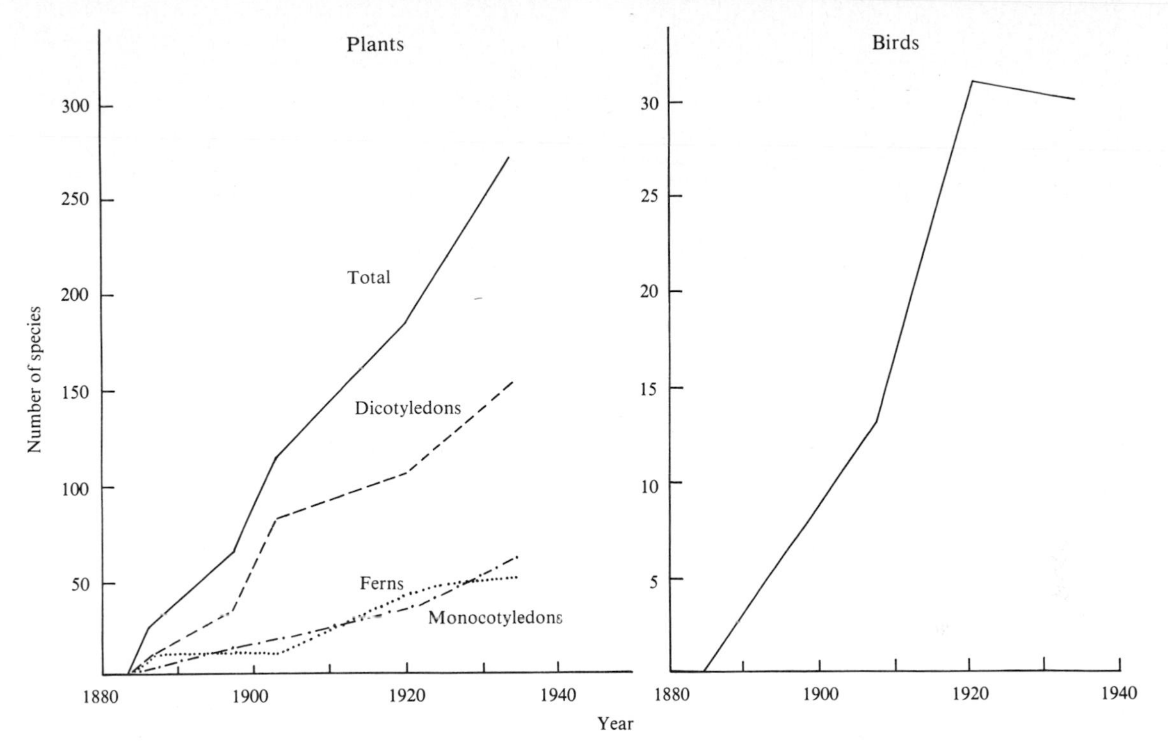

Figure 7.2
Rapid recolonization of the island of Krakatau by plants *(left)* and birds *(right)* that had dispersed successfully across the ocean. All life on the island was destroyed by a volcanic eruption in 1883, and several biological surveys recorded the colonizing species, which had probably traveled across at least 40 km of ocean. (After MacArthur and Wilson, 1967.)

have evidence, both from recent occurrences and from historical patterns, that other species have crossed equally long distances over continental areas (McAtee, 1947; Cruden, 1966).

Long-distance dispersal has three important consequences for biogeography. First, it can be used judiciously to explain the wide and often discontinuous distributions of many taxa of animals, plants, and microbes. Second, it accounts in part for both similarities and differences in the biotas inhabiting similar environments from different geographic areas, because the ability to disperse successfully over distance and over habitat barriers varies greatly among different kinds of organisms. And finally, because chance plays such an important role in successful dispersal and establishment of colonists, it causes a certain degree of taxonomic randomness (or stochasticity) in the composition of biotas.

The extent to which organisms can disperse successfully depends primarily on their abilities to (1) travel over long distances, (2) withstand potentially unfavorable conditions during the passage, and (3) establish viable populations on arrival. We shall consider each of these factors more carefully.

Mechanisms of movement

Active movement. Organisms can disperse either actively, moving under their own power, or passively, being carried by a physical agent

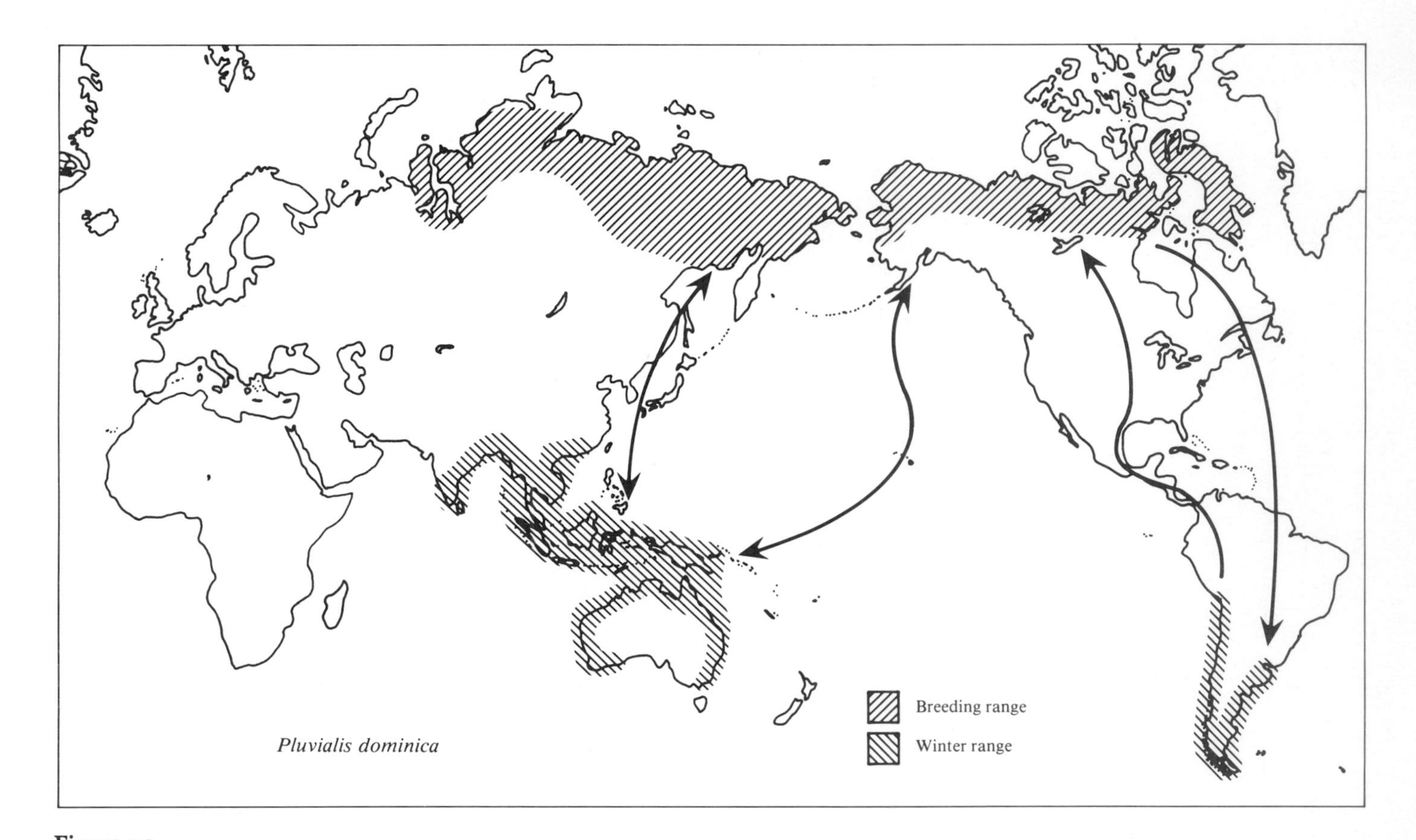

Figure 7.3
Migratory routes of the golden plover *(Pluvialis dominica)*, a shorebird that breeds in the Arctic and winters in temperate regions of the Southern Hemisphere. Every year these birds fly prodigious distances, often crossing huge expanses of open ocean. (Map compiled from various sources.)

such as wind or water or other organisms. Only a few relatively large animals have the capacity to travel long geographic distances under their own power. Of these the large, strong fliers, including many birds, bats, and large insects (e.g., dragonflies, some lepidopterans, beetles, and bugs) have the greatest capability for rapid, active long-distance dispersal. Many of these animals regularly travel hundreds or thousands of kilometers during seasonal migrations, which are a normal part of their annual life cycles. When stressed or aided by favorable winds, some of these same fliers can cover comparable distances during a single flight.

A few examples of normal migratory routes indicate the potential for dispersal by flight. The golden plover *(Pluvialis dominica)*, a medium-sized shorebird weighing about 150 g, breeds in the Arctic and winters in southern South America, southern Asia, Australia, and the islands of the Pacific. Migrating individuals regularly fly nonstop from Alaska to Hawaii, a distance of 4000 km (Figure 7.3) (Dorst, 1962). The ruby-throated hummingbird *(Archilochus colubris)*, at 3.5 g one of the smallest of all birds, regularly commutes twice a year nonstop across the Gulf of Mexico, a distance of 800 km, en route between its breeding grounds in the eastern United States and its wintering grounds in southern Mexico (Dorst, 1962). Every year a few individuals of bird species native to Europe are seen in eastern North America, usually following severe North Atlantic storms. Bats of the genus *Lasiurus* (body weight 10 to 35 g) migrate from northern North America to winter in the Neotropics (Findley and Jones, 1964). The species of this bat that is native to the Hawaiian Islands undoubtedly is derived from migrating individuals that went astray. Monarch butterflies *(Danaus plexipus)* and some dragonflies *(Anax)* migrate distances comparable to those flown by many songbirds, from southern Canada and the northern United States to the southern United States and central Mexico. Individual monarch butterflies may fly as much as 375 km in 4 days and 4000 km in their lifetimes (Urquhart, 1960; Brower, 1977).

An excellent example of successful long-distance dispersal and establishment of a highly mobile bird is provided by the cattle egret, *Bubulcus ibis* (Crosby, 1972). This small heron originally was native to Africa, where it inhabits tropical and subtropical grasslands in association with large herbivorous mammals and forages for insects and other small animals that are flushed by the grazing mammals. In the late 1800s cattle egrets colonized eastern South America, having dispersed under their own power across the South Atlantic Ocean. During the succeeding decades the immigrants thrived and spread throughout much of the New World, finding abundant food and habitat as a result of recent clearing of the tropical forests for grazing livestock. The cattle egret has expanded its breeding range northward to the southern United States and has colonized all of the major Caribbean islands (Figure 7.4).

A few animals, such as some of the larger mammals, reptiles, and fishes, may be able to disperse substantial distances by swimming or walking. Active dispersal by these means is generally less effective than by flight, because the animals are constrained to swim or walk through unfavorable intervening habitats whereas flying animals can simply vault barriers. Nevertheless, some large animals, especially aquatic ones such as many whales, sharks, predaceous fishes, and sea turtles, have wide and often discontinuous geographic distributions produced in part by active dispersal.

Certainly the presence of wings or legs does not automatically mean that such organisms will disperse great distances; in fact, many are not good dispersers. Two examples can be cited. In western North America there lives a butterfly *(Euphydryas editha)* whose local populations are often isolated and genetically differentiated. A 4-year study was conducted in a grassy area having three distinct populations only several hundred meters apart. This study

Figure 7.4
Colonization of the New World by the cattle egret, *Bubulcus ibis*. This heron crossed the South Atlantic from Africa under its own power, becoming established in northeastern South America by the late 1800s. From there it dispersed rapidly, and it is now one of the most widespread and abundant herons in the New World. (After Figure 11-19 [p. 353] from *Ecology and Field Biology,* second edition, by Robert Leo Smith. Copyright © 1974 by Robert Leo Smith. Reprinted by permission of Harper & Row, Publishers, Inc.)

revealed that only 3% of the 1048 marked butterflies moved from one population to another (Ehrlich, 1961, 1965). On the other side of the world, in Australia, certain species of morabine grasshoppers (Morabinae) live in exceedingly small areas, even on the particular gravestones in a cemetery, and their sedentary nature has promoted the genetic differentiation of local races and the formation of new species (White, 1973, 1978).

Passive dispersal. The vast majority of organisms disperse largely or solely by some passive means. Plants are excellent examples because only passive means are found. In any

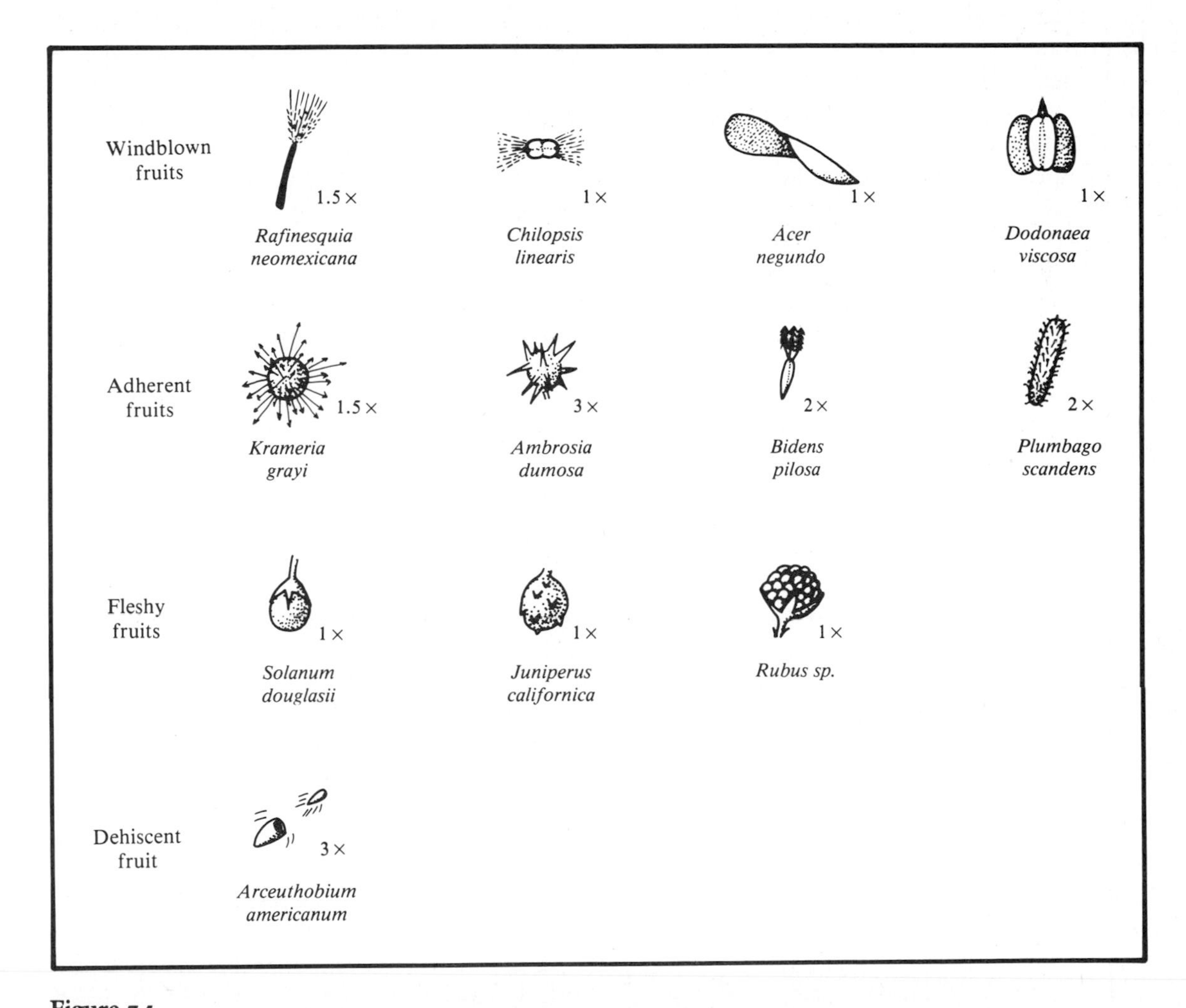

Figure 7.5
Variety of "fruits" from North American plants designed to enable seeds to disperse from the mother plant. Fruits that are carried by wind often bear hairlike wings *(Rafinesquia* and *Chilopsis)* or papery wings *(Acer* and *Dodonaea)*; those able to become attached to the bodies of animals have barbs *(Krameria* and *Bidens)*, sharp spines *(Ambrosia)*, or sticky glands *(Plumbago)*; those meant to be eaten and later defecated by birds and mammals are fleshy *(Solanum, Juniperus,* and *Rubus)*. In an unusual case, seeds of dwarf mistletoe *(Arceuthobium)* are explosively discharged from a fleshy base.

plant community, we can easily observe dispersal of diaspores (seeds, spores, fruits, or other propagating materials) from the mother plant. Wind carries seeds and fruits with attached wings, hairs, or inflated processes (Figure 7.5). Birds and mammals consume fleshy and dry diaspores, scattering some of them during feeding, distributing others later with their feces. The seeds and fruits of some species become attached to the feathers or the fur of animals and ride as hitchhikers until they are dislodged accidentally or by grooming. Some diaspores are explosively released for short distances, whereas others simply fall to the ground at the base of the plant. On the ground, ants, rodents, and birds compete for fallen seeds and may carry them considerable distances, in some cases many meters. Finally, flowing water, tides, and ocean currents may displace the seeds and fruits and carry them long distances (Stebbins, 1971a; Pijl, 1972).

Some of the aforementioned means of passive dispersal by plants or animals are obviously more effective than others. Aerial movement by winds, by birds, and in the tropics, by bats is especially successful in moving disseminules over long distances. Of course, the distance passively traveled in a generation for a spore, seed, or other disseminule depends on the velocity, distance, and direction moved by the dispersal agent. Animal transporters may be territorial residents, whose home ranges are fairly limited, or they may be nomadic or migratory, leaving their original habitats and covering great distances. Wind tends to have local effects except in violent storms, during which organisms can be widely disseminated. Jet streams are also alleged to be important dispersing air currents, especially for tiny spores and invertebrates (Gressitt, 1963; Clagg, 1966). Such so-called aerial plankton also includes tiny mites, spiders, and insects that do not necessarily have special stages adapted for dispersal.

Passive mechanisms for dispersal are not necessarily completely haphazard or random. Often certain stages of the life cycle are specialized for the journey. Some organisms produce special structures to facilitate their movement and are resistant to destruction by the physical conditions and biological enemies they are likely to encounter while crossing areas of unsuitable habitat.

The structural features that promote long-distance or waif dispersal have long fascinated naturalists. For bacteria, yeasts, fungi, mosses, and ferns, dispersal is effected by tiny spores that not only are readily transported by wind and water but also are highly resistant to extreme physical environments. They are light enough to float in water or remain aloft in the air for long periods. Many of the lower invertebrates also have small disseminules. Often, fertilized eggs or other life history stages encyst to form thick-walled structures that are metabolically inactive and capable of withstanding desiccation and wide ranges of temperature. Groups such as some protozoans, rotifers, tardigrades, worms, and crustaceans disperse over long distances by such means. For example, the brine shrimp, *Artemia,* lives in highly saline pools and lakes throughout the world. When these bodies of water dry up, the brine shrimp survive for months or years as encysted fertilized eggs until the pools refill, when they resume their life cycle. These dry eggs may be purchased at pet stores, and one need only add them to salt water to rear a new generation of *Artemia*. The tiny encysted eggs are also picked up by the wind, however, and dispersed great distances. It is apparently because of such long-distance dispersal that *Artemia* is found in isolated localities such as the Great Salt Lake in the western United States and the Dead Sea in Israel.

For larger dispersal units, the structures are obviously less buoyant and therefore require surface features to keep them aloft in the wind currents. We mentioned earlier the aerial adaptations of plants (Figure 7.5); some invertebrates have equally innovative designs, including gossamer parachutes in certain spiders (Aroneae, Arachnida) and long dorsal filaments in mealybugs (Hemiptera).

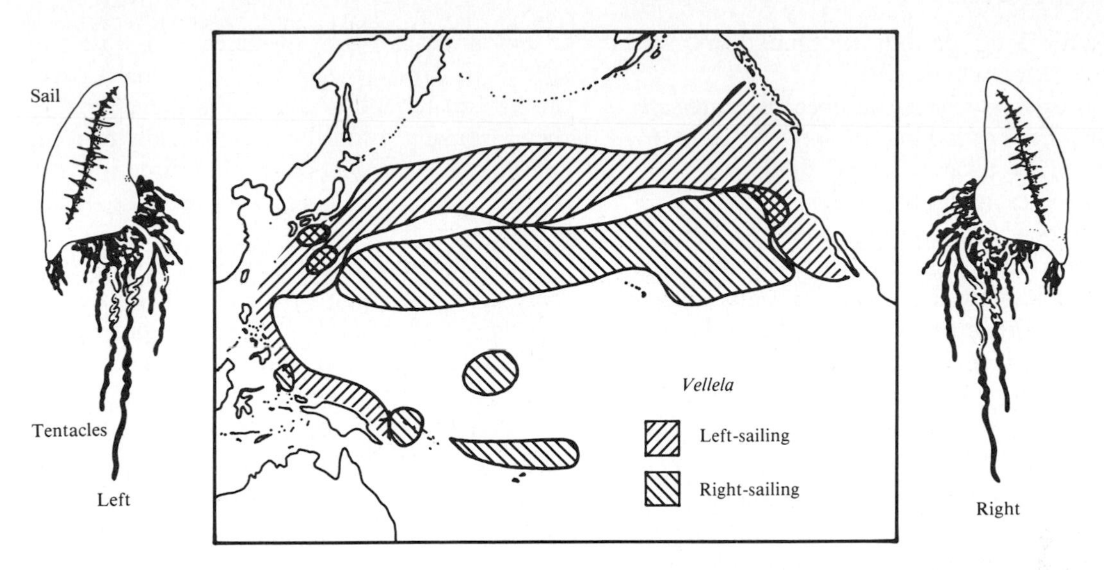

Figure 7.6
Coelenterates *Physalia* (Portuguese man-of-war) and *Velella* possess floats oriented with respect to their bodies so they tend to sail either to the right or to the left. Each form is restricted to certain regions of the ocean, and it is thought that by sailing in the wind these animals tend to counteract the tendency to be carried in ocean currents and thus remain within a limited area. (After Savilov, 1961.)

Most marine organisms have juvenile stages that are planktonic, i.e., free-living near the surface of the water. Tiny animal eggs and larvae, fish embryos, algal cells, and the like drift passively in ocean currents. Distances traveled by small versus large forms may differ, but all are controlled ultimately by the gyral and more local circulation patterns of the oceans (Chapter 2). Strong currents may move forms quickly from one locality to another in a definite direction, and the wind that drives ocean circulation causes dispersal of unusual species such as the adult jellyfish, the Portuguese man-of-war (*Physalia physalis,* Cnidaria), which is equipped with "sails" that are oriented to carry the individuals in somewhat predictable directions (Figure 7.6). The larvae of this species drift in the plankton.

Although terrestrial vertebrates are generally too large to be carried far by the wind, some are transported for surprising distances by water, often as passengers on mats of drifting vegetation or other debris, called rafts. Rafts have been recorded with entire trees and carrying a large variety of organisms (Ridley, 1930; Carlquist, 1965). Rodents and especially lizards apparently have managed to colonize many isolated oceanic islands by these means. Such large rafts are washed out to sea from large tropical rivers and may not break up for extremely long distances. Although the chance of a particular raft going ashore on a given island is admittedly very small, over sufficiently long periods of evolutionary time there is a high probability of such an event occurring. It is by the chance occurrence of such individually unlikely episodes of passive long-distance dispersal that oceanic islands and many other isolated habitats acquire much of their distinctive biotas.

A less haphazard means of dispersal by seawater is found in certain plant species, especially those of beaches and mangrove swamps, which have fruits, seeds, or vegetative parts that float in salt water and remain viable when submerged for long periods of time (Carlquist, 1965, 1974; MacArthur, 1972).

A surprisingly large number of organisms are transported over long distances by other organisms. This process is called phoresy, and many of the examples are obvious. Parasites are carried to new areas by their hosts. As Europeans explored and colonized the earth, bacterial and viral diseases were spread. In some isolated areas, populations of native peoples had never been exposed to these microbial pathogens and evolved no immunity to them. Consequently, diseases such as smallpox and measles had devastating effects when introduced among American Indians and Pacific Islanders (McNeill, 1976; Marks and Beatty, 1976). Some plants have sticky or barbed seeds or fruits that adhere to the integuments of mobile animals, such as birds and mammals (Figure 7.5). Aquatic birds may transport small invertebrates in their feathers or trapped in mud on their feet between widely separated lakes and ponds. Other seeds, usually those found in sweet, fleshy fruits, are adapted to pass through the digestive tracts of animals and, being resistant to the digestive juices, are still viable when dropped in the feces. In fact, the germination of certain plant seeds is enhanced when exposed to animal digestive chemicals. Internal transport is extremely common and important not only for colonizing isolated oceanic islands but also for dispersing in continental regions, transferring diaspores from one habitat to another. Birds are the usual agents, but mammals and reptiles also serve as useful internal seed dispersers.

A few animals are deliberate hitchhikers. Colwell (1973, 1979) and his associates have studied tiny mites that live in flowers. These mites do not have a highly resistant stage in their life cycle and seem poorly adapted for long-distance dispersal. However, the mites occur only in flowers regularly visited and pollinated by hummingbirds, and they are transported between flowers on the birds' bills. They crawl into the nasal openings where the protected microclimate enables them to survive and to be carried for long distances. At least two species of these specialized flower mites occur on the Greater and Lesser Antilles. Presumably, they originally colonized these islands from the continental mainland of tropical America, riding hundreds of kilometers on the bills of dispersing hummingbirds.

It is also interesting that some species have more than one type of disseminule: juveniles and eggs versus adults, vegetative propagules versus seeds and fruits, and sometimes even different forms of the same life stage. Examples of the latter can be found in both plants and animals. In plants, some seeds may be lighter and equipped with special structures for wind dispersal, whereas others on the same plant may be heavy and unlikely to move far from the parent (Figure 7.7). More drastic heteromorphic cases are found in insects, in which winged and apterous forms are produced in a single population or at different times of the year (Udvardy, 1969) (see also Chapter 12).

Once again we must point out that accounts of passive dispersal tend to highlight taxa with peculiar and superior properties that enable them to travel over significant distances and barriers. Emphasizing such one-step colonizers draws attention away from an equal or greater number of examples of taxa that migrate very slowly. Of course, each species disperses individually, but its successful colonization of distant areas may be facilitated if it travels in concert with other organisms and especially with other members of the community that facilitate its survival and reproduction.

One can observe such coordinated migrations on practically any shifting boundary between plant communities. For example, in some regions of North America temperate deciduous forest is gradually invading bordering grassland by casting out its propagules along the margin. One or a few pioneer species may take hold first, then others as shading and other changes favor forest forms, until eventually a stand of trees develops. Forest-dwelling animals migrate into the area as suitable habitats become avail-

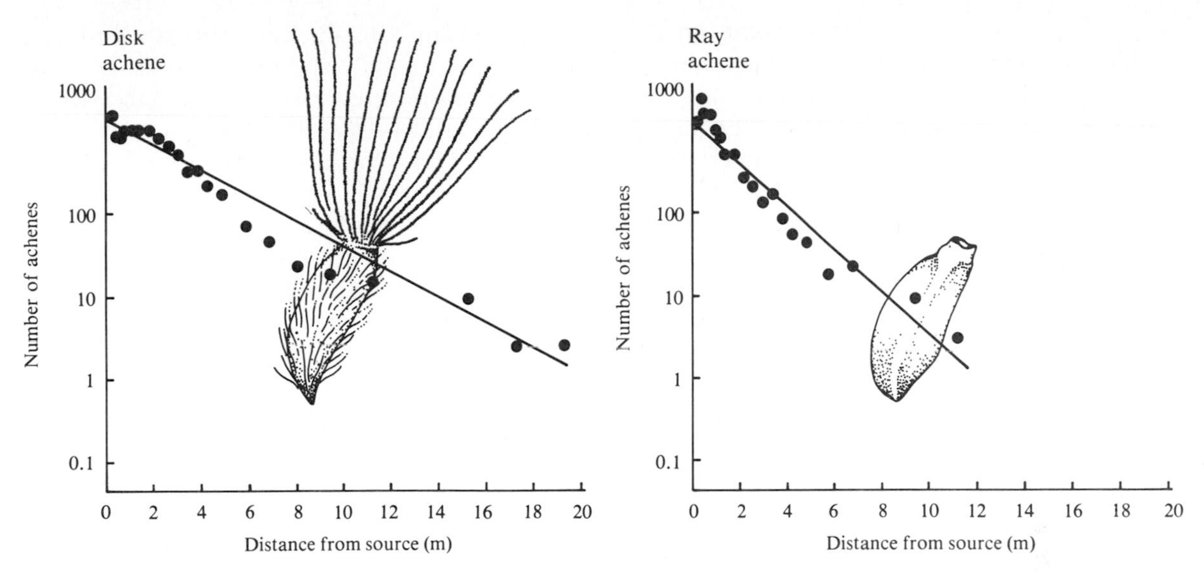

Figure 7.7
Seed dimorphism in the composite *Heterotheca latifolia* affects dispersal distance. This plant produces two kinds of fruits (achenes): disk achenes, which have an attached parachute-like structure that causes them to be carried by wind some distance from the parent, and ray achenes, which lack dispersal structures but have a thicker fruit wall and can survive for longer periods in the soil. (Courtesy L. Venable.)

able to provide proper food and cover. Meanwhile, grassland species of plants that need full sunlight cannot compete in shaded habitats. They are eliminated and replaced by shade-tolerant herbs and shrubs. This new assemblage of plants may also be temporary, a sere, yielding to another slowly migrating front until eventually a climax community is established. Although the details of succession are different for each plant community, range expansion of one community at the expense of another is a common phenomenon (Horn, 1974; Daubenmire, 1978; Maarel, 1979).

It may be stated (although it is a gross generalization) that in the majority of plant species seed dispersal is essentially a random process and most diaspores fall in the vicinity of the mother plant. An individual may end up less than a meter or several thousand meters from its parent, but the probability of traveling a relatively long distance is slight. Figure 7.1 shows a commonly observed pattern in herbaceous plants. The disseminules of this species are windblown achenes, but even so the majority of them land within 5 meters of the mother plant. This curve is decidedly leptokurtic, i.e., sharply peaked and having a long tail. The average distance traveled by tree seeds is greater but the pattern is similar. Aggressive and pioneer species also frequently show flatter curves, with greater numbers reaching distant areas. The shape of the curve, as well as the number of diaspores produced, will greatly influence the probability that propagules can cross distance barriers. On the other hand, the immigration of diaspores from distant areas has an important impact on population structure and the adaptation of these plants to local environments.

Even if plants and invertebrates do show highly leptokurtic dispersal curves, significant long-distance dispersal may still occur. If enough disseminules are produced, there is a small but significant probability that a few of them will be carried long distances by unforeseen events. One or two far-traveling disseminules may be extremely important biogeographic events, even if they occur less frequently than once per million times, because in most species of plants and invertebrates a population may produce many millions of offspring in each generation, and it only takes one or two successful dispersers to cause large range expansions.

The nature of barriers

Successful long-distance dispersal usually requires that organisms, while moving from one habitable area to another, survive for significant periods of time in environments very different from their usual sites. These unusual environments constitute physical and biological barriers that must be crossed by successful colonists. The effectiveness of such barriers in preventing dispersal depends not only on the nature of the environment but also on the characteristics of the organisms themselves. The latter, in particular, can vary from one taxonomic group to another so that particular barriers may not affect all residents of a habitat equally. Two examples will illustrate this point.

Most freshwater zooplankton have resistant stages that facilitate long-distance dispersal. Consequently, the same species are found in widely separated localities wherever environmental conditions are similar, such as in the cold temperate lakes of northern North America, Europe, and Asia. On the other hand, the fishes inhabiting these same lakes appear to be completely unable to disperse across terrestrial and oceanic barriers. Consequently, the same fish species inhabit only those lakes that have been connected by fresh water sometime in the past. Many isolated alpine lakes, which have never had such connections, support a diverse plankton fauna but have no fish at all (unless they have been introduced recently). Similarly, the isolated mountains of the southwestern United States are essentially islands of cool, moist forest in a sea of hot, dry desert. The deserts represent virtually no barrier for birds and bats, which rapidly and repeatedly fly over the deserts to colonize the montane forests. On the other hand, for small terrestrial mammals, reptiles, and amphibians, which must disperse much more slowly on foot, the deserts represent virtually absolute barriers and these taxa colonized the mountains only when they were connected by bridges of suitable forest habitat during the Pleistocene (J. Brown, 1978).

Because the effectiveness of a particular kind of barrier for a given taxon depends on both the physical and biotic environment provided by the barrier and the biological characteristics of the organism, it is difficult to make sweeping generalizations about the nature of barriers. It is usually true, however, that organisms that inhabit temporary or highly fluctuating environments are much more tolerant of extreme or unusual physical and biotic conditions than species that are confined to permanent or stable habitats. Plants and animals from temporary or fluctuating environments are also likely to have resistant life history stages, which not only enable them to survive periods of unfavorable conditions but also can serve as effective dispersal agents. Consequently, organisms from temporary or fluctuating environments are likely to be better dispersers and less limited by any kind of barrier than species from more permanent or constant habitats.

Janzen (1967) made this point in an insightful paper that addressed the more restricted distributions of species in mountainous areas in the tropics than in regions of comparable topographic relief in the temperate zones. He pointed out that mountain passes of a given elevation were effectively higher in the tropics than in the temperate zones. Because temperate environments experience great seasonal temperature variation, lowland plants and animals must be able to tolerate winter conditions that are very similar to those found at higher elevations in summer (Figure 7.8). In contrast, because of the great thermal constancy of the tropics, a lowland species would never experi-

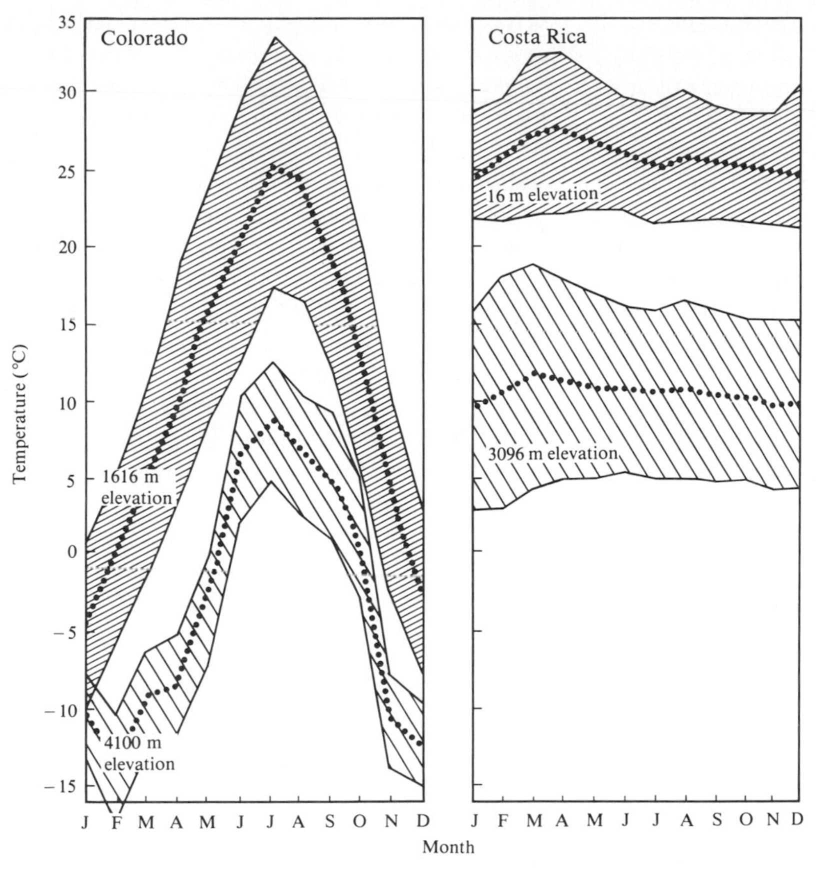

Figure 7.8
A given change in elevation tends to be a greater barrier to dispersal in the tropics than at higher latitudes, for the reason shown in these temperature profiles. In tropical regions sites separated by several thousand meters of elevation usually experience no overlap in temperature, whereas in the temperate zones winter temperatures at low elevations broadly overlap summer temperatures at much higher sites. (Reprinted from "Why mountain passes are higher in the tropics," *American Naturalist* 101:233-249, by D.H. Janzen, by permission of the University of Chicago Press. Copyright © 1967 by the University of Chicago.)

ence and would not be adapted to withstand the temperature regimes it would have to encounter if it were to travel over a pass from one lowland area to another.

A comparable example is provided by the pupfish *Cyprinodon variegatus,* an extremely euryhaline and eurythermal species that inhabits estuaries, tidal flats, and mangrove swamps along the eastern coast of North America from Cape Cod to Yucatán. This little fish has managed to cross hundreds of kilometers of ocean to colonize comparable habitats on many of the Caribbean islands. On the other hand, strictly freshwater fishes, such as the sunfishes *(Lepomis)* and basses *(Micropterus),* which are so abundant and diverse in the rivers, lakes, and streams of eastern North America are not native to the Caribbean, although successful introduction of many species by humans indicates that the saltwater barriers, rather than lack of suitable habitats on the islands, had prevented their colonization.

Probably the most severe barriers are presented by physical environments so far outside the range normally encountered that organisms cannot survive long enough to disperse across them. The vast majority of aquatic organisms live either in the oceans or in fresh water and cannot physiologically regulate their water and salt balance sufficiently to survive more than brief exposure to the other. Very few freshwater fishes have successfully colonized across the oceans. The same is true of most amphibians, which have permeable skins and are quite intolerant of exposure to salt water. Many terrestrial plants, too, are unable to withstand prolonged exposure to seawater at any stage of their life, including the seed. Hnatiuk (1979), however, reported that 56 of 69 terrestrial plant species native to Aldabra Atoll in the western Indian Ocean tolerated total immersion of their seeds in seawater for 8 weeks with no inhibition of germination. This tolerance would not only facilitate survival of these species on tiny islands subject to wave splash during heavy storms, it would also greatly increase the probability of these species dispersing across ocean barriers, thus perhaps explaining how many of them managed to colonize the islands in the first place.

Environmental temperature regimes can also serve as major barriers. We have already mentioned the mountain passes in the tropics. A similar situation is provided by the tropics themselves, which form a band of high temperature around the equator, isolating the cooler temperate and arctic areas toward either pole. For many cold-adapted organisms, both terrestrial and aquatic, these warm tropical climates have been a major barrier to dispersal. Some groups, such as the Nearctic avian family Alcidae (auks, puffins, guillemots, and murres) are good dispersers within their climatic zone and are well distributed around the world in one hemisphere, but they have not managed to cross the topics to colonize the other hemisphere (Figure 7.9), which is inhabited by forms of convergently similar morphology and ecology, such as the penguins (Spheniscidae) and diving petrels (Pelecanoididae).

Dispersing organisms not only must be able to physiologically tolerate the physical stresses imposed by the environments they traverse, they must also survive the biological hazards. Just as predation and competition can limit the distributions of species, they can also potentially prevent successful dispersal. One would expect biotic interactions to be particularly important components of barriers for these animals that neither move very rapidly nor disperse at a resistant stage. Although it is likely that competition and predation limit dispersal of such organisms, we are not aware of any well-documented examples.

Surprising as it may seem, psychological barriers appear to play a major role in preventing the long-distance dispersal of some organisms. Most organisms appear to have some mechanisms of habitat selection, the ability to recognize and respond appropriately to favorable environments. In some animals these traits are so well developed that they strongly inhibit most active dispersal. For example, some bird species that seem perfectly capable of flying long distances are apparently unwilling to cross certain kinds of barriers. Willis (1974) describes species of antbirds (Formicariidae) that have become extinct on Barro Colorado Island and not recolonized, even though they would have to fly only a few hundred meters across Lake Gatun. MacArthur et al. (1972), Diamond (1975b), and others document groups of strong-flying tropical birds, such as the New World cotingas and toucans and the Old World barbets and pittas, which are repeatedly absent or poorly represented on oceanic islands, even when good habitat appears to be abundant. Often species restricted to virgin rain forest are particularly sedentary, whereas other, even closely related species characteristic of second

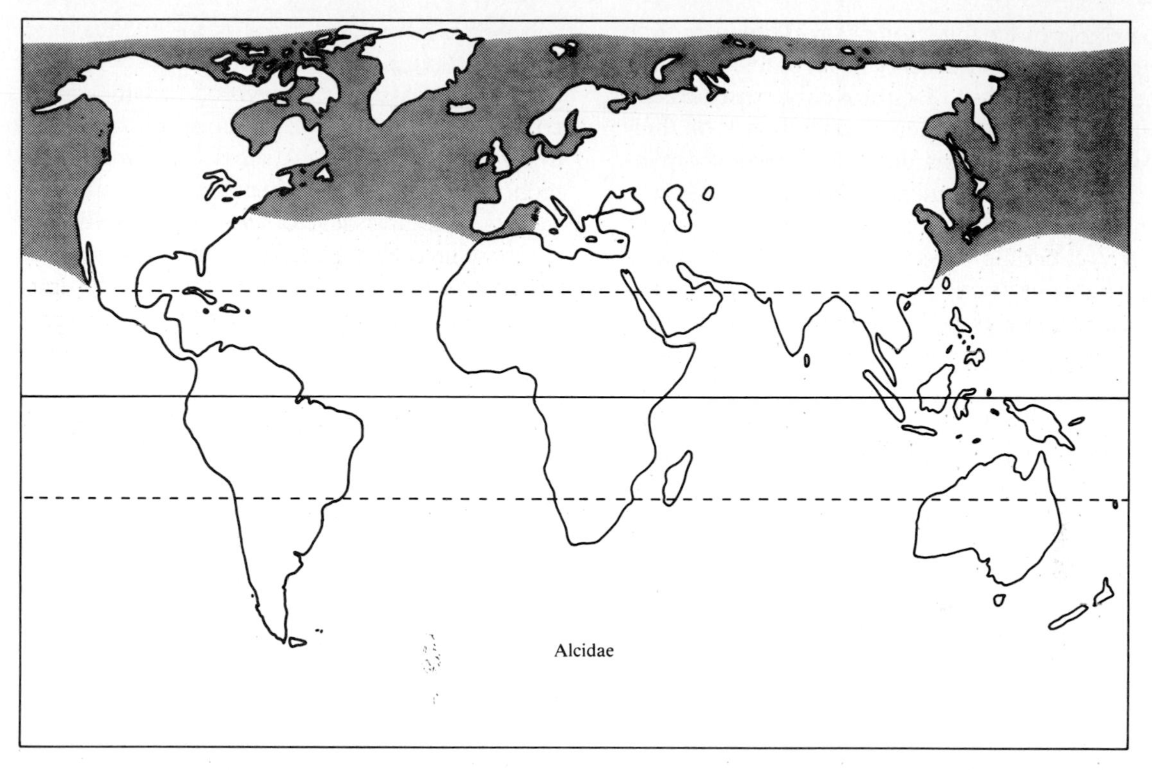

Figure 7.9
Avian family Alcidae (auks, puffins, and murres) is restricted to the cooler regions of the Northern Hemisphere, although all of these birds are winged and most are strong fliers. The tropics appear to constitute a major barrier to dispersal of this group, although the presence of distantly related but ecologically convergent seabirds in the Antarctic region may also prevent their colonizing the Southern Hemisphere. (After Shuntov, 1974.)

growth habitats are good dispersers. This pattern suggests that species adapted for successional habitats, which must continually disperse to new environments as their habitat patches succeed and become unsuitable, are much more likely to strike out across barriers and hence to disperse successfully for long distances than species that are restricted to very stable environments.

Establishing a colony

If dispersal is to be of biogeographic significance, an organism not only must be able to travel long distances and cross geographic and habitat barriers but also it must be able to establish a viable population after its arrival in a new area. As Carlquist (1965) has stated, "Getting there is half the problem." Obviously, the successful colonist is one that survives and reproduces. Several factors that may play little role in dispersal per se may be of great importance in determining the fate of a disperser once its journey is over.

Habitat selection. Habitat selection is often important in enabling the organism to find a habitat or a microenvironment in which it can survive and reproduce. All organisms exhibit some form of habitat selection. Highly mobile organisms actively seek out favorable environments, which they recognize either instinctively or from having learned characteristics of their place of origin. Many passively dispersed organisms, such as the planktonic larvae of sessile marine invertebrates, will not settle unless they perceive certain sensory cues that indicate a substrate is suitable for establishment. Even the seeds and spores of plants and the cysts of invertebrates have some capacity for habitat selection. Although these resistant structures are completely passively dispersed, certain specific conditions of temperature, moisture, light, and other factors are nevertheless usually required to break the diapause (dormant stage) and initiate growth and development. Unless cues indicating favorable conditions are received, the diapause remains unbroken and the resistant stage is capable of further dispersal.

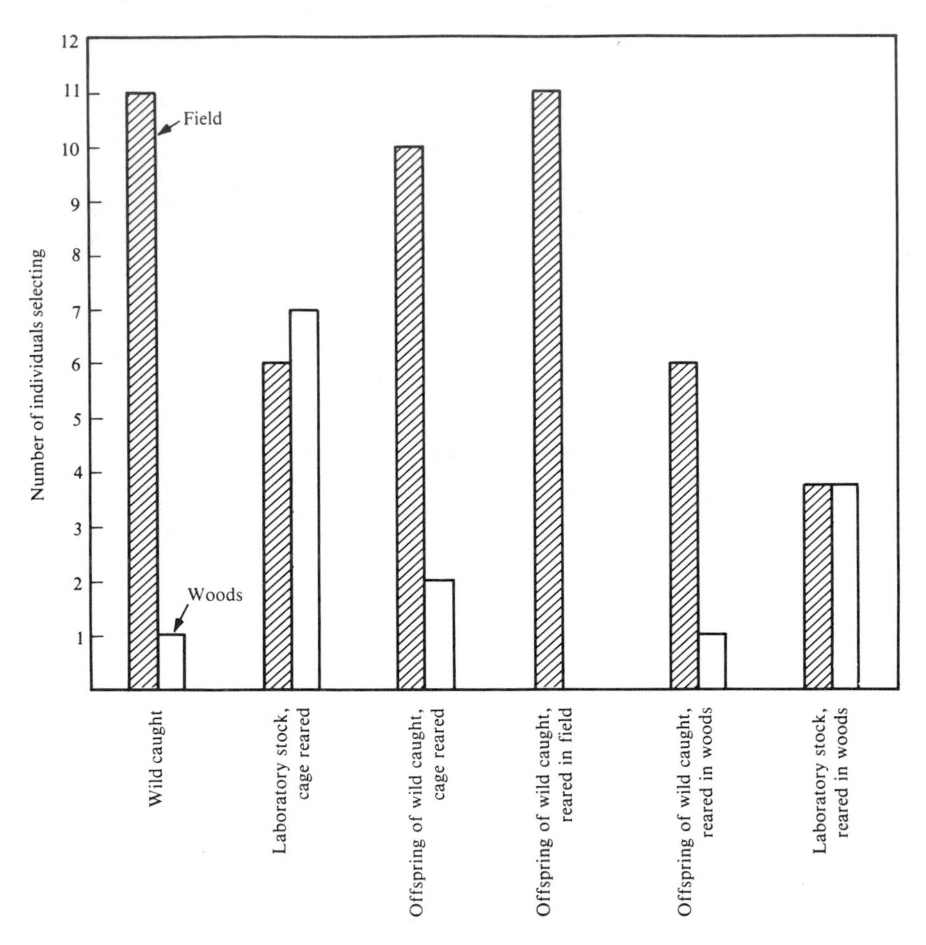

Figure 7.10
Habitat selection by the field race of the deer mouse, *Peromyscus maniculatus bairdi,* depends on both inherited ability and early experience. Field-caught individuals and their offspring, especially if they were exposed to field habitat at the time of weaning, spent most of their time in the field end of a large enclosure that covered a field-forest boundary. Many generations in the laboratory or early experience in the woods diminishes this preference. (After Wecker, 1963.)

Wecker (1963, 1964; see also Harris, 1952) performed a classic study of habitat selection in deer mice *(Peromyscus maniculatus)*. He worked on two subspecies of this species that inhabit the north central and northeastern United States: *Peromyscus maniculatus gracilis,* a form primarily restricted to mature, mesic forests, and *P. maniculatus bairdi,* which inhabits grasslands. The latter subspecies is of particular interest because it is one of those animals that originally inhabited the prairies of central North America but extended its range eastward and colonized agricultural cropland and old fields as the forests were cleared by Europeans (Hooper, 1942; Baker, 1968). In a series of experiments in which he reared mice in different environments and then tested them in a large outdoor enclosure that was half old field and half forest, Wecker showed that each subspecies tended to wander and explore until it found its appropriate habitat, whereupon it would establish residence (Figure 7.10). The mice use genetically inherited information, supplemented by early experience in their natal habitat, to select the correct environment. Because forest and field habitats form a temporally fluctuating patchy mosaic throughout the northeastern United States, this precise habitat selection probably is of major importance in the successful dispersal of deer mice into these successional habitats.

What constitutes a propagule? In addition to finding the right kind of habitat on arrival, successful colonists must also be able to reproduce and establish a viable new population. The size and composition of the dispersing unit are of critical importance in this regard. MacArthur and Wilson (1967) and others have used the term *propagule* to define the unit necessary to found a new colony. The question of what constitutes a propagule is a major problem for many higher plants and animals that reproduce sexually, because reproduction normally requires the participation of two individuals of different sexes or mating types. Since long-distance dispersal is a rare event, it is relatively unlikely that potential mates will arrive sufficiently close to each other in space and time so as to find each other and reproduce. On the other hand, asexually reproducing organisms are often successful colonists, because a single individual can found an entire population by fission, budding, or some other means of asexual reproduction. Although many microbes and lower plants and animals are capable of sexual reproduction, many of them can also reproduce asexually. Thus, for example, many of the crustaceans that comprise the freshwater zooplankton can go through many generations of asexual reproduction after emerging from encysted zygotes (which are produced by a sexual generation that occurs periodically during the life cycle). Many plants also are capable of asexual reproduction by budding from either the roots or shoots.

Even among groups of normally obligately sexual higher plants and animals there are species with reproductive patterns that make them much better colonists than others. A large number of plant species are both hermaphroditic and self-compatible; i.e., the same individual has flower parts of both sexes and can produce viable seeds by self-fertilization. This is true of many species, such as tumbleweed *(Salsola iberica),* that we normally think of as weeds. Dandelion *(Taraxacum),* another common weed, is apomictic; i.e., it can produce viable seeds from unfertilized ovules. Other weedy plant species, including Bermuda grass *(Cynodon),* are capable of extensive asexual reproduction by means of tillers or rhizomes. In all of these kinds of weeds a single seed can serve as a propagule and not only travel long distances across major barriers but also found a population in a new site. A few species of fishes, amphibians, and reptiles are parthenogenetic, and the entire population consists solely of females that produce female offspring asexually (White, 1973; Cuellar, 1977). Some fish can also change their sex, depending on the sex of other members of their local population. Although these reproductive patterns do not necessarily mean that the organisms will be good colonists, they appear to facilitate successful dispersal of some species. Thus, for example, at least three genera of gecko lizards have parthenogenetic species that are distributed primarily on oceanic islands (Cuellar, 1977). For these animals it only takes a single individual, drifting ashore on a floating log or arriving in an islander's canoe, to establish an entire new population.

In asexual species, the unit required for successful colonization of a new area is a single individual, whereas in sexual species, it is at least two individuals of different mating types. Some sexual species have reproductive cycles and social behaviors that facilitate simultaneous colonization by several individuals. In animals with internal fertilization, a propagule can consist of a single female with stored sperm or developing embryos. In plants in which several seeds are normally retained and dispersed in a fruit, this can be the unit of dispersal even in self-incompatible, obligately sexual species. Some birds, which have virtually none of the characteristics listed above, are good dispersers and successful colonists not only because they can fly long distances, but also because they normally travel in flocks, which obviously increases the chances of establishment of viable populations in new locations. Another widely distributed vertebrate in the tropical Pacific is the fruit pigeon *(Ducula),* a medium-sized, highly social bird that has been observed traveling in flocks far out to sea (Diamond, 1975b).

Even if a viable propagule arrives in an apparently suitable environment, successful colonization is not assured. Both stochastic (random) and predictable processes play a large role in determining its fate. When small numbers of individuals are involved, as is almost always the case in long-distance dispersal, chance becomes very important in determining the course of events. Colonies may fail because a freak storm or an unlikely predator kills just one or two individuals.

Because long-distance dispersal almost inevitably means organisms colonizing habitats that are not exactly like those from which they originated, successful colonists must be able to survive physical stresses and biological hazards to which they may not be adapted. Organisms from highly fluctuating and unpredictable environments probably tend to be good dispersers not only because they can tolerate stressful conditions encountered en route but also because they are well prepared to meet the unknown physical and biotic challenges they must face after they arrive. The biological hazards posed by competitors and predators (including parasites and diseases) should not be underestimated. A common pattern in dispersal is that organisms from large, continuous areas with diverse ecological communities, such as productive continental habitats, are relatively successful in colonizing small, isolated habitats with few species, such as oceanic islands. On the other hand, insular species rarely successfully invade mainland habitats. Presumably this is because the insular organisms are not adapted to cope with the threats posed by a diverse array of competitors and predators. This point will be developed further in Chapters 16 and 18.

Dispersal routes

So far our discussion has concentrated on long-distance waif dispersal across major geographic and ecological barriers. Such events are spectacular and undoubtedly extremely important in accounting for the distributions of many taxa, especially those species inhabiting oceanic islands or having almost cosmopolitan distributions. It is likely, however, that the vast majority of dispersal events are of another type: rather than crossing existing barriers, a species simply shifts its range. Sometimes peripheral populations may be able to adapt to conditions at the margin of the range and colonize new areas. More often, species probably follow changing environmental conditions. They expand into new regions as the environment becomes favorable and their previous range contracts as conditions deteriorate.

Examples of such movements abound. Throughout the temperate Northern Hemisphere plants and animals moved southward as Pleistocene glaciers advanced and the climate cooled and then retreated northward again with the return of warmer conditions (Chapter 14) (see also Figure 4.10). Most freshwater fishes are not only intolerant of salt water, they are also incapable of crossing land barriers. Nevertheless, many groups have spread over much of the world by moving from one body of water to another as stream capture and other geologic events have created aquatic connections.

Corridors. Dispersal or range shift in response to changing environmental conditions differs from waif dispersal across barriers in that many taxa are likely to use the same routes and to exhibit geographic ranges that reflect historical patterns of environmental change. Long-distance waif dispersal is a highly stochastic process in which the probability of each species crossing a barrier tends to be independent of the dispersal of other taxa. In contrast, when the climate changes or previously isolated landmasses or water masses come into contact, many species are affected similarly, and they may disperse almost synchronously along the same routes. Historical dispersal along such routes provides the most parsimonious explanation for many patterns of similar distributions in diverse taxonomic groups.

For land organisms, those that must use land during part of their life cycle, the easiest explanation to account for past biotic movement is the demonstration of past land connections. The same functional explanation, namely the continuity of suitable habitat, likewise applies to marine and freshwater organisms. To account for large-scale changes in the land and sea topography of the earth, four mechanisms have been proposed: (1) raised and lowered sea levels, (2) ancient sunken continents, (3) raised and sunken land bridges and island arcs, and (4) broad connections of continents now drifted apart. Even when land connections are present, however, they can affect the movement of organisms in different ways.

The term *corridor* was proposed for a route that permits the spread of many or most taxa from one region to another (Simpson, 1936, 1940). A corridor therefore allows a taxonomically balanced assemblage of plants and animals to cross a narrow region from one large source area to another, so that both areas obtain elements that are representative of the other. The corridor does not selectively discriminate against any form and must therefore provide an environment similar to the two source areas; otherwise such exchange would not occur.

Simpson's term was devised to describe a situation in which two different source areas come in close enough proximity to exchange biotas. Udvardy (1969) has adopted a different concept of corridor, also used by other authors. In this concept, a corridor is a broad band of continuous habitat, e.g., a biome. Organisms restricted to the habitat can have wide ranges and can disperse across great distances without having to cross major environmental barriers. Udvardy's example is the transcontinental temperate deciduous forest that extended uninterrupted from western Europe to eastern China during part of the Cenozoic. Although this once continuous pathway has since become fragmented, China and France still share a number of bird species.

Although most authors prefer to maintain Simpson's definition of corridor as an unselective bridge between two biotas, Udvardy has described a very common situation. We can find many examples in the fossil record. The ancient Tethys Sea, a marine system, extended all the way from the Orient and Malaysia to westernmost Europe, separating Africa from Eurasia. For nearly 500 million years the Tethys Sea served as a highway for many marine organisms, including benthic and pelagic forms of deep and shallow water, to disperse across the 120° of longitude (Adams and Ager, 1967). The exchange of taxa within the Tethys was not always uniform from one end to the other but it was effective over long periods, and this exchange was disrupted about 60 million years ago when Africa became connected with Asia through Arabia and when India drifted northward to join Asia. Now the Mediterranean Sea has a vastly different fauna than the Indian Ocean.

Do broad, continuous vegetational zones on continents (biomes) facilitate free dispersal between distant regions? The answer to this question has to be a highly qualified yes. Certain species are highly mobile or vagile, so individuals may traverse great distances over a lifetime. These species may be able to use habitats as a passageway from one place to another. Examples of this are the large carnivorous and herbivorous mammals, such as the moose *(Alces alces)* and the wolf *(Canis lupus)*, which are distributed around the world in northern coniferous forest habitats. In contrast, small mammals are quite sedentary, with small home ranges; therefore peripheral populations may become relatively isolated from those in the center and may become highly differentiated. A case in point is that of the small mammals of these same northern coniferous forests (Banfield, 1974). Careful examination of distribution maps reveals that many genera have two or more species in the forests (Figure 4.1) and widespread species have several, even many, differentiated

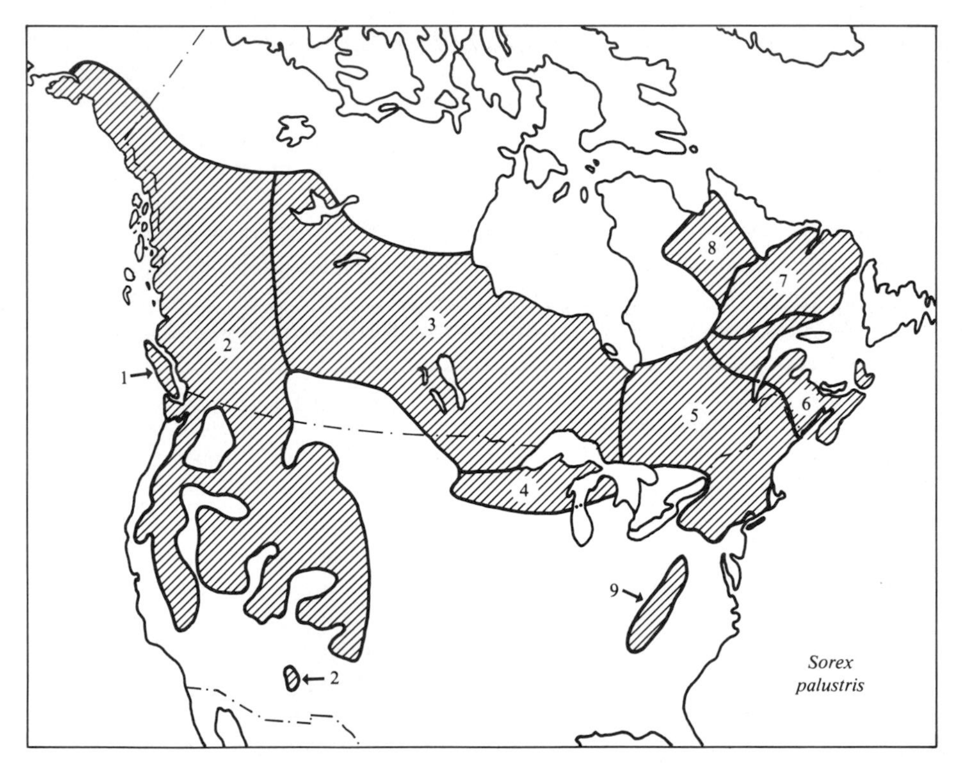

Figure 7.11
The distribution of the North American water shrew, *Sorex palustris,* a species confined to the northern part of the continent that is covered primarily by coniferous forest vegetation. Despite the fact that both this habitat and the distribution of the shrew are relatively continuous, *S. palustris* has differentiated into numerous recognized subspecies. (After Hall, 1981.)

subspecies (Figure 7.11). In addition, over half of the widespread species have broad distributions in other biomes. Similar patterns of differentiation within the forest belt can be demonstrated even for aerial animals, e.g., birds and insects (Kendeigh, 1974), which might be expected to be broadly distributed because of their potential for wide dispersal. From these observations we conclude that the expansion of a biome does permit some forms to extend their ranges, but continuous habitat does not ensure migration of many, even a majority of the species, from opposite ends. Moreover, by treating present-day biomes as continuous "corridors," we overlook important historical events that have caused the differentiation of species.

The same observation applies to marine systems. Many genera, in fact a great number of vertebrate and invertebrate species, have Indo-Pacific distributions extending from the Red Sea and eastern Africa to far eastward in the western Pacific Basin (Figure 7.12). This is not to say that each geographic area has the same fauna, because the eastern and western margins especially are extremely depauperate in comparison to regions around the islands of southeast Asia. The world's highest diversity of species for marine bivalves, hermatypic corals, coral reef fishes, and echinoderms occurs in Pacific waters around the Philippines.

Many ancient connections have been suggested for northern and southern continents and for the Old and New World, respectively. Where continents were connected but have since drifted apart, the broad regions of attachment should have allowed a balanced biota to be distributed throughout. The fossil record supports this prediction in cases where ade-

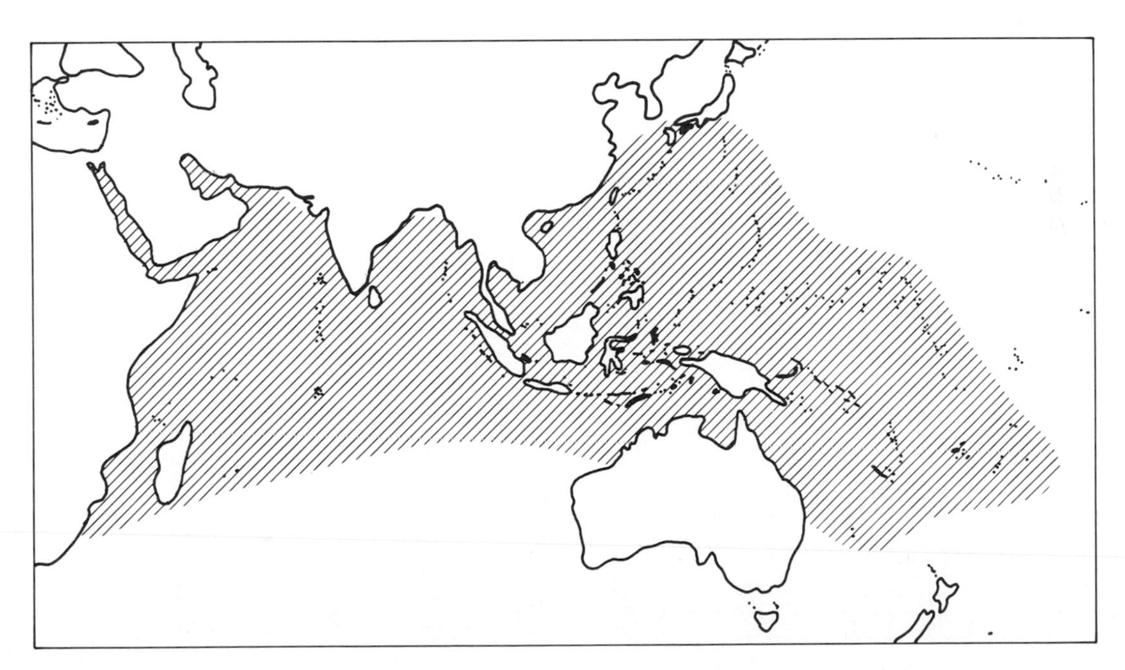

Figure 7.12
Distribution of the giant clam family Tridaenidae. Like other Indo-Pacific taxa this shallow water group has a wide range from the eastern coast of Africa to the islands of Melanesia and Polynesia, but it does not reach the eastern Pacific or the Mediterranean Sea. (After Rosewater, 1965.)

quate data are available on both the geologic connections and the distributions of organisms. The Bering Land Bridge between Alaska and Siberia was a two-way passageway for plants and animals throughout the Cenozoic and during the Pleistocene, but this probably was a corridor only during the first half of the Cenozoic, when a mild climate prevailed. Here, as in other land bridges on poleward margins, cold periods and breaks in the bridge must have stopped many forms from crossing.

Filters. If at least some barrier to dispersal exists between regions, the term *filter* is used. As the name implies, a filter is more restrictive to dispersal than a corridor and blocks passage of certain forms while allowing those able to tolerate conditions of the barrier to migrate freely. Following such dispersal, biotas on the two sides of a filter are generally fairly balanced or harmonic, sharing many of the same taxonomic groups, although some taxa will be conspicuously absent in each.

The Arabian subcontinent is a harsh filter that permits certain mammals, reptiles, nonpasserine ground birds, invertebrates, xerophytic plants, and other groups to maintain fairly continuous distributions between northern Africa and Central Asia; however, organisms that need abundant water, such as freshwater fishes, most amphibians, and forest dwellers in many groups, are stopped by the deserts. This has not always been true. As mentioned above, for millions of years Arabia and Asia were widely separated by the Tethys Sea, a formidable barrier to land inhabitants. When the Arabian Peninsula became emergent and contacted Persia, the region intermittently served as a benign land bridge for movements of Asian taxa into Africa and vice versa, e.g., in the Upper Miocene, when many African forms appeared in Asia for the first time (Cooke, 1972). Thus a filter of today may have been more like a corridor or a formidable barrier in the past and may become either in the future.

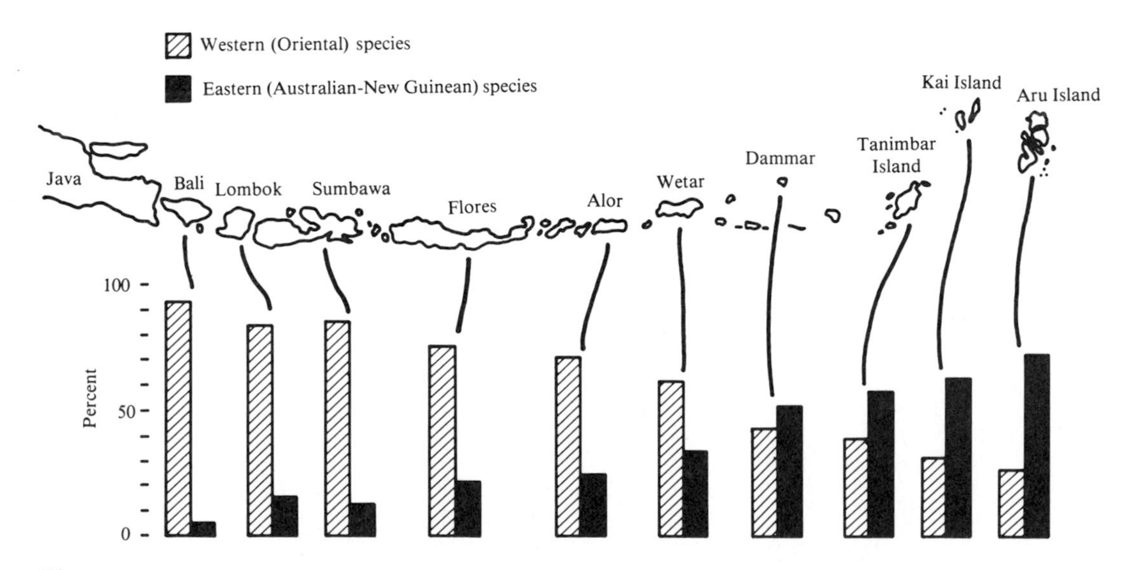

Figure 7.13
Lesser Sunda Islands between Java and New Guinea serve as a two-way filter for the reptilian faunas of southeastern Asia and Australia. This diagram quantifies the decline in Oriental species and the increase in Australian species going from west to east down the island chain. (After Carlquist, 1965.)

Filters may be produced by abiotic or biotic factors. Often they exhibit discontinuities of habitats, as in an island arc, so that some long-distance dispersal is required. The greater the barrier, the harsher the filter. Filters are generally easy to identify because the number of species in a taxon decreases gradually or abruptly from the source area. Figure 7.13 depicts an example observed in the islands between Java and New Guinea. Reptiles of Oriental origin proportionately decrease eastward, where Australasian groups are dominant, and, of course, the same decreasing trend is found for Australasian groups progressing westward. This is an example of a two-way filter bridge often found between two large source areas.

Another two-way filter bridge is the Isthmus of Panama. When the isthmus emerged 4 to 6 million years ago, fusing Central America with Colombia, organisms from both sides of the barrier were given an opportunity to migrate, resulting in a major biotic exchange. Most groups of mammals from southern North America appeared quickly in South America, and those of northern South America ventured northward (Figure 7.14). One assumes that other terrestrial groups behaved similarly, e.g., large-seeded vascular plants, but the data are not so complete. Nonetheless, to this day some groups have not been able to cross the isthmus. Consider animals that cannot tolerate seawater, the primary division freshwater fishes. As seen

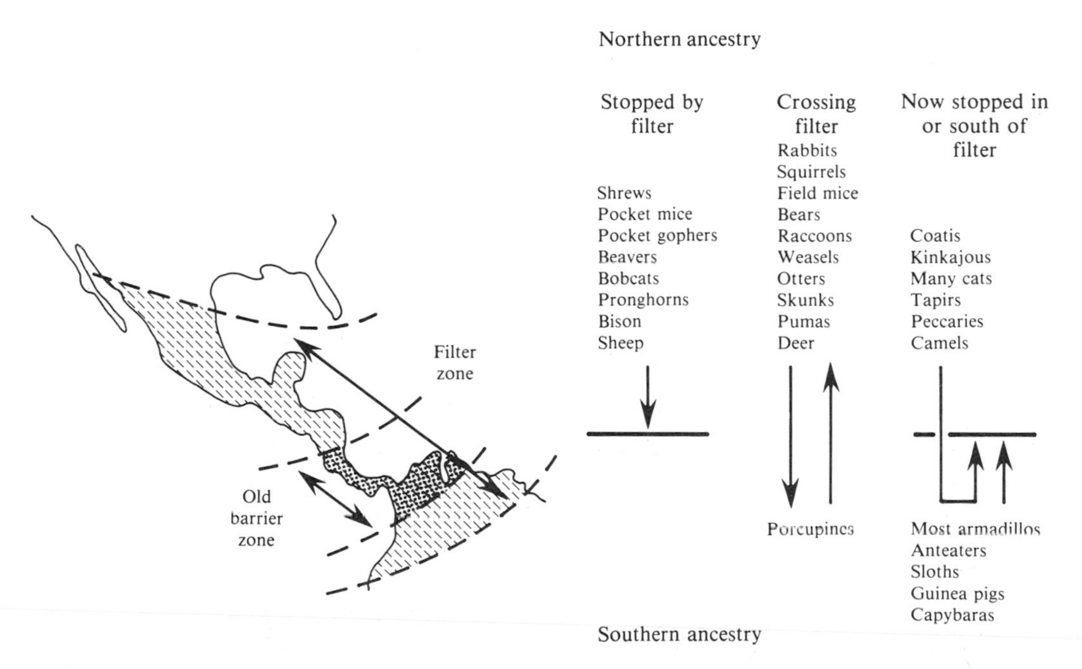

Figure 7.14
Central America has acted as a filter, allowing extensive interchange of some mammalian groups between North and South America but preventing the passage of other taxa. This diagram shows those groups of northern ancestry that both were stopped by and passed through the filter, the same groups for taxa of southern ancestry, and a final group, originally of northern ancestry, that invaded and survived in South America but became extinct in North America. (From *The Geography of Evolution* by G.G. Simpson. Copyright © 1965 by Chilton Book Co., Radnor, Pennsylvania.)

in Figure 7.15, only a few families of North American fishes occur in Central America and South America, and the reverse is likewise observed. Why have freshwater fishes been stopped? The answer lies in the topography of the area. Central America has a volcanic backbone, so the rivers flow predominantly toward the sea and not along the land axis. Consequently, there are no continuous waterways of fresh water from Nicaragua to Colombia. The only fishes able to pass through the isthmus are those slightly tolerant of seawater and those with special adaptations for overland dispersal, such as the South American bloodsucking catfishes (Trichomycteridae). Some of the Central American fishes may also be relicts dating back to periods when earlier versions of Central America had connections with the con-

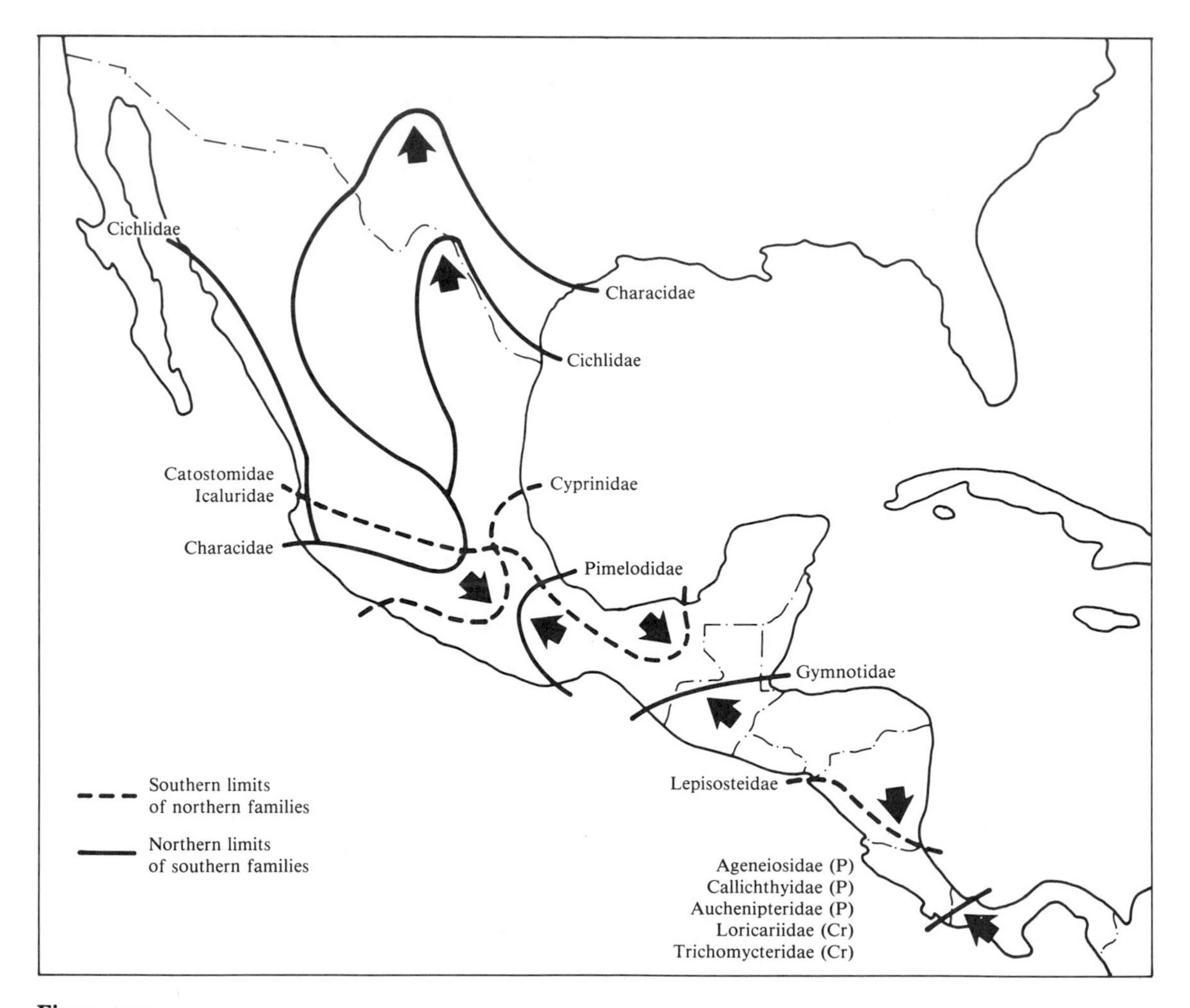

Figure 7.15
Distributional limits of certain families of freshwater fishes in Central America. For these groups this region has been a harsh filter. Only two species of primary division fishes of South American origin have reached the United States, and primary division forms from North America have dispersed only as far south as Costa Rica *(arrows)*. (After Miller, 1966.)

tinents (Miller, 1966; Rosen, 1975; Bussing, 1976).

Other kinds of routes. The term *sweepstakes* was coined by Simpson to describe chance dispersal from one locality to another across major barriers. In a sweepstakes, many individuals enter the contest but only a handful of lucky ones win prizes. In the natural world, many propagules continually disperse from established populations into new areas, but only a small fraction of these seeds, spores, juveniles, or adults is ever successful in founding a new population. As pointed out earlier, the greater the distance and the barrier the less chance of arriving at a new locality, and only those species or groups with features permitting long-distance jumps and tolerating physiological hardships have any chance of arriving in a remote area. Given enough time, however, colonization by some forms will occur. However, even when sweepstakes dispersal occurs and organisms cross barriers independently of each other, they may still use the same routes. Insular biotas, for example, usually have taxonomic affinities with the organisms inhabiting the nearest continent or other nearby landmass that served as a source of propagules. Most of the species inhabiting the islands of the South Pacific have Oriental and to a lesser extent Australasian affinities, and the number of taxa shared with these landmasses decreases with increasing distance out into the Pacific (Wilson, 1959; 1961; Solem, 1981) (see Figure 7.16).

When a landmass is shifted from one place

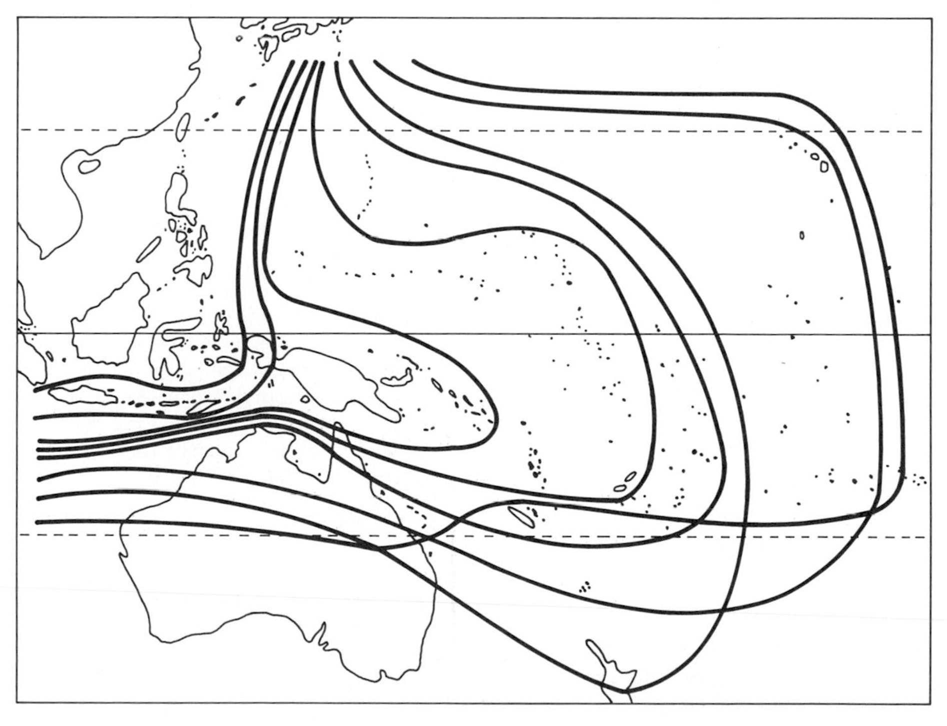

Figure 7.16
Limits of the distributions of eight different families of land snails in Australasia and the South Pacific. Each of these groups originated in southeastern Asia and has spread southward and eastward to a different extent. (After Solem, 1981.)

to another by seafloor spreading, it carries onboard a biota. This biota can be transferred directly without major barriers. India, which moved from southern Africa to Asia, is used as a prime example and has been called a "Noah's Ark" by McKenna (1973) because an assemblage of organisms was deposited en masse in a new environment. There are other parcels of land, in southern Europe, western North and South America, and eastern Asia, that may have been useful arks. These areas also carry fossil beds associated with their former locations.

Given that we accept the principle of actualism (Chapter 1), then any route of biotic exchange must be considered that we can observe operating today in nature. However, as noted earlier, controversy arises over the question of which means of dispersal is invoked for which cases, and excessive devotion of authors to particular mechanisms has given each a bad reputation. Vast numbers of transoceanic land bridges, now submerged, have been proposed on an ad hoc basis to account for the peculiarly disjunct distributions of many different taxa. Other proposals made exclusive use of Holarctic land bridges, continental drift models with totally unrealistic dates, stories of long-distance dispersal too outrageous to be believed, and so forth. Now that continental drift is accepted, there is a tendency to search for ancient land connections to account for many disjunctions.

Biologists have been eager to help the earth scientists find evidence of past land connections and physical barriers. Looking back with 20/20 hindsight, there was a time earlier in this century when biogeographers could have made substantial contributions to establish the occurrence, nature, and timing of continental drift, especially if their understanding of the phylogeny of organisms had been more advanced. But this did not happen, perhaps because biogeographers were too conservative. Now they must design their reconstructions with a realistic picture of the earth's history in mind. However, biogeography still can make some original contributions in these areas, especially by identifying barriers other than those created by drifting continents. Biological distributions generally cannot be used as proof of land connections in lieu of evidence from the rocks, but the analyses can corroborate or contradict geologic data and can provide valuable clues to past events overlooked by geologists. The analysis of routes for dispersal is best done separately by geologists and biogeographers, with both keeping watchful eyes for conflicting results. Meanwhile, the burden of determining positions of landforms and dates of changes still falls on geologists, and the reconstructions of the biotas, paleoecology, and paleoclimatology are projects for biogeographers.

Selected references

What is dispersal?

W.H. Brown et al. (1919); Carlquist (1965, 1974, 1981); Dammermann (1948); Docters van Leeuwen (1936); Gaines and McClenaghan (1980); Horton (1974a); Krebs (1978); MacArthur (1972); MacArthur and Wilson (1967); Maguire (1963); Platnick and Nelson (1978); Savile (1956); A.T. Smith (1974); Udvardy (1969).

Mechanisms of movement

R.R. Baker (1978, 1981); Berg (1975); Boer (1970); Bramwell (1979); Briggs (1974); Burtt (1929); Carlquist (1965, 1966a, 1966b, 1974, 1981); Colwell (1973, 1979); Crosby (1972); Cruden (1966); Darlington (1938, 1943); Dingle (1980); Dorst (1962, 1974); Falla (1960); Fosberg (1963); Gauthreaux (1980a); Gunn et al. (1976); Guppy (1906); Horton (1974a); C. Johnson (1963, 1969); F. Jones (1968); Leggett (1977); McAtee (1947); Mackie (1974); Maguire (1963); Pijl (1972); V. Proctor (1968); Rabinowitz and Rapp (1980, 1981); Ridley (1930); Scheltema (1971, 1977); Southwood (1962); B. Taylor (1954); Udvardy (1969); Urquhart (1960); Werner (1975); Wickens (1979); M. Williamson (1981).

The nature of barriers

G. Bartholomew (1958); J. Brown (1971b, 1978); Carlquist (1965, 1966a, 1974, 1981); Diamond (1975b); Ehrlich (1961, 1965); Gauthreaux (1980a, 1980b); Grinnell (1914); Hnatiuk (1979); J. Jackson (1974); Janzen (1967); MacArthur and Wilson (1967); A.T. Smith (1974); Southwood (1962); Vermeij (1978).

Establishing a colony

Carlquist (1965, 1966c, 1974, 1981); Cuellar (1977); Cuellar and Kluge (1972); Diamond (1975); V. Harris (1952); Hilden (1965); Klopfer and Hailman (1965); MacArthur (1972); MacArthur and Wilson (1967); A. Schoener (1974a); Soumalainen (1962); Southwood (1962); Terborgh et al. (1978); Wecker (1963, 1964); M. Williamson (1981).

Dispersal routes

C.G. Adams and Ager (1967); Bussing (1976); Carlquist (1965); H. Cooke (1972); C. Cox et al. (1980); Hopkins (1959, 1967, 1979); Howden (1974, 1981); C. Johnson and Bowden (1973); Kendeigh (1974); Marshall et al. (1982); McKenna (1973); R.R. Miller (1958, 1966); Por (1971); Rosen (1978); Savile (1956); G. Simpson (1936, 1940, 1950, 1952b, 1965, 1980b); G. Smith (1978, 1981); Salem (1981); Steenis (1962); Stoddart (1981); Terborgh (1973a); Udvardy (1969); Vermeij (1978); D. Walker (1973); S. Webb (1976); Wenner and Johnson (1980); E. Wilson (1959, 1961).

Chapter 8 Endemism, Provincialism, and Disjunction

The most pervasive feature of geographic distributions is that they are limited. No species is completely cosmopolitan, and most species and genera, and even many orders and families, are confined to restricted regions such as a single continent or ocean. For example, the entire avian family Furnariidae (ovenbirds), with 56 genera and 214 species, is confined (endemic) to South and Central America plus several neighboring islands. Several marine fish families, including the catfish eels (Plotosidae) and archerfishes (Toxotidae), are confined to the Indo-Pacific region. In South and Central America there are over 50 families and subfamilies of flowering plants that occur nowhere else (Table 13.5), many of which are small taxa, but there are also a number of very large families, e.g., the cacti (Cactaceae) and bromeliads (Bromeliaceae), that would be endemic to the New World if one species had not crossed the ocean recently by natural means (long-distance dispersal) and become established in Africa.

Endemics are not distributed randomly but tend to be found most abundantly in selected regions. Some regions, such as Australia, southern Africa, Madagascar, and New Caledonia, contain a large percentage of endemic species and numerous endemic higher taxa, whereas other regions such as Europe, northern North America, and the southern Atlantic Ocean share much of their biotas with other areas. Distantly related taxa of plants and animals tend to show patterns of endemism not only in the same ocean or on the same continent or island but also in the same localities of those regions. The coincident distributions of endemics, which we term *provincialism,* often do not correspond precisely to the present boundary of continents and oceans, and they certainly do not always coincide with known limits of abiotic and biotic environments. As pointed out in the previous chapter, human introductions demonstrate that species can thrive in regions far from their native habitats, as evidenced by the European rabbits and New World cacti that have become serious pests in Australia. Species of North American bass and trout have been introduced throughout the world to develop important sport fisheries in freshwater habitats. These successful introductions help to emphasize the unique influences of historical events in determining where organisms occur today. In many cases, the spread of taxa from regions in which they evolved has been blocked by barriers to dispersal; and in other cases, one species of a once widespread group may have persisted in a limited area after the representatives in other regions have become extinct.

Disjunct distributions provide a dramatic exception to this general pattern of provincialism. Disjunctions are those cases in which two or more closely related taxa live today in widely separated regions, being absent from intervening areas. Often these forms inhabit similar environments, e.g., rain forests on each landmass, but this is not always true. For example, the southern beech *(Nothofagus)* grows in wet cool temperate forests in South America, New Zealand, New Caledonia, New Guinea, New Britain, and Australia (Figure 13.10); but camels occur in the deserts and steppes of Asia and northern Africa and in montane habitats in South America. Disjunct distributions certainly must reflect major historical events: the organisms were dispersed long distances over geographic barriers, they were carried to distant sites aboard crustal plates as they drifted apart, or they are the surviving remnants of a once widespread taxon.

In this chapter we consider general patterns of endemism, provincialism, and disjunction, as well as some of the conceptual and methodological problems in analyzing and interpreting them. Methods for actually reconstructing historical biogeographic events by using endemics and disjunctions are introduced in Chapter 9.

Endemism

Because the term *endemic* simply means occurring nowhere else, organisms can be endemic to geographic regions on a variety of spatial scales and at different taxonomic levels. Taxonomic categories are hierarchical, so the distributions of lower taxa within a higher taxon are also organized in a hierarchical fashion. Just as an order contains a nested set of families, genera, and species that represent, it is hoped, the historical branching pattern of a single, or monophyletic, evolutionary lineage, so the geographic range of an order has nested within its boundaries the ranges of all of its families, genera, and species in a cumulative series. For this reason, the lowest taxonomic categories, species and genera, tend to be more narrowly endemic than the higher taxa, such as families and order, of which they are members. Figure 8.1 provides an example. The rodent family Heteromyidae, containing the pocket mice and their relatives, is endemic to western North America and northernmost South America. Each of the five genera of heteromyids have more restricted ranges, with the genus *Microdipodops* (kangaroo mice), for example, occurring only in the Great Basin of the western United States. The genus *Dipodomys* (kangaroo rats) has a much wider distribution, and its taxa vary greatly in their coverage, from *D. ordii,* which occurs in most of the desert regions, to *D. ingens,* which is endemic to an area of a few thousand square kilometers in the vicinity of Morro Bay on the California coast.

Many species and genera, and even some families and orders, are entirely restricted to tiny islands or equally small patches of habitat. The entire population of the Devil's Hole pupfish (*Cyprinodon diabolis,* Miller, 1948) numbers less than 600 individuals, which are confined to a spring pool measuring 20 by 3 meters in the Mojave Desert just east of Death Valley. Also in California, on the islands off the coast of southern California, are some remarkable plant endemics. On the tiny island of San Clemente lives the only population of the distinctive shrub *Munzothamnus* (Asteraceae); in one canyon of the island of Santa Catalina lives a population of four plants of a mountain mahogany, *Cercocarpus traskae* (Rosaceae), its only known locality. Also on Santa Catalina, San Clemente, Santa Rosa, and Santa Cruz islands occurs the Catalina ironwood, *Lyonothamnus* (Rosaceae), an elegant evergreen tree whose fossils are known from the mainland, including some from Death Valley. Even flying organisms can have narrow ranges. The todies (Todidae), for example, are a small family of birds entirely restricted to a few West Indian islands.

Cosmopolitanism. In contrast to such narrowly endemic organisms are the cosmopolitan taxa, which are widely distributed throughout the world. Relatively few species, genera, and families are truly cosmopolitan, but exceptional examples include the peregrine falcon *(Falco peregrinus),* the very broad and diverse plant genus *Senecio* (groundsel), and the bat family Vespertilionidae (Figure 12.11), which have achieved their broad distributions by natural means. Numerous minute animals and plants, such as protozoans, algae, and fungi, have extremely broad ranges because the organisms or their tiny disseminules are broadcast widely by the action of water or wind. In addition, there

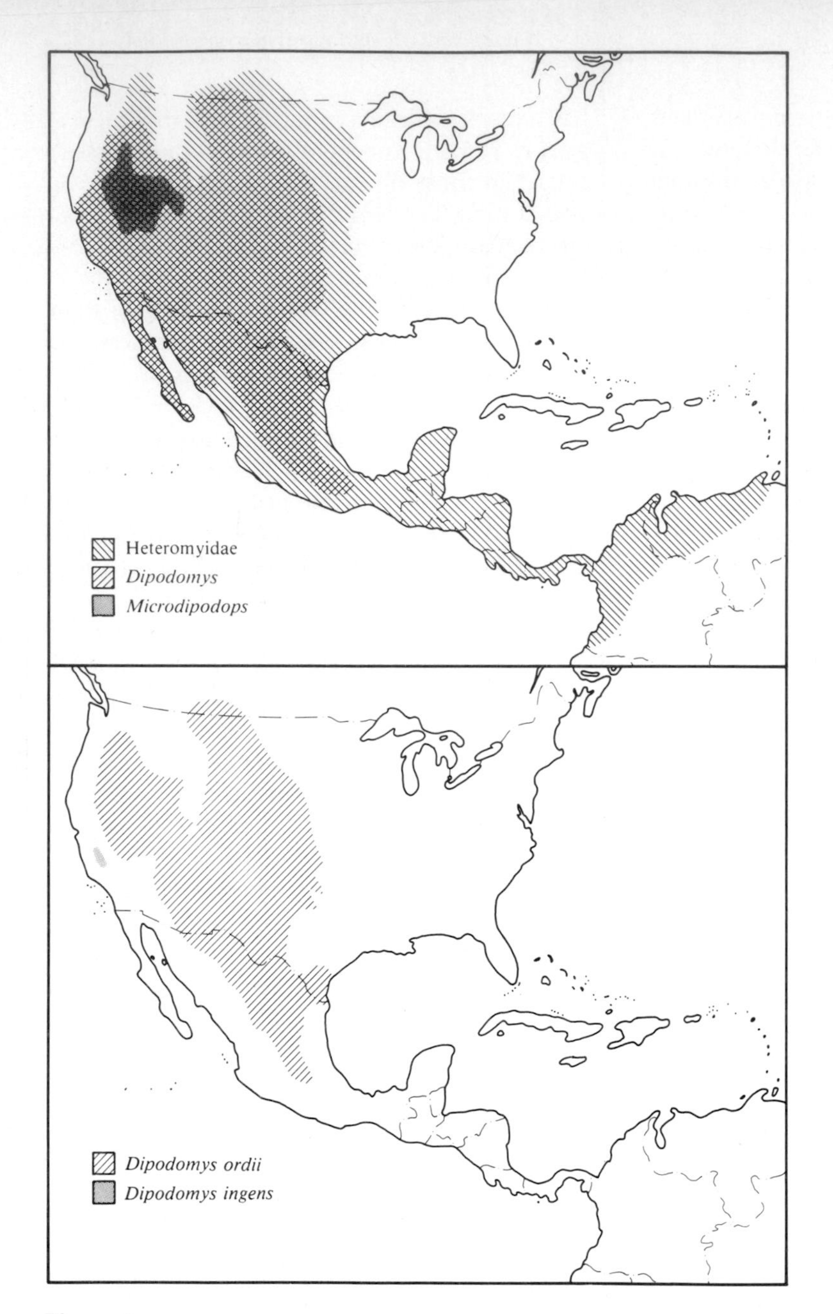

Figure 8.1
Hierarchical patterns of endemism in the rodent family Heteromyidae. This family, which includes the pocket mice and kangaroo rats, is restricted essentially to western North America, although it reaches northernmost South America. The family contains five genera, which are restricted to regions of varying size within this area. For example, the kangaroo rat genus, *Dipodomys,* is found in most arid and semiarid habitats in the western part of the continent, whereas the kangaroo mouse genus, *Microdipodops,* is endemic to the Great Basin Desert. Similarly, at the species level among the several species of kangaroo rats *D. ordii* has a very large geographic range, whereas *D. ingens* is restricted to a small portion of the San Joaquin Valley in California. (After Hall, 1981.)

is now a sizable list of plant and animal species associated with humans, our foods, pests, and diseases, which have intentionally or inadvertently been introduced into even the most remote habitable places. In the sea, where an organism can, theoretically, swim around the world unimpeded, relatively few species or genera are found in all the oceans. Most notable exceptions are some of the large predators and also organisms that have been transported by ships to distant ports. Hence we can generalize on the cosmopolitan distributions by stating that such ranges are rare, limited to less than 1% of all families. These families have at least some genera and species with very broad ecological tolerances, great capacities for dispersing long distances with or without human assistance, and in many cases, very general requirements for completing the life cycle. On the other hand, many orders and most classes are essentially cosmopolitan because the ecological diversity within these higher taxa is usually broad enough to include forms that can exist in most terrestrial or aquatic habitats, and also because these groups are old enough to have had historical opportunities to colonize most parts of the world.

Types of endemics. The origins and ages of endemics are indicated by a variety of terms. An autochthonous endemic is one that has differentiated in situ, where it is found today, whereas an allochthonous endemic has evolved its characters elsewhere and merely survives in its current area. Prime examples of allochthonous endemics are relicts or epibiotics, species that were once widespread but are now confined to a very small region.

There are two kinds of relicts, taxonomic and biogeographic. Taxonomic relicts are the sole survivors of once diverse taxonomic groups, whereas biogeographic relicts are the narrowly endemic descendents of once widespread taxa. Often the two categories coincide, especially for organisms called living fossils. For example, the tuatara *(Sphenodon punctatus)* is a primitive reptile superficially resembling certain lizards. This animal, which inhabits some small islands of New Zealand, is the only surviving species of the order Rhynchocephalia that was widespread on Mesozoic continents. Similarly, the coelacanth *(Latimeria),* known only from the deep waters off western Madagascar and the Comoro Islands, is the only living member of the lobe-finned fishes, the crossopterygians, (Figure 8.2), which were widespread in freshwater habitats as well as in oceans and shallow epicontinental seas in the Paleozoic and which gave rise to the amphibians. An example of a plant relict is the ginkgo (*Ginkgo biloba,* Ginkgoales), a gymnosperm narrowly endemic to eastern China, which is the survivor of a group that was fairly diversified in the Mesozoic.

For categorizing endemics by age, the terms

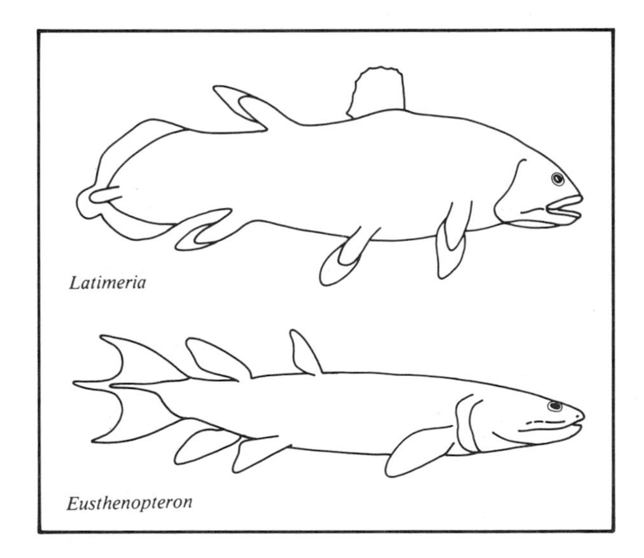

Figure 8.2
The living coelacanth, *Latimeria, (above)* is known only from relatively deep Indian Ocean waters near Madagascar. It is apparently the sole surviving member of the once diverse group of crossopterygian, or lobe-finned, fishes. This group, represented here by the late Devonian form *Eusthenopteron (below),* was geographically widespread and apparently occurred primarily in freshwater habitats. Crossopterygians are noteworthy for colonizing land and giving rise to amphibians.

paleoendemic and *neoendemic* are used to identify old and recently formed species, respectively. One can see immediately that using such terms requires an author to make judgments, usually subjective ones, on the origins of endemics. Easiest to label are the most recent endemics, those of Quaternary age. As pointed out in Chapter 14, the Pleistocene was a time of great geologic, climatic, and biogeographic change. Within just the last 10,000 years many species ranges have shifted dramatically in response to the warming of the climate and retreating of the ice sheets. Many once widespread distributions have contracted, so that now only small isolated populations are found, especially in the cool, moist mountains. On the other hand, boreal and tundra plants have reinvaded glaciated regions and in many cases developed as new polyploid species (Löve and Löve, 1963).

The restriction of a taxon to a particular geographic range is the consequence of both historical events and ecological processes. Ecological explanations must be invoked to explain the present limits of an endemic range. Abiotic and biotic limiting factors (Chapter 3) prevent the species from expanding at the periphery of its range and thus also determine the nature of barriers that must be overcome if it is to disperse to distant areas. On the other hand, historical events must be invoked to explain how this taxon became confined to its present range and to reconstruct the geographic origin, spread, and contraction of the taxon, over time. As one looks for historical explanations, such as barriers formed by drifting continents, changing sea levels, and glaciation, it must also be kept in mind that there were highly stochastic events going on at the same time, processes such as long-distance dispersal and extinction of small populations. All this occurred as the expansion and contraction of ranges caused new combinations of organisms to come in contact and have biotic interactions. Thus investigators search for and hope to find satisfying, simplistic explanations for the origins of endemics, but what they expect to find eventually is a very complex picture of how endemics evolved in time and space, to be explained in part by past geologic events and in part (maybe never fully explained) in terms of ecological processes.

Provincialism

Terrestrial biogeographic regions. When the ranges of organisms are examined closely, it is seen that endemic forms are neither randomly nor uniformly distributed across the earth but instead are clumped in particular regions. Three patterns are apparent. First, the most closely related species tend to have overlapping or adjacent ranges within restricted parts of continents or oceans. Second, completely unrelated higher taxa, for example, those of certain plant and animal orders and classes, often show similar patterns of endemism. Third, a significant portion of orders and families and some genera have markedly disjunct ranges, with species living in widely separated areas on different continents or islands. The first two patterns make it possible for biogeographers to identify circumscribed regions of the earth's surface that share common biotas that are taxonomically distinct from neighboring areas. The third pattern encourages us to search for historical explanations for how these organisms came to occupy their present positions in those distinct regions.

Provincialism was one of the first general features of land plant and animal distributions noted by such famous nineteenth century workers as the phytogeographers Schouw (1823) and de Candolle (1855) and the zoogeographers Sclater (1858) and Wallace (1876). As soon as biologists traveled to different continents, they were impressed by the differences in the biotas on the various landmasses. A goal of early classifications was to identify centers of origin, cradles where biotas were created, and those historical barriers that blocked the exchange of organisms between adjacent regions. As reviewed by Udvardy (1969) and Nelson and Platnick (1981), much effort has been devoted to this endeavor. The result was a division of the earth into a hierarchy of regions reflecting patterns of faunal and floral similarities. In order of decreasing size, the common subdivisions are usually referred to as realms or kingdoms, regions, subregions, provinces, and districts.

Regions and subregions can often be used to describe a great many distributions. One example is Australia, where the terrestrial fauna is typified by large marsupials rather than placental mammals and the flora has many species of myrtles (Myrtaceae) and proteads (Proteaceae). Zoogeographers recognize two subregions, the central arid and semiarid two thirds dubbed the Eremaean or Eyrean Subregion, named from an important basin in the eastern portion, and the moister fringe to the north, east, and south (Figure 8.3). The northern portion, called the Torresian Province, is a warm tropical belt that contains groups with close affinities to animals and plants in New Guinea and sometimes also southeast Asia. In general, these organisms were once distributed across the present Torres Strait from Australia to New Guinea when sea level was much lower. The southeastern portion of Australia, including Victoria, Tasmania, and some of the surrounding islands, called the Bassian Province, is inhabited by animals and plants adapted to cooler, temperate climates. The other section, called Westralia, includes mostly the southwestern corner where great numbers of endemic forms reside. For climatic and probably historical reasons many of the cool, temperate disjuncts between the southern portions of the southern continents reside today in the Bassian Province, and many of the

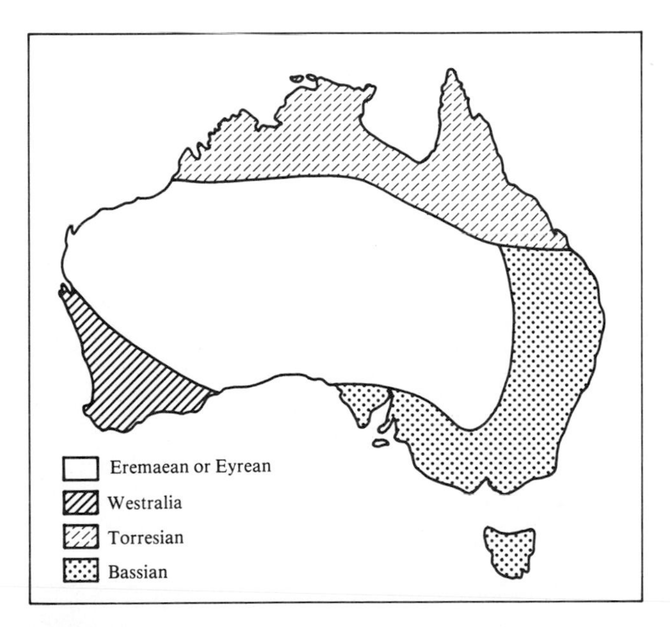

Figure 8.3
Biogeographic subregions and provinces of Australia. There are two subregions: the great arid and semiarid Eyrean or Eremaean Subregion in the central part of the continent, and the strip of moist, usually forested habitats around the northern, eastern, and southern periphery, which are designated the Torresian, Bassian, and Westralian provinces, respectively. The isolated Westralian Province experiences a mediterranean-type climate and has many endemics, including several plant taxa with closest relatives in South Africa, but many of these taxa also occur in semiarid habitats in South Australia as well.

South African–Australian taxa reside in Westralia.

The Australian Region is one of six large units recognized by Sclater to describe the world distributions of bird families and genera (Fig. 1.3). These were essentially the same six zoogeographic regions later accepted by Wallace in his major treatise on zoogeography. In the Northern Hemisphere, landmasses north of the tropical zone are called the Holarctic, being composed of the Nearctic (North America) and Palearctic regions (Eurasia and northernmost Africa). The remaining regions are primarily tropical ones, the Neotropical (Central and South America and the West Indies), the Ethiopian (Africa south of the Sahara, and Madagascar), and the Oriental Region (southeast Asia and the adjacent continental islands). The faunas of oceanic islands in the Pacific basin are anomalous in this classification because they contain a small number of taxa from adjacent continents and have relatively few unique groups.

The majority of articles and books on the distributions of organisms now use these regions as fundamental descriptors of animal distributions, and it has to be said that this classification has become one of the primary empirical foundations of biogeography. This is not only a tribute to the early zoogeographers, who were working with incomplete collections and primitive phylogenetic classifications, but also a testimony to the clarity and generality of the distributional patterns of terrestrial organisms.

The Sclater-Wallace classification also works reasonably well for plants, but there are a number of significant differences (Figure 8.4). Dis-

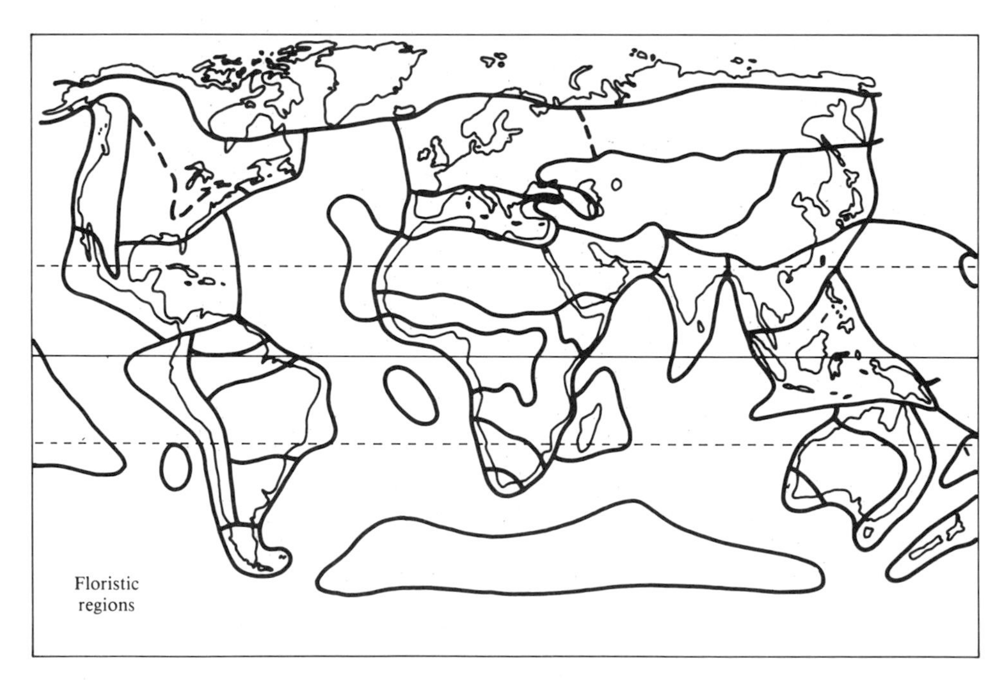

Figure 8.4
Division of the world into regions and provinces based on the distribution of plants. The major regions correspond well with those delineated by Sclater and Wallace (Figure 1.3) to describe basic patterns of animal distributions on the landmasses, but phytogeographers recognize subregions and provinces that reflect climatic regimes and other physical conditions that do not have such strong influence on most animal groups. (After Good, 1974.)

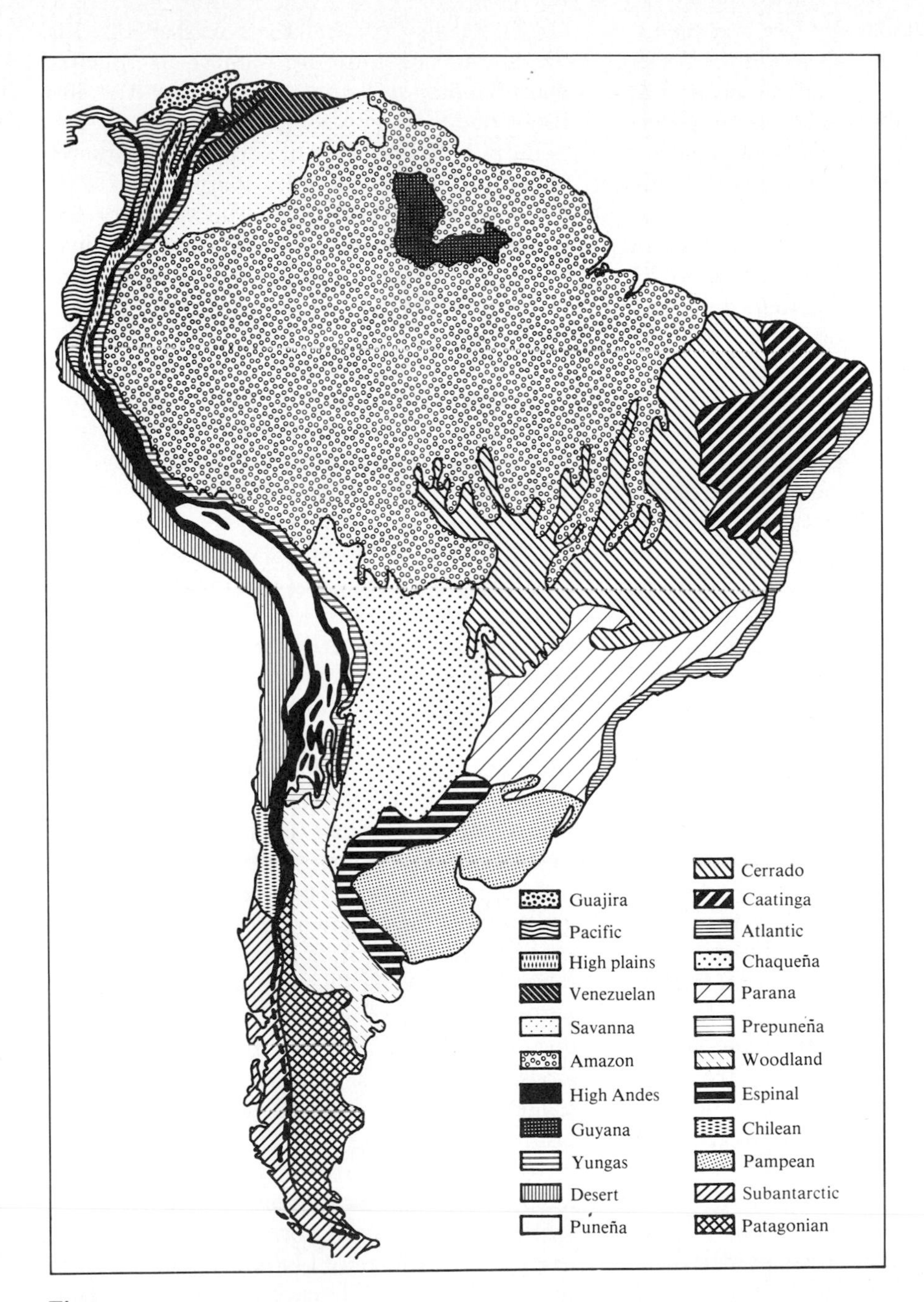

Figure 8.5
Biogeographic provinces of South America. The delineation of these regions is based primarily on the distributions of plants, and it largely reflects the relationship between distinctive vegetation types and climate. (After Cabrera and Willink, 1973.)

regarding for the moment the fact that phytogeographers translate the provinces of vertebrates as regions for plants, and the districts as provinces, there are still some important differences that reflect a basic dichotomy between animals and plants. Vegetation types are tightly restricted by abiotic factors, such as temperature and rainfall, so that regions of endemism are more clearly defined on the continents by climatic and other physical barriers than they are for animals, which are often able to overcome climatic barriers by physiological and behavioral adaptations. Hence the small tip of South Africa, which has a mediterranean-type climate, itself constitutes a botanical region with 90% specific endemism and an exceedingly rich flora (Dyer, 1975; Goldblatt, 1978), whereas animal groups of South Africa generally are not restricted to that zone and are, in some groups not very diverse (Werger and van Bruggen, 1978). South America has been divided into numerous provinces based primarily on the distributions of certain plant species that are in turn related to particular combinations of soil, temperature, and precipitation (Figure 8.5). However, Cabrera and Willink (1973), who defined these provinces, pointed out that certain animal species are largely restricted to or exceptionally abundant in these regions.

For historical biogeography, it is important to realize that regions and provinces have been shaped by both ecological and historical factors. Continental taxa have been stranded on selected landmasses not only because the range is limited by abiotic and biotic factors (Chapter 3) but also because the barriers isolated them on that landmass millions of years ago. To avoid complicating our general discussions here, we have placed lists of endemics for various taxa in Unit Three. When one examines these lists, the fact that the continents are centers of endemism will become quite evident. However, this observation by itself is of limited interest, because even randomly selected large areas would be expected to contain endemics, given the condition that no family can occur everywhere. For example, Beadle (1981) has observed that Australia has about one fifth the number of endemic plant families as South America, but it is also about one fifth the size. A more exciting question can be asked: To what extent is the history of the taxon or taxa under consideration a reflection of the history of the landmasses or water masses in which these organisms now live, and can distributions of organisms be used to reconstruct the history of geographic changes on the earth's surface? Thus the historical biogeographer is more deeply concerned about the geographic and taxonomic affinities of endemic and disjunct groups than about how many different taxa coexist within different regions, which is a problem for ecological biogeographers (Unit Four).

The recognition of formal biogeographic provinces implies that the biota within each region is more homogeneous than between adjacent areas. Unfortunately, most biogeographic units have not been quantitatively analyzed to show that they can, indeed, be treated as relatively homogeneous assemblages. At present investigators must be content to follow conventional wisdom for subdividing the biosphere. One of the observations that is still encouraging to an acceptance of the concept of provincialism is the rapid turnover of taxa between areas. Biogeographers have drawn lines to define fairly precisely the limits of regional biotas. Although these biogeographic lines are usually derived for one taxon at a time, they provide additional documentation of the discreteness of the regions. The limited distribution of endemics suggests a history of local origin and limited dispersal.

A second important characteristic of the major biogeographic regions is that they describe general distribution patterns that hold true for many different kinds of organisms. Despite tremendous differences in biological traits that influence habitat requirements and dispersal capabilities, such diverse kinds of organisms as angiosperms, freshwater fishes, earthworms, lizards, and birds all exhibit worldwide distributions that correspond remarkably well to the realms and regions recognized by Wallace in 1876. This is perhaps the best evidence that these distributions reflect historical events.

The third characteristic of biogeographic regions that suggests the influence of historical events is the environmental nature of both the regions themselves and the boundaries between them. All of the regions offer an extremely diverse range of terrestrial environments. For example, all have mountains more than 3000 m in elevation, all except the Nearctic and Palearctic contain well-developed lowland tropical rain forest, and all except the Oriental Region have large areas of desert habitats. If the distributions of organisms were limited solely by ecological factors, we would see a very different pattern of provincialism. Taxa would be found around the world in areas of similar climate and topography, rather than occupying a diversity of environmental regimes within a restricted area. In fact, cosmopolitan families are very uncommon, and even the circumtropical distributions, such as those of the palms (Arecaceae) (Figure 13.1) and the trogons (Trogonidae) (Figure 12.7), which are clearly limited to the tropics and near tropics by cold temperatures, still have intrafamilial endemism on the various southern landmasses (Moore, 1973). Thus even though about 10% of the families of angiosperms and vertebrates are circumtropical, a historical explanation is still required to explain the patterns of endemism within these families.

In general, the boundaries between biogeographic regions on the continents are geographic formations that pose major barriers to dispersal of land organisms. Often the barrier is seawater, which marks the limits of both continents and biogeographic regions; but in some cases the barriers are such things as mountain ranges and deserts, which do not correspond closely to the geologic boundaries of crustal plates (Chapter 5). For example, as discussed in some detail in Chapter 18, the present boundary between the Nearctic and Neotropical regions occurs in southern Mexico at the northern limit of tropical habitat and also in the West Indies rather than abruptly at the Isthmus of Panama, the narrowest and most recent connection between North and South America. Similarly, the continent of Africa is itself divided into two categories, the Ethiopian and Palearctic zones, by the Sahara Desert. The Oriental Region is actually a composite consisting of landmasses of Asian origin and India, which once was part of Gondwanaland and attached to what is now the Ethiopian Region. These cases indicate once again the difficulty in separating historical from ecological causes for distributional patterns—although the local origins and radiations of taxa are historical events, ecological limiting factors have played primary roles in blocking dispersal and preserving historical patterns.

Biogeographic lines. As noted in the previous section, boundaries between two major regions or provinces are often called biogeographic lines. The recognition of such discrete limits implies that there is now or has been in the past a major barrier to biotic exchange between adjacent territories. The most famous of these boundaries is Wallace's line between southeast Asia and Australasia. Wallace's line, extending between Borneo and Celebes, and Bali and Lombok, marks the approximate center of this region in which especially the fauna changes. Seven different lines (Figure 8.6) have been suggested to account for various zoogeographic discontinuities (Simpson, 1977). The one closest to southeast Asia is Huxley's line, which demarks the Sunda Shelf and includes Java, Sumatra, and numerous smaller islands of western Indonesia as part of the Oriental Region. The line furthest east is Lydekker's line, which demarks the Australian Region at the Sahul Shelf just east of New Guinea. The shelves were largely emergent at times during the Pleistocene, permitting landlocked organisms to reach their edges, where they subsequently became stranded as islands were formed by rising sea levels. Groups of organisms with the poorest capacity for long-distance dispersal drop out

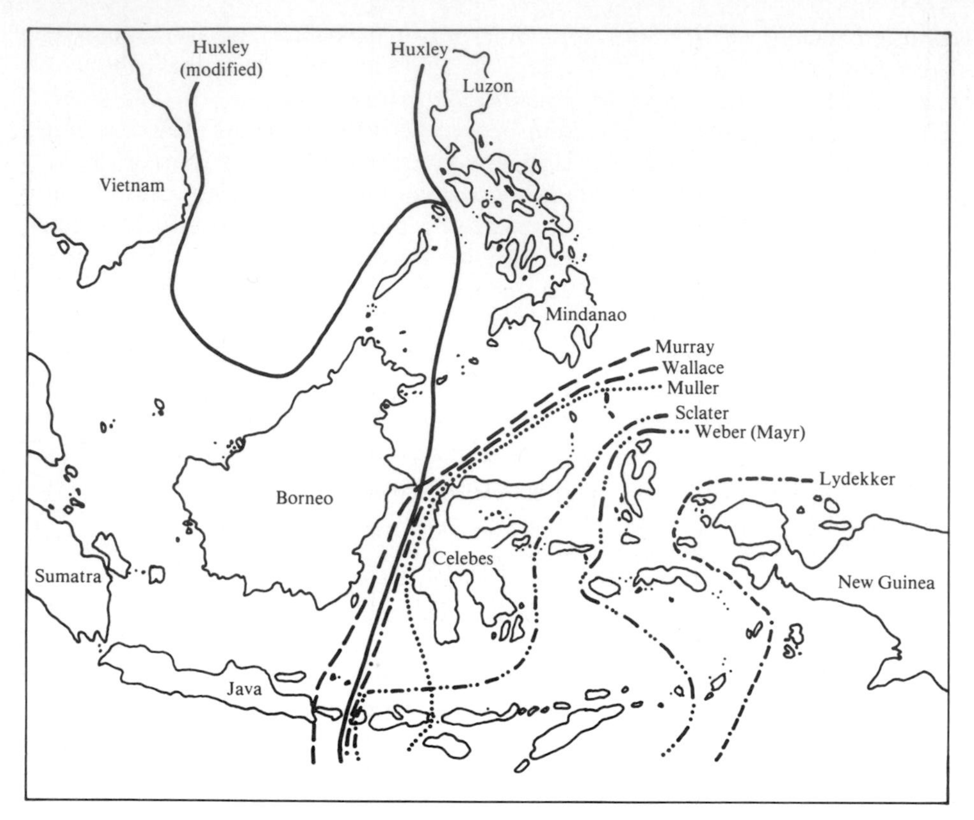

Figure 8.6
Various lines drawn by zoogeographers to define the boundaries between the Oriental and Australasian biotas. The multiplicity of lines reflects the fact that different taxa have managed to penetrate different distances from their continent of origin into the islands of the East Indies. The only two lines that appear to have general significance are Lydekker's and Huxley's (modified), which coincide with the limits of the continental shelves and consequently indicate regions that were largely above water during past periods of lower sea levels. (After Simpson, 1977.)

at these lines from both the Oriental and Australian sides although the bats, birds, butterflies, and plants extend beyond, many even penetrating the opposite region. Between Lydekker's line and Huxley's line is a zone of transition. The further subdivision of this transitional zone by additional lines no longer serves a useful function (Simpson, 1977).

Other lines have been proposed between biotas. In the Old World the Palearctic fauna of mediterranean Africa is extremely different from the Ethiopian fauna of mainland Africa south of the Sahara, even though the Sahara serves as a wide transition zone for some of the species. However, a mere 5000 years ago beasts such as hippopotamuses and lions lived in northernmost Africa, making these divisions less pronounced.

In the New World the line separating the Nearctic and Neotropical elements is neither very distinct nor does it correspond well with the ancient seawater barrier that separated the two continents during the Cretaceous and most of the Cenozoic. Rather the present boundary

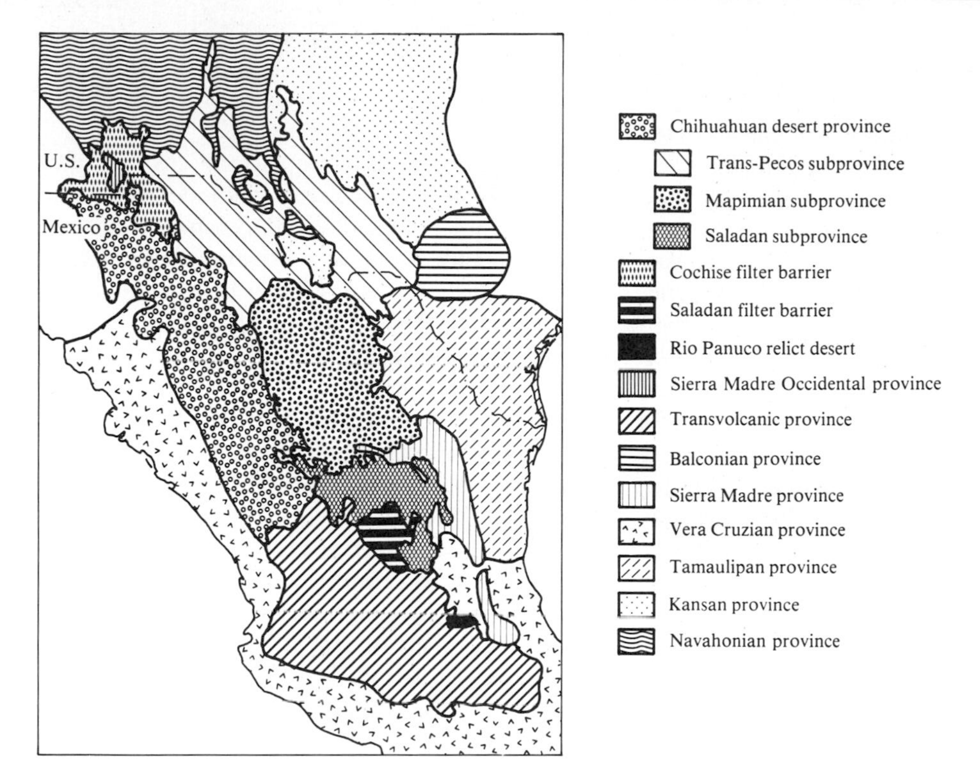

Figure 8.7
Biogeographic provinces and subprovinces of south central North America, based on quantitative analyses of biotic similarities among the reptiles and amphibians. Such regions defined on the basis of the distribution of animal groups usually reflect barriers to dispersal to a greater extent and climatic regimes to a lesser extent than provinces based on plant distributions. (After Morafka, 1977.)

between the regions seems to be influenced primarily by climatic and habitat factors. The flora and fauna of original North American and South American ancestry mix over a broad region in Middle America and are closely integrated even at the level of local communities. Generally we think that the neotropical flora extends northward to Oaxaca and Veracruz in southern Mexico, but of course there is a gradual decline of species from South America as one proceeds northward through Central America. Groups with dozens of species in Amazonia often have but one or two species in Guatemala and Chiapas. Meanwhile, the temperate forests of Mexico extend southward in the high mountains. First sweet gum *(Liquidambar)* drops out and then the pines *(Pinus)* in northernmost Nicaragua, while walnuts *(Juglans)*, willows *(Salix)*, and oaks *(Quercus)* range into the Andean Cordillera. Along with these woody plants, temperate understory species, such as many ferns, have disjunct distributions in high-elevation forests along the backbone of Central America (Goméz, 1982) and Mexico (Miranda and Sharp, 1950). Thus any lines that are drawn tend to follow elevational contours or drainage basins (for aquatic organisms) rather than political or latitudinal boundaries.

In addition to dividing the earth into major biogeographic regions, many specialists have attempted to understand the meaning of provincialism on a smaller scale. Figure 8.7, for example, shows a number of biotic provinces and subprovinces of northern Mexico and the adjacent United States as determined by the distributions of the reptiles and amphibians (Morafka, 1977). If these analyses are made with careful analytical procedures, a worker can not only determine which areas have a homogeneous set of species and which do not but also begin to define the ecological factors that limit the range of that assemblage. On the other hand, when analyses are made on scales smaller than the region, the results for historical biogeography appear much less general. The provinces described for the above North American reptiles and amphibians are probably not the same provinces for other groups of animals and plants even though they may be very similar to provinces of mammals (Hagmeir, 1966). The apparent reason for this discrepancy is that each taxon has different biological attributes and as a consequence its local distribution patterns are limited by different ecological tolerances and requirements (Chapter 3). Although the effects of the earth's most important barriers have similarly limited the dispersal of many different groups of terrestrial organisms, resulting in general, clearly defined biogeographic regions, the effects of more modest barriers operating on a smaller scale probably have not had such general effects.

Classifying islands. Since Wallace (1876, 1880), biologists have tried to classify islands as either continental or oceanic (see box on p. 237). It is generally assumed that continental islands were once part of the adjacent mainland and originally contained a virtually complete continental biota. Some species may have become extinct, however, and even recolonized since the island became isolated. Consequently, continental islands have harmonic biotas; i.e., the assemblages are taxonomically and ecologically very similar to those on the mainland, although they may lack some continental taxa. These islands typically are included in the same faunal or floral province as the nearby mainland from which the biotas were derived. Most islands with harmonic biotas are part of continental shelves and many have been connected to the mainland as recently as the Pleistocene, when sea levels were lower.

In contrast, oceanic islands, which have arisen as distant, isolated volcanoes from the ocean floor, have received colonists only via long-distance dispersal. Because long-distance or waif dispersal is a chancy process, it is not surprising that the composition of the biota is somewhat unpredictable. However, this stochastic aspect cannot obscure two important broad-scale deterministic patterns, known to all biologists. First, the recent successful colonists (as judged by their degree of differentiation from continental relatives) of distant islands are highly biased in favor of forms that possess certain traits that promote long-distance dispersal and successful colonization (Carlquist, 1965, 1974; Lack, 1976). As pointed out elsewhere (Chapter 16), island forms may lose their adaptations for long-distance dispersal as they undergo further evolution after establishment (Carlquist, 1966a, 1966b, 1966c). Second, the nearest landmass will usually contribute the greatest number of colonists to an island, but if an island is equidistant from two source areas, colonization from each does not necessarily have to be equal. Some areas are better sources of colonists because they contain more species or species better suited for the insular environment or because they lie in the path of a strong dspersal force, such as a prevailing wind, a strong ocean current, or an avian migratory flyway (see Figure 12.3). Given that a number of variables influence the origins of species for each island, it is not surprising that distant islands generally are not easily classified into regions and that phytogeographers especially have tended to recognize each as a discrete province or district when endemism is extremely high.

For years island biologists have known that some now-distant islands have continental (sialic or andesitic) rocks. With the acceptance of plate tectonics, they have finally had to face

Some biogeographically interesting islands classified according to their modes of origin

FULLY OCEANIC

Totally volcanic islands of fairly recent origin that have emerged from the ocean floor and have never been connected to any continent by a land bridge.

Midoceanic Island chains or clusters formed from hot spots (HS) *or along fracture zones* (FZ) *within an oceanic plate*

Austral-Cook Island Chain (HS or FZ); Carolines (HS); Clipperton Island (FZ); Galápagos Islands (FZ, Carnegie Ridge); Hawaiian Islands (HS); Kodiac-Bowie Island chain (HS); Marquesas (HS); Society-Phoenix Island chain (HS, but some contribution by the Tonga Trench).

Island arcs formed in association with trenches

Aleutians (may have been part of the Bering Land Bridge in the Cenozoic); Lesser Antilles; Lesser Sunda Islands; Marianas; New Hebrides; Ryukyus (may have been associated with neighboring islands); Solomons; Tonga and Kermadec.

Islands formed at presently spreading midoceanic ridges

Ascension Island; Azores (some islands have continental rocks); Faeroes; Gough Island; Tristan da Cunha.

CONTINENTAL ISLANDS

Formed as part of a continent and since have been separated from the landmass. Some of these have added oceanic material since they were formed.

Islands permanently separated from the mainland since the split was initiated (final separation in parentheses)

Greater Antilles (80 million years BP); Kerguelen Island (Upper Cretaceous); Madagascar (*ca.* 100 million years BP); New Caledonia (*ca.* 50 million years BP); New Zealand (80 to 90 million years BP); Seychelles (65 million years BP); South Georgia (45 million years BP).

Island groups with connections of some islands with the mainland but not others

Canary Islands.

Islands most recently connected with some mainland in the Pleistocene by land or an ice sheet

British Isles; Ceylon; Falklands; Greater Sunda Islands; Japan and Sakhalin; Newfoundland and Greenland; New Guinea (with Australia); Philippines; Taiwan; Tasmania.

the reality that these islands cannot be classified as either strictly oceanic or continental because their role has changed as organisms arrived by either transoceanic dispersal or via land bridges, respectively. Instead, workers now must make inferences about which groups are relicts, remnants dating from when the island was part of a continent, and which have colonized across water barriers since its isolation. New Caledonia is a prime target for such discriminatory studies.

As workers reexamine the biogeography of the larger islands, such as Madagascar and the Greater Antilles, that were known to have been involved in continental drift, it is good to reconsider also the interpretations of the origins of the biotas of other islands, no matter how distant, to make certain that some groups have not survived since the island was closely associated with a continental landmass. In most cases the original conclusions—that these islands are truly oceanic, have never been connected to large landmasses, and must have received their entire biotas by long-distance, over-water dispersal—will undoubtedly stand after reexamination. However, the goal must be to avoid dogmatism and to seek out and test alternative ideas.

Aquatic regions and provinces. Numerous biogeographers have attempted to define regions and provinces in the oceans. Some of the classifications, however, are really ecological characterizations based on such criteria as water temperature, depth, and substrate. Such a classification often emphasizes vertical rather than horizontal division of the three-dimensional marine realm, and understandably so, because there is much less distinctiveness among the marine biotas in different oceans than among terrestrial organisms on different continents. Not only are the oceans more interconnected than the continents, but also many marine forms have life history stages that are widely dispersed. Most marine animals are cosmopolitan at the familial level and many even at the generic level. Consequently, it is difficult to detect patterns in the distribution of marine taxa that clearly reflect the histories of the present water masses.

Part of the problem in reconstructing the history of oceanic faunas is a result of the very long and complicated histories of most oceans. By studying fossil marine invertebrates through the Paleozoic and Mesozoic we can see that in some taxa during certain periods there was very little provincialism. For example, in the Silurian, many genera of marine brachiopods, gastropods (Boucot and Johnson, 1973), and graptolites (Berry, 1973) had nearly cosmopolitan distributions, but the majority of phyla have some period during which distinctive assemblages occurred in one early ocean or another, such as the Tethys (Hallam, 1973a; Gray and Boucot, 1979). This accordion effect—ranges expanding and contracting to produce alternating stages of provincialism and cosmopolitanism—causes us to be concerned as to how present-day marine distributions can be used reliably for reconstructing anything besides the most recent historical events in the oceans, those of the Cenozoic.

Some history is preserved, however, in the distributions of certain taxa, especially those with limited dispersal, which often exhibit significant provincialism. The most obvious horizontal pattern is latitudinal variation in species diversity and composition, but this appears to be primarily the result of environmental temperature and related abiotic limiting factors rather than a legacy of historical events. Warm tropical oceans have served as significant barriers to dispersal to some cold-adapted temperate and arctic groups, such as the penguins (Sphenisciformes; Chapter 10), and the high latitude seas represent centers of endemism and speciation. Other polar groups do not exhibit such provincialism because they have frequently dispersed across the equator in deep water where temperatures are also very cold.

The boundaries of marine provinces are most sharply defined along continental coastlines. These are identified by rapid turnover in the species composition. However, such turnovers from a province to its northern or southern neighbor usually involve changes in individual species but not major changes in the higher taxa.

Some shallow water and coral reef organisms show significant differentiation in different regions. Coral reef fish faunas are quite variable in composition, so that certain regions in the Indo-Pacific and the Caribbean not only are centers of high species diversity but also contain a number of endemic genera (Briggs, 1974). There is also significant endemism at the generic level among the corals and molluscs. In the case of these invertebrates, fossils indicate that many genera were widely distributed in the Mesozoic, which means that narrowly restricted forms are often relicts of once circumtropical forms (Vermeij, 1978). Recent information suggests that many marine organisms do not disperse so far as was once believed, so the relationship between marine biogeography and historical tectonic events seems a fruitful area for future research.

One expects to find provincialism in freshwater systems comparable to terrestrial ones; and for landlocked groups such as obligate freshwater fishes and molluscs, one can clearly identify patterns of endemism that correspond to the Sclater-Wallace biogeographic classification (Chapter 10). However, the situation is different for most freshwater plant families and even genera and species, which tend to have broad ranges. Some of the species have nearly cosmopolitan distributions, especially those that are weedy, like duckweeds (Lemnaceae), some aquatic ferns *(Azolla, Salvinia, Marsilea),* water milfoil *(Myriophyllum),* and hornwort *(Ceratophyllum)*. Freshwater plants achieved these broad ranges with the aid of wading birds, which carry the seeds or plantlets from one pond or lake to another. As a consequence, provinces are easier to delimit for freshwater animals than for plants.

Quantifying similarity among biotas. The early biogeographers and many of their successors defined biogeographic provinces and regions subjectively. This does not necessarily mean that their classifications are unreliable; in fact, the human brain has an exceptional capacity for recognizing pattern. The biogeographic regions first defined subjectively by Sclater almost certainly summarize real patterns that could be defined objectively by modern quantitative analyses. During the last two decades quantitative techniques have been applied increasingly to systematics and biogeography in order to make the process of classifying organisms and biogeographic regions more rigorous, objective, and repeatable.

In principle, quantitative techniques are simple. The items to be classified are described using objective criteria. In the case of biogeographic studies, the data usually consist of a complete list of the appropriate taxa that occur at a specific site or within a given area. Several mathematical indexes can then be used to quantify the similarity between each pair of biotas. Some of the most commonly used similarity indexes are shown below. These all can be computed from simple presence-absence data, but

Simple coefficients used by various authors to estimate biotic similarities

A, Absent in both units compared; *C,* Present in both units; N_1, Total present in the first unit; N_2, Total present in the second unit (when the first unit contains the fewer taxa). (After Cheatham and Hazel, 1969.)

Jaccard

$$\frac{C}{N_1 + N_2 - C}$$

Simple matchings

$$\frac{C + A}{N_1 + N_2 - C + A}$$

Dice

$$\frac{2C}{N_1 + N_2}$$

First Kulczynski

$$\frac{C}{N_1 + N_2 - 2C}$$

Second Kulczynski

$$\frac{C(N_1 + N_2)}{2(N_1 N_2)}$$

Otsuka

$$\frac{C}{\sqrt{N_1 N_2}}$$

Correlation ratio

$$\frac{C^2}{N_1 N_2}$$

Simpson

$$\frac{C}{N_1}$$

Braun-Blanquet

$$\frac{C}{N_2}$$

Fager

$$\frac{C}{\sqrt{N_1 N_2}} - \frac{1}{2\sqrt{N_2}}$$

they differ primarily in the extent to which they incorporate taxa that are present in both regions, in the range of values they can assume, and how they behave mathematically (that is, how variations in presence-absence patterns are combined to produce a single number). The Jaccard and Simpson similarity indexes are the two that have probably been most frequently used in biogeographic analyses. Similarity values for each pair of biotas can conveniently be expressed in matrix form (Table 8.1).

Once the similarity between each pair of biotas has been computed a quantitative clustering method is used to divide the biotas into groups that reflect a hierarchy of distinctiveness. Again, a number of different clustering techniques are available; because these differ in their procedures for making mathematical computations, they can give somewhat different results. Once the biotas have been clustered, all that remains is to decide what levels of similarity to use for designating different biogeographic ranks, such as provinces and subprovinces, so the products can be plotted with exact boundaries on a map.

Given a suitable data set, the similarity indexes can be calculated and the clustering procedure can be performed rapidly using a computer. However, despite the advantages of having objective classifications and the apparent ease with which they can be obtained, there have been relatively few cases in which such quantitative methods have been applied to analyze whole biotas. Virtually all of these have been made for a single taxon within a very limited geographic region. For example, Kikkawa and Pearse (1969) divided Australia into 10 provinces on the basis of the avian distributions, and Connor and Simberloff (1978) have analyzed similarities among the avifaunas and the floras of the various Galápagos Islands. These cases demonstrate the practicalities of applying quantitative methods to biogeographic problems but are too limited in taxonomic scope to detect any general patterns. One of the most ambitious early attempts to use these methods was the analysis by Holloway and Jardine (1968) of the distributions of butterflies, birds, and bats in Indo-Malesia (Figure 8.8). This quantified overall similarities of significant

Table 8.1
Matrix of similarity coefficients (Simpson index) between the mammal faunas of various regions*
(After Flessa et al., 1979. Courtesy The Geological Society of America.)

	North America	West Indies	South America	Africa	Madagascar	Eurasia	SE Asian islands	Philippines	New Guinea	Australia
North America		40	55	8	9	19	8	9	6	6
West Indies	67		33	11	9	11	7	7	9	11
South America	81	73		3	7	4	7	6	4	3
Africa	31	27	25		30	25	21	27	17	12
Madagascar	38	27	35	65		32	26	22	22	17
Eurasia	48	27	36	80	69		75	64	25	14
SE Asian islands	37	20	32	82	63	92		73	30	18
Philippines	40	20	32	88	50	96	100		26	18
New Guinea	36	21	36	64	50	64	79	64		46
Australia	22	20	22	67	38	50	61	50	93	

*Values above the diagonal are similarities at the generic level; values below the diagonal are similarities at the familial level.

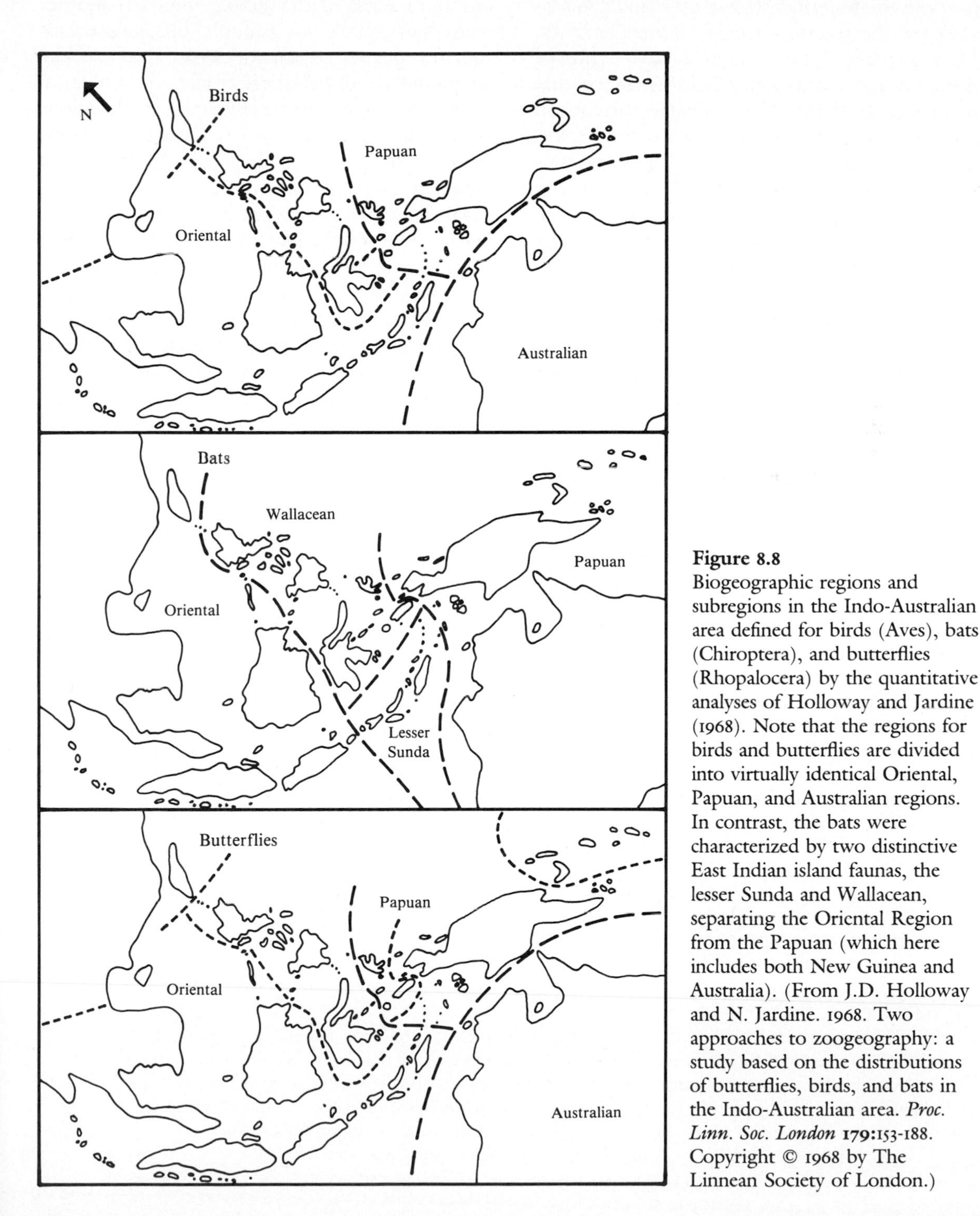

Figure 8.8
Biogeographic regions and subregions in the Indo-Australian area defined for birds (Aves), bats (Chiroptera), and butterflies (Rhopalocera) by the quantitative analyses of Holloway and Jardine (1968). Note that the regions for birds and butterflies are divided into virtually identical Oriental, Papuan, and Australian regions. In contrast, the bats were characterized by two distinctive East Indian island faunas, the lesser Sunda and Wallacean, separating the Oriental Region from the Papuan (which here includes both New Guinea and Australia). (From J.D. Holloway and N. Jardine. 1968. Two approaches to zoogeography: a study based on the distributions of butterflies, birds, and bats in the Indo-Australian area. *Proc. Linn. Soc. London* **179**:153-188. Copyright © 1968 by The Linnean Society of London.)

portions of the land faunas of islands, all assessed at the species level. From these analyses, Holloway and Jardine inferred past trends of dispersal and speciation. They concluded that birds and butterflies have similar patterns in this region but bat distributions are different. Their data also indicated that flying organisms invaded eastward (outward from Asia) in more than one wave. Using the original data of Holloway and Jardine, Nelson and Platnick (1981) have shown how alternative approaches in historical biogeography are possible if the original data are presented in such objective ways.

Although not denying the value of sophisticated analyses such as the ones just described or other methods for quantifying similarities using information theory (e.g., MacArthur et al., 1966; Pielou, 1979), many biogeographers still find some of the simple similarity indexes extremely useful. For example, the Simpson index was devised to quantify in an objective way the influence of historical events and ecological factors on the provincialism of distributions (Simpson, 1940, 1943a, 1953). Using this index, one can infer that the lower the percentage of similarity, presumably the longer the overland distance or the strength of an ecological barrier. Flessa et al. (1979; Flessa, 1980, 1981) have used similarity indexes, including the Simpson index, to analyze the distributions of mammalian genera and families among the continents. These analyses show that the similarity of biotas is best correlated with the kind and extent of present barriers between them and that the Simpson index is a fairly accurate estimator of the overland distance between them (Figure 8.9).

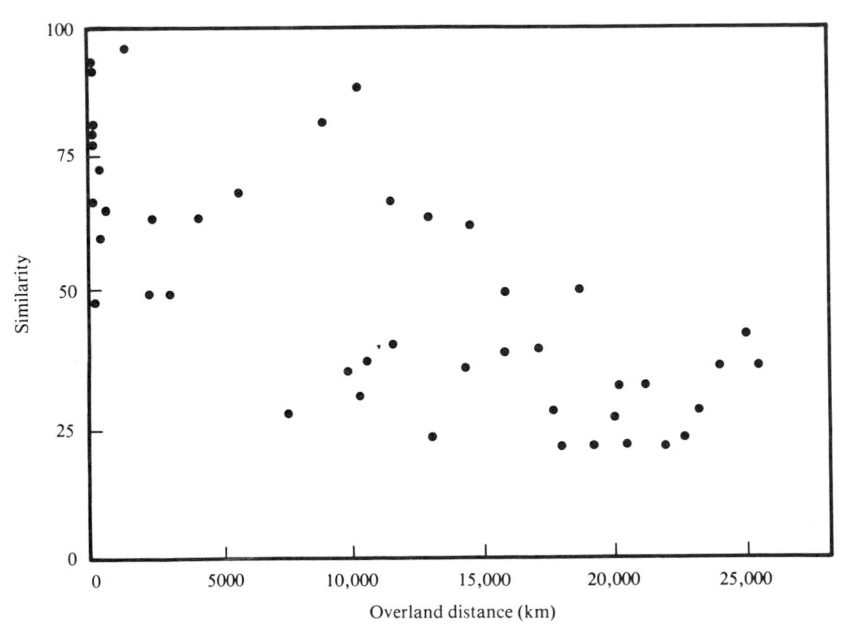

Figure 8.9
Relationship between faunal similarity, using the Simpson index applied to families of terrestrial mammals, and overland distance separating pairs of biogeographic regions. This analysis shows that the existing land bridges and water barriers between landmasses are clearly reflected in the distinctiveness of mammal faunas. Compare this figure to Figure 18.2, which is based on a similar analysis that used the Jaccard index applied to genera of terrestrial mammals. (From K.W. Flessa, 1980. *Bioscience* 30[8]:518-523. Copyright 1980 by the American Institute of Biological Sciences.)

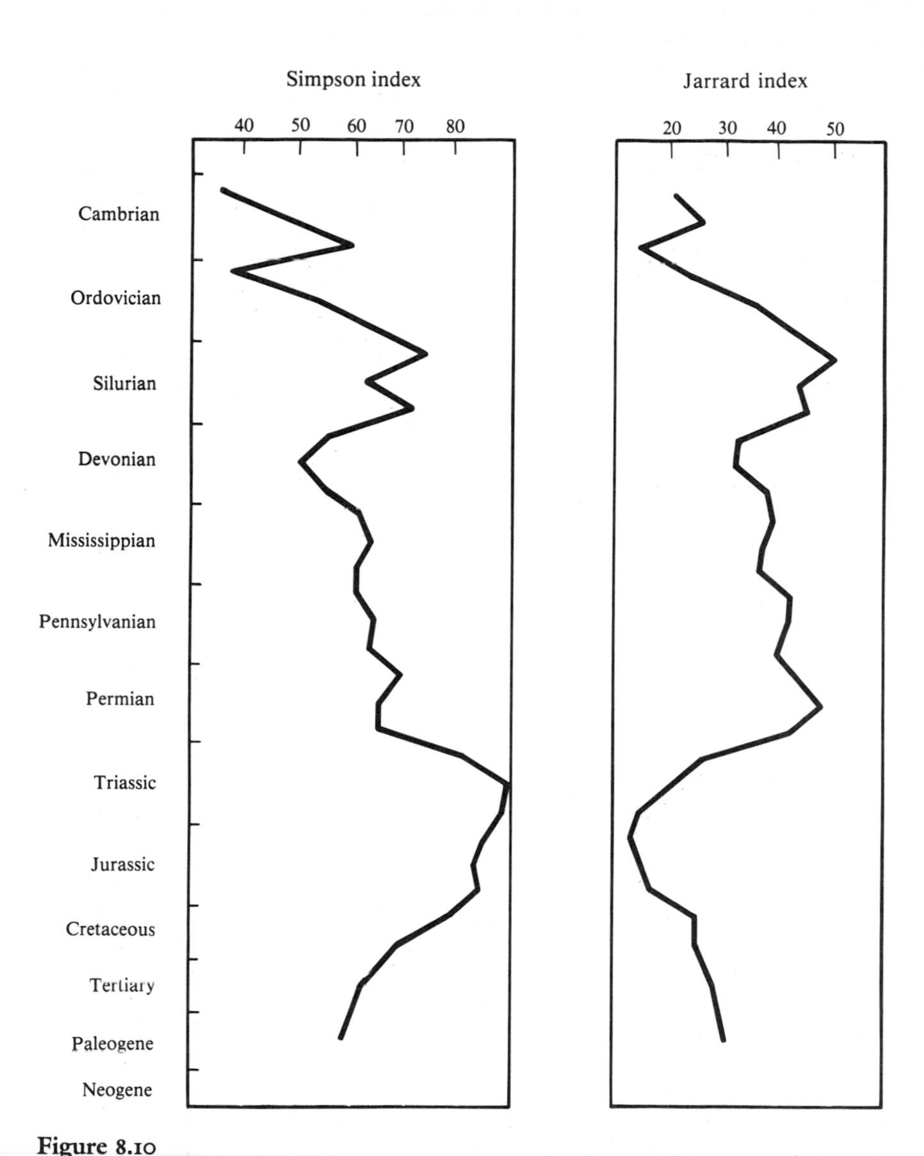

Figure 8.10
Graph comparing the Simpson and Jarrard coefficients of similarity calculated for exactly the same data for shallow-water marine invertebrates on opposite sides of the North Atlantic during the Mesozoic and Cenozoic. Note that the two coefficients give similar results from the Carboniferous through the Permian but suggest opposite trends from the Triassic to the late Tertiary. This distressing inconsistency indicates the importance of understanding the mathematical properties of the different coefficients and emphasizes the hazards of relying on a single index of similarity. (After Flessa, K.W., and J. M. Miyazaki. 1978. Geol. Soc. Am. Bull. **89**:467-477. Courtesy The Geological Society of America.)

Flessa's results suggest little need to invoke historical changes in the positions of landmasses to explain the present distributions of mammals. This is not surprising. Most of the radiation of the class Mammalia that gave rise to extant forms occurred since the Eocene, after the breakup of Pangea.

Although workers in biogeography must use quantitative analyses to evaluate provincialism, they should also be aware that choosing a similarity index or a more complicated method can be a difficult matter. Different conclusions can result from using different indexes. For example, by reanalyzing a paper by Fallaw (1977), Flessa and Miyazaki (1978) show how the Simpson and Jaccard indexes give contradictory estimates of similarity for trans–North Atlantic invertebrates throughout the Mesozoic and Cenozoic (Figure 8.10). This discrepancy is caused by large differences in the numbers of genera between the eastern and western areas. Fallaw (1978) and Simpson (1943, 1947a, 1960) defend the use of the Simpson index over the Jaccard index for certain types of data; but notwithstanding, one must exercise caution in choosing similarity indexes and in interpreting the resulting analyses.

Disjunctions

Disjunctions are those distributions in which two or more closely related taxa live today in widely separated areas. Since the early nineteenth century, one of the chief goals of biogeography has been to explain such distributions. Phytogeographers especially have devoted considerable effort to documenting the major and minor patterns of disjunctions (e.g., Wood, 1972; Thorne, 1972; Good, 1974) and the most remarkable individual examples. Workers have always hoped that knowledge of disjunctions would help to unravel the history of life on earth and to determine how biotas have been assembled through time. Within the last 10 years, disjunctions have become the central theme in vicariance biogeography (Chapter 9), which hopes to use an objective method of analyzing disjunctions for reconstructing sequences of major and minor historical events on earth.

There are innumerable cases of disjunct ranges at all taxonomic levels, and in fact nearly all species, genera, and families of animals and plants have discontinuous ranges at some scale. As a consequence, the term *disjunction* typically is used where the distance separating two populations of a species or two closely related species is substantial. For example, it is not surprising that the beach tree koa, *Acacia koa* (Fabaceae), which occurs on the Hawaiian Islands, has a discontinuous range because island populations are separated by a seawater barrier, but it is remarkable that the closest living species to *A. koa* is *A. heterophylla* of Mauritius Island in the Indian Ocean (Carlquist, 1965, 1974). Likewise, one can speak of the highly discontinuous range of storax (*Styrax officinalis,* Styracaceae) in California, but we think that its occurrence in eastern North America as well as the Mediterranean Region deserves special recognition.

Most discussions on disjunction concentrate on those examples above the species level as well as below the level of the order, a policy we have followed in this book. The reasons behind this choice are of course arbitrary but have some logical merit. First, discontinuities in the range of a single species by and large reflect the importance of both ecological factors and stochastic events in determining distributions; for example, environmental limiting factors prevent the species from being distributed continuously and many isolated areas have been colonized by waif dispersal. Second, disjunctions within a species by and large are considered to be very recently produced, as during the Holocene, with occasional exceptions like *Styrax,* so they have limited value in historical reconstructions. However, remember that groups that are of little interest in historical biogeography may be very informative in ecological biogeography. At the other extreme, orders and higher taxa have rarely been used in historical reconstructions because their distributions are too broad and evolutionary histories too long to focus in on special issues. One prominent exception are the ratite birds and tinamous (Struthiformes) (see Chapter 11).

There is definitely a risk in considering as disjuncts forms that appear to be the closest living pair of species but that are really very distantly related. If the morphological differences are some relatively simple developmental changes between two forms, such as different sizes and shapes of flowers or leaves, colors of pelage or plumage, or modifications in mouth hooks or other feeding apparatus, one can envision how rapid divergence of the two daughter populations could have proceeded in response to drift and selection (Chapter 6) when isolated by a barrier. However, in the case of two very different forms, which require great changes in design and undoubtedly long series of speciation events with many side branches and dead ends, one has to be aware that there is much more evolutionary history in the lineage. Here the missing pieces could have serious effects on simplistic conclusions that are based only on the known forms.

No one person could in a lifetime, even with modern record-keeping devices, simply list all the naturally occurring disjunctions of plants and animals from the genus to the suborder. To do this the worker would want disjunction to be strictly defined as those two taxa that are more closely related to each other than to any other living species (sister taxa). Although phylogenetic knowledge today is much greater than it has been before, there are still huge gaps in our knowledge of the relationships of disjunct taxa. Moreover, collections of tropical organisms are often poor, so judgments on the pattern of disjunction and the relationships of the taxa are always subject to future revision. Instead, biogeographers rely on short regional reviews, such as those for the Southern Hemisphere land areas (Fittkau et al., 1968, 1969; Battistini and Richard-Vindard, 1972; Williams, 1974; Mani, 1974; Kuschel, 1975; Kunkel, 1976; Werger and van Bruggen, 1978; Keast, 1981; Gressitt, 1982) that show where major patterns of relationships can be found in the various classes of organisms. Such summaries are invaluable to the advanced as well as the beginning biogeographer.

Causes of disjunction. Two different events can give rise to disjunct ranges. The most useful for reconstructing the past sequence of land connections would be an event in which a range was at one time relatively continuous but became subdivided by the formation of a geologic barrier or by an environmental barrier that caused intervening populations to become extinct. The second type of event is one in which a species has dispersed over long distances to colonize a distant locality that may or may not be similar to the original one. In many cases a habitat type is and always was disjunct, and closely related organisms that never lived anywhere else had to colonize them via some long-distance phenomenon, for example passive dispersal of certain plant groups between patches of a rare soil type.

So obvious are these different classes that we hardly need to provide extensive lists of examples. The fossil record confirms the existence in the past of numerous broad ranges whereas today the group is confined to one area or two or more disjunct areas in pieces of the former range (Hallam, 1973; Gray and Boucot, 1979). One such example is the sourgum or blackgum *Nyssa* (Nyssaceae), a genus of tree once widespread in the Northern Hemisphere but now confined to eastern North America, Central America, and southeast Asia (Figure 8.11). Throughout Unit Two we have introduced examples of taxa fitting this category. Equally obvious are the cases of plants and animals that have been carried long distances by passive means to colonize distant oceanic or habitat islands (Chapter 7 and 16) or under their own power were able to migrate from one patch to another as part of their life history strategy (Chapters 10 and 12). What concerns historical

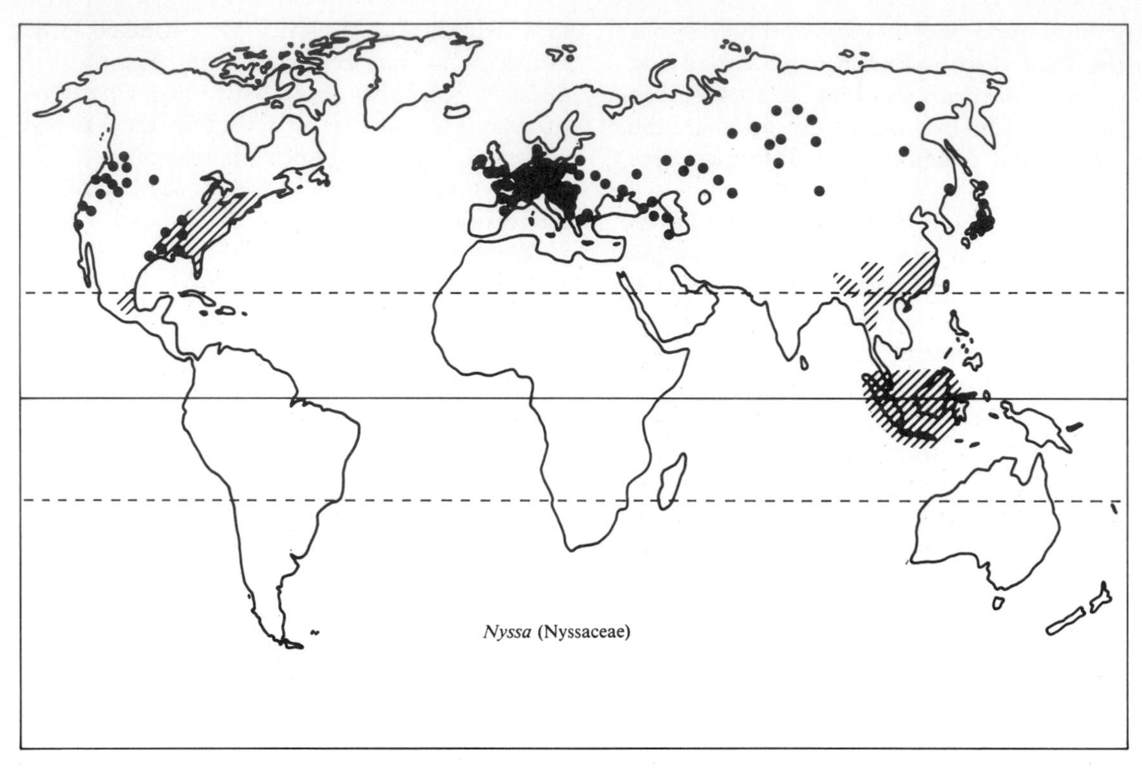

Figure 8.11
Distribution of extant *(shaded)* and extinct *(dots)* populations of the sourgums or blackgums, the tree genus *Nyssa* (Nyssaceae). Fossils show that this taxon was widely distributed across Europe, Asia, and North America during the Tertiary, but it is now confined to eastern North America (including Mexico and Guatemala) and southeast Asia (including the islands of the East Indies out to Wallace's Line). (After Wood, 1972.)

biogeographers is not that both phenomena have occurred and still do occur but rather the need to develop analytical methods that can treat the two causes of disjunctions as alternative hypotheses and can distinguish objectively between them.

The following six chapters deal mainly with disjunctions and how they can be interpreted in historical biogeographic reconstructions. What one discovers as the disjunctions of taxa are identified is that practically every conceivable disjunction occurs. This includes taxa that are disjunct between areas of the same climate and vegetation type, such as around the tropics in the major areas of rain forest; pole to pole, pole to mountaintop, desert to desert, as well as the unexpected ones diagonally across broad oceans, for example, Australia and western North America, or on opposite sides of the world, for example, South America and Madagascar. One remarkable plant disjunction is the little known Datiscaceae, which occurs in dryland habitats in western North America and Turkey as perennial herbs and then as huge tropical trees in tropical Wallacea. Our purpose here is merely to identify the diverse kinds of disjunctions that are known to occur and to offer some thoughts, although no final interpretations, on how these distributions may be explained. In all likelihood, the interpretations of these disjunctions will continue to shift as long as humans study them, even though the groups have not changed their distribution patterns. The critical issue is this: What method should be employed to use disjunctions in the reconstruction of historical events?

Selected references

Battistini and Richard-Vindard (1972); Cain (1944); Candolle (1855); Darlington (1957); Duellman (1979); Ekman (1953); Fittkau et al. (1968, 1969); Florin (1963); Good (1974); Graham (1972, 1973); Gressitt (1982); Hubbs (1974); Illies (1974); Keast (1977a, 1981); Keast et al. (1972); Kunkel (1976); Kuschel (1975); Mani (1974); Müller (1974); G. Nelson (1978); G. Nelson and Platnick (1981); Swainson (1835); Werger and van Bruggen (1978); W. Williams (1974).

Endemism

Beadle (1981); Berra (1981); Bond (1979); Goldblatt (1978); M. Johnson and Raven (1973); Löve and Löve (1963); J. S. Nelson (1976).

Provincialism

Ager (1963); Brenan (1978); Briggs (1974); Bussing (1976); Cheetham and Hazel (1969); George (1962); Goméz (1982); Gray and Boucot (1979); Hagmeir (1966); Hagmeir and Stults (1964); Hallam (1973a, 1973b); Hedgpeth (1957, 1979); Holloway and Jardine (1968); M. Horn and Allen (1978); Hughes (1973); Kremp (1974); MacArthur et al. (1966); Mayr (1944a); Middlemiss and Rawson (1971); Pielou (1979); H. Raven (1935); Schmidt (1954); Schouw (1823); Sclater (1858); G. Simpson (1940, 1960, 1977, 1980a); J. Valentine (1966, 1973); Vermeij (1978); A. Wallace (1860, 1876, 1880).

Disjunctions

Balgooy (1966, 1969, 1971); Bramwell (1979); Carlquist (1965, 1970, 1974); Constance (1963); Fernald (1925); Fryxell (1967); Kikkawa and Williams (1971); Lack (1976); Leroy (1978); Li (1952); G. Nelson and Rosen (1981); P. Raven (1963, 1972); Thorne (1972); D. Valentine (1972); Wood (1972).

Chapter 9 Reconstructing Biogeographic Histories

The realization that historical tectonic events have caused changes in the positions and relationships of landmasses and water masses over geologic time, together with recent advances in evolutionary systematics, has stimulated new interest and recent advances in historical biogeography. If the climate of the earth and the positions of continents had remained fixed over time, this would drastically limit the kinds of explanations that could be advanced for the worldwide patterns of endemism, provincialism, and disjunction. For example, terrestrial disjunctions between the Old and the New Worlds could be explained only by either long-distance dispersal across present barriers or as relicts of distributions that passed through the Northern Hemisphere, where saltwater barriers have not always existed. The dynamic nature of the earth's history, however, makes possible other explanations, such as transport on drifting continents and changes in the nature of barriers as a result of climatic shifts.

Traditionally, the central problem of historical biogeography has been to work out the evolutionary histories of each taxon and to map these histories on the changing surface of the earth. The corollary to this, which is now the avenue of active research, is determining whether the distributions of taxa can be used to reconstruct the exact sequence of geologic and geographic events of the earth, a study done in parallel with such studies in the earth sciences. Both require not only that the reconstruction of the geologic and climatic history of the earth be accurate but also that workers have reliable methods for recapturing the histories of evolutionary lineages and relating these to geography. Present knowledge of the physical history of the earth itself has been summarized in Chapter 5, and the gross details of the continental movements are well understood. Unfortunately, these are not the only events that biogeographers need to know. Ranges of biotas are subdivided by barriers other than deep oceans, e.g., glaciers, epicontinental seas, wide rivers, mountain ranges, and deserts. Geologists can tell us that such structures were present, but they cannot tell us how effective they were in limiting ranges or how such barriers did or did not initiate taxonomic differentiation. These are the issues biogeographers must face head on.

Developing systems of classification has been a human pastime for centuries. The original goals were to facilitate identification, to provide a convenient, practical means for communication, and to store information for retrieval. In the twentieth century major changes have been made in order to portray the true genetic relationships of taxa, such as kinship (genealogy), so as to provide a basis for learning more about evolutionary trends. Thus most contemporary systematists not only strive to make a phylogenetic classification system that accurately represents the evolution of all taxa, but they also attempt to obtain a system that can be tested in a scientifically rigorous fashion.

Having a scientifically sound phylogenetic system is the foundation of historical biogeography. One basic goal is to explain disjunctions: Why do two very closely related organisms today live in widely separated areas, in some cases on opposite sides of the globe? If the taxa discussed actually are not really close relatives, then any analysis is not only meaningless but also wrong and misleading. In addition, having an accurate phylogenetic classification enables historical biogeographers to use new analytical approaches to address old issues that were never resolved in any satisfactory manner. Our intention in this chapter is to consider some of the traditional issues and controversies and to introduce some recently developed analytical methods for resolving some of these issues.

Use of the fossil record

The task of reconstructing the evolutionary histories of organisms has by long tradition been worked on by two groups of scientists. On the one hand, systematists or taxonomists have been attempting to classify organisms into hierarchical groups (taxa) that summarize patterns of similarity. Because speciation is a branching process, the phylogeny of a group, i.e., its evolutionary ancestry, is inherently hierarchical and thus provides a logical basis for a hierarchical classification. Furthermore, phylogenetically related organisms share common ancestors, so they are also likely to share traits possessed by the ancestor. Consequently, a phylogenetic classification that reconstructs the pattern of divergence of taxa from a common ancestor should be a powerful predictive statement about similarity in biological attributes. The pair of species that was formed by the most recent speciation event should share more traits with each other than either does with a third species produced by an earlier dichotomy. Recognizing this, the systematist can use sets of shared characteristics of living forms to reconstruct the evolution of the organisms.

On the other hand, paleontologists have attempted to determine the evolutionary histories of organisms directly from the fossil record. Although this might seem to be a much more straightforward approach, it has its own set of practical and logical problems. For one thing, the known fossil record is woefully incomplete for most groups and its interpretation is filled with pitfalls. Only a minute fraction of all organisms and species that ever lived have been found as fossils. Animals lacking hard tissues and plants without durable chemicals in their cell walls are disproportionately and poorly represented in the fossil record because they are readily decomposed. Of those organisms that achieved fossilization, many have been destroyed by erosion or tectonic processes and many lie in deep, unexamined strata. The take-home message is a disappointing one: the fossil record is biased in favor of easily preserved organisms, geographically uneven, and still greatly unexplored. As if these problems are not enough, the fossils that have been preserved may have been "side branches" of the evolutionary tree that became extinct without giving rise to living species. Such fossils, whether or not they are evolutionarily or biogeographically important, still have to be classified into a hierarchical phylogenetic system on the basis of their known features, which are comparatively few. Hence we should not be surprised to learn that many biologists have a tendency to distrust the systematic conclusions drawn from the fossil record.

Despite these problems, fossils are extremely useful in biogeography and systematics. They provide minimum ages for taxonomic lineages (Table 9.1) and the minimum ages for occupation of an area by a particular taxon. However, no one can insist that the first fossil indicates the actual date of origin for the group. Another value of fossils is to document extinctions from an area and of major groups of organisms. Studies of oldest to youngest fossils in a particular taxon, e.g., mammals, reptiles, or fishes, have aided in determining primitive features and evolutionary trends. For many groups these evolutionary progressions could not have been ordered and relationships properly as-

Table 9.1
Oldest known fossils of selected taxa*

Taxon	Oldest known undisputed fossil	Age (10^6 yr)	Recent fossil review	Number of extant species*
Vascular plants (Tracheophyta)	*Cooksonia*	Pridolian, U. Silurian (405)	Banks (1975)	232,000
Ginkgoads (Ginkgoales)	*Sphaerobaiera*	U. Permian (240)	Bierhorst (1971)	1
Angiosperms (Annonopsida)	*Retimonocolpites* (and others)	Barremian, L. Cretaceous (117)	Doyle (1978)	220,000
Sunflower family (Asteraceae)	*Echitricolporites*	L. Miocene (20)	Muller (1970)	21,000
Crabs (Brachyura)	*Eocarcinus*	L. Jurassic (180)	Glaessner (1969)	4500
Insects (Insecta)	Protorthoptera	Namurian, L. Pennsylvanian (320)	Carpenter (1977)	800,000
Lampreys (Petromyzones)	*Mayomyzon*	M. Pennsylvanian (290)	Hardisty (1979)	31
Anurans (Salienta)	*Vieraella*	L. Jurassic (180)	Estes and Reig (1973)	2500
Turtles (Testudines)	*Proganochelys*	U. Triassic (200)	Gaffney (1975)	230
Birds (Aves)	*Archaeopteryx*	U. Jurassic (150)	Feduccia (1980)	8600
Mammals (Mammalia)	*Kuehneotherium* (and others)	U. Triassic (200)	Lillegraven et al. (1979)	4100
Monotremes (Prototheria)	*Obdurodon*	Miocene (20)	Griffiths (1978)	3
Bats (Chiroptera)	*Icaronycteris*	L. Eocene (50)	Jones and Genoways (1970)	860

*Number of extant species cannot be predicted from age of oldest fossil or projected age of group.

sessed by the study of only extant species. Fossils also provide a conservative record of the diversity of past life, such as the origin and radiation of early land plants and rise of early angiosperms. A major value of fossils is that some sequences are so extensive over long periods as to permit paleoclimatic reconstructions and to provide information on the immigration of forms from adjacent landmasses. Examples are the strata of vascular plants (leaves and pollen) and mammals in the temperate Northern Hemisphere over the last 50 million years. This information provides clues on past migration routes and on the composition of the assemblages that moved. Finally, paleontologists use fossils to study basic patterns of speciation and macroevolution (Chapter 6).

Every few years someone makes startling claims that fossils of a certain taxon have been found that are much older than any existing evidence. These are serious issues for biogeographers and evolutionists and often spark controversy until they are rechecked and either corroborated or invalidated. In the past, scientists eager to be up-to-date frequently cited these new discoveries as "fact" before they could be rechecked; consequently, major dating mistakes appear in the literature, and even in textbooks, and they take a long time to be erased. Here is a classic example. In 1969 Tidwell et al. reported a palm fossil, a modern-looking form, from the Jurassic shale of Utah. Most botanists had concluded that angiosperms first arose in the Lower Cretaceous; thus finding a palm in the Jurassic, especially a modern form, was counter to those conclusions but supportive of others (e.g., Axelrod, 1952) predicting the origin of angiosperms in the Paleozoic. The fossil report was soon included in general textbooks as evidence for early angiosperm origin. Shortly thereafter, a team of eminent geologists and paleobotanists (Scott et al., 1972) revisited the site and provided substantial evidence that the fossil was of Eocene age but had been transported down into older strata. This prompted the retraction of the original report (Tidwell et. al., 1972); however, citation of the "Pre-Cretaceous" angiosperm from Utah continued for some time.

Another famous episode was the discovery of *Amphitherium* of the Middle Jurassic Stonesfield fauna of Oxfordshire, England. When the beast was described in 1838, it was classified as a mammal. This raised an outcry because mammals were then believed to be much more recent in age, and the animal had an unusual jaw. However, it is now recognized that at least two lineages of mammals lived in Europe as early as the Upper Triassic, much longer ago than *Amphitherium* (Lillegraven et al., 1979).

Organisms are sometimes buried in places where they did not physically live. Two contemporary observations from western North America provide insight into how this could occur. It is not uncommon to find decaying specimens of freshwater fishes floating along the shores of the Great Salt Lake in Utah. These species do not live in the lake. Occasionally, they are carried by floods into the saline lake, causing their death and burial there. Nearby, one can place out pollen traps in the desert flats of Nevada and return to find high numbers of coniferous pollen grains in the sample. The coniferous pollen has blown in from mountain populations many kilometers away.

Each investigator must evaluate the fossil record and decide how, if at all, the evidence can be used. Given warnings on the potential errors in dating, identification, and interpretation of fossils, the biogeographer learns to check fossil reports and consult with current experts before using them in preparing distributional scenarios. In cases in which the fossil record is "good," paleontological data are meaningful tools to discriminate between alternative hypotheses; where it is meager, fossils raise more questions than they answer. Therefore the problem confronting the historical biogeographer is to know how fossils can be used in a scientifically sound rather than a whimsical manner. Authors are now trying to define when fossils can be used objectively in historical biogeography and paleontology. For example, Patterson (1981a, 1981b) has asked whether fossils can be used objectively to contradict inferences made from the pattern of present-day taxa. The answer is not a simple yes or no. Fossils play a valid role in documenting extinctions by indicating additional areas that must be analyzed. Fossils are also valuable for providing minimum ages for occupation of an area, which sometimes permits investigators to choose between different geologic events to account for geographic distributions. On the other hand, Patterson questions whether fossils can properly be used, as some have attempted, to identify directions of dispersal from source areas.

Criteria for determining center of origin

In attempting to reconstruct history, it seems logical to begin at the beginning. Since the time of Alphonse de Candolle (1855), historical biogeographers have tried to determine the precise birthplace of each taxon. Two general purposes prompted this search for centers of origin. First, authors wished to know whether certain geographic regions have been more important than others as cradles of new forms and, consequently, to know the underlying factors that promote the evolutionary innovations that are responsible for the success of new lines. Second, to understand how biotas have been assembled, authors needed to plot the routes that organisms followed as they dispersed and differentiated around the world. These goals are laudable, but the methods used to infer the location of the birthplaces have led to endless controversy.

Biologists have repeatedly tried to develop simple rules for determining the center of origin for any taxon. For example, Adams (1902, 1909) listed 10 criteria that could help identify such centers. Some authors have insisted categorically on using a single criterion to the exclusion of others. For instance, Matthew (1915) believed that centers of mammal origin were in the Holarctic, where new, evolutionarily more successful forms arose and eventually supplanted the original ones, forcing them southward to peripheral habitats and into the Southern Hemisphere. Thus the center of origin is where the derived forms reside. This is in direct opposition to the criterion that the center of origin is where the primitive forms live today. The latter is sometimes called the progression rule (Hennig, 1966). According to this rule, an ancestral population remains at or near the point of origin and derived forms disperse outward. In actuality, either explanation might be correct for different groups, because displacement of one form by another depends in part on how they speciate, disperse, and interact with their biotic and abiotic environments.

A thorough evaluation was made by the phytogeographer Stanley Cain (1944), who showed that none of the 13 criteria in use at that time could independently be trusted for demonstrating center of origin (see box). For example, some authors claimed that the center of origin should be where the greatest number of species of that group reside. This is, of course, invalid if the majority of forms inhabit a region of secondary radiation, as do the heaths of South Africa *(Erica),* with 605 species in the Cape region (Baker, 1967), or the hundreds of species of *Drosophila* on the Hawaiian Islands

Criteria used and abused for indicating center of origin of a taxon

(After Cain, 1944.)

1. Location of greatest differentiation of a type (greatest number of species)
2. Location of dominance or greatest abundance of individuals (most successful area)
3. Location of synthetic or closely related forms (primitive and closely related forms)
4. Location of maximum size of individuals
5. Location of greatest productiveness and relative stability (of crops)
6. Continuity and convergence of lines of dispersal (lines of migration that converge on a single point)
7. Location of least dependence on a restricted habitat (generalist)
8. Continuity and directness of individual variation or modifications radiating from the center of origin along highways of dispersal (clines)
9. Direction indicated by geographic affinities (e.g., all Southern Hemisphere)
10. Direction indicated by the annual migration routes of birds
11. Direction indicated by seasonal appearance (i.e., seasonal preferences are historically conserved)
12. Increase in the number of dominant genes toward the centers of origin
13. Center indicated by the concentricity of progressive equiformal areas (i.e., numerous groups are concentrated in centers, and numbers decrease gradually outward)

(Chapter 12). Is the location of a primitive form or the earliest fossil an absolute criterion? Not necessarily, said Cain, because primitive forms often survive in isolated regions, containing few competing species and located far from their original ranges. Examples of such relicts are described in Chapter 8.

To give the reader a flavor for a center of origin narrative, we can examine one case in which numerous criteria yield a similar conclusion. Consider for a moment the sea snakes (Hydrophiidae) of the Indo-Pacific region, a group that has no fossil record. Sea snakes are venomous marine predators closely related to the Elapidae, the cobras, kraits, coral snakes, and mambas. Over 50 species are known from tropical and subtropical coastal habitats along reefs in the western Pacific and Indian oceans, including some brackish inlets and rivers, and several species have entered and adapted to freshwater habitats in the Philippines and the Solomons (Dunson, 1975). The species with the widest distribution, including the range of the entire family, is *Pelamis platurus* (Figure 9.1), which tolerates the coolest water temperatures. *Pelamis* is a pelagic form found predominantly in narrow ocean slicks, where these snakes feed on a variety of prey. By following ocean slicks,

Figure 9.1
Two species of sea snakes. *Pelamis platurus (above),* the most widespread and pelagic species in the family. This individiual was washed ashore on a Mexican beach during a storm. *Laticauda laticauda (below),* is representative of a distinctive group often classified as a separate tribe or subfamily. This specimen was photographed near shore on the island of Fiji. (Photographs by R. E. Brown [*above*] and W. A. Dunson [*below*].)

Pelamis has colonized coastal western America from the Mexico–United States border south to Ecuador, and there is a real threat that the species could invade the Atlantic and the Mediterranean if a sealevel canal is ever built through Central America.

Determining a center of origin for sea snakes is made easier because they possess several useful attributes: (1) cold water (below 20° C) has been an absolute barrier to them, containing sea snakes within the Indo-Pacific Region, (2) sea snakes appear to be a young family (of Tertiary age), (3) closest relatives to Hydrophiidae overlap with them in distribution, and (4) lines of morphological specialization appear to be relatively unbroken, thus eliminating problems of tracing geographic patterns obscured by major extinctions.

A logical conclusion is that Australia-Papua was the center of origin (Cogger, 1975). This region retains a majority of the total species (over 30). Here one finds the putative elapid ancestors *(Rhinoplocephalus* and *Drepanodontis),* swamp-inhabiting Australian snakes, as well as all the hydrophiids regarded as being primitive (McDowell, 1969, 1972, 1974). *Ephalophis,* the putative primitive taxon, is a notably weak swimmer, like the elapids (Dunson, 1975). Most genera of Hydrophiidae are classified in the *Hydrophis* group, and in this assemblage most are Australian. Genera that occur northwest of Australia often have at least one representative in the Australian coral reefs as well. Finally, the genus *Laticauda* (Figure 9.1), which appears to be an early offshoot of the family (a separate tribe or subfamily), occurs around Papua. This reconstruction is not universally accepted, however; see Voris (1977) for a somewhat different interpretation of the same systematic and geographic data.

Narratives like the above imply perfect knowledge of the history of a group, although this is never the case. Therefore a narrative like this is not a useful basis to begin any scientific reconstruction of the geographic history of any taxon. Moreover, how does one go about rigorously testing the reconstruction or the assumptions? A scenario like the one presented is possible because the taxon probably was confined largely to a single area, and in this sense it is relatively uninteresting unless part of a general pattern. Rather than starting with an assumption about the center of origin for a group, current workers prefer first to trace the geographic spread and fragmentation of the group through time and not to look for a specific, static locality for the group's origin (Croizat et al., 1974). The sea snake narrative becomes broad and interesting if other groups have followed a similar pattern of differentiation. In this way, researchers learn whether centers of high diversity with many primitive forms are in any way related to the original distribution of the lineage or represent instead a secondary center of radiation or a relictual center of survivorship.

The relationship between biogeography and systematics

The importance of phylogeny. It is difficult to draw any important conclusions about the biogeographic history of any taxon unless the phylogenetic relationships of the organisms are known. To study phylogeny one must make three basic assumptions: (1) evolution has occurred, (2) patterns of inheritance exist, and (3) at least some of the features may be used to show relationships between taxa and to determine evolutionary sequences. Thus phylogeneticists believe that a fairly sophisticated evolutionary classification is possible if we accumulate enough information and interpret it correctly.

There was a time, in the 1960s and early 1970s, when many systematists argued the merits of purely phenetic classifications (e.g., Sneath and Sokal, 1973). These pheneticists suggested that because the complete phylogenetic history of a group could never be known with certainty, it is more useful and objective to base classifications solely on the phenotypic similarities and differences among contemporary forms rather than trying to reflect evolutionary relationships. This approach has lost favor in recent years with the advent of cladistic methods of phylogenetic reconstruction. This is fortunate for historical biogeography, because the phylogenetic and biogeographic histories are intimately interrelated. Biogeographers need to know which two taxa are most recently related, not which two look most alike. Moreover, a purely phenetic classification makes no assumptions or predictions about the past, whereas a phylogeny is a hypothesis that makes testable predictions about the historical relationships among both groups of organisms and biogeographic regions.

The literature abounds with examples in which poor understanding of systematics has clouded biogeographic relationships. In the earlier studies, especially, workers noted species in a particular genus or family that were anomalous in the overall biogeographic pattern. After careful examination, the taxa were split and reassigned as different phylogenetic groups. One case involved lizards of the family Helodermatidae. The genus *Heloderma,* containing the gila monster and Mexican beaded lizard, occurs mostly in Mexico. These venomous beasts were once classified with the earless lizards of Borneo, *Lanthanotus,* raising the problem of accounting for a very widely disjunct distribution. Later, researchers learned that *Lanthanotus* (now Lanthanotidae), is anatomically very close to *Varanus* (Varanidae), the monitor lizards that are common in the East Indies and other regions of the Old World tropics. This reclassification immediately dispensed with the question of how the wide disjunction had been achieved.

A similar case can be found in the porcupines. There are eight genera of these strange rodents. The four Old World genera range from Africa to southeast Asia, with one form, *Hystrix,* reaching southern Europe. Three of the New World genera are confined to South and Central America, but the other genus, *Erethizon,* is broadly distributed across North America from Mexico to northern Canada and Alaska. One might suspect that these porcupines are all closely related to each other, because in addition to having hairs modified to form long, stout, sharp protective quills, all of them possess the unusual hystricognath arrangement of jaw muscles in which one major branch passes through a large opening in the bony orbit surrounding the eye to attach to the front part of the skull. If porcupines were close relatives one might expect that, regardless of whether they originated in the New or Old World, they dispersed between the two regions, either across the Bering Land Bridge or by a Gondwanaland connection. For example, one prediction would be that the North American genus might be an intermediate in its relationships between the South American and Asian forms and that fossils of additional intermediate species might be found in northern Asia. Although this is a plausible scenario, it is incorrect. The New and Old World porcupines are not closest relatives at all. In fact, each of them is more closely related to other ratlike rodents on the same continents where they reside than they are to each other. The defensive quills and the jaw musculature are apparently the result of convergent evolution. Furthermore, the North American porcupine appears to have split off from its South American ancestors and dispersed northward quite recently, sometime since the completion of the Panamanian Land Bridge. Because the porcupines cannot be considered a single lineage, i.e., a monophyletic group, there is no need to invoke any kind of historical explanation to account for the disjunction between the New and Old World forms.

Even when we are using a taxon that is most certainly monophyletic, it is still imperative to know how the various elements of the taxon are phylogenetically related. Erroneous conclusions can be drawn when the evolutionary relationships within a taxon are not well understood. For example, many authors have used very general distribution maps of families to

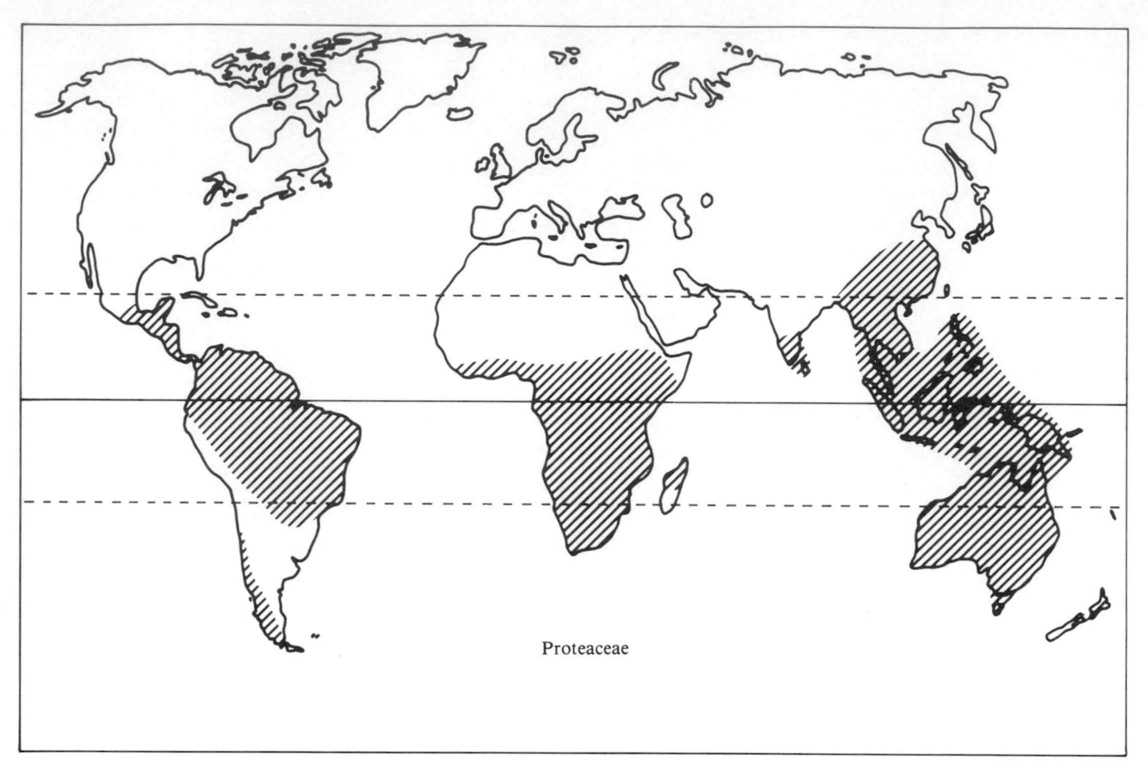

Figure 9.2
World distribution of the plant family Proteaceae. This group found on all southern continents and also on Madagascar and New Zealand but barely reaches the Northern Hemisphere. Hence, this family would be a likely candidate to have occurred originally on Gondwanaland and to have been carried to most of its present locations by continental drift. (Map adapted from Johnson and Briggs, 1975.)

demonstrate that closely related species occur on opposite sides of a major barrier, such as an ocean. Closer examination of the phylogenetic relationships reveals that the taxa opposite the barrier are not, in fact, very close relatives within the family. One case is the plant family Proteaceae, which is present on all southern continents (Figure 9.2) and is therefore a candidate to be a member of the original Gondwanaland biota. A phylogenetic reconstruction of this family provides a more complex picture (Figure 9.3). Genera in southern Africa and in South America are highly specialized members of two different subfamilies, Proteoideae and Grevilleoideae, respectively, so the genera are not closely related at all; actually each is more closely related to those in Australasia (Johnson and Briggs, 1975). The only intrafamilial taxon that is present both in the Neotropical and the Ethiopian region is the tribe Macadamieae of the subfamily Grevilleoideae, a highly specialized tribe that also occurs in southern Asia. On the other hand, the other genera of Grevilleoideae in the New World are closely related to species in southern Australasia and undoubtedly were part of the biota when Australasia and South America were joined by Antarctica.

Many other examples of this sort could be

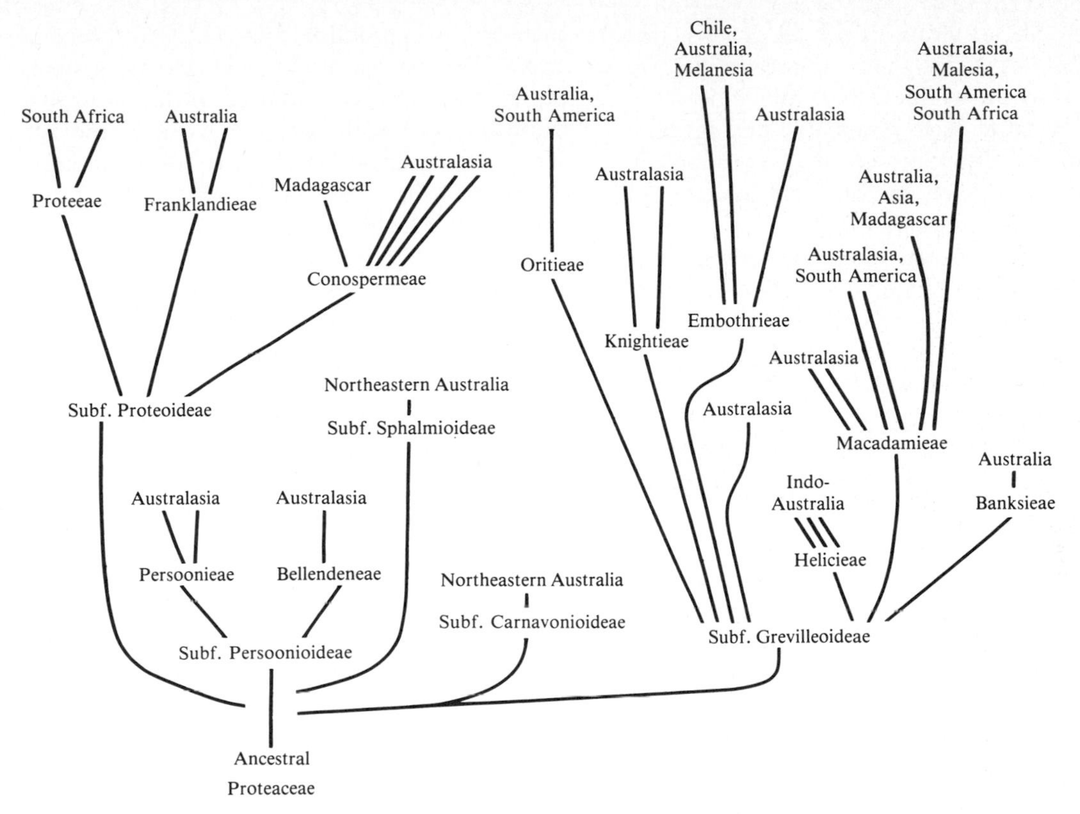

Figure 9.3
Phylogenetic model for evolution of the plant family Proteaceae, showing the relationships among the different subtribes. The number of lines arising from each tribe (e.g., Conospermae or Macadamieae) are the subtribes. (Simplified from Johnson and Briggs, 1975.)

mentioned, but these should be sufficient to demonstrate that it is extremely difficult even to frame reasonable biogeographic hypotheses in the absence of an accurate representation of phylogenetic relationships. This is a problem, because systematics itself is a developing discipline. New methods of inferring relationships are still being worked out and classifications are continually being revised. In the past, biogeographic information was often used by systematists in classification schemes as part of the process of elucidating relationships, but it is important to work out relationships without involving distributions and thereby to avoid circular reasoning when biogeographic analyses are made. It is easy to get trapped into erecting a group of superficially similar organisms because they occur together within a region and then to explain the pattern of endemism as a reflection of a common evolutionary history.

As systematists make efforts to revise the classifications of some groups and avoid past problems, distribution patterns that once seemed easy to explain may raise intriguing new problems for biogeographers. Although it might seem logical in theory for historical biogeographers to wait until taxonomists have finished their task and have produced accurate phylogenetic classifications for all groups, this would be impractical and counterproductive. For one thing, it will be a long time indeed before we begin to have reasonable phylogenetic classifications for many of the more poorly known groups, such as kinds of insects and lower plants. More importantly, it is essential that there be continual interplay between systematics and biogeography; in fact, if biogeography can describe distributional patterns, the patterns can be used to test the validity of existing phylogenetic reconstructions as new classifications are produced.

Determining which taxa are most closely related

Taxonomy, the classification of organisms on the basis of their similarities to and differences from each other, is one of the oldest disciplines in biology. Over 2000 years ago Aristotle produced the first major classification. By the middle of the eighteenth century, after a long series of conceptual advances by European biologists, Linnaeus made the pioneering steps in developing the kind of hierarchical classification that we use today. However, it was only in the late 1800s, after the advent of Darwinism and the general acceptance of the fact of organic evolution, that authors began to work toward a hierarchical system that would reflect as perfectly as possible the phylogenetic relationships of all taxa.

It is important for the beginning biogeography student to appreciate not only where systematics is headed today, but also how far it has come over the last 100 years. A mere 50 years ago the comparative morphology of animals and plants was just starting to play a crucial role in the development of good evolutionary classifications. Comparative studies between extant and extinct forms led to many of the currently held views on what is primitive versus what is derived, what features are most useful in identifying relationships, and so forth. In fact, the first really useful evolutionary classification of mammals was published by G.G. Simpson in 1945. The first avowedly phylogenetic systems for angiosperms were formed in the 1950s, first published in widely accepted form in the late 1960s, and critically revised only recently (Thorne, 1976a, 1981; Takhtajan, 1980; Dahlgren, 1980; Dahlgren et al., 1981; Cronquist, 1981). Although in both plant and animal systematics it is easy to lose sight of the fact that the field is still in its infancy, the contributions to date have been enormous. Whether or not early workers interpreted everything properly is not as critical as the foundation they laid for the exciting and fast-moving field of modern systematics.

The development of present-day systematics has occurred largely since 1950. Like any vigorous field, it is a discipline of changing ideas, competing viewpoints, and clashing personalities. Phylogenetics especially is an area of active debate because the methods are diverse and authors disagree on the assumptions used with each method. There are vigorous debates over how many characteristics should be used, what kinds weighted in what ways, whether taxa must be strictly monophyletic, whether fossils can be used reliably or designated as ancestors, whether a phylogeny should be all-inclusive or be restricted to certain types of data, and how methods can be more formalized. As you will see, none of these issues is trivial. All are germane to the discussion on how to make evolutionary classifications logically rigorous and objectively quantitative.

To clarify the scope and methods of systematics, authors have recognized two general approaches to the field (Cracraft, 1974b; Eldredge and Cracraft, 1980; Nelson and Platnick, 1981; Wiley, 1981). The traditional approach is called evolutionary systematics; a newer, formalized method that was started by W. Hennig (1950, 1966, 1979) is called phylogenetic systematics or cladistics. There are, of course, not only two extreme, polarized schools of thought but also a variety of intermediate approaches that use some tenets of both.

Evolutionary systematics. Traditional phylogenetic classifications, such as those still in use for most taxa, often were devised by the systematist examining the traits of the organisms under consideration and then weighting these subjectively depending on whether the systematist judged them to be either primitive or highly derived, conservative or subject to rapid change. The resultant classifications often attempted to summarize simultaneously two aspects of the history of evolutionary lines: (1) the actual historical branching pattern of phylogenetic lineages, called clades or phylads (cladogenesis); and (2) the grade of evolutionary development attained by a particular taxon (anagenesis). Although there is no one methodological protocol, evolutionary systematics, using data from comparative studies, determines the primitive and derived character states of each character and then arranges the taxa so that the taxon with the most primitive features is placed at the trunk of an evolutionary tree and those with highly derived features are arranged on the branch tips. The worker chooses at the start (a priori) the features that are most important and relies on these heavily to define the categories and to draw the phylogeny. Fossils or extant species are placed into the diagram at branching points wherever they appear to fit the general reconstruction and are chronologically consistent.

Figure 9.4 shows how a phylogeny of living vertebrates might be portrayed by a worker using the methodology of evolutionary systematics. Some groups of organisms are indicated as the probable ancestors of other groups if fossil forms are at the proper level of structural specialization (i.e., the appropriate grade) to have

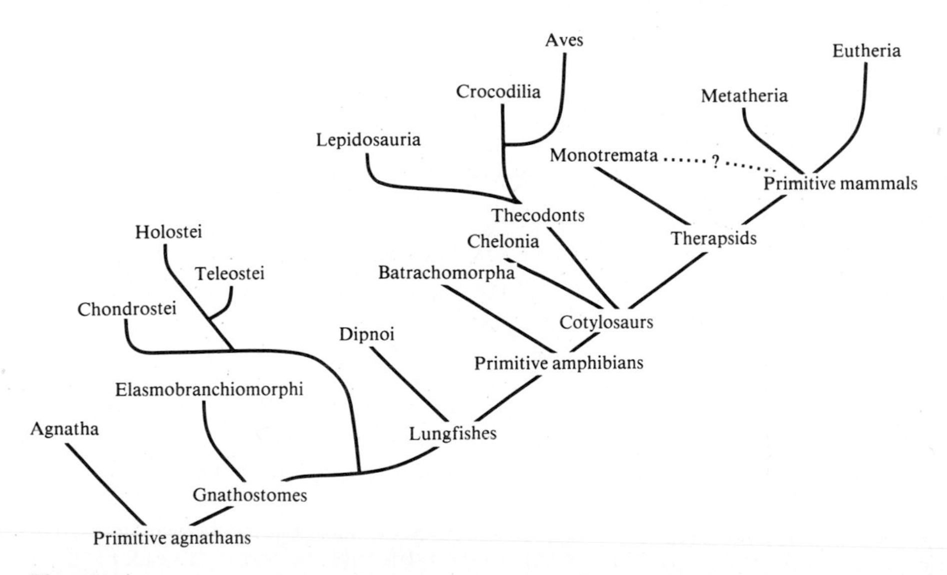

Figure 9.4
Phylogenetic model for the evolution of living vertebrates drawn using traditional methods of evolutionary systematics. Putative fossil ancestors are placed on the diagram to help visualize their characteristics. These ancestral groups often were quite diverse, so from this diagram alone one cannot determine which particular family has been designated as the probable ancestor. Two lines leading to the Monotremata indicate the two opposing views on their origin with most authorities favoring a separate (from other mammals) origin from the therapsids. (Compiled from several sources.)

given rise to the next advanced group. Thus by following through the diagram, one follows the presumed evolution of structural specialization of certain important vertebrate features such as the skeleton, jaw, circulatory system, and so forth; and from the fossils inserted as ancestral forms one receives a general image of what an ancestor may have looked like.

Recently, practices in evolutionary systematics have been severely criticized. In many cases, such broad statements about relationships have permitted taxa to be classified together that are not derived from the same ancestral form, i.e., they are not strictly monophyletic. Fossils have often been overliterally interpreted as ancestor-descendent lineages, and many fossils have had to be demoted from their ranks as ancestors when they were discovered to be far too specialized to have given rise to any group of contemporary organisms. The methods are also culpable because authors often did not fully describe the criteria used for drawing phylogenies. Many early authors, and even some recent ones, seem to have employed more art and intuition than objective science, a confidence game in which one is asked to trust the author's experience in drawing the lines a certain length and the glyphs in a certain size and shape and placing the branches in the right places. Finally, one often cannot transcribe all the information in the phylogenetic diagram into a printed classification of the taxa, nor can the reverse be done by the nonspecialist. Clearly then, there are numerous areas in evolutionary systematics in which resulting classifications cannot be tested directly, even though many of the phylogenetic conclusions may be absolutely correct.

Phylogenetic systematics or cladistics. Both logic and empirical data demonstrate that it can be difficult, if not impossible, to summarize both cladogenesis and anagenesis perfectly in the same classification, because the acquisition of major new morphological features and other evolutionary innovations may not correspond with the patterns of speciation or the origins of higher taxa that determine the branching of phylogenetic lines. A classic example is provided by the crocodiles. The order Crocodilia, which includes the crocodiles, alligators, caymans, and gavials, has been traditionally classified as a group of reptiles because its representatives share many features, such as scales, many simple teeth, and quadrupedal locomotion, with other extant reptilian groups. However, comparative studies reveal that the crocodiles are derived from an ancient group of reptiles, the archosaurs. Archosaurs not only gave rise to crocodiles and some extinct reptilian groups like the thecodonts and pterosaurs but also led to the birds. Thus workers have the unresolvable dilemma that a classification based on grades would keep the crocodiles with the

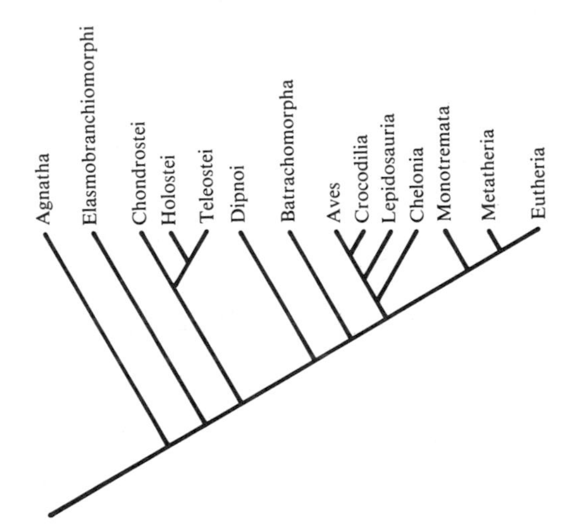

Figure 9.5
Cladistic model expressing phylogenetic relationships among living vertebrates. In this diagram representatives of living or fossil groups are not placed at the branching points to represent ancestral forms. Instead, the shared derived characteristics of the ancestor representing each node can be hypothesized precisely. Note that in this diagram the birds and crocodiles are shown as sharing a more recent common ancestor than either does with any other group. (After Nelson, 1969b.)

reptiles, whereas one based on clades would realign them with the birds (Figure 9.5). This is, of course, an extreme example; and often the evolution of clades and grades are relatively congruent, so they can be summarized reasonably well in a single classification.

The practical problems raised by cases such as the crocodiles point out the logical problems inherent in attempting to construct classifications based on both clades and grades. To escape this problem, an alternative approach was advocated by Hennig to represent accurately only genealogy, i.e., the branching of lineages, and this approach appears to have been gaining favor. There are several reasons for the popularity of a cladistic approach. The first is the logical consistency of basing the hierarchical system of classification on the similarly hierarchical branching pattern of evolutionary lineages. Here the only problem is the trivial one of deciding what levels of the phylogenetic branching to designate arbitrarily as corresponding to the discrete taxonomic levels of genera, families, orders, classes, and so on. On the other hand, a classification based on evolutionary grades must always be a logically unsatisfactory compromise between phylogenetic and phenetic relationships.

Another major advantage of the cladistic approach is that Hennig and others have developed precise methods for deducing phylogenetic relationships based on the characteristics of shared forms. Cladists are distrustful of the traditional, often ad hoc approaches because the methods are not uniform. Like any method of reconstructing history, this one is only as valid as its assumptions, but these are clearly stated and open to rigorous empirical tests. The principal assumptions of phylogenetic systematics are the following: (1) speciation is allopatric (Chapter 6) in the majority of cases; (2) the features analyzed are homologous, i.e., descended from the same ancestors and the same ancestral genes; (3) parallel evolution of individual characters and suites of characters is rare; and (4) organisms with derived characteristics generally do not give rise to taxa with more primitive ones. Given that these assumptions are correct for any group under consideration, the method appears logically sound.

Cladistic analysis includes the following chronological steps. First, the taxa to be classified are clearly defined so that the worker knows how many species are involved and what their characteristics are. For each character, e.g., number of vertebrae or the nature of the crowns on a molar, different forms, called character states, are defined and a transformation series is determined. A transformation series is merely a linear statement representing the evolutionary history of a feature from a primitive condition called a plesiomorphic character state, to a highly derived condition called an apomorphic character state.The plesiomorphic condition is identified through comparisons with outgroups, closely related taxa that presumably share the same ancestry and hence the same ancestral traits. These character states are scored for each taxon, i.e., plesiomorphic or apomorphic, and then the taxa are grouped into categories based on the pattern of shared derived characteristics, called synapomorphies. Traits that are shared by all members of a set, i.e., the primitive condition for the set, do not provide any information for assessing relationships within the group. On the other hand, characters that are shared by some members of the group but not by others are used to define relationships within the group. This assumes that different traits arose independently of each other, so the relative relationships among taxa are reflected by the proportion of the derived character states that they share. Because both the phylogeny and the classification must be hierarchical, the pattern of branching can be reconstructed by examining the distribution of specific traits among taxa.

The product of these analyses is a diagram, called a cladogram, that gives the best (most parsimonious) organization of the taxa based on their synapomorphies. Figure 9.6 shows a simple example with three of the possible cases

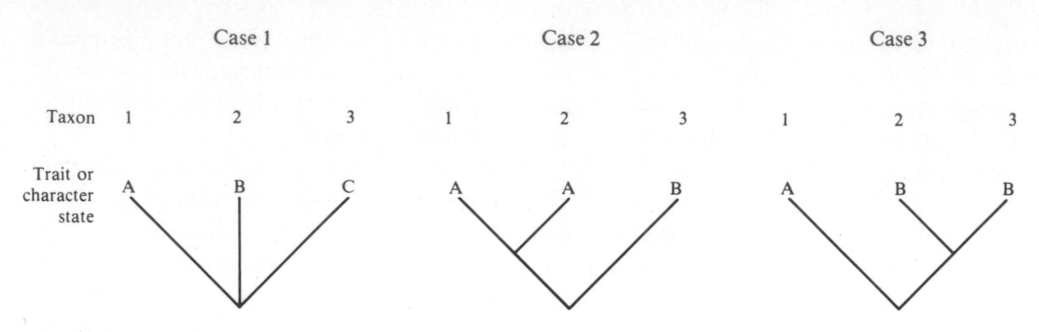

Figure 9.6
Hypothetical cladograms expressing possible phylogenetic relationships among three taxa. In case 1 there is no basis for assuming that any two are most closely related, whereas in cases 2 and 3, two of the three taxa share the same derived trait and are hypothesized to be more closely related to each other than to the third taxon.

for expressing the relationships for three taxa. In case 1 none of the three taxa share derived traits, so there is no basis for assuming that any pair is more closely related than any other. In case 2, taxa 1 and 2 have trait *A*, and it is assumed they branched off from a common ancestor that also possessed trait *A*. This also means that they diverged more recently than taxon 3, which does not share trait *A*. Case 3 is similar to case 2 except that taxa 2 and 3 are more closely related by virtue of sharing derived trait *B*. The cladogram serves as a phylogenetic hypothesis of the taxa studied, to be checked by the analysis of different characters or additional taxa to see whether they falsify the predictions made.

There are 13 possible combinations for the relationships of three taxa, including hypotheses that show the species as ancestors of the others, and as you would expect, the number of possible combinations for higher numbers of taxa gets very large indeed. Therefore one of the principal goals of cladistic analysis is to reduce the number of reasonable choices, getting rid of those that are totally inconsistent and focusing on those that are likely. Then, as new characters become available for analysis, the remaining choices can be judged on the basis of how well each predicts the new results. Ideally, cladograms should be dichotomous statements of synapomorphy so that a definite test can be made at each node (branching junction) on the diagram. However, this is not a necessary requirement for a cladogram, which only hypothesizes the relationships to the level of sophistication of the data used. A node may have three or more emerging branches because the data cannot discriminate between alternative choices by the methods used; or the multiple branches may reveal that the process of differentiation in those taxa cannot be reduced to a simple dichotomous diagram no matter how the data are analyzed.

The final operation in cladistic analysis is to produce the phylogenetic classification. In cladistics this is simple and straightforward because the diagram can be translated directly into a Linnaean system of subordination and phylogenetic sequences. Adjacent branches on the cladogram are called sister taxa, and a cladogram is merely a hierarchical statement of hypothesized sister taxa. Therefore ranks of taxa are decided often by the number of dichotomies, in which each node can be a taxonomic level higher than that of its branches, and so forth, although the naming of levels has considerable flexibility.

Cladograms and their phylogenetic classifi-

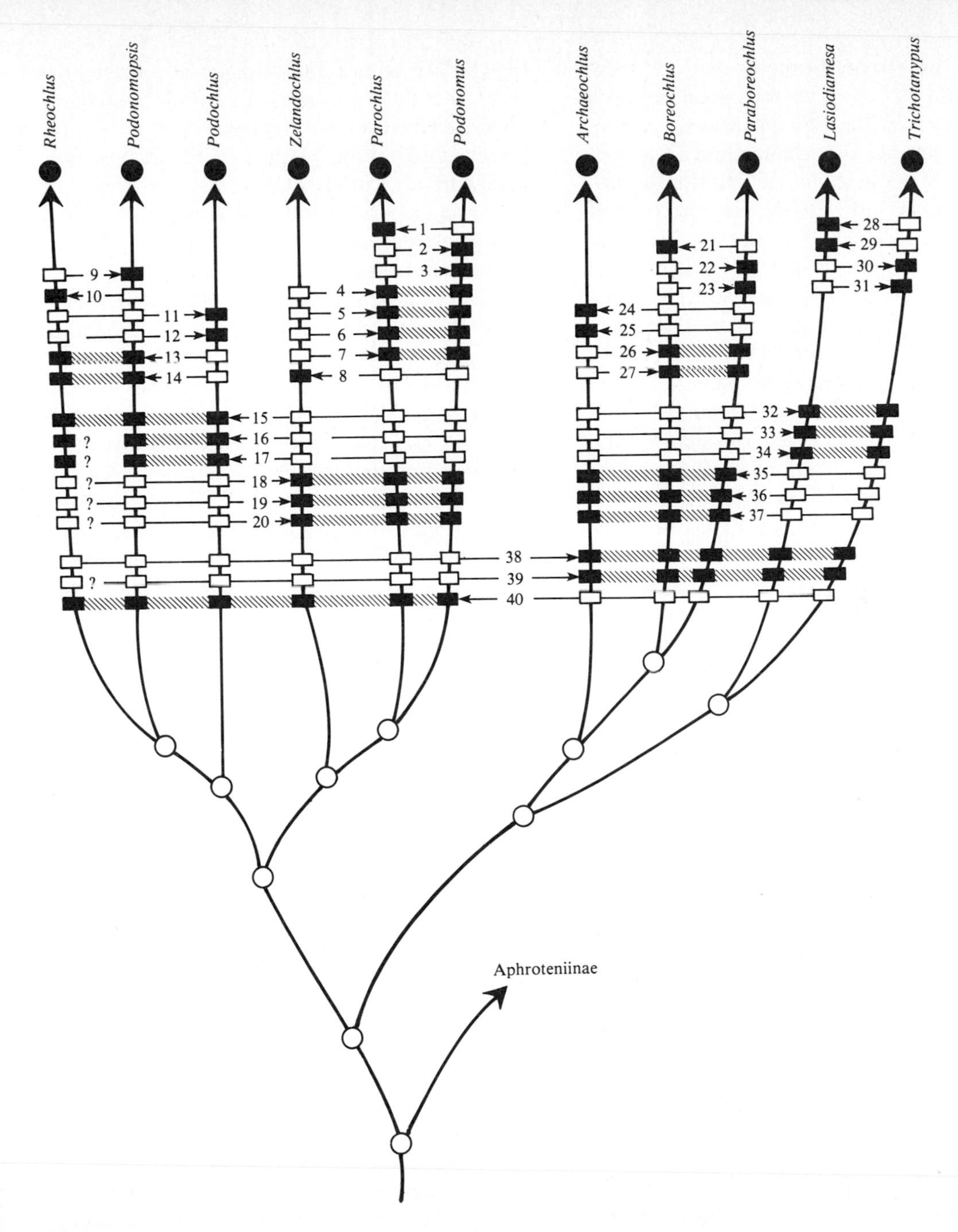

Figure 9.7
Brundin's (1965) hypothesis for the intergeneric relationships of chironomid midges in the subfamily Podonominae. Each bar connecting solid rectangles is a derived character state that defines that taxon and no other one. For example, six genera all share derived feature 40 whereas five others share derived features 38 and 39; and in the latter clade, *Archaeochlus, Boreochlus,* and *Paraboreochlus* share derived features 35 to 37 whereas *Lasiodiamesa* and *Trichotanypus* share derived features 32 to 34.

cations can be tested, both by studying new characters and by repeating the procedure on a larger number of taxa. Discrepancies in the results would indicate that the method is not robust, probably because a small, statistically unreliable number of characters determined critical branch points. For example, because parallel evolution, the origin of similar forms from a dissimilar common ancestor, is notoriously difficult to recognize, many characters should be used in cases in which it is suspected. Otherwise the nodes might be determined by only one or two characters, which could be the ones involved in parallel evolution. Perhaps the most powerful test is a check on the congruence of a cladogram based only on extant organisms with one including the known fossil history of the group, even though there are technical problems in doing this (Patterson, 1981a, 1981b).

An example of a real cladogram, showing reconstructed relationships among chironomid midges (Brundin, 1965, 1966, 1967) appears in Figure 9.7. The solid bars that connect adjacent branches in the diagram indicate the shared derived character states used to determine the nodes. In the original paper the bars are coded to show that each represents a different trait.

Brundin's work on midges still represents one of the few thorough cladistic analysis of a diverse taxonomic group. Much of the work in cladistics continues to be of a rather theoretical nature. Critics of the cladistic approach have challenged many of its assumptions and methods. Many of these criticisms are well founded, and proponents of cladistics have responded by trying to revise the methods and work out the bugs while trying to maintain the logical rigor and objectivity that Hennig's original approach seemed to promise. The details and history of these developments are beyond the scope of this book (but see Eldredge and Cracraft, 1980; Nelson and Platnick, 1981; Wiley, 1981). Suffice it to say that cladistics offers a promising but still largely untried method of reconstructing phylogenetic relationships. Regardless of whether or not the cladistic approach provides the final answers, historical biogeographers need accurate phylogenies to interpret present distributions, especially the disjunctions of apparently closely related taxa, in terms of past events. It is hardly surprising that vicariance biogeography, a promising new method for reconstructing distributional histories, is derived from and intimately based on cladistics, a currently favored method for reconstructing phylogenetic lineages.

Attacks on cladistics are legendary. Some critics note that parallelism is common in certain taxa and that primitive features can reappear in a lineage (a reversible character state of a transformation series). For example, in *Drosophila* parallelism is common (Throckmorton, 1965, 1968). Mayr (1974) attacked the fundamental tenet of cladistics that closest relatives share more derived genes than more distant ones. There are also some problems with assuming allopatric speciation as the only form of geographic differentiation (Chapter 6), and interspecific hybridization, especially in plants, has raised some problems. One by one these issues have been addressed (e.g., Funk, 1981) and adjustments made in the methods to accommodate such problems and to test the assumptions.

Major assaults have also fallen on the cladistic classification, including complaints that one loses much information about the evolution of grades and the time frame of evolution. In general, however, the information conveyed in a well-developed cladistic hypothesis is as useful as traditional ones, if not more so, when properly executed. Many authors have also been reluctant to accept taxa with adaptively incongruous forms such as birds, dinosaurs, and crocodiles and to split up traditional groups such as reptiles that are useful research and teaching categories. Finally, a major objection was the initial insistence on a dichotomous classification, which in groups with numerous branches would result in a highly inflationary system. This has been answered and solved by a process of sequencing whereby numerous taxa can be listed in a classification at a single rank. In a situation, for example, in which five families occur in a single suborder without superfamilial categories, each taxon is the sister of the next one on the list, thus permitting a cladogram to be drawn from the classification. Hence solutions to rectify most complaints about cladistic methods and classifications have been proposed by authors trying to work out the "bugs" and still keep the method objective in all aspects (Eldredge and Cracraft, 1980; Nelson and Platnick, 1981; Wiley, 1981). In reality, systematists eventually want to know which two species were most recently produced from a common ancestor, and cladistics or some future form of it may be the key to that answer. Moreover, biogeographers need to have that type of information to interpret disjunctions.

Vicariance biogeography

What can the distributions of organisms tell us about the history of the earth? Why are certain taxa widely disjunct, and how were these disjunctions achieved? These major questions being asked today in historical biogeography are not fundamentally different than those asked a century ago (see box, p. 11), but the methods used now are more refined and the answers are often considerably different.

There are three primary reasons why the explanations have changed, sometimes drastically, even within the last 30 years. First, the phylogenetic relationships of most taxa had not been satisfactorily determined and, as in any discipline, a certain volume of descriptive work is required before major syntheses can be produced. Second, much of the early discussion about land connections between now-distant continents concerned the distribution patterns of groups too recent in origin to answer important geologic questions. We can appreciate, of course, why the large familiar organisms, such as birds and mammals, and small, well-classified taxa were studied first, but we suspect, for example, that most of these groups originated after the separation of Africa and South America. Finally, but not least, much of the thinking of the period had been conditioned by the permanence theory, the idea that the earth has always had a stable configuration of continents and ocean basins.

This last point, namely the intransigence of leading figures to recognize or consider alternative explanations, has been cited as *the* principal factor holding back scientific discoveries in historical biogeography. Not denying that this occurred, we must keep in mind that prior to the mid-1960s biological evidence for the past union of southern continents was meager, and other than groups like glossopterid gymnosperms, *Nothofagus,* certain freshwater fishes, ratite birds, and a few others, a case for a common history for many forms within the Southern Hemisphere was equivocal. In fact, nearly all of the phylogenetic models for southern groups have undergone major revisions since the mid-1960s. Earlier classifications could not have been used in any type of Popperian methodology because they did not meet the current standards of phylogenetic systematics. Whereas many writers underestimated the importance of Mesozoic land connections for the dispersal of life and to account for disjunctions, most syntheses were neither entirely wrong nor completely right. Certain themes, such as the importance of the long geographic isolation of Australia and South America, have been substantially corroborated. Meanwhile, many of the examples originally used to demonstrate continental connections in the Southern Hemisphere have been discredited and replaced by examples from other groups or at higher taxonomic levels.

Given the background of historical biogeography, in which interesting ideas were disregarded because they were in conflict with the accepted geologic descriptions of earth's history, numerous biogeographers now ask whether historical analyses can be made in a manner that allows our hypotheses to remain objective without depending on the current geologic model for approval. In other words, historical biogeography should develop in parallel with and complement but not be subservient to another discipline, such as plate tectonics. Under these conditions, it is hoped that biogeography can contribute new knowledge on the history of the earth and evaluate independently the conclusions of other fields.

A historical perspective. Cladistic systematics has had a major impact on historical biogeography as investigators have realized that similar techniques can be used to reconstruct the geographic histories as well as the phylogenetic relationships of taxonomic groups. This approach, often referred to as vicariance biogeography, has been popularized only within the last decade, although some of its beginnings can be traced to the writings of the phytogeographer Leon Croizat (1952, 1958, 1960, 1964).

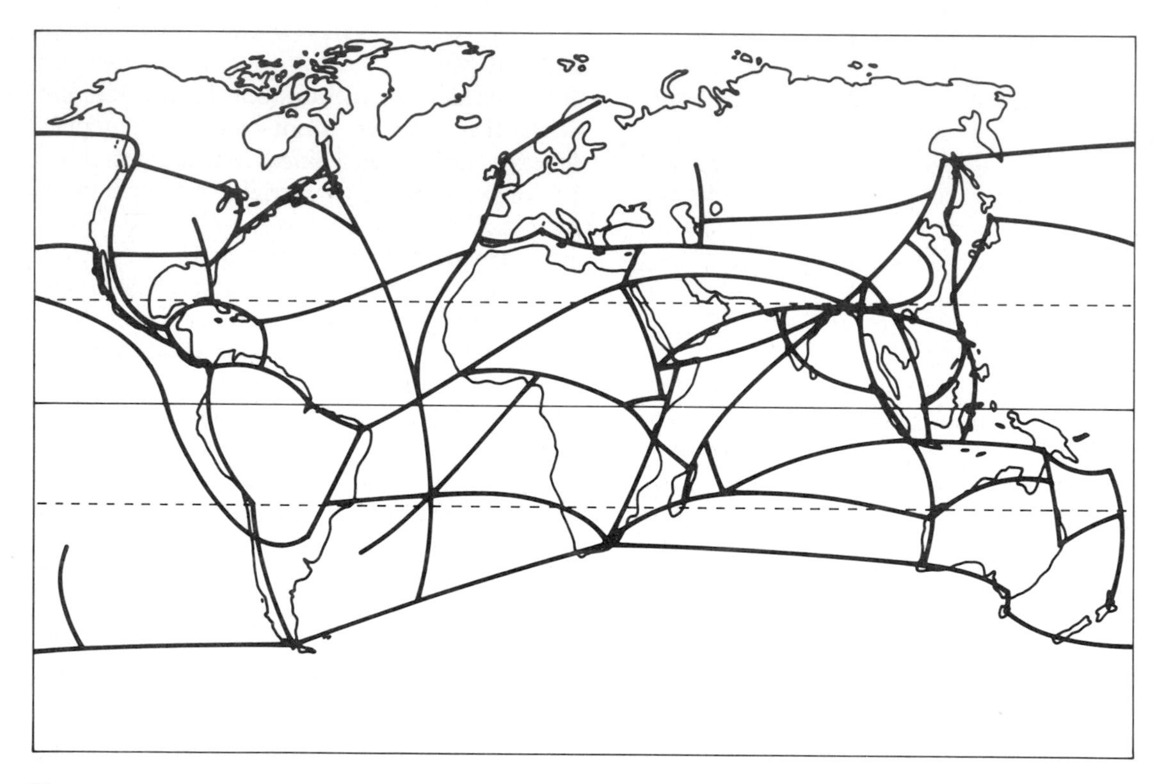

Figure 9.8
Composite drawing of the major and many moderately important "tracks" hypothesized by Croizat (1952, 1958, 1960, 1964) to show hypothesized distributions of ancestral land biotas connecting distant regions. These tracks are derived by plotting distributions of endemic species in several taxa and then drawing lines to connect regions that share disjunct endemics in different groups. For example, tracks radiating out from Madagascar show that its endemics are most closely related to other endemic taxa in South America, tropical eastern Africa, tropical India, Melanesia, and western Australia. Some of the tracks mimic known land connections, such as the Bering Land Bridge and the Mesozoic connection between Africa and South America, whereas others connect continents with islands and islands with each other. In general, the tracks avoid desert regions and open oceans, and instead follow the backbones of principal mountain ranges and island arcs.

Like many phytogeographers before him, Croizat recognized that the present limited distributions of narrowly endemic, disjunct taxa are the historical relicts of entire, integrated biotas. Unlike previous workers, however, Croizat suggested that all organisms with similar disjunctions, whether they have the capacity for long-distance dispersal or are conservative in the dispersal modes, should be treated as having equal importance in analyses.

Using disjunctions from all the continents, Croizat attempted to demonstrate that each disjunct pattern was, in fact, the result of fragmentation of an entire biota that formerly connected those regions. Positions of narrowly endemic species were plotted on a map and lines, called tracks, were drawn connecting sister taxa. Where numerous tracks were superimposed, Croizat labeled these generalized or multiple tracks (Figure 9.8). The generalized track presumably represented what was left of an entire biota that formerly was continuous but subsequently fragmented. By plotting all the generalized tracks on a map, one could theoretically recreate how regional biotas developed in time and space. Croizat rejected dispersal scenarios to account for these coincidences in distribution because he reasoned that "dispersal must be orderly and continuous in time and space" (1958). He suggested that we should not trust our senses concerning dispersal, because long-distance dispersal is not and has not been a major force, even for the colonization of distant oceanic islands.

When first published, Croizat's ideas received little favorable response. For one thing, the basic concept of historical subdivision and isolation of former biotas was hardly new. For another, some of Croizat's own ideas and examples were so extreme or ludicrous that many could not take his work seriously. Especially distressing was Croizat's insistence that long-distance dispersal had to be dismissed for explaining disjunctions. His books had many technical and phylogenetic errors. In addition, with all the disjunction information in hand, Croizat categorically denounced Wegenerism (Chapter 5) and sought different geologic mechanisms for explaining amphioceanic disjunctions. Only much later did he reverse his stand.

As noted above, the essence of the vicariance concept was a venerable theme in biogeography. Systematists had long noted that certain geographically isolated regions contained sets of related species in many different groups, suggesting that the areas were once part of a continuous biota. Two persons are especially noteworthy in advancing this concept. From 1840-1878 Asa Gray, an eminent American botanist, published a series of taxonomic papers on the similarities between plant species in North America and Japan. Since Linnaeus, authors had noted strong similarities between these areas, but Gray made intensive studies to show that there are an unexpectedly high humber of species pairs and "identical species" shared between these very distant regions. Gray proposed that a continuous flora existed across the northern Northern Hemisphere through the Bering Strait, which was subsequently broken up by Pleistocene glaciation leaving disjunct populations of those species stranded on different continents. Through time some of these populations differentiated as new species and others remained essentially unchanged. These conclusions were later reworked by Li (1952) and Wood (1971, 1972) to the point that most are now judged as pairs of species, morphologically distinct and reproductively isolated.

In 1908, David Starr Jordan, a renowned ichthyologist, published a taxonomic analysis of organisms living on the two sides of the isthmus of Panama. He described the morphological changes in groups that have occurred since the land barrier formed about 4 to 5 million years ago, and severed a marine connection and the continuous marine fauna between the Gulf of Mexico and the Pacific Ocean. A great many of the organisms in the two areas are now morphologically distinguishable, usually as separate species that he called geminate or twin species.

The coastal fishes show this pattern admirably with over 100 pairs, and so do the crabs, echinoderms, and nearly every successful littoral marine group in the area (Ekman, 1953). Vermeij (1978) lists about 250 pairs of geminate species (vicariants) of molluscs on the opposing coasts of the barrier. Among the seven species of sea turtles, four *(Dermochelys coriacea, Eretmochelys imbricata, Caretta caretta,* and *Chelonia mydas)* have different subspecies in the two areas, but most pelagic and abyssal forms, including, for example, most epipelagic fishes, do not show speciation caused by the land barrier (Parin, 1970).

Notwithstanding these earlier studies, Croizat had begun to develop formalized methods for inferring past distributions from present ones, a facet that was missing from historical biogeography. In addition, it was important that Croizat insisted that such methods should be applied first without regard for the dispersal properties of the taxa under consideration. Croizat was reacting to the many traditional accounts in which the listener is asked to trust the senses and to interpret the past dispersal of organisms in light of the present-day properties of those species or families. The audience must tolerate untestable assumptions that certain organisms crossed vast stretches of ocean or land via long-distance dispersal. These narratives include long discourses about singular organisms caught alive in places where they normally do not live and breed making the point that they do occasionally disperse long distances and, by chance and given enough time, are likely to land successfully on some distant shore and survive. Even though most of these narratives are probably correct, Croizat and many later vicariance biogeographers dismiss these dispersal scenarios as metaphysical and untestable and therefore invalid.

It remained for others, especially G. Nelson, D. Rosen, and N. Platnick from the American Museum of Natural History, to overhaul the original methods, tie these to good phylogenetic hypotheses, and then interject vicariance concepts into the mainstream of current biogeographic thought. The evolution of the vicariance methods of Croizat to the development of a sound methodological construct was closely tied to cladistic studies and was built on basically the same premises.

The method. Although vicariance biogeography remains an area of extremely active research and rapidly changing techniques, the current state of the art is summarized in a number of recent publications, perhaps most clearly by Platnick and Nelson (1978), Nelson and Platnick (1981), and Wiley (1981). Platnick and Nelson note that historical explanations for disjunct distributions of related organisms fall into two classes: dispersal explanations (dispersal biogeography), in which the organisms are assumed to have migrated across preexisting barriers, and vicariance explanations (vicariance biogeography), in which the formation of barriers fragmented the ranges of once continuously distributed taxa. Extant distributions are usually inadequate to determine whether a given barrier was in existence before or after the migration of a particular taxon to its present disjunct areas. This is particularly true if the taxon occurs in only two isolated regions. If, however, the organisms inhabit three or more disjunct areas, the techniques of cladistics can be used to determine the relative degree of relationship among them (Figure 9.9). If one can assume that the rate of divergence for populations is proportional to the time of their isolation, then the phylogenetic relationships also indicate the age of separation of the disjunct groups. Thus the cladistic phylogeny not only provides a phylogenetic hypothesis about the historical relationships among taxa but also constitutes a biogeographic hypothesis about the historical relationships among geographic localities. Both hypothetical reconstructions reflect the hierarchical branching of a taxonomic lineage in time and space. The hypothetical geographic reconstruction is called an area cladogram because of its precise logical analogy to a phylogenetic cladogram.

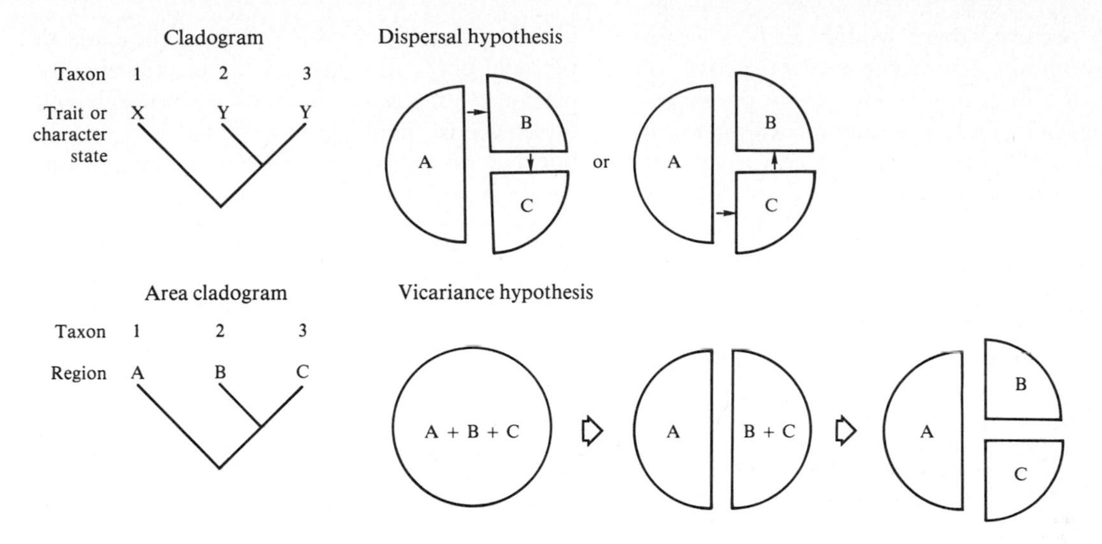

Figure 9.9
Hypothetical cladogram showing presumed phylogenetic relationships among three taxa, and corresponding dispersal and vicariance hypotheses to account for their distribution. Note that even without any extinctions and multiple colonizations there are two dispersal hypotheses that account equally well for the known data, but only a single area cladogram and vicariance hypothesis.

Figure 9.9 shows a hypothetical example. Three taxa are distributed in three disjunct areas. The arrangement of the diagram indicates the relationships among the taxa as revealed by cladistic analysis. Taxa *2* and *3* are more similar to each other than either is to taxon *1*. From this one can infer that taxa *2* and *3* share a more recent common ancestor and a more recent common geographic range than either does with taxon *1*. One can now advance two kinds of biogeographic hypotheses that are consistent with these relationships. A dispersal hypothesis would have an ancestral population inhabiting area *A*, with a propagule dispersing to area *B* and then, after some additional time, another propagule dispersing to area *C*. Because taxa *2* and *3* are symmetrically related to each other, the colonization order could also have been $A \rightarrow C \rightarrow B$, but dispersal from either *B* or *C* to *A* is inconsistent with the phylogenetic diagram. A vicariance hypothesis would have the ancestral taxon inhabiting all three areas, *A*, *B*, and *C*, followed by the formation of a barrier to isolate *B* and *C* (still interconnected) from *A*, and finally by the formation of a barrier to isolate *B* from *C*.

How would we test these hypotheses? Platnick and Nelson (1978) and others point out that for a single taxonomic group, dispersal hypotheses are extremely difficult to falsify, even with information from the fossil record. This is especially true if one allows for the possibility of repeated episodes of colonization and extinction; then almost any pattern of distribution of fossil forms is possible. On the other hand, a vicariance hypothesis is easier to falsify because it makes explicit predictions about historical connections among areas. These predictions may be compared with data on the geologic and climatic history of the earth, the distributions of extinct forms of the taxon being considered, and the extant distributions of other taxa that occur in the same areas at the same time. In the case of our hypothetical example in Figure 9.9, clear geologic evidence that the barriers between the areas were in existence before the taxa occurred there would falsify a vicariance hypothesis. The occurrence of a fossil of taxon *2* or *3* in area *A* would falsify the vicariance area cladogram, as would a fossil of *3* in *B* or of *2* in *C*. One could also construct area cladograms for other extant groups with endemic taxa in areas *A*, *B*, and *C*. If the two groups were in existence at the same time one would expect them to respond similarly to sequential fragmentation of their ranges, so they should, if they have differentiated in each area and have not experienced extinction in those areas, exhibit similar area cladograms. Of course, there is some probability that area cladograms will or will not be congruent by chance alone. However, these probabilities can be determined because they depend on the number of areas and taxa being considered. As the number of areas and especially the number of independent taxonomic groups included in the analysis increases, the probability of observing congruent cladograms simply by chance diminishes rapidly. Calculating these probabilities may not be as simple as it may seem, but at least precise probabilities can someday be produced so that patterns can be tested in a statistical framework (Simberloff et al., 1981a, 1981b).

Current status. At present it is difficult to evaluate the substantive impact of the vicariance approach on our understanding of the biogeographic history of different kinds of organisms. Certainly the emphasis on logical rigor and quantitative analysis represents a major advancement, whatever the ultimate fate of the methods themselves. In the first place, this interjection of a new method has forced historical biogeographers to take more care in describing disjunctions by making certain that the phylogenetic relationships of each unit are carefully studied and discussed. Second, workers will be discouraged from publishing idle speculations on the causes of particular disjunctions without some in-depth comparisons with disjunctions in other taxa occupying similar areas. Finally, it has taught biogeographers to evaluate both vicariance and dispersal explanations rather than just choosing one or the other on the basis of personal bias and using it without critical comparisons. Vicariance methods do not rule out explanations invoking long-distance dispersal, but they do discourage using such explanations before evaluating alternative hypotheses.

Whether vicariance methods result in lasting contributions or not will depend not only on whether the assumptions and cladograms survive empirical tests and prove to be robust and sound but also on whether the task of constructing and testing taxonomic and area cladograms can be made a truly practical endeavor for systematics and biogeography. At present, the development of the science is still mostly in the theoretical stage. If vicariance biogeography is to have an impact more lasting than a heuristic call for objectivity and quantification, however, those who do the actual work of revising taxonomic groups and working out their geographic histories will have to adopt the methods of cladistics and vicariance biogeography. Application of these methods to specific taxa and geographic regions will have to provide new insights into the past. It is simply too soon to tell whether or not this is likely to happen.

There have already been several case studies in vicariance biogeography, but so far these have been like demonstration models of new automobiles or other consumer products. They look untarnished, impressive, and useful, but only time will tell if they are economical and durable or if they are to be replaced by new and better models even before they have a chance to be widely used. Recent case studies such as those of Brundin (1965, 1966, 1967) (Figure 9.7 and Chapter 12) on chironomid midges (Chironomidae) and Rosen (1975, 1978) on the organisms of the West Indies are promising, if only because they have served as successful demonstration models. At least it is now known that the methods can be practically applied to real situations and that the resulting cladograms are reasonable biogeographic hypotheses. However, if vicariance biogeography is to maintain its momentum, it must now pass beyond the theoretical stage and begin demonstrating its practical value by producing lasting empirical contributions. Hence it is somewhat discomforting to find that in cases in which the area cladograms for a number of different life forms have been compared for the temperate Southern Hemisphere, several patterns and not a single general pattern are revealed (Humphries, 1981).

Some persistent issues in historical biogeography

In the previous section we attempted to provide an objective overview of vicariance biogeography: what it is and where it hopes to go. This does not mean that all workers in historical biogeography have accepted vicariance methods as the best or only approach, nor does this imply that all who recognize the value of vicariance methods fully agree with the interpretations of multiarea cladograms.

Stochastic elements in historical biogeography. Although vicariance methods offer much promise for resolving certain kinds of problems in historical biogeography, they are, at least in their present stage of development, inadequate to provide satisfying answers to other kinds of questions. It seems clear that long-distance dispersal hypotheses must account for many disjunct distributions. Investigators know this must be true, for example, for the inhabitants of oceanic islands, and it is confirmed by the experience of the recolonization of Krakatau following its eruption (see Chapters 7 and 15) and in other similar cases. Although it may be, as Platnick and Nelson (1978) have pointed out, that stochastic elements of colonization and extinction make it difficult or impossible to distinguish between alternative dispersal hypotheses, it does not mean that we should stop talking about them. Many of the organisms on an oceanic island group, such as the Hawaiian Islands or the Galápagos Archipelago, provide two-area statements: the island endemic is the sister taxon of a form on the mainland, and all the close relatives of that mainland species also reside in the same mainland area. For example, over 90% of all the plants on the Galápagos Islands occur also in the nearby lowlands of Ecuador or the neighboring Andean Cordillera (Porter, 1979). Workers in vicariance biogeography would assert that such two-area statements are uninformative; and as we have just discussed, this is true from a logical viewpoint because one cannot determine the time that a disjunction occurred with only two areas. However, comparing the biological properties of island and mainland forms, island biologists have been able to provide a convincing reconstruction for the origin of the biota via long-distance dispersal. It is unlikely that the few multi-area cladograms that do apply to the Galápagos organisms will show a different interpretation or that any pattern arising out of the few cases could be distinguished from what one would expect by chance alone.

Although vicariance biogeographers are asking why taxa occur in particular areas and not others, other historical biogeographers would like to understand why a taxon is absent. So far, scientific methodology has not been developed that deals with the highly stochastic elements of extinction and long-distance dispersal in historical analyses. Results of these partially stochastic processes are likely to obscure to varying degrees the kinds of highly deterministic patterns that might otherwise be expected to result from the general spread and fragmentation of ranges under vicariance models. Even though these questions are difficult to answer, they are interesting and should not be restricted from our study.

Consider, for example, the history of life on the southern continents. Several different taxa have endemic forms in Africa, South America, and Australia that are well suited for developing and testing vicariance models. Area cladograms for these groups can be prepared and tested by comparing them with each other, the distributions of fossil forms, and current geologic data on continental drift. Even if a single, parsimonious reconstruction can be obtained, however, it cannot account for the absence of most of these taxa from other Gondwanaland regions: Antarctica, India, Madagascar, and New Zealand. These absences are interesting, and reasonable hypotheses can be advanced to account for them. The obvious one is that the ancestral forms were present throughout Gondwanaland, including those portions where the taxa are missing today. Once the ancient supercontinent became fragmented, some of the isolated populations became extinct: on Antarctica, because the continent drifted into polar latitudes of unsuitable climate; on India, because the biota was eliminated by competition with diverse Oriental forms after the Indian Plate collided with southern Asia; and on Madagascar and New Zealand, because these were relatively small areas that supported limited populations that consequently had high probabilities of stochastic extinction. These hypotheses are interesting, but they will be difficult to test, and in the meantime authors should feel free to discuss them even though they do not fit into the restricted view of historical biogeography as defined by vicariance methodology.

Assuming that common patterns have common causes. Another source of strong disagreement between vicariance biogeographers and those who propose dispersal scenarios is the assumption of vicariance biogeography that common patterns result from common causes (McDowall, 1978b). In this case, a shared ancestral biota is used to account for coincident distributions with coincident patterns of differentiation. This is based on two assumptions: that all elements with similar disjunct distributions must be of the same age in the regions presently occupied and that all these elements dispersed along identical pathways. Both of these premises are difficult to test. Suppose, for example, the extant pattern is Eurasia–Africa–South America. There are at least three simple and plausible interpretations for explaining the disjunction between Africa and South America: (1) an ancestral group was shared when the continents were connected and divergence followed the split, (2) the organisms crossed the water gap after the continents started to drift apart, or (3) the organisms dispersed through adjacent landmasses, such as Antarctica or the northern

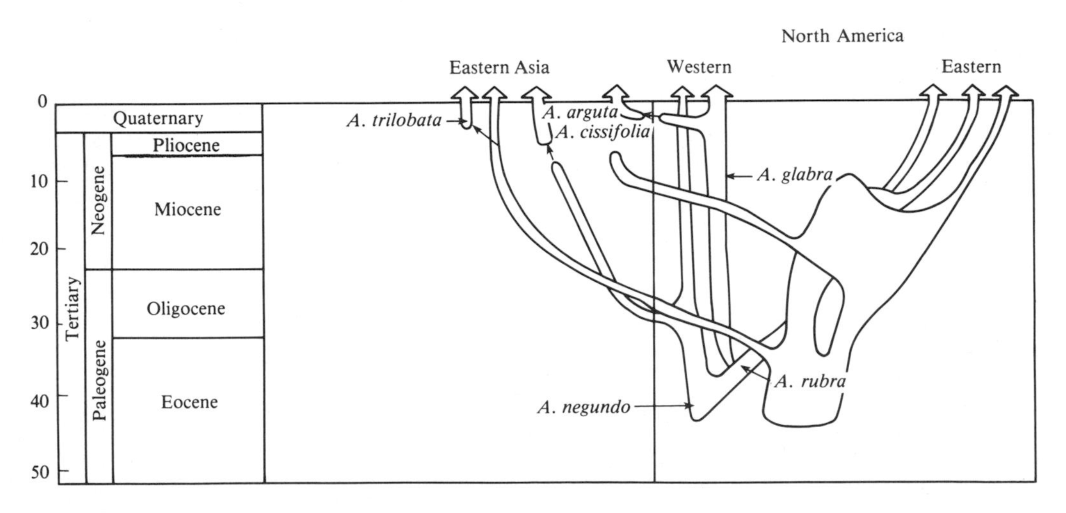

Figure 9.10
Phylogenetic and biogeographic reconstruction for the maple genus, *Acer,* by Wolfe (1981). This reconstruction hypothesizes several independent colonizations of Eastern Asia from North America to account for the present distribution of the various groups.

continents, any time after the breakup of the continents and left no extant representatives in the intervening areas. The phylogeny for all three explanations could look the same. For simplicity, neatness, and predictability, scientists seek general patterns and search for single causes, but nature is not always so simple. One must at least acknowledge that some of the elements of a pattern may not have been achieved in the same manner as the others.

One study that sheds light on this matter is an analysis of disjunctions in the maples (*Acer,* Aceraceae) of the Northern Hemisphere (Wolfe, 1981). Maples have many disjunct species groups with three area disjunctions between parts of North America and Asia. Fortunately, these plants left amazingly good fossil records of leaves and fruits, so one can also trace their range extensions through time. Figure 9.10 shows how some sections of *Acer* achieved vicariance in the Eocene, some in the Oligocene, some in the Miocene, and some more recently. Although one might disagree with the manner in which the study was done on the grounds that cladistic methods were not used to construct the phylogeny or that one cannot trust the fossil record for an accurate record of events, certainly this study shows how extant taxa could be disjunct in the same areas but have very different histories.

Stability of ranges: Tertiary vegetation of North America. Vicariance hypotheses are based on the assumption that geographic distributions of both individual taxa and entire biotas have relatively simple histories of expansion and fragmentation. If instead their histories have been characterized by repeated episodes of migration, isolation, and extinction, one would have the same difficulties reconstructing and testing vicariance explanations as one does for dispersal scenarios when allowing for repeated colonization and extinction events. Of course, vicariance methods do not assume that ranges have been exactly stable but, unless the changes have been sufficiently conservative to preserve the effects of major geologic and climatic events in the distributions of present taxa, the approach is based on an incorrect assumption and can be expected to yield misleading results.

As workers search for ways to reconstruct the geographic positions of past biotas from the positions of present-day fragments they are also confronted with the fact that most distributions are dynamic, not stable. This can be observed directly on a small scale in plant and animal species in many communities, but there is also an extensive record of large displacements of ranges during the Quaternary (Chapter 14). One can wonder, therefore, whether present distributions and the simple reconstructions of their history might be obscuring some very complex and interesting events.

One of the best cases with which to analyze the relationship between a simple vicariance-type model and a complex one is the Tertiary vegetational history of North America. North America has an excellent series of paleofloras, and these have been used by various authors to reconstruct the vegetational history of the continent. However, they also can be used to clarify the geologic and climatic changes across the Northern Hemisphere that should have affected many other groups.

Whereas Alaska today has mostly cold tundra and cool coniferous forest biomes, Wolfe (1969a, 1971, 1972, 1975, 1978, 1979) has shown that Alaska had paratropical rain forest in the Paleogene. This means that the area had abundant precipitation, no pronounced dry season, a mean annual temperature of 20° to 25° C, and a general lack of frost and freezing temperatures (Chapter 13). Such forests included plant groups with undivided (entire) leaf margins similar to present-day evergreen rain forest trees. The earliest Tertiary forests for which fossils are preserved were dominated by such groups as laurels (Lauraceae), menisperms (Menispermaceae), nutmegs (Myristicaceae), witch hazels (Hamamelidaceae), and walnuts (Juglandaceae) but also included were palms (Arecaceae) and a dipterocarp (Dipterocarpaceae). These are all groups with Asian relatives. At least in the Middle Eocene, when a firm land bridge connected Siberia and Alaska, tropical and subtropical plants and animals could probably have migrated between Asia and western North America without any major climatic barrier.

Figure 9.11 traces the history of Alaskan temperatures (estimated from vegetational analysis of leaves) through the Tertiary. In the latest Eocene there occurred a marked cooling period when deciduous species became more common, followed by a warming trend and more tropical vegetation in the Lower Oligocene. The flora changed again in the Upper Oligocene, having a lower species diversity that included more broad-leaved deciduous taxa. By the Middle Miocene a rich flora had been reestablished, but most of the trees were broad-leaved deciduous species mixed with a small group of evergreen dicotyledons and narrow-leaved conifers. At the end of the Miocene, Alaska had coniferous forests and a cool temperate climate, and in the Pliocene the mean temperatures decreased toward the glacial conditions of the Quaternary (Chapter 14). Today Alaska has not only ex-

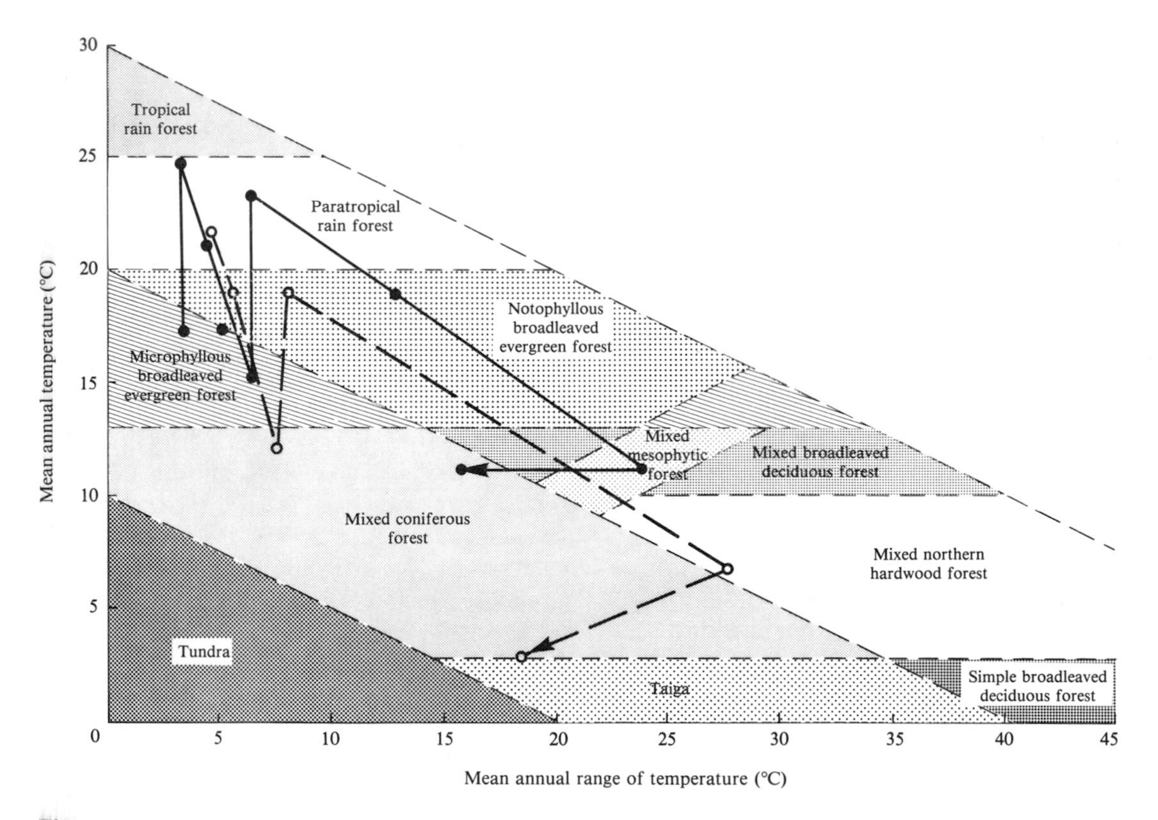

Figure 9.11
Relationship between climate (here expressed as mean annual temperature and precipitation) and vegetation types comprising humid and mesic forests in the Northern Hemisphere. By comparing characteristics of fossil leaves from the Pacific Northwest and southern Alaska the climatic history of those regions can be inferred *(arrows)*. The solid line with solid circles indicates the Tertiary vegetational changes in the Pacific Northwest and the dashed line with open circles indicates the same for Southern Alaska, which has been consistently cooler. Major changes in the climate and vegetation of these regions occurred between the Eocene and the present. (After Wolfe, 1979.)

tremely cool winter temperatures but also strong seasonality, which probably was not the case during most earlier periods.

As one can imagine, when the forest types changed in Alaska, many plant groups did not become extinct but were simply displaced to other regions where the climate was suitable. Maps representing those movements are still incomplete. Nevertheless, it is known that as northern regions became colder, the laurel forests moved southward to where they are now in the southeastern United States and the highlands of Mexico, and the California bay (*Umbellularia californica,* Lauraceae) and some fan palms (Arecaceae) came to assume relictual distributions in the southwestern United States (Figure 13.2). The Tertiary history of wet forest types in Mexico is virtually unknown, and existing reconstructions of them are very speculative and inaccurate. Nevertheless, the list of Mexican taxa with closest relatives in the Orient is fairly long, and are all potentially attributable to historical exchanges across the Bering Land Bridge.

Reconstructing the history of dry vegetation types in North America has been more troublesome because the fossil records are less complete and their interpretations are taxonomically more difficult. Even when the taxa have been identified accurately, authors are still faced with the problem of determining whether the shifts in vegetation resulted from changes in precipitation or temperature or a combination of the two. Semiarid vegetation types, especially chaparral, existed in some form during the Miocene of California, and as the continent dried and developed strong seasonality in temperature and rainfall, semiarid and arid vegetation became dominant.

A widely cited reconstruction of the vegetational history of western North America is that of Axelrod (1958, 1975, 1979) (Figure 9.12). Axelrod treats the flora of North America as basically the interaction of three major and fairly discrete vegetational units that he calls geofloras, assemblages of plants that were relatively homogeneous and stable through time and space. At the beginning of the Cenozoic, the continent was occupied by broad-leaved evergreen forest (Neotropical-Tertiary Geoflora) in most of the south, a temperate and deciduous forest (Arcto-Tertiary Geoflora) in the north, and a sclerophyllous and microphyllous forest (Madro-Tertiary Geoflora) in the semiarid Southwest. As climate changed over the last 65 million years, these geofloras migrated northward and southward, and shifted elevations on mountains. Remnants of the Arcto-Tertiary Geoflora now remain not only in the north temperate zone but also in the Mexican highlands and the riparian deciduous woodlands in the arid Southwest. As indicated earlier, the tropical elements of the Neotropical-Tertiary Geoflora, including the laurels and palms, have retreated to tropical Mexico; and the Madro-Tertiary Geoflora, consisting of an assemblage of Mexican taxa, has invaded arid and semiarid western North America. Thus according to Axelrod the plant communities we see today are very similar to those of millions of years ago.

In contrast, Wolfe views the Tertiary as a dynamic continuum, one in which the species were distributed individually through time, not in stable blocks as preformed plant communities. This view corresponds well with our understanding of the organization of contemporary communities (Chapter 4). The Arcto-Tertiary Geoflora, which is so widely mentioned in textbooks and research articles, was never a discrete floral assemblage almost exactly comparable to what we see in certain parts of the continent today; rather, in the north these temperate deciduous species were originally mixed in with evergreen species, like palms and laurels, and the composition changed gradually so that evergreens disappeared and deciduous plants became more dominant as the climate got progressively cooler and more seasonal. Some of the so-called neotropical elements were actually imports from Asia. One could find several species living in the same community 40 million years ago, whereas now they occur in

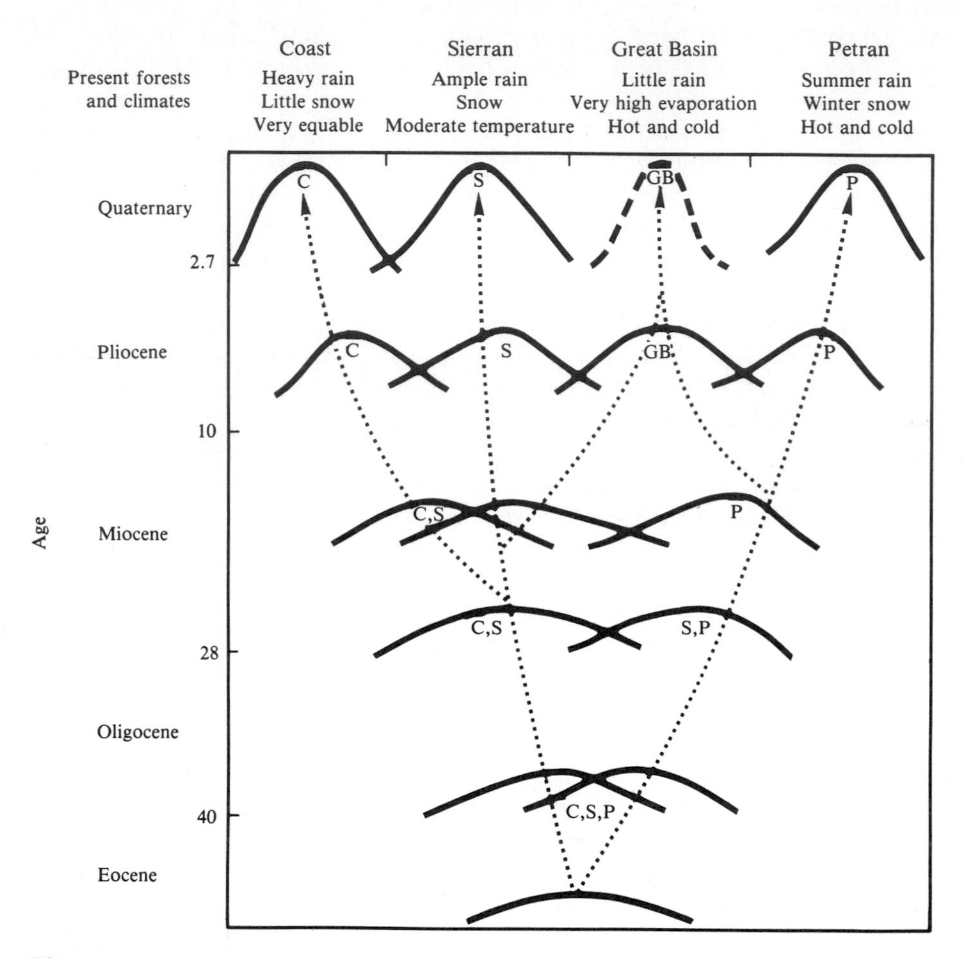

Figure 9.12
Axelrod's (1977) model to explain the derivation of contemporary vegetation types in western North America from earlier Cenozoic plant assemblages. The model hypothesizes that the vegetation of the Late Eocene, when the climate was equable, was a fairly homogeneous assemblage that included plants adapted for living in both wet and dry habitats. As the climate became more seasonal and temperate, those groups preadapted to different regimes of temperature and moisture assorted themselves into distinct vegetation types. Although occasionally a new community developed by acquiring forms from existing assemblages, in general the communities remained relatively stable and characteristic of particular habitats throughout the Cenozoic. Thus this model treats the dispersal of plants in time and space as the movement of entire, integrated communities rather than as the individualistic response of particular taxa.

widely divergent plant communities and regions. Wolfe (1979) concluded that one cannot appreciate the intricate patterns and processes that constitute the historical biogeography and paleoclimatology of North America by relying on oversimplified notions of geofloras.

Wolfe (1979) has also suggested that the presence of paratropical forests in the Eocene of Alaska is a strong indicator that the daylight-darkness regime was not the same as it is today because broad-leaved evergreens and palms, growing at 65° N latitude, could not have tolerated 6 months of darkness and 6 months of daylight. This reasoning has led Wolfe and others to conclude that the earth's axis was tilted much less or not at all at that time. If geoscientists can test this idea, this will be one case in which biogeography has made a substantial contribution to the physical sciences.

Selected references

Ball (1975, 1980); Eldredge and Cracraft (1980); Funk and Brooks (1981); Mayr (1969); G. Nelson (1969a, 1973); G. Nelson and Platnick (1981); G. Nelson and Rosen (1981); C. Patterson (1982); G. Simpson (1961a, 1965); Wiley (1981).

Use of the fossil record

Bierhorst (1971); Brodkorb (1971b); Carpenter (1977); Cracraft and Eldredge (1979); Doyle (1969, 1977, 1978); Eldredge (1979); Eldredge and Cracraft (1980); Estes and Reig (1973); Farris (1976); Feduccia (1980); Gaffney (1975); Gingerich (1976a); Hallam (1975b); M. Hecht et al. (1977); Lillegraven et al. (1979); Miles (1977); G. Nelson and Platnick (1981); Raup and Stanley (1978); Romer (1968); T. Schopf (1972); G. Simpson (1965, 1980a); Solem (1979b, 1981); Stanley (1979); Wolfe (1978).

Criteria for determining center of origin

Ball (1975); Cain (1944); Croizat et al. (1974); Darlington (1957); Hennig (1979); Matthew (1915).

The relationship between biogeography and systematics

Ciochon and Chiarelli (1980); L. Johnson and Briggs (1975); Sneath and Sokal (1973).

Determining which taxa are most closely related

Ball (1975, 1980); Blackwelder (1977); Bock (1973); Bonde (1977); Cracraft (1974b); Cracraft and Eldredge (1979); Cronquist (1981); Dahlgren (1980); Dahlgren et al. (1981); Eldredge and Cracraft (1980); Farris et al. (1970); Funk and Brooks (1981); Funk and Stuessy (1978); Gaffney (1975); Gingerich (1976a); Greenwood et al. (1973); G. Griffiths (1973); M. Hecht and Edwards (1977); M. Hecht et al. (1977); Hennig (1950, 1966, 1979); Hull (1970, 1974); Mayr (1942, 1963, 1969, 1970); G. Nelson and Rosen (1981); Platnick and Gaffney (1977, 1978b); P. Raven (1976); Romer (1968); Rosen et al. (1979); G. Simpson (1945, 1961a); Sokal (1975); Takhtajan (1969, 1980); Thorne (1968, 1976a, 1981); Wiley (1976, 1981).

Vicariance biogeography

Ball (1975); Brundin (1965, 1966, 1967); Carlquist (1981); Cracraft (1974c, 1975a, 1975b); Croizat (1952, 1958, 1960, 1964); Ekman (1953); Funk and Brooks (1981); Gressitt (1963); Humphries (1981); Keast et al. (1972); Li (1952); G. Nelson (1969a, 1973, 1974, 1978); G. Nelson and Platnick (1978, 1980, 1981); G. Nelson and Rosen (1981); C. Patterson (1981a, 1981b); Platnick (1976); Platnick and Nelson (1978); Rosen (1974, 1975, 1978, 1979); Rosen and Greenwood (1976); Simberloff et. al. (1981a, 1981b); Vermeij (1978); Wiley (1981); C. Wood (1971, 1972).

Some persistent issues in historical biogeography

Axelrod (1958, 1966, 1979); Axelrod and Bailey (1969); Bramwell (1979); Carlquist (1981); Craw (1979); Graham (1972); McDowall (1978b); Sauer (1969); Wolfe (1969a, 1969b, 1971, 1972, 1975, 1978, 1979, 1981).

UNIT THREE DISTRIBUTIONS OF TAXA IN SPACE AND TIME

To reconstruct past biogeographic events workers must have some basic information on the present and past distributions of animals and plants. Sources for this information include lists of species from each region, monographs on selected taxa, summaries of taxa endemic to a particular region, summaries of disjunctions, and new data from the original documented specimens in scientific collections. Because this data set is infinite, no one can possibly assimilate every significant pattern. Consequently, workers tend to choose familiar examples, although realizing that less spectacular plants and animals exhibit similar patterns. In the following chapters the examples selected illustrate both the major patterns between large regions on the globe and some of the problems or issues that should be considered in analyzing them.

Chapter 10 examines the distributions of aquatic animals, organisms that require a continuous water system to disperse from one region to another. A large percentage of aquatic animals are confined either to fresh water or to salt water and cannot tolerate the other medium at all because individuals cannot osmoregulate over a broad range of salinity. Organisms with narrow and conservative osmoregulatory properties are the best ones to use in discerning the histories of the landmasses or water masses because they can only disperse through continuous habitats. Hence obligate freshwater groups such as ostariophysan fishes, planaria, and mayflies aid in reconstructing the history of continental connections and show the effects from the breakup of Pangea. On the other hand, freshwater organisms that are able to tolerate salt water must be studied very critically to determine whether their disjunctions were the result of unique colonization across a saltwater barrier or were instead attributable to some past land connection with another landmass. Marine distributions pose many problems in interpretation because their ranges have been produced by a variety of dispersal and reproductive strategies and over a long period of time, during which oceans have been interconnected.

Those organisms that disperse through terrestrial systems exhibit a wide variety of mechanisms to aid in long-distance dispersal through the air or across water gaps, but by and large terrestrial animals are landlocked and must be studied also from the point of view that they may help us to understand the history of land connections. Chapter 11 presents the major patterns of distribution for many past and present taxa of tetrapods to show that some orders and families may be useful in future analytical studies, e.g., groups occurring in each of the southern landmasses. In cases in which there are abundant fossils, as in the land tortoises and mammals, workers are challenged to seek interpretations that are consistent with extinct as well as extant distribution records. Groups such as earthworms and land snails hold important information for reconstructing historical events and will be extremely valuable when their phylogenies are sufficiently known to identify multiarea disjunctions.

Chapter 12 attempts to show that many groups having the ability to fly still exhibit marked provincialism and interesting patterns of disjunction. As expected, some families have members with very wide ranges, especially if they have migratory behavior. Nonetheless, the zoogeographic regions that we recognize today, which were created by P. Sclater in the nineteenth century, were formulated originally from endemic patterns of birds and are supported by the distribution patterns of insects and bats. Each southern landmass has a long list of endemic families and subfamilies of aerial organisms, and sister taxa are now disjunct across wide stretches of ocean. Even in the Northern Hemisphere, where land connections existed throughout the Cenozoic, there are marked differences at the family level between the flying animals of North America and those in Europe or Asia. The myriad of examples of insect distributions are abstracted here by three case studies, that help to illustrate the range of patterns found in this the largest of all animal classes. The insects of Australia, perhaps 5% of the world's total, show the effects that long geographic isolation has on the directions of evolution. The freshwater chironomid midges, which have an aerial phase in the life cycle, exhibit vicariance patterns for the Southern Hemisphere and illustrate how selected groups may elucidate the sequence of historical events. Finally, the cosmopolitan genus *Drosophila,* which is well known phylogenetically, can contribute to our understanding of how complex distribution patterns can be at the local level and how they may be related to ecological as well as historical factors.

Historically, phytogeographers have paid close attention to the many factors limiting the distributions of plants, and from this has emerged a wide variety of methods to model or quantify precisely the relationships between climatic regimes and vegetation types. Each method differs in its ability to explain the distribution patterns of vegetation on each continent and to predict the type of vegetation that should be found in each local habitat, given climatic information or the plant forms that are present. Angiosperms dominate in most biomes. Angiosperms arose in the Mesozoic, at least by the Lower Cretaceous, and presumably from seed ferns in tropical latitudes; and since then most orders have achieved very wide ranges, expanding into all habitable situations. Some of the taxonomic differentiation appears attributable to the splitting of lineages during continental drift since the Mesozoic. High familial endemism on each of the southern landmasses demonstrates the role geographic isolation has played in the multiplication of angiosperms. Interpreting plant disjunctions as evidence of past land connections is complicated by the abilities of many groups to disperse over long distances, across wide oceans, even from pole to pole, and to colonize distant islands. Plants also have special properties, such as high frequency of polyploidy and specialization of different soil types, that provide great challenges for explaining speciation events within regions and hence for using plants in the reconstruction of local geologic events.

Perhaps the greatest challenge confronting historical biogeographers is to justify the historical reconstructions that are made using present disjunctions when most ranges are quite different than they were even 20,000 years ago. Chapter 14 summarizes the salient climatic events in the Quaternary that led to great shifts in the ranges of land and aquatic organisms. Expansions and contractions of distributions occurred in both the glaciated and the unglaciated Northern Hemisphere, where great ice sheets invaded middle latitudes; but major shifts also occurred in the tropics, where rain forests were probably restricted to small, isolated outposts, and in the mountains, where vegetation of the high elevations descended to lower elevations and permitted cooler forms to disperse along mountain ranges. Freshwater and saltwater lakes were well developed in areas that are now arid or semiarid, and the elimination of most habitats has caused great extinction while also producing a new set of species in geographically isolated habitats. The extinction of large mammals was so noticeable during the Pleistocene, especially at the end of the last glaciation, that workers debate whether this was caused by normal ecological processes or instead by the overkill of these beasts by human hunters. Finally, well-documented studies can show how the distributions of animals and plants are still undergoing shifts in ranges, suggesting that the dynamism of the Pleistocene was only an exaggerated phase of a continuous process. In light of this dynamism, workers must determine how present-day distributions can be used to reconstruct small-scale geographic changes on the earth's surface.

Chapter 10

Distribution Patterns of Aquatic Animals

Almost three fourths of the earth's surface is covered by water, which is the exclusive home of innumerable organisms. Many of these organisms cannot survive out of water, which is therefore the principal medium for dispersal. There are, of course, a few well-publicized cases in which adult aquatic animals occasionally disperse over land or through the air. For example, air-breathing catfishes (Clariidae) can migrate short distances overland from pond to pond using their ventral fins. More commonly, organisms may be unintentionally carried from one locality to another on objects, e.g., the transport of algae, freshwater snails, and numerous small invertebrates in the mud on birds' feet; and an even greater number have windborne, desiccation-resistant encysted disseminules (Chapter 7). We are mostly interested, however, in those taxa that must use a continuous water medium for dispersal.

Physiological tolerances of aquatic animals

Salinity. Living organisms must preserve relatively constant internal conditions in order to permit essential chemical and physical processes to proceed at the necessary rates for survival, growth, and reproduction. Although the degree of constancy varies between species and between life stages of the same species, limited ranges of temperature, pH, chemical composition, water content, and other factors must be maintained. To exist, organisms maintain a constant internal chemical state despite fluctuations in the external environment. This regulatory process, called homeostasis, is costly, and many organisms can maintain their internal homeostasis only by remaining in a relatively narrow range of environmental conditions.

For many years biogeographers have observed that some aquatic organisms live in seawater and not fresh water, some in fresh water but not seawater, and a few are able to switch regularly between both media. Those organisms tolerant of only a narrow range of salinity, such as species restricted to either fresh water or seawater, are termed stenohaline, whereas those that can live in a wide range of salt concentrations, such as the species inhabiting estuaries, are termed euryhaline. These niche specializations are the result of adaptations for maintaining homeostasis of internal conditions.

Water movement into and out of an organism is governed by basic physical and chemical laws. A brief lesson in chemical potential will illustrate the problems. Figure 10.1 shows a container of water subdivided by a semipermeable membrane. Small water molecules can move across this membrane, but salt ions cannot pass. On one side we have placed pure water (lacking electrolytes) and on the other water plus solutes. Assuming that both sides are under constant pressure and temperature, there will be a net movement of water by diffusion from chamber *A* (pure water) into chamber *B*. Diffusion no longer occurs when the system

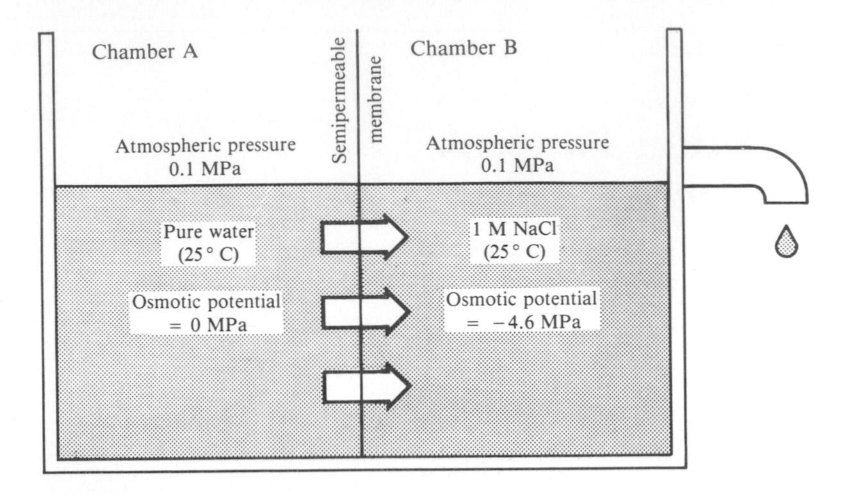

Figure 10.1
Movement of water *(arrows)* between compartments separated by a semipermeable membrane and containing solutions of different osmotic concentrations. In this case, after starting with equal amounts on both sides, water moves from chamber *A* to chamber *B* because the osmotic potential of the salt solution is less than that of pure water. Solution will build up in chamber *B*.

reaches equilibrium at the lowest free energy of the solutions.

One can immediately grasp the significance of this for aquatic organisms. If no regulatory system existed, an organism would gain or lose water to its environment by passive movement through its semipermeable cell membranes or skin whenever its internal salt concentration did not match that of the environment. Figure 10.2 shows two typical situations in which a marine fish is placed into salt water, called a hyperosmotic environment, and a freshwater fish is placed into a freshwater or hypoosmotic solution. Notice that the organism would become desiccated in salt water and in fresh water it would take up too much water. To prevent this, organisms must control the intake and exit of water and salts. When the ion concentration of the blood must be kept *lower* than that of the environment, water is taken up, and the accompanying salts are excreted by special chloride cells in the gills. Water is retained by a fairly impermeable skin and by reducing the amount excreted by the kidneys. In freshwater organisms the ion concentration of the blood must be kept *higher* than the solution; hence the kidneys must eliminate much water. However, this water loss would mean concomitant loss of body salts together with nitrogenous waste compounds. Therefore salts are reabsorbed from urine in the kidney tubules and also extracted from the water by special glands in the gills. Additional salts are injested from prey. In both cases, the regulation of a constant internal concentration different from the liquid environment requires the expenditure of considerable energy to move substances across membranes against the chemical potential gradient.

This description applies to fishes specifically. Different groups of aquatic organisms maintain homeostasis in somewhat different ways, although the physical principles are the same. They use a variety of compounds and electrolytes to adjust blood osmolality, and they have different excretion patterns, skin permeabilities, and tolerances to body fluids of different ion concentrations. Among such aquatic animals as crabs, amphibians, lampreys, sharks, and bony

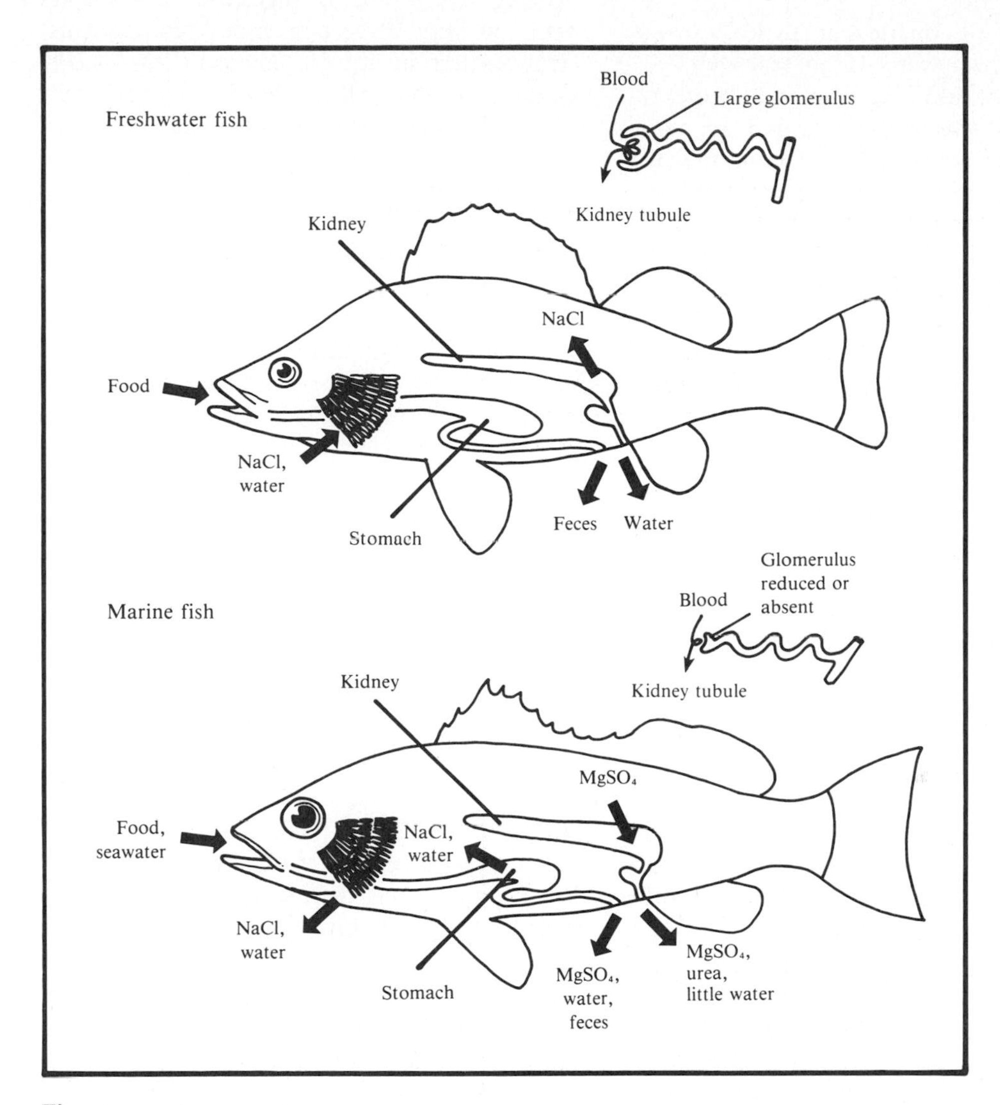

Figure 10.2
Osmoregulatory mechanisms used by fishes to maintain constant internal osmotic concentrations in fresh water *(above)* and salt water *(below)*. Freshwater fishes maintain osmotic balance in a dilute environment by minimizing water uptake, excreting a dilute urine, and absorbing salts through the gills. Marine fishes maintain osmotic balance in a concentrated environment by minimizing water loss through the skin and urine, and by excreting salts through the gills. The glomerulus is well developed in freshwater fishes to enable the kidneys to filter and then expel great quantities of water. (Modified from Webster and Webster, 1974.)

fishes, a broad range of osmoregulatory mechanisms can be found.

Over 90% of all aquatic animals live exclusively in either fresh water or in seawater because they cannot osmoregulate in the other medium. The vast majority of marine plant and animal classes and orders are osmoconformers, which means that they do not osmoregulate but instead maintain internal concentrations identical with seawater. However, this is not true of fishes, of which 58% are obligate marine forms and 33% are obligate freshwater ones. Other groups heavily represented with obligate freshwater species are algae, hydroid coelenterates, planaria, crustaceans, desmospongioid sponges, aquatic insects, and bivalve and gastropod molluscs.

Historical biogeography generally emphasizes how major differences in salinity, e.g., crossing from seawater into brackish or fresh water, affect large-scale distribution patterns. Most marine organisms can tolerate a small range of salinity concentrations. For example, in the world's oceans each of the large water masses (Chapter 2) has a characteristic temperature and salinity profile, but these small differences in salinity between adjacent water masses do not appear to limit most species. Even in such seas as the northern Gulf of California and eastern Mediterranean, where evaporation greatly exceeds freshwater input, producing concentrations of 40‰, distributions are not limited by salinity. Only in extremely hypersaline environments, such as the Great Salt Lake and the Dead Sea, which are several times more concentrated than seawater, is diversity greatly reduced. At high concentrations, salts greatly reduce primary productivity, interfere with the sexual reproduction of all taxa, and may be toxic for even the most hypersaline marine animals.

Temperature. It is easy to understand why water temperature is so important in the lives of aquatic animals because the majority are ectotherms, i.e., their body temperature is set by that of the medium. Temperature controls the rates of development, digestion, and movement, all very important fitness components. Temperatures in aquatic environments usually are much more constant than in terrestrial habitats, and aquatic organisms typically can tolerate only modest variations in temperature. For similar reasons, aquatic tropical and subtropical species are more stenothermal than their relatives of higher latitudes, and marine forms are more stenothermal than freshwater ones.

Knowledge of the effects of extreme water temperatures for limiting geographic ranges has been invaluable for reconstructing aquatic distribution patterns. In 1853 James D. Dana published the first major zonal classification of life in oceans based on temperature, using isocrymes, which are lines of mean minimum monthly sea-surface temperature (SST). Subsequent work has verified and improved these thermal analyses on extant faunas (Ekman, 1953; Briggs, 1974; Lehner, 1979), and the results have been extrapolated to explain past distributions in terms of historical changes in sea temperature.

Figure 10.3 depicts the Pacific coastline of western North America. On this map we have plotted the currently accepted marine coastal provinces and regions. Distributional boundaries of numerous marine groups correspond fairly closely to changes in sea-surface temperature, especially those in the coldest month (Lehner, 1979). The sea-surface temperature regimes are controlled primarily by the joint influences of climate and the temperature of water masses that flow into the region via currents. Climatic parameters other than temperature are less useful in predicting the faunal limits, especially as one proceeds poleward. The geographic ranges of particular species and the overall species diversity of tropical groups, e.g., the rocky shore fishes, correspond closely to the minimum winter sea-surface temperature of 20° C, the isocryme that occurs at 24° N and 3° S latitude (Ekman, 1953). These boundaries are

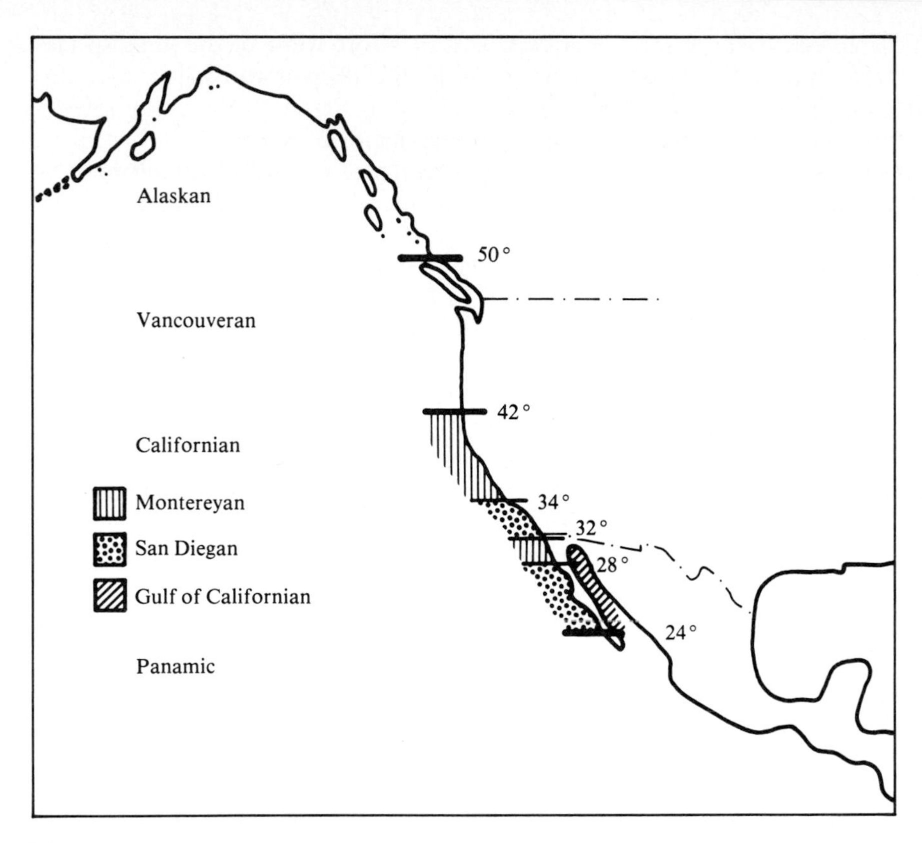

Figure 10.3
Map of the Pacific coast of western North America showing the marine provinces defined for shallow-water organisms and their latitudinal boundaries. (Modified from Southern California Coastal Water Research Project, 1973.)

set by the contact of warm tropical water with cold water from the temperate anticyclonic California and Peru currents.

Temperature is a primary determinant of latitudinal faunal changes, but other environmental factors also play important roles. Local topographic and hydrographic features may produce major faunal dissimilarities between adjacent zones, e.g., where freshwater input is great or where substrate changes markedly. A group such as rocky shore fishes have discontinuous ranges in which rocky habitats are interrupted by mangrove, estuary, coral reef, or sandy shoreline; in fact, authors have described a Central American faunal gap between 16° and 13° 1′ N latitude, where sandy but no rocky shores occur on the Pacific side. Depending on the vagility of the group, such breaks in habitat may act as formidable barriers to dispersal and limit the ranges of species.

Patterns along the eastern Pacific coastline are much more straightforward than those along other pole-to-pole transects. For example, the coasts of the Carolinas and Georgia in the western North Atlantic sometimes receive warm water from the Gulf Stream out of the

Gulf of Mexico and sometimes cold water from the north. Menzies et al. (1966) noted a sudden sea-surface temperature change in April from 11° to 19° C owing to the influx of tropical water. Such sudden shifts can be accompanied by rapid changes in distribution when warm-water pelagic organisms are carried north of their normal ranges. This is dramatically illustrated by the occurrence in the cold North Atlantic of huge warm-water eddies, spun off as gyres as the current proceeds northward and carrying entire biotas of tropical and subtropical organisms (Figure 2.13).

Life in deep bodies of water is likewise stratified along vertical thermal gradients. As pointed out in Chapter 2, incoming solar radiation is absorbed only in the uppermost layer, and from there water temperature diminishes with depth. Seasonal changes may influence surface temperatures, but at progressively deeper levels seasons are less pronounced. At abyssal depths, where neither light nor heat penetrates, water temperatures remain at or slightly below 4° C year-round. Abyssal temperatures may be even colder at the poles and in regions of downwelling.

Keeping this temperature gradient in mind, one might expect that the greatest diversity of animal life should occur in the warmest layer above the thermocline, in the photic zone where the primary producers reside, and decrease with temperature, leaving a depauperate fauna in the abyssal and hadal aphotic zones where phytoplankton cannot survive. The littoral and sublittoral zones, e.g., coral reefs, are indeed very rich in species, and the upper portion of the intertidal zone is mostly species poor because few organisms can withstand periods of such physical stress (Chapter 3).

Water at 4° C is much closer to the surface at the poles than in the tropics, 0.5 km versus over 3 km, respectively. Occurrence of frigid water close to the polar surface has been termed equatorial submergence (Ekman, 1953) or polar emergence (Menzies et al., 1973). Many aquatic organisms are adapted to live only in these cold waters, and therefore some of the same species that occur in the deepest tropical waters also reappear at shallow depths at the poles, following isothermal bands of water.

Oxygen content and other dissolved substances. Theoretically, oxygen is present at sufficient levels for animal life even at the coldest depths, where water occurs under tremendous pressure. In fact, the solubility of oxygen in water actually increases with decreased temperature, and at the same temperature is more soluble in fresh water than in salt water. The availability of oxygen, however, does not always correspond to patterns predicted from these simple physical considerations. At the surface, aeration produced by turbulence and the photosynthetic activity of algae may bring oxygen levels to near saturation, but in shallow, heavily populated bodies, nighttime respiration and high bacterial growth may rapidly deplete the oxygen supply after sunset. Lower depths, especially in deep tropical seas and lakes, may become oxygen deficient and remain essentially anaerobic because there is little exchange with aerated upper levels. Insufficient oxygen may serve as an ecological barrier to migration, at least on a short-term basis.

The presence or absence of other dissolved substances can greatly affect aquatic distributions. A deficiency in calcium limits the growth of animals with calcareous exoskeletons. On the other end of the scale, the excess of an ion can be toxic for an animal or for its food source. Such deficiencies and excesses of ions are commonly observed in freshwater areas, where surface runoff or local parent material can dramatically influence the ionic composition. Recent changes in hydrogen ion concentration or pH resulting from runoff of acid rain have made many freshwater lakes unsuitable for many of their original inhabitants, resulting in extinction of numerous populations. Although this problem was caused by human pollution, over longer time scales similar processes have created unusual microhabitats in which endemic species developed and have acted as important barriers to dispersal.

Substrate and form

Habitat specificity is the rule for most organisms, and aquatic animals show some of the clearest interrelationships between the physical properties of the habitat and the organism's form, feeding method, and strategy for avoiding predation. For example, an aquatic organism may be free living, epifaunal (living on a substrate), or infaunal (living within a substrate). Given that the organism lives on a substrate (for example, an adult clam or sea anemone), the species may be sessile (immobile) or capable of varying degrees of mobility. The extent of locomotion often depends on feeding behavior, for example, whether the organism is a browser, a filter feeder, a scavenger, or a predator. Some organisms can be buried, whereas others become permanently attached to firm or shifting substrates. Many workers believe that each architectural design reflects a compromise that enables the organism to obtain food while avoiding death by predation and stressful abiotic factors (Vermeij, 1978). The architectural design is also evolutionarily constrained by the morphological and physiological properties of the stock from which it evolved.

Habitat relationships of aquatic organisms are not considered very important for interpreting distributions by some workers in historical biogeography. On the other hand, paleobiogeographers and paleoecologists extensively use such data to reconstruct ancient habitats, identify the presence of shallow seas, locate ecological barriers, and thereby explain ancient distributions (Hallam, 1973b, 1975b; Gray and Boucot, 1979).

Dispersal and reproductive properties of aquatic animals

Distributions cannot be understood without some knowledge of life histories. For example, a marine alga, fish, mollusc, sponge, crab, and starfish may occupy the same coral reef, share similar preferences in physical conditions, but exhibit geographic distributions of different sizes and shapes (Figure 10.4). Why are some species wideranging and others narrowly endemic? These differences could reflect (1) the subtle ways that different species adapt to abiotic conditions, (2) the biotic interactions of each species with other species, (3) the age and history of the species, and (4) the manner in which each species reproduces and its offspring are disseminated. All of these can be important, but the last has proven particularly useful for accounting for the distributions of aquatic organisms.

Pelagic marine organisms. Pelagic organisms generally have no reliance on fixed substrates so that reproduction and dispersal are accomplished in open water. There are, of course, numerous exceptions, species that use the shore or sea bottom either for spawning, such as the whale shark (*Rhincodon typus,* Rhincodontidae) with demersal eggs, or for feeding, e.g., many adult eels (Anguillidae). Other oddities include sea turtles (Dermochelidae and Cheloniidae) and grunion (*Leuresthes,* Atherinidae), which lay eggs in beach sand above the waterline. In another vein, the truly pelagic sargassum fish (*Histrio histrio,* Antennariidae) attaches its eggs by filaments to floating mats of the brown alga *Sargassum.*

Most pelagic organisms, especially invertebrates and many algae, are planktonic forms that drift more or less passively with water masses. Although many plankton exhibit active daily vertical migrations, they are incapable of sufficient locomotion to affect geographic ranges significantly. Consequently, planktonic taxa can become isolated in individual water masses, promoting high endemism, as in small lantern fishes (Myctophidae).

In contrast, nektonic forms are the active swimmers of the pelagic environment. Distributions of the strongest swimmers, mostly vertebrates, are more often determined by the availability of food and the tolerance of abiotic

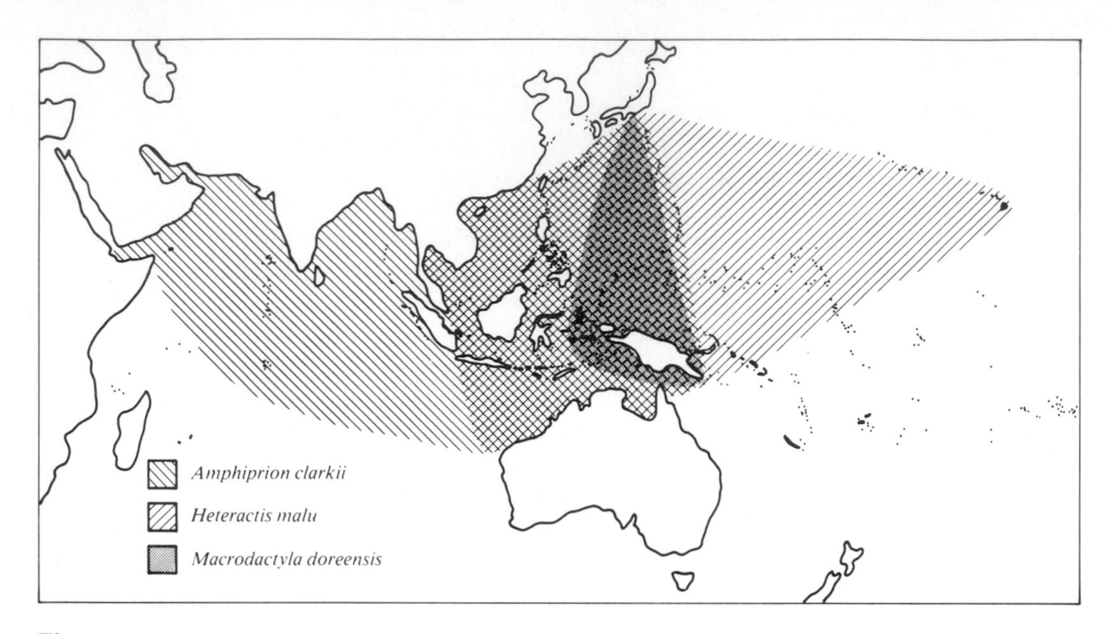

Figure 10.4
Distribution of the Indo-Pacific anemone fish *(Amphiprion clarkii)* and two sea anemones *(Heteractis malu* and *Macrodactyla doreensis)* that serve as hosts. *Amphiprion clarkii* is the only fish that is mutualistic with *H. malu,* but it is associated with other anemones, including *M. doreensis.* This shows that even some relatively specific mutualists need not have completely overlapping ranges, because some species (e.g., *H. malu*) may be able to survive without their partners, whereas others (e.g., *A. clarkii*) may be able to switch to use other species as mutualists. (Unpublished data courtesy D. Dunn, California Academy of Sciences.)

conditions, such as thermal regimes, than by direct effects of large-scale circulation patterns. Examples of very wide-ranging species can be found among cetaceans and sea turtles, distributions that have been greatly diminished by human hunting and interference.

Whether planktonic or nektonic, the offspring of most pelagic marine organisms are released into the open water and the young, starting as eggs, larvae, or fingerlings, join the plankton, where they are transported by currents stronger than their own locomotion. Forms with a pelagic stage therefore tend to have broad ranges that are not homogeneous from season to season or year to year. Brooding and other forms of parental care are uncommon in pelagic marine species, although there is an excellent exception in whales, which usually produce only one offspring per pregnancy.

Whales also provide good examples of migration within the continuous marine environment. Species with large individuals, such as the humpback whale *(Megaptera novaeangliae),* the blue whale *(Balaenoptera musculus),* the fin whale (*B. physalus,* Balaenopteridae), and the California gray whale (*Eschrichtius robustus,* Eschrichtiidae) undertake annual migration between polar and subtropical or tropical waters to complete their reproductive cycle (Figure 10.5). In summer these whales gorge themselves on krill or other crustaceans in the high latitudes where, as a result of the presence of oil-rich diatoms, prey populations are extremely high. As cold weather approaches and feeding areas become

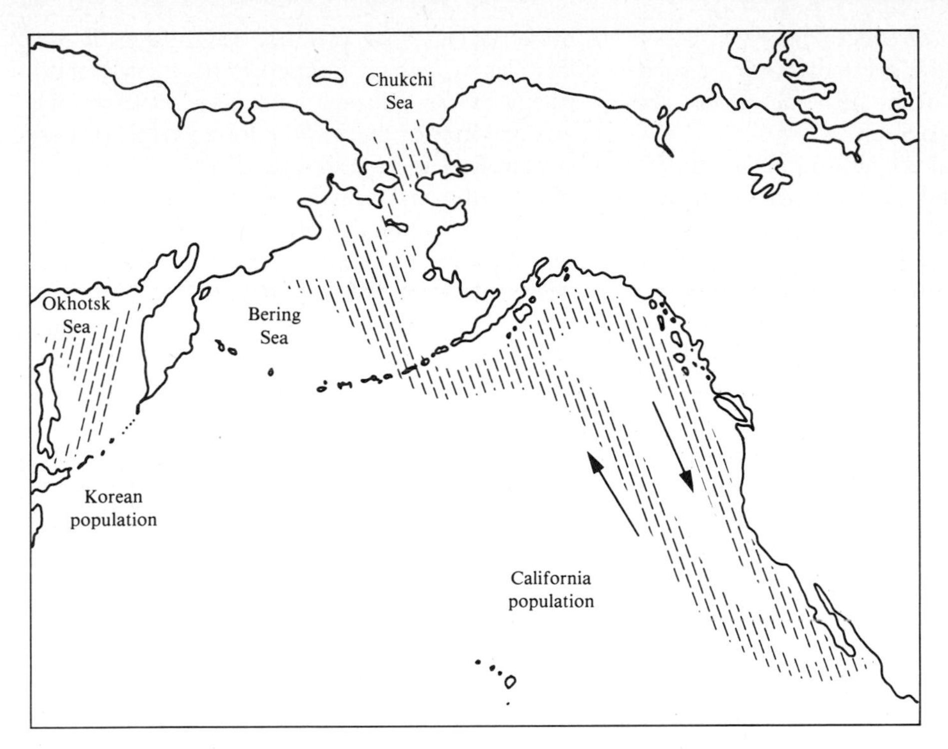

Figure 10.5
Annual migratory routes of the California gray whale *(Eschrichtius robustus)* between their summer feeding areas in cold arctic and subarctic waters and their winter breeding grounds in warm southern waters. This map is based on records from early in this century when the whales were still reasonably abundant. The Korean herd may now be extinct and may merely be stray migrating animals of the California population. The California population suffered a drastic decline as a result of hunting, but is now increasing. (After Walker, 1962.)

covered with ice, herds head for the tropics. There the females give birth to young, and mating also takes place in some species. Although tropical waters are rich in the number and variety of organisms present, they contain only about 1% of the polar prey biomass, and thus the whales live off their blubber reserves during the long migration and reduce energy expenditure by staying in warm waters. It is possible that whales tolerate this long fast in order to provide a warm environment for the protection and rapid early growth of their milk-drinking newborns, which have very low fat reserves.

Many other nektonic species show migration patterns in which adults live and feed in regions of very high productivity, called growing or fattening areas, but use inland or warmer waters (spawning areas) for breeding. The green turtle *(Chelonia mydas)* uses isolated beaches far from its feeding grounds for laying eggs. Many fishes, such as albacore and other tunas (family Scombridae), appear to migrate along specific routes between fattening and spawning areas (Jones, 1968; Parin, 1970).

Benthic marine organisms. At some stage of their life cycle, most marine species either become attached to bottom substrates or use these substrates for feeding or concealment. It is risky to generalize about the great diversity of such benthic forms, but some generalizations are noteworthy for biogeography.

In general, benthic organisms (most invertebrates and many fishes) have pelagic planktonic larvae that exhibit one of two strategies: (1) planktotrophic larvae, those that must feed on other plankton during their development, and (2) lecithotrophic larvae, those that live on large yolk reserves (Thorson, 1950). Planktotrophic larvae generally live in the plankton for a long time. In a minority of species, the eggs and larvae never leave the benthos because the lecithotrophic larvae develop in situ or while being brooded by an adult (Vance, 1973a, 1973b) or the demersal larvae may have a free-living benthic feeding period (Mileikovsky, 1971).

In tropical waters over 80% of bottom-dwelling marine invertebrates have planktotrophic larvae with long pelagic lives. Representatives, such as molluscan veligers, cnidarian and coral planulae, annelid trochophores, and others, develop from small eggs, so they must feed on minute plankton to survive. These larvae are carried long distances for days, sometimes months, before they settle to start new populations, thus promoting a wide distribution of the genotypes. During their pelagic phase the larvae tend to move toward the light (positively phototaxic), allowing them to rise to the more productive, warmer, and faster moving upper layers. As the time for metamorphosis approaches, those larvae that have survived predation at the surface become sensitive to gravity (positively geotaxic) and repelled by (negatively phototaxic) or unresponsive to light, descend in the water column, and settle on the proper substrate. Occasional observations indicate that pelagic phases can be extended when an appropriate substrate is not present. A large proportion of the planktonic larvae die before metamorphosis, and it is not surprising that vast numbers are produced.

Progressing into cooler waters, one finds more species that produce lecithotrophic pelagic larvae, which frequently have fairly short pelagic phases. In polar and deep seawaters the vast majority of species lack long pelagic phases, and there is an increased tendency for brooding of offspring and other types of concealment so that the swimming phase is brief or totally absent. A case in point is the arrowworm *Eukrohnia hamata* (Chaetognatha) at Point Barrow, Alaska, which broods eggs and young. Other taxa showing considerable parental care for offspring include copepods, amphipods, and some brachiopods.

Thorson's (1950) excellent summary noted that poor nutrition of a larva slows growth, prolongs larval life, and consequently prolongs exposure of the young to predators. In cold water, where development can be very slow, life histories are selected to permit rapid development at near zero temperatures. This is precisely the result predicted by mathematical models (Vance, 1973a, 1973b): planktotrophy is favored when planktotrophic development time is not substantially longer than lecithotrophic ones and when planktonic predation is not substantially greater than benthic predation; and for lecithotrophy, benthic development is favored when developmental time is long or planktonic predation is much greater than benthic predation, or both. Differences in mode of reproduction and development are considered to be one of the main reasons for the restricted distributions of benthic marine species (Thorson, 1950).

Most benthic marine fishes generally have pelagic eggs (Breder and Rosen, 1966), but some recent studies of fish communities have found that certain species of shallow water communities have nonpelagic eggs, even in the tropics. In the Hawaiian Islands, Leis and Miller (1976) observed nine families of inshore reef species with brooded or demersal eggs, and Lehner (1979) and Thomson and Gilligan (1983) have noted that primary residents of rocky intertidal communities appear to have demersal, sticky eggs, adherent to protected rock faces. These species exhibit little daily migration, highly specific microhabitats, and high degrees of territoriality. Families living in estuaries and embayments also often have demersal eggs. Limited dispersal should help lower losses of eggs and young of species that occupy habitats limited in area and very patchy in occurrence.

Marine insects. Of perhaps 800,000 species, only several hundred kinds of insects are able to live in salt water. They are derivatives of freshwater or terrestrial groups and have special adaptations to avoid respiratory and osmoregulatory complications. An excellent recent summary (Cheng, 1976) shows that most species occur along shorelines and in estuarine habitats where they feed in the water but lay their eggs above the waterline. The only forms inhabiting the open sea are five tropical species of surface-living water striders (*Halobates,* Gerridae, Hemiptera), predaceous sucker-feeders that probably lay eggs on flotsam. Consequently, these particular species have extremely wide distributions (Figure 10.6).

Freshwater organisms. Freshwater habitats constitute a minute fraction of the world's water but provide homes for about 10% of all aquatic species. Whereas marine habitats have fairly constant conditions year-round except along the shoreline, freshwater habitats may exhibit marked variability. Small and shallow water bodies experience major annual temperature

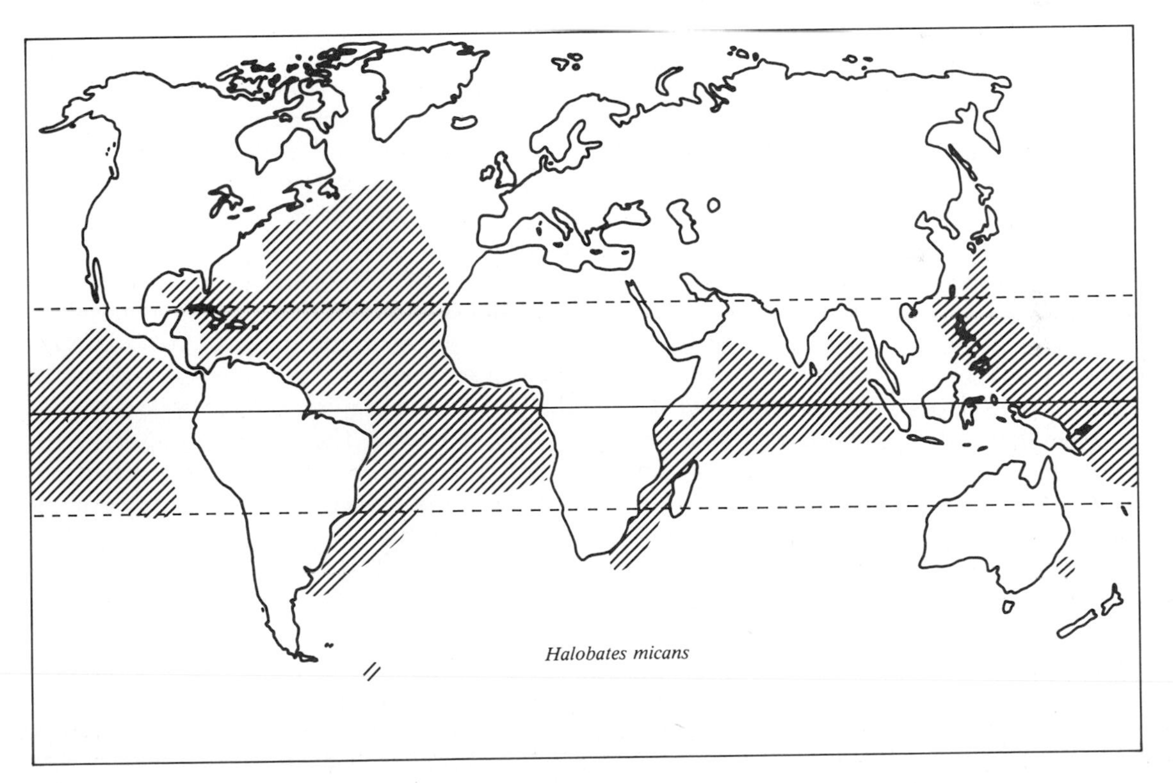

Figure 10.6
Circumtropical distribution of the marine water strider, *Halobates micans*. This is a completely pelagic insect. Adults and juveniles live on the surface tension of the water, and eggs are attached to floating debris, such as seabird feathers and bits of vegetation. (After Cheng, 1974.)

fluctuations, some even freezing in winter and others drying up in summer. Fresh water is also more subject to influence from runoff from the land. Thus for the invertebrate phyla common to marine and freshwater habitats, similar modes of reproduction usually are observed, but those living in stressful freshwater environments may have dormant or resistant stages added to the life cycle.

Of the 7500 species of freshwater fishes, many spend most or part of the life cycle close to the bottom, e.g., having demersal eggs adhesive to rocks or vegetation, and many others bear live young or exhibit parental care. Eggs sitting around are food for predators, and eggs have osmoregulatory difficulties in fresh water.

Over 25,000 species of insects, including both free-living and parasitic forms, spend most of their lives in fresh water. Usually at least the eggs and larvae are aquatic, whereas the adult may be winged and partly or entirely terrestrial. Flight is used in courtship, in mating, and sometimes in dispersal. Consequently, aquatic insects, like amphibians, can have two different media for dispersal; but aerial dispersal does not necessarily confer great migratory power. In fact, in a group like the mayflies (Ephemeroptera) flight is very ineffective for long-distance dispersal (Edmunds, 1972, 1975, 1981).

Snails are important members of many freshwater habitats. Eggs and juveniles are often so small that they hitchhike from one locality to another on larger animals (Chapter 7). In addition, many widely distributed species are ovoviviparous or parthenogenetic or both, and some so-called freshwater taxa are euryhaline, tolerating brackish water, and have free-swimming veliger larvae. Dispersal of many species has been promoted by human shipping activities that unintentionally give adults, protected by their shells, long rides to far-off ports and islands.

Diadromous fishes. Diadromous fishes must live in both seawater and fresh water to complete the life cycle. Those that live in the sea but breed in fresh water, such as most lampreys (Hyperoartii), marine sturgeons (*Acipenser,* Acipenseridae), and salmon (Salmonidae), are called anadromous, whereas those living in fresh water and breeding at sea, such as many eels (Anguillidae), are called catadromous. The importance of these life cycles in biogeographic interpretations is obvious: their life cycles enable these exceptional fishes to cross both freshwater and marine barriers. Most workers have been careful not to use these groups as indicators of past land or water connections.

Distribution patterns of freshwater fishes

George Myers (1938) proposed an ecological classification of freshwater fishes based on their intrinsic abilities to tolerate seawater. Those totally intolerant of salinity are called primary division freshwater fishes, those slightly tolerant of but not preferring seawater are secondary division freshwater fishes, and those able to osmoregulate and switch from one medium to another are peripheral freshwater fishes. Noting that primary division fishes are theoretically landlocked, biogeographers have attempted to use their continental distributions as indicators of past land connections. Unless there is fossil evidence to the contrary, it is assumed the common ancestor was strictly freshwater if all extant descendents are also strictly freshwater.

Ostariophysi. The largest group of freshwater fishes is the Ostariophysi (5500 species). Authors debate whether this is an order or superorder and how its lesser subunits should be classified, but all accept it as a monophyletic taxon. These fishes share the unique and specialized Weberian apparatus, a modification of the anterior vertebrae to connect the air bladder with the inner ear, presumably an improvement in hearing. Over 95% of the species are primary division forms, and their distributions serve to illustrate many biogeographic patterns.

The distributions of the ostariophysan families are summarized in the box on p. 295. Simply stated, most families are endemic to only

Distributions of families in the Ostariophysi

Based on the subordinal classification of Fink and Fink (1981). Families with marine forms (m) are noted.

SUPERORDER OSTARIOPHYSI

Series Anotophysi

Order Gonorynchiformes
- Chanidae (m); Indo-Pacific
- Kneriidae; Africa
- Phractolaemidae; Africa
- Gonorynchidae (m); Indo-Pacific

Series Otophysi

Subseries Cypriniphysi

Order Cypriniformes
- Cyprinidae; North America, Africa, Eurasia
- Gyrinocheilidae; southeast Asia
- Psilorhynchidae; southeast Asia
- Catastomidae; eastern Asia, North America
- Homalopteridae; southeast Asia
- Cobitidae; Eurasia, Northern Africa

Subseries Characiphysi

Order Characiformes
- Characidae; South America to Texas, Africa
- Erythrinidae; South America
- Ctenoluciidae; South and Central America
- Hepsetidae; Africa
- Cynodontidae; South America
- Lebiasinidae; South America
- Parodontidae; South America
- Gasteropelecidae; South America to Panama
- Prochilodontidae; South America
- Curimatidae; South America
- Anostomidae; South America
- Hemiodontidae; South America
- Chilodontidae; South America
- Distichodontidae; Africa
- Citharinidae (including Ichthyboridae); Africa

Order Siluriformes

Suborder Siluroidei
- Diplomystidae; South America
- Ictaluridae; North America
- Bagridae; Asia, Africa
- Cranoglanididae; Asia
- Siluridae; Eurasia
- Schilbeidae; Africa, India
- Pangasiidae; southeast Asia
- Amblycipitidae; southeast Asia
- Amphilidae; Africa
- Akysidae; southern Asia
- Sisoridae; southern and western Asia
- Clariidae; Africa, Asia
- Heteropneustidae; southeast Asia
- Chacidae; southeast Asia
- Olyridae; southeast Asia
- Malapteruridae; Africa
- Mochokidae; Africa
- Ariidae (m); pantropical
- Doradidae; South America
- Auchenipteridae; South America to Panama
- Aspredinidae; South America
- Plotosidae (m); Indo-Pacific, Australia
- Pimelodidae; South and Central America
- Ageneiosidae; South America to Panama
- Hypophthalmidae; South America
- Helogeneidae; South America
- Cetopsidae; South America
- Trichomycteridae; South America to Costa Rica
- Callichthyidae; South America to Panama
- Loricariidae; South America to Costa Rica
- Astroblepidae; South America to Panama

Suborder Gymnotoidei
- Gymnotidae; South and Central America
- Electrophoridae; South America
- Apteronotidae; South America
- Rhamphichthyidae; South America

one zoogeographic region, and closely related families often show provincialism. However, there are some noteworthy exceptions, disjunctions that span major saltwater barriers. Most noticeable are the characoids (Characiformes), including the piranhas and tetras, which occur predominantly in South America, recently entering Central America; but about 200 species are endemic to Africa (Géry, 1977). The African disjuncts include Citharinidae and Distichodontidae, which are now believed to be the most primitive extant forms of the order (Fink and Fink, 1981), the African tetras (Alestinae of Characidae), and the monotypic Hepsetidae. This split range strongly suggests a formerly continuous distribution in West Gondwanaland, one in which families had already begun to diverge prior to the separation of South America and Africa.

The catfishes (Siluroidei), which are highly specialized forms lacking scales, occur on all continents but are very rich and diverse in Asia and South America. The most primitive living siluroid is believed to be *Diplomystes* of southern South America (Fink and Fink, 1981). Lacking crucial phylogenetic information on this suborder, we can only surmise now that siluroids, like the characoids, were present in West Gondwanaland in the middle Mesozoic. Evidence for this comes from several observations. First, the occurrence of primitive-looking species in South America suggests that the taxon arrived at an early date. Second, the endemic families of catfishes in South America show marked diversification, as if radiation occurred in long isolation, and only a few of these families have managed to enter Central America following the recent establishment of the Panamanian Land Bridge (since 4 million years BP). Third, a derived group of siluroids is the suborder of knifefishes and ostariophysan eels (Gymnotoidei), including the electric eel *(Electrophorus electricus)*, which are also South American families. Finally, the catfishes of North America seem to be most closely related to Asian forms rather than the neotropical ones. All this suggests that the South American siluroids and gymnotoids have evolved in isolation for a long period without interchange with North America, so, presumably, the stock could have been present in South America from an initial Gondwanaland biota. Species that have reached Australasia and Madagascar belong to Plotosidae and Ariidae, essentially marine families that evolved secondarily from freshwater ancestors of tropical Asia.

The order Cypriniformes includes six families, that are especially well represented in southeast Asia and that have dispersed mainly through the northern continents. Three families are endemic in the Oriental region, the Cobitidae occur in northern Africa as well as in Eurasia, and Catostomidae are temperate and arctic fishes of eastern Asia and North America. The largest fish family is Cyprinidae, the minnows, whose species are mostly absent from the Southern Hemisphere; and the African cyprinid lineages are apparently recent derivatives of fairly modern Eurasian stocks.

Here then are several closely related groups of freshwater fishes: one is clearly produced from a West Gondwanaland biota, one is certainly Laurasian with recent entry into Africa, and one is primarily Old World, dispersing from Eurasia into North America by land bridges and across Wallace's line to Australasia by sea-going forms. The South American forms have been radiating in isolation for a long time (Figure 10.7). This is not the scenario once proposed by P.J. Darlington (1957) (Figure 10.8) to explain these same distributions but prior to the general acceptance of continental drift. Darlington chose an Oriental origin for Ostariophysi followed by dispersal through northern continents into the southern landmasses. Thus according to that reconstruction amphi-Atlantic disjunctions do not demonstrate direct faunal exchanges of freshwater fishes.

Lungfishes and osteoglossomorphs. There are two other noteworthy groups of primary division freshwater fishes, the lungfishes (Lepidoseriformes or Dipnoi) and the osteoglosso-

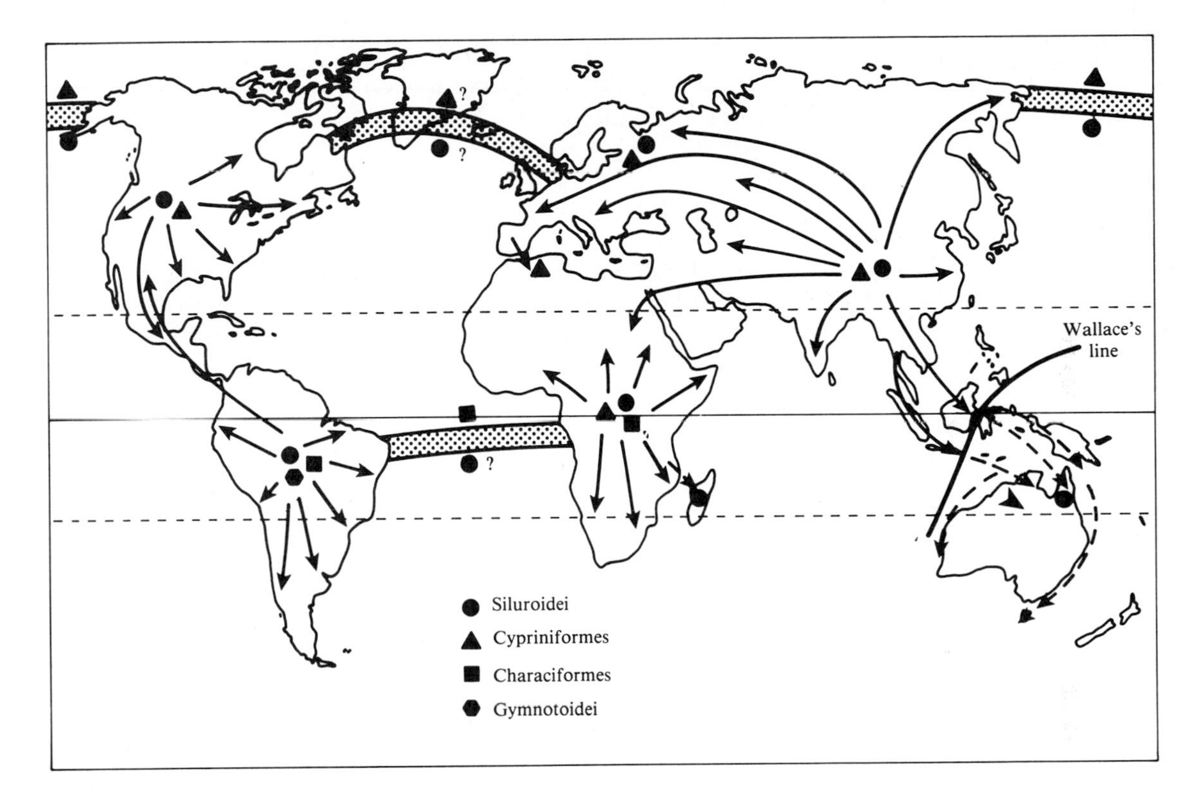

Figure 10.7
Model summarizing the views of various authors on the history of dispersal and differentiation of the Otophysi. The orders and suborders of these fishes inhabiting each continent are shown. Stippled bridges indicate where past land connections (either emergent land bridges or predrift continental connections) enabled groups to move between the Old and the New World. Solid arrows indicate probable routes of dispersal within and between continents that are now connected. Dashed arrows show dispersal of saltwater-tolerant forms across marine barriers, especially across Wallace's line to Australasia. Limited interchange has occurred between North and South America since the formation of the Panamanian Land Bridge. This model does not consider the possible effect of a drifting India in transporting ostariophysans between Gondwanaland and Asia.

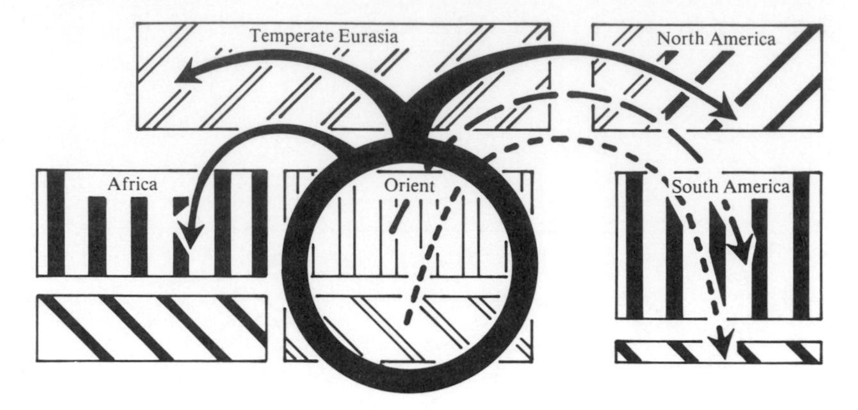

Figure 10.8
Darlington's (1957) model to explain the distribution of primary division freshwater fishes. This model supposes that the groups originated in the Oriental region, subsequently spread via land and freshwater connections to Africa, Eurasia, and North America, and finally colonized South America from North America via the Panamanian Land Bridge. This model, proposed prior to the general acceptance of continental drift, should be compared to the current interpretation diagrammed in Figure 10.7.

morphs (Osteoglossomorpha), that have Southern Hemisphere distributions. Lungfishes (Figure 10.9) should show Gondwanaland relationships because they belong to an ancient group of fishes, presumably the stock that gave rise to tetrapods. Indeed, the South American *Lepidosiren* (Lepidosirenidae) is probably the vicariant of the African *Protopterus* (Protopteridae). In contrast, the Australian lungfish (*Neoceratodus,* Ceratodontidae) is the sole surviving member of a family that was wideranging in the Mesozoic (Figure 10.9) and that also had marine relatives. Thus the lungfish reconstruction cannot be treated simply as a tripartite vicariance sequence leaving a living lungfish on each southern landmass.

Osteoglossomorpha is a small group, including the very large and graceful arapaima and arawana sometimes seen in large public aquaria. In the family Osteoglossidae, *Arapaima* of eastern South America is apparently most closely related to *Heterotis* of West Africa, whereas the two species of *Osteoglossum* of South America are the close relatives of *Scleropages* not only of northern Australia and New Guinea but also in Southeast Asia. Other living osteoglossomorphs reside in Africa (Notopteridae, Pantodontidae, and Mormyriformes), but there are fossils of this superorder found in the Northern Hemisphere. Hence in the osteoglossomorphs, as with lungfishes, we suspect that early forms were part of the Gondwanaland biota, but the evidence needed to reconstruct their biogeographic histories is still equivocal.

Secondary division freshwater fishes. Given the above problems in interpreting primary division freshwater fishes, imagine the controversies surrounding the distributions of secondary division fishes, some species of which can tolerate seawater. Take first the simple example of the predatory gars (Lepisosteidae), found in freshwater and occasionally brackish water of eastern North America to Lake Nicaragua, with an endemic species in fresh water on Cuba and the Isle of Pines. Because gars are sometimes captured at sea, a classic interpretation of the Cuban gar has been that they arrived by crossing a short marine barrier (Darlington, 1957). A recent monograph on gars by Wiley (1976) of-

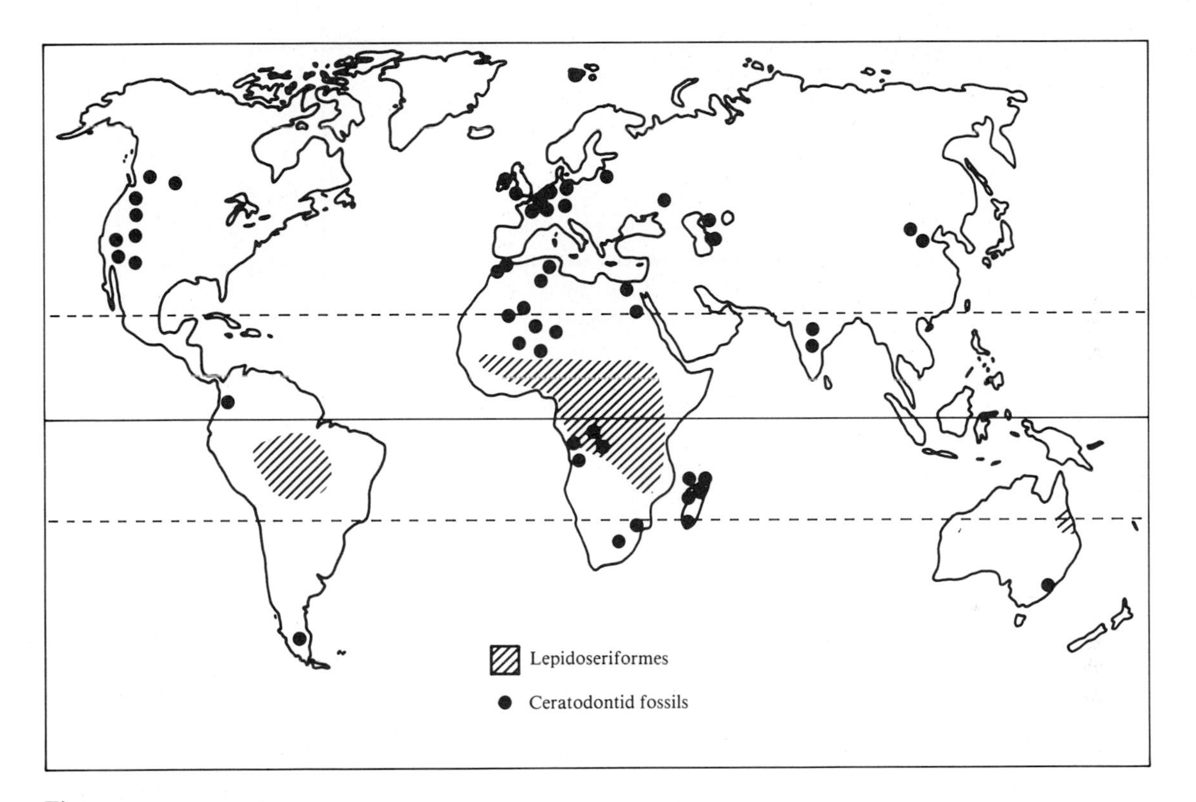

Figure 10.9
Present distribution *(hatching)* of the lungfishes (Lepidoseriformes), showing how they are confined to southern continents, and the distribution of fossil representatives of the family Ceratondontidae, showing how they were widely distributed (even in the Northern Hemisphere) in the Mesozoic. The sole surviving species in this family, *Neoceratodus fosteri,* lives in Queensland, Australia. This shows how dramatically important fossil discoveries can potentially change historical reconstructions of geographic distributions. (After Sterba, 1966, and Keast, 1977a.)

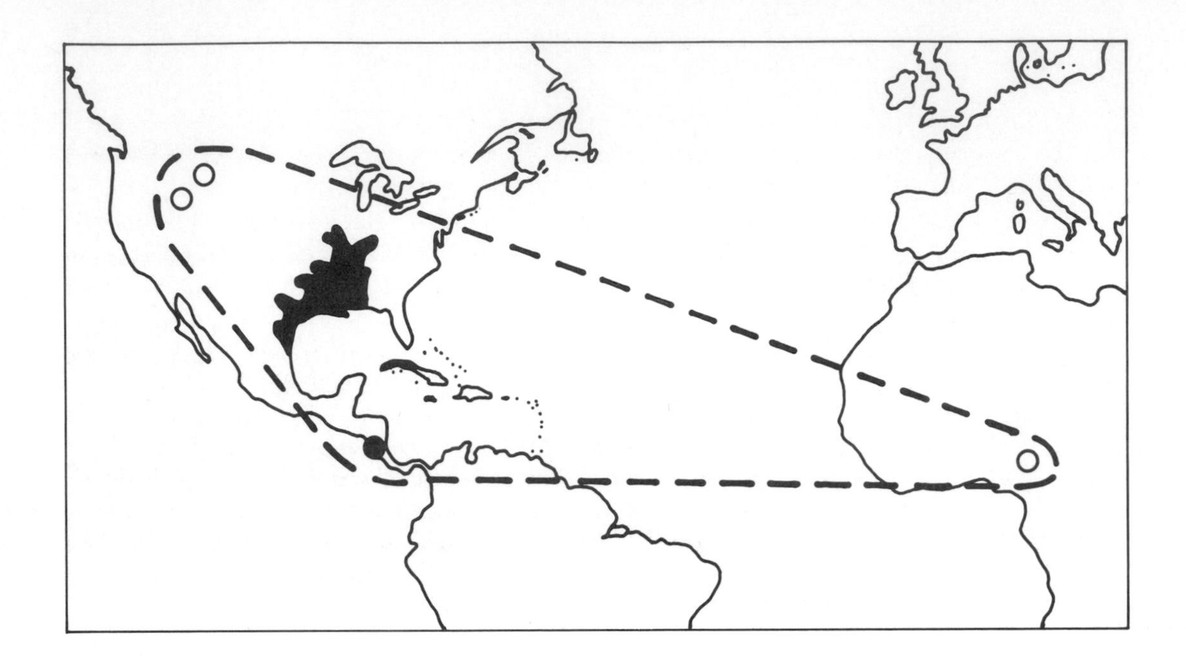

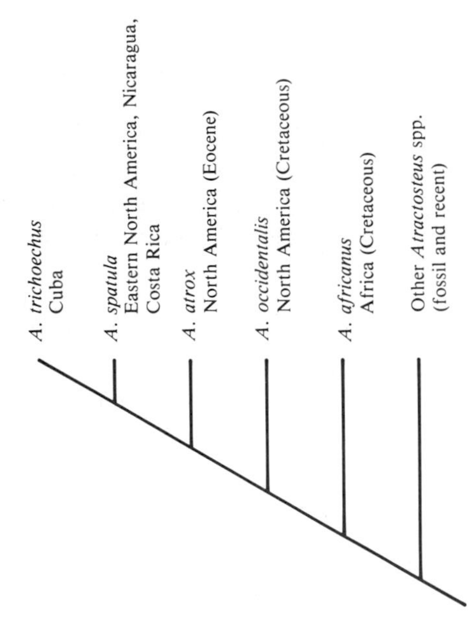

Figure 10.10
Distribution of the living *(solid)* and fossil *(open circles)* forms of the *Atractosteus* species group of gars (Lepisosteidae) and cladogram reconstructing the phylogeny of this group of fishes. Since gars are somewhat tolerant of salt water it was traditionally assumed that they had reached such distant outposts as Cuba by crossing marine barriers, but an alternative interpretation based on ancient land connections seems to be consistent with their phylogenetic relationships. (After Wiley, 1976.)

fered another plausible explanation. Wiley recognized two genera, each having fossil and living forms that each have a disjunction between eastern North America and the Old World (Figure 10.10). *Atractosteus tristoechus* of Cuba is presumably the sister taxon of *A. spatula,* the alligator gar of the United States and Middle America, several fossil species from North America, and the fossil *A. africanus* of central Africa. This permits the gar distribution to be a remnant of an older land connection, first when North America was adjacent to Africa in the Mesozoic and then when what is now Cuba once occupied a position in Central America. Meanwhile, the gar captured occasionally in the Gulf of Mexico is believed to be a separate saltwater species, and its phylogenetic relationships must be understood before the Cuban disjunction can be fully assessed.

The largest family of secondary division fishes in Cichlidae (700 species), which includes many familiar aquarium fishes such as discus and oscars. Members of this family can often tolerate brackish water but are unlikely candidates for long ocean voyages because they tend to have highly specialized modes of courtship, egg laying, and parental care. Most species occur in the East African rift valley (Chapter 6) but they have also radiated in South America and occur on Madagascar and the Indian subcontinent. Cichlids are now dispersing northward through Central America (Figure 7.15). This is therefore a pattern reminiscent of Pangea. In contrast, suborder Cyprinodontoidei (500 species) is a Northern Hemisphere group now moving southward. Both Cyprinodontoidei and Cichlidae have disjunct taxa in the Greater Antilles that have been interpreted as products of long-distance dispersal (Darlington, 1957) or as landmass vicariants from Central America (Rosen, 1975).

The nature of freshwater fishes on islands. In order to assess the conclusions that freshwater fish distributions are largely determined by their ability to tolerate salinity, one can turn to oceanic islands to see whether the predictions hold true. Wherever the entire freshwater ichthyofauna consists of only peripheral forms, we can feel fairly confident that the locality has been isolated by marine barriers, because peripheral fishes do not demonstrate land connections. An example is Fiji, a volcanic island group born in the Eocene. Its native freshwater fish assemblage consists of less than 20 genera, of which most are marine species from the East Indies that occasionally enter rivers or estuaries (Fowler, 1959). In fact, all oceanic islands lack primary and secondary division fishes. However, keep in mind that very few small islands have aquatic habitats.

The strength of conclusions on fish distributions can be tested for consistency by analyzing the isolated Australasian Plate. Only the lungfish and the osteoglossid *Scleropages* are possible candidates as primary division taxa, and their origins need clarification. Cyprinidae extend eastward from Asia as far as the Lesser Sunda Islands (*Rasbora* and *Puntius*), and other Asian groups of freshwater fishes have not penetrated eastward beyond Wallace's line. In New Guinea and northern Australia the most interesting taxa are the endemic rainbow fishes (Melanotaeniidae) and blue-eyes (Pseudomugilidae). Coastal species of these two families can live in brackish water and the families have evolved from marine ancestors, presumably the Atherinidae. Other species occurring in fresh water are described as "entering rivers and estuaries" (Munro, 1967), showing that these are marine groups invading fresh water, some of which ultimately have evolved as obligately freshwater species such as *Clupeoides papuensis* (Clupeidae) and a freshwater eel, *Anguilla interioris* (Anguillidae), of New Guinea.

In southern Australasia, notably New Zealand and Tasmania, several different families predominate. McDowall (1978a) has analyzed in great detail the 27 indigenous species of New Zealand. Four native species spend a portion of their lives at sea, including a lamprey (Geotriidae), an eel (Anguillidae), and two galaxiids (Galaxiidae). The remaining 23 species are en-

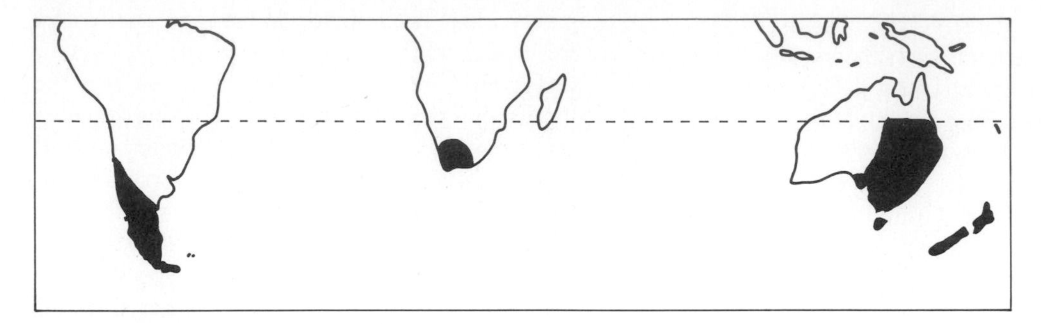

Figure 10.11
Distribution of the fishes in the superfamily Galaxioidea (families Galaxiidae, Retropinnidae, Aplochitonidae, and Prototroctidae). These peripheral fishes are restricted to nontropical waters in the Southern Hemisphere. The Galaxiidae are the only family to occur on each of the major landmasses, but the Aplochitonidae are disjunct between Tasmania and southern South America. (After Berra, 1981.)

demic but closely related fishes of Tasmania and southern Australia, or they are upland freshwater derivatives of coastal forms. Most interesting is the superfamily Galaxioidea, a favorite subject because the taxon is restricted to southern Gondwanaland (Figure 10.11). Of these the Southern Hemisphere smelts (Retropinnidae) are endemic to Australia and New Zealand but are sometimes caught at sea in shallow water. Galaxiidae have a broader distribution, including South Africa, and diadromous species have been caught 700 km from land. The eggs and larvae of the coastal species are commonly collected at sea. Another interesting species is *Gobiomorphus cotidianus* (Eleotridae), which has both lake and seagoing populations.

The southern continental affinities of New Zealand and Australian fishes do not necessarily indicate that they arrived when land connections existed with the rest of Gondwanaland. Given that the fishes are tolerant of salt water and could have dispersed across oceanic barriers, their restriction to southern temperate latitudes may simply require an ecological, not historical answer. McDowall suggested that the original colonizers of New Zealand arrived mostly from the sea, carried by eastward currents from Australia and Tasmania. A few taxa may have been present when the island was once part of a southern landmass with Antarctica and Australia. On the other hand, Croizat et al. (1974), Rosen (1974), and Ball (1975) prefer to interpret such vicariance events as remnants of a southern land connection, discounting long-distance dispersal across the sea, a conclusion that McDowall (1978b) flatly rejected. Regardless of who wins the disagreement, the fish fauna of Australasia is consistent with our expectations based on its long history of isolation from other continents: a diminutive, taxonomically biased, and distributionally biased sample of freshwater fishes.

Distribution patterns of other freshwater animals

After wading through the introductory discourse on fishes, it should come as no surprise that similar distribution patterns and problems of interpretation can be observed in other freshwater taxa. Some selected examples will illustrate this conclusion.

Freshwater planaria. Ball (1975) used a new phylogenetic classification of planaria (Dugesiidae) to reevaluate the distribution of this group. Because these animals are intolerant of

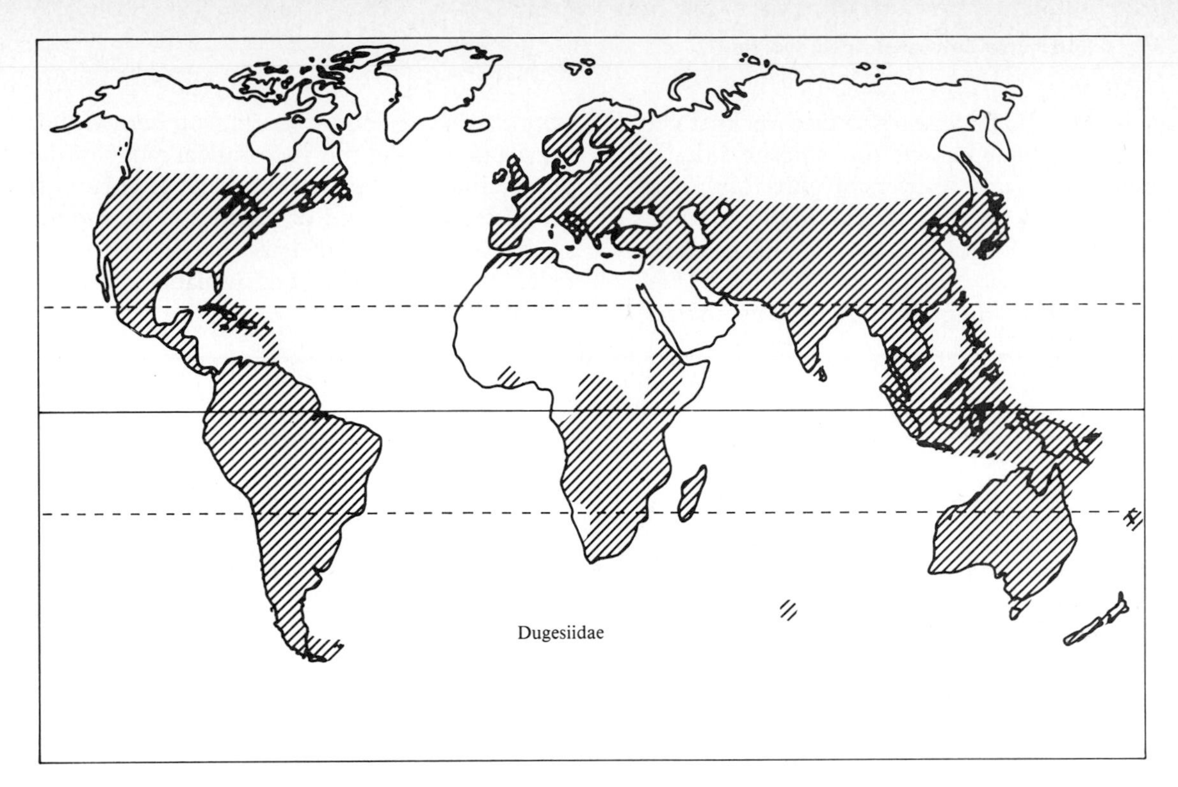

Figure 10.12
Distribution of freshwater planaria of the family Dugesiidae. (After Ball, 1975.)

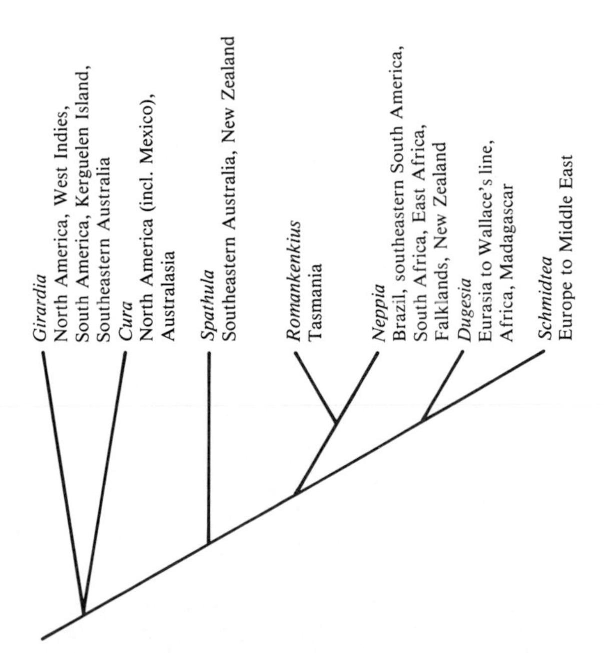

Figure 10.13
Cladogram showing hypothesized relationships among the seven genera in the family Dugesiidae. Ball preferred to explain the distribution as having originated in Gondwanaland and subsequently dispersed northward. According to this scenario *Cura* in North America and *Dugesia* and *Schmidtea* in Eurasia are fairly late arrivals from the south, not vicariant taxa dating back to the separation of Laurasia.

salt water, it is interesting to discover that certain genera are vicariant on opposite sides of present-day marine barriers (Figure 10.12). Ball has presented a model to account for disjunctions of austral taxa and the dissimilarities between taxa in the Nearctic and Palearctic (Figure 10.13). According to his cladistic hypothesis, early clades were all present in southern Gondwanaland, making this the original range of the group. Initial diversification of Dugesiidae was completed by the start of the Mesozoic, and the spread and subsequent isolation of each taxon reflect the land connections available in Pangea and its breakup.

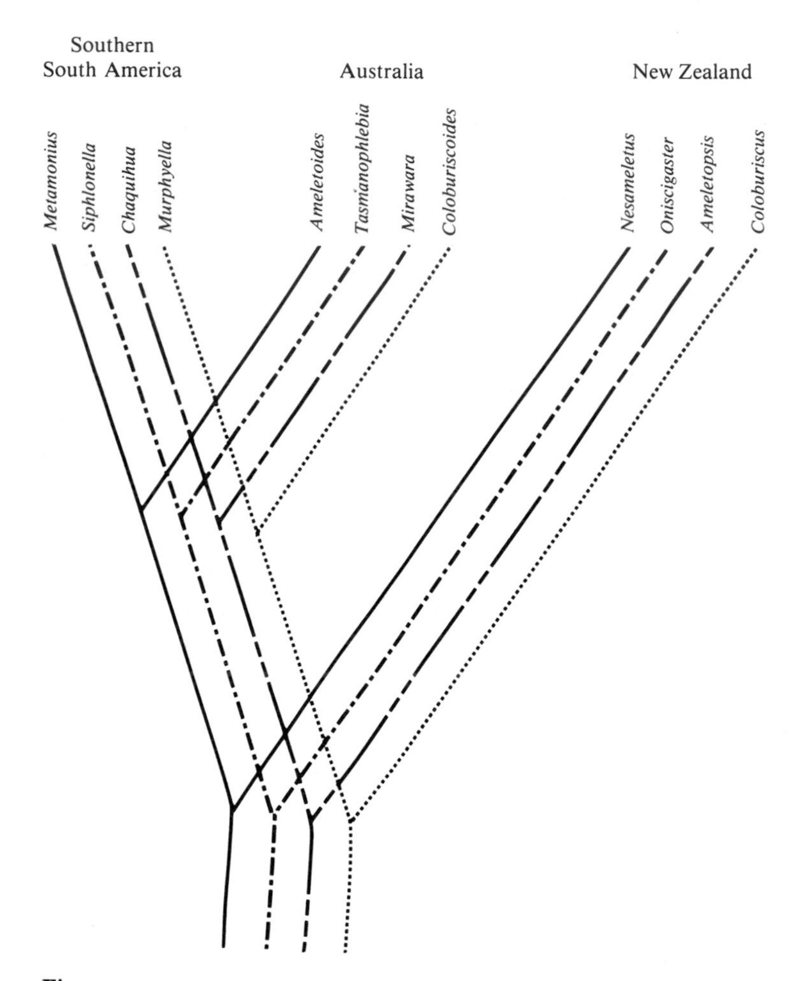

Figure 10.14
Phylogenetic relationships among four taxa of temperate mayflies (Ephemeroptera) of the Southern Hemisphere as worked out by Edmunds (1981). There are four different groups, each having representatives in South America, New Zealand, and Australia. Note that the geographic pattern of differentiation in each clade (e.g., the solid line) mimics the pattern of the others, indicating that the sequence of geologic events producing the genera in one clade was the same for all clades. In all cases the South American and Australian forms are more closely related to each other than either is to the New Zealand representative. Edmunds suggests that this is a vicariant pattern caused by the breakup of Gondwanaland. (Modified from Edmunds, 1981.)

Mayflies. Of all the groups of freshwater insects, the mayflies (Ephemeroptera) are of greatest interest because they have a history of conservative dispersal (Edmunds, 1972). Most of their lives are spent in water as eggs and larvae, interrupted by a very brief mating flight of adults, which are rapidly killed by desiccation. In addition, only a few families have species on oceanic islands, supporting the notion that the majority of forms are landlocked. Because the fossil record of mayflies apparently goes back to the Carboniferous, the vicariant events of this group could mimic the splitting of the continents and subcontinents.

Two aspects of mayfly distribution are extremely interesting. First, four monophyletic clades with a southern disjunction between South America, New Zealand, and Australia have identical three-area cladograms (Figure 10.14). The New Zealand generic split occurred earlier than the split between South America and Australia (Edmunds, 1975, 1981). This sequence is the same as geologists have proposed for the breakup of these southern landmasses: New Zealand first and Australia–South America second (Chapter 5). A second major pattern of interest is that several mayfly sister taxa are found on opposite sides of the South Atlantic in Africa and South America (Edmunds, 1975), particularly in the tropical groups. This implies a tropical connection before these genera had differentiated from a common ancestor.

Molluscs. Attempts to elucidate the distributions of freshwater molluscs are greatly handicapped by our poor understanding of their phylogenetic relationships, not only among the families but also among the genera. So much structural convergence has occurred in snail evolution that classification schemes have been modified repeatedly. Phylogenetic diagrams, especially cladograms, are rarely published and are unavailable for the families considered most interesting by historical biogeographers. Moreover, the richest nonmarine molluscan fauna, which occurs in the Neotropics, is poorly known.

Recognizing these limitations, we can nonetheless take a glimpse at some freshwater patterns in the Southern Hemisphere. A focal point for discussion is the well-studied malacofauna of Africa (Brown, 1978, 1979, 1980; Bruggen, 1977). Africa is species poor, having only 326 known species, of which a portion have been introduced recently from Europe. The low number may reflect the fact that most inland waters have not persisted very long (Brown, 1980). Because living Mollusca include many old families, one would expect to find numerous interesting patterns of disjunction between Southern Hemisphere landmasses. Instead, no family of the freshwater Afrotropical fauna is endemic, although many of the genera and species are; of the indigenous species, most share phylogenetic affinities with groups in tropical Asia. There are a few taxa that may have belonged to an original Gondwanaland biota. One interesting disjunction occurs with South African species of *Tomichia* and its relatives. These were formerly assigned to the predominantly northern family Hydropiidae but now are reclassified in the family Pomatiopsidae, which also occurs in southern South America and Australasia. Several other families residing in the Cape Region of Africa need to be examined for possible Australian relationships. Interestingly, molluscs of Madagascar share a number of taxonomic similarities to those in New Caledonia, whereas both are very different from those of Sri Lanka, which is geographically much closer to either than Madagascar and New Caledonia are to each other (Starmühlner, 1979).

In temperate Australasia, at least, the freshwater mussels of the family Hyrridae and the little river snails (Hydrobiidae) seem to have sister taxa in southern South America.

Salamanders. Treating amphibians as a group of aquatic or as land organisms is, of course, misleading because at least part of the life cycle is usually conducted in both habitats. Among amphibians, salamanders (order Caudata or Urodela) are most closely tied to water,

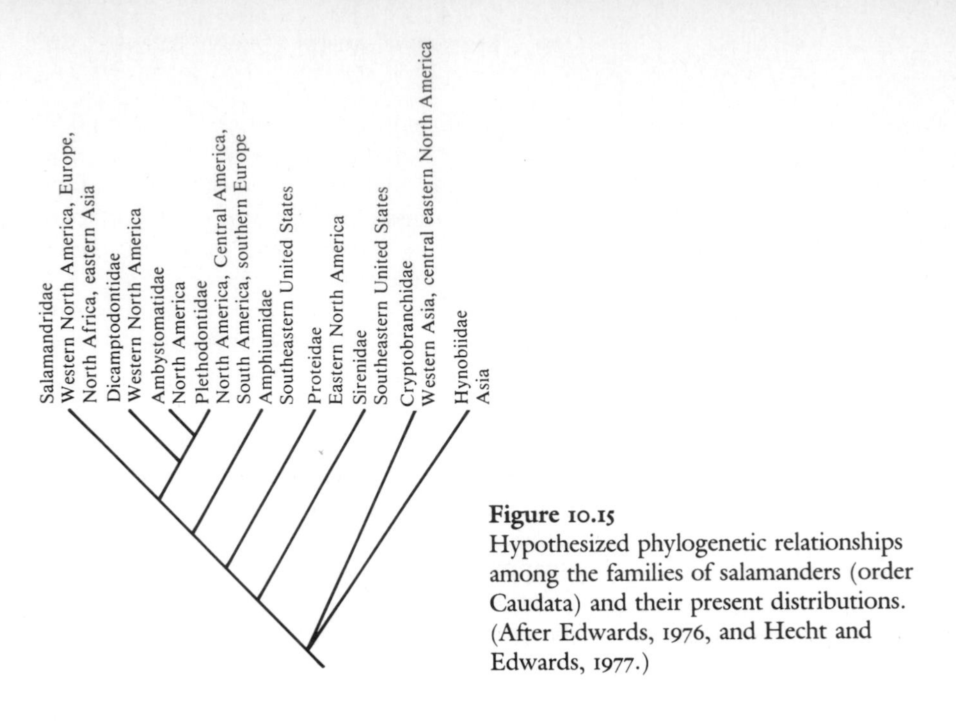

Figure 10.15
Hypothesized phylogenetic relationships among the families of salamanders (order Caudata) and their present distributions. (After Edwards, 1976, and Hecht and Edwards, 1977.)

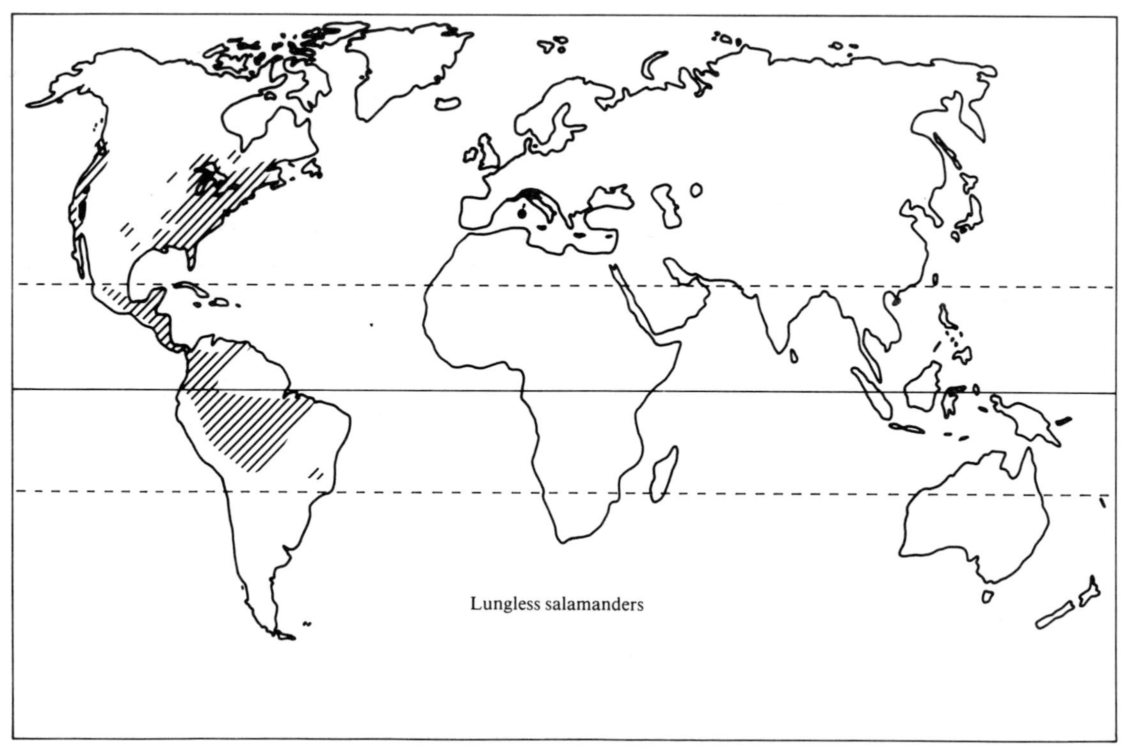

Figure 10.16
Distribution of the lungless salamanders (Plethodontidae), showing many wide disjunctions in an old and presumably once widely distributed group. (After Wake, 1966.)

some forms being entirely aquatic and some rarely leaving water; but there are a number of interesting taxa living and breeding in trees (arboreal) or in burrows (fossorial).

The phylogenetic and distributional history of salamanders can only be inferred from living species because the fossil record of these organisms is very poor. Using a recent phylogenetic hypothesis of Edwards (1976) (Figure 10.15), we can see that all the groups with primitive features (that are fully aquatic) and nearly all derived families occur exclusively in the Northern Hemisphere. They do not occur on oceanic islands, confirming other evidence that they are intolerant of seawater.

For this holarctic order, the most interesting disjunctions occur in the modern family of lungless salamanders (Plethodontidae) (Figure 10.16), which has more than half of the total species (Wake, 1966). Several items are noteworthy. The family has disjunctions between mesic areas in eastern and western North America with relict populations in the central United States, showing that this family once occurred throughout forested regions of North America. In addition, the genus *Hydromantes* has a remarkable disjunction between California and Italy, France, and Sardinia in the Old World, indicating a continuation of its mesic distribution within the Holarctic. Only the supragenus *Bolitoglossa* (126 species) occurs in the Neotropics to Brazil and Bolivia, but it misses the West Indies, implying a recent invasion of South America. Finally, the genus *Ensatina* is famous for its interesting ring of races or subspecies that encircles the Central Valley of California (Stebbins, 1949).

Thus here is a taxon, the entire order of salamanders, that has used northern migration routes, exhibits only one currently successful invasion of the tropics (Wake, 1970), and has experienced much contraction of its range in the north, leaving numerous disjunct populations.

Freshwater turtles. Much biogeographic research is done on taxa for which there is little or no fossil record; hence the only avenue for reconstructing the history of the group is to analyze the disjunctions of living forms. How accurate are such reconstructions? The principal way of assessing this is to compare distributions of living forms with the known fossil record to determine similarities and difference. One group we can analyze are the freshwater turtles, which show major inconsistencies between extant versus fossil distributions.

The box on p. 308 summarizes some of the remarkable distributions of living and fossil turtles. The majority of families have undergone a major range contraction and do not occur in areas where they once lived. For example, the snapping turtles (Chelydridae) inhabit the New World but formerly occurred in Europe. More remarkable is the New Guinea plateless river turtle (*Carettochelys insculpta,* Carettochelyidae), now a narrow endemic of southern New Guinea and northern Australia but whose relatives once lived in Eurasia in the Pliocene.

The side-necked turtles (infraorder Pleurodira) now are restricted to the Southern Hemisphere, suggesting an ancient Gondwanaland distribution, but many fossil records occur in the Northern Hemisphere. In the Pelomedusidae, *Podocnemis* occurs in South America and Madagascar (this genus was once very widespread), *Pelusios* in Africa, and *Pelomedusa* in Africa and Madagascar. Complementary to this West Gondwanaland range, the sister taxon Chelidae has its genera split between South America and Australia–New Guinea. In fact, these are the only freshwater turtles in Australia. The only other turtles in New Guinea, with the exception of the relict *Carettochelys* mentioned above, are at least partially marine species, e.g., the seagoing *Pelochelys bibroni* (Trionychidae), which crossed Wallace's line from southeast Asia.

In contrast, the majority of turtle families and the three principle superfamilies of freshwater and terrestrial species strongly show a Laurasian pattern, often with a relatively recent invasion into the Neotropics. The widespread emydines (Emydidae) and the soft-shelled tur-

A classification of extant and fossil(†) turtle families, order Testudines

Classification includes the distributions of extant families (after Gaffney, 1975).

Order Testudines
Suborder Proganochelydia
Proganochelyidae†

Suborder Casichelydia
Infraorder Pleurodira (side-necked turtles)
Pelomedusidae, South America, Africa, Madagascar
Chelidae; South America, Australia, New Guinea
Infraorder Cryptodira
Parorder Paracryptodira
Superfamily Baenoidae
Glyptopsidae†
Baenidae†
Parorder Eucryptodira
Superfamily Trionychoidae
Kinosternidae; North, Central, and South America
Dermatemydidae; Central America, Mexico
Carettochelyidae; New Guinea
Trionychidae;. eastern Asia to New Guinea, Africa, including the Mediterranean, North America
Superfamily Chelonioidea (marine turtles)
Plesiochelyidae†
Protostegidae†
Toxochelyidae†
Dermochelyidae; tropical seas
Cheloniidae; tropical seas to Newfoundland and Scotland
Superfamily Testudinoidea
Chelydridae (incl. Platysternidae); North, Central, and South America, southeastern Asia
Emydidae; North, Central, and South America, West Indies, Eurasia, northern Africa
Testudinidae; North, Central, and South America, West Indies, Galápagos Islands, Eurasia, Africa, Madagascar, Seychelles Islands

tles (Trionychidae) have very closely related genera in North America and eastern Asia; and where African genera are present, they are related to Asian forms. These observations support a conclusion that eastern Asia has served as an important source area for the dispersal of turtles in the Northern Hemisphere. However, there are so many fossils to be considered in any historical reconstruction that details of early vicariance and dispersal events of Testudines are not at all understood.

As mentioned above, freshwater turtles are probably barred from crossing open ocean. For example, of the four genera in the widespread New World family Kinosternidae, only *Kinosternon* occurs on islands, and it seems fairly certain that these islands, such as Trinidad and Long Island, recently had continental connections. *Kinosternon* is absent from the Antilles. In the family Emydidae, which includes most living freshwater species as well as semiaquatic and terrestrial forms, the species are not capable of long ocean voyages but can live in brackish water. For example, the diamondback terrapin, *Malaclemys terrapin,* occurs along the coastline of the eastern United States, and the terrapins of *Pseudomys* occur in the Bahamas, the Cayman Islands, and the Greater Antilles. Moreover, the terrestrial *Terrapene* (Emydidae) has an endemic species in the Greater Antilles, and Rosen (1978) has proposed that these turtles were present on the landmass when the Greater Antilles were located where Central America is today (Chapter 5).

Interpreting marine distributions

Because oceans are vast interconnected regions one expects marine organisms to have fairly continuous and broad distributions. There are, of course, many thousands of marine examples of a species or a genus that occurs in an area as large or larger than most species ranges on continents. This is particularly common for species of the open oceans and invertebrate species having planktotrophic larvae. Boundaries of these individuals' distributions are set by geologic barriers such as a landmass; by climatic barriers, such as cold temperature or the intersection of two water masses of different quality and direction; by biotic factors, such as low productivity; or by the absence of a particular substrate, such as shallow sandy bottoms or rocky coastlines. Biogeographers, especially those working with fossil marine faunas, have made great use of taxa with broad ranges for identifying ancient marine provinces (Middlemiss et al., 1971; Hallam, 1973b; Gray and Boucot, 1979).

One should not assume, however, that marine distributions are all necessarily very broad; on the contrary, there are many species with narrow ranges. Those organisms with relatively restricted distributions tend to be restricted to shallow water and often to particular substrates. Such species are fairly common along north-south trending continental coastlines, in the tropics on large coral reefs, and in water bodies that are somewhat isolated from the influence of large water masses. Along western North America hundreds of molluscs have fairly short linear ranges along the coastline. Much of this endemism is undoubtedly related to the narrow habitat preferences of each species; hence the populations can only occur where favorable patches of habitat are present.

Proximate ecological factors are easy to hypothesize but difficult to demonstrate to account for the richness of groups on tropical coral reefs, such as the Great Barrier Reef of Australia, in the Philippines, and in the Caribbean region. In such cases, a small patch of habitat may have dozens of closely related species, many of them living right next to each other. For these cases we still need both historical explanations for where the species came from and how they achieved their present ranges and ecological studies of resource use and community organization to account for their coexistence. Much of the speciation on these coral reefs probably occurred during the Cenozoic, if not mostly within the last 2 million years, as the coral reefs were subdivided into smaller, isolated patches in which differentiation proceeded. Subsequently, the ranges of these isolates could have become sympatric as the habitat expanded. This cycle of differentiation in isolation followed by sympatry could have occurred frequently, resulting in the development of extremely diverse communities.

Reconstructing the sequence of events giving rise to marine endemics is perhaps the biggest frontier left in historical biogeography. To date, most dependable reconstructions using endemics have been fairly straightforward. For example, when Panama became emergent land 4 million years ago, it severed the ranges of a marine system that was continuous from the eastern Pacific Ocean to the Gulf of Mexico. Since then, the sister populations of most coastal and some open water forms have differentiated into separate but usually morphologically very similar races, subspecies, and species (Ekman, 1953; Vermeij, 1978). However, when workers turn their attention to complex regions, such as the Philippines with its great numbers of coexisting species, no simple historical picture has yet emerged. It would be extremely desirable to use biotic distributions of narrow endemics in a place such as the Philippines to unravel the history of the islands and explain how the islands and their barrier reefs were connected in the past. Unfortunately, if workers are going to use vicariance methods to address these issues, carefully prepared phylogenies of the groups must be produced. This has never been an easy task for marine taxa, because their morphological traits often are highly specialized for particular functions and ecological niches. For example, many characteristics of mollusc shells appear to be adaptations for avoiding specific predators (Vermeij, 1978).

Pinnipeds and penguins. The crowd-pleasing seals and sea lions are morphologically adapted for aquatic life but return to land each year to breed. There are 34 extant species belonging to two independent evolutionary lines. The true or earless seals and elephant seals, the Phocoidea, are presumably most closely related to otters (Mustelidae); and the eared seals, the Otarioidea, which include sea lions, fur seals, and the walrus, show strong similarities to bearlike carnivores (Ursidae). Hence although all of these mammals are called pinnipeds and most books treat them as a single order, Pinnipedia, present evidence favors our treatment of these animals as two convergent clades.

The reason we insert them here as aquatic examples is because their historical biogeography is better than that of other aquatic vertebrates because there is a fairly extensive fossil record. A comprehensive study of both fossil and recent forms (Repenning, 1976; Repenning et al., 1979) clearly demonstrates that we could never understand the complex history of pinnipeds merely by studying extant distributions. The fossils provide a rich and worthwhile data set that cannot be ignored. The study also shows, however, that no fossil record is ever as complete as is needed to recapture the complete history.

The most useful reconstruction of pinniped history is available for the eared seals (Otarioidea) and is illustrated in Figure 10.17. Comparative studies suggest that the extinct *Enaliarctos mealsi,* which lived in the northern Pacific region about 22 million years BP, was very close to the ancestral stock of otarioids. Whereas present-day species are sexually dimorphic and rock breeders, *Enaliarctos* was a small coastal form showing no sexual dimorphism or rock breeding attributes. Subsequent increase in body size, development of sexual dimorphism, and modification of the ear gave rise to the pelagic feeding strategy and harem breeding of derived forms. The now extinct Desmatophocidae were very abundant in the north Pacific in the Miocene and especially from 14 to 12 million years BP and they gave rise to the Otariidae, the sea lions and fur seals, by diversifying in that region. Another clade, the bottom-feeding walruses (Odobenidae), appeared first in the fossil record about 14 million years BP in the Pacific, but by 8 million years BP walruses had invaded the Caribbean, where they became greatly diversified. The modern walrus *(Odobenus rosmarus)* apparently is derived from a Caribbean form that moved northward with the warming of the Gulf Stream in the Pliocene, entered the north Atlantic, and finally returned to the north Pacific via the arctic route only in the Upper Pleistocene. If this scenario is correct, today the sympatry of *Odobenus* with other Otariidae in the north Pacific is a secondary occurrence.

Figure 10.16 also depicts the hypothesized evolutionary history of Phocoidea, which includes a single family, Phocidae. The fossil record of this family is less valuable because at 15 million years BP one already finds the two extant subfamilies, the monachinoid seals (Monachinae) and the harbor seals and relatives (Phocinae). Fossils of the monachines appear first from the Atlantic coast of the United States and subsequently in western and eastern Europe as well as the Caribbean. Therefore this lineage was differentiating in the Atlantic province while otarioids were differentiating in the north Pacific. By 5 million years BP just prior to the formation of the Panamanian Land Bridge, monachinoid seals undoubtedly occurred in the eastern Pacific because the closure of Central America produced two present-day populations, the West Indian monk seal *(Monachus tropicalis)* and the Hawaiian monk seal *(M. schauinslandi)*. The other extant species of monk seal, *M. monachus,* formed as an eastern

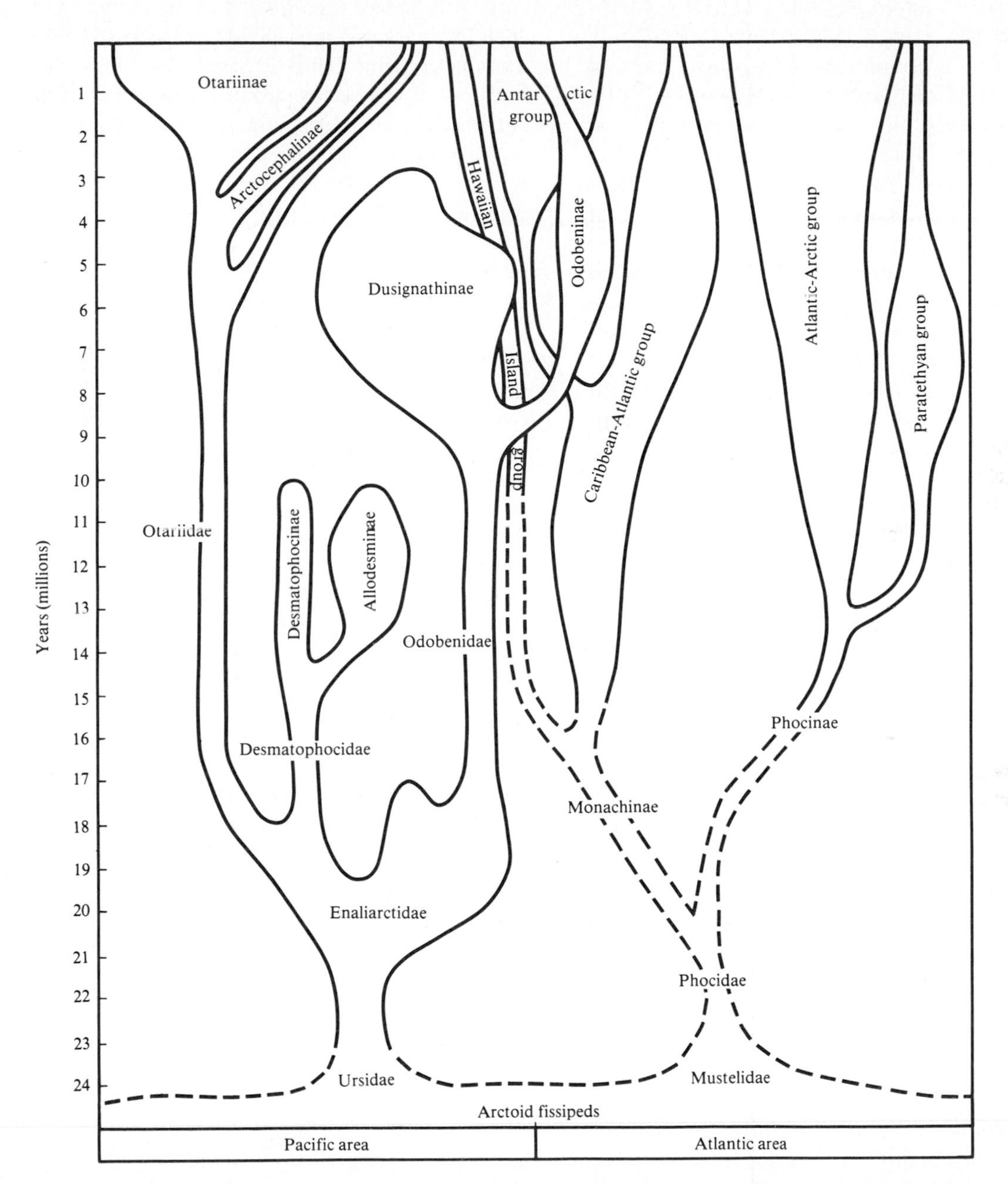

Figure 10.17
Model of the phylogenetic and biogeographic history of the seals and sea lions. Once considered to be a monophyletic order, Pinnipedia, these marine mammals are represented here as comprising two different groups: the Otarioidea, which evolved from bearlike ancestors in the Pacific, and the Phocoidea, which differentiated from weasel-like ancestors in the Atlantic. (After Repenning et al., 1979.)

isolate in the eastern Atlantic Ocean, the Mediterranean Sea, and the Black Sea.

There is little information on how most of the genera of seals and sea lions living in polar waters achieved their present ranges and how they differentiated into a variety of new genera and species. For example, for the harbor seals it is simple to envision how the Baikal seal of Lake Baikal *(Phoca sibirica)* became stranded from other populations and differentiated in isolation since 300,000 years BP, but it is more difficult to explain the origin of the four monospecific genera *(Cystophora, Erignathus, Halichoerus,* and *Histriphoca)* that live in the northern waters in and around Eurasia and North America. During the periods that these phocine genera were differentiating in the northern latitudes, four monachine genera were differentiating around the Antarctic. One of these, the genus of elephant seals *(Mirounga),* has an interesting disjunction, with one circumtemperate species in the Southern Hemisphere and one in the waters of California and Baja California in the Northern Hemisphere. Then, to complicate matters even further, one must explain the disjunctions in the Otariidae. Most genera live in the northern Pacific region, but the fur seals *(Arctocephalus)* live mostly in widely disjunct locations in the Southern Hemisphere (Australia, Kerguelen Island, New Zealand, Juan Fernández Island, temperate South America, the Galápagos Islands, and the Cape Region of South Africa). Here is a case in which the fossil record gives no specific answers, and authors are forced to rely on dispersal scenarios. Dispersal explanations are quite believable for seals and sea lions, which are good navigators and strong long-distance swimmers, but workers are always stuck with the scholarly problem that it is difficult to obtain hard evidence to support such explanations.

The same type of equivocal statement can be made about the penguins. Penguins (order Sphenisciformes) are flightless marine birds that occur only in the Southern Hemisphere, typically from south temperate latitudes to the high Antarctic, where they inhabit some of the same areas as the otarid pinnipeds just discussed. The most interesting genus is *Spheniscus,* which has two endemic species around southern South America *(S. magellanicus* and *S. humboldti),* on two islands of the Galápagos Archipelago *(S. mendiculus),* and on the islands off southern and southwestern South Africa *(S. demersus).* Because these penguins are strong pelagic swimmers and are migratory, an easy interpretation is fairly recent dispersal through the sea (Simpson, 1976). However, Cracraft (1974c) suggested that *Spheniscus* has a vicariant distribution linked to the breakup of western Gondwanaland. The fossil record does not help to resolve this issue, because the oldest fossil penguins have been found in the Paleogene of New Zealand and Antarctica, suggesting that the group may only be Cenozoic in age. Workers must continue to search for new data, especially additional fossils, and for ways to distinguish between the alternative hypotheses that can be advanced to explain these distributions.

Selected references

Physiological tolerances of aquatic animals

Bentley (1971); Briggs (1974); Ekman (1953); Gordon et al. (1982); Hoar and Randall (1969); Hochachka and Somero (1973); M. Horn and Allen (1978); Hubbs (1948); G. Hutchinson (1957, 1967); Kinne (1975); Lehner (1979); Maloiy (1979); Menzies et al. (1973); Menzies et al. (1966); Parin (1970); Perkins (1974); Saunders (1971); Schmidt-Nielsen (1975); Thomson and Lehner (1976); J. Valentine (1973).

Substrate and form

J. Gray and Boucot (1979); J. Gray et al. (1981); J. Valentine (1973); Vermeij (1974, 1978).

Dispersal and reproductive properties of aquatic animals

Abele and Walters (1979); Breder and Rosen (1966); A. Brown (1977); Cheng (1976); Edmunds (1972, 1975, 1981); Thomson and Gilligan (1983); Hynes (1970); F. Jones (1968); Leggett (1977); Lehner (1979); Leis and Miller (1976); Macan (1973); Matthews (1980); Mileikovsky (1971); Norris (1966); Parin (1970); Perkins (1974); Popham (1961); Rex (1981); Rice and Wolman (1971); Scheltema (1978); Slijper (1962); Steele (1974); G. Thorson (1950); Vance (1973a, 1973b).

Distribution patterns of freshwater fishes

Berra (1981); C. Bond (1979); Briggs (1974, 1979); Cracraft (1974c, 1975a, 1975b); Croizat et al. (1974); Darlington (1957, 1965); Ekman (1953); Fink and Fink (1981); Fowler (1959); Fryer and Iles (1972); Géry (1969, 1977); Greenwood (1974); Greenwood et al. (1973); Jubb (1967); Keast (1977a); McDowall (1978a, 1978b); McDowall and Whitaker (1975); Miles (1977); R. Miller (1958, 1966); Munro (1967); Myers (1938); Naiman and Soltz (1981); J.S. Nelson (1976); Roberts (1973); Rosen (1974, 1975, 1979); Rosen and Greenwood (1976); G. Smith (1981); Sterba (1966); T. Thorson (1976); Wheeler (1977); Wiley (1976).

Distribution patterns of other freshwater animals

Ball (1975); D. Brown (1978, 1979, 1980); Bruggen (1977); Brundin (1966, 1967, 1972); Edmunds (1972, 1975, 1981); J. Edwards (1976); Goin et al. (1978); M. Hecht and Edwards (1977); Pritchard (1979); Rosen (1978); B. Smith (1979); Starmühlner (1979); R. Stebbins (1949); Wake (1966, 1970); Weatherley (1967).

Interpreting marine distributions

Briggs (1974); Brinton (1962); Ekman (1953); New Zealand Department of Scientific and Industrial Research (1979); Repenning (1976); Repenning et al. (1979); G. Simpson (1976b).

Chapter 11

Distribution Patterns of Terrestrial Animals

For assessing the importance of aquatic barriers between landmasses, investigators naturally turn to animals like ourselves that, without aids, must move on solid land. Biogeographers have long observed that taxa with the capacity for long-distance dispersal are the groups represented on distant oceanic islands (Carlquist, 1965, 1974). On the other hand, investigators prefer to use groups with conservative dispersal properties and life history parameters for reconstructing the histories of continental faunas.

Abilities of land creatures to cross water barriers

A small fraction of the large terrestrial animals are also able to swim short distances, and another small but interesting fraction of small to tiny land creatures have ways of crossing water and other habitat barriers at some time in their life cycle, e.g., by hitching rides on other animals or being trapped on flotsam.

Stories about swimming terrestrial vertebrates have appeared repeatedly in the literature and, as absurd as some reports may seem, we cannot deny that these occasionally occur. A well-documented case is the elephant, which enjoys being in water. However, who would think that an elephant would be found in the ocean? Odd as it may seem, actual reports are well authenticated. For example, a cow and her calf not only were photographed voluntarily swimming at sea in fairly deep water off the coast of Sri Lanka but were also followed and timed, showing that they made a return trip (Johnson, 1978, 1980, 1981). Johnson and others believe that elephants have been able to traverse narrow straits between islands and mainlands in search of new food supplies and therefore are not reliable indicators of a solid land connection. Other large vertebrates, such as tigers and terrestrial snakes, have been spotted over a kilometer at sea, but in many of these cases the organisms were probably not there by choice, having been washed out to sea during a torrential storm.

No biogeographer today wants to believe that swimming was the means used by organisms to cross wide oceans; in fact, transoceanic swimming of terrestrial organisms is discounted, except during the early stages of seafloor spreading when distances between two landmasses were extremely narrow. A more plausible means for achieving dispersal over water is by rafting. These rafts are relatively firm assemblages of vegetation washed out to sea by tides and storms, sometimes including whole trees with roots and attached soil and carrying onboard a variety of terrestrial creatures, especially small rodents, lizards, snakes, snails, and soil creatures (Carlquist, 1965). These rafts have been reported drifting hundreds of kilometers out at sea, and although they eventually break apart, some will and do come to rest on a distant parcel of land with live colonizers. A gravid female, a mated pair, or a parthenogenetic form would be required to found a new population.

On continents and large islands, water barriers pose a less permanent barrier to dispersal simply because organisms can often migrate around the obstruction or can cross directly by swimming or floating. Of course, many of these barriers are more ephemeral than the barriers created by oceans and large epeiric seas. This does not mean, however, that water cannot terminate distributions of terrestrial species. For example, Grinnell (1914) discovered that totally different forms of pocket gophers (Geomyidae) live on opposite sides of the Mississippi River in central North America. Many vertebrates, such as primates in South America (Figure 11.1), and invertebrates have ranges terminated abruptly by wide rivers.

For the majority of terrestrial animals having long-distance dispersal capabilities, one very successful method has been passive dispersal by being carried by larger creatures, especially externally on the feathers or the feet of flying birds or the fur of mammals. Of course, internal or external parasites travel with their hosts.

It is no wonder therefore that terrestrial groups with mainland-island disjunctions often show obvious means for crossing water barriers, especially those carried by birds and the

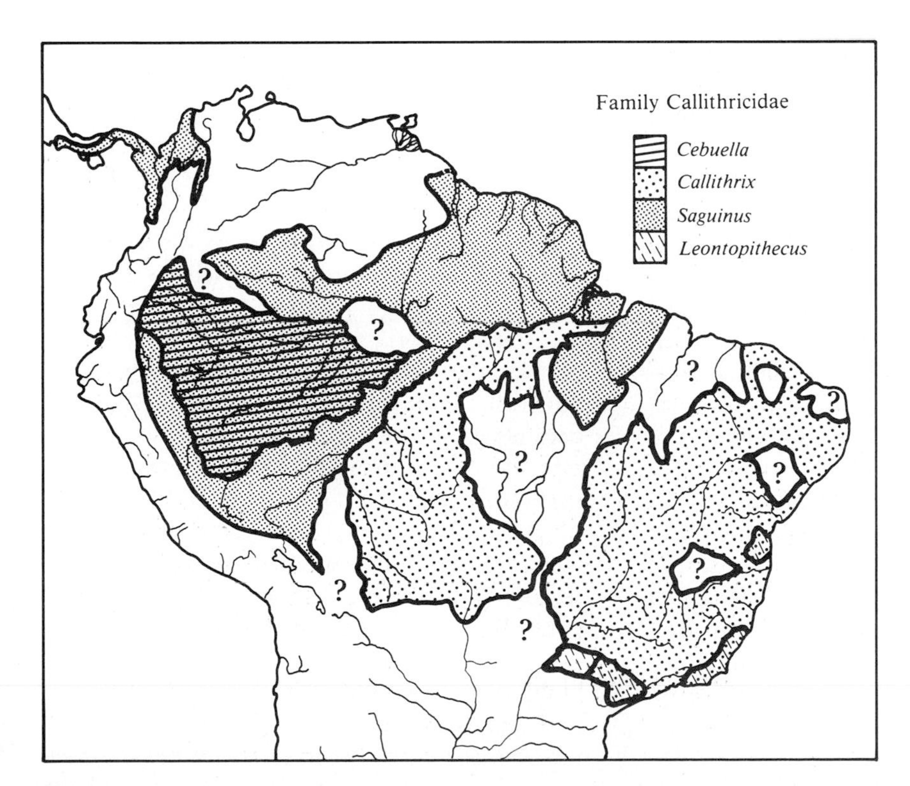

Figure 11.1
Distribution of New World monkeys of the family Callithricidae. Note that many of the taxa have adjacent but nonoverlapping geographic ranges that reflect the roles of major rivers in limiting their distributions. (Reprinted from *Living New World Monkeys* by P. Hershkovitz, by permission of The University of Chicago Press. Copyright © 1977 by the University of Chicago.)

wind. If land organisms can cross wide oceanic gaps, then these same means of dispersal must be used with even greater frequency on mainlands. Although investigators cannot presently assess the significance of long-distance dispersal for terrestrial animals on continents, they tend to consider that it occurs but is often not necessary to invoke compared with the greater effects of direct overland migration.

Mesozoic land vertebrates

Before we describe the extant distributions of interesting terrestrial animals, we should, for historical purposes, explore the patterns of some ancient creatures to see what they might show about past land connections. Alfred Wegener (Chapter 5) and his advocates believed that the distributions of these extinct beasts would corroborate his theory of continental drift. One example was *Mesosaurus,* a small aquatic reptile known from the Permian of southern Brazil and South Africa. For years opponents rejected this example because they thought the organism could have traversed the South Atlantic, even though it apparently lived in estuarine habitats. At the same time, authors were not ashamed to use as evidence of a northern land connection similarly wide disjunctions in the Northern Hemisphere, like the pelycosaur *(Edaphosaurus)* and labyrinthodont amphibians, shared between the Lower Permian of Texas and France. Although the Permo-Carboniferous fossil beds are plentiful in the Northern Hemisphere, in general records of terrestrial tetrapods are scanty, except in eastern and southern Africa. The Upper Permian land reptiles of South Africa show close affinities to Laurasian cotylosaurs and pareiosaurians in northern Russia (Colbert, 1979).

It is not until the Triassic, especially the Lower Triassic, that tetrapods provide convincing evidence of Gondwanaland unity. As discussed in Chapter 5 (Table 5.2), *Lystrosaurus* and several other distinctive reptiles and amphibians occurred in southern Africa, India, and Antarctica; the overlying *Cynognathus* zone, another assemblage of tetrapods, also has a widespread occurrence in Gondwanaland. Africa, South America, India, Antarctica, and Australia were certainly closely joined in the Lower Triassic to show such strong faunal similarities.

Beginning in the Upper Triassic, dinosaurs became widespread and dominated terrestrial communities. Nonetheless, in places like South America other tetrapod lineages coexisted with them. Small mammal-like reptiles were beginning to disappear as mammals were evolving (Lillegraven et al., 1979). In the Upper Triassic similar tetrapods occurred in such widely separated regions as Rhodesia *(Syntarsus)* and North America *(Coelophysis);* Argentina, western China, and western North America (therapsids); or China, Europe, and Africa (prosauropods). Tritylodonts were present on all continents (Clemens, 1982).

Dinosaurs dominated the world landscapes in the Jurassic and Cretaceous. Unfortunately, there are few fossil collection sites from the Jurassic of the Southern Hemisphere, the best site being in Argentina, where there is a record of dinosaurs and two frogs assigned to Ascaphidae. The Upper Jurassic and Lower Cretaceous still show a worldwide fauna, but three major, new families of dinosaurs evolved in the Cretaceous of the Northern Hemisphere. Dinosaurs became extinct at the end of the Mesozoic in a mass extinction that also occurred in many marine clades.

A number of authors have analyzed these faunas to determine whether any periods or any regions showed provincialism during the Mesozoic. Based on current fossil records, Cox (1973a, 1973b) has shown that all landmasses had highly similar terrestrial and semiaquatic tetrapod faunas in the Triassic (Table 11.1), indicating the existence of a fairly continuous supercontinent. The smallest percentage of similarity was found between India and Asia, although both were individually extremely similar to Europe and to Africa. Having closer faunal similarities between India and southern Gondwan-

Table 11.1
Estimations of faunal similarities of terrestrial tetrapod families occurring on Triassic continental areas

The values below were obtained by dividing the number of shared families in two areas by the total number of families in the area with the fewest families and multiplying by 100, i.e., the Simpson index (Chapter 8). Many pairwise comparisons show highly similar tetrapod faunas. (Reprinted with permission from C.B. Cox, 1973. The distribution of Triassic terrestrial tetrapod families, pp. 369-371. *In* D.H. Tarling and S.K. Runcorn [eds.], Implications of continental drift to the earth sciences, vol. 1. Copyright © Academic Press, Ltd.)

	Europe	Asia	South America	Africa	India
North America	87.5%	44%	56%	75%	59%
	Europe	80%	68%	57.5%	81%
		Asia	49%	89%	41%
			South America	74%	56%
				Africa	75%

aland is, of course, consistent with its position in the continental drift model. In the Triassic, there are northern and southern trends; for example, the *Lystrosaurus* and *Cynognathus* zones show strong Gondwanaland distributions, whereas the labyrinthodont *Metaposaurus* and the crocodile-like phytosaurs were in Laurasia. Later, in the Triassic, the Manda Formation from eastern Africa and the Santa Maria Formation of Brazil have similar rhynchosaurs, cynodonts, dicynodonts, and thecodonts, presumably showing a Gondwanaland bridge. However, Romer (1975) has pointed out that these forms also occurred in northeastern North America and Europe.

Charig (1973a) has presented a useful review of known Jurassic and Cretaceous dinosaur distributions, by continent and by family, which indicates that some families had almost cosmopolitan distributions. In other words, the data now available do not show vicariance events caused by the sequential breakup of Pangea in the Mesozoic. The trends can be readily spotted, some north versus south tendencies, but exceptions and wide-ranging families lead to the conclusion that there is much more to learn about the fossil distributions of tetrapods before biogeographers can understand what land connections were important to these organisms. Until many more dinosaurs have been unearthed in South America and Australia, a global reconstruction of even that relatively well-known group is not attainable.

The same reason, lack of fossils in southern continents, can be cited as an insurmountable barrier to our current understanding of Mesozoic mammals (Lillegraven et al., 1979; Clemens, 1982). There is a strong possibility that mammals occurred in South America in the Jurassic, but investigators have no fossils to prove this. The occurrence of mammals in South America and Australasia prior to the breakup of Gondwanaland is, of course, crucial for our mammalian discussion later—where did the endemic groups on those continents come from?

Terrestrial amphibians and reptiles

No class of tetrapods is entirely terrestrial because all have some fully aquatic species.

Some of the aquatic groups have been mentioned in Chapter 10, so here we will review some interesting distributions of amphibians and reptiles that are predominantly terrestrial.

Regional endemism and areal relationships of anurans. A group that is both terrestrial and aquatic in all senses is Anura, the frogs and toads. The richest area for anurans, including the greatest diversity of species as well as the most families and subfamilies, is the Neotropical region (see box). In fact, over one quarter of all known anurans occur in the small region of Central America and Colombia. Workers have had great difficulties in ascertaining relationships of anuran families especially because of morphological convergence. It has only been within the last 20 years that progress has been made in elucidating how these organisms dispersed between continents.

As phylogenetic relationships of the families have been revealed, many of them have indicated a Gondwanaland pattern. The most ob-

Classification of Recent families of frogs and toads (order Anura)

This classification is similar to one presented by Duellman (1975) but for convenience recognizing four families (Ascaphidae, Allophrynidae, Heleophrynidae, and Pelodryadidae) that are submerged by Duellman. Distributions of the families are based on Savage (1973) and Goin et al. (1978).

ORDER ANURA

Suborder Xenanura

Superfamily Discoglossoidea
- Leiopelmatidae; New Zealand
- Ascaphidae; Western North America
- Discoglossidae; Eurasia, northern Africa, the Philippines

Superfamily Pipoidea
- Pipidae; South America to Panama, Africa
- Rhinophrynidae; Mexico, Central America

Superfamily Pelobatoidea
- Pelobatidae; North America, Eurasia to Wallace's line
- Pelodytidae; southwestern Europe, southwestern Asia

Suborder Neobatrachia

Superfamily Bufonoidea
- Myobatrachidae; Australia, New Guinea
- Leptodactylidae; North America, Central America, South America, West Indies
- Bufonidae; cosmopolitan
- Brachycephalidae; Brazil
- Rhinodermatidae; Chile, Argentina
- Allophrynidae; northern South America
- Heleophrynidae; southern Africa
- Pelodryadidae; Australia, New Guinea to other parts of Melanesia (does not cross Wallace's line)
- Dendrobatidae; Central America, South America
- Pseudidae; South America
- Hylidae; North America, Central America, South America, the West Indies, Eurasia, North Africa
- Centrolenidae; Central America, South America (mostly northern areas)

Superfamily Microhyloidea
- Microhylidae; circumtropical to Queensland, temperate eastern North America, temperate eastern Asia, southern Africa

Superfamily Ranoidea
- Sooglossidae; Seychelle Islands
- Ranidae; cosmopolitan
- Hyperoliidae; Africa, Madagascar, Seychelle Islands
- Rhacophoridae; southeast Africa, Madagascar, tropical Asia to Celebes

vious West Gondwanaland distributions are the aquatic frogs of the family Pipidae, with the subfamily Pipinae in tropical South America and the subfamily Xenopinae in Africa (Figure 11.2). Fossils of Pipidae have been found in the Cretaceous of Africa, South America, and Israel (Estes and Reig, 1973). The Mexican burrowing toad (Rhinophrynidae) of Middle America is the closest relative of Pipidae.

A cluster of endemic families in the Southern Hemisphere appears to be the product of the breakup of Gondwanaland. Most familiar of these are the leptodactylids (Leptodactylidae), the largest and most diversified family of South and Central America, which have arboreal, terrestrial, semiaquatic, and aquatic adult lifestyles and many different reproductive strategies. These frogs are closely related to three other small South American families, Brachycephalidae, Allophrynidae, and Pseudidae; to family Myobatrachidae of Australia and New Guinea; and to family Heleophrynidae of South Africa.

In the Old World one finds another interesting group of families presently occupying pieces of Gondwanaland, the frogs of the ranid alliance. Hyperoliidae with four subfamilies occur in Africa, Rhacophoridae with two subfamilies occur in Africa, Madagascar, and southern Asia to Japan and Celebes, and Sooglossidae (two monospecific genera) occur only on the Seychelles (Figure 11.3). This distribution

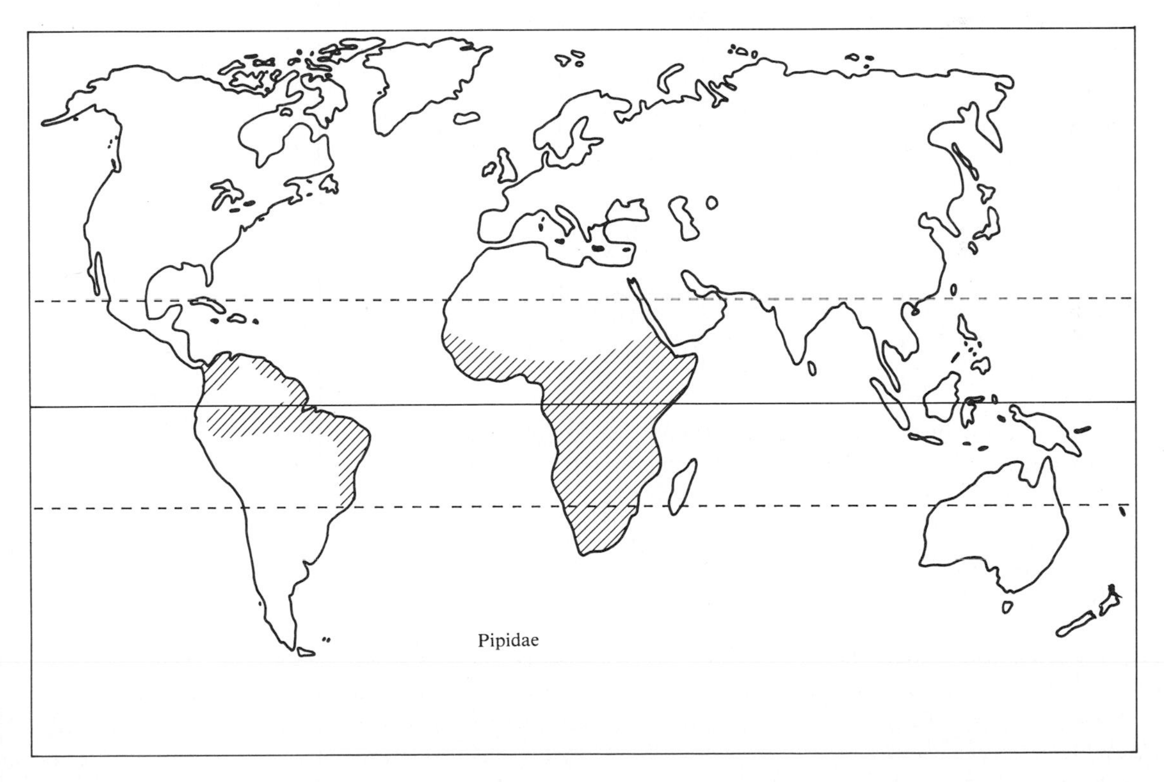

Figure 11.2
Distribution of clawed aquatic frogs (Pipidae). This family is comprised of two subfamilies, the Pipinae in tropical South America and the Xenopinae in tropical Africa. This suggests a vicariant pattern for a group that was once distributed in western Gondwanaland. (After Savage, 1973.)

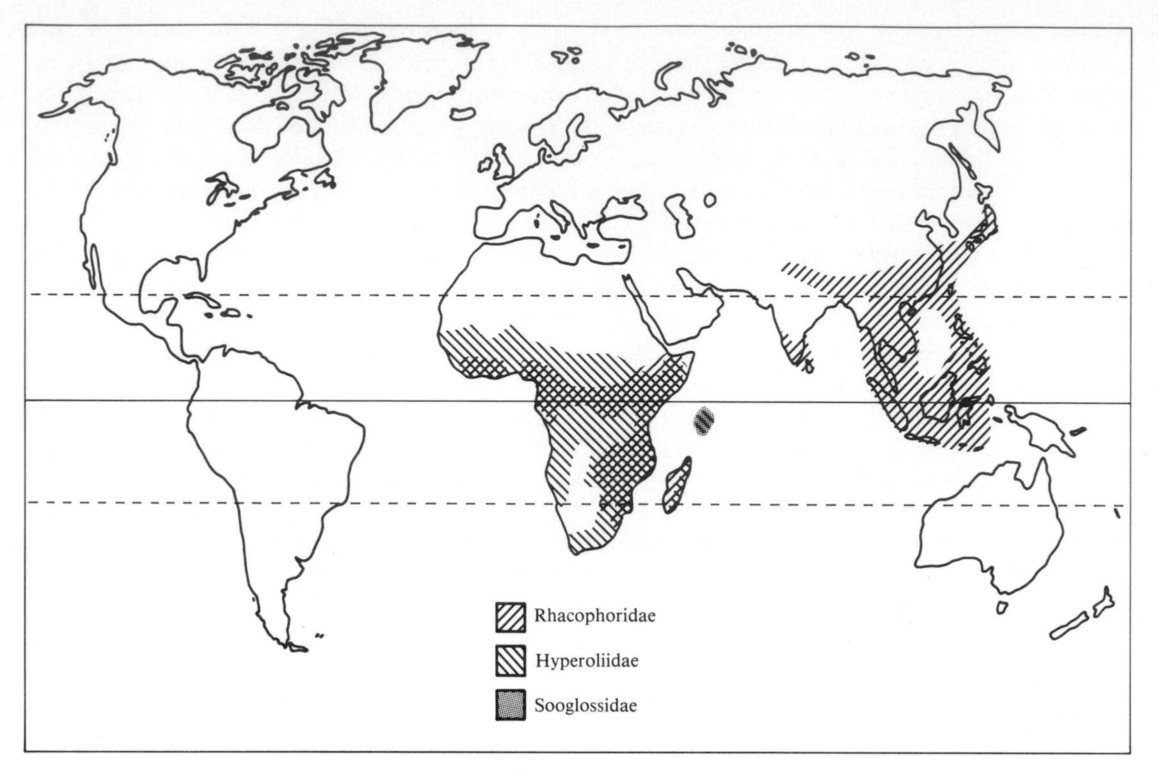

Figure 11.3
Distribution of frog families Hyperoliidae, Rhacophoridae, and Sooglossidae. These related groups of anurans could represent the relicts of a widespread taxon that occurred on Gondwanaland before formation of the Indian Ocean; alternatively they may be relicts of a broadly distributed paleotropical taxon. (After Savage, 1973, and Goin et al., 1978.)

could have resulted from the fragmentation of the Africa-Madagascar-India landmass during the formation of the Indian Ocean, but it also could represent in part a paleotropical distribution between Africa and Asia that was severed during the Cenozoic.

An Antarctic land connection between South America and Australia is certainly required to explain the distribution of certain hyloid taxa. Phyllomedusine frogs (subfamily Phyllomedusinae of Hylidae), which are usually treated as three arboreal genera, have some unusual and highly derived skin features that were recently discovered in certain species of *Litoria* (Pelodryadidae) in Australia (Bagnara and Ferris, 1975). Most of the tree frogs of the family Hylidae are found only in the New World, and the South American Centrolenidae and the highly toxic and brilliantly colored neotropical poison frogs (Dendrobatidae) are probably also derivatives of the South American hylids.

Before leaving the Southern Hemisphere, one more endemic, *Leiopelma,* must be mentioned. This endemic family of New Zealand (Leiopelmatidae) appears to be related to the monotypic Ascaphidae of western North America and to the family Discoglossidae (four genera), which has a fragmented Palearctic and Oriental distribution. Such a disjunction, with taxa at the distant corners of the globe, is hard to rationalize. However, fossils assigned to Ascaphidae have been found from the Upper Jurassic of Argentina, suggesting a possible origin of *Leiopelma* as a vicariant event with South America. This was followed by great reduction in the range of that cluster of families to their present-day highly relictual pattern.

Several extant families show only northern relationships. One popular example is found in the family Pelobatidae, in which the spadefoot toads are disjunct between North America *(Scaphiopus)* and Europe *(Pelobates)*. The little-known Xenosauridae has two genera, *Xenosaurus* in southern Mexico and Guatemala and *Shinisaurus* in southern China. *Scaphiopus* inhabits harsh desert habitats and has adjusted its life cycle to survive drought in burrows and breed in ephemeral pools following heavy rains.

Nearly cosmopolitan families are the toughest ones to consider because relationships of the genera are often obscured by uncritical maps and imprecise classifications. Two anuran cosmopolitan families help to illustrate the issue. The true frogs (Ranidae) are basically Holarctic and Ethiopian, but there was a recent invasion of South America and Queensland by the species in the highly specialized and successful genus *Rana*. In contrast, the true toads (Bufonidae) have a very similar overall distribution, but in South America there are two endemic genera plus the related monotypic family Rhinodermatidae. This type of taxonomic differentiation of toads in South America may require an entry of the stock there earlier than the land bridge formed in Panama. Investigators cannot yet say whether this invasion was from North America or via Africa directly.

The anurans certainly were present in the Mesozoic, and fossils are known from that time in West Gondwanaland (Estes and Reig, 1973), so one must consider the possible effects of the southern land connections on their distributions and how the breakup of Pangea influenced the origins of families and genera (Savage, 1973). A very different explanation was proposed by Darlington (1957), who stressed the importance of migration through the northern continents into the Southern Hemisphere, the same model that he proposed for freshwater fishes (Figure 10.8). However, anurans are by and large poor colonizers of islands and intolerant of seawater because they cannot osmoregulate (except for beasts like *Bufo marinus,* which can), so that land routes are generally required. By studying anurans in light of continental drift, one discovers that all groups endemic to Australasia could have come by way of some Gondwanaland connection; only a few highly specialized genera have managed to cross Wallace's line and invade New Guinea and Australia from southeast Asia in recent times.

Lizards and snakes. There are nearly 6000 extant species of lizards and snakes (order Squamata). The number of families and subfamilies of these terrestrial reptiles is much smaller than that of terrestrial mammals, even though the ages of both groups date from the early Mesozoic. This pattern of low familial diversity in squamates is greatly amplified by including fossil mammals and squamates because many mammalian families arose and disappeared in the last 100 million years. Nearly as many families of mammals have become extinct as are living today.

Kurtén (1969) was concerned about the relatively low number of major groups of reptiles as compared to mammals. He suggested that continental drift would have influenced the radiation of reptiles and mammals in different ways: reptiles evolved earlier, when all the landmasses were closely attached, whereas mammals radiated later in isolation on independent continents. Although this is partly true, such an explanation also ignores the observations that squamate evolution was also strongly influenced by continental drift and does indeed show the taxonomic effects of the breakup of Pangea.

Snakes do not show dramatic adaptive radiation morphologically because they have, by virtue of their highly specialized body form and perhaps by its relationship to food preferences and poikilothermy, been constrained from developing entirely new designs. On the other hand, mammals, by virtue of their generalized quadrupedal form, fast pedal locomotion, and homeothermy, have evolved new designs to exploit many different food resources and to adapt through great changes in shape and size to a wide variety of ecological niches. It is harder to explain why the adaptive radiation of lizards has been so limited. Perhaps lizards were restrained in their designs by competition, at first with their larger cousins, the dinosaurs, which showed that reptiles could achieve many life-styles, and later in the Tertiary by placental mammals, which usurped many roles for both large and small terrestrial vertebrates.

Because lizards and snakes existed in the Mesozoic, one can expect to find disjunctions that mimic the splitting of West Gondwanaland into Africa and South America. Indeed, this is what is found.

A vicariant event can be suggested to account for the origin of two closely related lizard families whose ranges are separated by the South Atlantic (Figure 11.4). The Old World lizards belong to the family Lacertidae, which occurs throughout Africa, Europe, and Asia but does not occur in Australasia or on Madagascar. Lacertids are predominantly carnivorous terrestrial lizards. Similar niches are often filled in the New World by the related Teiidae, which are particularly abundant in the Neotropics. Northward penetration of teiids into the United States is by the genus *Cnemidophorus*, which is well known for its bizarre modes of speciation,

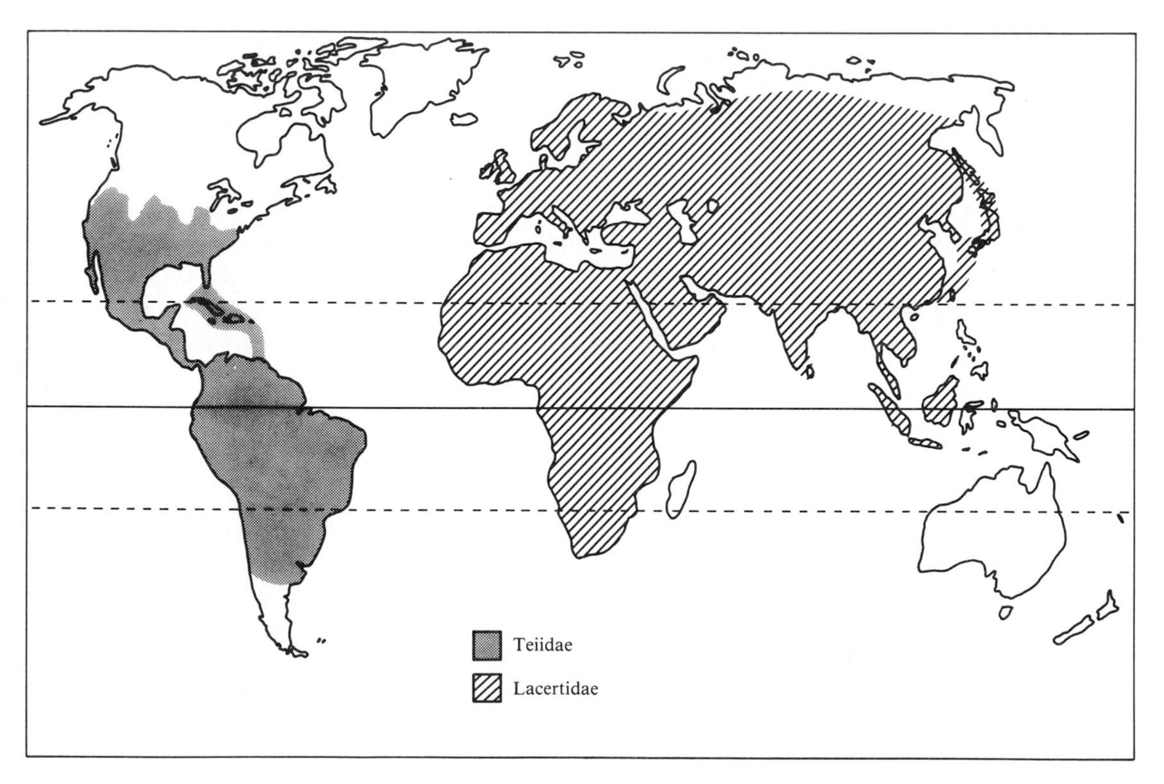

Figure 11.4
Geographic ranges of lizard families Teiidae and Lacertidae. These two families are closely related and ecologically similar. The early division between these New and Old World groups may date back to the breakup of West Gondwanaland and the formation of the South Atlantic. (After Goin et al., 1978.)

e.g., the occurrence of polyploidy and all-female species (Lowe et al., 1970). This genus appears to be of relatively recent lineage.

West Gondwanaland was certainly the birthplace of the infraorder Iguania (Figure 11.5). Iguanas (Iguanidae) are mostly New World lizards, being the largest family in the New World, where they have evolved a wide variety of life-styles. The agamids (Agamidae) are their Old World counterparts, occurring most abundantly in Africa but also reaching Australia. The chameleons (Chamaeleonidae), which are mostly present in Africa and Madagascar but range to southern Spain and the Indian subcontinent, are believed by many authors to be closely related to and probably derived from agamid stock. A Gondwanaland scenario for Iguania is made certain by two additional facts. First, there are two genera of Iguanidae *(Chalarodon* and *Oplurus)* endemic to Madagascar. Second, iguanidlike fossils have been found in the Upper Cretaceous and Paleocene of Brazil, indicating that the group is old enough to have been present around the time of continental movements in the Mesozoic.

There is also one monospecific genus of Iguanidae in the South Pacific, *Brachylophus,* on Fiji and Tonga. Therefore we find it remarkable that another taxon, the large boas (Boinae of Boidae) have disjunctions identical to these of the iguanas (Figure 11.6), with living species in the New World (five genera), in Madagascar

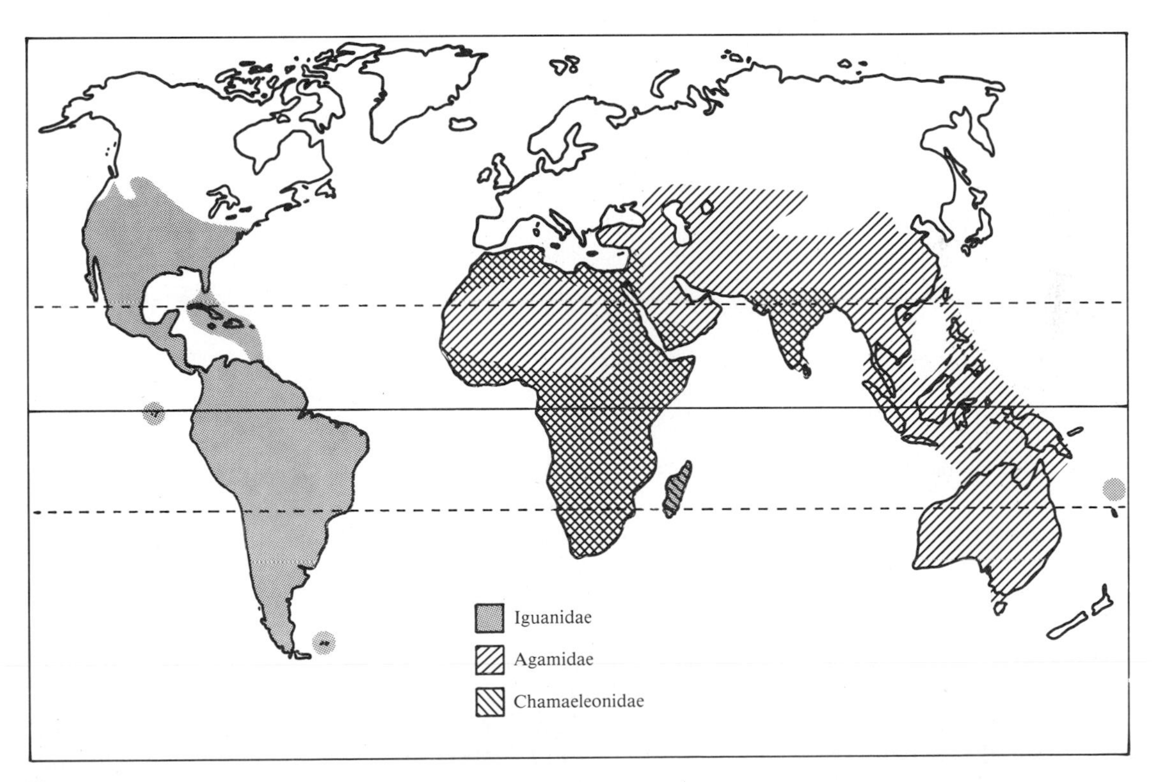

Figure 11.5
Distribution of three lizard families, Iguanidae, Agamidae, and Chamaeleonidae. These related groups, which form the infraorder Iguania, probably represent a vicariant pattern derived from ancestors that lived on West Gondwanaland. (After Goin et al., 1978.)

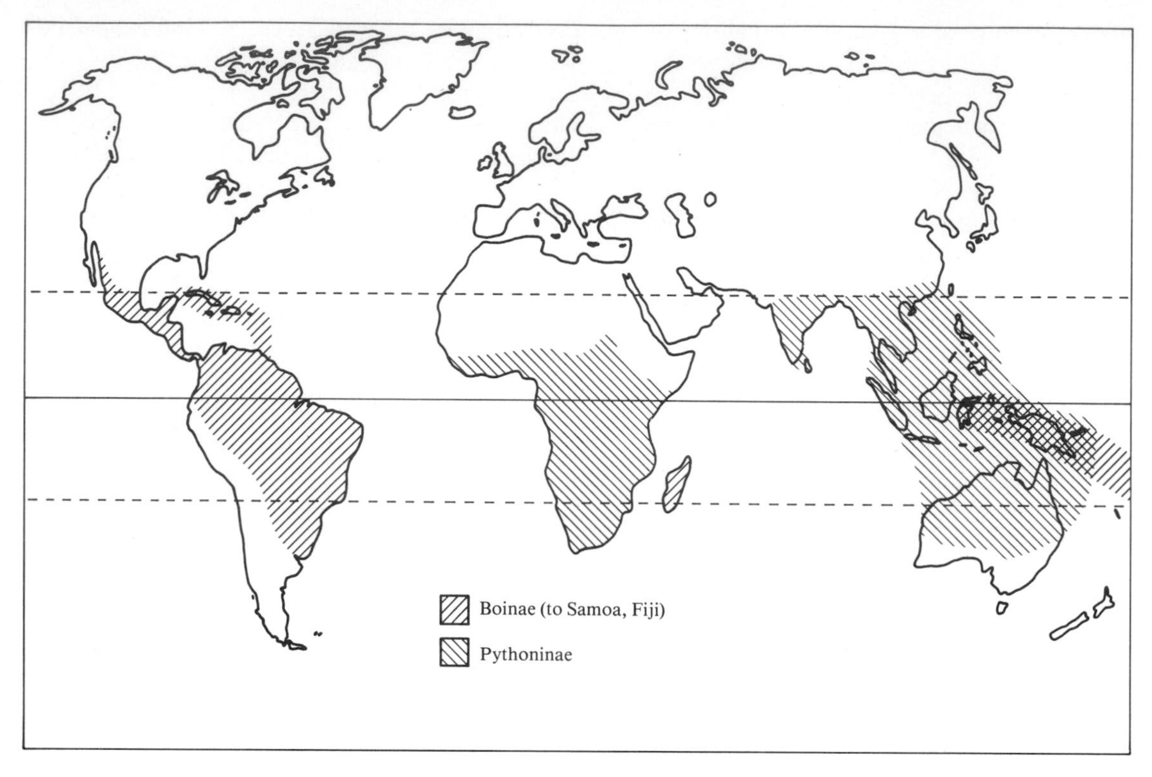

Figure 11.6
Geographic ranges of two subfamilies, Boinae and Pythoninae, of the snake family Boidae. Both the present and fossil distributions of these snakes are largely consistent with an explanation invoking an ancestral range that was fragmented by the breakup of Gondwanaland. Like the iguanid lizards, the Boinae have interesting outlying representatives occurring in Polynesia. (After Goin et al., 1978.)

(Acrantophis and *Sanzinia),* and a genus of three species in Polynesia *(Candoia)*. The Boinae are also known from fossils of Africa from the Upper Cretaceous and Eocene and from the Eocene of France and North America, making this a likely example for a distribution created by the breakup of Gondwanaland (Underwood, 1976). The boid snakes have some other interesting disjunctions, such as the rosy boas (Tropidophidae) in the Neotropics and Mauritius (two genera) and the separation of the New World boas from the pythons (Pythoninae of Boidae) (Figure 11.6), which occur in Africa, Eurasia, and Australasia. All this suggests that the iguanioids and boids were contemporaneous and experienced similar fates when the landmasses of Gondwanaland became isolated.

The primitive and bizarrely specialized worm lizards (Amphisbaenidae, 35 species) also show a tropical distribution consistent with an ancestral distribution in West Gondwanaland, as do the blind snakes (infraorder Scolecophida, Figure 11.7), consisting of the amphi-Atlantic families Typhlopidae and Leptotyphlopidae and the neotropical endemic Anomalepidae. The occurrence of these organisms in the West Indies requires considerable attention, however; it is known that the widespread Typhlopidae occur

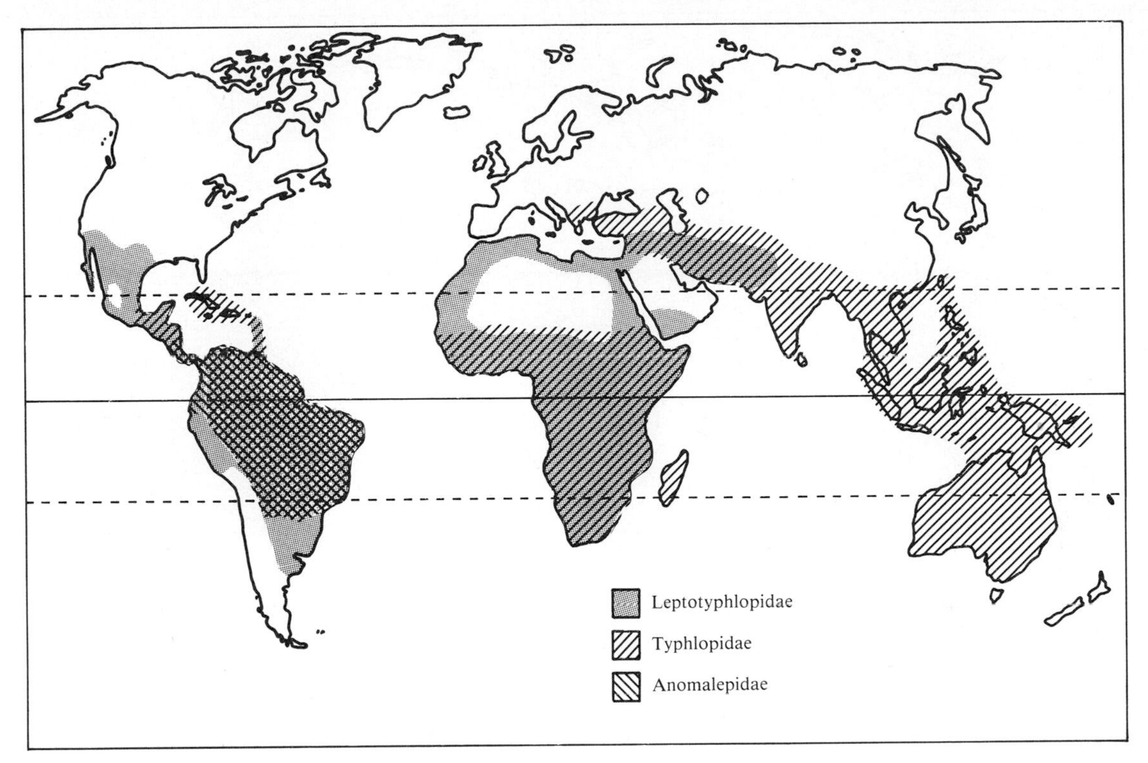

Figure 11.7
Distribution of three families, Typhlopidae, Leptotyphlopidae, and Anomalepidae, in the blind snake infraorder Scolecophida. Like several other groups of tropical reptiles and amphibians, these related taxa have a distribution suggesting that their ancestors may have occurred in West Gondwanaland. (After Goin et al., 1978.)

on some very recent oceanic islands and therefore may be capable of crossing short ocean barriers.

Much biogeographic research is needed on the truly cosmopolitan families Scincidae (skinks) (Figure 11.8) and Gekkonidae (geckos) and the large families of snakes (Colubridae and Elapidae). Geckos and skinks have long been noted for their excellent representation on oceanic islands (Carlquist, 1965) where no other land tetrapods occur, and they have been observed rafting at sea. This pattern notwithstanding, much of the continental distributions had to be achieved by land connections, e.g., to Australasia. In Australasia are the endemic subfamily Diplodactylinae of Gekkonidae, with closely similar forms in New Zealand and New Caledonia, and the endemic elapid snake radiation in Australia. The Greater Antilles also have some peculiar endemic snakes and lizards with life-styles not conducive to traversing ocean gaps. In other words, some of the taxa could have attained their present distributions recently by dispersal, whereas others undoubtedly have relied on migration across ancient land connections.

Numerous squamate phylads exhibit patterns requiring only dispersal through the

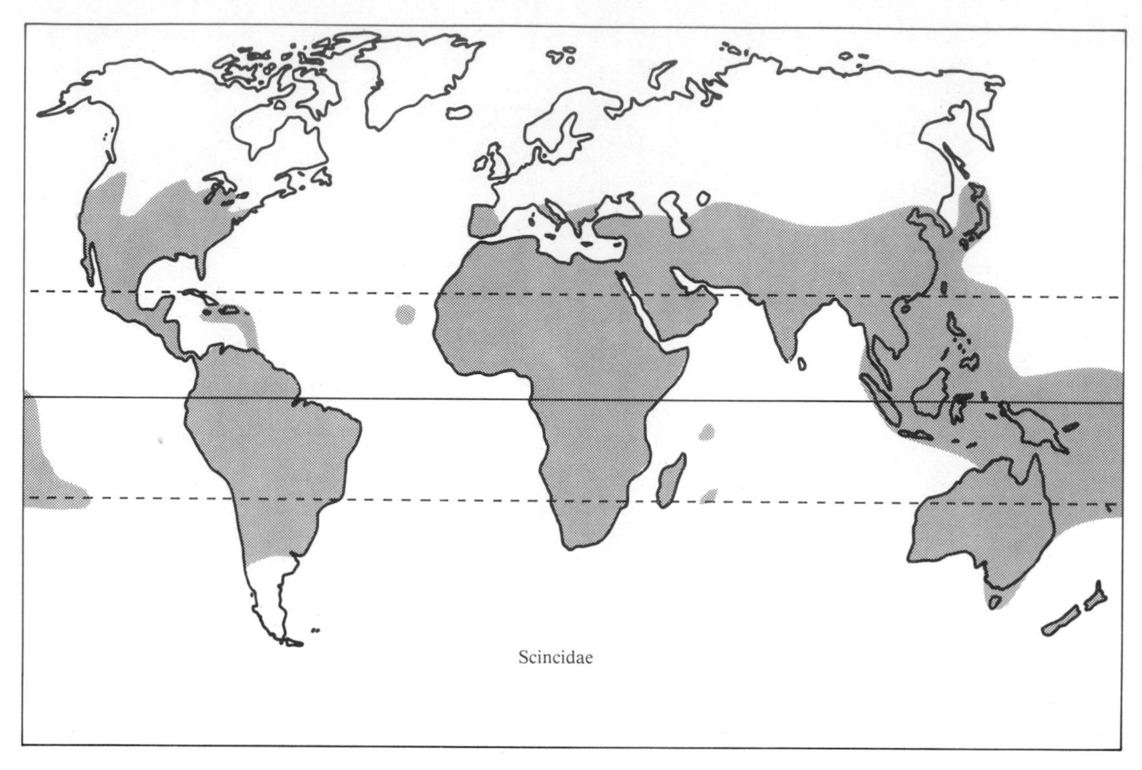

Figure 11.8
Distribution of the cosmopolitan lizard family Scincidae. Several other large reptilian families have similarly wide ranges, and further analysis of systematic and biogeographic relationships among their genera are required to work out their histories. (After Goin et al., 1978.)

Northern Hemisphere and from there southward. A perfect example of this is the family Crotalidae, the pit vipers, sometimes classified also as a subfamily of the true vipers, Viperidae. Figure 11.9 shows one reconstruction of the geographic evolution of this family, with the primitive forms occurring in Asia followed by dispersal across the Bering Land Bridge to North America, where rattlesnakes *(Crotalus)* have evolved as the principal groups of venomous snakes in the Nearctic region. Some of the most feared snakes of South America are the fer-de-lance *(Bothops)* and the bushmaster *(Lachesis),* which are considered to be derivatives of Tertiary crotalines of North America and hence immigrants to South America, probably following the formation of the Panamanian Land Bridge.

Land tortoises. The land tortoises (Testudinidae) are suspected of being derived from the Emydidae (Chapter 10), with which they share many morphological and cytological features (Auffenberg, 1971, 1974; Albrecht, 1976; Bickman, 1976). Because there are fossils of four distinctive genera in the Paleogene of the Northern Hemisphere, land tortoises were probably present in the Cretaceous as well. Now species occur on all continents except Australia.

Primitive features of Testudinidae are found in the subgenus *Manouria* of *Geochelone,* which may represent the stock from which all other

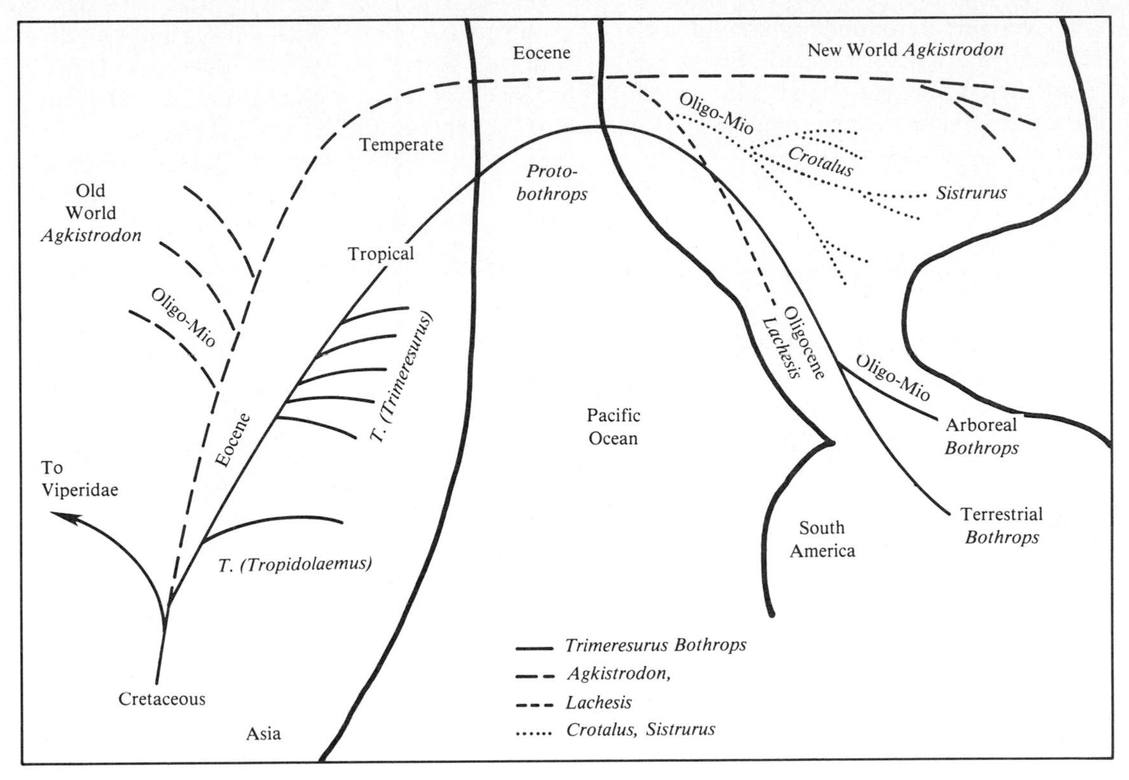

Figure 11.9
Reconstruction of the phylogenetic and biogeographic history of the Crotalidae, the family of venomous snakes that includes the rattlesnakes, copperhead, and fer-de-lance. This group apparently originated in the Old World during the Mesozoic, and two different representatives subsequently colonized the New World via the Bering Land Bridge in the Eocene. One lineage eventually reached South America. (After Brattstrom, 1964.)

subgenera of *Geochelone* and other genera of the family arose. In the New World, *Geochelone* gave rise to two genera, *Stylemys* (now extinct) and *Gopherus,* and the New World subgenus *Chelonoidis,* which appeared in South America in the Miocene before the Panamanian Land Bridge was formed. Now *Chelonoidis* occurs not only on the mainland but also in the West Indies and the Galápagos Islands. In the Galápagos Islands, *G. elephantopus* has 14 distinct forms on separate islands or parts of islands, some of which recently became extinct as a result of human predation. These tortoises helped spawn Darwin's evolutionary ideas.

In the Old World, Eurasian tortoises differentiated as a number of African taxa, including the dwarf monotypes *Pyxis* and *Acinixys* on Madagascar. Such occurrences in Madagascar prompt us to wonder whether land tortoises were present in Africa before the Middle Cretaceous, when Madagascar and mainland Africa were still united. To cast doubt on this hypothesis, Simpson (1943b) has noted that land tortoises are known to tolerate seawater for extended periods; Carr (1952) has observed tortoises swimming off the Florida coastline. Given that *G. forsteni* is a land tortoise that has crossed Wallace's line from southeast Asia, most biogeographers prefer to think that land tortoises arrived on numerous islands by swimming. Nonetheless, biogeographic reconstructions on this group are still very tenuous.

Ratites

The ratites are the gigantic flightless birds commonly exhibited in municipal zoos. Typically seen are ostriches *(Struthio camelus)* of Africa and emus *(Dromaius novae-hollandiae)* of Australia, but this group also includes the rheas (*Rhea* and *Pterocnemia*) of South America, the kiwis *(Apteryx)* of New Zealand, the cassowaries *(Casuarius)* of New Guinea and Queensland, and several extinct taxa, the moas (Dinornithidae) of New Zealand and the elephant birds (Aepyornithidae) of Madagascar. For years ornithologists have debated whether all these birds are related (i.e., a monophyletic group) or whether they have evolved large size and flightlessness independently several times from winged ancestors. If these families and genera are not related, they could have evolved in isolation on each southern landmass and be absent from northern continents only by coincidence; but if ratites are monophyletic, land connections are certainly required to disperse them across the Southern Hemisphere.

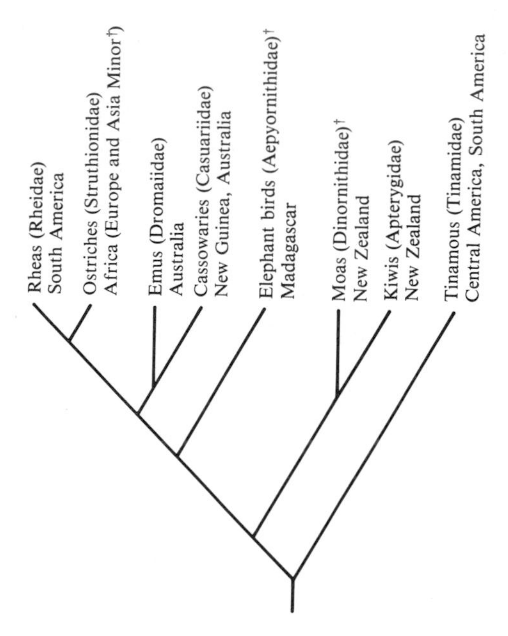

Figure 11.10
Phylogenetic relationships among the families of the ratites as inferred by the cladistic analysis of Cracraft (1974) and the present and fossil ranges of these giant flightless birds. Dagger(†) indicates extinct forms. The most parsimonious explanation of this pattern is one invoking fragmentation of an ancestral range on Gondwanaland, but fossil ostriches from the Northern Hemisphere indicate that other hypotheses cannot be discounted.

Cracraft (1974a) has provided a strong case for classifying these as a monophyletic group, the suborder Ratiti, being most closely related to the weak-flying tinamous (Tinamidae, suborder Tinami) of the Neotropics. He has presented a cladogram to postulate the relationships for these birds (Figure 11.10). Putative sister groups are the kiwis and extinct moas of New Zealand and the emus and cassowaries of Australia, respectively; but the ostriches and rheas, sometimes classified as subfamilies in the same family (Struthionidae), are separated by the South Atlantic. The simplest, most parsimonious explanation for these phylogenetic relationships is that ratites dispersed overland through Gondwanaland and differentiated subsequently in isolation on the various parcels.

Distributions of extant and extinct mammals

Mammals are splendid organisms for biogeographic study because they provide excellent examples of most of the important patterns and processes. Mammals occur in every habitable region, from hot to frigid, wet to arid, and arboreal to subterranean. This is facilitated by homeothermy, the ability to maintain a relatively constant and high body temperature from the energy produced during oxidative metabolism (endothermy) and aided by fur (which insulates against excessive heat loss) and mechanisms (such as sweating and panting) to prevent overheating through evaporative cooling. The natural ranges of species may be exceedingly wide or extremely narrow, including little or great amounts of genetic differentiation for local populations. Some groups, such as horses (Equidae) and proboscideans (Proboscidea), have extensive fossil records that enable scientists to reconstruct quite accurately their evolutionary and biogeographic histories. Others, like the pangolins and scaly anteaters (Pholidota), have no appreciable fossil record. Rodents (Rodentia) and marsupials (Metatheria) also illustrate the complexities of mammalian evolution with large numbers of species and numerous cases of parallel and convergent evolution on different continents.

Only by considering both extant and extinct taxa can one understand the distributions of terrestrial mammals. There are about 3000 terrestrial species (excluding here flying and marine forms), and of these over half are rodents, members of a single order (Rodentia). Fossil species of the remaining 15 orders of recent mammals and several extinct Tertiary orders probably collectively outnumber the living ones. The fossil history of a taxon may be rich and exceedingly informative, especially in those orders and families with only one or two extant species. Returning to the order Proboscidea, which includes the families of mastodons (Mammutidae), gomphotheres (Gomphotheriidae), stegodonts (Stegodontidae), and elephants and mammoths (Elephantidae), this fact is clearly illustrated. Elephantidae has 25 valid species in five genera (Maglio, 1973), but only the African elephant *(Loxodonta africana)* and the Indian elephant *(Elephas maximus)* survive today, and now they are greatly threatened by human activities. The situation was very different in the Paleogene, when the gomphotheres exhibited great morphological diversity and other families were starting. Whereas extant proboscideans are now confined to eastern Africa and southern Asia, fossils of proboscideans have been found throughout the Northern Hemisphere as well as on islands, such as off the coast of California, in the Mediterranean, on Japan, and in the East Indies as far east as Celebes. Maglio (1973; Maglio and Cooke, 1978) has proposed that the family Elephantidae originated in Africa and subsequently dispersed into Europe and Asia, and from Asia into North America via the Bering Land Bridge (Figure 11.11).

The elephant reconstruction shown in Figure 11.11 has included five aspects that are common in many reconstructions of mammalian orders and families: (1) exchange between Africa and Europe or Asia has been affected many times; (2) eastern Asia and western North America have exchanged taxa frequently since the early Cenozoic; (3) most terrestrial groups have not been able to cross Wallace's line and pass from Asia to Australia and New Guinea by island hopping; (4) young African clades, e.g., those that radiated in the Cenozoic, by and large do not occur on Madagascar because this island has been widely separated from the mainland since Middle Cretaceous times; and (5) many families had more extensive ranges in the past, so their fossils have been found in places that today would be totally uninhabitable for them, such as regions that are now tundra or hot deserts.

Mammals of the Northern Hemisphere. On geologic grounds, one expects mammalian families of northern North America to occur also in Eurasia and vice versa, and this is generally the rule. Of the 24 terrestrial families now native in the Nearctic, 17 occur or once occurred in the Palearctic region. The remainder are shared with South America or are endemic to North America (Table 11.2) without external fossils. For those families of the Palearctic, relationships with North America are similar, but the fauna also shows strong ties with mainland Africa and absolutely none with South America, except through North America. For example, the camels (Camelidae) are presently known in Asia, northern Africa, and western

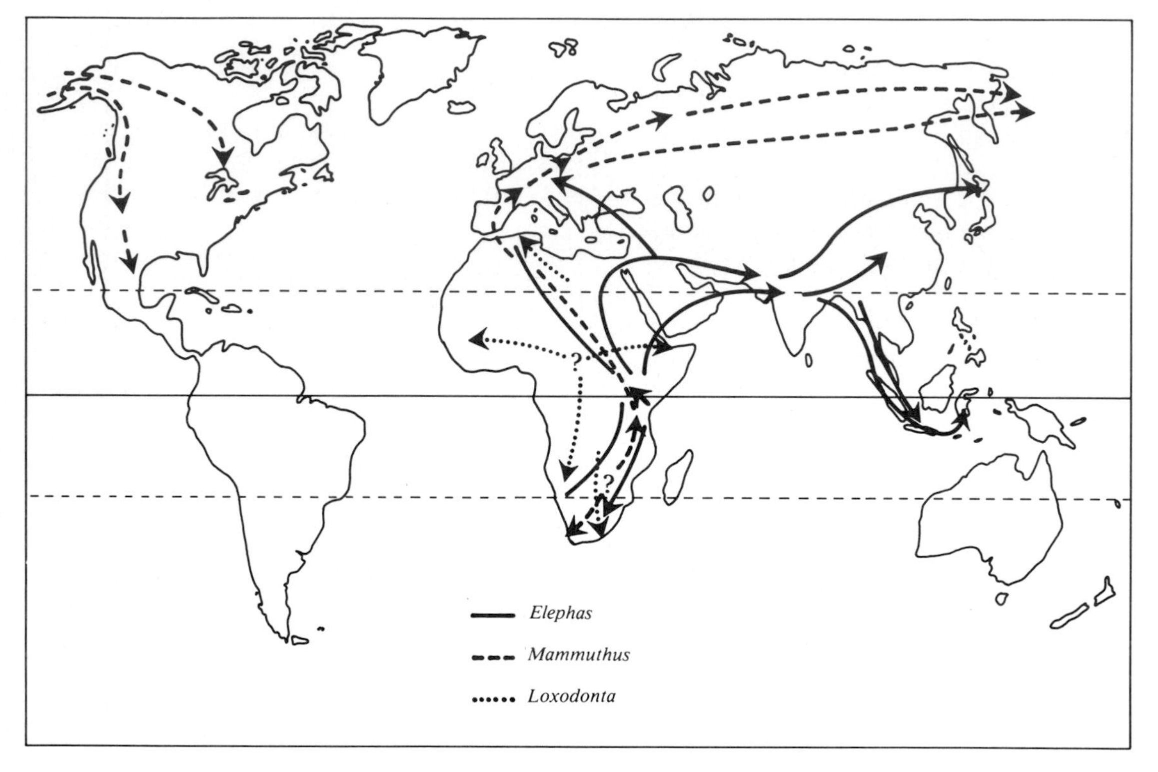

Figure 11.11
Proposed history of geographic expansion of three genera of elephants, *Loxodonta, Mammuthus,* and *Elephas*. All groups are thought to have originated in central Africa and then spread to other regions. Only two species, the African elephant *(Loxodonta africana)* and the Indian elephant *(Elephas maximus)*, survive today as relicts of these groups. (After Maglio, 1973.)

Table 11.2

Families of terrestrial mammals endemic to each zoogeographic region

Totally extinct groups are not listed, and numbers in parentheses (mostly after Honacki et al., 1982) indicate only species presumed living. In some cases, these families are called subfamilies by some authors.

Family	Order
Nearctic region	
Antilocapridae (pronghorn antelope; 1 sp.)	Artiodactyla
Aplodontidae (mountain beaver; 1 sp.)	Rodentia
Palearctic region	
Seleviniidae (dzhalman; 1 sp.)	Rodentia
Spalacidae (mole rats; 3 spp.)	Rodentia
Oriental region	
Cynocephalidae (flying lemurs or colugos; 2 spp.)	Dermoptera
Hylobatidae (gibbons and siamangs; 9 spp.)	Primates
Platacanthomyidae (dormice; 2 spp.)	Rodentia
Tarsiidae (tarsiers; 3 spp.)	Primates
Tupaiidae (tree shrews; 16 spp.)	Insectivora or Scandentia
Ethiopian region	
Mainland Africa	
Anomaluridae (scaly-tailed squirrels; 7 spp.)	Rodentia
Bathyergidae (mole rats; 9 spp.)	Rodentia
Chrysochloridae (golden moles; 18 spp.)	Insectivora or Lipotyphla
Ctenodactylidae (gundis; 5 spp.)	Rodentia
Galagidae s.s. (galagos; 8 spp.)	Primates
Giraffidae (giraffe and okapi; 2 spp.)	Artiodactyla
Hippopotamidae (hippopotamuses; 2 spp.)	Artiodactyla
Macroscelididae (elephant shrews; 15 spp.)*	Insectivora or Macroscelidea
Orycteropodidae (aardvark; 1 sp.)	Tubulidentata
Pedetidae (spring hare; 1 sp.)	Rodentia
Petromyidae (dassie rat; 1 sp.)	Rodentia
Potamogalidae (otter shrews; 3 spp.)	Insectivora or Lipotyphla
Procaviidae (hyraxes or dassies; 7 spp.)[†]	Hyracoidea
Protelidae (aardwolf; 1 sp.)	Carnivora
Thryonomyidae(cane rats; 2 spp.)	Rodentia
Madagascar	
Cheirogaleidae (mouse and dwarf lemurs; 7 spp.)	Primates
Daubentoniidae (aye-aye; 1 sp.)	Primates
Indriidae (indris; 4 spp.)	Primates
Lemuridae s.s. (lemurs; 7 spp.)	Primates
Lepilemuridae (weasel-lemurs; 2 spp.)	Primates
Tenrecidae (tenrecs; 30 spp.)	Insectivora or Lipotyphla

*Also in Palearctic Africa.
[†]Also in Arabia.
[‡]Organisms endemic to the Greater Antilles.
[§]This family has a Nearctic species.
[‖]One species reaches Celebes.

Continued.

Table 11.2—cont'd

Families of terrestrial mammals endemic to each zoogeographic region

Totally extinct groups are not listed, and numbers in parentheses (mostly after Honacki et al., 1982) indicate only species presumed living. In some cases, these families are called subfamilies by some authors.

Family	Order
Neotropical region	
Abrocomidae (chinchilla rats; 2 spp.)	Rodentia
Agoutidae (agoutis; 2 spp.)	Rodentia
Bradypodidae (including Choloepidae; tree sloths; 5 spp.)	Edentata
Caenolestidae (rat opossums; 7 spp.)	Marsupialia
Callimiconidae (callimico; 1 sp.)	Primates
Callithricidae (marmosets; 16 spp.)	Primates
Capromyidae (hutias; 13 spp.)[‡]	Rodentia
Caviidae (guinea pigs; 14 spp.)	Rodentia
Cebidae (New World monkeys; 6 spp.)	Primates
Chinchillidae (chinchillas; 31 spp.)	Rodentia
Dasypodidae (armadillos; 20 spp.)[§]	Edentata
Dasyproctidae (pacas and agoutis; 13 spp.)	Rodentia
Didelphidae (opossums; 76 spp.)[§]	Marsupialia
Dinomyidae (pacarana; 1 sp.)	Rodentia
Echinomyidae (spiny rats; 55 spp.)	Rodentia
Erethizontidae (New World porcupines; 12 spp.)[§]	Rodentia
Hydrochaeridae (capybara; 1 sp.)	Rodentia
Microbiotheriidae (colocolos; 1 sp.)	Marsupialia
Myocastoridae (nutria or coypu; 1 sp.)	Rodentia
Myrmecophagidae (anteaters; 4 spp.)	Edentata
Octodontidae (including Ctenomyidae, tuco-tucos; 41 spp.)	Rodentia
Solenodontidae (solenodons; 2 spp.)	Insectivora or Lipotyphla
Australian region	
Burramyidae (pygmy phalangers; 7 spp.)	Marsupialia
Dasyuridae (marsupial cats and mice; 50 spp.)	Marsupialia
Macropodidae (kangaroos and wallabies; 58 spp.)	Marsupialia
Myrmecobiidae (numbat; 1 sp.)	Marsupialia
Notoryctidae (marsupial mole; 1 sp.)	Marsupialia
Ornithorhynchidae (platypus; 1 sp.)	Monotremata
Peramelidae (bandicoots; 17 spp.)	Marsupialia
Petauridae (ringtails, gliding phalangers; 23 spp.)	Marsupialia
Phalangeridae (phalangers; 15 spp.)[‖]	Marsupialia
Phascolarctidae (koala; 1 sp.)	Marsupialia
Tachyglossidae (spiny anteaters; 2 spp.)	Monotremata
Tarsipedidae (honey possum; 1 sp.)	Marsupialia
Thylacinidae (marsupial wolf or Tasmanian tiger; 1 sp.)	Marsupialia
Thylacomyidae (rabbit-eared bandicoots; 2 spp.)	Marsupialia
Vombatidae (wombats; 3 spp.)	Marsupialia

South America, but fossil camels and llamas are abundant in North America until the Holocene. The tropical and subtropical Oriental region shares most taxa with tropical Africa and the rest of Eurasia. The most famous paleotropical distributions in tropical Asia and Africa are three primate families (Cercopithecidae, Lorisidae, and Pongidae), the pangolins (Manidae, Pholidota), and rhinoceroses (Rhinocerotidae). However, the Orient does share some groups with South America, e.g., the tapirs (Tapiridae) in Malaya and the Amazon, but once again fossil tapirs are known from North America and Europe.

Many authors have reviewed in great detail the exchange of placental mammals during the Cenozoic and Pleistocene between Eurasia and North America (e.g., Matthew, 1915; Simpson, 1947a, 1947b, 1965; Darlington, 1957; Savage, 1958; Kurtén, 1971; Kurtén and Anderson, 1980). In most accounts the Bering Land Bridge has been used, but McKenna (1972a, 1972b, 1972c) has also recommended consideration of a land bridge (the De Geer route) over the North Atlantic, which probably existed until the Paleocene. Although the fossil record of some groups is poor or nonexistent, by extrapolation we suspect that all extant orders probably were in existence by the Eocene, and some of them had originated before then. Hence a late Cretaceous

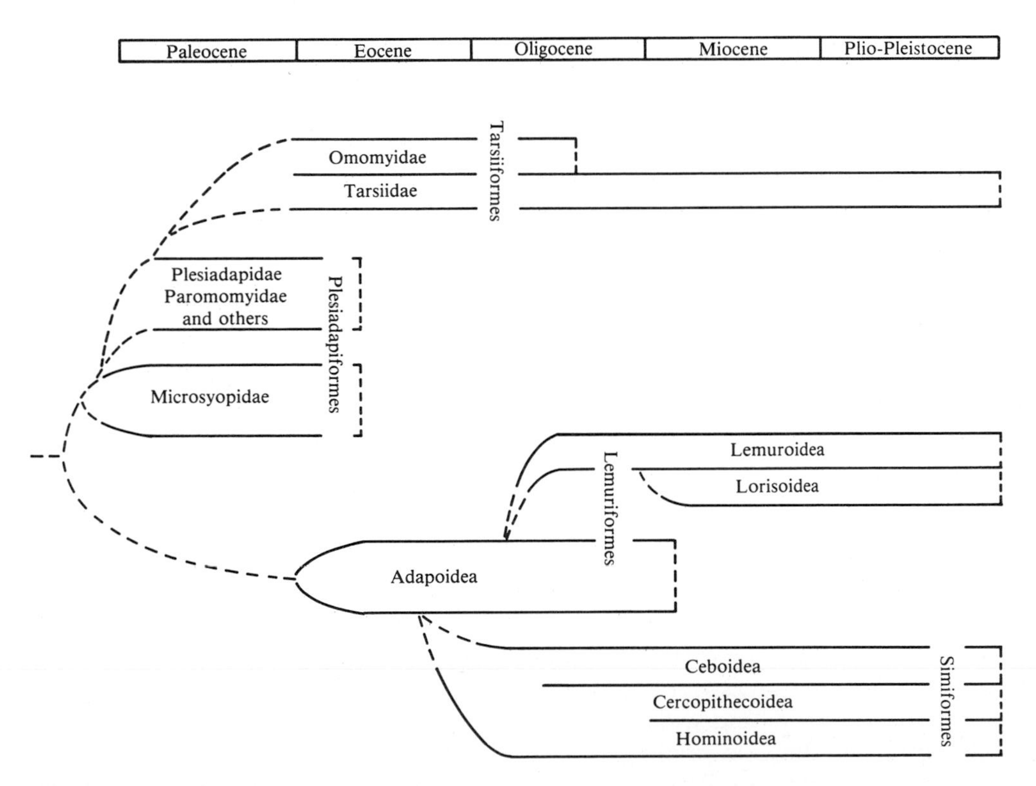

Figure 11.12
Reconstructed phylogeny of the mammalian order Primates by Gingerich (1976b). Although living representatives of this group occur in South America, southern Asia, and Africa (including Madagascar), the primitive group Plesiadapiformes was widely distributed in Europe and North America.

and Paleocene connection between western Europe and eastern North America must be evaluated. What we suspect, however, from fossils is that only primitive members of most orders were present in the Paleocene. For example, in the Primates, infraorder Plesiadapiformes, located near the base of the ordinal tree (Figure 11.12), lived in western Europe and western North America from the middle Paleocene through the Eocene (Gingerich, 1976b). From this group arose the infraorder Tarsiiformes, which retains many primitive traits and is currently confined to southern Asia. Most extant families of primates are alleged to be more recent in origin than Eocene, so only the Bering avenue was available for exchange of families, genera, and species between Eurasia and North America.

One way that investigators infer organisms crossed a land bridge is by comparing fossil strata on opposite sides. A group can be abundant on one side and then suddenly show up equally abundantly on the other side. Some groups, like Elephantidae, which have their oldest fossils and plesiomorphic features in the Old World, appear only in the Neogene of North America. In contrast, groups like horses (Equidae) and camels (Camelidae) appear to have originated in North America and then migrated to Asia. Before North American groups dispersed to Asia, they typically were widespread on the continent, occurring as far south as parts of Panama in the Miocene, which had horses and rhinoceroses. The camels are, as indicated earlier, now extinct in their homeland.

Mammals of Africa and South America. Africa is said to be a continent with a Pleistocene eutherian fauna. It is not the richest fauna per unit area but does include the largest survivors of groups that were prominent and widespread in the Old World until the end of the Ice Age. This fauna comprises almost 60 families of mammals (Honacki et al., 1982). Of the terrestrial families, 15 are endemic to the mainland, 6 to Madagascar, and several others are nearly endemic (Table 11.2). Several subfamilies are also endemic. Of the nonendemic families, their outside taxa are mainly in the Oriental region or are subtropical to temperate species that extend northward into Europe, groups that probably occurred even farther north at times in the past when climate was more equable.

The oldest known fossil mammals from Africa are found in the Upper Eocene of Egypt (Cooke, 1972; Maglio and Cooke, 1978), but all workers concede that mammalian lineages were well established millions of years earlier. The evolution of African mammals seems to have proceeded relatively independently of lineages in Eurasia over most of its history, but from time to time African clades suddenly appeared in Europe or Asia. One such episode was at the end of the Miocene and another occurred at the end of the Pliocene. In the Upper Pliocene some Asian taxa appeared rather suddenly in Africa. These data suggest that substantial land bridges connected the continents at those times. Cooke (1972) attempted to show that epeiric seas that separated Africa and Eurasia for much of their history were indeed reduced during those episodes.

Although investigators have no Cretaceous fossils of mammals in Africa, several types of indirect evidence suggest that a reasonably diverse fauna was present. One clue is the early diversification of proboscideans, hyraxes (Hyracoidea), and Embrithopoda (extinct order) in Africa. Another is that the insectivores (Insectivora), which are regarded as fairly primitive eutherian mammals, have a number of interesting endemic families in Africa. Finally, on Madagascar occur the five families of lemurs (Lemuroidea), primitive primates, and the tenracs (Tenracidae) that probably could not have colonized the island by traversing the wide and deep Mozambique Channel. A land bridge to Madagascar, i.e., before its departure from Gondwanaland, would necessitate a mid-Cretaceous presence in Africa or on the India-Madagascar landmass.

Probably the most reasonable explanation

for those African groups shared with the Oriental Region is that a fauna or series of faunas must have existed between the continents through Arabia during benign periods in the Tertiary. The pattern is well pronounced in many tropical groups, such as angiosperms, birds, and insects. An alternative explanation would be that India transported a placental fauna along with other elements from Africa to Asia. Much research is needed to determine which groups are old enough and have the right pattern to have departed from Africa en route to Asia versus those that might have dispersed as a continuous biota around the border of the Indian Ocean after India collided with Asia.

Another way of dating the age of mammal lineages in Africa is to compare its fauna with

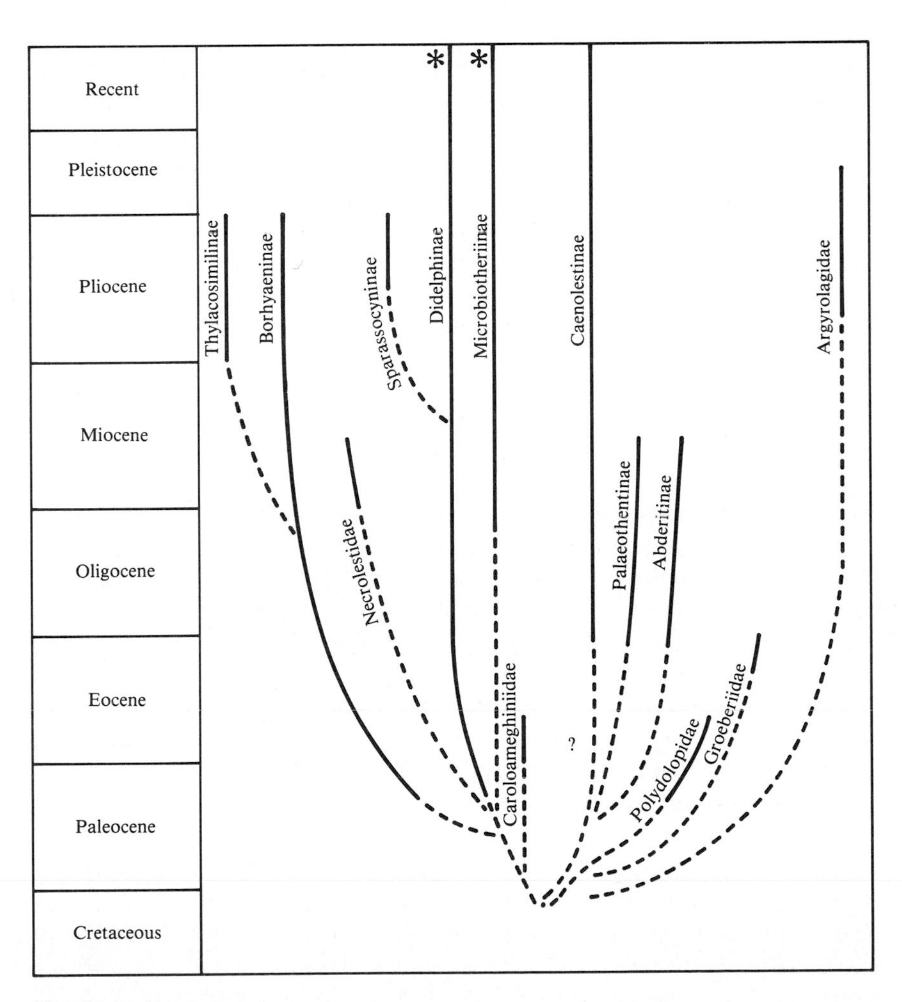

Figure 11.13
Phylogeny and fossil distribution of South American marsupials as reconstructed by Patterson and Pascual (1972). Marsupials apparently became isolated in South America and Australia with the breakup of Gondwanaland; during the early Tertiary they radiated on these two continents while placental mammals were diversifying on other landmasses.

that of South America; if strong relationships exist, then workers may learn something about the mammals present in the first half of the Cretaceous in West Gondwanaland. Unfortunately, workers have discovered that the South American mammalian fauna is much different than the African one in practically every category (Hershkovitz, 1972). To begin with, a small but significant portion of the neotropical fauna is composed of marsupials (60 spp., Kirsch, 1977; Kirsch and Calaby, 1977), which are totally absent in Africa. Of these, all but one species, the opossum *(Didelphis virginiana),* is endemic only to the Neotropics. An extensive fossil record in South America reveals a marsupial radiation in the Tertiary (Patterson and Pascual, 1972) (Figure 11.13), at a time when only placental radiations can be documented in Africa.

The Tertiary mammals of South America also included unique groups that experienced adaptive radiations: the edentates (Edentata or Xenarthra), including armadillos, glyptodonts, anteaters, and sloths, and large grazing animals called the South American ungulates (Pyrotheria, Litopterna, Xenungulata, Notoungulata, and Astrapotheria). The ungulate-like grazers of South America are believed to have been derived independently from condylarth-type animals (primitive grazers) rather than from the same stocks as the ungulates of Africa. During much of the Tertiary these endemic orders dominated the fauna of South America, becoming extinct sometime in the last 15 million years, especially since the establishment of the inter-American land bridge through Panama in the Pliocene (Figure 7.14). Some clades lasted much longer and even spread into North America. In fact, among the edentates the Shasta ground sloth *(Nothrotheriops)* disappeared in southern Arizona only about 10,000 years ago. The armadillo *(Dasypus novemcinctus)* is still expanding its range northward in the United States.

Patterns of radiation in isolation are demonstrated in several other early South American mammals, but most notably in the caviomorph rodents, represented by such divergent forms as guinea pigs, chinchillas, agoutis, and capybaras, which occupy a broad variety of ecological niches occupied by similar but unrelated rodents on other continents.

South America has rodents, primates, and anteaters, as does Africa, and from time to time authors have suggested that this fact is evidence that mammalian stocks of these forms occurred in West Gondwanaland before Africa and South America split apart (e.g., Ciochon and Chiarelli, 1980). Arguing against this thesis, most workers treat the caviomorph rodents (Caviomorpha) and the New World monkeys (Platyrrhini of superfamily Ceboidea) (Figure 11.12) as Tertiary arrivals from North America. The South American anteaters (Myrmecophagidae, Edentata) are not true vicariants of the Old World Pholidota but are instead highly specialized forms of South American edentates. Some confusion still exists on the relationships of New World and Old World porcupines (Erethizontidae and Hystricidae, respectively), and needs clarification.

A typical and traditional explanation for the origin of the entire South American mammal fauna interprets each lineage as having come from North America (Simpson, 1950, 1969a, 1980b; Patterson and Pascual, 1972). Marsupials and early condylarths would have been the initial colonists, followed by successful invasions by ancestors of caviomorph rodents, platyrrhine primates, and maybe others, and finally a large land invasion of North American families occurred after Panama emerged in the Upper Pliocene (Montehermosan). North American taxa quickly entered South America, outcompeting some of the native forms as they radiated into many new niches, so that now almost 40% of the mammalian fauna of South America consists of species that have arrived since the Pliocene. Meanwhile, South American taxa dispersed northward but did more poorly and have contributed little to the present northern fauna, presumably because they have been poorer competitors.

Although the general outline of events for the colonization and diversification of mammals in South America is probably correct, some of the details are presently being reexamined. For example, Clemens (1982) has reopened the controversy on how mammals originally colonized South America by suggesting that some of the original mammalian lineages of South America may have evolved from the ancestral forms of West Gondwanaland. At the other end of the spectrum, Marshall et al. (1982) place less faith in the assumption that competition was the primary force in the extinction episode during the Pliocene interchange. Authors sandwiched between these opposing views continue to debate whether the rodents and especially the primates (e.g., Ciochon and Chiarelli, 1980) dispersed from North America to South America during the Tertiary or were instead products from West Gondwanaland.

Mammals of Australasia. The Australian mammals clearly reflect a long history of isolation from the other continents (Figure 11.14). The fauna is dominated by marsupials of diverse morphologies and ecological preferences. Some of these provide a classical example of convergent evolution of similar life-styles (Chapter 18) (Carlquist, 1965), in which one monophyletic group has radiated adaptively to fill the niches of absent placental mammals, whereas in South America, metatherians, which were once diverse, have been mostly replaced by placental counterparts. There is considerable disagreement on the phylogenetic relationships of all marsupial families (Marshall, 1979; Patterson, 1981a), but most consider the Australian mammals to be the sister taxon of a South American group, so a southern migration along a land bridge is strongly favored (Tedford, 1974).

The other principal terrestrial component of Australian mammal fauna is the order Rodentia. More than 100 species of rodents (Muridae) are endemic to Australasia. These forms are clearly derived from ancestors that originated in southeast Asia and invaded Australasia by island hopping. Once again this group, which may represent numerous invasions, has converged to resemble representatives of several rodent families and subfamilies on the other continents. Also in Australia and New Guinea survive all three species of the world's most primitive mammals, the egg-laying monotremes (Prototheria) (Griffiths, 1978). The remaining putatively native terrestrial mammal, other than aboriginal humans, is the canid dingo *(Canis dingo)*, which almost certainly was brought to Australia by the aborigines.

Interpreting Australian patterns has always resulted in controversy. For years the field was dominated by scenarios in which the marsupials and placentals arrived from southeast Asia via island hopping. Phylogenetic analyses suggested that the murid rodents probably arrived over the last 10 million years by no less than seven separate invasions; and the marsupial radiation developed from some early Tertiary colonization. Throughout, however, some authors have attempted to move marsupials to Australia via the Antarctic connection with South America. When continental drift clearly demonstrated that this passage was available until the Lower Eocene and that the climate was warm enough for forest plants and animals, the southern route became acceptable. Authors (e.g., Patterson, 1981) still argue whether the primitive forms lived in South America or Australia and whether marsupials ever lived in Africa and therefore represent a broader Gondwanaland pattern. Moreover, no one has ever elucidated why placental mammals never made the southern trek across Antarctica. Definitive marsupial fossils in the Upper Cretaceous and lowermost Tertiary of Antarctica would provide convincing evidence for the role of Gondwanaland connections in the historical biogeography of Australian marsupials. Finding them in New Zealand, which presently has no native terrestrial mammals, would set the entry into Australia at 80 to 90 million years BP, when New Zealand drifted northward from Antarctica. Finally, the first of these tests has been realized with the discovery on March 7,

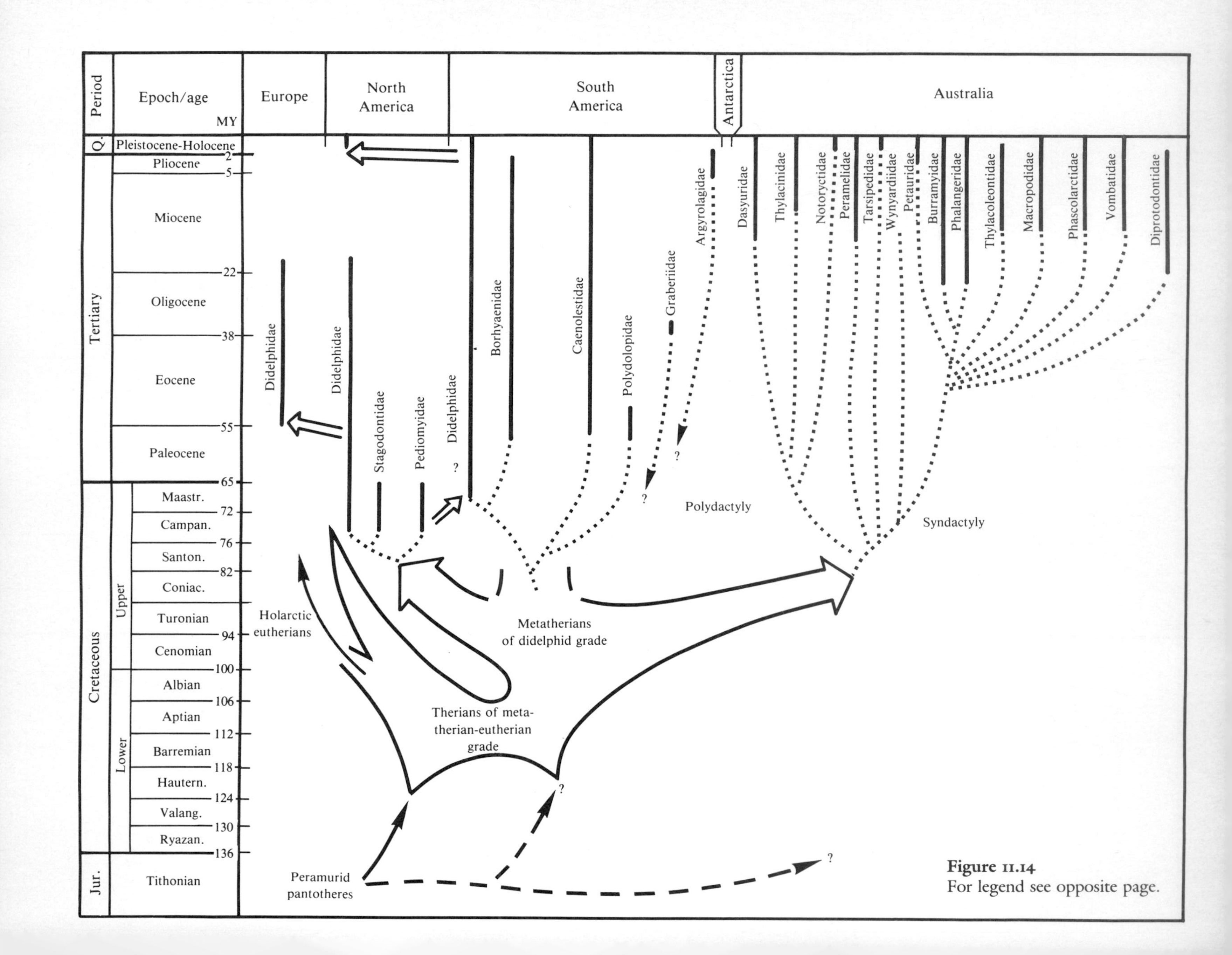

Figure 11.14
For legend see opposite page.

Figure 11.14
One of the numerous phylogenetic hypotheses on the origin of Australian marsupials from South American ancestors and the subsequent radiation of Australian taxa to fill ecological niches now occupied by placental mammals on other continents. (After Keast, A. 1977. In B. Stonehouse and D. Gilmore, editors: *The Biology of Marsupials*. Baltimore, 1977, University Park Press. By permission of Macmillan, London and Basingstoke.)

1982, of marsupial fossils from the Tertiary of Antarctica by M.P. Woodburne.

The mammals of New Guinea are closely related to those in Queensland, and marsupials gradually decrease in numbers proceeding westward. Rafting and swimming are considered reasonable methods for dispersal of the westernmost species that occur on islands that were not connected to New Guinea by land bridges.

Mammalian parasites. Some attention has been given to parasites of mammals, because parasites should be transferred with the host and therefore could show a similar distribution and a similar pattern of vicariance. In this vein, it was interesting to discover that obligate ectoparasitic lice (Amblycera, order Phthiraptera) of marsupials in South America (Trimenoponidae) are a sister group to the ones on Australian marsupials (Boopidae). A similar story can be found for the insect pests of the plant genus *Nothofagus* from the same regions (Schlinger, 1974). Other cases of such coincident distributions for host and parasite are known, and Hennig (1966) has used cladistics to show congruent relationships for certain parasites of perissodactyls.

The problem in pursuing this subject is that many parasite distributions do not parallel closely enough the phylogeny of hosts to be of any biogeographic value. Many parasites are opportunistic, at least over evolutionary time intervals, using a variety of hosts, e.g., different orders of homeotherms, and occasionally switching from one host taxon to another. Some groups of parasites, such as mallophagans (Phthiraptera) and cestodes (Cestoda), are more host-specific than others (Mayr, 1957; Traub, 1980). The value of host-specific parasites in historical biogeographic studies has been reviewed recently by Brooks et al. (1981), who have demonstrated some beautifully coincident distributions in sting rays and their cestode parasites.

Some patterns of slow-moving land invertebrates

Until the acceptance of continental drift in the 1960s, discussions and debates were focused mainly on vertebrate groups, in part because they are after all our closest relatives, and in part because they are smaller taxa and better known systematically and ecologically. As we have pointed out, however, these vertebrate analyses, as exemplified by the mammals, often used groups that could have been explained by either Gondwanaland or Laurasian routes and many of which were too young to reveal anything meaningful about the breakup of Pangea.

Megadrile earthworms. If any group holds potential for helping biogeographers understand past land connections, it might well be the earthworms, which are by and large totally intolerant of seawater and not at all adapted for long-distance dispersal. One of the first zoogeographic reconstructions that used continental drift was Michaelsen's (1922) treatment of earthworms, which was subsequently treated by Wegener as strong corroborative biological evidence supporting his theory.

The term *earthworm* is misleading because there are numerous aquatic species (see box). Not surprisingly, some freshwater species tend to have wider distributions and disjunctions than some of their fully terrestrial relatives. For example, species of the extratropical Holarctic genus *Bimastos* (Lumbricidae) are semiaquatic and now occur in such distant places as South Africa and the Hawaiian Islands. A few widespread species are euryhaline littoral forms, but the most cosmopolitan species are those that have been transported, even intentionally introduced around the world, such as the common earthworm of agriculture, *Lumbricus terrestris* (Lumbricidae).

Although the phylogenetic relationships of many earthworms are still very confused, a fairly useful tribal classification of suborder Lumbricina is available for analysis (see box). As indicated in this classification, some tribes or higher categories are restricted to a continent, some to closely adjacent landmasses, and some have widely disjunct distributions on two or more continents separated by wide oceans. Strong faunal links are observed between earthworms of Africa and South America in the family Alluroididae, the subfamilies Glossoscolecinae and Alminae (Glossoscolecidae), and the tribes Ocnerodrilini, Acanthodrilini, and Di-

A classification of suborder Lumbricina (megadrile earthworms)

Included are notes on the distributions and habits of these organisms. (From Jamieson, 1974.)

SUBORDER LUMBRICINA

Superfamily Alluroidoidea

Family Alluroididae (mostly aquatic)
- Subfamily Alluroidinae; Africa, Argentina
- Subfamily Syngenodrilinae; Kenya

Superfamily Lumbricoidea

Family Glossoscolecidae
- Subfamily Glossoscolecinae (terrestrial)
  - Tribe Glossoscolecini; tropical America
  - Tribe Microschaetini; South Africa
- Subfamily Kynotinae; Madagascar (one terrestrial species)
- Subfamily Hormogastrinae; southern Europe, northern Africa (terrestrial)
- Subfamily Alminae (aquatic)
  - Tribe Almini; tropical Africa and America, Indo-Malesia (*Glyphidrilus* crosses Wallace's line)
  - Tribe Criodrilini; southern Europe, the Middle East
- Subfamily Biwadrilinae; Japan (one aquatic species)
- Subfamily Sparganophilinae; North America and introduced into Europe (aquatic)

Family Lumbricidae; Eurasia, North America and islands, and introduced worldwide (terrestrial)

Superfamily Megascolecoidea

Family Megascolecidae (over 1000 species)
- Subfamily Ocnerodrilinae (terrestrial and aquatic)
  - Tribe Ocnerodrilini; South America to Baja California, the West Indies, Africa, and introduced elsewhere
  - Tribe Malabarini; southern India
- Subfamily Acanthodrilinae (terrestrial)
  - Tribe Acanthodrilini; Australasia, Africa, South America (especially southern parts), Madagascar, North America
  - Tribe Octochaetini; New Zealand, Indo-Malesia
  - Tribe Howascolecini; Madagascar, Indo-Malesia
- Subfamily Megascolecinae (terrestrial)
  - Tribe Dichogastrini; Australia, New Zealand, Indo-Malesia, China, the Americas
  - Tribe Perionychini; Australasia, Indo-Malesia, North America
  - Tribe Megascolecini; Australasia, Indo-Malesia, China, Japan

Family Eudrilidae; Africa (terrestrial)

chogastrini (Megascolecidae). Michaelsen had perceived this African-American relationship, but the evidence he used was of shared genera, nearly all cases of which are now rejected on systematic grounds (Jamieson, 1974). Even the genus *Wegeneriella* (Megascolecidae), named to honor Wegener and created for a presumed monophyletic group in Africa and South America, has been rejected. The undeniable relationships that describe the splitting of West Gondwanaland appear to be at levels higher than the genus. In mainland Africa the family Eudrilidae and subfamily Syngenodrilinae (Alluroididae) are endemic, whereas the highest endemic taxon in the New World is the tribe Glossoscolecini (Glossoscolecidae). Meanwhile, the tribes Perionychini and Megascolecini of the Megascolecidae are totally absent from West Gondwanaland, occurring in Australasia and the Northern Hemisphere.

Of the five families of earthworms in suborder Lumbricina, apparently only one, Megascolecidae, is indigenous to Australasia (Jamieson, 1974), where there is a high rate of generic endemism and 100% specific endemism (excluding a few euryhaline forms). Even within the Australian region each landmass has mostly endemic genera. For example, between Australia and New Guinea, the only known infrageneric affinity is *Pheretima,* which also ranges into southeast Asia. New Caledonia and New Zealand share the genus *Acanthodrilus.* Today, the strongest affinities for Australasia are with Indo-Malaysia, less so with the New World; almost nothing is shared with Africa. On the other hand, the group with the primitive features, tribe Ocnerodrilini, occurs in the pieces of West Gondwanaland; the origin of Australian megascolecids is almost certainly attributable to Gondwanaland. This may be why India and Australia have closely related genera—remnants of past Gondwanaland connections. Whether or not earthworms used Antarctica to migrate between South America and Australia remains to be determined.

Land snails. The outstanding feature of land snails (over 25,000 species in 72 families) is that they are, as a whole, very ineffective in crossing filter zones (Solem, 1959). This conclusion is clearly demonstrated by several lines of evidence. First, the majority of land snails cannot tolerate seawater at any stage of their life cycle, as adults, veligers, or eggs. Second, because these organisms are mostly intolerant of seawater, very few genera have colonized distant islands. This is clearly illustrated in Figure 7.16, in which Solem (1981) shows the attenuation of the Palearctic-Oriental land snail families across the saltwater filter from southeast Asia to Australia. Third, the majority of ranges of wide-ranging or highly disjunct individual species are attributable to human activity. Finally, in family after family, snail species on continents tend to have extremely narrow ranges and cannot even cross minor ecological barriers of substrate or dryness.

Having narrow ecological tolerances is especially pertinent to our discussions on the relationships between continents and deserves a vivid illustration. In the southwestern United States and adjacent Mexico is found the family Helminthoglyptidae, including about 160 species, of which over two thirds are endemic either to California or to Arizona. For example, 54 species of *Sonorella* (total 68 species) are endemic to Arizona and the genera *Helminthoglypta* (47 species) and *Mohavelix* (1 species) are endemic to California. In the desert mountains of southern Arizona, each peak can have a different species or form of *Sonorella,* and in a single species, *S. sabinoensis,* there are four distinct populations in the mountains surrounding Tucson (Bequaert and Miller, 1973).

A full and comprehensive description of land snail distribution patterns is not only too lengthy to present here but also not possible yet because many phylogenetic matters are still unresolved. Nevertheless, some recent reviews (Solem, 1976, 1979a, 1979b, 1981; van Bruggen, 1980) help us to appreciate that snails illustrate many interesting distributions.

Unlike many land groups featured in this

book, land snails have an old and fairly extensive fossil record, dating back to the Upper Paleozoic. Comparative studies reveal that molluscs evolved to live on land many different times; by the end of the Paleozoic, fossils of at least five of these lineages can be observed. Practically all the Paleozoic and Mesozoic fossils have been found in Europe and North America, and Europe had an exceedingly diverse malacofauna until the Miocene, when tropical climates deteriorated there and the organisms either migrated southward or became extinct.

Based on the abundance of modern families in the fossil beds of the Northern Hemisphere and the distribution patterns of many snail superfamilies, Solem (1979b) regards the Northern Hemisphere as the site for basic radiations of land molluscs. He suggested that Gondwanaland per se was never a major factor in their initial diversification. There are few intrafamilial and interfamilial disjunctions between tropical Africa and tropical South America, the most obvious ones with sister groups in these two regions being the following: acavoids (superfamily Acavacea), the Old World Pomatiasidae and the New World Chondropomidae, and possibly the Dorcasiidae of Africa and Strophocheilidae of South America, which may comprise a superfamily (both have Paleocene fossils in their respective regions).

Several taxa show a distribution suggesting southern Gondwanaland origins. Two endemic genera of Rhytididae in southern Africa have their putative rhytidid sister taxa in Australia and New Zealand; and the Charopoidae of southern Africa, South America, and Australasia also form a closely related group. Another vicariant event appears to be in Bulimulidae of southern South America and Australasia. Some of the other families occurring on the three southern continents need more systematic investigation before workers can evaluate the role of plate tectonics in their differentiation (e.g., the circumtropical Veronicellidae, and the Diplommatinidae and Cyclophoridae. Solem has also noted close affinities between land snails of the Mediterranean region and those of the Greater Antilles, especially in superfamily Achatinacea, which may be tied to the breakup of Pangea.

One of the surprising facts is the relatively low number of endemic families of land snails in South America, which otherwise has the richest overall biota. This may mean that relatively few snail families were present in South America prior to its departure from Gondwanaland. Places rich in endemic families are Europe, North America and the West Indies, and the oceanic islands of the Pacific. In the Pacific occur the insular endemics Achatinellidae, Endodontidae (s.s.), Partulidae, Amastridae, and part of the Poteriidae. Solem treats these as ancient relicts that have survived in Polynesia since the early Mesozoic by colonizing one island after another. This would have to be the means of continuing these clades, unless they have colonized the Pacific more recently, because the present-day islands are all Tertiary in age or younger. These types of dynamic historical events are generally lost to investigators forever.

Selected references

Abilities of land creatures to cross water barriers

Boer (1970); Carlquist (1965, 1966a, 1974); Clagg (1966); Gressitt (1970); Gressitt and Yoshimoto (1963); Grinnell (1914); D. Johnson (1978, 1980, 1981); G. Simpson (1952b); Solem (1959); Wenner and Johnson (1980).

Mesozoic land vertebrates

Alvarez et al. (1980); Charig (1973a); Clemens (1982); Cluver and Hotton (1979); Colbert (1971, 1972, 1973a, 1979, 1980); C. Cox (1973a, 1973b, 1974); Ganapathy (1980); Kalandadze (1974); Keast (1972a); Lillegraven et al. (1979); Molnar et al. (1981); Olson (1971); Panchen (1980); Romer (1975).

Terrestrial amphibians and reptiles

Blair et al. (1976); Blanc (1972); Brattstrom (1961); Carlquist (1965); Carr (1952); Cracraft (1974c, 1975a, 1975b); Cogger (1979); Darlington (1957); Duellman (1966, 1979); Edwards and Lofty (1977); Estes and Reig (1973); Gaffney (1975, 1977); Gans and Parsons (1969–1981); Goin et al. (1978); Kurtén (1969); Lowe et al. (1970); Marx and Rabbil (1970); Peabody and Savage (1958); Pritchard (1979); Rawlinson

(1974); J. Savage (1966, 1973); G. Simpson (1943b); Vial (1973).

Ratites

Cracraft (1974a, 1981).

Distributions of extant and extinct mammals

Bigalke (1972); Carlquist (1965); Charig (1973b); Ciochon and Chiarelli (1980); Clemens (1977, 1982); Cockrum (1962); H. Cooke (1972); Corbet (1978); Corbet and Hill (1980); Darlington (1957); Eisenberg (1981); Finerty (1980); Fooden (1972); Funk and Brooks (1981); Gaines and McClenaghan (1980); George (1962); Gingerich (1976a, 1976b, 1979); Gordon et al. (1977); M. Griffiths (1978); E. Hall (1981); M. Hecht et al. (1977); Hershkovitz (1972, 1977); Hoffstetter (1972); Honacki et al. (1982); Keast (1968, 1972a, 1972b, 1972c, 1977b); Keast et al. (1972); Kirsch (1977); Kirsch and Calaby (1977); Kurtén (1969, 1971, 1972); Kurtén and Anderson (1980); MacFadden (1980); McGowran (1973); McKenna (1972a, 1972b, 1973, 1975b); Maglio (1973); Maglio and Cooke (1978); Marshall (1979); Marshall et al. (1982); P.G. Martin (1977); Matthew (1915); Mayr (1952, 1957); Papavero (1977); B. Patterson and Pascual (1972); C. Patterson (1981a); D. Savage (1958); Schlinger (1974); G. Simpson (1945, 1947a, 1947b, 1950, 1953, 1956, 1961b, 1965, 1969, 1978, 1980a, 1980b); Szalay (1975); Tedford (1974); Traub (1980); Vaughan (1978); Vereshchagin (1967); E. Walker (1975); S. Webb (1976, 1977, 1978); Whitmore and Stewart (1965).

Some patterns of slow-moving land invertebrates

Backhuys (1975); Bequaert and Miller (1973); Bruggen (1980); Cameron and Redfern (1976); C. Edwards and Lofty (1977); Ferris et al. (1976); Jaeckel (1969); Jamieson (1974); Sims (1978); Solem (1959, 1979a, 1979b, 1981); Southwood (1962); Wallwork (1976).

Chapter 12

Distribution Patterns of Flying Animals

As children, our fascination with flying objects probably convinced us that having wings permits animals to fly in any direction for any distance. The capability for flight potentially permits long-distance dispersal over both land and water barriers and rapid moves over or through unfavorable regions. Of course, one can band birds and bats or use genetic markers for insects in order to follow their movements and thereby demonstrate the effectiveness of flight for dispersal. Other evidence for the vagility of winged animals is the high representation of birds and flying insects on remote oceanic islands and the almost cosmopolitan distributions of certain avian species, such as the barn owl *(Tyto alba)*, the osprey *(Pandion haliaetus)*, and the peregrine falcon *(Falco peregrinus)*, not to mention the sizable list of cosmopolitan winged genera.

However, our first lesson in the distribution of bird life of the continents is in direct conflict with this preconception. As Sclater pointed out (Figure 1.3), each major continental area has a peculiar bird fauna, with many species and higher taxa not found on any other landmass. Clearly then, all birds do not traverse distances like that traveled by the arctic tern *(Sterna paradisaea)*, which flies from pole to pole twice a year, an annual migration route of 40,000 km. In fact, many species of flying animals are narrowly endemic, and even groups that are carried passively by the wind are generally not cosmopolitan.

Aerial dispersal and migration

Having so many common examples of wide-ranging flying organisms, it is all too easy to avoid discussing those with narrow ranges. Although some of these can be explained simply as historical events—a population started in one locality and cannot leave because it lacks the ability for long-distance dispersal—generally the limited distributions of flying organisms are controlled by more subtle ecological processes. A common factor limiting aerial dispersal is some narrow ecological requirement, such as the distribution of a certain food resource or the proper habitat for raising offspring. For example, the oilbird of the Neotropics (Steatornithidae) is limited not only by the distribution of its chief food, certain palm fruits, but also by the availability of caves for colonial nesting. Another neotropical bird, the strange leaf-eating hoatzin (Opisthocomidae), cannot live outside the range of its food plant, an arum (Araceae).

Some flying insects, such as butterflies and moths, tend to have highly restricted distributions because their eggs and larvae can only develop on a particular host plant. An excellent example of narrow host-plant preferences is illustrated by the neotropical butterflies in the genus *Heliconius* (Nymphalidae). Each species lays its eggs on the leaves of a particular species of passion flower *(Passiflora)* (Gilbert and Singer, 1975; Turner, 1981). In fact, food specialists are the rule rather than the exception in many insect groups. This is a fundamental reason why there are so many species with fairly narrow ranges. Organisms with more general food requirements, such as nectar-feeding birds or nectivorous bats, which may visit dozens of unrelated flowers, are theoretically not as limited in range as those dependent on a single food item that is itself narrowly restricted.

Competition is another important reason for flying organisms to have fairly small ranges. Not only does competition influence the local distribution of a species but also it may operate at higher taxonomic levels for an entire continent. To illustrate this, we can examine some bird examples. The New World warblers (Parulidae) are replaced in Eurasia by the Old World warblers (Sylviidae), and the New World flycatchers (Tyrannidae) are replaced by Old World flycatchers (Muscicapinae). Australia also has its own set of warblers (Acanthi-

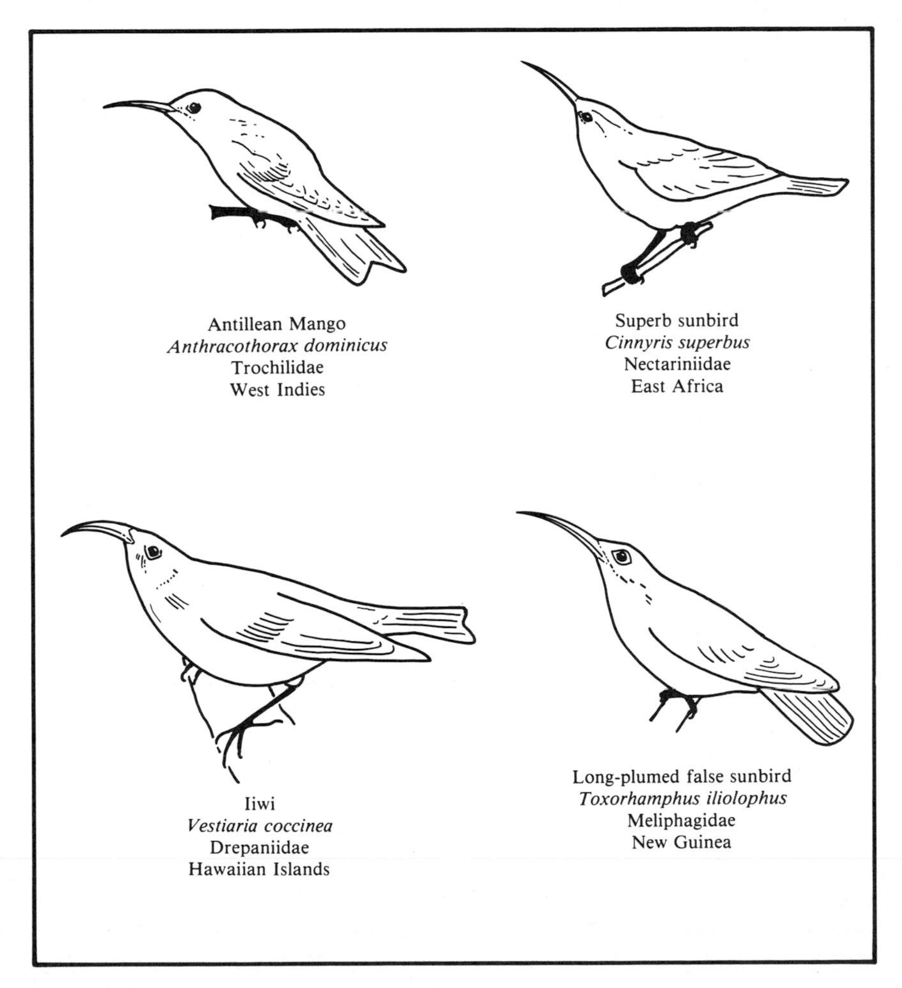

Figure 12.1
Drawings of four species of nectar-feeding birds representing four different families. In geographically isolated regions convergent evolution has produced birds of similar morphology (especially specializations in the bill and tongue) adapted for feeding on floral nectar.

zinae, Muscicapidae) independently evolved from those in the Northern Hemisphere. Other common examples are the nectar-feeding birds of the world (Figure 12.1), represented by the hummingbirds (Trochilidae) in the New World, the sunbirds (Nectariniidae) in the paleotropics, the honeyeaters in Australasia (Meliphagidae), and the honeycreepers (Drepanini or Drepaniidae) on the Hawaiian Islands. All of these have specialized bills and tongues to extract nectar from long tubular flowers.

Although flying organisms with narrow ranges often have poor abilities for long-distance dispersal, biogeographers have long realized that the potential for aerial dispersal can be diminished or lost after a population arrives at its present location. Classical examples are the flightless birds and insects on remote oceanic islands (Carlquist, 1965). On these islands the population accumulates those genes that favor its survival on the island at the expense of traits suitable for colonizing distant places. However, these same types of shifts toward flightlessness or poor dispersibility can also occur on mainlands wherever a flying organism becomes a specialist on a limited resource. When distribution maps are drawn for narrow endemics, the fact that this species may have come from a highly dispersive stock will be lost.

The importance of migration for flying organisms. Every spring, approximately on the 15th of March at Hinckley Pond in northern Ohio, turkey vultures *(Cathartes aura septentrionalis)* return to breed and raise their young. Where do they come from? The birds have traveled thousands of kilometers from warmer areas in South America where they had escaped the stressful and lean winter months in the north. This is an example of true migration (Chapter 7), in which an individual travels long distances but eventually returns to its place of origin. Flying species, especially birds, are famous for such return movements, but so are other vertebrates, such as the Pacific salmon (*Oncorhynchus* spp.) that swim upstream to lay eggs and die; the freshwater eels of western Europe and eastern North America (*Anguilla* spp.), which return to the Sargasso Sea in the South Atlantic for breeding en masse; and the migrating whales and sea turtles (Chapter 10).

In cases of true migration, it is easy to see that the entire process of spatial displacement is genetically fixed and intentional, and serves the adaptive function of transporting individuals from a place of dwindling resources or deteriorating climate to one more favorable for continuing the life cycle. The departure of migrants is typically cued by abiotic factors that are reliable predictors of seasonal change, so that in many species this departure is obviously the product of natural selection to maximize the chances of individuals to survive and leave offspring. Migration also enables populations to inhabit a much larger area than they could if they were sedentary. Many taxa, especially of birds, would not be able to inhabit temperate and arctic regions if they were unable to migrate to warmer regions for the winter. Migration may also facilitate the colonization of new regions. Migrating populations may gradually extend their ranges or they may occasionally be carried abruptly to new sites by unusually severe storms. Of course, true migration is specialized behavior that involves not only the use of the proper cues for timing and orienting movements but also special physiological adaptations permitting successful long-distance movement. This is much different than the generalized phenomenon of dispersal, which includes all acts of movement and displacement, including those that are accidental or random events without any specific directionality.

One can see immediately why authors would have difficulty in defining migration, because there is a continuum of patterns ranging from round-trip journeys between distant locations on a strict annual schedule to local and nomadic movements (see box on p. 347). Many insects, most notably locusts, are legendary for mass migration flights. However, the term *flight* is somewhat misleading because once the locusts are airborne the direction of movement is determined by the prevailing winds (Dingle, 1980). Many other insect groups are passively dispersed by prevailing winds, but some, such as aphids, can exercise some choice over when and where they will land. Whether one calls this migration or not is a matter of semantics, but certainly the dispersal is genetically coded, intentional, often directional, and adaptive, requiring specialized behavior and physiology.

In insects or other invertebrates that migrate on prevailing winds, an individual does not return to its exact place of origin because its life span is too short and the winds are not dependable enough to deliver an individual to an exact location. Some active flying is required for a precision return movement. A return movement is recorded for the monarch butterfly *(Danaus plexippus)* in North America, which starts to migrate in August from southern Canada toward the southeastern United States (Figure 12.2). For years naturalists thought the entire annual migration was completed by single individuals, as with birds, but now they suspect that members of a different generation return to northern latitudes the following spring. Nonetheless, the collective movements of these several generations have the same general biogeographic effect as the round trip of a single individual.

There are some polymorphic populations in which some individuals are morphologically, behaviorally, and physiologically adapted for long-distance flights, whereas others can make only short flights or do not fly at all (Harrison, 1980). Long-winged (macropterous), short-winged (brachypterous), and wingless (apterous) forms are known from many insect species or between closely related species. For example, the aquatic water striders of Europe (Gerridae) show many differences between and within species (Table 12.1). The percentage of individuals able to leave the habitat by flight may be greater in certain geographic regions, e.g., in cool, temperate climates; at certain times of the year, e.g., in the warmer months; in ephemeral habitats; under certain nutritional conditions,

Classes of migration for insects (After Johnson, 1969.)

Class I. Adults emigrate from a breeding site without returning; life span limited to within one season.
- A. Each generation relinquishes old habitats to find new ones; dispersal is in all directions.
- B. In some generations alate forms are produced that disperse by prevailing winds; in other generations a succession of apterous parthenogenetic forms may be produced.
- C. Forms migrate between different host plants in the same region, eventually returning to the original host.
- D. Individuals move from one breeding area to another as resources become limiting.
- E. Individuals disperse in a particular direction more or less independently of prevailing winds; they oviposit and die after reaching their destination, and return is by their offspring.

Class II. Adults emigrate and return to the point of origin; life span is limited to within one season.

Class III. Adults emigrate to hibernate or aestivate, and return flights are by the same individuals after imaginal diapause.
- A. Individuals migrate within the breeding area to hibernate or diapause and then undergo another, usually shorter, movement before breeding.
- B. Individuals migrate out of the breeding area for diapause and then return for breeding.
- C. Individuals disperse from the breeding area, aestivate a long distance from the breeding area, make another short flight before hibernating, and after hibernation return to the breeding area.

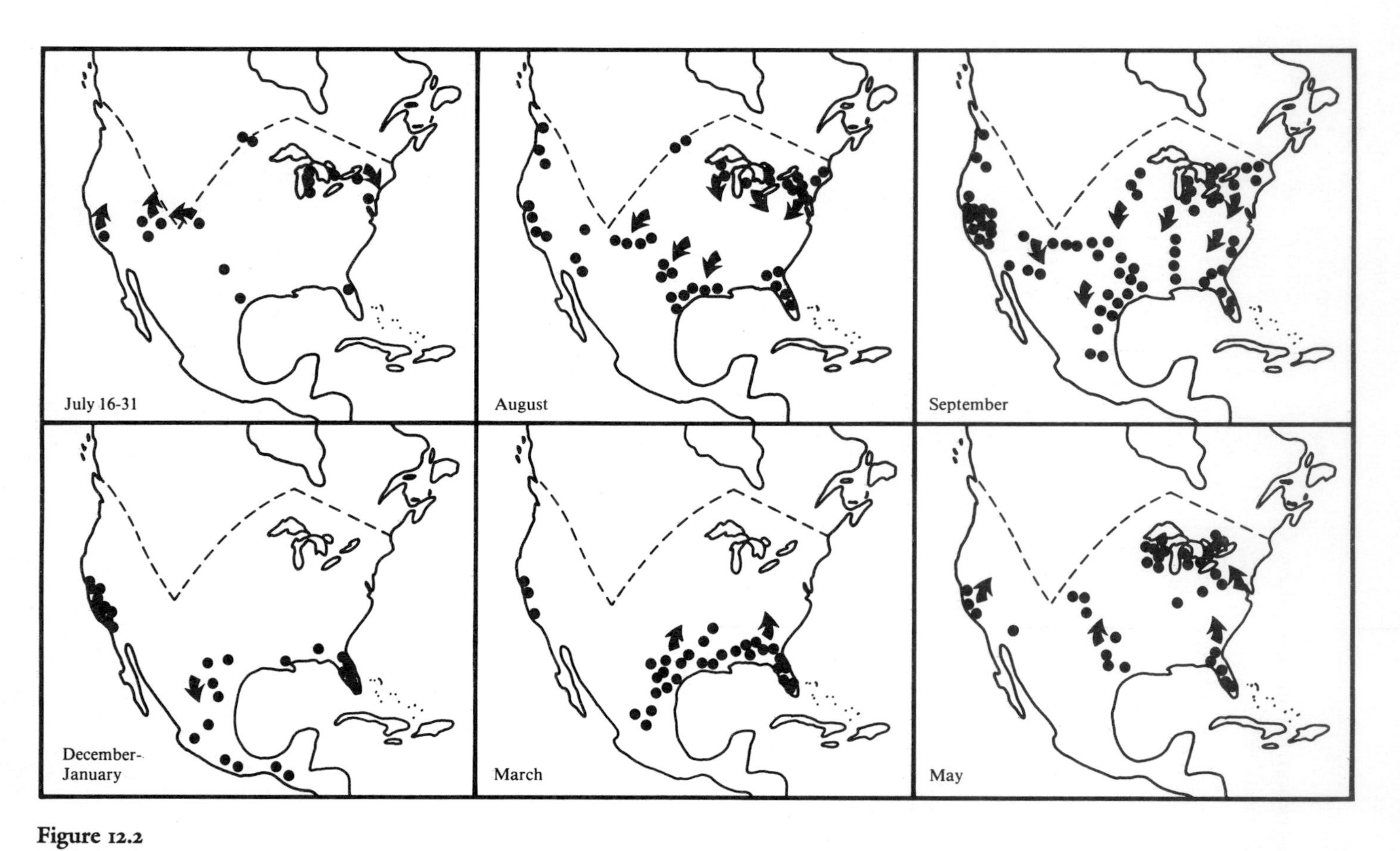

Figure 12.2
Annual migration pattern of the monarch butterfly *(Danaus plexippus)* in North America. Although single individuals may travel hundreds of kilometers, the entire migration south in the fall and north again in the spring is probably completed by different individuals in different generations. (After Baker, 1978.)

Table 12.1
Polymorphic life history strategies of water striders (Gerridae)
Species of northern and central Europe are arranged in an approximate progression from completely long-winged to completely short-winged populations. (After Dingle, 1980.)

Taxon	Strategy	Habitat
Gerris rufoscutellus	Monomorphic, long-winged, univoltine*	Temporary ponds
G. thoracicus (Central Europe)	Monomorphic, long-winged, multivoltine*	Ponds and gravel pits subject to drying in long warm summers
G. thoracicus (southern Finland)	Monomorphic, long-winged, long life, interreproductive flights; "supertramp" strategy	Small temporary, highly unproductive rock pools
G. thoracicus and *G. odontogaster* (Finland)	Seasonal dimorphism with long-winged winter (diapause) and short-winged summer generations	Temporary ponds, streams, ditches with cool summers
G. odontogaster (Central Europe)	Long-winged diapause generation of strong flyers; summer generation dimorphic	Temporary ponds, streams, ditches with long warm summers
G. argentatus and *G. paludum* (Finland)	Long-winged diapause generation and short-winged summer generation	Ponds, medium size lakes and streams with cool summers
G. argentatus and *G. paludum* (Central Europe)	Long-winged diapause generation and dimorphic summer generations	Ponds, medium size lakes and streams with long warm summers
G. lacustris (Finland)	Genetically dimorphic winter generation and short-winged summer generation	Rivers, lakes with cool summers
G. lacustris (Finland and Europe)	Long-winged diapause generation and short-winged summer generation	Ponds, medium size lakes, ditches, rivers
G. lacustris (Central Europe)	Genetically dimorphic summer and winter generations	Rivers, lakes with long warm summers
G. lateralis (Finland)	Genetically dimorphic, univoltine	Large lakes, rivers with cool summers
G. asper	Genetically dimorphic, probably bivoltine*	Large lakes, rivers with long warm summers
G. najas (Poland)	Genetically dimorphic	Streams with warm summers
G. lacustris (Southern Finland)	Monomorphic short-winged, univoltine	Isolated permanent ponds with cool summers
G. najas (Finland)	Monomorphic short-winged, univoltine	Isolated pools of permanent streams with cool summers
G. sphagnetorum	Monomorphic short-winged	Isolated bogs

*Breeds once (univoltine), twice (bivoltine), or many times (multivoltine) per year.

e.g., when food resources are depleted; or at particular phases of the reproductive cycle, e.g., when a female is gravid. Polymorphism of dispersal properties is probably an intermediate stage in the origin of fully migratory species, and it is often difficult to decide whether migration or long-distance dispersal is the proper descriptor of these behaviors.

Migration patterns of birds. No discussion on migration could be considered complete without some consideration of migratory birds. Seasonal migrations are favorite topics for historical biogeographers, because they could hold some information on the history of life on continents. Seasonal migrations include not only those with regular departure and arrival times in standard summer breeding areas and winter nonbreeding areas but also those with more irregular timetables, such as the species that wander widely (vagrants) or whose movements facultatively follow the seasonal rains over a continent. Some of the most widespread groups are also migratory, e.g., hawks (Accipitridae), pigeons (Columbidae), nightjars (Caprimulgidae), swifts (Apodidae), swallows (Hirundinidae), and vespertilionid (Vespertilionidae) and molossid bats (Molossidae).

Two seasonal migratory patterns, elevational and latitudinal, can be observed in any mountain range of the southwestern United States. Certain resident species (those that do not leave) such as the dark-eyed junco *(Junco hyemalis)*, Cassin's finch *(Carpodacus cassinii)*, and Steller's jay *(Cyanositta stelleri)*, live in the high montane forests during warm weather and descend the mountain to varying extents in the wintertime to take advantage of more abundant food and warmer temperatures. In the same forests live the broad-tailed hummingbird *(Selasphorus platycercus)*, violet-green swallow *(Tachycineta thalassina)*, and western tanager *(Piranga ludoviciana)*, which are highly migratory species that head southeast in the fall along the Sierra Madre Occidental to spend the winter in more equable regions of Mexico and Central America. In the spring the migrants return, but along the way the tanagers and swallows can be observed at desert oases and the hummingbirds stop to feed on the red tubular flowers in the foothills and mountains of Mexico, which open at the right time to attract the birds and use them as pollinators.

The route along the Sierra Madre Occidental from the north is one of the major flyways used by birds migrating from the temperate zone to the tropics (Figure 12.3). By such routes each tropical area annually receives a large influx of land and freshwater birds: the Neotropics, 147 species (of a total 3300 species in the fauna); India, 115 species (of 1200 species); subsaharan Africa, 118 species (of 1481 species); and southeast Asia, 142 (of 1198 species) (Karr, 1980). In some tropical areas (especially in the Southern Hemisphere) the migrants constitute only 1% of the bird fauna, but in other areas, especially those closest to the temperate zone, such as Mexico, Cuba, and Hispaniola, up to 50% of the winter assemblage may be composed of migrant species (Terborgh, 1980). The highest latitudes have small avifaunas, over 70% of which migrate; but proceeding southward, as the number of species in the breeding avifauna doubles, the percentage of migrants decreases by half (Slud, 1976). Islands tend to have lower percentages of migrants than comparable mainland localities.

Much of the research on bird migrations is not focused directly on biogeographic problems, but the salient patterns are pertinent to our historical treatment here. The best known patterns are those studied in depth for the Nearctic-Neotropics system, which was the subject of a recent symposium (Keast and Morton, 1980). To begin with, migrants should not be considered to be temperate species that invade the tropics; instead, most of these migratory birds had a tropical origin, they still live in their tropical habitats 6 or 7 months per year, and they have clearly defined niches in the tropical ecosystems. Many workers had originally assumed that migrants were intruders into the tropics, northern birds that had been forced to

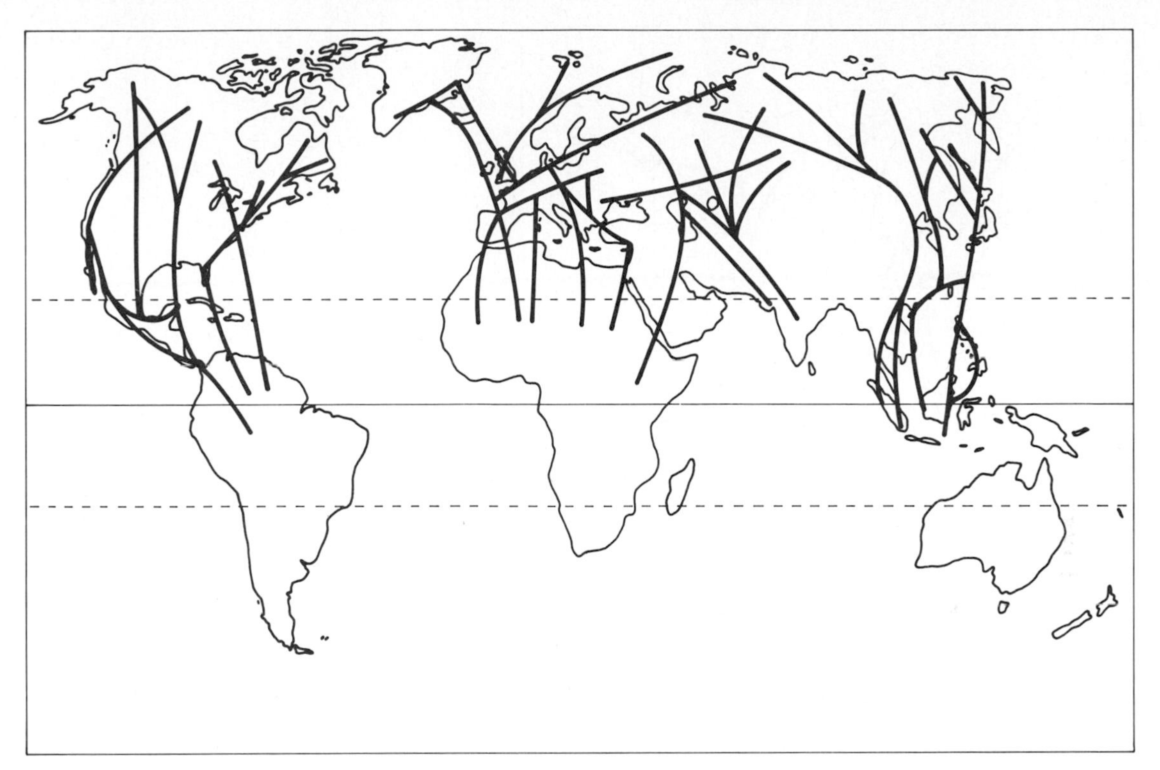

Figure 12.3
Major avian flyways of the world. Note that although there is extensive migration between temperate and subarctic breeding areas in the Northern Hemisphere and wintering grounds in the New and Old World tropics, there is little annual long-distance migration in the Southern Hemisphere. (After McClure, 1974, and Baker, 1978.)

fly south and compete with or displace tropical species when their seasonal environments became unsuitable. Studies now show that in the wintering area most migrants have different niches than the residents, and the residents do not completely fill the vacant niches after the migrants have left. Therefore one hypothesis to explain this pattern is that bird species from tropical climes have evolved an annual trip northward to produce their offspring in a habitat where food resources are seasonally abundant and interspecific competition is less. Even with losses during migration, more offspring can be produced by such seasonal movements than could have been produced in situ in the tropics.

The migratory routes between the Palearctic and the Old World tropics (Figure 12.3) have been clearly documented (Moreau, 1966, 1972; McClure, 1974), but so little is known about the tropical wintering areas that comparisons with the neotropical system cannot be made at this time. The situation in Africa may be quite different from that in South America because so much of Africa has a drier climate and less mountainous terrain.

A few words are needed about migration in Australasia, which is unusual in having numerous intracontinental but relatively few intercontinental movements. An example of a latitudinal migrant is the bronze cuckoo *(Chrysococcyx lucidus),* which breeds in southernmost Australia,

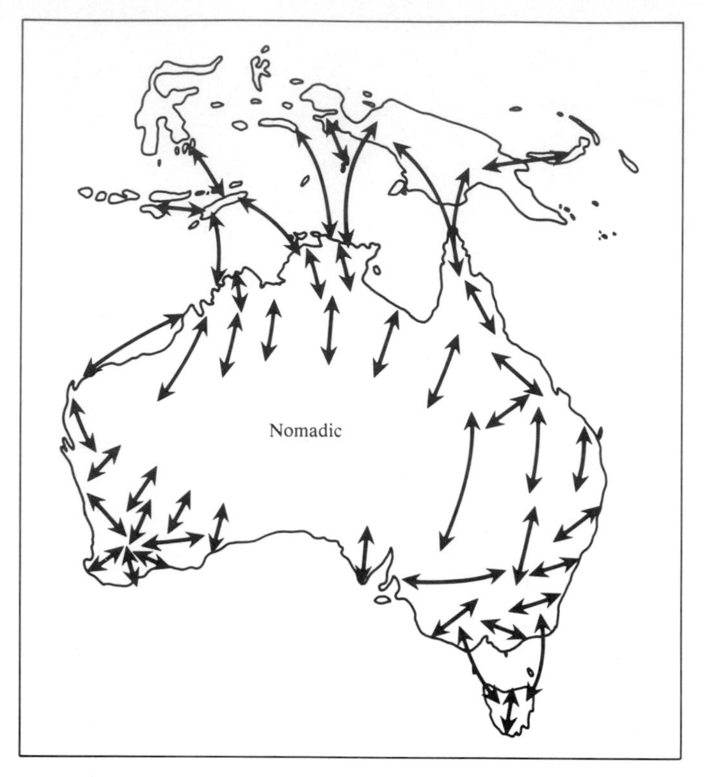

Figure 12.4
Primary migration routes of Australian birds. Note that there is much movement within the continent but much less migration to the islands of the East Indies and virtually none to southeastern Asia. (After Nix, 1977.)

Tasmania, and New Zealand but which moves to New Guinea and other equatorial islands for the winter. Such movements are the exception rather than the rule, however; much of the known migration of birds encompasses only small portions of Australia, such as within the southwestern corner, between Tasmania and Victoria, and between Queensland and New Guinea (Nix, 1977). Figure 12.4 shows the probable seasonal movements of Australian birds, illustrating how they track areas of seasonal productivity. In the interior arid deserts of Australia, where rainfall is extremely scarce, unpredictable, and erratic, the birds are nomadic, opportunistically moving into areas in response to plant and insect production.

Large-scale patterns of bird distributions

Seabirds. For a historical biogeographer, the first task in comparing avifaunas is to recognize the seabirds so they can be deleted from the analysis. As one might expect, many seabirds are wide ranging, found around the world. Examples include the tropicbirds (Phaethontidae), frigatebirds (Fregatidae), and many species of terns and gulls (Laridae). However, there are families with narrowly endemic forms. For example, many species of cormorants *(Phalacrocorax)* are narrowly endemic. The most notorious is the Galápagos flightless cormorant

(Nannopterum harrisi), whose fearlessness of humans and awkwardness on land were made famous by the evolutionary writings of Darwin. Rarely are seabird families restricted to a single hemisphere, exceptions being the northern auks (Alcidae) (Figure 7.9) and the southern diving petrels (Pelecanoididae) and the penguins (Spheniscidae; Chapter 10).

So little is known about the intergeneric relationships of seabirds and their long evolutionary history, which is not preserved well in the fossil record, that the historical biogeography of these organisms is mostly unknown. Investigators know that for strong fliers distance has not generally been a barrier to their dispersal around the world. For example, pelicans (Pelecanidae), anhingas and darters (Anhingidae), and skimmers (Rynchopidae) are widespread, occurring on all continents. However, these same groups illustrate a different pattern of distribution because in each family a single species occurs in each area, so that the species have nonoverlapping ranges. Ecologically this can be explained as competitive exclusion—only one species can successfully occupy the particular niche on a given coastline—but one is still left with the mystery of how the species were formed in the first place and whether the sequences of speciation events are related to any important changes in the physical conformation of the landmasses.

Land and freshwater birds of tropical America. In the region stretching from southern Mexico to the tip of South America and including the West Indies there can be found over 40% (about 3500 species) of all species of land and freshwater birds. In comparison, the rest of the New World has a depauperate avifauna, and the Neotropics have twice the number of species found in mainland Africa. Examining equivalent parcels of rain forest in South America and Africa, one finds that some 10 km^2 plots of neotropical forest, with over 300 resident species, have a richer avifauna than the entire equatorial African country of Liberia, with over 80,000 km^2 (Amadon, 1973). On the other hand, neotropical forests are not noticeably richer in bird families than other tropical regions, but instead those which are present have radiated to produce much greater species diversity. Understanding the origins and nature of the avifauna of the Neotropics is essential to reconstructing the biogeographic history of birds as a whole.

The box on p. 354 lists the endemic bird families and subfamilies of the Neotropics along with those having most of their species appearing in tropical America (Meyer de Schauensee, 1970; Haffer, 1974; Gruson, 1976). Disregarding a number of very small endemic families, such as the trumpeters (Psophiidae) of South America and the todies (Todidae) of the West Indies, consisting of slightly differentiated forms, the most remarkable aspect of neotropical endemism is the number of large families, especially in the order Passeriformes, e.g., the ovenbirds (Furnariidae) and antbirds (Formicariidae). The ovenbirds clearly demonstrate evolutionary diversification because these taxa have managed to fill most of the habitats and have convergently evolved basically the same morphological designs as unrelated small birds in Africa (Vaurie, 1980). As noted in earlier chapters, such adaptive radiation implies that the group was present and speciated before other competitive groups occurred there.

Ovenbirds and antbirds belong to a group of passerine birds called suboscines, and 90% of all suboscines are present in tropical America. In contrast, the largest group of living birds, the songbirds or oscines (over 4000 species) are proportionately less abundant than in other tropical regions. Suboscines certainly were isolated in South America during their period of major adaptive radiation, as shown by many South American genera that neither penetrate Central America nor occur yet in the West Indies (except the continental island of Trinidad), even though suitable habitats exist. In contrast, the common bird groups of Central America extend into northern South America; e.g., the

Bird families and subfamilies endemic or nearly so to South and Central America

NEOTROPICAL ENDEMICS (700 spp.)

Nonpasserines

Tinamidae (tinamous; 46 spp.)
Rheidae (rheas; 2 spp.)
Anhimidae (screamers; 3 spp.)
Opisthocomidae (hoatzin; 1 sp.)
Psophiidae (trumpeters; 3 spp.)
Cariamidae (seriemas; 2 spp.)
Eurypygidae (sunbittern; 1 sp.)
Thinocoridae (seedsnipe; 4 spp., also on the Falklands)
Steatornithidae (oilbird; 1 sp.)
Nyctibiidae (potoos; 5 spp.)*
Galbulidae (jacamars; 17 spp.)
Bucconidae (puffbirds; 32 spp.)
Ramphastidae (toucans; 41 spp.)
Todidae (todies; 5 spp., all West Indian)*
Momotidae (motmots; 8 spp.)

Suboscine passerines

Dendrocolaptidae (woodcreepers; 50 spp.)
Furnariiae (ovenbirds; 214 spp.)
Formicariidae (antbirds; 230 spp.)
Conopophagidae (gnateaters and antpitits; 8 spp.)
Rhinocryptidae (tapaculos; 29 spp.)
Pipridae (manakins; 53 spp.)
Phytotomidae (plantcutters; 3 spp.)
Oryruncidae (sharpbill; 1 sp.)

Oscine Passerines

Dulineae of Bombycillidae (palmchat; 1 sp.)*
Catamblyrhynchinae of Emberizidae (plush-capped finch; 1 sp.)

NEW WORLD ENDEMICS BEST DEVELOPED IN THE NEOTROPICS (1458 spp.)

Nonpasserines

Cathartidae (New World vultures and condors; 7 spp.)*
Cracidae (curassows; 44 spp.)
Aramidae (limpkin; 1 sp.)*
Trochilidae (hummingbirds; 315 spp.)*

Suboscine passerines

Cotingidae (cotingas and becards; 79 spp.)*
Tyrannidae (tyrant flycatchers; 384 spp.)*

Oscine passerines

Troglodytidae (wrens; 59 spp., except the widespread common wren, *Troglodytes troglodytes*)*
Icterinae of Emberizidae (blackbirds, New World orioles, and caciques; 92 spp.)*
Emberizinae of Emberizidae (New World sparrows and finches; 234 spp., with some exceptions)*
Thraupidae (tanagers; 233 spp., including Tersinidae)*

NEW WORLD ENDEMICS WITH RELATIVELY FEW SPECIES THAT BREED IN THE TROPICS (226 spp.)§

Miminae of Turdidae (New World thrashers and mockingbirds; 30 spp.)*
Vireonidae (vireos; 39 spp.)*
Parulidae (wood warblers; 120 spp.)*
Cardinalinae of Emberizidae (cardinals and grosbeaks; 37 spp.)*

---

*Occurs in the West Indies.
§Many species in these oscine families are migrants to tropical latitudes from the temperate zone.

majority of species in Belize also occur in South America (Russell, 1964).

Haffer (1974, 1978, 1981) has shown that patterns of distribution for resident land birds of the lowland neotropical forests are consistent with the model of allopatric speciation. Figure 12.5 shows the known distributions of six species of blue cotingas *(Cotinga)*, members of a superspecies. The three species in Central America are each geographically separated, and the Andes separate *C. nattererii* in western Colombia from the two taxa of the Amazonian basin, *C. cotinga* and *C. maynana*, which are narrowly parapatric and can hybridize in the zone of contact. The sixth species, *C. maculata*, is geographically isolated in the forests of southeastern Brazil. This distribution pattern is found in other superspecies of birds.

Amazonia has the richest avifauna in the world, consisting of no fewer than 930 species of resident land birds, of which at least 650 occur in lowland forests and 409 species are endemic. The richest area is adjacent to the eastern slopes of the Andes in eastern Peru and Bolivia, where 150 species are endemic. Naturally, such centers of endemism have attracted the attention of biogeographers seeking answers to the question of why speciation has been so high. Haffer has attributed much of this diversity to events during the Pleistocene. Before the Pleistocene, the lowland rain forest was expansive and fairly continuous, but during

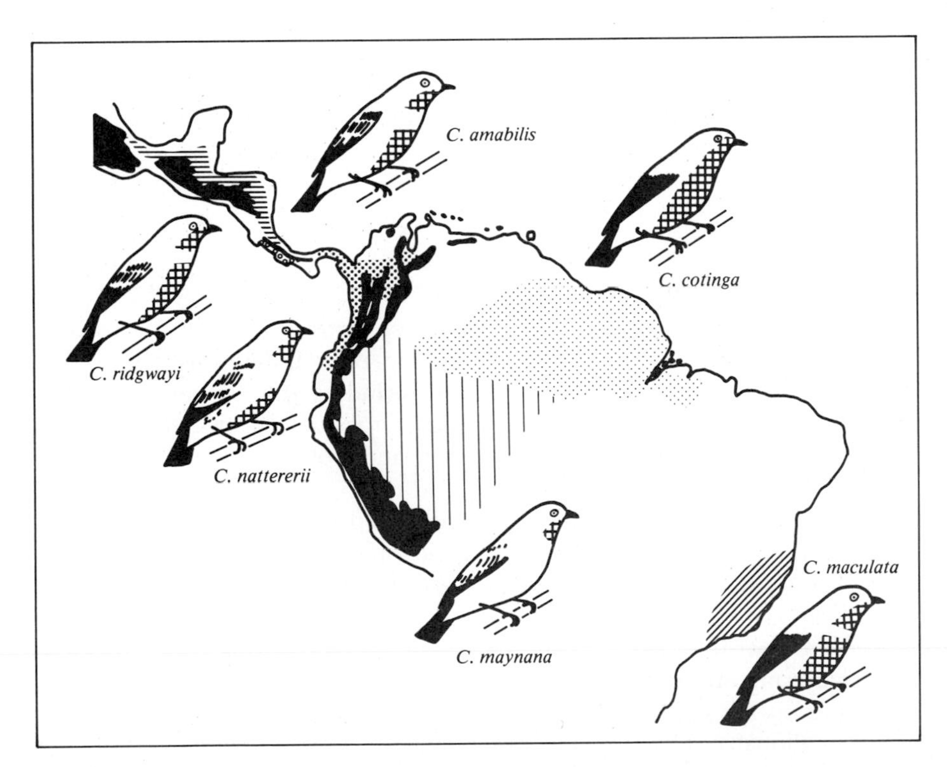

Figure 12.5
The geographic distribution of the blue cotingas *(Cotinga)* in the Neotropics. Each of the six species of these fruit-eating birds occupies a geographic range that is essentially allopatric, probably reflecting speciation when forest habitats were fragmented during the Pleistocene. (After Haffer, 1974.)

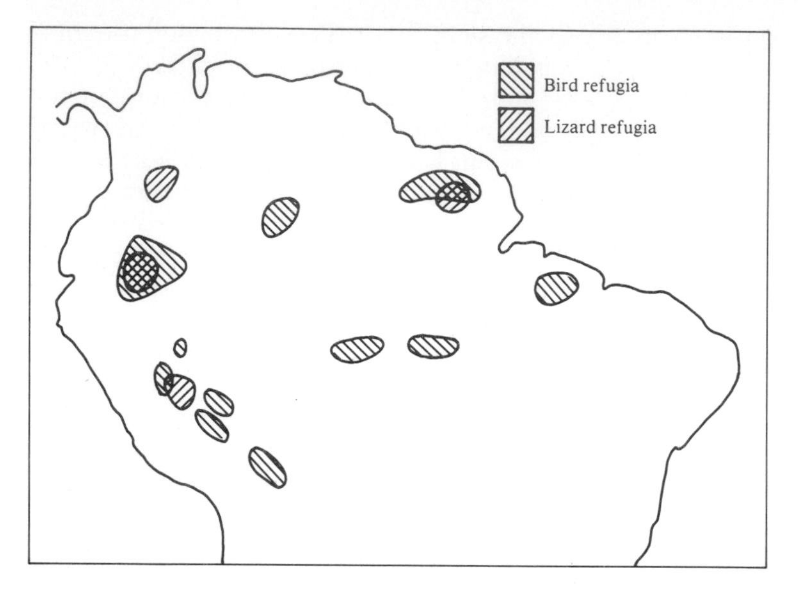

Figure 12.6
Centers of endemism for bird and lizard species in Amazonia. These centers are interpreted as marking the location of forest refugia, where isolated populations differentiated during the Pleistocene. Note the correspondence between the patterns for birds and lizards, which suggests that both groups were affected by the same historical biogeographic events. (After Haffer, 1969, and Vanzolini and Williams, 1970.)

dry climatic phases parcels of rain forest were isolated by more xeric savanna and shrub habitats. Speciation during isolation would have occurred in many groups within these pockets of rain forest. Then, with the reestablishment of a continuous lowland forest since the Pleistocene, bird populations would have expanded from those centers. By plotting the distributions of species in a superspecies or subspecies of a species occurring in the Amazon, Haffer has shown that there are centers of concentration of these taxa (Figure 12.6). Where populations from two centers now come in contact they either remain distinct or hybridize along the zone of secondary contact, depending on the extent to which they have diverged in allopatry. Haffer therefore interprets the high diversity in neotropical forests to recent subdivision of the vegetation type, which permitted multiplication of the original avian stocks. This phenomenon is also expressed in other groups of organisms (Chapter 14).

A number of families in mainland tropical America and the West Indies have pantropical distributions, including the nonpasserine cuckoos (Cuculidae), parrots (Psittacidae), barbets (Capitonidae), and trogons (Trogonidae) (Figure 12.7). The nearly cosmopolitan families of pigeons (Columbidae) and woodpeckers and piculets (Picidae) are also abundant in South America. Two groups, the trogons (38 species) and barbets (78 species) appear to have dispersed into the tropics through the Northern Hemisphere. Trogonidae are absent in Australasia and are most diverse in southeast Asia, especially Sumatra, and in tropical American highlands. The coppery-tailed trogon *(Trogon elegans)* enters the United States in the southeastern corner of Arizona. Fossils of Trogonidae appear from the Upper Eocene to the Mio-

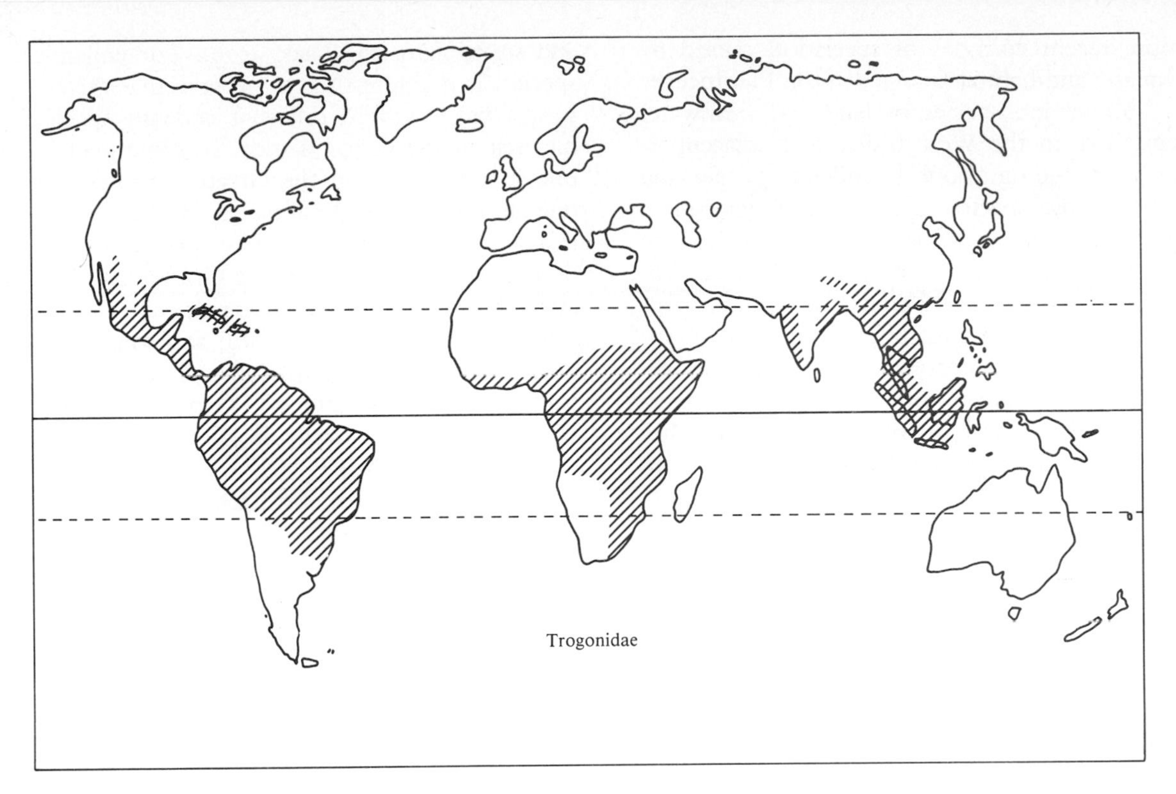

Figure 12.7
Pantropical distribution of the avian family Trogonidae. This pattern is typical of the wide distributions of several tropical bird families. At least in the case of the trogons, fossil evidence shows that this group of fruit-eating birds occurred well up into the Northern Hemisphere in the mid-Tertiary when climates probably were more equable.

cene of France, a time when northern climates were more tropical. Barbets have a similar extant range but with greatest diversity in Africa and southeast Asia. In contrast, Cuculidae, Psittacidae, and Columbidae are well represented in Australasia, even by endemic tribes or subfamilies, as well as other tropical regions, and Cracraft (1973) has suggested that these three families may be vicariants of Gondwanaland. So much phylogenetic work needs to be done on these families that any interpretations remain very tenuous.

As alluded to in these few introductory remarks, the avifauna of South America appear to have had a long and complex history. Some of the taxa have southern continent affinities, suggesting that their ancestors were either shared with or derived from Africa when the two continents were still in close proximity. Many of the endemic families fall into this category, and they have radiated during the geographic isolation of South America. Other families, especially those of oscines, have strong historical ties with North America, from which most of the South American lineages apparently have been derived. The diversity of many groups, such as the hummingbirds (Trochilidae, over 300 species), is undoubtedly related to the origin of diverse habitats forming with the rise of the Andean Cordillera in the Miocene and to much more recent episodes of speciation caused by climatic and habitat changes in the Pleistocene.

About 300 species of land and freshwater birds live in the West Indies and adjacent islands, including about 31 endemic genera and 125 endemic species. This number may seem surprising because some of the islands are very close to points on the mainland: Florida, Yucatan, Honduras, and Venezuela. Endemic genera are most numerous in hummingbirds (Trochilidae), wood warblers (Parulidae), mockingbirds (Miminae of Turdidae), tanagers (Thraupidae), and sparrows and bullfinches (Emberizinae of Emberizidae). Eight genera are endemic to Cuba (Garrido and Montaña, 1975) and six to Jamaica (Lack, 1976). For endemic species, it is interesting to observe that certain genera have evolved different endemic species on each of the major islands. In some genera, one species occurs in the Greater Antilles and one in the Lesser Antilles, as in the pewees (*Contopus*); some have separate species on each of several large islands in the Greater Antilles, as in the emerald hummingbird genus *Chlorostilbon* or the todies (Todidae); and some have a number of species restricted to individual islands in both the Greater and Lesser Antilles, as shown most dramatically in the parrot genus *Amazona* (Figure 12.8).

The fine treatment by Lack (1976) on bio-

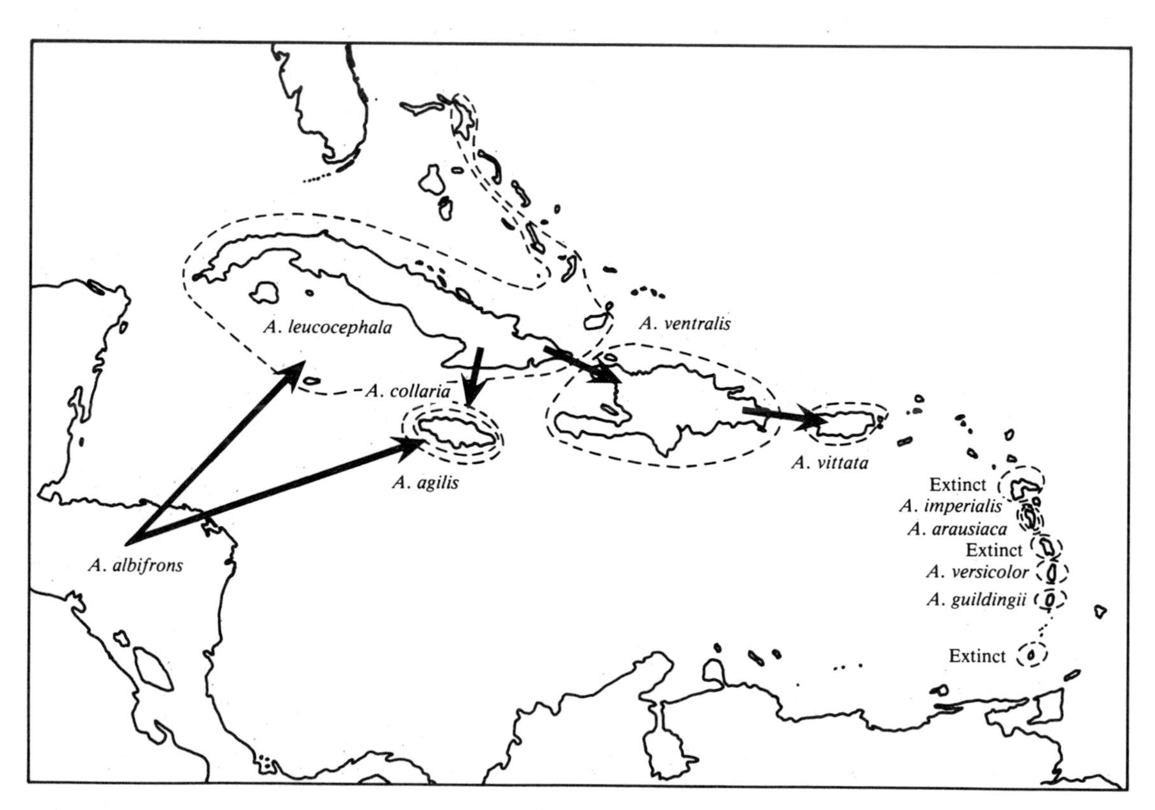

Figure 12.8
Distribution of the parrot genus *Amazona* in the West Indies, showing the hypothesized colonization events giving rise to the endemic species on different islands of the Greater Antilles. (After D. Lack, 1976. *Island biology illustrated by the land birds of Jamaica.* Studies in Ecology 3. University of Calfornia Press.)

geographic patterns of West Indian birds is a model study for those who seek to interpret historical biogeography from a dispersalist and ecological viewpoint. The resident land birds of the Greater Antilles share strong similarities with the avifauna of tropical North America (southern Mexico to Panama). Of the 66 species of land residents on Jamaica, 25 species also occur on the mainland nearby and at least 20 others have their closest relatives (presumably once the ancestors) in Central America, supporting a conclusion that most Jamaican birds could have colonized from the closest mainland sources. Lack discussed a small list of species that could have been derived from South America.

Looking at the patterns within the Greater Antilles, one sees that only half of the species on Jamaica occur on nearby Cuba and Hispaniola. There are many differences in the avifaunas of these three large and close islands even though they all have about the same number of resident species. A fairly elaborate dispersal type of explanation can be produced to account for these differences. For example, in Figure 12.8 the arrows illustrate an interpretation that a Honduran parrot (aff. *Amazona albifrons*) spread to Cuba to form *A. leucocephala,* which in turn colonized Jamaica to become *A. collaria* and colonized Hispaniola to become *A. ventralis*. *A. ventralis* gave rise to the diminutive *A. vittata* on Puerto Rico. Meanwhile, a separate colonization of Jamaica from the Honduran population produced *A. agilis,* leaving two species on that island. This scenario is based on analysis of only external features, and a well-documented phylogenetic scheme would be preferred. It is also possible that the speciation events just summarized could have been produced by a vicariance model (land connections); however, the dispersal scenario is appealing for Jamaica, which was completely submerged for a period prior to 15 million years BP (Robinson and Lewis, 1971), long after the islands of the Greater Antilles had achieved their present positions (Chapter 5).

Birds of Africa and Eurasia. For convenience we will consider here the gross patterns of Africa and Eurasia together. This includes three zoogeographic regions, the Palearctic, Oriental, and Ethiopian, which are linked because many groups of temperate birds migrate into the tropical zones of the other two regions.

Because temperate birds migrate to the tropics, it is not surprising that temperate Eurasia has no endemic families. Because birds tend to migrate north and south, it is also understandable that many groups that are abundant in Eurasia are essentially absent from the New World. Old World families barely represented in North America include the Old World warblers (Sylviidae; 6 of 339 species), nuthatches (Sittidae; 3 of 21 species), penduline tits (Remizidae; 1 of 9 species), long-tailed tits (Aegithalidae; 1 of 7 species), wren-tits (Timaliidae; 1 of 252 species), dippers (Cinclidae; 1 of 4 species), shrikes (Laniidae; 2 of 70 species), and larks (Alaudidae; 2 of 76 species). The North American representatives are often more closely related to species in eastern Asia than in Europe. There are also several large New World groups, such as the tyrant flycatchers (Tyrannidae), wood warblers (Parulidae), and emberizine finches (Emberizinae of Emberizidae), which are absent or greatly reduced in Eurasia. Given that different groups are dominant in the New World and in Eurasia, one can conclude that there has been only limited exchange of bird groups since the origin of early passerine clades and what little exchange did occur must have passed primarily through Beringia.

The tropical forests of southeast Asia have a rich avifauna; but unlike other tropical areas the Oriental region has only one endemic bird family, the leafbirds (Irenidae, 14 spp.). Other Oriental groups extend to Australasia or into the western Pacific basin (frogmouths, Podargidae; flowerpeckers, Dicaeidae; and crested swifts, Hemiprocnidae), and the parrotbills (Paradoxornithinae of Muscicapidae) range northward into temperate Asia. However, by far the strongest taxonomic affinities are with Africa.

Some interesting families shared with tropical Africa are the sunbirds (Nectariniidae), bulbils (Pycnonotidae), bee-eaters (Meropidae), honey-guides (Indicatoridae), shrikes (Laniidae), hornbills (Bucerotidae), and broadbills (Eurylaimidae) as well as closely related species or genera of raptors (Accipitridae and Falconidae), the pantropical groups discussed earlier (Columbidae, Cuculidae, Psittacidae, and Capitonidae), and the paleotropical waxbills (Estrildidae). Many groups of ground birds, e.g., pheasants (Phasianidae), have radiated in southeast Asia. A great number of bird species migrate between the Orient and temperate Asia.

That mediterranean Africa is part of the Palearctic region is amply demonstrated by comparing the avifauna of Morocco with that of southern Spain directly across the Straits of Gibraltar. Of the 132 species of land birds in the Atlas Mountains, all but 13 occur in southern Spain, and only 8 species that breed in southern Spain do not occur also in Morocco (Moreau, 1966). Therefore at the species level the short water gap between them would not appear to be very restrictive; but it turns out that 26 of the nondesert passerines of Morocco are represented by subspecies distinguishable from the Spanish populations.

The Sahara Desert is the northern boundary of the Ethiopian region. It includes about 6 million square kilometers of barren landscape that receives less than 10 cm of annual rainfall. Seventy species are listed from the desert area, but many of these are residents of the oases. The true deserticolous birds number only 25 species, including a widespread falcon *(Falco concolor)*, 5 species of ground birds, and 19 species of passerines. All desert species are wide ranging and lack any distinctive subspecies within the desert; some of these range all the way to India, Central Asia, and the Gobi Desert (Moreau, 1966). Although the latter distributions show Palearctic exchange, other species are more closely related to groups south of the Sahara. In Lower Egypt, for example, the avifauna is most diverse along the Nile and consists of 35 species of Palearctic birds, 22 species of groups typical of mainland Africa south of the Sahara, and 18 species that are widespread in both regions, especially water birds. Thus the Sahara appears to be a zone of transition for birds, a blending of dry land and water birds as well as predators and scavengers from the north and the south.

Moreau recognizes about 1400 species of land and freshwater birds living south of the Sahara on mainland Africa, of which 900 species are passerines. The majority of nonpasserines and over half of the passerines live in nonforest habitats, unlike the situation in the Neotropics, where the greatest diversity occurs in the forests. Dominant oscine taxa are the starlings (Sturnidae), flycatchers (Muscicapidae), warblers (Sylviidae), sunbirds (Nectariniidae), goldfinches (Carduelini of Emberizidae), and waxbills and weavers (Ploceidae). The box on p. 361 shows the groups endemic or nearly so to this region. The families restricted to Africa are all small taxa, a striking contrast to some of the large neotropical families. In fact, when the avifaunas of the Congo and Amazonia are compared, there are relatively few taxonomic relationships at the family level, and these are the wide-ranging families (Amadon, 1973). Consequently, most authors favor the conclusion that Africa and South America have not exchanged bird taxa since the formation of the South Atlantic, at least not in recent times. This is best illustrated by the dominance of suboscines in South America and oscines in Africa. Ratites (Chapter 11) and several nonpasserine pantropical families (Psittacidae, Columbidae, and Cuculidae) are mainland groups that could have dispersed across West Gondwanaland when no major ocean barrier existed between them.

Madagascar's avifauna, which is strongly African in character, is noteworthy not for its richness but rather for being impoverished and unbalanced compared with that of the mainland. This huge island has only 129 breeding land bird species, somewhat low for a tropical

Bird families and subfamilies endemic or nearly so to mainland Africa and Madagascar (M)

Nonpasserines

Struthionidae (ostrich; 1 sp.)
Aepyornithidae (elephant birds; extinct in recent times)
Balaenicipitidae (whale-headed stork; 1 sp.)
Scopidae (hammerhead stork; 1 sp.)
Sagittariidae (secretary bird; 1 sp.)
Mesitornithidae (Madagascar rail; 1 sp.) (M)
Musophagidae (turacos; 22 spp.)
Coliidae (colies and mousebirds; 6 spp.)
Indicatoridae (honeyguides; 14 spp.)*
Leptosomatidae (courol; 1 sp.) (M)
Brachypteraciidae (ground roller; 1 sp.) (M)
Atelorninae of Coraciidae (ground rollers; 2 spp.)
Phoeniculidae (wood hoopoes; 8 spp.)

Passerines

Philepittidae (asities; 4 spp.) (M)
Malaconotinae of Laniidae (bush shrikes; 32 spp.)
Vangidae (vanga-shrikes; 12 spp.) (M)
Hypositinae of Salpornithidae (Madagascar nuthatch; 1 sp.) (M)
Ploceidae (weavers and parasitic viduines; 150 spp.)*

*Some species occur in Eurasia.

island of its size (Chapter 15), but of these over two thirds of the species and one third of the genera are endemic, greater than for Africa itself. Endemic families include the rail-like Mesitornithidae, the suboscine Philepittidae, the vanga-shrikes (Vangidae), the courol (Leptosomatidae), and a close relative of the rollers of Africa, Brachypteraciidae.

Considering the short distance across the Mozambique channel to the mainland, it is remarkable that the island lacks such Ethiopian birds as pipits, barbets, hornbills, turacos, orioles, cranes, colies, wood hoopoes, finches, shrikes, and helmet-shrikes, even though apparently suitable habitats are abundant on Madagascar. In the tropical forest there is a poor complement of fruit eaters and seed eaters, and eagles and vultures are nearly absent. The typical genus has only one to several species (average of 1.7 species per genus), and the largest genus, *Coua,* has only 10 species. The most interesting case of adaptive radiation on Madagascar is that of the 12 species of vanga-shrikes, whose differences are so great that ornithologists classify them as 10 distinct genera! However, in general most Madagascan species are not greatly different from their close relatives in eastern Africa, although on the island they tend to have broader niches and to occupy a wider range of habitats. Authors have wondered whether the low species numbers for the avifauna and for each genus is a longstanding historical pattern or the result of recent extinctions. One clear case of recent extinction is the family of elephant birds (Aepyornithidae), an apparent Gondwanaland relict (Chapter 11) and the largest of all known birds, which were probably hunted to extinction by humans. However, it is much less likely that humans eliminated the many kinds of small flying birds that are conspicuous by their absence on Madagascar, so some climatic cause would have to be hypothesized.

The easiest way to account for the flying birds of Madagascar is to treat most of them as colonists that arrived from the African mainland over the many millions of years since separation. However, such an explanation may be too simplistic. For example, the occurrence of the endemic suboscine Philepittidae suggests that the ancestors of all suboscines were present across West Gondwanaland; Philepittidae has the primitive features for the entire group (Cracraft, 1981), so its disjunction with the other families in the Neotropics suggests that there was exchange between Africa and South America. If this was the case, they have become extinct on the mainland. Whether parrots, cuckoos, pigeons, and suboscines had to have a solid land bridge to disperse between the two continents or whether they could have crossed a narrow South Atlantic Ocean cannot be answered directly, but it is likely that some of these clades were part of the Cretaceous biota across the landmass of West Gondwanaland.

Birds of Australasia. Turning to Australasia, we once again encounter peculiar elements in the biota, this time in an avifauna of about 1600 species of which more than 1000 species are endemic to the Australian Plate and several of the neighboring islands, e.g., the Moluccas, Solomons, Bismarcks, and Sundas. Relatively few species migrate annually to Australasia because birds of temperate eastern Asia winter mostly in tropical southeast Asia, but 37 Palearctic migrants, mostly shorebirds, are recorded from New Guinea (Rand and Gilliard, 1967). Conversely, very few Australasian land birds migrate northward to Asia in the winter (Figure 12.4), so its avifauna is still geographically isolated, as it has been for millions of years.

Taxa that are restricted or nearly so to Australasia are shown in the following box. Always paramount in discussions are the flightless ratites, the emus, cassowaries, kiwis, and extinct moas, which signify for most biogeographers that certain groups of birds arrived in Australia and New Zealand via a southern land connection (Chapter 11). If land bridges provided opportunities for migration of ratites, then other old groups may have also arrived in a similar fashion. One candidate is the family of megapodes, whose closest relatives appear to be the Opisthocomidae and Cracidae of South America (Figure 12.9). Other nonpasserine groups using Gondwanaland connections could have been some of the pantropical families mentioned earlier, especially if their sister groups are in South America. The kagu (Rhynochetidae) of New Caledonia seems to be closely related (Figure 12.10) to the South American trumpeters (Psophiidae), seriemas (Cariami-

Bird families and subfamilies endemic or nearly so to Australasia

Nonpasserines

Casuariidae (cassowaries; 3 spp.)
Dromaiidae (emu; 1 sp.)
Apterygidae (kiwis; 3 spp.)
Megapodiidae (megapodes; 12 spp.)*
Pedionomidae (plains wanderer; 1 sp.)
Rhynochetidae (kagu; 1 sp.)
Aegothelidae (owlet-nightjars; 7 spp.)
Gourinae of Columbidae (crowned pigeons; 3 spp.)
Loriidae (lories and lorikeets; 52 spp.)
Cacatuidae (cockatoos; 18 spp.)
Micropsittinae of Psittacidae (pygmy parrots; 6 spp.)
Nestorinae of Psittacidae (kakas and keas; 3 spp.)
Strigopinae of Psittacidae (kakapo; 1 sp.)

Passerines

Acanthisittidae (New Zealand wrens; 3 spp.)
Malurinae of Muscicapidae (wren warblers; 29 spp.)
Orthonychinae of Muscicapidae (logrunners; 20 spp.)*
Acanthizinae of Muscicapidae (Australian warblers; 59 spp.)*
Pachycephalinae of Muscicapidae (whistlers and shrike thrushes; 48 spp.)*
Epthianurinae of Muscicapidae (Australian chats; 5 spp.)
Meliphagidae (honeyeaters; 169 spp.)*
Menuridae (lyrebirds; 2 spp.)
Atrichornithidae (scrubbirds; 2 spp.)
Climacteridae (Australian tree creepers; 6 spp.)
Callaeidae (wattlebirds; 3 spp.)
Grallinidae (magpie larks; 4 spp.)
Artamidae (wood swallows; 10 spp.)*
Craticidae (bell magpies; 4 spp.)
Ptilonorhynchidae (bowerbirds; 19 spp.)
Paradisaeidae (birds of paradise; 40 spp.)

*At least one species occurs in mainland southeast Asia.

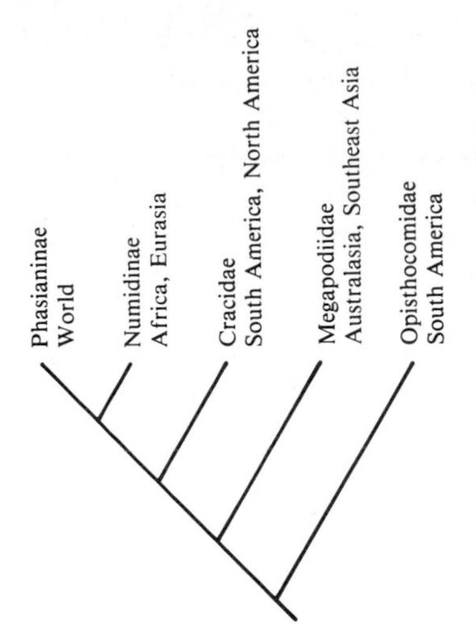

Figure 12.9
Hypothesized phylogenetic relationships of the Australian mound nest–building megapodes and other groups of gallinaceous birds on other southern continents. Although these birds might have been distributed via Gondwanaland routes, other explanations are possible. (After Cracraft, 1980.)

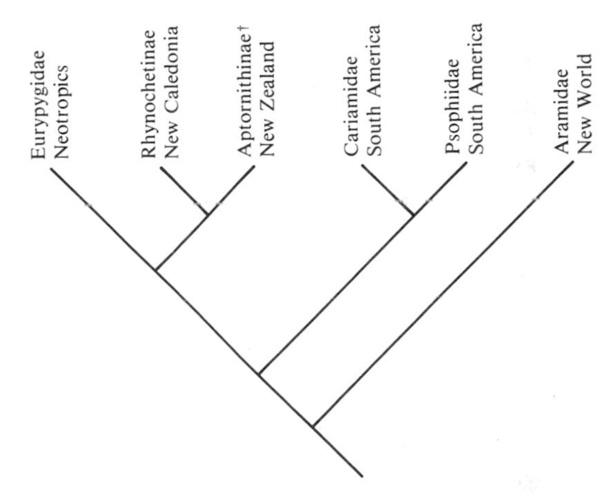

Figure 12.10
Reconstructed phylogenetic relationships of the kagu (Rhynochetidae) to other bird families of the Southern Hemisphere. The kagu, a strange, heronlike bird endemic to New Caledonia, has long perplexed systematists and biogeographers trying to reconstruct its history. Here Cracraft (1981) suggests its closest relatives are South American groups, suggesting a vicariant pattern dating back to the fragmentation of Gondwanaland.

dae), and sunbittern (Eurypygidae), and possibly to the recently extinct *Aptornis* of New Zealand (Cracraft, 1981).

The remaining families, and in particular the oscines, show the strongest phylogenetic relationships with the Oriental region. For example, the muscicapids and meliphagids have experienced great adaptive radiations on the Australian Plate (Keast, 1977c) following their arrival from some Asian source. Another interesting radiation has been the crowlike birds, including the bell magpies, bowerbirds, lyrebirds, birds of paradise, magpie larks, and wattlebirds (Cracraft, 1981). Much research is needed to determine when these taxa may have arrived in Australasia and to evaluate whether Wallacea (the continental islands of southeast Asia) or India served as stepping-stones between Asia and Australia when the two continents were widely separated in the Cretaceous and early Tertiary. It is hoped, of course, that some fossil evidence will be found to help in these reconstructions. Unfortunately, the fossil record of birds is very poor from the Cretaceous and Lower Tertiary (Brodkorb, 1971b; Feduccia, 1980), and interpreting these fossils is beset with many problems because assigning these to extant families is troublesome and because many families have fossils occurring outside their present ranges.

Bats

Bats (order Chiroptera) live in practically every habitable land area on earth; in fact, the family Vespertilionidae is cosmopolitan (Figure 12.11), occurring from the Arctic Circle to the tips of the southern landmasses and even on such remote islands as the Hawaiian Archipelago and the Azores. Many species of high latitudes have been banded and recaptured to show that individuals migrate long distances, either between breeding roosts and winter hibernacula or from subarctic and temperate habitats to more tropical regions in nonhibernating species. One of the documented cases of long individual migration occurs in the insectivorous guano bat *(Tadarida brasiliensis)*, which makes seasonal movements between southern Colorado and Mazatlan in Sinaloa, Mexico, a distance of more than 2000 km (Cockrum, 1969). Seasonal migration of nectar-feeding bats such as *Leptonycteris* between southern Arizona and warmer areas of Mexico follow the seasonal flowering patterns of specialized bat-pollinated plants, which do not occur farther north than southern Arizona. Some bats of wet tropical re-

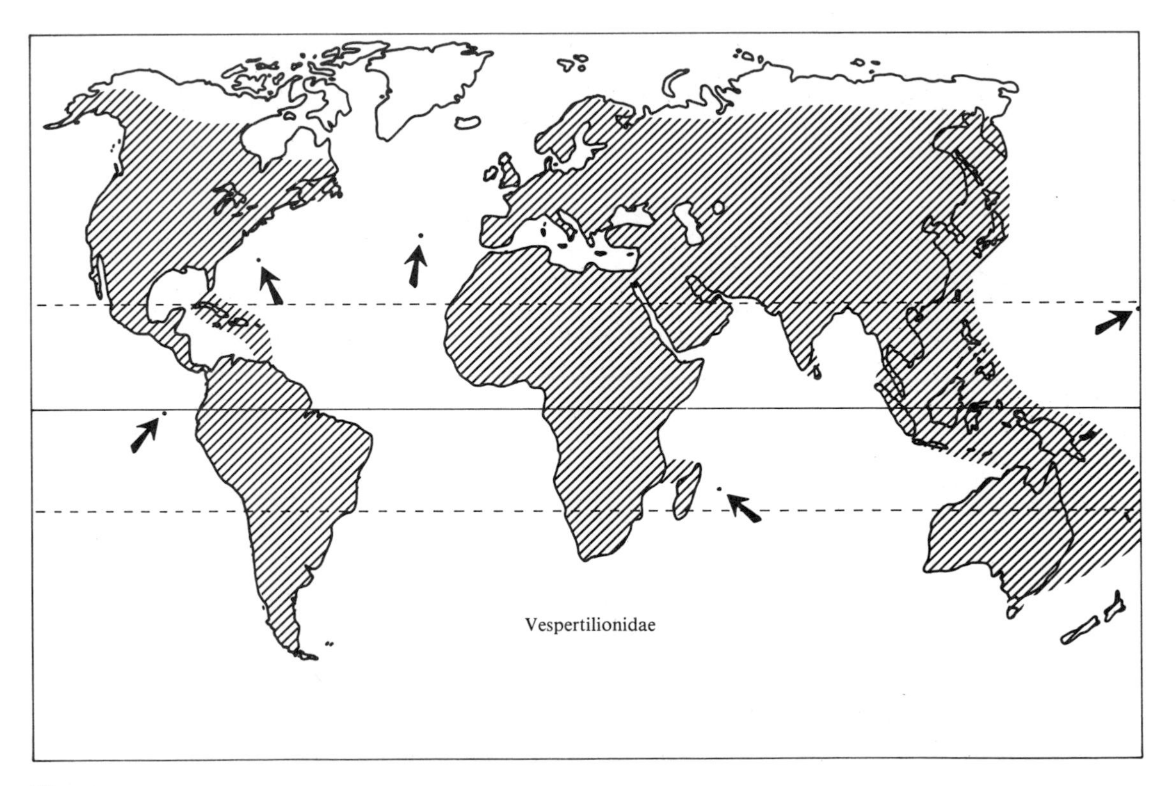

Figure 12.11
Distribution of the cosmopolitan bat family Vespertilionidae. Several other groups of flying animals have similarly wide distributions. Note, however, that this distribution is not truly worldwide; these insectivorous bats are absent from some oceanic islands and from perennially cold regions. (After Koopman and Jones, 1970.)

gions appear to lack migratory behavior, but there are those that do migrate; a great deal of research is needed to understand the distributions of these bats.

Chiropterologists recognize two suborders of bats, which could have arisen as separate evolutionary events. The Megachiroptera (150 species) include a single large family (Pteropodidae) of fruit-eating bats and flying foxes of the Old World and the western Pacific. The suborder Microchiroptera consists of 17 extant families (700 species), seven of which have only one or two species. Phylogenetic relationships within and between these families are poorly known and are complicated by parallel and convergent evolution of many features. The oldest bat fossils, which have features intermediate between the Megachiroptera and Microchiroptera, have been found in Eocene beds of Europe and North America, when they had equable subtropical to tropical climates. Because most bats are essentially tropical, except for members of three highly derived families, a tropical origin for bats is suspected.

The Megachiroptera are often quite large and strong flyers. Their distributions today fol-

Present distributions of extant bat families in suborder Microchiroptera
(Mostly after Koopman, 1970; Koopman and Jones, 1970.)

SUPERFAMILY EMBALLONUROIDEA

Rhinopomatidae (mouse-tailed bats; 2 spp.); northern Africa east through southern Asia to Sumatra
Emballonuridae (sheath-tailed bats and ghost bats; 44 spp.); pantropical, including Arabia and western Pacific islands to Samoa
Craseonycteridae (hog-nosed bat; 1 sp.); Thailand

SUPERFAMILY RHINOLOPHOIDEA

Nycteridae (hispid bats; 13 spp.); Africa south of the Sahara, Madagascar, and Malesia
Megadermatidae (false-vampire bats and big-eared bats; 5 spp.); Old World tropics
Rhinolophidae (horseshoe bats; 130 spp.); Old World except temperate Asia, New Zealand, and south and western Australia

SUPERFAMILY PHYLLOSTOMOIDEA

Noctilionidae (bulldog bats; 2 spp.); Neotropics (including the West Indies) to Sinaloa, Mexico
Mormoopidae (ghost-faced and mustached bats; 8 spp.); Neotropics to southern Texas and Cuba
Phyllostomidae (leaf-nosed bats; 150 spp.); Neotropics to southwestern United States

SUPERFAMILY VESPERTILIONOIDEA

Natalidae (funnel-eared bats; 4 spp.); Neotropics
Furipteridae (smoky bats; 2 spp.); northern South America, Panama, and Trinidad
Thyropteridae (disc-winged bats; 2 spp.); Neotropics
Myzopodidae (sucker-footed bat; 1 sp.); Madagascar
Vespertilionidae (common or vespertilionid bats; 280 spp.); cosmopolitan
Mystacinidae (short-tailed bat; 1 sp.); New Zealand
Molossidae (free-tailed bats; 82 spp.); nearly cosmopolitan but missing much of the Holarctic

low the occurrence of Old World tropical forests, which produce the fleshy fruits for their diets. These bats appear to have spread from southeast Asia to Australia and on to Samoa, and pteropodids have colonized nearly every forested island in the southwestern Pacific basin. This suggests that they would do well in the Neotropics; their restriction to the Eastern Hemisphere favors their origin and differentiation in the Old World much after South America separated from Africa. Holloway and Jardine (1968) have made statistical analyses of bat faunas in Indo-Malaysia and Melanesia that suggest, not surprisingly, that Australasian bats arrived from the Orient in more than one invasion.

The principal foods of Microchroptera are flying insects. Exceptions include some predatory species that feed on vertebrates (Megadermatidae and Phyllostomidae) and nectar and fruit-eaters (Phyllostomatidae). Therefore it is not at all surprising that the insectivorous Microchiroptera and the frugivorous Megachiroptera coexist in the Old World by using different food resources nor is it surprising that Microchiroptera adapted to frugivory independently in the Neotropics.

Koopman (1970; Koopman and Jones, 1970) recognized four superfamilies of microchiropteran bats (see box on p. 365). The superfamily Rhinolophoidea is Old World, Phyllostomoidae is New World, Emballonuroidea is divided between the Old and New World with most species in the Old World, and Vespertilionoidea includes two widespread families, three endemic in the Neotropics, one endemic in Madagascar, and one endemic in New Zealand. Much phylogenetic work is needed before these disjunctions can be properly interpreted.

Koopman's (1968, 1975) studies of Antillean bats serve to illustrate a common biotic pattern in the West Indies. Bats of the southernmost Antilles are those that have invaded from South America, and as one travels northward through the Lesser Antilles, southern species drop out. Groups endemic to the Greater Antilles or of northern origin invade southward along the Lesser Antilles and also drop out along the chain. Hence location within a chain of islands can have a major influence on the affinities of the fauna. Terborgh (1981) (Figure 12.12) has graphically shown this countercurrent pattern for breeding land birds of the Lesser Antilles and the Virgin Islands (part of the Greater Antilles).

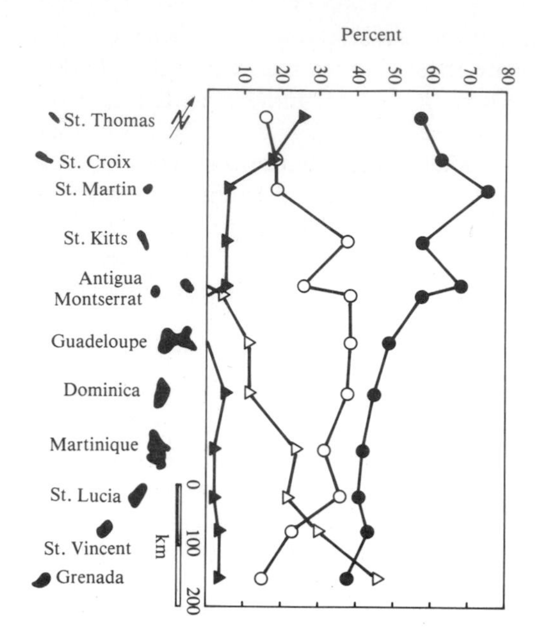

Figure 12.12
Countercurrent distributions of resident land birds of the Lesser Antilles and Virgin Islands, showing the decrease of South American species and the increase of species from the Greater Antilles from south to north along the island chain. Widespread and endemic species show no such pattern. (After Terborgh, 1981.)

Some patterns of insects

Class Insecta has more known living species than all other plants and animals combined. Some entomologists predict the current figure (800,000 species) could be doubled by investigating thoroughly just the insects of South America. Faunas of other regions are less diverse and better collected and monographed. Most biogeography books give brief treatment to insects because of the overwhelming task of merely classifying these beasts and mapping their distributions. There has as yet been little attempt to synthesize all the existing information to describe general insect patterns and to suggest the processes that produce them. Even the well-collected, photogenic butterflies (superfamily Papilionoidea, Lepidoptera), consisting of about 10,000 species, have not been thoroughly analyzed on a global scale. Within the five butterfly families, investigators scarcely understand the relationships of the genera, never mind the tribes and families. Phylogenetic models are greatly complicated in butterflies, as in many insect orders, by parallelism and convergence, especially mimicry (Turner, 1981). The magnitude of the problem is underscored by the fact that 25 unrelated species can mimic a single unpalatable model in neotropical forests.

Because of the difficulties of treating insects as a whole, we focus here on three interesting examples that illustrate some of the patterns observed in insects: the Australian insect fauna, which provides insights into the long history of a diverse continental biota; the biogeographic history of the chironomid midges, organisms that show clear relationships around the southern end of the world; and the genus *Drosophila,* which permits a more detailed view of the Tertiary history of diversification within a monophyletic group.

The insects of Australia. There are 55,000 species of insects presently known from Australia. These represent a fair cross section of the insect orders and belong to 560 families (C.S.I.R.O., 1970). Almost one third of these species are in huge cosmopolitan families that each have more than 1000 species in Australia. In this way only, the insect fauna of Australia is fairly typical. In other ways insect distributions of this continent illuminate the historical events shaping the fauna.

First, Australia is rich with endemics at all taxonomic levels from species to families. Perhaps 80% of all species are restricted to Australia and its neighboring islands, often narrowly so, and in many distinctive Australian groups generic endemism is also very high. An example is the subfamily Melolontinae of scarabaenid beetles (Scarabaenidae, Coleoptera), in which only 2 genera of 80 have ranges outside Australia proper. Listing endemic families and subfamilies would make a lengthy table. Such broad-scale endemism, in the majority of orders, indicates once again that Australia has experienced a long period of isolation sufficient to promote major genetic changes in the taxa present.

In Chapter 11 we discussed the marsupials of Australia and briefly mentioned that the group has produced a radiation of designs to fill many ecological niches normally occupied by placental families on other continents. Similar adaptive radiation and convergent evolution characterizes many groups of Australian insects. Several examples from the order Hymenoptera will suffice. The most common groups of bees in Eurasia are mostly absent from Australia. Instead, the most primitive family of bees, Colletidae (850 species in Australia) has radiated to fill almost all the typical bee niches. Gall-forming, a niche normally dominated by Cynipidae in Eurasia, has been filled by the Chalcidoidea (2800 species) in Australia. Lastly, many of the leaf-mining and leaf-feeding niches have been assumed in Australia by the Pergidae (136 species) instead of a number of families in Eurasia.

Groups like Pegidae have another similarity with the marsupial example, because their sister groups are both in South America. In order after order, one can find examples of interesting endemic Australian insect genera or species groups whose closest relatives reside in southern South America, New Zealand, or both. In addition to the mostly aquatic mayflies (Ephemeroptera, Chapter 10) and the chironomid midges and their relatives (Nematocera, following section), some other examples include the

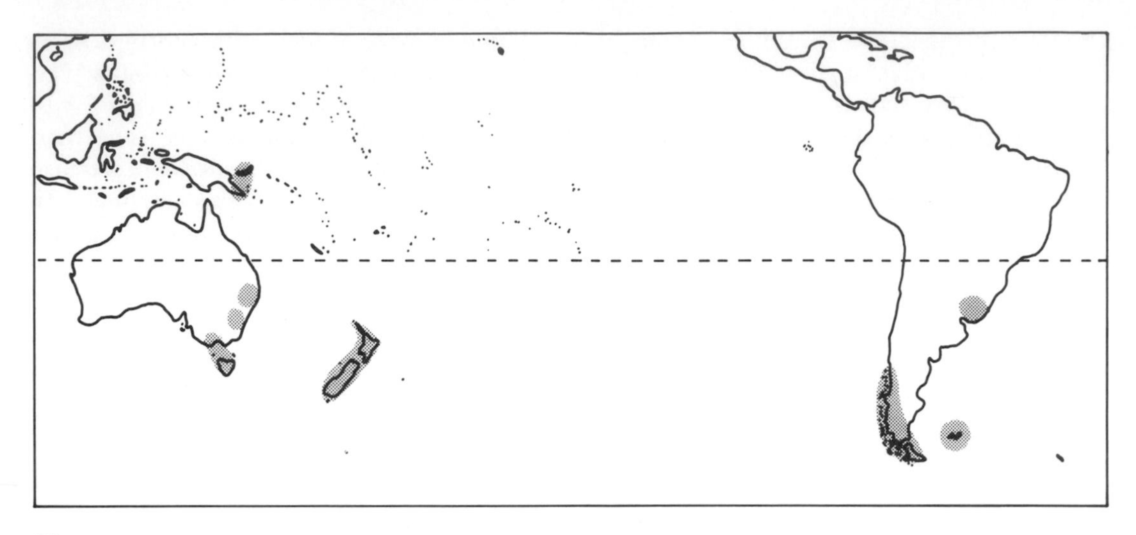

Figure 12.13
Disjunct distribution of south temperate beetles of the tribe Migadopini of the family Carabidae. This and similar patterns in other insect groups suggests an ancestral form occurring on Gondwanaland, but alternative explanations based on long-distance dispersal are tenable. (After Darlington, 1965.)

stoneflies (Plecoptera), with three of four Australian families shared with New Zealand and southern South America; Nannochoristidae (Mecoptera) and *Archichauliodes* and its relatives (Corydalidae, Megaloptera); Belidae (Coleoptera), with six genera in Australia, one in New Zealand, and two in Chile; Migadopinae (Figure 12.13), Broscinae, Psydrinae, and Trechinae (Carabidae, Coleoptera), with close relatives in New Zealand, Chile, and Patagonia (Darlington, 1965); and lacewings (Neuroptera) of Stilbopterygidae (Australia and South America) and the primitive subfamilies of Osmylidae (Australia, New Zealand, and South America). In other words, a great many cool temperate groups have these southern continent temperate disjunctions. A smaller number of taxa, e.g., Corydalidae, Carabidae, and Meinertellidae, have relatives also in southern Africa and more distant places. In general, these south temperate distributions appear to reflect historical Gondwanaland connections through Antarctica, although alternative explanations (e.g., widespread dispersal but inability to cross the tropics) are possible for some groups.

The characteristic tropical families of the Oriental region, especially Malaysia, are primarily found in tropical and subtropical northern Australia and New Guinea. In many ways they appear to be recent arrivals, although some taxa have expanded southward to colonize the drier habitats of Australia.

As with the monotremes and marsupials, Australia is also the home of some endemic, primitive families for certain insect orders. Heading the list is *Mastotermes,* the most primitive living termite (Isoptera), surviving today only in Australia but known from Tertiary fossils in other parts of the world. Australia also has other primitive termites (Termopsidae), the most primitive family of bees (Colletidae), archaic dragonflies (zygopteran Odonata), and some fairly primitive Neuroptera, Lepidoptera, Hymenoptera. This evidence suggests that Australia had an extensive insect fauna while it was still part of Gondwanaland.

If we are interpreting this general record correctly, the same themes emerge as before with aquatic and terrestrial animals (Chapters 10 and 11): Australia began the Mesozoic with a Gondwanaland biota; continued to share elements, especially temperate ones, with South America until the Lower Eocene; and finally in the Neogene received major waves of new colonists by groups capable of long-distance dispersal over the short ocean gaps from southeast Asia. Additional temperate elements may have arrived via long-distance dispersal between Australia and southern Africa or South America. Finally, although the continents remained in isolation during the Tertiary, groups present from early Gondwanaland connections or early colonists were able to differentiate and assume varied ecological roles.

Chironomid midges. The aquatic larvae of chironomid midges (Chironomidae, Nematocera) are important freshwater predators on small invertebrates in temperate, slow moving streams and ponds in both the Northern and the Southern Hemisphere. These larvae, which live in free floating or attached tubes, may drift downstream with the currents, but the gnatlike adults can be strong fliers and return upstream. The midge life cycle is therefore reminiscent of the mayflies (Ephemeroptera; Chapter 10), although midges usually fly for longer periods. There are also some midges that live in highly saline habitats.

A broad and comprehensive study of chironomid midges by Brundin (1965, 1966, 1967) was the first attempt to use both cladistic and vicariance methods to quantify biogeographic patterns for a large taxon, and also for one well developed in the temperate Southern Hemisphere. Figure 12.14 shows the multiarea cladogram of the midges in the subfamilies Podonominae and Aphroteniinae as constructed by Brundin. What impresses us is that the genera in one southern landmass, e.g., in New Zealand, are not haphazardly related to taxa in all the other southern areas but are always most closely related to South American forms. Similarly, Australian taxa are most closely related also to South American forms and not to those in nearby New Zealand. A conclusion that explains this pattern is a vicariance one: a midge fauna in southern Gondwanaland was sequentially broken up and the phylogeny reflects the splits between each of the landmasses. This type of analysis not only reinforces our contention that Australia, New Zealand, Africa, and South America were all once connected with Antarctica but also it goes a step further to identify the barriers that promoted taxonomic diversification.

***Drosophila:* a case study.** More genetic research has been done on *Drosophila* (vinegar or fruit flies; Drosophilidae, Diptera) than any other eukaryotic animal. One tends to think of *Drosophila* in terms of a few species cited repeatedly in genetic papers, but in fact the genus is a very large one, is well understood phylogenetically, and illustrates well several small-scale biogeographic patterns and processes.

Any student who has taken a course in genetics knows that drosophilids have large polytene salivary chromosomes with conspicuous bands. By studying karyological changes in chromosome number, fusion of parts, positions of heterochromatin, and inversion of portions of the chromosomes, investigators have been able to piece together much of the evolutionary history of the various species groups and individual species. If the precise origin of one species from another can be known by tracing the evolution of unique chromosomal mutations, the phylogeny can be used to unravel the history of speciation through time.

Drosophila and several closely related genera in its subfamily include about 2000 known species. Figure 12.15 gives the current phylogenetic hypothesis for major species groups within this monophyletic taxon. Some of the radiations are small but the majority are extremely large. For example, the Hawaiian drosophilids probably include 600 to 700 species (Carson et al., 1970), although normally they are lumped under the headings *Scaptomyza* and "Hawaiian *Drosophila.*"

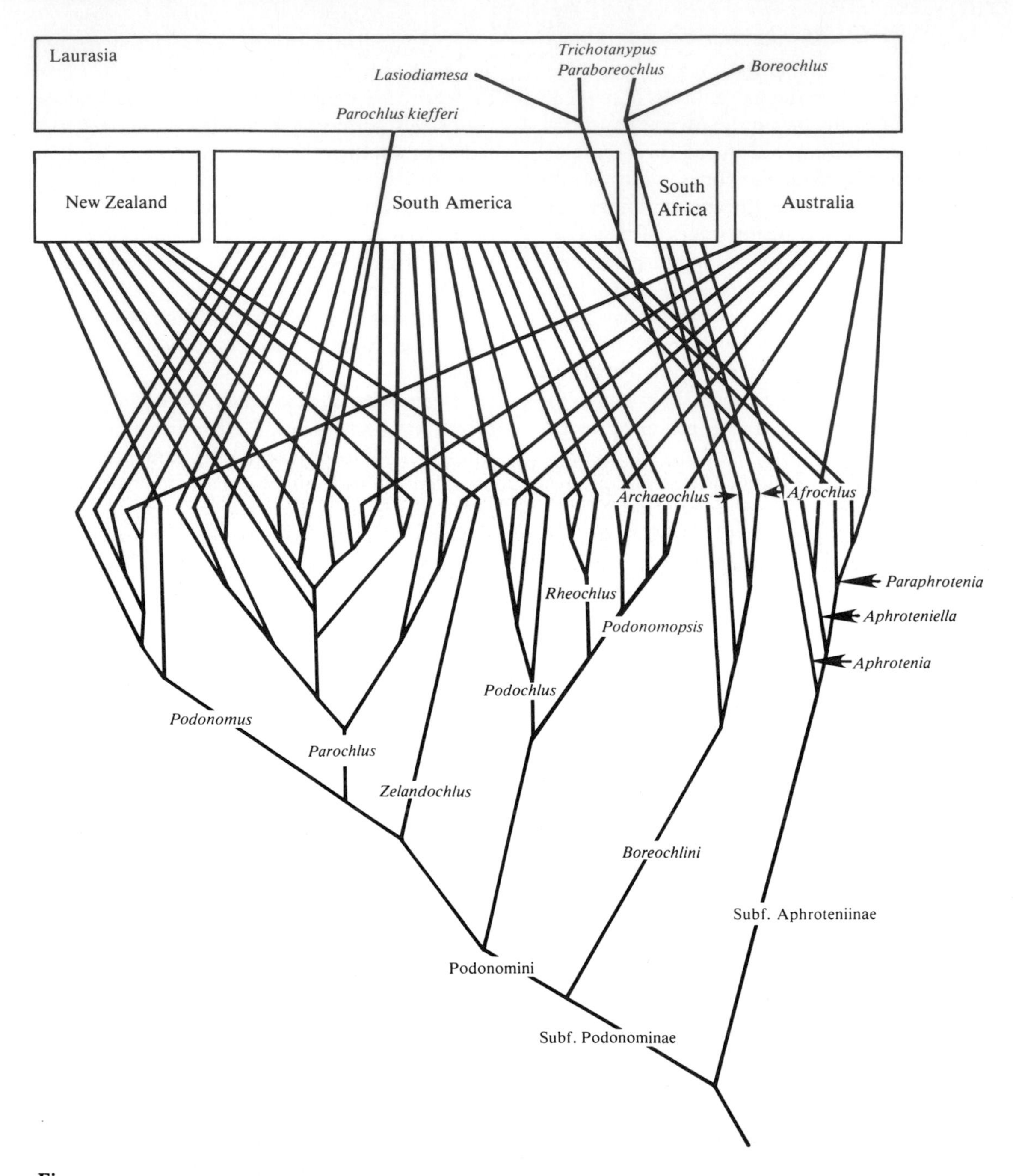

Figure 12.14
Multiarea cladogram for related groups of chironomid midges, Podonominae and Aphroteniinae, from the systematic and biogeographic analyses of Brundin (1965). This cladogram shows that representatives of different lineages usually have representatives on several southern landmasses but few taxa on the northern continents. This suggests a vicariant pattern dating back to the breakup of Gondwanaland.

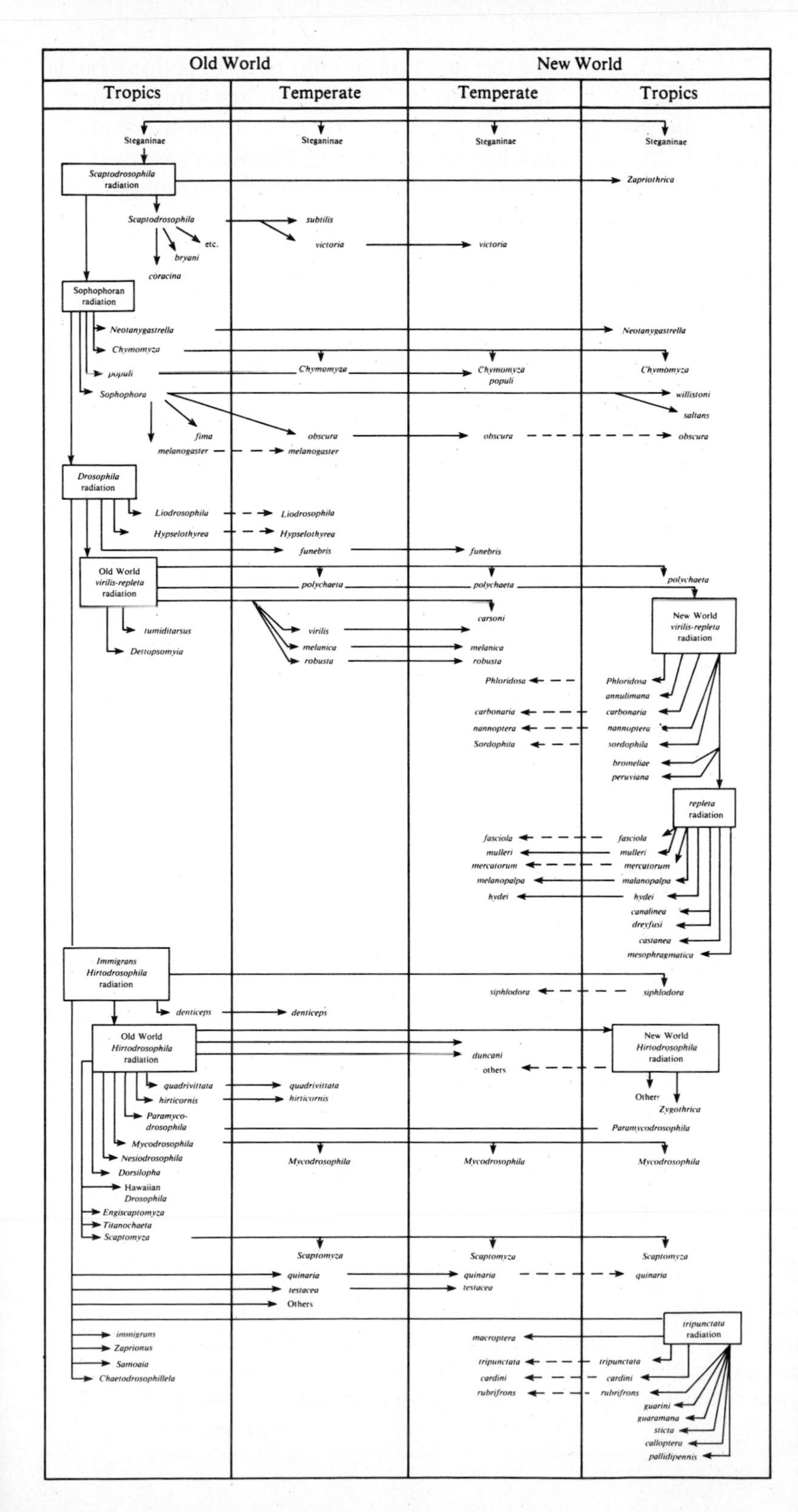

Figure 12.15
Hypothesized phylogeny of the fruit fly genus *Drosophila* and derived forms. The amazing radiation of this group involved repeated invasions of the New World from the Old World and repeated migrations between the tropics and temperate zones. (After Throckmorton, 1975.)

Of all the species groups, the picture-winged Hawaiian flies give the clearest story of speciation. As indicated in Chapter 5, the Hawaiian Islands have formed as a narrow chain from a hot spot in the mantle. Islands have emerged at the present locality of the hot spot for at least 80 million years. These extend from the Aleutian Trench (the Emperor Seamounts) southward to an elbow in the central Pacific and then southeastward to the youngest emergent island of Hawaii, less than 1 million years old and located above the hot spot. The hot spot has never been closer to North America than it is today, and its position relative to Asia has probably remained stable during the same period.

The ancestors of the Hawaiian drosophilids probably arrived originally on emergent islands that have since become submerged, so they would have dispersed to new islands as they formed at the hot spot. As deduced from the karyotypic phylogeny, Figure 12.16 shows the proposed dispersal events from one island to another accounting for new radiations on islands. Starting from Kauai (age 5 million years), a minimum of 22 interisland colonizations are needed to account for the picture-winged species on the Maui complex (including Maui, Molokai, and Lanai), Oahu, and Hawaii. The Maui complex of islands has been fused and separated at least twice in their short history by the lowering of sea levels, which adds another interesting dimension to the speciation picture. Remaining to be explained are the timing and mechanisms of the many additional speciation events occurring within each island and the ecological mechanisms that enable so many species to coexist, even using the same apparent food resource. For example, nearly 50 species of *Drosophila* (leaf-mining larvae) have been reared from leaves of *Cheirodendron* (Carson et al., 1970). Many species that coexist in a *Cheirodendron* forest, even within a single decaying leaf, are sister species that probably diverged while populations were sympatric.

Having a reliable phylogeny of *Drosophila*

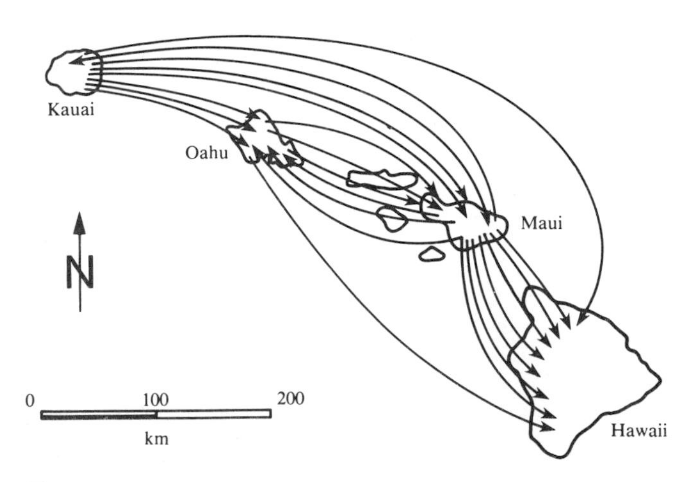

Figure 12.16
Hypothesized scenario to account for the phylogeny and biogeography of picture-winged Hawaiian *Drosophila*. Arrows show the minimum number of interisland dispersal events necessary to account for the distribution of the contemporary species. Note that most of the dispersal occurred from the older islands at the western end of the chain to the more newly formed eastern islands. (After Carson et al., 1970.)

permits workers to speculate not only on the geography of evolution but also on the evolution of food preferences and the mechanisms of speciation. Throckmorton (1975) suggested that the drosophilids probably began by exploiting slowly fermenting leaves and other fleshy plant structures on the floor of humid tropical forests in Asia. Although long-distance dispersal over oceans is theoretically possible (i.e., the colonization of Hawaii), the movements of clades from the tropics in the Old World to the New World is coincident with the occurrence of the tropical Cenozoic land bridge in the Northern Hemisphere, which permitted entry to the New World from Asia. Speciation and radiation would have accompanied both entry into new areas and exploitation of new types of food resources, especially those rich in carbohydrates, such as fermenting fruits. As climates cooled in the late Cenozoic, the Bering Land Bridge became temperate and would have permitted temperate lineages to be exchanged between Asia and North America; Asia may have received back specialized species of North American clades. This type of back and forth exchange of taxa appears to have been commonplace in Beringia for plants (Chapter 9) and vertebrates as well as for invertebrates (Ross, 1958).

Perhaps the clearest examples of the relationship between food resource specialization and *Drosophila* distribution and evolution comes from the so-called cactophilic *Drosophila,* formerly of the arid and semiarid regions of the New World but now occurring in semiarid Australia and elsewhere (Barker and Starmer, 1982). Representatives of several different species groups have independently invaded desert habitats and specialized to live and breed in the juicy rotting stems of cacti, feeding on the yeasts and bacteria that live in soft rots of decomposing cacti. Certain species of the *D. repleta* complex are host specific for certain cactus species (Fellows and Heed, 1972; Heed, 1982), such as *Drosophila pachea* (*nannoptera* group), which requires a sterol from senita cactus *(Lophocereus schottii)* to complete development. Other species of *Drosophila* are excluded from feeding in senita by the presence of toxic alkaloids (Kircher et al., 1967). Other species of the *nannoptera* groups and some in the *repleta* group are intimately associated with the columnar cacti of North America (Gibson and Horak, 1978), and there is growing evidence that host specificity of these drosophilids closely follows the evolution of their host plants, a subtle but important historical event. As a corollary to this, numerous investigators are interested in the ability of some flies that live in columnar cacti, where fermentation occurs, to live for long periods without food in certain types of alcoholic atmospheres, presumably using the alcohols as carbon sources. The ability to use alcohol is credited to the presence of the enzyme alcohol dehydrogenase, which permits the organism to break down alcohol. Here may be a case in which the product of a single gene (ADH locus) with its modifiers can potentially have an important adaptive function in natural habitats (Starmer et al., 1977).

Studying distributions of *Drosophila* makes one appreciate that simple historical explanations are rarely applicable for groups currently involved in extensive and rapid speciation events. For example, in the New World *repleta* complex (76 spp.) there are many narrow endemics, species with wide disjunctions, and weedy forms introduced around the world; generalists and extreme specialists in food preferences and nest sites; and easily identified forms as well as numerous cryptic species with phenotypes so similar that they often cannot be properly identified even by experts. Although some of the distributions of closest relatives do conform to simple allopatric models of speciation, the majority do not. Biogeographers must attempt to unravel the obscure patterns because they carry different but no less interesting data on the history of biotas. Complex patterns such as these are undoubtedly the rule rather than the exception in many types of organisms, particularly insects and plants.

Selected references

Aerial dispersal and migration

R.R. Baker (1978, 1981); Benson et al. (1975); K. Brown (1976); Carlquist (1965); Darlington (1943); Dingle (1980); Dorst (1974); Emlen (1969); Emsley (1965); Fretwell (1980); Galler et al. (1972); Gauthreaux (1980a, 1980b); L. Gilbert (1975); L. Gilbert and Singer (1975); Gwinner (1977); Harrison (1980); C. Johnson (1969); C. Johnson and Bowden (1973); Karr (1980); Keast and Morton (1980); McClure (1974); Moreau (1966, 1972); Nix (1976); Schmidt-Koenig (1975); Slud (1976); Terborgh (1980); J.T. Turner (1981).

Large-scale patterns of bird distributions

Amadon (1950, 1973); J. Bond (1971); Brodkorb (1971a, 1971b); Cody (1973a); Cracraft (1973, 1977, 1980, 1981); Dorst (1972); Feduccia (1980); Garrido and Montaña (1975); Gruson (1976); Haffer (1974, 1978, 1981); B. Hall and Moreau (1970); Harris (1962); Howard and Moore (1980); T. Howell (1969); Johnsgard (1981); Keast (1961, 1976); Lack (1976); Mani (1968); Mayr (1944); Meyer de Schauensee (1966); Moreau (1966, 1972); J.B. Nelson (1978); S. Olson and Feduccia (1980); Palmer (1962, 1976); A.L. Rand and Gilliard (1967); Russell (1964); Sclater (1858); Serventy (1960); Short (1970); S. Simpson and Cracraft (1981); Terborgh et al. (1978); Vaurie (1965, 1980); F. Vuilleumier (1975).

Bats

Holloway and Jardine (1968); Koopman (1970, 1975, 1976, 1978); Koopman and Jones (1970).

Some patterns of insects

Barker and Starmer (1982); Brundin (1965, 1966, 1967); Carson (1970, 1975, 1981); Carson et al. (1970); Carson and Kaneshiro (1976); C.S.I.R.O. (1970); H. Daly et al. (1978); Darlington (1965); Erwin et al. (1979); Fellows and Heed (1972); Gressitt (1974); Heed (1982); Hennig (1979); Holloway (1979); Michener (1979); Ross (1958); Starmer et al. (1980); Throckmorton (1975); Zimmerman (1948, 1958).

Chapter 13 Distribution Patterns of Plants

Even though vascular plants have a simpler structural organization than that found in higher animals, their distributions are no less complex. Plants are of course stationary organisms, but many can and do change their distributions in very short periods of time. Although each species has its own unique distribution, repetitive patterns are common. Some of these distributions reflect past land connections and climates, and others indicate ecological limits imposed by the present environment. Thus it is an important matter to determine which groups have been most influenced by historical events.

Factors limiting growth and reproductive success of plants

If dispersal issues (Chapter 7) can be set aside for the moment, the other critical factors limiting distributions of vascular plants can be considered. During the year plants must achieve some net carbon gain and some growth in order to survive and reproduce. The regimes of light, temperature, and water availability control the physiological performance of these photoautotrophs.

We can use water as an example. Not only does water serve as the medium for biochemical reactions and diffusion of solutes, including their transfer from the soil to the shoot, but also green plants split water to yield energy in the light reaction of photosynthesis, in which hydrogen is removed and ultimately incorporated into carbohydrates. A consequence of the opening of the stomates for the uptake of carbon dioxide is a simultaneous loss of water vapor in large quantities, called transpiration. This seemingly wasteful loss of water also is advantageous in hot, sunny habitats where otherwise damaging heat loads of leaves are reduced by evaporative cooling. Water is also the necessary substance for cell enlargement (vacuolation), and cell turgor and hydration of cell walls provide much of the support for a plant. In other words, a green plant cannot live in an environment where its basic water requirements are not met.

Distributions of plants, like animals, are limited by both abiotic and biotic factors (Chapter 3). Unlike most animals, however, plants are stationary and therefore must tolerate the full range of physical conditions in their local environment. Not unexpectedly, plants native to a given climatic regime have growth and reproductive characteristics suitable for that same clime. Some of the important observed patterns are as follows.

1. Plant species, even the local populations, often have adaptations to ensure that germination occurs at the most favorable time of the year for seedling establishment. Water must be present to trigger germination, and temperature and photoperiod are usually important in breaking dormancy. At cold latitudes and elevations seeds germinate in warm seasons to avoid freezing of tender organs, whereas in warm deserts, where water is scarce but freezing is infrequent, germination coincides with

rainy seasons. Consequently, a desert locality like Tucson, Arizona, which receives heavy precipitation in both winter and summer, has two germination peaks. In wet years many small, narrow-leaved annuals carpet the ground in the winter, and a different set of taller annuals appear in the hot summer. In these same plant communities, seeds of many woody perennials, germinate mostly following summer rains and therefore escape the early summer heat and drought, but even so many never live through the first year. Under the constant temperature and moisture of the humid tropics, germination can occur year-round, and many seeds have no dormancy period. In most latitudes the weeds typically have the ability to germinate without a dormancy period but also can tolerate long periods of dormancy in the soil.

2. Each species is adapted to live in a portion of the available light spectrum, e.g., full sun, partial sun, shade, deep shade, and even deeply submersed in water as an aquatic species (phanerogam). Structural features and biochemical adaptations used to harvest light maximally and safely in full sun are inappropriate in deep shade, and vice versa.

3. Plant species often show preferences for particular soil types. Soils differ in water-holding capacity, concentrations of soluble ions, pH, compactness, and concentrations of dissolved gases, such as oxygen and carbon dioxide. A deficiency or excess of one substance (e.g., nitrogen or chloride, respectively) severely inhibits certain plants from living there but, if not too extreme, may favor others. A variance in pH values determines the elements that are present in an available form to the plants. Therefore even though the climate of a region may appear suitable, a species may be excluded by the structure and chemistry of soils that are not compatible with the species' physiological requirements.

Many plants are known to need certain fungal mycorrhizae associated with their roots to become established and flourish, and soil factors may affect distributions of the vascular plant indirectly by controlling the distribution of the mycorrhizal organisms.

4. Each species has a particular regime for flower and fruit production. The timing of reproduction is selected to correspond to the presence of pollinating and dispersal agents and is also coordinated with a time in the plant's life cycle when enough water and photosynthate are available to produce these expensive structures. Photoperiod and temperature are common cues used to initiate flowering.

5. The presence or absence of a species may be related to its ability to withstand periodic disturbances or catastrophies, such as burning or flooding.

6. Interspecific plant-plant and plant-animal interactions are believed to be responsible for the absence of some species in apparently favorable areas. Whether the reason is competition, predation (including disease), or mutualism, the effect is a lowering of fitness for that species so that, ultimately, that species cannot leave progeny.

A fairly small number of plant species are present only when another plant species is present. Common examples are the parasites that are specific on a single host and whose range therefore is a subset of the host, such as dwarf mistletoes *(Arceuthobium)* on the stems of certain conifers (Hawksworth and Wiens, 1972) or the root parasite *Epifagus* on beech *(Fagus)* of North America. In another vein, in the southwestern United States the range of *Mohavea confertiflora* (Scrophulariaceae) lies within the broader range of *Mentzelia involucrata* (Loacaceae) because the *Mohavea* flowers are mimics that offer no rewards. *Mohavea* flowers are pollinated by extra male bees of the genus *Xeralictus,* the *Mentzelia* pollinator, and therefore can only exist where surplus males are available to visit its flowers.

Plant distributions are rarely controlled by a single physical or biotic parameter but instead are mediated by complex interactions with climates, soils, and other organisms. For example, plants of lowland tropical rain forests do not experience cold and therefore are greatly intolerant of prolonged cold. Vegetative growth is relatively continuous except during dry spells because a year-round equable clime prevails; plants are day-neutral in response to photoperiod because little seasonal variation in day length occurs. When investigators find many members of these same taxa in the fossil record, having similar morphologies to all of their tropical descendents, they must conclude that the paleoclimate was warm and fairly similar to that inhabited by comparable extant vegetation types.

Such reasoning can be applied to explain the past and present distribution of palms. The distribution of Arecaceae (Figure 13.1) shows that this family is pantropical, largely restricted between the Tropic of Cancer and the Tropic of Capricorn, with only a few genera on each continent extending into the subtropics and neighboring temperate areas. The limiting factor to this distribution is cold; most palms cannot tolerate prolonged temperatures below 10° C. An exception is *Washingtonia filifera,* now distributed poleward to 34° N latitude in California and Arizona (Figure 13.2) in isolated, moist, and protected canyon oases of the desert mountain ranges. The fossil record indicates that in the Eocene palms were widely distributed in the Northern Hemisphere, occurring in the Pacific Northwest, Maryland, Utah, and even England. This indicates that 40 to 50 million years ago the tropics and subtropics were

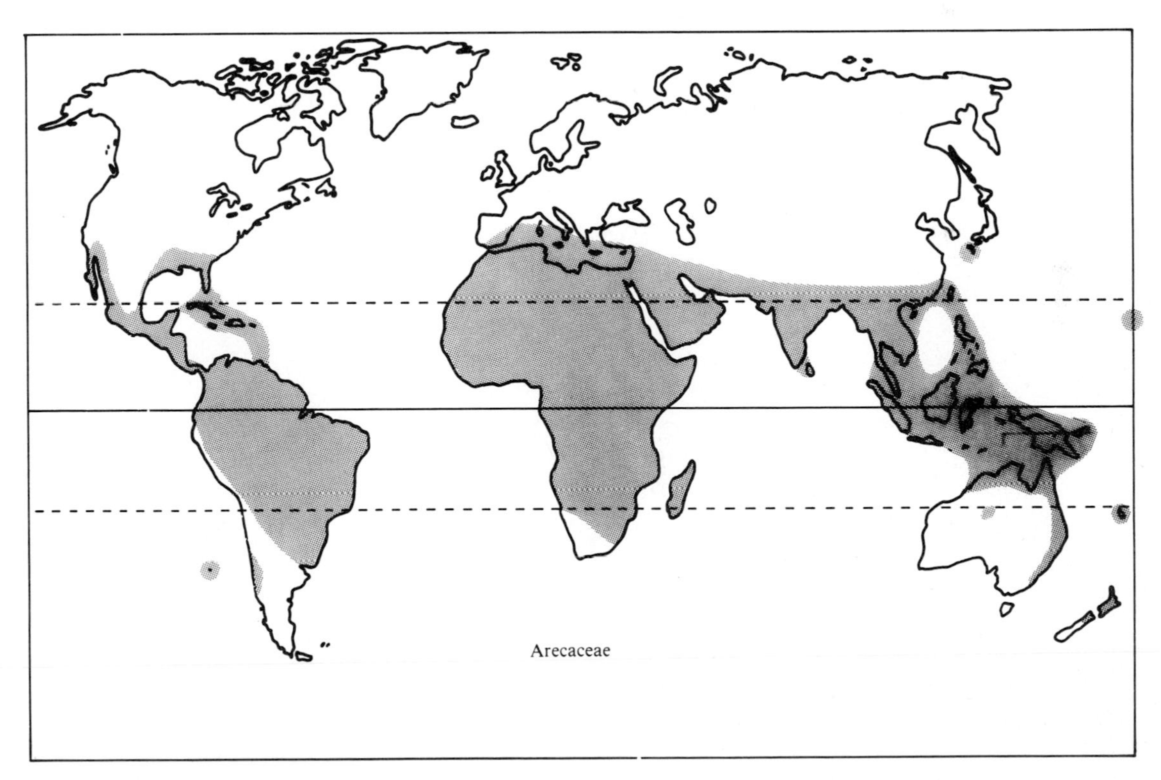

Figure 13.1
Pantropical distribution of the palm family, Arecaceae. Primary factor limiting the distribution of these plants is low temperatures. Although a few palms can tolerate freezing, none can withstand extreme cold. (After Good, 1974, modified using data in Moore, 1973.)

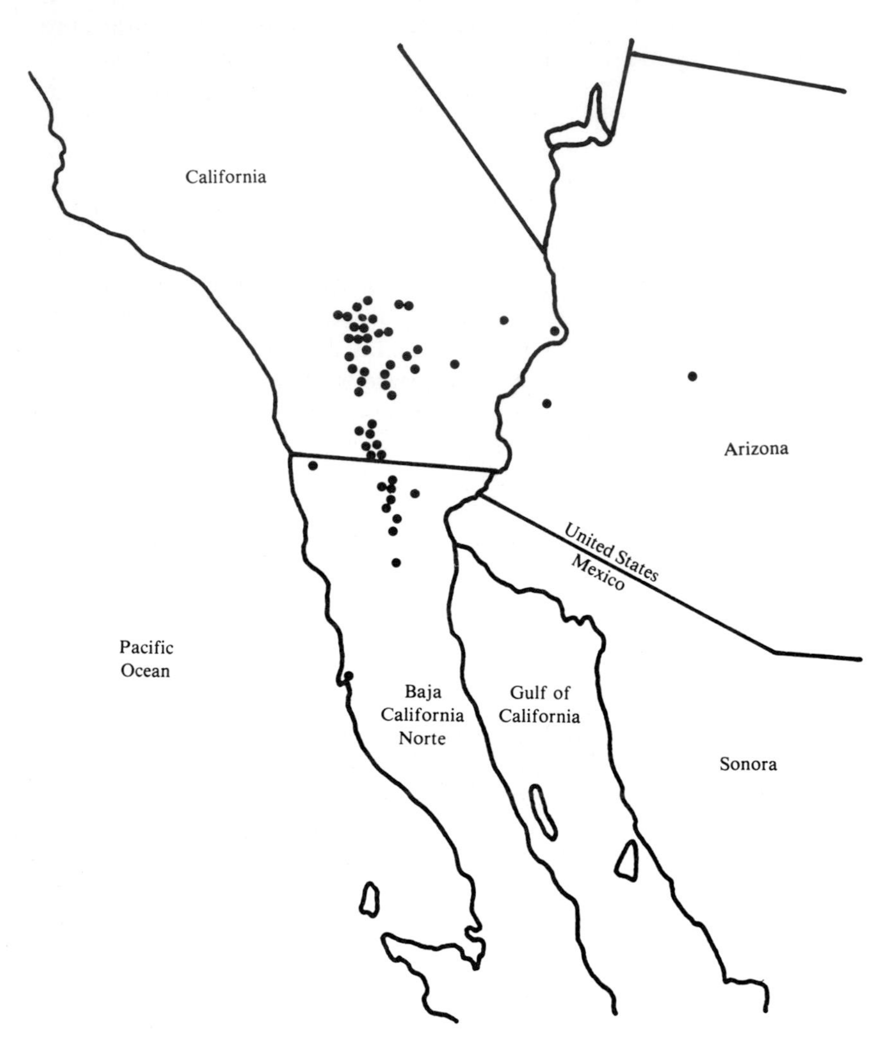

Figure 13.2
Reported distribution of populations of the California fan palm, *Washingtonia filifera*. This species occurs at springs and other mesic sites in the deserts of California, Arizona, and northern Mexico. These localities represent the northern limit of the distribution of palms in western North America. Palms may no longer be present in some of these locations. (After Brown et al., 1976.)

much broader. There is also corroborative evidence: other tropical plants and warm water corals occurred in the same latitudes at the same time. The spotty distribution of northern palms represents the breakup of a more continuous distribution, now shrunken because of decreased moisture and minimum annual temperatures.

An investigator quickly discovers that many years of exhaustive research are required to explain all the major factors limiting the distribution of a single species. By developing a fully mechanistic explanation for one species, an investigator can lose sight of broad patterns, although as the example with palms shows, physiological ecology can explain why certain taxa are disjunct or narrowly endemic today. In time these data will help investigators to predict and quantify paleoclimates and in this manner aid in reconstructing the history of disjunctions.

Relationships of vegetation type to climate

Whereas an investigator can describe species distributions one at a time, a different and highly rewarding endeavor has been to quantify the relationships between climate and whole vegetation types, such as chaparral and lowland tropical rain forest. This approach was started by von Humboldt at the beginning of the nineteenth century, and it has been a constant challenge to biogeographers.

In this age of great computers and precise instruments for monitoring climate, one might expect to find a single comprehensive model representing how climate and vegetation are related. Nothing could be further from the truth. The principal stumbling block for modeling climate versus vegetation is the problem of determining which climatic parameters should be used. The complexity of this issue is illustrated best by one common in-depth climate diagram (Figure 13.3). This graph gives data on climatic variables of temperature and precipitation for a single site. Some values are means, some maxima, some minima, and some measure the duration of a stressful or benign factor. Humidity, wind velocities, soil-water moisture, and solar irradiance are important as well but are not included. Which set of these climatic factors gives the best fit for defining all types of vegetation around the world? No one seems to have a crystal ball.

The Köppen classification. In their first exposure to patterns of climate and vegetation, students learn Köppen's classification (Table 13.1). This system uses letters to symbolize overall temperature regime and seasonal rainfall patterns of a region. For some continents zonal belts of vegetation are roughly coincident with climatic provinces, although on other landmasses they are not. Most biogeographers consider the Köppen classification too crude for quantitative purposes.

Thornthwaite's climatic water budget. Two localities may receive identical rainfall but one would have more water available for plant growth. For example, in a hot region with low relative humidity, water will evaporate quickly from the soil surface and be lost through plant evapotranspiration. Thornthwaite (1948) devised a way of compensating for the temperature effects so he could determine whether the water provided by the soil to the plant is severely limiting at certain times of the year. Thornthwaite used monthly values of precipitation, potential evaporation, actual evapotranspiration, and soil-water surpluses or deficiencies to determine the monthly amount of water available for plant growth at any locality. This system thus identifies the months that have the most effective precipitation.

Figure 13.4 shows three stations in California with different annual water balances. Richardson's Grove in northern California has a wet, cool, maritime climate with great water surpluses in the cool months and a mild deficit during the summer, which is not hot. At the other extreme, Palm Springs has scanty rainfall,

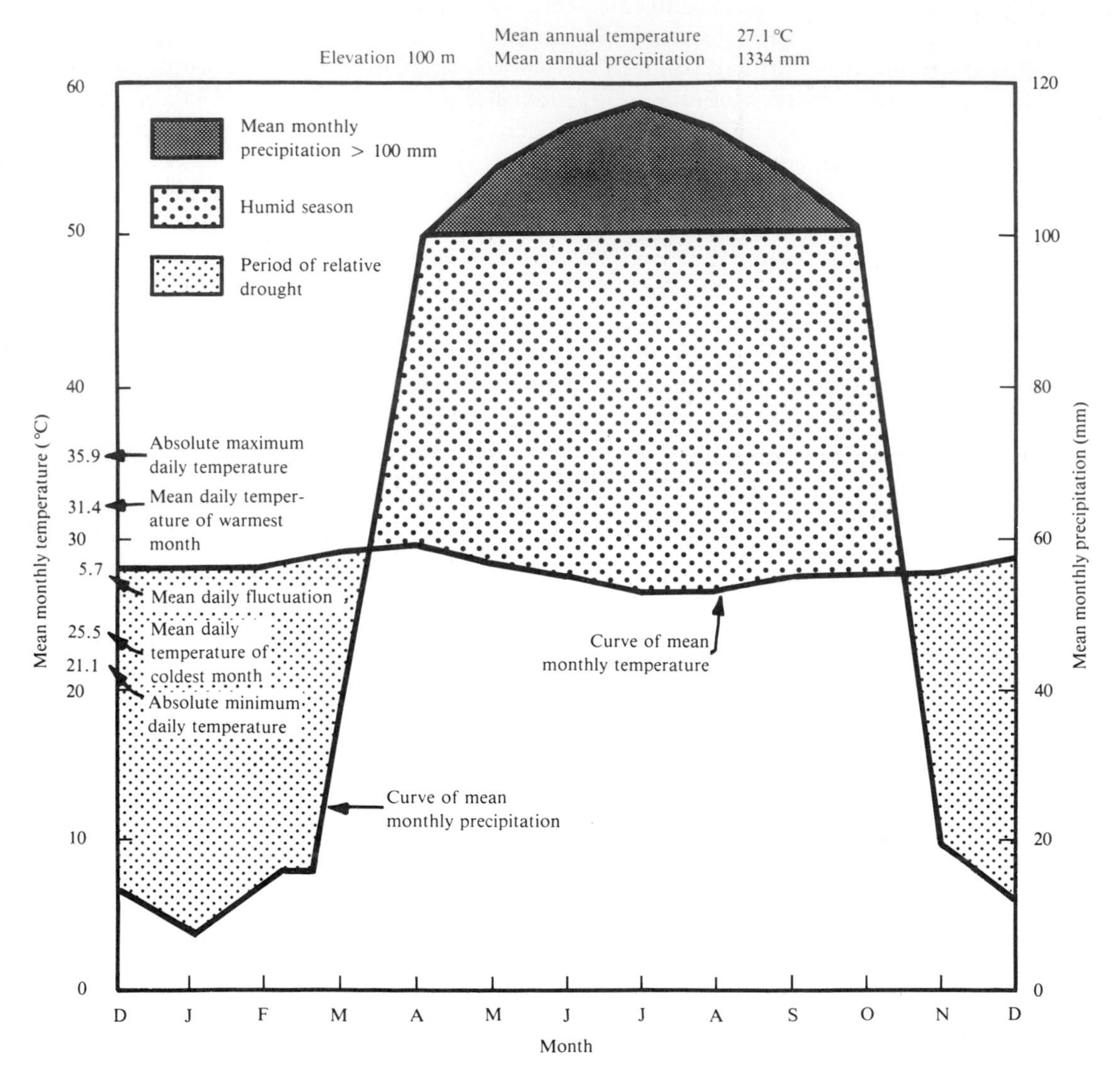

Figure 13.3
Climate diagram for Calabozo, Venezuela, showing the kinds of data that can be expressed on a single graph. This tropical site experiences a uniform temperature year-round with only slight daily fluctuation (5.7° C); the warmest and coldest temperatures ever recorded are 35.9° and 21.1° C, respectively. The curve for the mean monthly precipitation shows a pronounced wet season with slightly cooler temperatures lasting from March to October and a dry season extending from November to February. Climate diagrams of this format facilitate comparisons between regions because differences in temperature and precipitation regimes make for visual contrasts. (After Walter, 1979.)

Table 13.1

The principal climatic types recognized in the Köppen classification

Capital letters denote the overall temperature and precipitation regimes of an area, and lowercase letters indicate when the rainy season or a drought occurs and the temperature characteristics during the summer or winter. Examples of the vegetation associated with each main climatic type are noted.

Köppen type	Description	Climax vegetation type	Selected locations
Af	Always hot, always moist	Tropical rain forest	Amazonia, Congo, Borneo
Am	Always hot, seasonally excessively moist	Monsoon forest	Burma and Thailand
Aw	Always hot, seasonally droughty	Savanna	Eastern Brazil
BSh	Semiarid, hot	Thorn scrub	Mali, Somalia
BSk	Semiarid, cool	Desert scrub	Colorado Plateau
BWh	Arid, hot	Arid desert scrub*	Sahara, Saudi Arabia
BWk	Arid, cool or cold	Arid desert scrub	Tarim Basin
Cfa	Mild winter, always moist with long, hot summer	Subtropical broad-leaved evergreen forest	Southeastern United States, eastern China
Cfb	Mild winter, always moist with short, warm summer	Temperate rain forest	Pacific Northwest of North America
Cfc	Mild winter, always moist with very short, cool summer	Heath	Iceland
Csa	Mild moist winter with long, hot, droughty summer	Chaparral	Coastal southern California
Csb	Mild moist winter with short, warm, droughty summer	Fynbos	South Africa
Cwa	Mild winter, moist summer with long, hot summer	Subtropical thorn forest	Northern India
Cwb	Mild winter, moist summer with short, warm summer	Subtropical evergreen forest	Natal
Dfa	Severe winter, always moist with long, hot summer	Prairie grassland	Iowa
Dfb	Severe winter, always moist with short, warm summer	Boreal forest	Northern United States
Dfc	Severe winter, always moist with very short, cool summer	Boreal forest	Hudson Bay

*Vast regions of BWh are totally barren, devoid of woody plants.

Continued.

Table 13.1—cont'd
The principal climatic types recognized in the Köppen classification

Köppen type	Description	Climax vegetation type	Selected location
Dfd	Severe winter, always moist with short summer and excessively cold winter	Boreal forest	Central U.S.S.R.
Dwa	Severe winter, moist summer with long, hot summer	Boreal forest	Northern China
Dwb	Severe winter, moist summer with short, warm summer	Boreal forest	Southeastern Siberia
Dwc	Severe winter, moist summer with very short, warm summer	Boreal forest	Central Siberia
Dwd	Severe winter, moist summer with short summer and excessively cold winter	Boreal forest	Northeastern Siberia
EF	Polar climate in which plant growth is impossible		Greenland
ET	Polar climate with very short period of plant growth	Tundra	Northern Alaska
H	Unspecified mountain climates		

Figure 13.4
Climate graphs using the Thornthwaite system show pronounced differences in temperature and precipitation regimes among three sites in California. Richardson's Grove has a cool, mesic climate, Palm Springs shows the dramatic temperature fluctuations and large water deficit typical of a desert, and San Diego has a warm maritime climate. (After Major, 1977.)

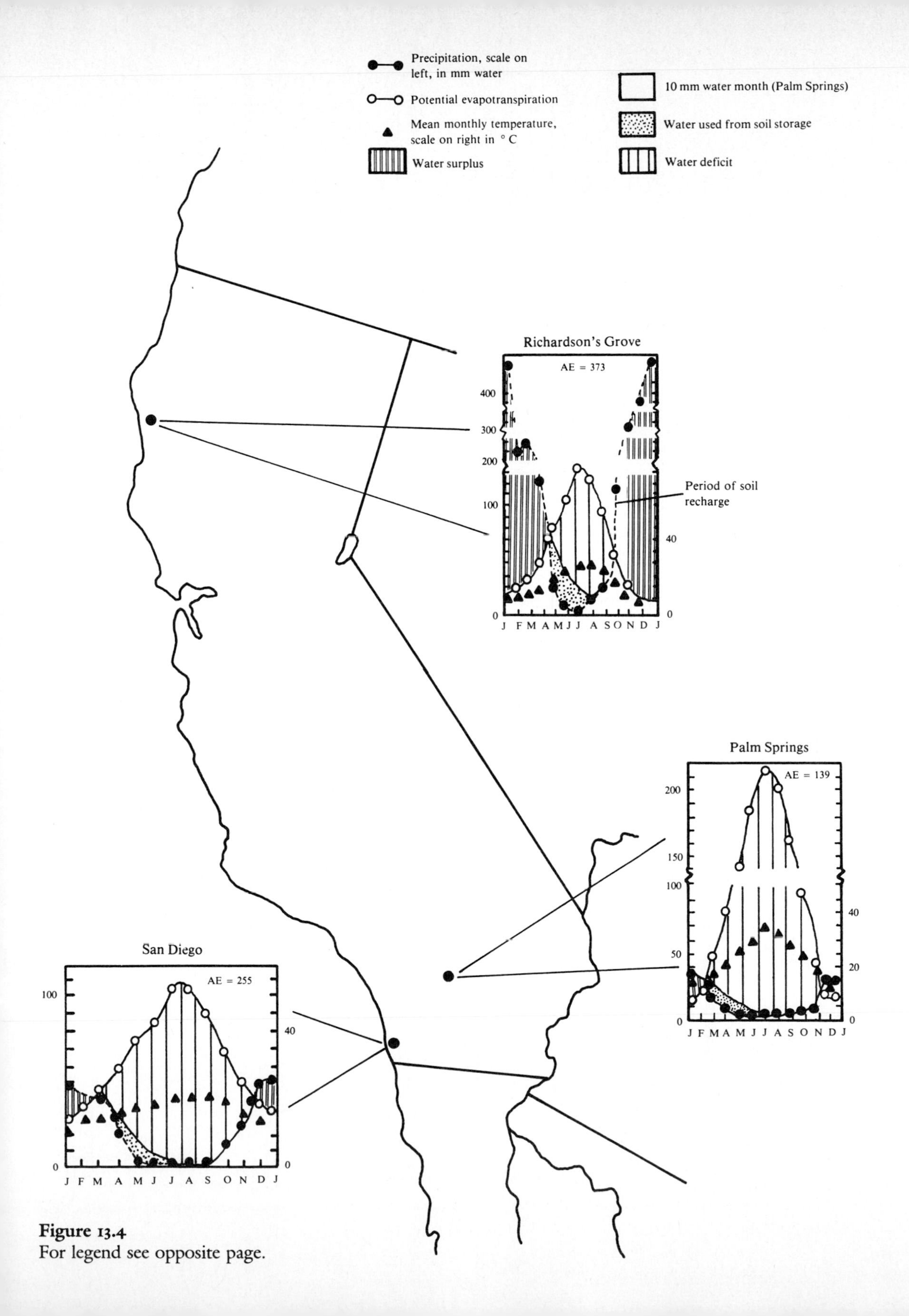

Figure 13.4
For legend see opposite page.

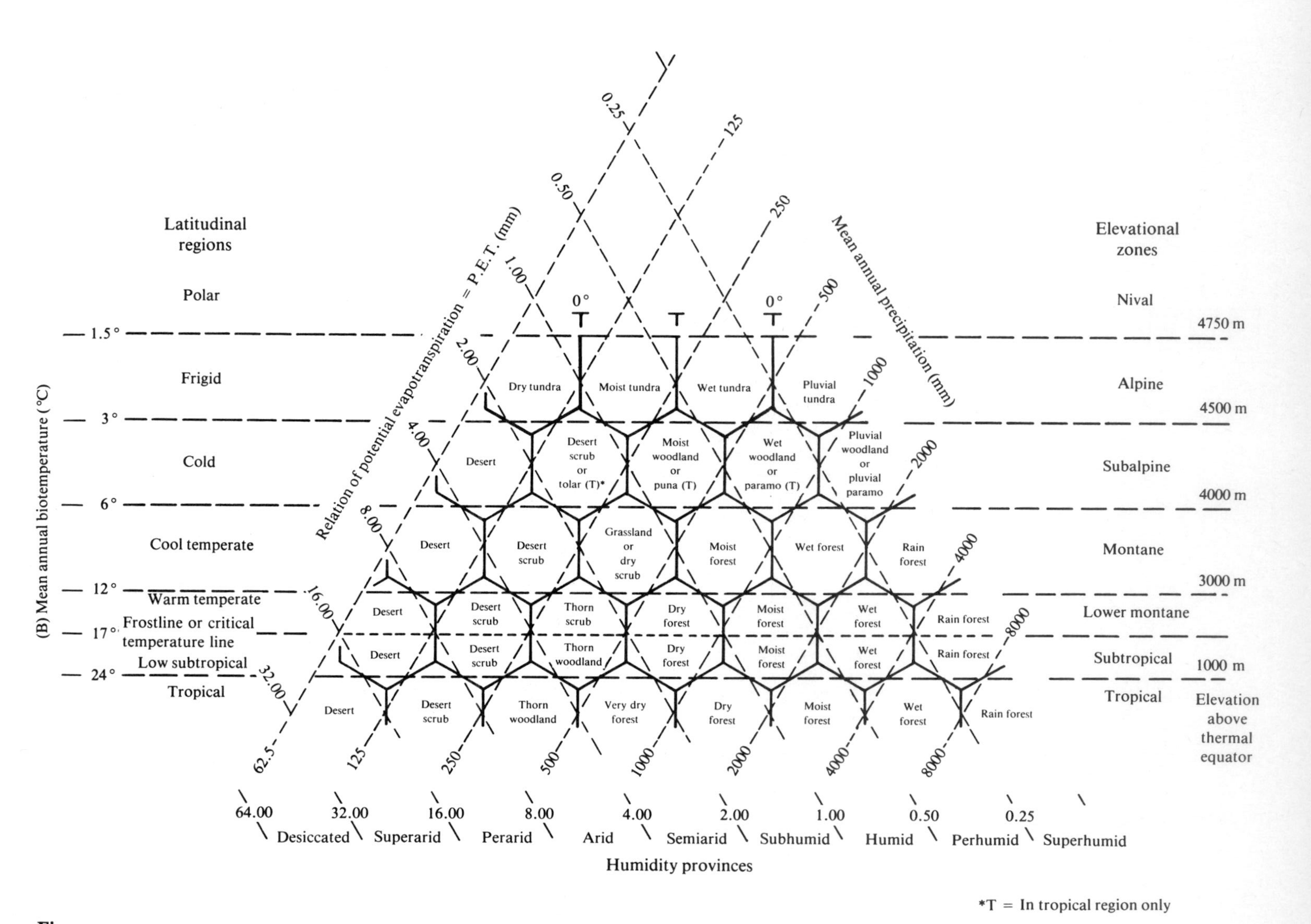

Figure 13.5
For legend see opposite page.

mostly during the winter months, and a long, hot, dry summer and fall. The former climate supports a lush redwood forest, whereas the latter is a sparse desert scrub. Intermediate is the climate of San Diego, with cool temperature and some water surplus in the winter and hot, dry summers, supporting chaparral vegetation.

As recent reviews have indicated (Major, 1977; Schulze and McGee, 1978), there are various problems in properly calculating the water budget of a locality, and on some continents Thornthwaite's calculations are not reliable for plotting vegetation on a local scale. Nevertheless, this system does relate climate and vegetation fairly well on a subcontinental level.

The Holdridge life zone system. Holdridge (1947; Holdridge et al., 1971) introduced a model to classify climates so that each life zone is equivalent to a biome type. His system used primarily three long-term climatic parameters: average total annual precipitation, mean annual biotemperature, and potential evapotranspiration ratio. Biotemperature was originally defined by Holdridge as the mean of temperatures with the substitution of zero for all unit-period values (e.g., hours) below 0° C. He reasoned that all subzero periods could be eliminated from calculations because plants are physiologically inactive at such low temperatures. For tropical climates Holdridge also now substitutes zero for values above 30° C because at such high temperatures plants presumably experience no net carbon gain. The potential evapotranspiration ratio is a theoretical value calculated from the amount of water needed for evapotranspiration divided by the precipitation moisture. Biotemperature is used to calculate the ratio, which is positively correlated with humidity. The overall humidity of the locality and the frost line or critical temperature line are given because these are meaningful factors controlling plant life.

Figure 13.5 is a chartlike representation of the model for the global life zones proposed by Holdridge. This shows the primary relationship among the life zones, implying that a shift in environmental parameters, e.g., drier or wetter, should result in a different vegetation as predicted by the diagram.

The Holdridge model has not been tested in enough habitats to determine its full value, but at least one study on southern Africa has shown that the model does not accurately mimic vegetation patterns in seasonally dry climates (Schulze and McGee, 1978). Moreover, numerous vegetation types are not accounted for in the model. The shortcomings of the model may be that it uses annual temperature and precipitation values, which misrepresent the important seasonal factors influencing the growth of plants, especially in semiarid habitats. The concept of biotemperature is also subject to criticism and needs careful evaluation by ecophysiologists.

Life forms of Raunkiaer and phytosociology. Phytosociology is defined as the study of sociological relationships among plant species, which depend on how plants coexist and modify the habitat. The approach is traceable to J. Paczoski, who wrote on the subject before the turn of the century, but the field was enhanced by the work of Raunkiaer (1934). Raunkiaer

Figure 13.5
Holdridge's (1947) system for classifying vegetation types in relation to climate. This superficially complex graph actually uses only three kinds of data: mean annual precipitation, potential evapotranspiration (P.E.T.), and mean annual biotemperature. These are graphed in relation to each other so as to form cells that indicate a particular combination of climatic conditions and an associated vegetation type. (See text for further explanation.) (After Holdridge, L.R.: Science 105:367-368, 1947.)

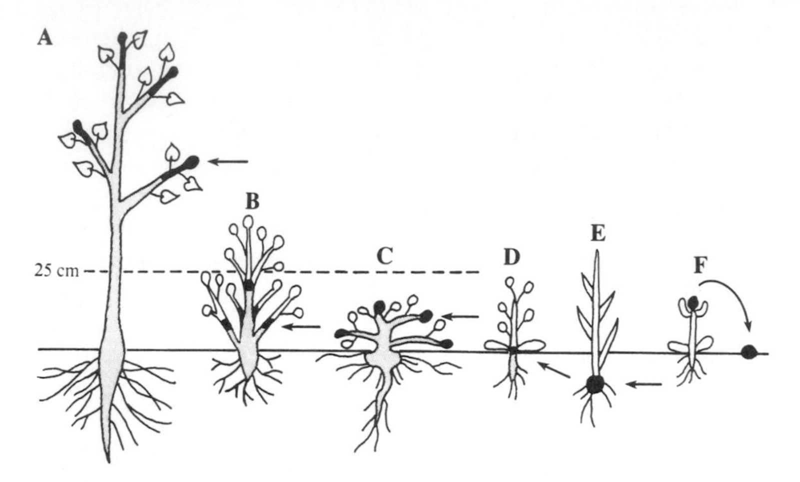

Figure 13.6
Five major life forms defined by Raunkiaer (1934). Perennating tissues are shown in black, woody tissues are lightly shaded, and deciduous tissues are unshaded. The life forms are, **A,** Phanerophyte, tree or shrub with buds greater than 25 cm above the ground; **B** and **C,** Chamaephytes, semishrubs or subshrubs with buds less than 25 cm above the ground; **D,** Hemicryptophyte, perennial herb with bud at ground surface; **E,** Geophyte, perennial herb with perennating organ below ground surface; and **F,** Therophyte, annual plant that survives unfavorable periods as a seed.

proposed a morphological classification of plants, called life forms, which depends on the location of the perennating tissues (Figure 13.6). Originally six major life forms were recognized, although several authors have proposed additional categories (Orshan, 1953; Böcher and Lyshede, 1968).

A number of European workers have used these categories to quantify the form and structure of natural communities, comparing the percentages of each life form. Because life forms are direct adaptations to local conditions, one should be able to compare regions in a statistical manner without regard for the names of the species. Communities convergent in life forms should have similar climates, and the closer the morphologies, the more similar the conditions that the plants experience.

This form of phytosociological analysis has never been popular in North America, but even so approaches such as those of Braun-Blanquet (1965) are recognized and have been recently analyzed (Whittaker, 1978b).

Physiognomic forest types of the Northern Hemisphere. Wolfe (1978, 1979) classified humid and mesic forest biomes of the Northern Hemisphere, using temperature regimes as the variable that controls climax vegetation. He assumed that precipitation, irradiation, and edaphic factors are not limiting. Forest types of eastern Asia were plotted on a graph with temperature axes, and the limits of each forest type correspond exceedingly well with mean annual temperature, mean annual temperature range, and mean temperatures of the coldest and warmest month (Figure 9.11). Wolfe considered the temperature parameters used to be very important for plant life (productivity and surviving stressful periods), although they may only approximate the factor that actually limits each distribution. For example, coldest mean month could signify the critical minimum one-time temperature that a plant can survive, the maximum length of time of cold a plant can withstand, or limits placed on its photosynthetic capabilities.

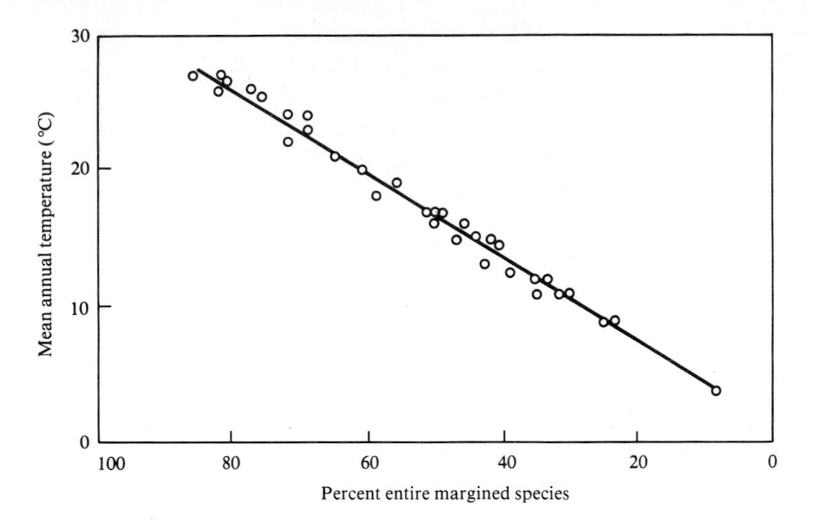

Figure 13.7
Correlation between the percent of species in a flora with entire-margined leaves and the mean annual temperature is very precise. Wolfe (1978) has used this as a basis for estimating past climates and vegetation types in different regions of the Northern Hemisphere.

Wolfe also found that forest types of North America, Europe, and parts of Australasia also mimic the thermal patterns of eastern Asia. These analyses led him to develop a new classification of physiognomic types, applicable throughout the Northern Hemisphere, and new definitions of climates. In his scheme, he recognized six temperature classes based on mean annual temperature: tropical (>25° C), paratropical (20° to 25° C), subtropical (13° to 20 ° C), temperate (10° to 13° C), paratemperate (3° to 10° C), and subtemperate (<3° C). When montane forests were studied, the traditional values of 18° C and 6° C cold mean month temperatures for dividing tropical from subtropical and subtropical from temperate, respectively, became unreliable. Instead, Wolfe suggested that the subtropical-temperate boundary is the 1° C cold month isotherm.

Since 1971 Wolfe has stressed the close correlation between the percentage of entire leaves in a forest and the mean annual temperature (Figure 13.7). The index of a forest, either present or past, could be used as an indicator of temperature regime. This leaf-shape method can be criticized (Plumstead, 1973; Pielou, 1979) because forest types of the Southern Hemisphere do not always fit the prediction. However, Wolfe has advocated its use only in the Northern Hemisphere until comprehensive analyses like his have been performed for the southern forests, especially in seasonally dry climates and on special soils, where sclerophylly is common. A synthesis like Wolfe's could help immensely in reconstructing the paleoclimates of terrestrial plant communities, permitting workers to put real temperature values into their discussions.

The origin of angiosperms

Nearly everyone who writes about distribution patterns of seed plants feels obligated at some time to speculate on the early history of angiosperms: their age, relatedness to other seed plants, probable center of origin, and early migration routes. In Darwin's day the answers to such inquiries were obscure, but gradually authors have acquired data that answer some of the questions and suggest probable explanations for some puzzling aspects of plant distribution.

From the outset, dating the origin of angiosperms was plagued by great gaps in knowledge and by misinformation. Scientists were impressed initially by the great diversity of fossil leaves in the Cretaceous. These were given names of, and inferred to represent, extant genera of angiosperms. If so many distinctive genera existed in the Cretaceous, certainly the origin of flowering plants had to significantly predate that period. Numerous authors proposed dates of Triassic or older to accommodate this interpretation, e.g., Axelrod (1952, 1960, 1970) and Croizat (1958), who insisted that extant families were demonstrably present early in the Cretaceous and cited putative pre-Cretaceous fossil angiosperms.

Beginning in the 1960s, the idea of a Triassic or earlier age for angiosperms lost considerable ground. Painstaking investigations by many workers provided powerful evidence of a more recent origin. Their findings are summarized as follows: (1) there are no reputable pre-Cretaceous angiospermous fossils, (2) the current oldest positive angiospermous pollen grains and leaves occur in the Barremian of the Lower Cretaceous, (3) simple Barremian pollen grains occur in fossil localities around the world at low paleolatitudes, (4) throughout the remainder of the Cretaceous there was a buildup in the diversity and specialization of pollen grains, leaves, and woods, (5) recognizable extant genera appear only in the upper half of the Cretaceous, and (6) the increase in diversity and specialization recorded by Cretaceous fossils mimics the best phylogenetic systems of classification for extant forms. Most botanists today accept the idea that the evolution of the group and its fossil record are reasonably coincident on a broad level; the so-called explosive evolution of angiosperms in fact covered 60 million years of diversification in the Cretaceous, a period equal to the time since then, in which enormous change also has occurred.

What are the ancestors of angiosperms? This question has been asked so many times that a thorough review of the literature here is impossible. It will suffice to say that nearly all orders of extant and extinct gymnosperms have been suggested as likely candidates. Only the seed ferns (Pteridospermales) have remained a viable option, however, as considerable evidence accumulated denying angiospermous origin from conifers, gnetophytes, ginkgoads, cycads, glossopterids, taxads, and cycadeoids. Mesozoic seed ferns have the prerequisite architectural design and anatomical peculiarities to give rise to angiosperms, and they were also the ancestors for fossil Caytoniales, which are closely similar to angiosperms and have also been suggested as their ancestors (Stebbins, 1974). The key to understanding lies in a comparison with Jurassic seed ferns, which has never been accomplished because of an inadequate fossil record. Finding an Upper Jurassic seed fern with all or nearly all of the structural angiospermous features would be strong evidence that flowering plants arose in the lowermost Cretaceous. Unfortunately, the absolute criterion for defining angiospermy is double fertilization, the fusion of two sets of nuclei to form the zygote and the endosperm, respectively, a feature that is not preserved in the fossil record.

Perhaps some analysis of angiospermy is required, inasmuch as its innovations are what have promoted the dominance and dispersal of the group. Double fertilization was extremely important. In known groups of cycads (and presumably in seed ferns), nutrition for the embryo is produced long before an embryo is formed. Therefore the plant invests significant resources in a structure before fertilization occurs, a costly risk. In angiosperms, endosperm or another storage tissue is started only when fertilization has been effected, an investment to ensure development and germination of *that* embryo. Shifting to double fertilization carries three additional benefits. First, the period for production of a seed is shortened from two years to a single growing season, giving flexibility to the life histories of plants. Second, because production of the embryo and its nutritive tissues are synchronous, a broad spectrum of new seed designs could be attempted, thus promoting greater variation in seed size and shape with obvious advantages for dispersal. Third, plants could develop new ways to store materials with the embryo, in harmony with dispersal design, to ensure germination of the seed.

Another key invention was the flower, which may have been developing in seed fern lines. The flower is a means to exploit animals as pollinating agents, which in turn vastly changes the reproductive structure of a plant population by allowing some control over the gene flow. In comparison, other gymnosperms use mainly wind pollination of unisexual flowers and obligate outcrossing, with great wastage of pollen. The potential for speciation was naturally increased many fold by manipulating what sort of animal visits the flower and how broadly genes will disperse. Moreover, flowers offer opportunities for inbreeding strategies. Our current evidence says that primitive angiosperms had hermaphroditic flowers, not unisexual ones.

Innovations in vegetative structure were no less important in the successful domination of the world by angiosperms. From the beginning angiosperms have shown remarkable plasticity in the growth of leaves and shoots. This plasticity, coupled with opportunities for herbaceous as well as woody growth habits, provided the necessary physiological flexibility to enter all types of habitats and outcompete those plants with slower growth rates and lower fitness. Angiosperms also evolved more efficient vascular tissues for moving water and carbohydrates. The angiosperm vessel is a vertical conduit without crosswalls, which moves water faster than the tracheid system of gymnosperms, which has periodic crosswalls that impede water flow. Vessels are considered a major improvement for water conductance (Carlquist, 1975). The angiospermous stock apparently lacked vessels but their descendents developed them very early; if current phylogenetic schemes are correct, vessels evolved independently in angiosperms at least five times, including a minimum of once in monocotyledons and four times in dicotyledons. Vegetative and reproductive parts also achieved new and more varied secondary metabolism to either discourage or encourage visitation by animals. Herbivore pressure is considered an important cause of plant diversity in the tropics (Janzen, 1970; Raven, 1977).

Searching for the birthplace of angiosperms has been a favorite goal of phytogeographers and, as usual, many localities have been suggested. Fortunately, most authors currently realize that no evidence can provide unequivocal proof of a birthplace, and it is not terribly important to assign a definite center of origin. Instead, authors now prefer to identify the probable region of initial radiation. Two regions, southeast Asia, sometimes including Australasia (Smith, 1967, 1970, 1973; Takhtajan, 1969; Thorne, 1963, 1976b, 1978), and the interior of West Gondwanaland (Raven and Axelrod, 1974), are popular candidates.

Tropical Asia is a region of great floral richness, not only in the total number of species but also in the total number of tropical plant families and genera. In addition, the tropical montane forests especially are homes for dicotyledons with very primitive features. The list for southeast Asia (including the East Indies as far as Borneo and excluding India) is very impressive, including 19 of 32 families and subfamilies of Annonales, all 11 taxa of ancient Hamamelidales, and other major old lineages of Berberidales, Theales, and Cornales. For many of these tropical families and orders (and others) the genera with the plesiomorphic features occur in the Orient. Nearby on the Australasian Plate are found additional taxa with numerous primitive and ancient features, e.g., on New Caledonia, rich in vesselless Winteraceae. Phylogenetic lines (with morphological specialization) appear to radiate from southeast Asia into the

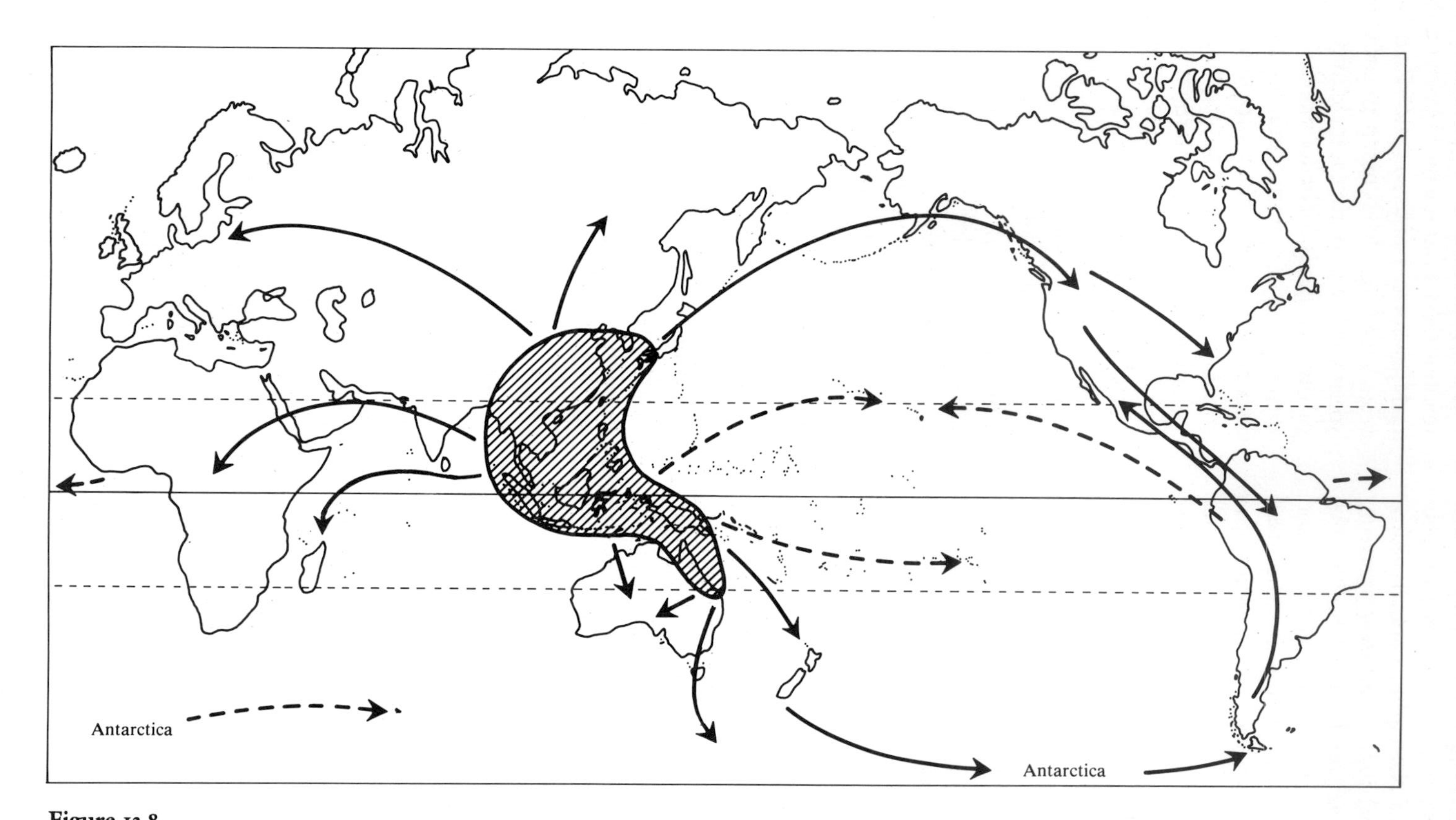

Figure 13.8
Hypothesized routes of dispersal of early angiosperms from their presumed center of origin in southeast Asia. (After Smith, 1970.)

adjacent landmasses (Figure 13.8), suggesting that taxa migrated and, in the process, diversified. In comparison, primitive-looking taxa are more poorly represented on other landmasses. All temperate families are derived from tropical ones.

Although these are correct observations, there are justifiable reasons to question the interpretations because this could be a region of secondary radiation, not the initial one. Instead of calling southeast Asia the cradle for developing angiosperms, authors such as Stebbins (1974) and Raven and Axelrod (1974) consider this a museum, merely a center of survivorship. They still recognize, however, that southeast Asia and Australasia were important source areas; from Asia migration has proceeded into adjacent landmasses such as Europe, Africa, and eventually North America.

The explanation that Gondwanaland was the initial center of radiation is based on completely different evidence. Raven and Axelrod (1974) concluded that significant migration of angiosperms took place between Africa and South America before the wide Atlantic Ocean formed. This means that a diversified flora was present in West Gondwanaland prior to their drifting apart. The present southeast Asia (including India) area is relatively new, a composite resulting from the drifting together of several unrelated geologic blocks; therefore the modern home of primitive angiosperms cannot be the ancient region for early differentiation.

To appreciate this way of looking at an old problem, one could produce a cartoon of the Lower Cretaceous world showing how primitive suborders of extant dicotyledons are positioned. In this model, a Gondwanaland origin permits more direct access to the Australasian plate. In the southeast Asia model (Figure 13.8), angiosperms had to reach Australia from Asia or vice versa by one of four means: (1) over great distances of waters at a period when angiosperms had poor dispersal properties, (2) the long way around through Africa or South America via Antarctica, (3) from some intermediary, such as a fragment (drifting block) of Asia, or (4) late in time (to Australia) as the two landmasses approached each other.

Proponents of an initial Gondwanaland radiation bolster their case with the following arguments.

1. Surviving species of primitive groups in Asia are often polyploids, meaning that they may have arisen like weeds in open habitats. The evidence provided for this is indirect and equivocal, based on the invasion properties of polyploids.

2. West Gondwanaland was strategically located for accommodating migration to all regions where primitive groups are found. Of course, there is no reason to assume a priori that an initial radiation had to be centrally located.

3. Angiosperms first radiated into the dry interior of West Gondwanaland because the arid and semiarid environment would have promoted the evolution of their important features: vessels, closed carpels, insect-mediated pollination systems, abundant secondary metabolites, and drought-resistant leaves. Moreover, speciation is promoted in dry habitats, especially ecotonal or border areas. All aspects of this statement are worth studying, but they are highly speculative when applied to historical phenomena. Initially, the conclusion of aridity in central Gondwanaland has never been properly substantiated. Even if the interior was dry, there is no reason to conclude that angiospermous features arose in very dry climates, at least not deserts or semiarid habitats in the sense of physical geographers. Vessels for efficient water movement, closed carpels for protection of ovules, insect pollination for economic reproduction, secondary compounds against predation, and leaf adaptations for maximizing photosynthesis all have a real value in practically any habitat. The presence of vessels in nonflowering desert plants, such as *Selaginella* and Gnetales, is questionable evidence on this matter. It

is more interesting that extant angiosperms lacking vessels or having very primitive ones cannot and do not live in exceedingly dry habitats (Carlquist, 1975).

Work is continuing to settle these issues. When scientists explore fully the progression of the angiospermous fossil record on each continent, from shales and coal balls of the Cretaceous, then some of the questions raised here can be answered.

Major disjunct ranges of plants

Phytogeographers have been repeatedly fascinated by those taxa that exhibit major disjunctions in range over vast distances. The widest of these, gaps of intercontinental or equivalent size, have been summarized by Thorne (1972), who used the most up-to-date systematic interpretations of the genera involved (see box), in this way erasing many persistent errors in older

Major intercontinental disjunctions in the geographic ranges of seed plants (After Thorne, 1972.)

EURASIAN–NORTH AMERICAN
- Arctic
  - Circum-Arctic
  - Beringian-Arctic
  - Amphi-Atlantic-Arctic
- Boreal
  - Circum-Boreal
  - Beringian-Boreal
  - Amphi-Atlantic-Boreal
- Temperate
  - Circum-North Temperate
  - North and South Temperate
  - Fragmentary North Temperate
    - Amphi-Atlantic Temperate
    - Mediterranean-American
    - Eurasian–eastern and western American
    - Eurasian–eastern American–Mexican
    - Asian–eastern and western American
    - Eurasian–eastern American
    - Asian–eastern American
    - Eurasian–western American
    - Asian–Mexican highland
  - Wide intracontinental disjuncts and Epibiotics

AMPHI-PACIFIC TROPICAL

PANTROPICAL

AFRICAN-EURASIAN (-PACIFIC)
- African-Mediterranean
- African-Eurasian
- African-Eurasian-Malesian
- African-Eurasian-Pacific
- African-Eurasian-Australasian
- Indian Ocean-Eurasian (-Pacific)

AMPHI-INDIAN OCEAN

ASIAN-PACIFIC
- Asian-Papuan
- Asian-Papuan-Melanesian
- Asian–Papuan–Pacific Basin
- Asian-Papuan-Australasian

PACIFIC OCEAN

PACIFIC-INDIAN-ATLANTIC OCEANS

AMERICAN-AFRICAN

NORTH AMERICAN–SOUTH AMERICAN
- Tropical
- Temperate
- Bipolar

SOUTH AMERICAN–AUSTRALASIAN
- South American-Australasian
- South American-Australasian-Asian
- South American-Australasian-Madagascan

TEMPERATE SOUTH AMERICAN–ASIAN

CIRCUM-SOUTH TEMPERATE

CIRCUM-ANTARCTIC

SUBCOSMOPOLITAN

ANOMALOUS

accounts. A fun project is to draw these infrageneric disjunctions on a world map to show that plant examples can be found to mimic all possible land connections between adjacent or opposing landmasses. Of course, many of these historical connections have been proposed, but Thorne presented a conservative case for interpreting such disjunctions; he said that some are reliable indicators of past land connections and some are not.

To understand the problems in interpreting the meaning of disjunctions, let us examine selected patterns identified by Thorne.

South American–Australasian. Continuous and disjunct ranges of plant genera and subtribes between Australasia and southern South America may provide information about southern migration routes of the past. The most intriguing examples are the forest plants whose seeds are not readily dispersed. Heading this list are the Southern Hemisphere beeches, *Nothofagus* (Fagaceae), the dominants in many south temperate forests. There are three easily recognized species groups within *Nothofagus,* each with distinctive pollen grains. The fossil record reveals that all three groups were widespread in southernmost latitudes (Figure 13.9), which accounts for their current distribution pattern (Figure 13.10), as they occur now in temperate eastern Australia, Tasmania, New Guinea, New Britain, New Caledonia, New Zealand, and southern South America. Because

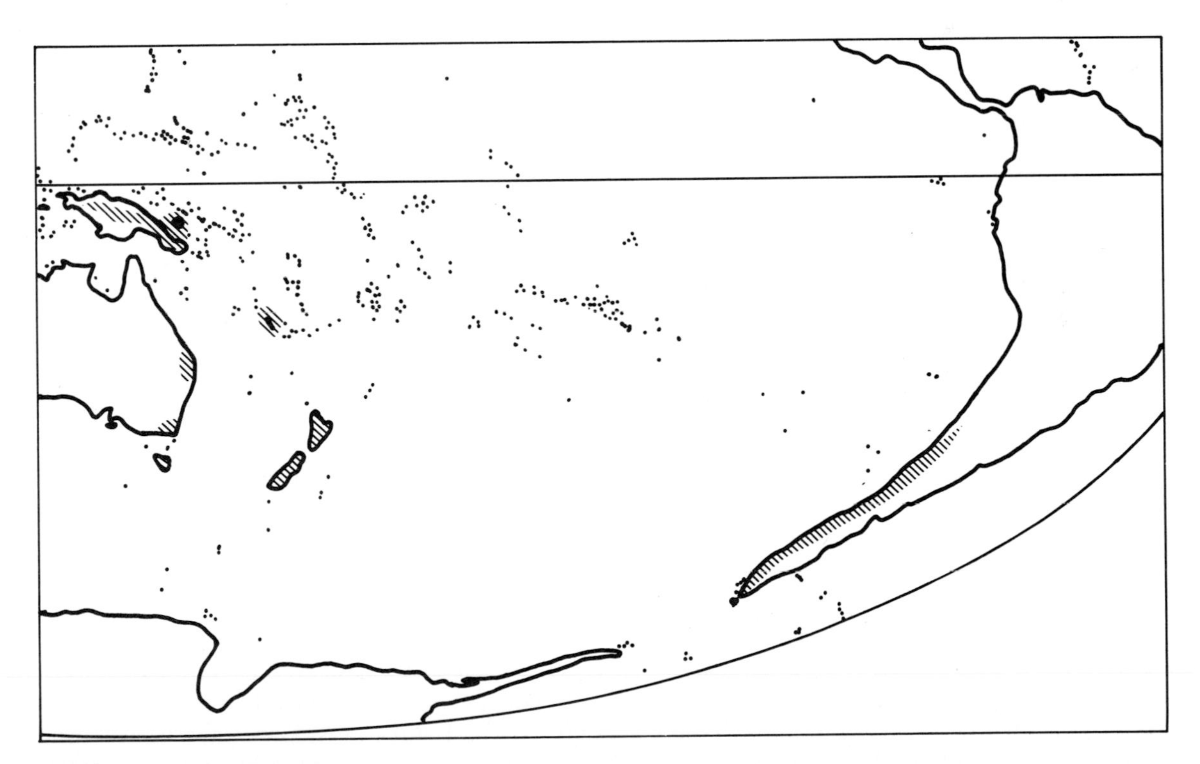

Figure 13.9
Current distribution of the Southern Hemisphere beech, *Nothofagus,* showing widely disjunct populations on several southern landmasses. Most of these are commonly thought to reflect the connection between South America and Australasia via Antarctica that persisted until the early Tertiary. (After Thorne, 1973.)

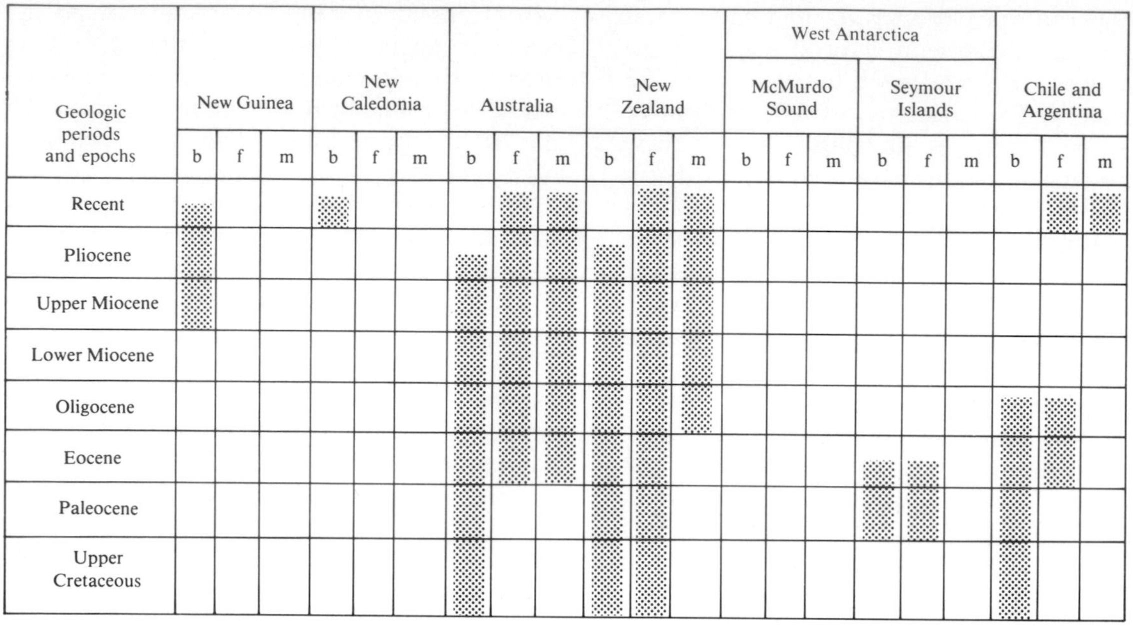

Geologic periods and epochs	New Guinea			New Caledonia			Australia			New Zealand			West Antarctica: McMurdo Sound			West Antarctica: Seymour Islands			Chile and Argentina		
	b	f	m	b	f	m	b	f	m	b	f	m	b	f	m	b	f	m	b	f	m
Recent	■			■				■	■		■	■								■	■
Pliocene	■						■	■	■	■	■	■									
Upper Miocene	■						■	■	■	■	■	■									
Lower Miocene							■	■	■	■	■	■									
Oligocene							■	■	■	■	■	■							■	■	
Eocene							■	■	■	■	■					■	■		■	■	
Paleocene							■			■	■					■	■		■		
Upper Cretaceous							■			■	■								■		

Figure 13.10
Occurrence of fossil pollen grains of the Southern Hemisphere beech, *Nothofagus* (Fagaceae), at various localities on the southern landmasses. There are three basic types of pollen grains in the genus: *brassii (b)*, *fusca (f)*, and *menziesii (m)*. All three types now occur or formerly occurred in Australia, New Zealand, and South America. The existence of *brassii* and *fusca* types in the Upper Cretaceous and Lower Paleogene of the above regions and Antarctica supports the view that *Nothofagus* forests occurred in Antarctica in the early Tertiary. (After Schlinger, 1974. Reproduced, with permission, from the Annual Review of Entomology, vol. 19. Copyright © 1974 by Annual Reviews, Inc.)

this genus does not successfully disperse across wide water gaps today, and now that geologists have shown that Australasia and South America were joined by Antarctica until the early Tertiary, phytogeographers overwhelmingly accept *Nothofagus* as strong evidence to document the effect of the Antarctic Land Bridge.

Recently Humphries (1981) attempted to use an analytical approach to determine whether *Nothofagus* can be helpful in elucidating the history of southern land connections. Do the southern beeches and cooccurring groups of animals and plants show congruent vicariant events, such that these sequences of taxonomic differentiation can be used to document the sequential breakup of the landmasses? Humphries concluded that the present reconstruction for *Nothofagus* does not share a single, unique sequence of vicariance with other four- and five-area cladograms, but instead there are several general patterns exhibited in the 20 taxa analyzed. Therefore at present there is no reason to treat *Nothofagus* as an example more scientifically superior to any other taxon. Much research must be done before a single best model can emerge, and in doing this research workers must fully evaluate an axiom of the analytical approach, that the common history of differentiation for many taxa is caused by the same geologic events.

If biogeographers must obtain a large set of south temperate disjunctions to apply analytical

Noteworthy genera of seed plants occurring in the forests of Australasia and South America

These genera, along with *Nothofagus,* may have been part of the early Cenozoic biota across Antarctica. Several taxa (e.g., *Weinmannia* and *Podocarpus*) also occur in Middle America, Africa, and Asia and were formerly widespread.

AUSTRALASIA	SOUTH AMERICA
Cupressaceae	
Diselma	*Fitzroya*
Libocedrus-Papuacedrus	*Austrocedrus-Pilgerodendron*
Podocarpaceae	
Podocarpus	*Podocarpus*
Dacrydium	*Dacrycarpus*
Acmopyle (fossils in Antarctica and India but not found yet in Australasia)	*Acmopyle*
Microcachrys	*Saxegothaea*
Dacrycarpus (fossils in Antarctica, New Zealand, Tasmania, Australia, and India)*	*Dacrycarpus*
Araucariaceae	
Araucaria	*Araucaria*
Monimiaceae	
Laurelia	*Laurelia*

AUSTRALASIA	SOUTH AMERICA
Winteraceae	
Tasmannia	*Drimys*
Lauraceae	
Beilschmiedia	*Beilschmiedia*
Cryptocarya	*Cryptocarya*
Cunoniaceae	
Weinmannia	*Weinmannia*
Geissois	*Belangera*
Saxifragaceae	
Griselinia	*Griselinia*
Proteaceae	
Orites	*Orites*
Lomatia	*Lomatia*
Oreocallis	*Oreocallis*
Kermadecia-Sleumerodendron-Bleasdalea	*Gevuina-Euplassa*
Loranthaceae	
Elytrantheae	Elytrantheae
Muellerina-Illeostylus-Tupeia	*Lepidoceras*

*Only known as fossils in Australasia (localities in parentheses).
†Extant species are found on other continents.

methods, then the best place to look for information would be among the other plants of the forest understory and floor that live in association with each other and often with *Nothofagus,* especially those forest genera having diaspores that are not adapted to cross great water gaps. Among the gymnosperms and old groups of angiosperms, the following Australasian–South American vicariants (see box).

Numerous additional taxa have the South American–Australasian distribution, but unfortunately we cannot feel as confident that all were part of the same ancient flora. Many of these not only appear to be of more recent origin but they also have seeds capable of long-distance dispersal, suggesting that they may have attained their present disjunct distributions by colonizing distant sites after the southern continents had separated. In some of the wet, cold habitats of Tasmania and New Zealand live numerous cushion plants, e.g., *Donatia* and *Phyllachne* (Stylidiaceae), *Drapetes*

(Thymelaeaceae), and *Gaimardia* (Centrolepidaceae), which reappear in similar habitats in Chile. In this case the distributions of at least *Donatia* and *Phyllachne* are believed to be fairly old because these taxa are morphologically very divergent within their family. Growing with these species are other disjunct genera, such as *Oreobolus* (Cyperaceae) and *Astelia* (Asteraceae), which also occur in Malesia and the Hawaiian Islands. These species have fleshy fruits, apparently transported by birds. Other plants of these southernmost latitudes have extended distributions in Asia, the Pacific Islands, and South America, e.g., *Coriaria, Fuchsia, Gaultheria, Gunnera,* and *Vaccinium,* each with baccate (berrylike) fruits; whereas *Oreomyrrhis* (Apiaceae), *Carpha* (Cyperaceae), *Hebe* and *Ourisia* (Scrophulariaceae), *Colobanthus* (Caryphyllaceae), and *Montia* (Portulacaceae) have small to minute seeds that are easily transported in mud on the feet of wading birds. Seeds of *Acaena* (Rosaceae) and *Uncinia* (Cyperaceae) commonly adhere to bird feathers, and the spores of ferns are like dust and easily carried by the strong prevailing westerlies from Australasia to South America and the distant Pacific Islands. In contrast, shoreline plants like *Tetragonia* and *Carpobrotus* (Aizoaceae) and *Sophora* (Fabaceae) have fruits that float on seawater, which they can tolerate physiologically. Thus although one could argue on the basis of phylogenetic relationships that all these taxa were once part of a single shared flora, the ease with which some of these taxa disperse and the presence of outlying disjunctions make us cautious not to overgeneralize.

American-African. Comparing the floras of Africa and the New World is an enormous task, and it is not at all surprising that the first good in-depth syntheses have been done fairly recently, beginning with Thorne (1973). Thorne integrated knowledge of distributions of plant groups with information on their habitats and means of dispersal. His findings may be summarized briefly as follows.

1. The floras share many taxa, including ranks from families to species. Shared species, which constitute only 0.6% of the total flora, indicate little about the historical biological relationships of these two continents because they are mostly weedy, shoreline, or aquatic plants recently introduced into the flora via long-distance dispersal and human intervention.

2. Africa and South America do not share a substantial number of families, those that are restricted either to the Old World or the New World. This decreases greatly their overall floristic similarities.

3. One third of the shared families (82) are those that have apparently penetrated southward from the Northern Hemisphere, and therefore African and South American representatives are not closest relatives (vicariants).

4. Several families held in common across the Atlantic Ocean, such as the Proteaceae (Figure 9.2), are misleading disjunctions because the taxa in South America and Africa are each related to forms in southeast Asia and Australasia, so they are disjunct but not vicariant across the Atlantic.

5. Of the tropical families shared between Africa and South America, some that are diverse on one continent are barely represented on the other. For example, there are seven American families that have only a single disjunct species in Africa: Bromeliaceae (2330 species; *Pitcairnia feliciana*), Cactaceae (1550 species; *Rhipsalis baccifera,* polymorphic in Madagascar), Loasaceae (280 species; *Kissenia,* probably one polymorphic species), Vochysiaceae (200 species; *Erismadelphus exsul*), Rapateaceae (80 species; *Maschalocephalus dinklagei),* Houmiriaceae (50 species; *Sacoglottis gabonensis),* and Mayacaceae (9 species; *Mayaca baumii)*. Each of these cases is clearly explained by one long-distance jump from South America to Africa. Several other African-Neotropical families probably have crossed the Atlantic via long-distance dispersal. Many important genera on one continent are absent or nearly so from the other, and many of these have easily dispersed diaspores.

6. The remaining core of families are pantropical. Of these, one finds a stronger phylogenetic relationship between South America and tropical Asia and between Africa and tropical Asia than between South America and Africa. Families common to both continents probably colonized them after the breakup of Gondwanaland, mostly invading each southern continent independently from the Asian tropics through the northern routes.

Raven and Axelrod (1974) used Thorne's (1968) phylogenetic classification to reevaluate Thorne's 1973 conclusions, but they presented a vastly different interpretation. These authors accepted in large part many conclusions made by Thorne (points 1 to 5), but they attempted to demonstrate that most distributions of tropical families and orders of angiosperms shared by Africa and South America do fit a vicariance model. The following are some of the pertinent points they used in developing their case.

1. Angiosperms had come into existence at least by the Lower Cretaceous, when Africa and South America were still united as a single landmass, West Gondwanaland. West Gondwanaland was, in all likelihood, the center of origin for angiosperms.

2. West Gondwanaland began to split by seafloor spreading in the Lower Cretaceous, but the two continents were still close enough to exchange diaspores until the Tertiary. During the Cretaceous, tropical groups could have migrated fairly easily between Africa and South America.

3. Throughout the Cretaceous North America was more distant from South America than the latter was from Africa, making biotic exchange more likely between the two southern continents than from the north.

4. A significant number of pantropical families, subfamilies, tribes, and genera show vicariance between Africa and South America even though they may have Asian relatives.

5. Judging from plant fossil records, many extant tropical families, even genera, probably had originated by the Paleocene and therefore were probably present in West Gondwanaland and distributed over its various parts.

6. Numerous groups that once lived on mainland Africa have become extinct, so investigators cannot properly read the historical relationships based solely on the flora of mainland Africa today. Presence of these taxa on Madagascar, which separated from mainland Africa in the Cretaceous, is evidence of what probably lived on the mainland. In this sense, the ancient flora of Africa is in part preserved on Madagascar.

7. If angiosperm groups were present in West Gondwanaland before it split up, many ancestors of modern orders, families, and even genera probably were transferred between Africa and South America more or less directly. This diminishes the need to explain the origin of a South American flora by invasion from the north, although some of this certainly occurred.

Today phytogeographers either have accepted one of these two interpretations or are caught somewhere between them. Thorne (1978) himself has acknowledged that he may have underestimated the importance of direct overland migration in West Gondwanaland and overemphasized long-distance dispersal. However, he thinks that most families, tribes, and genera originated late, in or after the Cretaceous, and immigration from the north and long-distance dispersal across the widening Atlantic Ocean were major factors in angiosperm colonization to and from South America. Only the oldest orders and families, having the appropriate disjunct distributions, poor means of dispersal, and very early fossil records, should be considered good candidates for a vicariant explanation based on direct overland migration within Gondwanaland.

North American–South American. Of special interest are those groups that span the tropics in the New World, taxa that occur in North and South America but are essentially absent in tropical latitudes. Three amphitropical patterns are recognized (Raven, 1963): bipolar disjuncts, occurring in the high latitude, subtemperate climates; temperate disjuncts, mostly intrageneric splits between southwestern South America and western North America (and occasionally eastern areas); and desert disjuncts, with members of the same genera in the horse latitude deserts of North and South America.

Because of the great distances involved, the bipolar disjunction is a remarkable pattern, recorded for 30 species or species pairs of vascular plants. Recent authors have recognized these as clear cases of fairly recent long-distance dispersal by migratory birds. The plants at the two poles are often not even distinguishable as different species; they are frequently self-compatible or apomictic, their diaspores are readily adapted for bird transport, and they live in habitats used by these migratory birds.

When discussing bipolar patterns, one example is widely cited. Empetraceae, the crowberries, is an extremely small family with a bizarre distribution (Figure 13.11). *Ceratiola* is endemic to the coastal plain of southeastern United States, *Corema* occurs on the eastern coastline of Canada and the United States as well as in Iberia and the Azores, and *Empetrum*

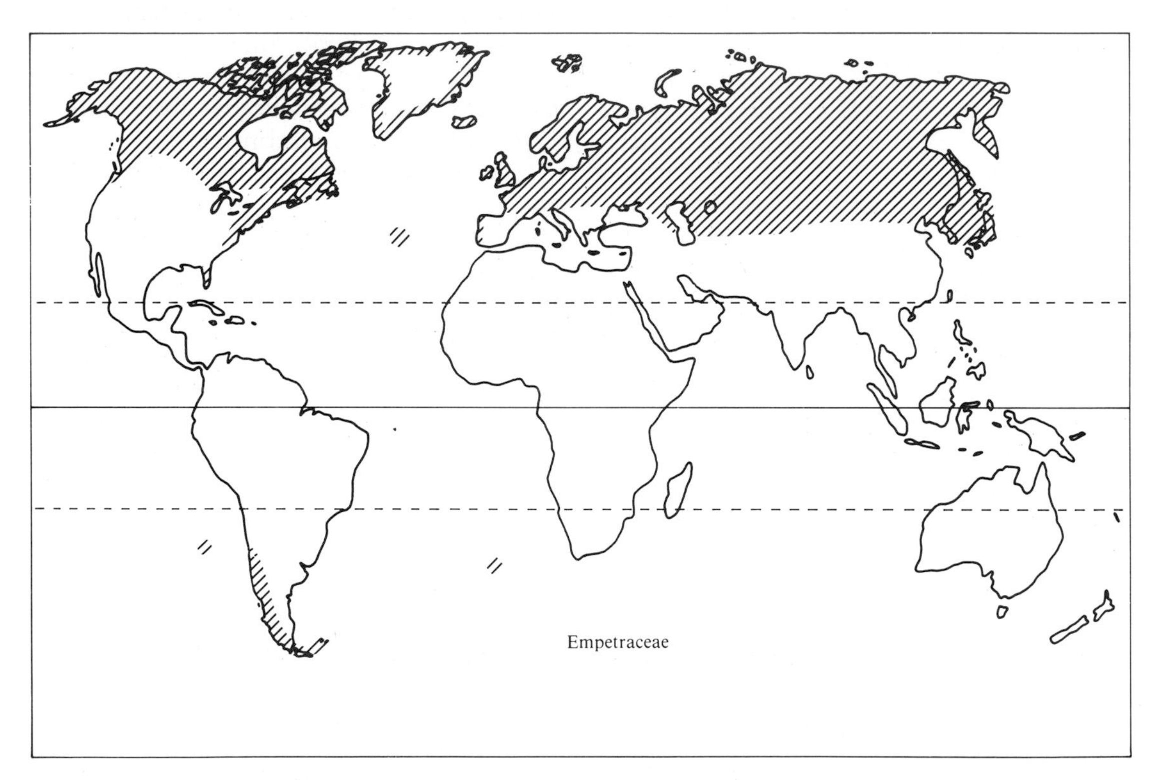

Figure 13.11
Approximate geographic range of the crowberry family (Empetraceae) showing a widely disjunct distribution. The genus *Empetrum* is widespread in the Northern Hemisphere but also occurs in southern South America and on several southern islands. Other genera occur in Portugal, western Spain, the Azores, and Florida. (After Thorne, 1972, and other sources.)

has a bipolar disjunction. The normally dioecious and diploid *E. nigrum* with black baccate fruits has a wide cicumboreal distribution in the Northern Hemisphere. Occasionally found in this northern area are populations with red fruits *(=E. kardokovii)* or polygamous or monecious individuals and black, red, or purple fruits *(=E. eamesii)*. In the Southern Hemisphere occurs the plant called *E. rubrum,* a dioecious diploid with red fruits, possibly only a variety of *E. nigrum,* with an Atlantic outpost population on Tristan da Cuhna and Gough Island having mostly monoecious flowers (Moore et al., 1970). At least the *Empetrum* distribution is easily attributed to bird dispersal, even though Croizat (1958) attached profound significance to these disjunctions to suggest land connections.

Temperate disjunctions are recorded for about 130 species, species pairs, and species complexes of angiosperms, and the phylogenetic relationships of these have been diligently pursued so the puzzle can be solved. Distributions shown in Figure 13.12 are typical of the general disjunctions observed between North and South America.

Such distributions could have come about by four means: (1) remnants of a past continuous flora (true vicariants), (2) single events of long-distance dispersal mediated by migratory birds, (3) long-distance dispersal probably mediated by migratory birds but using intermediate stations in the equatorial mountains as stepping-stones, or (4) apparent disjuncts that are not actually close relatives but rather have evolved independently on the two continents. The first interpretation is strongly rejected; these groups are minor elements of their respective biotas, and neither the dominant plants and animals nor the pathogens of these vicariants show a vicariance pattern. Moreover, there is no paleontological evidence of a past continuous temperate flora through the equatorial region. Many of the disjuncts live in habitats visited by migratory birds, and they are self-compatible, sometimes autogamous, so a new population could be founded by only one plant, even without a suitable pollinator. Many of the taxa are not montane species and cannot live in the cool habitats of the montane tropics; however, some of the taxa do have montane populations along the Andean Cordillera. In other words, of the 130 disjunct pairs, some seem likely to have migrated in steps whereas others must have made the migration in one jump.

Desert disjunctions are interpreted in a similar way, invoking single long-distance events and stepping-stone migration (Raven, 1963; Hunziker et al., 1972; Solbrig, 1972). Prominent genera of shrubs and subshrubs of North American deserts reappear in Chile and Argentina (see box on p. 401), but some also occur in northwestern South America and semiarid regions of Venezuela. Deserticolous species of *Prosopis* do not appear to be true vicariants but have evolved independently from different ancestors in the two continents.

The best studied desert disjunction is that of the creosotebush, *Larrea* (Mabry et al., 1977). *Larrea tridentata* is widespread in the deserts of North America, even having a chromosomal race in the Chihuahuan (diploid), Sonoran (tetraploid), and Mojave deserts (hexaploid), but its four related species all occur in southern South America. Most authors interpret *Larrea* as a southern group that has migrated northward, perhaps along Andean stepping-stones and probably in fairly recent times.

An opposite example is *Mentzelia* (Loasaceae), whose plesiomorphic features occur in North American species. Three sections of this genus have outlying representatives in South America: species of section *Mentzelia* are all North American except *M. grisebachii* of northern South America; *Trachyphyton* are desert annuals in the Mojave Desert of North America but reappear in the monte of Argentina; and *Bartonia* are North American, but *M. albescens* occurs in both Texas and northern Mexico and in Argentina and Chile, in both places growing with *Larrea. Mentzelia albescens* is weedy and

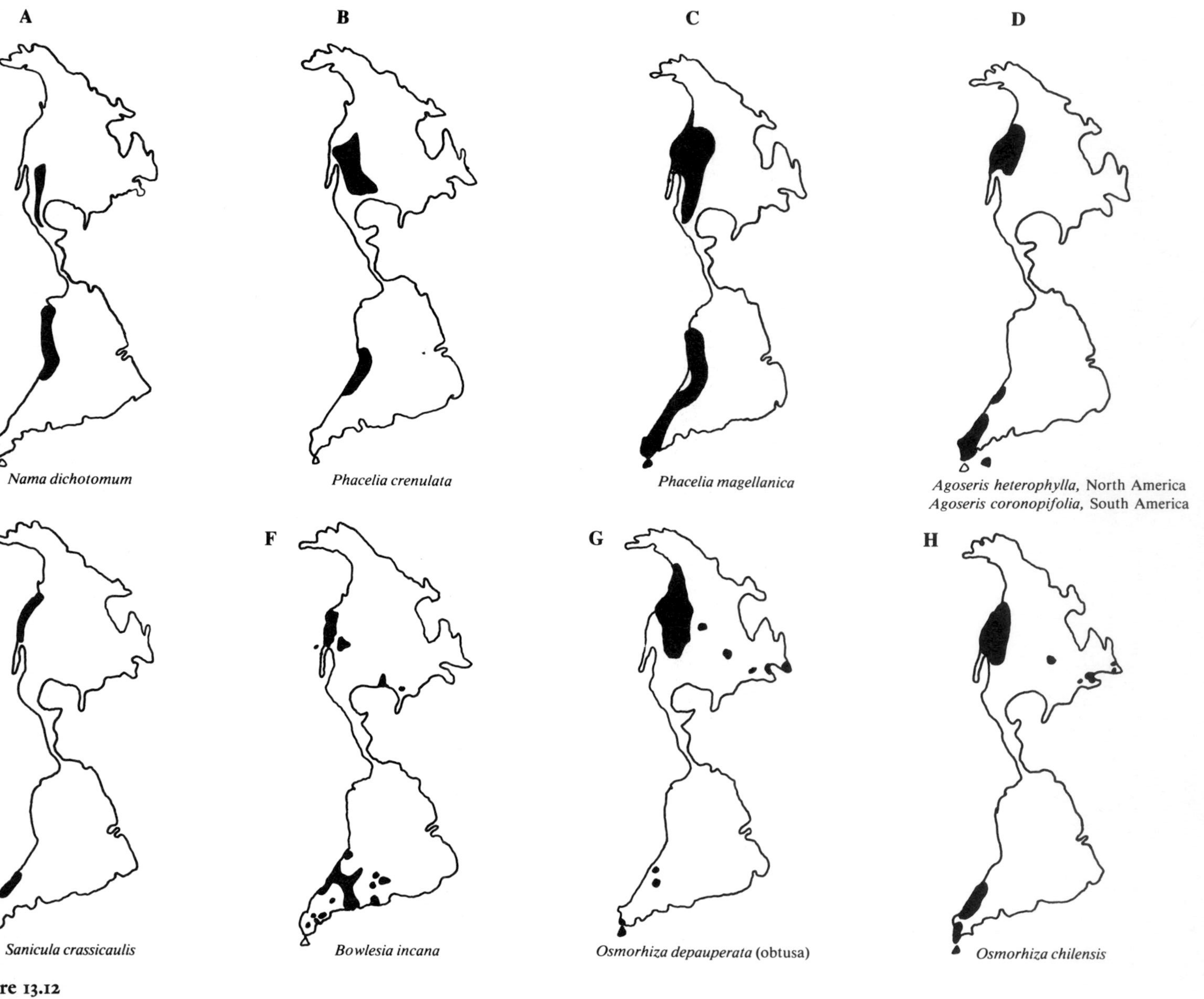

Figure 13.12
For legend see opposite page.

Figure 13.12
Several amphitropical disjuncts: **A,** *Nama dichotomum,* **B,** *Phacelia crenulata,* and **C,** *Phacelia magellanica* of the Hydrophyllaceae. **D,** *Agoseris heterophylla* (North America) and *A. coronopifolia* (South America) of the Asteraceae. **E,** *Sanicula crassicaulis,* **F,** *Bowlesia incana,* **G,** *Osmorhiza depauperata,* and **H,** *O. chilensis* of the Apiaceae. These are representative of an even greater number of plants with conspecific populations or at least very closely related species in temperate North and South America. (After Chambers, 1963, Constance, 1963, and Heckard, 1963.)

Some noteworthy genera of angiosperms occurring in desert regions of western North America (to Texas) and southern South America

Some of these genera have the same species or closely related species in the two areas, but in others populations also appear in isolated semiarid areas, e.g., in the Andean Cordillera or the tropical arid forest of Venezuela.

Aloysia (Verbenaceae)
Ambrosia (Asteraceae)
Amsinckia (Boraginaceae)
Andropogon (Poaceae)
Atamisquea (Capparaceae)
Ayenia (Sterculiaceae)
Baccharis (Asteraceae)
Blepharidacne (Poaceae)
Bothriochloa (Poaceae)
Bouteloua (Poaceae)
Calandrinia (Portulacaceae)
Cassia (Fabaceae)
Celtis (Ulmaceae)
Cercidium (Fabaceae)
Condalia (Rhamnaceae)
Cottea (Poaceae)
Cressa (Convolvulaceae)
Cryptantha (Boraginaceae)
Desmanthus (Fabaceae)
Digitaria (Poaceae)
Dissanthelium (Poaceae)
Dodonaea (Sapindaceae)
Encelia (Asteraceae)
Enneapogon (Poaceae)
Eragrostis (Poaceae)
Errazurizia (Fabeceae)
Evolvulus (Convolvulaceae)
Fagonia (Zygophyllaceae)
Flourensia (Asteraceae)
Frankenia (Frankeniaceae)
Gaillardia (Asteraceae)
Gilia (Polemoniaceae)
Glandularia (Verbenaceae)
Gochnatia (Asteraceae)
Gutierrezia (Asteraceae)
Hedeoma (Lamiaceae)
Helietta (Rutaceae)
Heliotropium (Boraginaceae)
Hoffmanseggia (Fabaceae)
Hymenoxys (Asteraceae)
Jatropha (Euphorbiaceae)
Koeberlinia (Capparaceae)
Larrea (Zygophyllaceae)
Leptochloa (Poaceae)
Lippia (Verbenaceae)
Lycium (Solanaceae)
Lycurus (Poaceae)
Malvastrum (Malvaceae)
Maytenus (Celastraceae)
Menodora (Oleaceae)
Mentzelia (Loacaceae)
Mimosa (Fabaceae)
Muhlenbergia (Poaceae)
Pappaphorum (Poaceae)
Pectis (Asteraceae)
Phragmites (Poaceae)
Polygonum (Polygonaceae)
Portulaca (Portulacaceae)
Proboscidea (Martyniaceae)
Prosopis (Fabaceae)
Salvia (Lamiaceae)
Schkuhria (Asteraceae)
Scleropogon (Poaceae)
Senecio (Asteraceae)
Setaria (Poaceae)
Sida (Malvaceae)
Stipa (Poaceae)
Tessario (Asteraceae)
Tribulus (Zygophyllaceae)
Trichloris (Poaceae)
Vallesia (Rubiaceae)
Verbena (Verbenaceae)
Verbesina (Asteraceae)
Willkommia (Poaceae)
Ziziphus (Rhamnaceae)

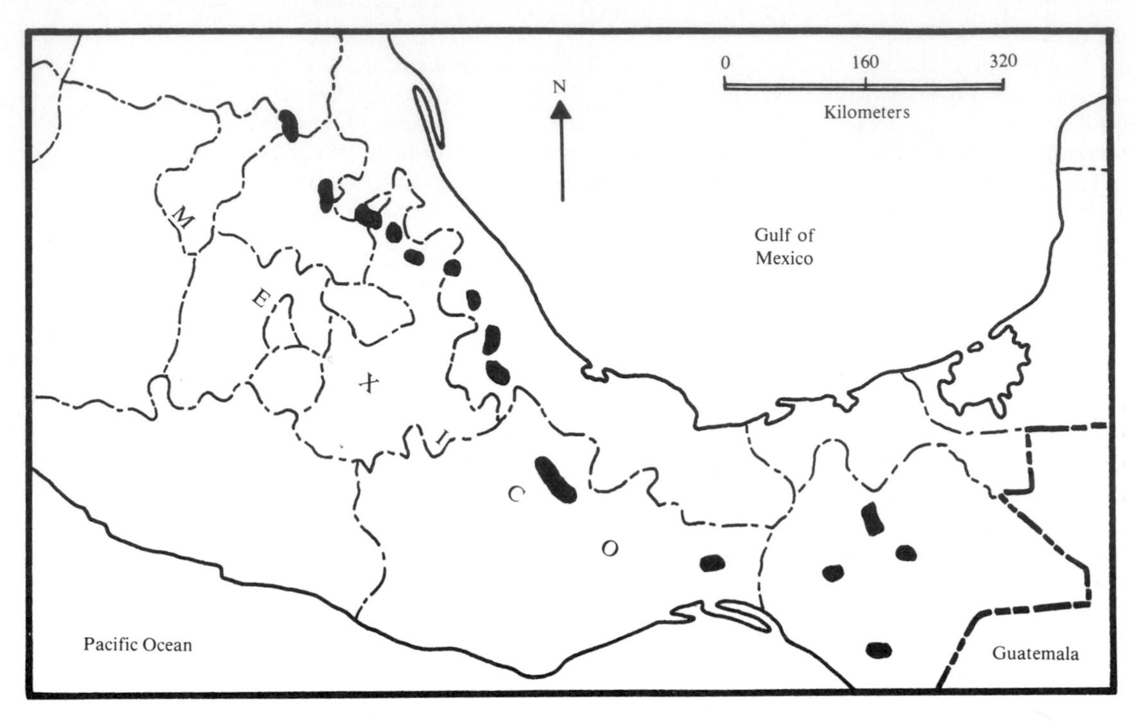

Figure 13.13
Discontinuous distribution of the sweet gum *(Liquidambar)* in Mexico. This tree is widespread in the deciduous forests of the eastern United States, and it extends southward to Guatemala as a series of isolated populations in the cool climate at high elevations. *Liquidambar* is found in all of these sites, but other forms typical of temperate deciduous forests also occur in these areas. (From F. Miranda and A.J. Sharp. Characteristics of the vegetation in certain temperate regions of eastern Mexico. *Ecology* 31[3]:313-333.)

autogamous. Thus two disjunct groups, whose phylogenies are well understood, grow together although their origins and directions of migration are exactly opposite. This is an exceedingly common theme in many broad plant disjunctions. Data like these do not negate a conclusion that for many other disjunctions the phylogeny and migration of plant clades appear to be entirely unidirectional, as in the historical development of the flora of the Galápagos Islands (Porter, 1979), but this makes us wary not to use plant distribution maps uncritically.

As Thorne (1972) has noted, almost all taxa have discontinuous distributions, and we emphasize mainly those with large gaps. However, intracontinental plant disjunctions are equally interesting, especially as they relate to the processes of speciation. In the Sierra Madre Oriental, Sierra Madre Occidental, and Sierra Madre de Chiapas of tropical and subtropical Mexico there are isolated local regions with temperate deciduous forest, characterized by sweet gum, *Liquidambar* (Figure 13.13). Many of the associated species of vascular plants are also those found in the summer-green deciduous forests of the eastern United States (see box on p. 403). In addition, there are numerous other disjunctions between the eastern United States and these temperate Mexican forests (Miranda and Sharp, 1950), and a small number of

Some species of vascular plants occurring in sweet gum *(Liquidambar)* temperate deciduous forests

Species listed occur in Hidalgo and Pueblo, Mexico, and also (or related species) in similar forests of the eastern United States. Some of these have populations in western North America or even close ties with species in the Orient. (After Miranda and Sharp, 1950.)

Acer negundo (box elder)
Berchemia scandens (supplejack)
Carpinus carolinianas (ironwood)
Carya mexicana (hickory)
Cercis canadensis (redbud)
Chimaphila maculata (spotted wintergreen)
C. umbellata (pipsissewa)
Cystopteris fragilis (bladder fern)
Drosera brevifolia (sundew)
Epifagus virginianas (beech drops)
Equisetum laevigatum (horsetail)
Fagus mexicana (beech)
Gelsemium sempervirens (yellow jessamine)
Gratiola virginiana (hedge hyssop)
Hypericum hypericoides (Saint Andrew's cross)
H. mutilum (Saint John's-wort)
Leucothoë mexicana (leucothoe)
Liquidambar macrophylla (sweet gum)
Lycopodium clavatum (running pine)
L. complanatum (ground cedar)
Lobelia cardinalis (cardinal flower)
Mitchella repens (partridgeberry)
Nyssa sylvatica (blackgum)
Osmunda cinnamomea (cinnamon fern)
O. regalis (royal fern)
Ostrya virginiana (hophornbeam)
Parthenocissus quinquefolia (Virginia creeper)
Prunella vulgaris (self-heal)
Prunus serotina (black cherry)
Rhus trilobata (skunkbush)
Sambucus mexicana (elderberry)
Smilax bona-nox (greenbrier)
S. glauca (greenbrier)
S. hispida (greenbrier)
Tovara virginiana (Virginia knotweed)
Toxicodendron radicans (poison ivy)
Triodanis perfoliata (Venus' looking glass)
Xolisma ferruginea (staggerbush)

moss species have similar distributions (Delgadillo M., 1979). The current interpretation, supported by the fossil record, is that a forest corridor of temperate elements existed from the eastern United States through southern Texas and into northern Mexico during the mid-Tertiary. Therefore these Mexican outposts are true vicariants, remnants of this paleoflora. In the middle latitudes some of these same genera, with others, are disjunct between eastern and western North America, eastern Asia, and western Europe, and a rich fossil record shows that here too a broad forest belt once existed (Wolfe, 1971; Graham, 1972).

Regional endemism and patterns of speciation in plants

In the southern tip of Africa exists one of the richest floras in the world. Thousands of plant species in the Cape Region and nearby deserts and semiarid grasslands are endemics. Four angiospermous families (sensu Thorne, 1981), three of these in suborder Brunineae (order Pittosporales), and three other families and two subfamilies extend only a short distance northward into neighboring provinces and countries. *Stangeria eriopus* of the monotypic cycad family Stangeriaceae is endemic to coastal districts in open grassveld and forest of Natal and the Cape Province, while on the Atlantic side in the coastal fog desert of Angola and Namibia lives yet another gymnospermous oddity, *Welwitschia bainesii* (Welwitschiaceae).

Turning to lower taxonomic levels, patterns of high endemism are equally intriguing. Of the more than 1900 plant genera and 18,500 species indigenous to southern Africa (Dyer, 1975; Goldblatt, 1978), 150 genera have 25 or more species in southern Africa and about 20 have more than 100 species. Tremendous speciation has occurred in certain families, producing enormous numbers of endemic species: irises (Iridaceae; 800 species) and lilies (Liliaceae and Amaryllidaceae; 1100 species); legumes (Fabaceae; 1100 species) and daisies (Asteraceae; 1800 species); stem and leaf succulents of ice plants and living stones (Aizoaceae; approximately 1800 species), carrion flowers (Asclepiadaceae), Geraniaceae, Crassulaceae, and Euphorbiaceae; the incredibly beautiful Proteaceae (325 species); and even some lesser known families like Restionaceae (300 species), Rutaceae (300 species), and Santalaceae. Most remarkable are the heaths of South Africa (Ericaceae), some 23 genera of them, including 800 species. *Erica* itself has about 600 species in the Cape region, and in one small district around Caledon in the extreme southwestern tip, an area of 5000 km^2, reside 220 species, nearly all very narrowly endemic (Baker and Oliver, 1967). At the other extreme, the world's largest plant genus, *Astragalus* (nearly 2000 species) has but a single species present in southern Africa. These facts of distribution can only stir excitement in the curious observer. What are the factors controlling regional distinctiveness and richness in plants? Some of these issues are discussed in Chapters 17 and 18.

In the search for disjunctions, workers tend to look past the fact that many plant taxa, even some families, are fairly narrowly endemic. For example, there are almost 80 families and subfamilies of angiosperms endemic or nearly so to the Neotropics (see box on p. 405) and 55 endemic to Africa and its neighboring islands (see box on p. 406). Tribes, subtribes, and genera are frequently much more restricted. Hence in the Asteraceae, the largest plant family, whose tribes are often cosmopolitan, half the genera are restricted to a small geographic area, such as only a portion of a continent or an island. For example, the five genera of subtribe Zinniinae are North American, and in subtribe Madiinae 11 genera are restricted to either California or western North America (Stuessy, 1977). Interestingly, the genus *Madia* is a temperate disjunct between North and South America, and the remaining three genera (*Argyroxiphium, Dubautia,* and *Wilkesia*) of Madiinae, which are endemic to the Hawaiian Islands, certainly have an American origin (Carlquist, 1965).

An exceptional degree of regional endemism typically suggests some combination of unusual ecological conditions and geographic isolation for enough time to permit new forms to evolve, speciate, and become the characteristic taxa of the flora. Isolation may be produced as an island formed distantly from a mainland, such as the Hawaiian Islands and St. Helena; as a parcel of land once connected to a landmass and now separated by water, such as Cuba and probably New Caledonia; or as a segment of a continent with long-term physiognomic and climatic barriers limiting biotic exchanges with adjacent floras, such as the Cape Region of South Africa or southwestern Australia. All these examples boast 90% specific endemism. Another feature of high endemism is environmental heterogeneity, great habitat diversity over short distances, often caused by dissected mountainous landscapes with isolated valleys and peaks, mosaics of soil types, and in some regions, patchy semiarid habitats. These factors encourage plant populations to differentiate and specialize for particular microhabitats. Under such environmental conditions, genetic drift and natural selection can result in rapid speciation and adaptive radiation to produce high numbers of endemic plant forms, especially in places like semiarid western Peru, southern Africa, southwestern Australia, California, and the Galápagos Islands.

Endemism in Australasia. No phytogeographer can resist the temptation to discuss, at least briefly, patterns of endemism in Australasia, especially Australia and New Caledonia. In addition to the summary in various monographs on the islands of Australasia (Williams, 1974; Kuschel, 1975; Keast, 1981; Gressitt, 1982), there are some very useful lists of the endemic plant groups for Australia (Burbridge, 1960; Beard, 1977; Beadle, 1981); and nearly all discussions on endemism in Australasia have made some attempts to explain how the Australian

Families and subfamilies of angiosperms endemic or nearly so to the Neotropical region

Aextoxiacaceae
Agdestioideae of Phytolaccaceae
Alstroemerioideae of Liliaceae
Alvaradoroideae of Simaroubaceae
Asteranthoideae of Lecythidaceae
Bixaceae
Brunelliaceae
Calyceraceae
Caryocaraceae
Catopherioideae of Lamiaceae
Columellioideae of Saxifragaceae
Cyclanthoideae of Cyclanthaceae
Cyrillaceae
Dictyolomatoideae of Rutaceae
Duckeodendroideae of Solanaceae
Eremolepidaceae
Goetzeaceae
Gomortegaceae
Goupioideae of Celastraceae
Gyrocarpoideae of Hernandiaceae (except *Gyrocarpus americanus*)
Halophytaceae
Henriquezioideae of Rubiaceae
Houmiriaceae*
Lactoridaceae (Juan Fernández Island)
Lecythidoideae of Lecythidaceae
Ledocarpaceae
Lennoaceae
Leonioideae of Violaceae
Lithophytoideae of Verbenaceae
Lophophytoideae of Balanophoraceae
Luzuriagoideae of Liliaceae
Malesherbiaceae
Marcgraviaceae
Mayacaceae*
Misodendraceae
Morkillioideae of Zygophyllaceae
Neotessmannioideae of Tiliaceae
Nolanoideae of Solanaceae
Pakaraimoideae of Dipterocarpaceae
Pellicieroideae of Theaceae
Peridiscaceae
Phytelephantoideae of Arecaceae
Picramnioideae of Simaroubaceae
Plocospermatoideae of Loganiaceae
Quillajeoideae of Rosaceae
Quiinaceae
Rapateaceae*
Rhabdodendroideae of Rutaceae
Siparunoideae of Monimiaceae
Stegnospermataceae
Styloceratoideae of Buxaceae
Tetralicoideae of Tiliaceae
Theophrastoideae of Myrsinaceae
Thurnioideae of Juncaceae
Tovarioideae of Capparaceae
Tropaeolaceae
Vellozioideae of Velloziaceae
Vivianiaceae
Vochysiaceae*

*Taxa with African taxa.

Families and subfamilies of angiosperms endemic or nearly so to the Ethiopian region

Island endemics are noted in parentheses.

Achariaceae
Asteropeioideae of Theaceae (Madagascar)
Barbeuiaceae (Madagascar)
Barbeyaceae
Brexioideae of Saxifragaceae
Capunonianthoideae of Meliaceae (Madagascar)
Caryotophoroideae of Aizoaceae
Curtisioideae of Cornaceae
Didiereaceae (Madagascar)
Didymeleaceae (Madagascar)
Diegodendroideae of Ochnaceae (Madagascar)
Dioncophyllaceae
Dirachmoideae of Geraniaceae (Socotra)
Foetidioideae of Lecythidaceae
Geissolomataceae
Geosiridoideae of Iridaceae (Madagascar)
Gilgiodaphnoideae of Thymelaeaceae
Glossocalycoideae of Monimiaceae
Greyiaceae
Grubbiaceae
Hoplostigmataceae
Huaceae
Hydrostachyaceae
Hymenogynoideae of Aizoaceae
Jollydoroideae of Connaraceae
Kirkioideae of Simaroubaceae
Medusagynaceae (Seychelles)
Medusandraceae
Melianthaceae
Mesembryanthemoideae of Aizoaceae
Monotoideae of Dipterocarpaceae
Myrothamnaceae
Mystropetaloideae of Balanophoraceae
Neomangenotioideae of Meliaceae
Oliniaceae
Penaeaceae
Pentadiplandroideae of Capparaceae
Psiloxyloideae of Myrtaceae (Mascarenes)
Ptaeroxylaceae
Quivisianthoideae of Meliaceae
Retzioideae of Loganiaceae
Roridulaceae
Ruschioideae of Aizoaceae
Sarcolaenaceae (Madagascar)
Sarcophytoideae of Balanophoraceae
Scytopetalaceae
Sphaerosepalaceae (Madagascar)
Strephonematoideae of Combretaceae
Stilboideae of Verbenaceae
Wellstedioideae of Verbenaceae (Socotra)

Table 13.2

Families and subfamilies of angiosperms endemic or nearly so to Australasia

Included in Australasia are Australia, Tasmania, New Guinea, New Caledonia, and New Zealand. The closest extant relatives are mostly those proposed by Thorne (1981).

Taxon	Distribution outside Australasia	Hypothesized closest relative and its distribution
Akaniaceae	Argentina (fossils)	Sapindaceae of Australasia to southeast Asia
Alseuosmioideae of Saxifragaceae		Saxifragaceae of Australasia to southeast Asia
Amborellaceae		Chloranthaceae of Malesia
Anarthrioideae of Restionaceae		Restionaceae of Australia
Austrobaileyaceae		Other Laurineae of Malesia
Balanopaceae	Fiji	Daphniphyllaceae of Malesia and Buxaceae of the world
Baueraceae		Cunoniaceae of Australasia
Brunonioideae of Goodeniaceae		Goodeniaceae of Australia
Byblidaceae		Other Pittosporineae of Australasia
Carnavonioideae of Proteaceae		Proteaceae of Australasia
Cephalotaceae		Saxifragaceae of Australasia and Malesia
Chamaelaucioideae of Myrtaceae		Myrtaceae of Australasia
Chloanthoideae of Verbenaceae		Verbenaceae of Australasia and Malesia
Corynocarpaceae	New Hebrides	Cunonineae of Australasia and Malesia
Dactylanthoideae of Balanophoraceae		Balanophoraceae of Australasia
Dampieroideae of Goodeniaceae		Goodeniaceae of Australasia
Davidsoniaceae	Fiji	Cunonineae of Australasia to southeast Asia
Ecdiocoleaceae		Restionaceae of Australia
Emblingioideae of Sapindaceae		Unknown Sapindaceae
Eremosynoideae of Saxifragaceae		Saxifragaceae of tropical Asia
Eupomatiaceae		Magnoliaceae and Himantandraceae of Australasia to southeast Asia
Gyrostemonaceae		Dodonaeoideae of Australia
Hectorelloideae of Portulacaceae		Portulacoideae of the Southern Hemisphere
Himantandraceae	Malesia	Magnoliaceae of Malesia and southeast Asia
Hydalellaceae		Uncertain
Idiospermoideae of Calycanthaceae		Calycanthaceae of the Orient
Micrairoideae of Poaceae		Unknown Poaceae
Oncothecaceae		Theineae of Australasia and Malesia
Paracryphiaceae		Theineae of Australasia and Malesia
Persoonioideae of Proteaceae		Proteaceae of Australasia
Phellinaceae		Theineae of Australasia and Malesia
Prostantheroideae of Lamiaceae		Unknown Lamiaceae
Sphalmioideae of Proteaceae		Proteaceae of Australasia
Sphenostemonaceae		Theineae of Australasia and Malesia
Stackhousiaceae	Philippines	Celastraceae of Malesia and southeast Asia
Strasburgeriaceae		Unknown Theales
Stylobasioideae of Sapindaceae		Dodonaeoideae of Australia
Tetracarpoideae of Saxifragaceae		Unknown Saxifragaceae
Tremandraceae		Pittosporineae of Australasia
Trimeniaceae	Fiji, Marquesas	Chloranthaceae and Monimiaceae of Malesia
Wittsteinioideae of Epacridaceae		Epacridaceae of Australasia
Xanthorrhoeoideae of Liliaceae		Probably Dracaenoideae of southeast Asia

flora was assembled and which groups have lived and radiated in Australia for fairly long periods of time.

There are over 40 families and subfamilies of angiosperms endemic or nearly so to Australasia (Table 13.2). As with the Ethiopian region (see box on p. 406), these Australasian endemics have very few species, and a number of them are monotypic families. One question that may be asked is whether these Australasian endemics are products of vicariance from the departure of Australasia from Africa, Madagascar, India, and South America. Table 13.2 notes the names and distributions of the hypothesized sister taxa of Australasian endemic families and subfamilies (Thorne, 1981); in most cases the most closely related taxon also occurs in the Australian region, and half are also native to tropical southeast Asia, including the large continental islands. Hence one suspects that this set of Australian endemics arose mostly by internal events on the Australasian plate itself, although the separation of Australia from India may have also played some role in the origin of Australian forms. There are, however, 18 angiospermous families that are endemic to or have maximum development on the three southern continents (Beadle, 1981). As discussed earlier in this chapter (p. 395), some very old angiospermous families, such as Proteaceae, Monimiaceae, Lauraceae, Winteraceae, Cunoniaceae, and Saxifragaceae, have disjunctions between Australasia and southern South America that are most easily explained by the presence of an Antarctic land bridge between them until the Eocene. The occurrence in Madagascar of some typically Australian taxa of the Proteaceae, Winteraceae, and Cunoniaceae (to name the best examples) may signify that Madagascar was still part of that landmass when these families were in existence.

Most difficult to deal with are the families that have endemic species in both southern Australia and southern Africa, such as Restionaceae, Aizoaceae, Philesiaceae, Pittosporaceae, and Hypoxidaceae, which might have been transferred between continents via Gondwanaland connections but which also could represent either cases of long-distance dispersal or broader distributions that have been restricted to the southern latitudes in fairly recent times. Unfortunately, most workers in vicariance biogeography have concentrated on the relationships between Australasia and South America, and the analytical approach has not been used to evaluate whether a general pattern emerges for the landmass of Africa, Madagascar, India, and Australia–New Guinea.

More than 75% of the species and 32% of the genera in Australia are endemic, but there are in addition about 100 genera that are nearly endemic to Australia and most certainly originated there. Excellent examples are *Eucalyptus* (about 450 species) and *Melaleuca* (140 species), two large Australian genera of the myrtle family (Myrtaceae), which dominate many vegetation types with the proteads (Proteaceae) and wattles (*Acacia*, Fabaceae) on the continent. In fact, Myrtaceae (1300 species) is the largest family in Australia, comprising 10% of the flora. This illustrates how certain plant families have radiated in isolation on Australia, a story that parallels the radiations of marsupials (Chapter 11), birds, and insects (Chapter 12) in the absence of groups more typical of the other continents. In fact, many of the groups characteristic of Eurasia and Africa occur only in the northeastern corner of Australia, in the lowland tropical forests of Queensland, and here they appear to be recent immigrants from southeast Asia via New Guinea. Pollen records show now that these rain forests in Queensland were eucalypt woodland less than 10,000 years ago (Flenley, 1979a).

One of the outstanding cases of an endemic flora is the southwestern corner of Australia, which has been summarized in an evolutionary manner by Carlquist (1974). So distinctive is this region that phytogeographers classify it as a floristic province, at the same level as the endemic flora of South Africa (Good, 1974), which has similar climatic properties but very different families composing the flora. The genera of shrubs and herbs especially have differentiated into many species, many of which have narrow ranges and are restricted to special soil types or other microhabitats. Researchers have just begun to unravel the historical and ecological reasons why so many species (about half the total of Australia) have come to reside in such a relatively small region of the earth.

As mentioned in Chapter 8, certain islands pose difficulties for interpreting present-day distributions. One of the most perplexing is New Caledonia, a small landmass with continental rocks. Nestled between and equidistant from Queensland, New Guinea, and New Zealand, New Caledonia has one of the most remarkable endemic floras, including about 75% endemism of species, high generic endemism, and at least five endemic families, as well as many groups with primitive features for angiosperms (Thorne, 1965, 1969). Nonendemic species mostly occupy disturbed habitats or are some of the common species of the closest islands. The fauna of the island is extremely disharmonic. Living on New Caledonia are no native mammals or amphibians, but seven species of bats, one freshwater fish (*Neogalaxias,* a close relative of the secondary division *Galaxias* of Australasia), some interesting lizards, some earthworms and land snails, and so forth (Holloway, 1979). A terrestrial snake occurs on the nearby Loyalty Islands but has not been found yet on New Caledonia. Thus the fauna is unbalanced, with some important groups totally absent.

Without adequate geologic information to describe the origin of New Caledonia, biogeographers have sought a synthesis based on the distributions of its endemic organisms. When the phylogenetic affinities of the plants are evaluated, strong similarities are found with the floras of Queensland and New Guinea but fairly weak ones with New Zealand. On the other hand, the strange and endemic kagu bird (Rhynochetidae) of New Caledonia, the mayflies, certain cerambycid beetles, and some moths (Holloway, 1979) have closer phylogenetic ties to New Zealand. To which, if any, of these landmasses was New Caledonia connected at one time by a land bridge? Much more phylogenetic and geologic research is needed to answer this stimulating question.

Special soil types. Plant endemism is especially affected by soil type because roots are stationary and therefore the plant must be able to tolerate the range of physiological conditions imposed by the local soil-microclimate interaction. In semiarid regions like southern Africa and southwestern United States and Mexico, the patchiness of soil types has played a noticeable role in the distribution and evolution of many plant groups. Botanists suspect that plant-soil characteristics are developed in ways similar to the way plants colonize chemically harsh environments, like the generally toxic tailings of zinc mines (Antonovics and Bradshaw, 1970; Antonovics et al., 1971; Antonovics, 1971). Rare phenotypes of a species are able to grow on the bare and toxic surface. This phenotype increases in frequency on the tailings, so that its gene frequencies become different than the parental stock. As local populations specialize to grow on the unusual soil, they gradually lose their ability to compete with other plants, so they cannot successfully grow and reproduce on the adjacent normal soil. Eventually the mine population could become a race and then a new endemic species. Evidence from populations that colonized tailings of mines within historic times suggest that such evolution can be surprisingly rapid.

This scenario is likely for unusual substrates, such as serpentine, gypsum, limestone, strongly acid, saline, sodic, and selenium soils, as well as soils with high concentrations of the toxic heavy metals, but less extreme versions are conceivable for soils differing only in water-holding capacity or pH. The indirect evidence for this is the diversity of soil preferences, particularly among members of the same genus. On the other hand, other families are remarkably conservative, e.g., Ericaceae on acid soils and Chenopodiaceae often on highly saline or alkaline soils.

Directional selection in plants apparently can result in rapid evolution of soil preferences and probably also of pollination systems. Thus the location of such wide or narrow endemics often will not resemble an idealized model of allopatric speciation but will instead mimic the distribution of that agent controlling differentiation. Special soils occur irregularly and are often greatly disjunct, so the species living on that soil type probably never had a continuous range because the habitats have not been continuous. Long-distance dispersal would have to overcome the distances between sites. Endemics on gypsum outcrops in Mexico are excellent examples of plants that never had continuous distributions (Powell and Turner, 1977).

Plant polyploidy. Another aspect of plant speciation influences how plant distributions can be used for interpreting historical patterns. As noted in Chapter 6, polyploidy is exceedingly common in vascular plants—probably 50% of all species have some polyploidy ancestry—and the majority of these are allopolyploids, the hybrid of two markedly different taxa. Under these circumstances reticulate evolution with repeated branching and joining of lineages is considered common, e.g., in certain families and genera of California plants such as oaks *(Quercus)*, California lilac *(Ceanothus)*, columbines *(Aquilegia)*, larkspurs *(Delphinium)*, and sages *(Salvia)*, to mention just a few (Grant, 1963, 1971; Stebbins, 1971b). The allopatric model of speciation is simply inappropriate to represent these phylogenetic relationships, although they can be represented in a cladistic model.

In this century much research has been devoted to identifying the distribution patterns of polyploids, although interest in the subject peaked in the 1960s, when workers sought but failed to find consistent patterns. Polyploids may differ from diploids in the following ways (Johnson et al., 1965): (1) flowering plants, especially monocotyledons, of high latitudes and at high elevations tend to be polyploids, particularly those occupying recently glaciated areas or cold, wet soils; (2) many polyploids tend to do best in open, pioneering, or highly disturbed habitats; and (3) in cold climates many polyploids have some form of apomixis and are often autogamous. As the data accumulated, it became increasingly difficult to defend any one explanation, although the significance of polyploidy for wider distribution is believed to be the greater genetic variability conferred by the possession of two different genomes (Pielou, 1979). Exceptions to these generalizations abound, and earlier conclusions were challenged by showing that frequency of polyploidy was also related to the taxa involved, there being higher polyploidy in certain families and in perennials. Also, many widespread diploids are known, indicating that polyploidy is not required for broad distributions of a species, and some polyploids have extremely narrow ranges. Polyploids are equally successful as weeds as are diploids, but polyploidy is generally low among annuals.

Although polyploidy does show geographic trends, e.g., higher frequencies in arctic areas, polyploidy in high latitudes cannot be easily separated from other adaptations, such as apomixis, self-compatibility or autogamy, and physiological and morphological specializations that appear to facilitate establishment, growth, and reproduction under stressful conditions. Polyploidy appears to be part of an overall colonizing strategy, and it is certainly effective as an isolating mechanism, permitting populations with a new ecological combination to achieve a different distribution than their parents (Lewis, 1979).

Distribution patterns of nonvascular plants

Relatively few biogeographic discussions mention the distribution patterns of algae and bryophytes. The phylogenetic relationships of nonvascular plant species, genera, and even families are incompletely understood, and we expect few historical patterns to appear because the orders are very old and the taxa have seemingly easy means of dispersal by resistant spores. However, biogeographers have been finding that many distributions are fairly conservative. In the brown algae (Phaeophyta), order Fucales has several widespread genera, but a large number of narrow endemics occur in Australasia (Nizamuddin, 1970). The kelps (order Laminariales) show different patterns. There are 16 genera that have primarily northern distributions, but three of these have amphitropical disjunctions. On the other hand, kelps in *Ecklonia* and *Lessonia* predominantly occur in the Southern Hemisphere, but each has disjuncts occurring around Korea and Japan (Dzitzer, 1975). Whereas local occurrence probably reflects types of substrates and predation pressure (Lubchenco, 1980; Souza et al., 1981), broad distributional limits of brown algae often appear to be limited by high temperature (with different maximum temperatures for different groups), which stops the growth or kills the gametophytes and may suppress sexual development. In contrast, red algae (Rhodophyta) are mostly tropical and subtropical forms, with the reverse set of abiotic requirements.

Schuster's paper of 1969 on bryophytes was one of the first attempts to interpret plant patterns in light of continental drift theory. Species of Jungermanniales have strong phylogenetic relationships between Australasia and South America. Since then many biogeographic accounts have demonstrated that bryophytes can be used to interpret some past events (e.g., Schofield and Crum, 1972; Delgadillo M., 1979), and recently distribution patterns of lichens have attracted attention (Sheard, 1977). However, in many groups investigators are still awaiting better collections of lower plants and their taxonomic monographs.

Selected references

Cain (1944); Good (1974); Polunin (1960); Wulff (1943).

Factors limiting growth and reproductive success of plants

Bannister (1976); Barbour et al. (1980); Berg (1975); Billings (1952); Daubenmire (1978); Eyre (1968); Grime (1979); J. Harper (1977); Larcher (1980); Levitt (1980); Nobel (1974); Solbrig (1980); G. Stebbins (1971b, 1974); Strain and Billings (1974); Whittaker (1975).

Relationships of vegetation type to climate

Beals (1969); Braun-Blanquet (1965); Daubenmire (1943, 1978); DeLaebenfels (1977); Hare (1954); Holdridge (1947); Holdridge et al. (1971); Köppen (1900); Köppen and Geiger (1936); Major (1963, 1977); Mather (1978); Penman (1956); Pielou (1979); Plumstead (1973); Raunkiaer (1934); Schnell (1970–1971); Schulze and McGee (1978); Thornthwaite (1931, 1948); Thornthwaite and Hare (1955); Thornthwaite and Mather (1955, 1957); Walter (1971, 1973, 1979); Walter et al. (1975); Walter and Lieth (1967); Whittaker (1978b); Wolfe (1971, 1978, 1979).

The origin of angiosperms

Axelrod (1952, 1960, 1970); Bakker (1978); Beck (1976); Bierhorst (1971); Carlquist (1975); Croizat (1958); Cronquist (1981); Dahlgren (1980); Dahlgren et al. (1981); Doyle (1969, 1977, 1978); Hughes (1976); J. Muller (1970); R. Raven and Axelrod (1974); A.C. Smith (1967, 1970, 1973); G. Stebbins (1974); Takhtajan (1969, 1980); Thorne (1963, 1976a, 1976b, 1981).

Major disjunct ranges of plants

Axelrod (1972); Balgooy (1966, 1971); Beard (1977); Carlquist (1981); Constance (1963); Darlington (1965); Florin (1963); Fryxell (1967); Goldblatt (1978); Gómez (1982); Graham (1972, 1973); Humphries (1981); Leroy (1978); Li (1952); Melville (1981); Meyer (1966); Miranda and Sharp (1950); D. Porter (1979); Quezel (1978); P. Raven (1963, 1972); P. Raven and Axelrod (1972, 1974, 1975); Schlinger (1974); Solbrig (1972); Steenis (1962); Thorne (1972, 1973, 1976a, 1976b, 1978); D. Valentine (1972); Vester (1940); Wolfe (1975, 1978); Wood (1971, 1972).

Regional endemism and patterns of speciation in plants

Antonovics (1971); Antonovics and Bradshaw (1970); Antonovics et al. (1971); Axelrod and Raven (1982); Beadle (1966, 1981); Bradshaw (1971); Burbridge (1960); Carlquist (1965, 1974); Dyer (1975); Goldblatt (1978); V. Grant (1963, 1971); Harlan and DeWet (1975); Hopper (1979); A. Johnson and Packer (1965); A. Johnson et al. (1965); M. Johnson and Raven (1973); Kruckeberg (1954, 1969); H. Lewis (1966); W. Lewis (1979); Pielou (1979); Powell and Turner (1977); P. Raven (1964); G. Stebbins (1971b, 1974); Thorne (1969, 1976a).

Distribution patterns of nonvascular plants

Bisby (1943); Delgadillo M. (1979); S. Jeffrey (1981); Lubchenco (1980); New Zealand Department of Scientific and Industrial Research (1979); Nizamuddin (1970); Schofield and Crum (1972); Schuster (1969); Sheard (1977); Wagner (1972).

Chapter 14 Quaternary Events

Throughout this book we have included observations on biogeographic events occurring in the Quaternary, i.e., within the last 1.7 million years (Haq et al., 1977). By devoting a full chapter to the Quaternary we are admitting that these events had profound effects on the distributions of plants and animals; but by leaving the discussion of these events for the last chapter in the unit on historical biogeography, we are stating that such changes have not erased the patterns resulting from earlier events. Quaternary changes have mostly involved shifts in distributions within the continents, causing some groups to diminish in importance and some to become extinct, while others assumed prominence by moving into new or recently vacated habitats. In the Holocene (the last 11,000 years), human activity has greatly interfered with more natural changes, sometimes by destroying previous climax vegetation types. This frustrates efforts to understand historical patterns. In places with a long history of dense human populations and intensive agriculture, alterations in vegetation have been so extensive that the most reliable information comes from the fossil record.

Pleistocene glaciation

Any discussion of the Quaternary logically begins with Pleistocene glaciation. There is nothing unique about this event, because large-scale glacial episodes have occurred numerous times in the last billion years (Hambrey and Harland, 1981); but the latest episodes of glaciation are so well documented that one may hope to understand their causes, geographic distributions, and effects on world climate as well as their effects on the past and present distributions of organisms.

The extent of glaciation. From the Miocene until the beginning of the Quaternary, global temperatures generally decreased, and in the Pleistocene there followed alternating periods of cold and warm climates. Investigators can estimate these temperature changes by several qualitative methods (Chapter 13) or more quantitatively by plotting oxygen isotope ratios in marine fossils. Most marine exoskeletons are composed of calcite ($CaCO_3$), made by combining water and carbon dioxide. In water, the two common isotopes of oxygen are ^{16}O and the heavier ^{18}O. During periods of high temperature, more water that is composed of ^{16}O is preferentially evaporated from the ocean surface than of the heavier isotope. As global water temperatures fall and glaciers form, the $H_2{}^{16}O$ is bound up as ice and the ratio of ^{18}O to ^{16}O in the ocean increases. Consequently, during colder periods organisms incorporate ^{18}O at a progressively higher rate. Hence by determining $^{18}O/^{16}O$ ratios for marine cores, one can determine how the temperature of the water has changed over time (Figure 14.1).

Broadly generalizing, a variety of evidence shows that there were four periods of very cold temperatures for the Northern Hemisphere in the Pleistocene, the first starting around 600,000 years BP. At these times, glaciers in the Arctic (present in the Pliocene) expanded into the lowlands and mountains of middle latitudes in Eurasia and North America and during the

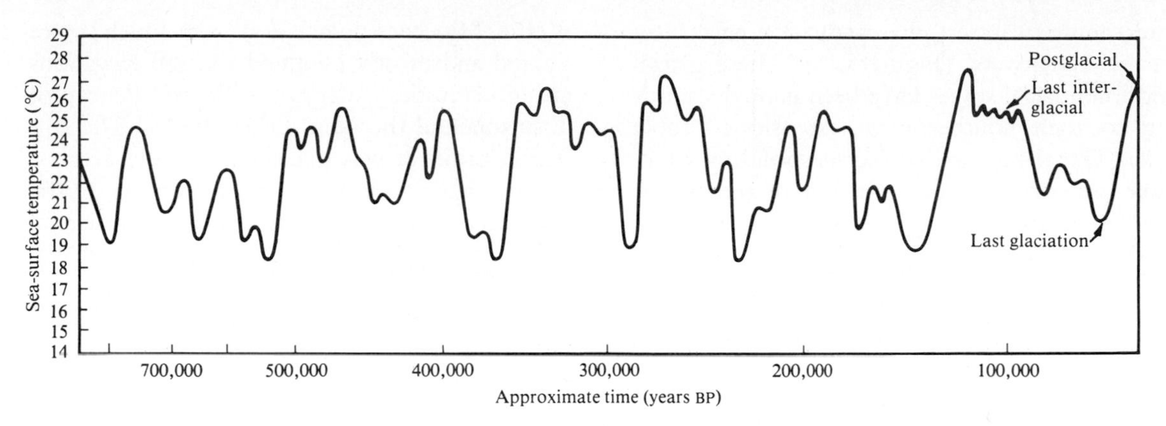

Figure 14.1
Sea-surface temperature variations at a typical tropical locality over the last 800,000 years. This pattern has been reconstructed from the $^{18}O/^{16}O$ isotope values obtained from materials in the skeletons of surface-dwelling Foraminifera deposited in deep ocean sediments. Warmer temperatures indicate interglacial periods, but not all colder water temperatures correspond to glacial episodes.

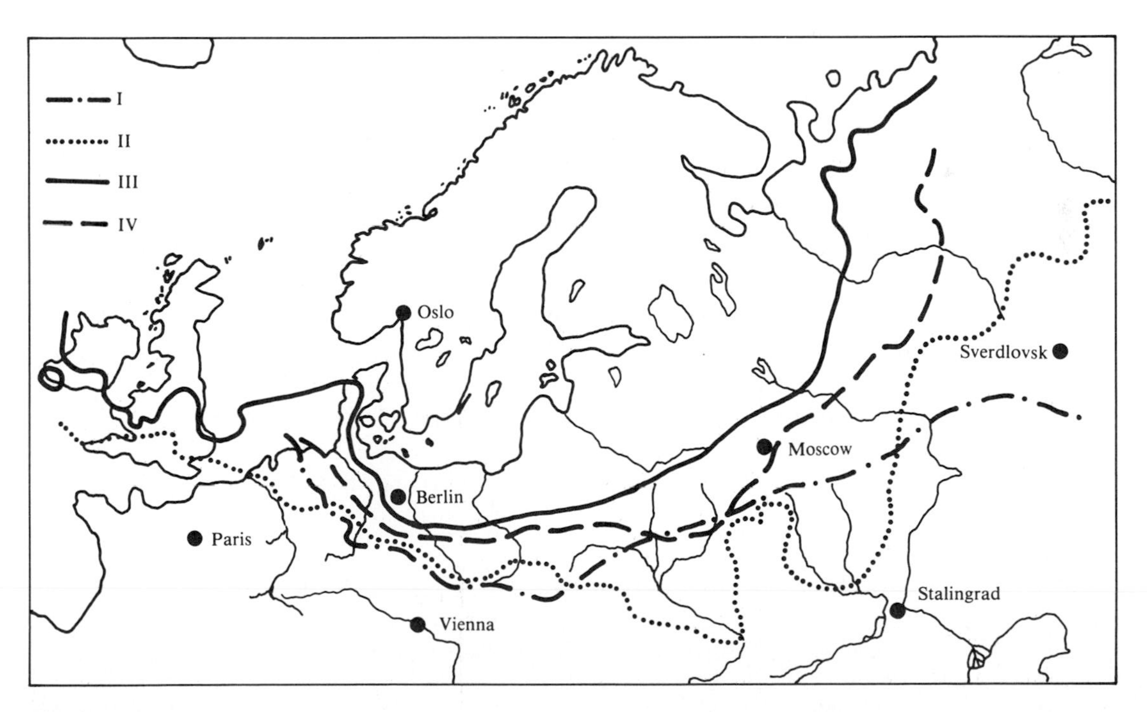

Figure 14.2
Limits of the continental ice sheets in northern Europe at various times during the Pleistocene. (After West, 1977.) The lines (I to IV) indicate the southernmost limit of the four major glacial ice sheets (oldest to youngest).

intervening warmer phases the ice sheets retreated northward (Figure 14.2). These glacial and interglacial stages have been named according to their southernmost extensions (Table 14.1). The names and the corresponding timetable are convenient to use, but of themselves they give us only approximate time ranges when land was available for colonization. For any given locality one must consult specific paleofloras to obtain exact information on when the glacier disappeared and to provide information on the local climate. Moreover, some areas were covered by glaciers almost continually during the Quaternary (e.g., Greenland, which still has a huge ice sheet), some during each of the four separate intervals, some fewer times, and others not at all. Furthermore, when oxygen isotope profiles like Figure 14.1 are studied closely, they show that temperatures fluctuated wildly during the Pleistocene, even during the glacial and interglacial episodes, warning one not to overgeneralize about the movements of plants and animals during these fluctuating climates.

Most authors discuss northern glaciation episodes because over 80% of the glacial ice occurred there. In the Southern Hemisphere (except in Antarctica, where glaciers began to form in the Miocene), glaciation was much more limited and mostly confined to high elevations at high latitudes, such as the Central Plateau of Tasmania and the New Zealand Alps. The Andean Cordillera was glaciated, but the greatest ice coverage was in Chile and Argentina. In the Pleistocene mainland Australia was unglaciated except for a small mountain range in Victoria, and Africa lacked glaciation except in the Palearctic Atlas Mountains of the extreme northwestern corner and in the high eastern mountains. None of the glacial areas in the Southern Hemisphere shows the four discrete glacial maxima of the northern latitudes (Flint, 1971).

Nonglaciated areas in northern latitudes. Because each successive glacial stage not only denuded the landscape but also scrubbed away most existing records of previous biotas, historical biogeographers studying the Pleistocene are left with two basic questions: Where did the organisms go after they were displaced, and how were the biotas reestablished? Obviously, this is the clearest case that exists for massive immigration of organisms into new continental areas. For example, in much of northern North America within the last 18,000 years (the glacial maximum of the Wisconsin), organisms have invaded totally barren habitats, formed new

Table 14.1
Sequence of glacial and interglacial stages in various regions of the Northern Hemisphere
The sequence described has occurred during the last 600,000 years (youngest to oldest). (After West, 1977.)

	North America	European Alps	British Isles	Northern Europe	U.S.S.R. (European)
Fourth glaciation	Wisconsin	Würm	Devensian	Mecklenburgian	Waldai
Interglacial	Sangamon	Riss/Würm	Ipswichian	Neudeckian	Mikulino,
Third glaciation	Illinoian	Riss		Polandian	Moscow
Interglacial	Yarmouth	Mindel/Riss		Helvetian	Odintzovo
Second glaciation	Kansan	Mindel	Wolstonian	Saxonian	Dnieper
Interglacial	Aftonian	Günz/Mindel	Hoxnian	Norfolkian	Lichwin
First glaciation	Nebraskan	Günz	Anglian	Scanian	Oka

communities, and in many cases been replaced again by other species, which are thriving today.

During glacial periods when northern regions were covered with ice, most organisms obviously moved to more southerly latitudes if they did not become extinct; but reconstructing the patterns of displacement and resettlement has not been simple. Many fairly simplified maps have been published to represent the shifting distributions of Pleistocene vegetation. Figure 14.3 serves to illustrate their common features. First, during glacial maxima biomes were shifted from 10° to as much as 20° latitude south from where they are located today (compare with Figure 4.12). Second, the biomes, i.e., tundra, subtemperate boreal forest or grassland, and temperate forest or semiarid scrub or grassland (Chapter 4), occupied the same relative positions north-to-south as they do today. This is because the climatic belts of the earth (Chapter 2), then as now, created zonal patterns of vegetation, although the zones were compressed during glacial episodes. Finally, north-south mountain ranges permitted arctic and boreal taxa to extend their ranges far southward in cool situations.

Two primary types of evidence allow workers to reconstruct vegetation maps for the Pleistocene; the analysis of pollen floras throughout

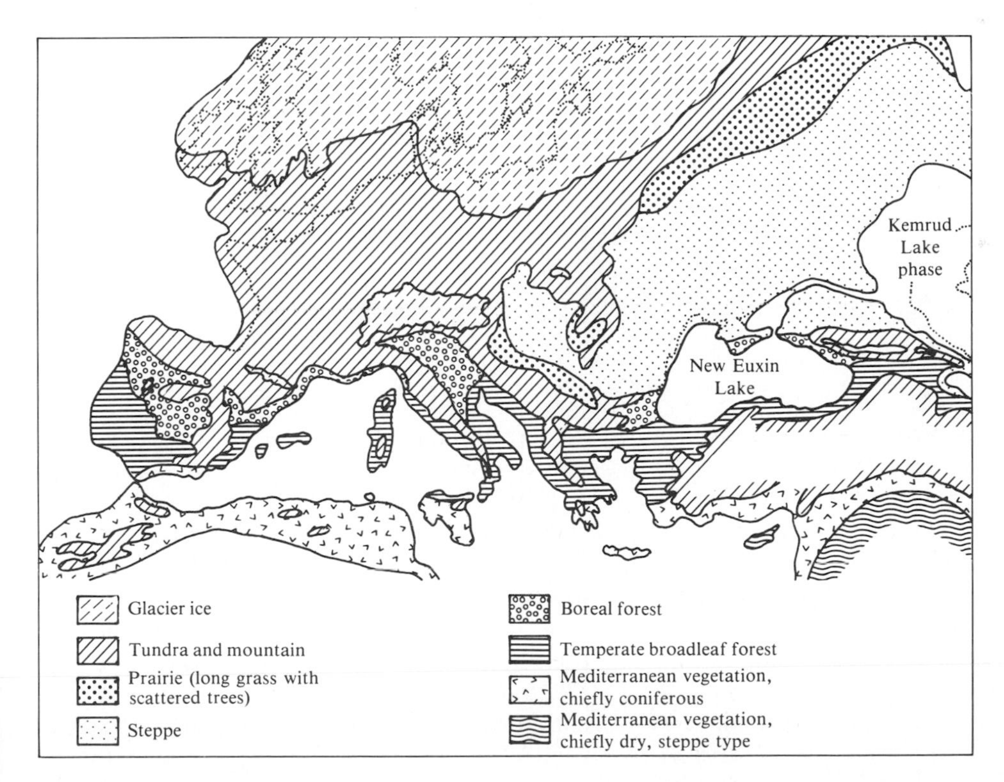

Figure 14.3
Reconstructed zones of vegetation in Europe during the Würm glacial maximum (18,000 years BP). Major vegetation types were shifted southward of their present locations by 10° to 20° of latitude, and the precursors of the Black Sea and the Caspian Sea were interconnected. (After Flint, 1971.)

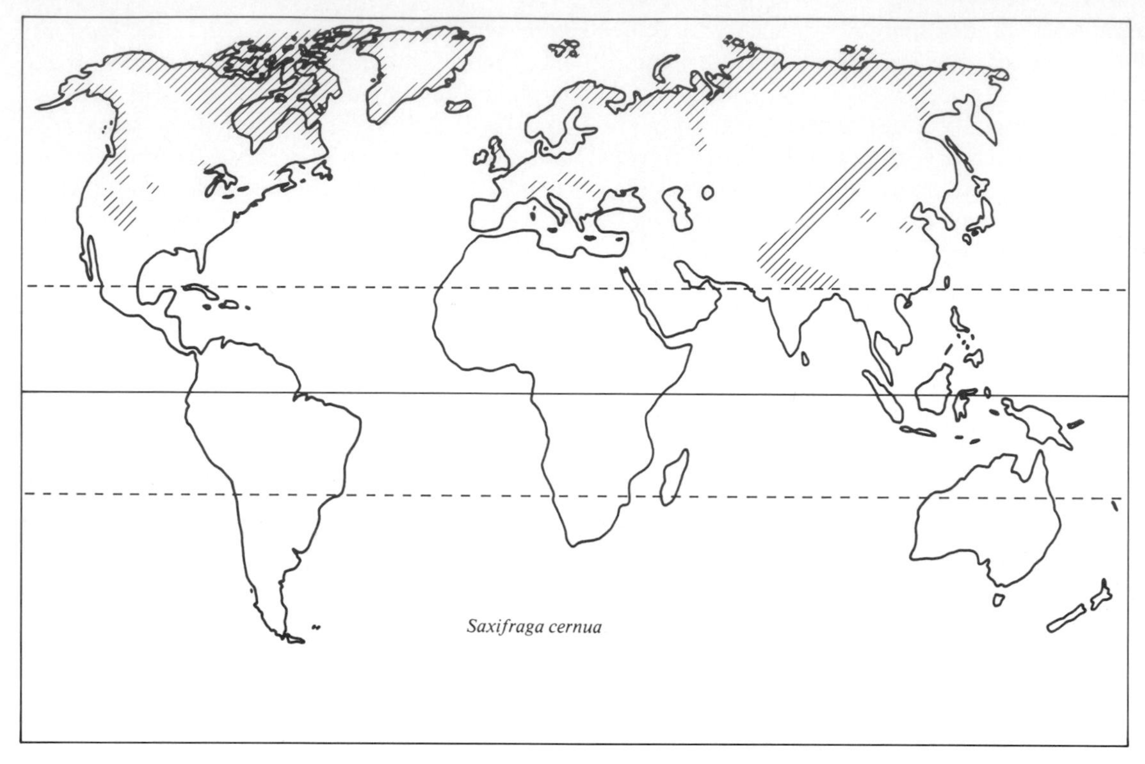

Figure 14.4
Disjunct distribution of the arctic plant *Saxifraga cernua* (Saxifragaceae) in northern and mountainous regions of the Northern Hemisphere. Like many plant and animal species, the range of this form shifted southward during glacial periods. Isolated relict populations have survived in cool climates at higher elevations since the glaciers retreated and the climate warmed. (After Hultén, 1964.)

the region and the occurrence today of disjunct arctic (Figure 14.4) or subtemperate boreal taxa in the southern mountains. It can be shown by fossil pollen that the typical vegetation of southern regions during the glacial periods were the same genera that now have relict populations in the mountains.

The problem of any reconstruction of late Wisconsin floristics is that the vegetation near the ice sheet was undoubtedly unstable and shifting continuously as the climate and the position of the glacial front changed. The data are not sufficiently precise to map the complex and dynamic relationships of these events on a very fine temporal or spatial scale.

Returning to the maps of the glacial and interglacial stages one can see that all of northern Europe (including here European U.S.S.R.) (Figure 14.2) and nearly all of the northern three quarters of North America had sheets of ice during the last glacial maximum. In contrast, much of Siberia and northern Alaska show no geologic evidence of glaciation in the Pleistocene. The reasons why these areas apparently remained unglaciated are still not known completely. Although it is known that biotas migrated generally southward before the advancing glacial front, plant and animal species of the Arctic could have persisted in places such as the Yukon and Siberia. Such refugia could

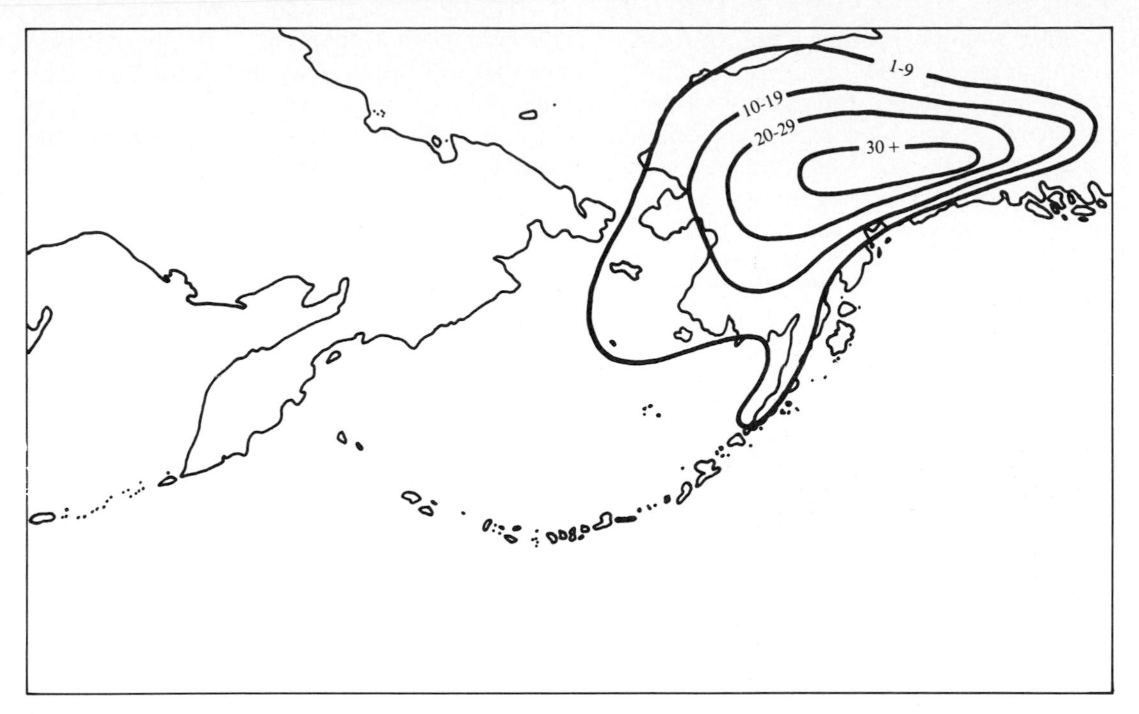

Figure 14.5
High frequency of endemic species in central Alaska was used by Hultén (1937) to advance the idea that a large part of Beringia remained unglaciated and served as a refugium for arctic forms during the Pleistocene. Each contour indicates the number of narrow endemics present in that region.

have served as a source of recolonization during interglacial times.

This thesis, that Beringia was a Pleistocene refugium, was clearly presented by the famous phytogeographer, E. Hultén (1937) (Figure 14.5) based on the study of plant distributions. Hultén's work is a classic, one of the best efforts to document exact plant distributions on a broad scale. It should be required study for all biogeographers because his exact distribution maps stimulated a creative synthesis.

Although Hultén's idea was ahead of its time, several recent lines of evidence help to confirm that Beringia was a vegetated region during the Wisconsin glacial maximum. First, Colinvaux (1981) and others have analyzed fossil pollen records and shown that treeless tundra occurred in western and central Alaska at the height of the Wisconsin glacial. Second, Hopkins and Smith (1981) have rather strong evidence for the occurrence of deciduous dicotyledonous trees and larch (*Larix,* Pinaceae) in the Yukon at this same time. Finally, Weber et al. (1981) have excavated late Pleistocene mammalian fossils (age from 40,000 years) in the interior of Alaska, finding many animals common then in open vegetation such as tundra and steppe.

A popular topic in Quaternary phytogeography is the nunatak. As mentioned earlier (Chapter 8), ***nunatak*** is a special term for a refugium that persists within an ice sheet. Nunataks could possibly develop on the southern continental margins of a glacier, as isolated pockets along a mountain range, e.g., on the tops of the highest peaks, along a coastline where bluffs are too steep for glaciers, or between adjacent ice sheets. Geologists have identified several nunataks. In North America the most famous nonglaciated region within the Laurentide ice sheet was the so-called "driftless area," in southern Wisconsin and adjacent Illinois and Iowa (Figure 14.6, *A*), an elliptical area that was bypassed by the glacial front. The Laurentide ice sheet covered most of northeastern and north central North America, but the Cordilleran glacier complex occurred in westernmost Canada, extending from the Canadian Rockies and adjacent southern Alaska (through the Aleutians) to the coast, and including small

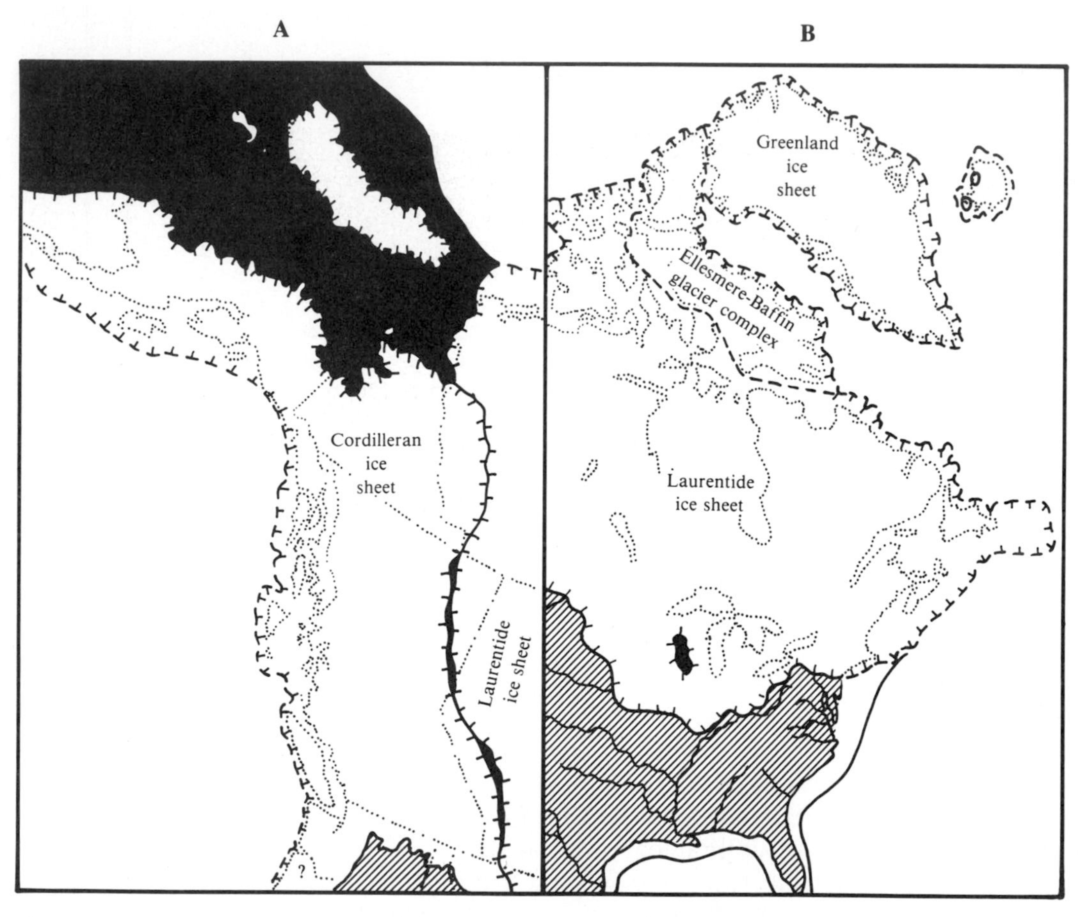

Figure 14.6
Several nunataks, isolated patches of unglaciated land surrounded by continental glaciers, have been proposed by various authors. Among the most widely accepted nunataks are; **A,** a series of unglaciated areas *(black)* between the Laurentide and Cordilleran ice sheets in western North America and, **B,** the "driftless area" *(black)* surrounded by the Laurentide ice sheet in what is now Wisconsin and Illinois. (After Flint, 1971.)

portions of northern Washington, Idaho, and Montana. A number of narrow nunataks separated these large expanses of ice (Figure 14.6, *B*).

Numerous investigators have used plant distributions as evidence for other northern nunataks, but the response to these, especially of geologists, has not always been favorable. Nunataks are alleged to have occurred where one finds small, isolated pockets of arctic-alpine vegetation that are especially rich in species, including rare and endemic forms; these same regions occur in the east and in the west. Examples are two alleged nunataks in Norway, which contain some interesting disjunctions. Here is found the Lapland rosebay (*Rhododendron lapponicum*), which occurs in Greenland and in the arctic and alpine regions of the New World and Asia but not elsewhere in Europe. Pielou (1979) has discussed the many problems in interpreting these as nunataks. The most serious is that it is difficult to demonstrate that these were not areas of recolonization following the retreat of the ice sheets at the end of the Pleistocene.

Sea level changes in the Quaternary. Pleistocene glaciation had great effects on sea levels. At glacial maxima in the Northern Hemisphere nearly 29% of all land was covered with thick glaciers, and sea ice also occurred in both polar regions. The consequence of this was removal from the ocean of a large volume of water, probably exceeding 50 million km^3 of ice (in excess of present-day ice sheets in Antarctica and Greenland). Removing that much water caused sea levels to drop substantially.

Biogeographic and geologic authorities differ on how much the sea level was lowered during the Pleistocene. Some biogeographers have used values as high as 200 m because this would be sufficient to connect all the major continental islands of southeast Asia out to Wallace's line. In contrast, Quaternary geologists find values ranging from 80 to 110 m more in keeping with calculations based on ice volume estimates and the various submerged geologic structures, such as benches, terraces, deltas, shelves, bars, and fossilized terrestrial mammals located within 100 m of the ocean's surface. However, even a lowering of the sea level by 100 m exposes a considerable area of the continental shelf and would allow land bridges to form between major land areas, e.g., between Australia and both New Guinea (Torres Strait) and Tasmania (Bass Strait); between Siberia and Alaska (much of the Bering Strait); and from mainland Asia out to many islands.

Two geologic phenomena complicate our simple estimates of where land bridges may have formed. First, a well-documented geologic event is isostasy, which means that continents or parts of continents are floating on a viscid layer of the mantle, the asthenosphere. The heavier the block, the deeper it will float in the asthenosphere. During glaciation, the presence of heavy glaciers on landmasses causes them to downwarp; when ice is removed, the land block rebounds. This uplifting (called updoming or upwarping) after glaciation is substantial, e.g., over 300 m for some land areas in the high northern latitudes. Hence when one cites a value of sea level change for a glaciated region, one must also adjust the value according to the amount the land has rebounded. In tropical regions this is not a concern, but even here local uplifting or subsidence could have occurred at any time in the past and may substantially change the length and width of local land bridges. Consequently, changes in sea level must be assessed cautiously. Biological evidence may often be more reliable than crude geologic estimates.

Some Pleistocene events in the tropics

The lowland tropical rain forests (Chapter 4) are awe-inspiring structures, teeming with an amazing diversity of life. Traditionally such places have been regarded by many investigators as stable refuges that persisted virtually unchanged for many millions of years while cli-

Figure 14.7
Pleistocene forest refugia in Amazonia that have been identified by Prance (1981) because they are centers of endemism for woody angiosperm plants. During glacial episodes most of the Amazon Basin was grassland or savanna habitat, and tropical rain forest was restricted to isolated patches as shown here. These refugia were important centers of differentiation and speciation of tropical forest biota. Compare this reconstruction based on plant endemism to the proposed refugia for birds and lizards in Figure 12.6.

mates became harsh and fluctuating elsewhere. However, two recent lines of evidence suggest that these rain forests have experienced marked changes in climate, so that within the last 40,000 years, the rain forests have varied greatly in distribution, sometimes being much more restricted than at present.

Neotropical Pleistocene refugia. In the Amazon Basin, the most extensive continuous rain forest in the world today, there are known to be centers of high endemism for both plants and animals. For example, Haffer (1969, 1974, 1978, 1981) has identified six principal centers in which about 150 species of birds are narrowly restricted (Figure 12.6). Frequently these centers are inhabited by subspecies of the same species or similar species of the same superspecies. This pattern is repeated in families of reptiles, amphibians (Vanzolini and Williams, 1970; Dixon, 1979, Lynch, 1979), woody plants (Vuillemier, 1970; Prance, 1973, 1978, 1979), and butterflies (Brown, 1976, 1977), although a greater number of centers can be identified for plants and butterflies.

The insular patterns of endemism in many different groups in generally homogeneous habitats have suggested to Haffer and others that these centers were refugia, islands of lowland forest that were left isolated by habitat changes during glacial periods. A worldwide lowering of temperature would theoretically cause all biomes to shift to lower elevations following the appropriate temperature and precipitation regime. There is also evidence to suggest that many lowland tropical areas experienced drier as well as cooler climates during glacial periods. Vegetation better able to tolerate low temperatures and more arid conditions would have replaced many tropical lowland forest sites, leaving isolated stands in the warmest and wettest regions. Such fragmentation into insular habitats would have been a vicariance event, permitting isolated populations to differentiate as new species.

This model predicts that when lowland rain forest expanded and reformed into continuous forest across Amazonia, organisms from the various refugia likewise expanded their ranges and came in contact with related populations from adjacent centers. Not only do we observe such zones of secondary contact, but in insects, lizards, and birds these zones are regions of frequent introgressive hybridization between the closely related forms. Complete reproductive isolation between many close relatives apparently has not been achieved during their most recent isolation.

The evidence in favor of and against this explanation for high speciation in Amazonia was reviewed in a symposium (Prance, 1982). The pattern is not without some exceptions, opponents, and complicating examples, and the genealogical relationships of all taxa involved are only beginning to be scrupulously worked out using vicariance methods and tests for congruence. Moreover, there is evidence of similar climatic and habitat changes on the tropical islands of the West Indies (Pregill and Olson, 1981), similar patterns and processes not obvious in tropical regions of other continents, making investigators wonder why the Neotropics are so different. Nevertheless, the model has led to important emphasis on Pleistocene effects on climate and distribution in nonglaciated areas.

Palynological evidence for tropical shifts. Whereas the centers of endemism imply widespread changes in climate and vegetation, a more direct way of assessing this is to study changes in pollen records. For equatorial latitudes, the small but growing list of tropical pollen records has been summarized by Flenley (1979), with interesting results. For a baseline of comparison, Flenley reviewed the relationship of pollen rain to extant tropical vegetation along gradients of elevation and aridity. These data clearly indicate that in equatorial rain forest habitats all around the world the zone of tropical lowland forest has shifted its elevational distribution substantially over the last 30,000 years. In Figure 14.8, pollen profiles of a Colombian mountain in South America diagra-

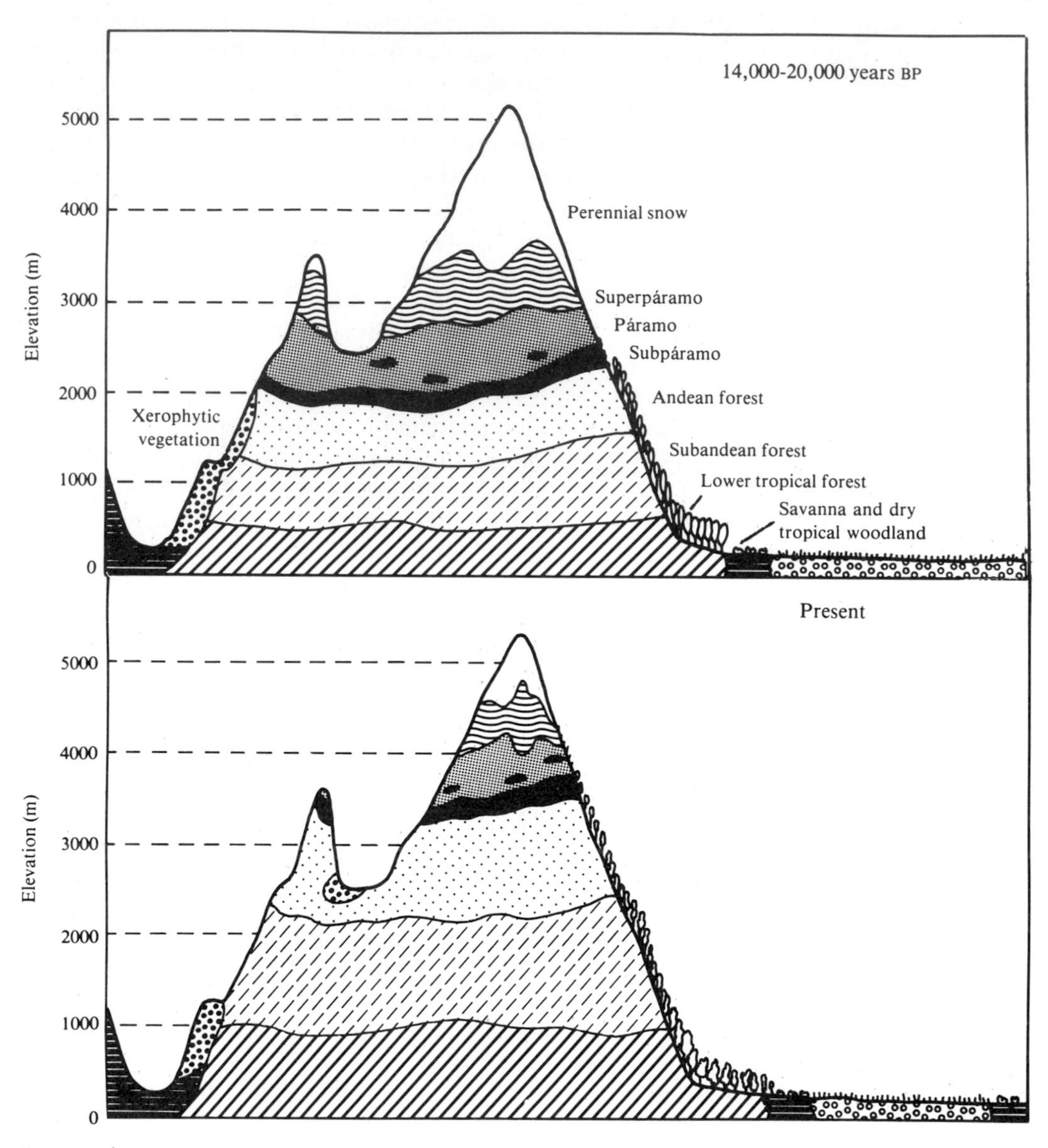

Figure 14.8
Reconstruction of the vegetation zones in the Eastern Cordillera of the Andes in Colombia during the last glacial maximum *(above)* and at present *(below)*. This is based primarily on the distribution of fossil pollen. Note that the lower zones became narrower as they shifted to lower elevations during the cooler climate of the glacial period. (After Flenley, 1979a.)

matically show how each of the vegetation types and the snow line have shifted upward since the last glacial maximum. The lower tropical sub-Andean and Andean forest types (including the tropical rain forest) have much wider elevational amplitude now than they did 14,000 years ago. Subparamo has remained about the same, but paramo and superparamo are elevationally compressed. Because the upper vegetation types occur closer to the narrow mountain peaks, the total area covered by these biomes has fluctuated greatly, causing extinctions when populations were isolated in restricted areas.

Figure 14.8 also helps to demonstrate how downward shifts of high elevation vegetation have created avenues of dispersal; such a lowering would allow a plant species, previously isolated on individual peaks, to cross ridges and migrate along mountain ranges. This mechanism is precisely the one invoked by many authors (e.g., Simpson-Vuillemier, 1971) to account for the intracontinental dispersal of vegetation types. When followed by isolation it

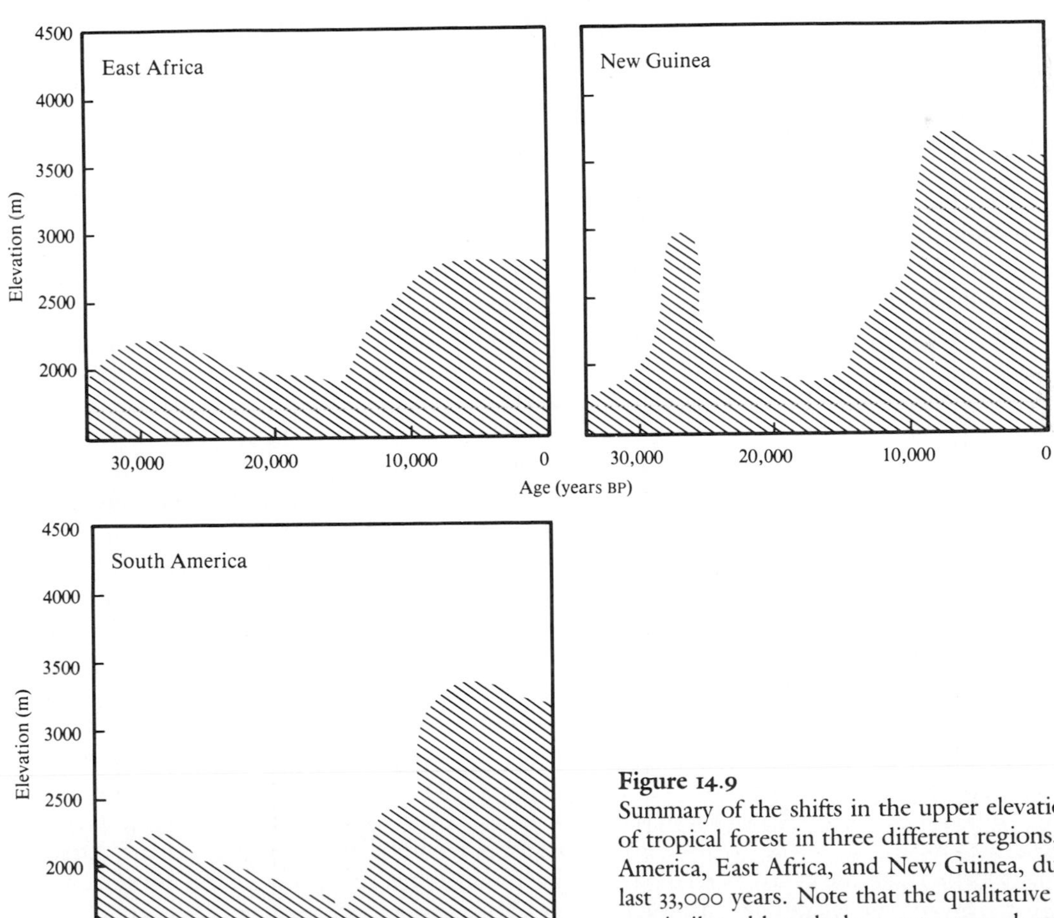

Figure 14.9
Summary of the shifts in the upper elevational limit of tropical forest in three different regions, South America, East Africa, and New Guinea, during the last 33,000 years. Note that the qualitative patterns are similar, although the exact upper elevational boundary was somewhat different in each region. (After Flenley, 1979a.)

results in taxonomic disjunctions (Simpson, 1975, 1979; Vuillemier and Simberloff, 1980; Vuillemier, 1980).

Figure 14.9 tracks the probable upper elevational limit of subtropical forest types in three equatorial regions. Each graph suggests several drastic elevational shifts for forests during the last 34,000 years. The curves are not identical, but in each case the upper elevational limit of forest began to decrease gradually beginning about 29,000 to 27,000 years BP but then reversed and increased sharply around 16,000 to 15,000 years BP. Flenley also noted that in Africa, South America, Indo-Malaya, and New Guinea, montane vegetational zones were depressed by about the same amounts (1000 to 1500 m) at roughly the same times. However, Queensland exhibited an exceptional and rapid change from sclerophyllous woodland to rain forest between the late Pleistocene and 6000 years BP without a corresponding shift in nearby New Guinea. The latter point helps illustrate how difficult it is to extrapolate trends in terrestrial paleoclimatology and to understand vegetational history merely by studying present-day vegetation.

Late Quaternary in desert regions

Because deserts in the Northern Hemisphere are located mostly south of 40° N latitude (Chapters 2 and 4), they were not glaciated by ice sheets or mountain glaciers. This does not mean, however, that such nonglaciated regions were not affected by Pleistocene events. In fact, one would be rather surprised if it were possible to visit our desert regions 10,000 to 15,000 years ago, because they had substantially different climates and vegetation.

Arid-basin pluvial lakes. During the times of glacial maxima the climates of many unglaciated temperate periods were cooler and wetter than those of today. These are called pluvial periods. Large freshwater or saline lakes formed on the continents because temperatures were cooler and because precipitation was higher and evaporation rates lower.

Few biogeographers appreciate the size and number of pluvial lakes in places that now have desert climates. A splendid area to study this is the southwestern United States. This desert region has a basin-and-range topography, which means many low elevational flat areas (basins) interrupted by isolated mountain ranges. During pluvial times, many of these basins filled with water (Figure 14.10). The largest water body was Lake Bonneville (not entirely a pluvial lake) in Utah and parts of Nevada and Idaho. At times it contained fresh water and drained northward into the Snake and Columbia rivers. In the Middle Wisconsin this lake was 330 m deep and had an area exceeding 50,000 km^2, slightly smaller than present-day Lake Michigan. The present Great Salt Lake is a small remnant of Lake Bonneville. Nevada, southern Oregon, eastern California, southeastern Arizona, and southwestern New Mexico had numerous large and small lakes, most of which had evaporated dry by 10,000 years ago. One of the lake basins, Death Valley, the lowest elevation in North America (−93 m), contains now perhaps the most extreme desert on the continent.

Pluvial lakes were also present in what are now deserts on other continents. Their remnants, saline lakes and dry lake beds, are abundant in the Atacama Desert of northern Chile, the monte of Argentina, many areas in interior Australia, the region of the Dead Sea in the Middle East (ancient Lake Lisan), the Kalahari Desert of southern Africa, and many places in arid and semiarid Asia (Flint, 1971). Perhaps the most remarkable is Lake Chad. This relatively small lake (16,000 km^2) was once 950 km long and covered over 300,000 km^2 (the present size of the Caspian Sea), including a significant portion of the southern Sahara Desert. Lake Chad was full from 22,000 to 8500 years BP. The western portion of the Sahara Desert was still relatively mesic until 5000 years ago.

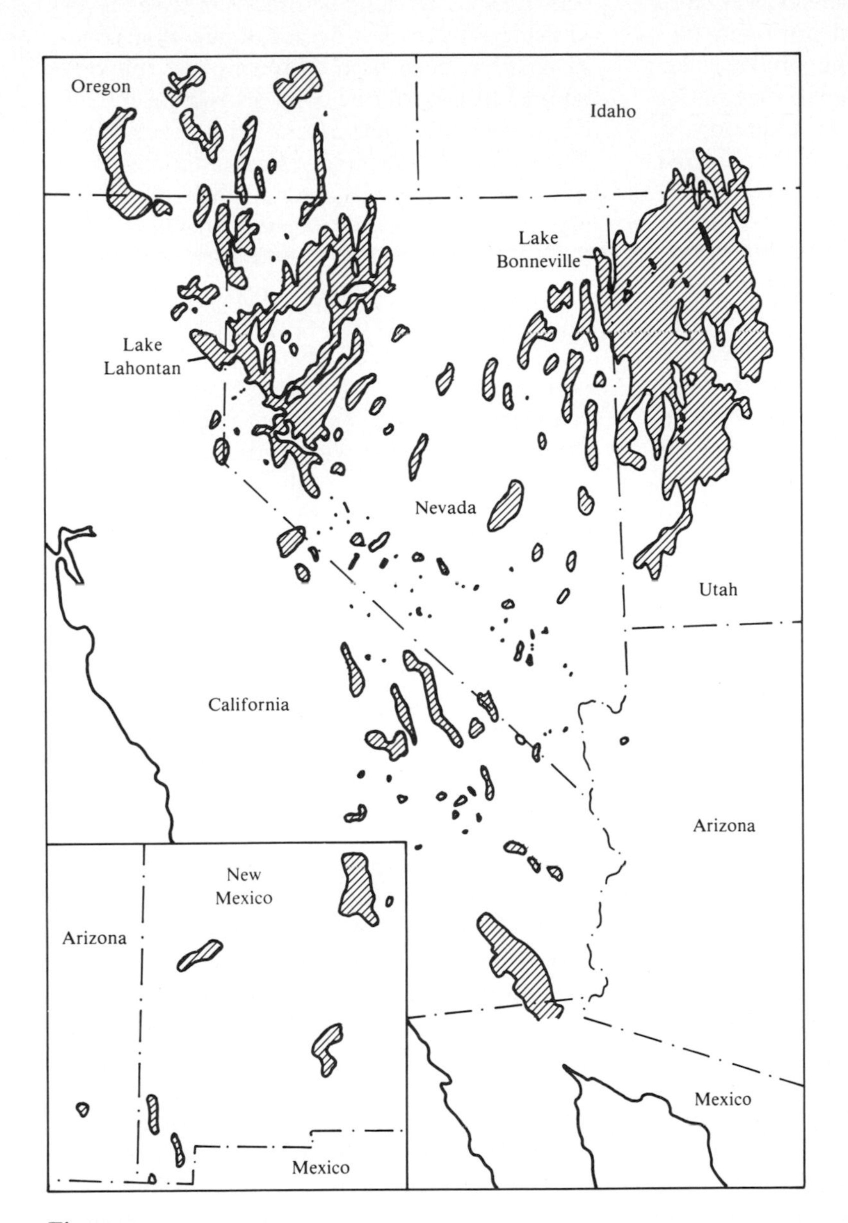

Figure 14.10
Distribution of pluvial lakes in western North America during the last (Wisconsin) glacial maximum. During the glacial periods most of the arid region of the continent experienced a wetter and cooler climate, and lakes and marshes filled what are now desert valleys. Relict populations of fishes and other aquatic organisms that were much more widely distributed during glacial episodes still survive in isolated springs that provide permanent aquatic habitat in this area. (After Flint, 1971.)

The disappearance of the pluvial lakes since the Pleistocene has had several profound biogeographic effects. It caused the wholesale extinction of many plants and animals living in or around these bodies of water. In addition, the dissection of large lakes into smaller isolated units has led to the vicariant speciation of surviving forms, as with the pupfishes *(Cyprinodon)* of the southwestern United States (Miller, 1961b, Smith, 1981). Although these climatic and habitat changes are most easy to visualize for regions that were once moist and highly productive but recently have been converted into arid deserts, similar shifts have occurred in other habitat types. For example, forest biomes were much more widespread in these desert regions 10,000 to 20,000 years ago but have since retreated to the more mesic mountains. Organisms inhabiting the forests, such as land snails (Chapter 11), were stranded on isolated ranges and prevented from migrating between neighboring mountains by the inhospitable desert lowlands. So in a broader sense, the disappearance of the pluvial lakes is an indicator of major climatic and vegetational shifts around the world.

Evolution of plant communities in the southwestern United States. Because dry deserts are notoriously poor environments for fossilization of either pollen or plant and animal megafossils, it has been difficult to reconstruct exactly the biotic history of deserts. Is a particular desert old or young? Did deserts move closer to the equator during glacial and pluvial periods or stay where they were, but with greatly contracted ranges? Are the desert communities seen today the same ones that existed before and during the Quaternary? Where did the desert vegetation survive during the mesic pluvial periods? Many of these questions remain largely unanswered, but they are the focus of much current research.

Traditionally, the history of deserts has been inferred mainly from theoretical paleoclimatological models reinforced wherever possible by fossils, especially the pollen records from the sediments of pluvial lakes. Neither provides a very reliable account of the small-scale history of deserts. In the last decade, a new type of fossil data has been used to reconstruct the vegetational history in one complex region, the desert zones of semiarid and arid southwestern United States. Packrats *(Neotoma),* are abundant rodents in xeric habitats, where they hoard plant materials as large caches that are sometimes protected in caves and under rock ledges. The remains of these caches become solid structures, called middens, and persist for thousands of years if kept dry. They are excellent sources of megafossils, because they provide a relatively complete and often quite continuous sample of the plants growing within 100 m of the den during its occupation. By collecting packrat middens at different elevations and locations and dating the materials by ^{14}C methods, the shifts in elevation and composition of vegetation types can be reconstructed and past climatic regimes can be deduced.

Results from the packrat middens are just beginning to be synthesized (Van Devender, 1977; Van Devender and Spaulding, 1979; Wells, 1979), but general trends are apparent. It is clear that there have been major changes in the vegetation at all elevations throughout the arid Southwest within the last 20,000 years. In general climates were substantially cooler and wetter during the last (Wisconsin) glacial maximum. During this period vegetation zones were displaced as much as 500 to 1000 m below their present limits. The change to the present configuration of climate and vegetation types occurred primarily within the last 8000 to 12,000 years. In general, the packrat middens reveal a history consistent with pollen records and geologic studies of pluvial lakes, but they provide a much more detailed picture of vegetation changes. It is now clear that plant communities did not simply move as entire entities up and down the mountains. Rather they changed dramatically in composition. For example, during the last full-glacial period (21,000 to 15,000 BP) most plant species in the Grand Canyon occurred 600 to 1000 m lower than at the present time. This indicates a cooler, wetter climate (Cole, 1982). At the top of the Grand Canyon, the plant community there did not resemble that found at higher elevations in the same region today. In fact, the community contained several species that are presently found together in northeastern Nevada and northwestern Utah at least 500 km to the north (Cole, 1982). The packrat midden studies reinforce the idea, presented in Chapter 4 and discussed earlier in this chapter, that plant communities are not discrete associations in either space or time. Individual species are relatively independent entities that have shifted their ranges over great distances and have coexisted with different combinations of species at times in the past when climate and perhaps other conditions differed substantially from those at present.

There are two major problems in synthesizing data obtained from packrat middens. First, as usual, current data are sufficient to say that belts of vegetation have shifted over the last 30,000 years, but much information is lacking on the latitudinal movements. For example, 15,000 years ago the lowest point in the present-day Chihuahuan Desert was a juniper woodland (Van Devender, 1977). What happened to desert taxa? Did the present flora assemble by and large following the last glaciation? Second, whereas all authors agree that the elevational and latitudinal ranges of plants were strongly affected by changing climates, they cannot agree on which climatic parameters were changed: total precipitation, annual rainfall pattern, temperature range or extremes, and so forth (Wells, 1979). Out of necessity because data are still scanty, reconstructions emphasize the shifts of key taxa. The field awaits broad and quantitative regional models like those produced by Fritts (1976) from tree ring data.

Extinction of large mammals

On display in American natural history museums are skeletons and models of many large mammals that dominated American faunas during the Pleistocene but are now extinct (Figure 14.11). Gone from North America are most of the large herbivores, such as mastodons, mammoths, camels, llamas, horses, tapirs, ground sloths, and cave bears, as well as species of artiodactyls related to contemporary deer, bison, and pronghorn antelope (Martin and Wright, 1967; Kurtén and Anderson, 1980). Gone also are many large predators that hunted those herbivores, including hyenas, dire wolves and other canids, saber-toothed tigers, and even lions. So many fossils have been unearthed from Pleistocene beds that paleontologists cannot help but be impressed by the disappearance of the North American megafauna from the fossil record at the end of the Quaternary. Extinction on other continents was more limited, with the possible exception of Australia. On the African mainland such mass extinctions apparently did not occur.

The demise of large mammals in North America prompted workers in the early nineteenth century to treat this as a major and sudden episode of mammalian extinction and to search for a specific cause for the wholesale destruction of the megafauna. Were the extinctions sudden or gradual? Did small animals and plants become extinct at the same time? Were Pleistocene extinctions caused by climatic and geologic changes or did intensive hunting by humans result in the extirpation of these large beasts? These are questions that must be answered to understand faunal change and biogeography in the Quaternary.

The prehistoric overkill hypothesis. The prehistoric or Pleistocene overkill hypothesis states that humans were responsible for the massive extirpation of large herbivorous mammals (over 50 kg) and the dependent carnivores after the Wisconsinan glaciers had retreated. This is an old hypothesis but is also the one most clearly presented (Martin, 1967, 1973) as a straightforward explanation that has potentially falsifiable assumptions and predictions. Let us consider this in detail for comparisons with alternative explanations.

Figure 14.11
Drawings of some of the large mammals that comprised the Pleistocene megafauna of North America. It is hard to imagine that these and many other species were widespread and abundant only 20,000 years ago.

The model suggests that a population of aggressive and skillful hunters entered North America in the latest Wisconsinan time by crossing Beringia from Asia. Once these hunters colonized America, they spread southward and eastward through North America and into South America, killing the big game as they went (Figure 14.12). The native animals lacked adequate defensive behaviors to outwit or elude their new predators. Abundant food supplies obtained from these hunts permitted human populations to remain high and in constant need of new and massive food sources. Behind this trail of carnage, there were no more waves of mammalian immigrants from Asia to replace those that became extinct. Most large mammals that survived were those that had spread to the New World from the Old World since the evolution of Pleistocene humans, so they were presumably already adapted to avoid human hunters.

The evidence favoring this scenario is of several types. First, fossil evidence shows that prehistoric humans and large mammals coexisted in the Americas and that the people hunted the extinct herbivores. Hunting arrow points in carcasses and signs of massive animal kills are clearly demonstrated. Second, late Wisconsinan extinctions in North America were nonrandom in that many more large to very large mammals than smaller forms became ex-

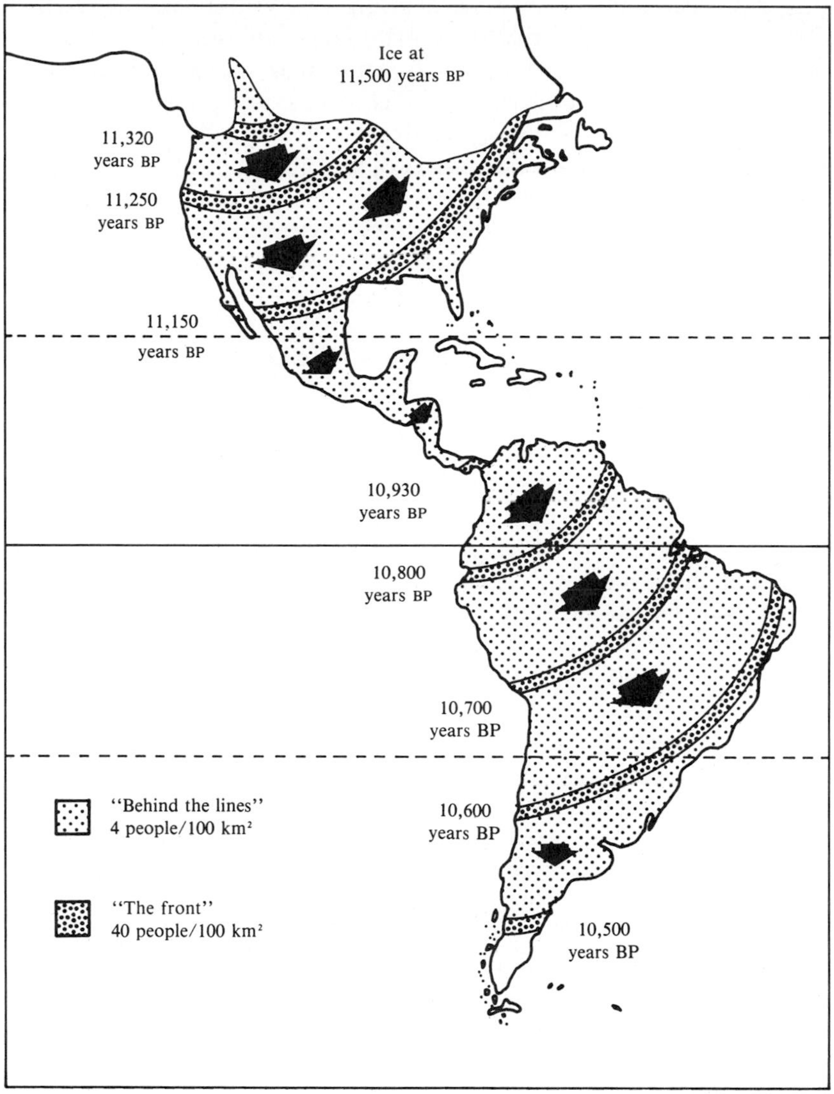

Figure 14.12
Temporal sequence of advancing populations of big game hunters reconstructed by Martin (1973) is suggested to correlate well with the progressive extinction of the large Pleistocene mammal species. According to Martin's scenario, sophisticated hunters crossed the Bering Strait and expanded southward, maintaining a relatively dense front population subsisting on large mammals. (After P.S. Martin, 1973, "The discovery of America," *Science* **179**:960-974. Copyright © 1973 by the American Association for the Advancement of Science.)

tinct in the period from 12,000 to 10,000 years BP. Third, extinction of large mammals appears to have begun in the north and proceeded rapidly and systematically southward (Figure 14.12). Finally, when the dates of the last known occurrence of a species are compared with a computer simulation of southward human migration (assuming high human population densities), the two appear to coincide rather closely (Mosiman and Martin, 1975).

The Pleistocene overkill model can be cast into doubt and falsified by showing that many different types of animals and plants became extinct at the same time, that extinctions were under way before humans arrived, that aggressive human hunters coexisted with large mammals for long periods, that human populations were never at high densities, or that comparable extinctions on other continents did not correspond with an invasion of hunting peoples. Eurasian and African faunas did not show the same degree of mass extinction from human hunting but show instead long periods of coexistence; the magnitude and causes of extinctions of large Australian mammals remain to be fully investigated. However, it may be possible to demonstrate that human invaders on long-colonized islands in the Holocene eliminated large prey. Therefore the extinction of the elephant birds of Madagascar (Aepyornithidae) and the moas on New Zealand (Diornithidae) would be consistent with the model.

Alternative explanations for Pleistocene extinction of mammals. As with any controversial issue in biogeography, there are several alternative explanations to account for Pleistocene extinctions of mammals. The overkill hypothesis, if correct, casts a rather brutal and disparaging picture of the early human pioneers to North America. Some authors feel these colonists have been given a "bum rap"—they may have been instrumental in reducing prey population sizes, but extinction was already proceeding in response to climatic shifts at the end of the Ice Age. They point, for example, to other groups of organisms, such as birds, which also experienced high rates of extinction at the same time (Grayson, 1977).

One puzzling observation has always been that the North American fauna did not disappear rapidly until *after* the glaciers retreated. Hence late Pleistocene extinctions cannot be related directly to the ice, cold climate, or any other catastrophic geologic event, such as a flood. Nevertheless, many workers contend that climatic changes were the direct cause, either through increased aridity (Guilday, 1967) or by decreased equability (Slaughter, 1967; Axelrod, 1967).

An excellent discussion of Pleistocene extinction by Kurtén and Anderson (1980) indicates why paleontologists generally prefer a climatic-based explanation. Quaternary extinction of mammals was not restricted to the period of 12,000 to 10,000 years BP but was part of a fairly continuous episode in the latest Cenozoic (Table 14.2). The Pliocene Blancan extinction (mostly between 3.3 and 2.4 million years BP) resulted in the disappearance from North America of at least 125 mammalian species, of which three fourths were animals smaller than 1 kg in weight (Figure 14.16). During that time, aridity increased, grasslands replaced forests, and many forest dwellers and browsers died out. Following that depletion in the fauna, surviving grazers and rodents underwent evolutionary radiation, and the small carnivores also increased in abundance. In the Irvingtonian extinctions of the Pleistocene (1.8 to 0.7 million years BP) only 89 taxa disappeared, 80% of which were small to medium-sized. Extinction during the glacial period (Rancholabrean, since 0.7 million years BP) was fairly low and constant until the Wisconsin, when many small, medium, and large animals disappeared. However, as stated earlier, more large mammals became extinct in the late Wisconsin than during other episodes.

Explanations for the declining numbers and reduced sizes of large mammal populations in the late Wisconsinan stage include changes in the species composition of grasslands in re-

Table 14.2
Extinction of North American mammals during the last 4 million years
Duration of periods in million years BP. (After Kurtén and Anderson, 1980.)

Animal size	Blancan (3.5-1.8)	Irvingtonian (1.8-0.7)	Rancholabrean (0.7-0.01)	Extinction total	Surviving species
Small (1-907 g)	97	55	29	181	166
Medium (908 g-181 kg)	31	25	33	89	50
Large (182-1730 kg)	5	12	35	52	16
Very large (over 1730 kg)	1	2	6	9	1

sponse to hotter, drier summers and cooler winters; increased competition from small animals; loss or reduction of suitable habitats; low reproductive rates that prevented rapid recovery from population reductions; the appearance of new predators, diseases, and parasites; and so forth. In other words, many combinations of ecological or evolutionary factors might be invoked to explain why large mammals decreased in abundance as the Pleistocene ended, and hunting by early human colonists may merely have delivered the final and fatal blow to many species. Unfortunately, this explanation is difficult to evaluate, even if it is correct.

Regardless of its cause, the elimination of the majority of large mammal species has been one of the most important events in the relatively recent history of the North American biota. Not only did mammals themselves experience reduced populations and geographic ranges ending in extinction, their disappearances had potentially important effects on other species. The herbivores that did survive have been faced with fewer potential competitors, and the surviving carnivores have had to make do with fewer prey. Other dependent species, parasites and mutualists, have either switched to new associates or become extinct. Finally, it is known from fossil feces (coprolites) that the extinct herbivores consumed large quantities of certain extant species. To what extent has release from such predation contributed to the shifts in plant species ranges and changes in distribution of vegetation types that are known to have occurred within the last 20,000 years? It is perhaps as important to investigate these questions as it is to solve the riddle of the extinction of the Pleistocene megafauna.

Biogeographic dynamism in recent centuries

Throughout this chapter we have emphasized the fluctuating nature of distributions, especially over the last 600,000 years. These dynamic changes are characteristic of biomes and communities as well as of individual species (Udvardy, 1969). In the northern latitudes biotas have been repeatedly displaced by shifting Quaternary climates and habitats. In the process, countless organisms became extinct, although many others survived and even speciated. For example, in plants many reinvaders of glaciated regions have been new polyploids.

At a single location, such as central eastern England (East Anglia), fossil records show that the vegetation changed from tundra to subarctic forest to boreal coniferous forest to temperate deciduous forest and back again during a single interglacial stage (West, 1977). Areal dynamism also characterized the nonglaciated regions of the world, as lowland forests and deserts expanded and contracted in relationship to other biomes, as well as up and down mountains. Meanwhile, oceans and pluvial lakes have also changed in size and temperature, affecting aquatic distributions and matters of speciation and extinction. However, such dynamism has not occurred only on such a large temporal scale that it can be assessed only from fossil evidence. In part because modern humans have had such great impact on their environment, major changes in the distributions of plants and animals have been documented in recorded human history.

Photographic studies of vegetational change. One of the simplest but most graphic ways to demonstrate vegetational changes is to keep a photographic record of selected locations over intervals of time. Although this has been done many times on a local scale, there are also a few excellent broad-scale studies. Two examples come from North America. In the 1930s Homer E. Shantz took a series of photographs of representative natural vegetation in the western United States. These sites, together with the locations of other early photographs, were revisited by Hastings and Turner in the 1960s and rephotographed at exactly the same spot and in the same direction. In a book entitled *The Changing Mile* (1965), they reveal the dramatic changes in desert communities for many sites in the Southwest (Chapter 3) in an interval of less than a century. They attributed these shifts to changes in climate and increases in human activity, especially grazing by domestic livestock. A less well-known study was that of Phillips (1963) on vegetational turnover in grassland communities. Prairie communities have mostly disappeared not only because they have been converted into huge grain and livestock farms but also because fire, which maintained the original species composition, has been totally suppressed.

With the advent of weather satellites came a new technology for documenting vegetational change throughout the world. Infrared imagery permits the identification of different types of vegetation from several hundred miles above the earth's surface. One of the uses of such photographs is to show that deserts of the world are increasing in size, a phenomenon known as desertification. Desertification is especially easy to observe in northern Africa and southern Asia, where steppe and thorn scrub are becoming barren wastelands. A primary cause for desertification is human activity, most notably the rearing of more domestic animals than can be supported on a sustained basis by the dry climates and scanty vegetation of those regions. These animals not only can denude the landscape but also alter the soil and use vast quantities of water, all of which adversely effect native vegetation. Recent years of drought and human use of firewood, particularly in Africa, has intensified these problems.

One does not need photographic evidence to show how the lowland tropical rain forests are being destroyed by lumbering activities. Lumbering and slash-and-burn agriculture have systematically and progressively resulted in the replacement of primary rain forest by secondary species and undoubtedly caused the unrecorded extinction of thousands of undescribed native plant and animal species. These range reduction extinctions confound attempts to apply historical and ecological biogeography in the tropical zone. Moreover, when primeval rain forest is removed, the relatively infertile lateritic soils may become baked hard, severely inhibiting both the reestablishment of native forests and the continued use of the soil for productive agriculture. Finally, many climatologists worry that the rapid elimination of tropical forests will cause major changes in global precipitation and temperature, further threatening the capacity of the biosphere to sustain life.

Shifts in ranges of animal species. Another type of well-documented dynamism includes the shifts in ranges of individual species. In Chapter 3 we reviewed cases such as the opposum *(Didelphis virginiana)* and the house sparrow *(Passer domesticus)* that have expanded their ranges in response to human activities that have simultaneously caused the reduction and extinction of other species. There are several different ways that these shifts come about: (1) as a result of changes in habitats or food sources that promote or limit the distribution of certain animals, (2) as a result of unintentional transport by humans (anthropochory) to new areas where propagules became established and spread, and (3) as a result of the intentional transport by humans of species and their release or inadvertent escape elsewhere. Often the spread of species cannot be categorized as caused entirely either by human activities or natural causes because both are so intimately related today. For example, migratory bird species, such as the redwing blackbird *(Agelaius phoeniceus),* have shifted both their breeding and wintering sites to exploit habitats and food resources that have been destroyed by modern humans. Udvardy (1969) discusses a number of examples of shifting distributions attributable to natural or human causes or both.

It is not hard to imagine how humans enabled species to colonize new regions, especially via rapid transport of species on ships from one port to another. This is especially true for freshwater molluscs, some of which now are widely distributed on oceanic islands.

A special case of rapid transport is that of the beetles native to Europe but also occurring in eastern North America. Lindroth (1957) studied these and concluded that many had been transported in ship ballast, loaded at the coast of Europe and unloaded when the ships reached North American ports of call. The species that made that journey and became established exhibit the following characteristics: (1) they live on the ground, (2) they are tolerant of dry conditions, (3) they prefer open and disturbed habitats, (4) they are not dependent on special kinds of food, i.e., they are polyphagous, (5) they are flightless, and (6) they have parthenogenetic reproduction. In other words, those capable of successful ballast transport have many features of good colonists, but they lack wings so they have not been able to disperse on their own power. A similar pattern can be observed in weedy plants, which have short generation times, high seed production, the capacity for autogamy or parthenogenesis, and minimal nutritional requirements for growth and reproduction.

Eurasia in particular has contributed many species of animals and weedy plants to the biota of North America, whereas the number of successful reverse introductions have been relatively small. Lindroth (1957) proposed that the European forms have succeeded in North America because they have landed where there are open niches, a suitable climate, but few natural enemies; but the American species have come in contact with filled niches and highly competitive species because the Old World has long been subjected to agriculture and other kinds of human disturbance. American species have therefore been unable to compete with the existing European forms. Thus the dispersal capacity of an organism is only one of several processes controlling the dynamism of colonizing forms (Chapters 3, 4, and 7).

The message of Quaternary dynamism for historical biogeography. Although dynamism is usually treated very briefly in historical biogeography, it can create major problems in interpreting disjunctions of plants and animals. Earlier in this chapter we accepted the commonly held belief that disjunctions carry information on the migration of organisms during glacial maxima, and we cited examples of relict populations in the mountains of southern Europe and eastern North America that show how the northern biomes had escaped inhospitable northern glacial climates by moving southward along the mountain ranges. Now we can legitimately ask, however, whether this dynamism violates an assumption of vicariance methods, namely that endemics are located today in the exact places where they differentiated. If dynamism is the rule rather than the exception in nonglaciated as well as glaciated areas, the precise location of an extant continental species may give a misleading picture of how a vicariance event actually occurred. Taxa may now exist where there was a refugium or where they have migrated rather than where they originally differentiated in time and space.

The strength in the method of vicariance biogeography lies in finding taxa that have identical patterns of multiarea disjunctions, which mimic the phylogenetic splitting of the clades involved. As discussed in Chapter 13, one can already find examples, such as the maples *(Acer)* in the Northern Hemisphere, in which the current positions of endemics on different continental areas do not reflect the origins of the taxa, resulting in a potentially incorrect reconstruction of the history of the multiarea patterns. Consider now the Pleistocene dynamism in these distributions within North America. The northward and southward movements produced shifts of 10° to 20° of latitude numerous times in the last 600,000 years. It is unlikely that a formalized method will be able to convert distributions on previously glaciated land, some of which are narrow endemics, into a testable framework. Accurate geologic and biogeographic reconstructions of the Northern Hemisphere north of the tropical zones may be attained more reliably by direct analysis of the geologic, paleoclimatic, and fossil evidence than by precise cladistic studies of extant taxa.

Turning to nonglaciated regions, especially in the tropics and subtropics, one can anticipate that many of these endemics might better fit the assumptions of vicariance biogeography because the shifts probably did not force the organisms completely out of their original ranges. If the rain forest organisms retreated to small, isolated forest refugia, these may (but not necessarily) have been located within both their original and their current ranges. Likewise, elevational shifts could also mimic models of allopatric speciation. However, one characteristic of these events might be that numerous isolated populations would have been produced simultaneously. Over a short period of time a widespread lowland biota could have been isolated in many mountains separated by strong barriers to dispersal. A vicariance hypothesis generally prefers a dichotomously branched cladogram, which would not properly represent the temporal and spatial pattern of the geologic and biogeographic events even if the genealogical relationships are correctly presented.

In summary, one would expect formalized vicariance methodology to be most successful for investigating intercontinental disjunctions, slightly less successful for wide disjunctions within a landmass (especially in unglaciated regions), and very difficult to interpret for small-scale disjunctions or regions that have experienced much recent climatic and habitat change. Narrow disjunctions have generally been produced in the last portion of the Quaternary. Moreover, those taxa that have speciated greatly within the last million years in concert with the fluctuating environment are the ones for which systematists have had difficulty in reconstructing phylogenies (e.g., annual plants, land snails, butterflies, *Drosophila,* freshwater fishes). As historical biogeographers continue to investigate how distributional patterns relate to geologic and climatic changes, it will be important to heed the message of Quaternary dynamism and avoid applying models with unsupportable assumptions.

Selected references

Butzer (1976); Cushing and Wright (1967); Davis (1969); Haq et al. (1977); Imbrie and Kipp (1971).

Pleistocene glaciation

Ager (1963); Anderton et al. (1979); Bousfield and Thomas (1975); Colinvaux (1964, 1981); Davis (1976); Emiliani (1966); Fernald (1925); Flint (1971); Frenzel (1968); Frost (1977); Goodwin (1975); Gjaeveroll (1963); Hambrey and Harland (1981); A. Hecht et al. (1979); A. Hecht and Imbrie (1979); Holland (1981); Hopkins (1959, 1967, 1979, 1980); Hopkins and Smith (1981); Hultén (1937, 1958, 1963a, 1963b, 1968); Ives (1974); Lindsey (1981); Löve and Löve (1963); D. Murray (1981); H. Richards (1971); Thunell (1979); D. Walker and West (1970); W. Watts (1980); Weber et al. (1981); West (1977); Woillard (1978).

Some Pleistocene events in the tropics

Ashton (1969); K. Brown (1976, 1982); Dixon (1979); Flenley (1979a, 1979b); Frost (1977); Haffer (1969, 1974, 1978, 1981); A. Hamilton (1976); Lynch (1979); S. Porter (1979); Prance (1973, 1978, 1981, 1982); Pregill and Olson (1981); B. Simpson and Haffer (1978); Sowunmi (1981); Street (1981); Vanzolini and Williams (1970); B. Vuillemier (1971); F.

Vuillemier (1973, 1980); P. Webster and Streten (1978).

Late Quaternary in the desert regions

Bernabo and Webb (1977); Brakenridge (1978); Burke (1976); Butzer (1978); Butzer et al. (1978); La Marche (1978); Livingstone (1975); R.R. Miller (1961b); Van Devender (1977); Van Devender and Spaulding (1979); Wells (1976, 1978, 1979); Wells and Berger (1967).

Extinction of large mammals

Axelrod (1967); Grayson (1977, 1979); Guilday (1967); Hester (1967); Kurtén (1971, 1972); Kurtén and Anderson (1980); Malie (1972); P.S. Martin (1966, 1967, 1973); P.S. Martin and Wright (1967); R. Martin and Webb (1974); Mosiman and Martin (1975); Slaughter (1967); S. Webb (1974).

Biogeographic dynamism in recent centuries

H.G. Baker and Stebbins (1965); Boyd and Nenneley (1964); Brice (1978); Fritts et al. (1979); Glantz (1976); Hastings and Turner (1965); Horton (1974a); T. Hutchinson and Havas (1978); Lindroth (1957); Lintz and Simonett (1976); Paylore (1976); W. Phillips (1963); Port and Thompson (1980); Porter (1974); Sabins (1978); Schanda (1976); Shantz and Turner (1958); Swain and Davis (1978); Udvardy (1969); Veblen et al. (1981); Whitney and Adams (1980).

UNIT FOUR ECOLOGICAL BIOGEOGRAPHY

Up to this point we have been concerned primarily with the roles played by ecological processes and historical events in determining the present and past distributions of particular species or higher taxa. In the last four chapters we shall attempt to account for geographic patterns found in the number and ecological characteristics of species. Although many of the patterns are well documented, most hypotheses to account for them have yet to be tested rigorously. The occurrence of similar patterns in distantly related biotas in widely separated regions suggests that the underlying mechanisms are largely ecological, but a complete understanding will have to include an integrated analysis of the roles for both ecological processes and historical events.

Chapter 15 examines MacArthur and Wilson's equilibrium theory of island biogeography, which stimulated major resurgence in research on insular and ecological biogeography beginning in the 1960s. This theory attempts to account for variation in species diversity among the islands of an archipelago or other restricted region without considering their taxonomic labels. It suggests that the number of species represents an equilibrium between opposing rates of colonization and extinction, which are affected primarily by island isolation and size, respectively. Despite many studies supporting predictions about variation in insular species diversity with island size and isolation, the theory has proven deceptively difficult to test rigorously. The necessary measurements of species turnover are difficult to obtain for permanent resident populations on real oceanic islands. Until unambiguous data become available, the validity of the theory must be questioned.

There can be no doubt, however, that MacArthur and Wilson's theory stimulated research that greatly increased knowledge of the factors influencing the distribution and ecology of insular organisms. Some of these findings are considered in Chapter 16. MacArthur and Wilson's approach has been applied to investigate the determinants of species richness in isolated habitats such as mountaintops, lakes, and caves as well as on real islands. Several of these studies show that some of the insular biotas are not in equilibrium, but are instead gradually increasing or decreasing in diversity following Pleistocene perturbations. Variation among species in dispersal ability and susceptibility to extinction importantly influences the kinds of species found on islands and in insular habitats. Variation in the numbers and kinds of coexisting species causes differences in competition and predation among islands and between islands and the mainland, and these in turn result in variation in niche characteristics and population densities. Interspecific interactions also appear to play an important role in the evolution of insular species, causing them to differentiate in ways that ultimately doom them to extinction. The result is a taxon cycle in which populations colonize islands and consequently evolve in a setting in which competitive pressures are reduced; eventually these species are replaced by more recent colonists with better competitive abilities.

Chapter 17 considers patterns of species diversity on continents and in the oceans. Undoubtedly the most striking of these is the rapid increase in species richness in most taxa from the poles to the equator, but diversity also varies in regular ways along gradients of elevation and moisture in terrestrial environments and along gradients of water depth in aquatic environments. The causes of these patterns have long been debated by ecologists and biogeographers. Diversity is clearly influenced by historical perturbations, productivity, harshness, climatic stability, habitat heterogeneity, and biotic interactions, but not all of these factors may be important primary causes. We suggest that there are compelling reasons to favor equilibrial as opposed to historical explanations and that productivity and harshness may be the two most important factors causing the major geographic patterns of species richness.

The final chapter attempts to bring together many of the themes developed earlier in the book and to consider the interacting effects of historical events and ecological processes on the distributions of species and the composition of biotas. Closely related species tend to occur in the same general region and often to exhibit overlapping geographic ranges. Despite this trend, it is often possible to detect the influence of interspecific competition in preventing cooccurrence of ecologically similar species. Distantly related species also interact, especially when long isolated biotas come into contact, and the outcomes of these interactions vary. Sometimes the two biotas remain distinct, sometimes there is extensive interchange, and sometimes one almost replaces another. Although it is extremely difficult to find unequivocal evidence for competition among such distantly related taxa, the results of these contacts suggest that some equilibrial level of diversity tends to be maintained within large biogeographic regions, which in turn implies that organisms tend to evolve in isolation to fill most of the available niches. This is supported by the dramatic examples of convergence of particular species and of entire biotas to fill similar ecological roles in geographically isolated regions that have similar abiotic environments.

Chapter 15 The Equilibrium Theory of Island Biogeography

Islands have always had a great influence on biogeography, far out of proportion to the tiny fraction of the earth's surface that they cover. The reason for this is straightforward. Islands and other insular habitats, such as mountaintops, springs, lakes, and caves, represent replicated natural experiments. They often occur in archipelagos, clusters of tens or hundreds of individual, isolated microcosms that are generally similar to each other but that differ in enough ways so that it is possible to determine the factors that determine the distribution of species. In these natural experiments, as in artificial manipulations, some factors are held constant among replicates while others are allowed to vary. In this way the effects of each factor on the system can be assessed. Islands have an important advantage over artificial manipulations in that they were established sufficiently long ago that there has been time for long-term evolutionary responses to the variables. All one needs to do is to gather the data and interpret them correctly, but this has not been an easy task. Basic data on insular biotas are still being compiled, and their interpretation has often been the subject of controversy. However, the hard work of gathering and analyzing data and the debating and testing of alternative methods and ideas, are essential to scientific progress.

Historical background

The impact of islands on scientific thought began in earnest in the early nineteenth century, when various European nations undertook to explore, map, and study the world they had recently discovered. Naturalists frequently accompanied the voyages of exploration. The best of these naturalists, especially Wallace, Darwin, and Hooker, not only described and collected specimens of what they found but also they noted patterns in nature and sought explanations for them. Some of the clearest patterns were apparent among the replicated natural experiments provided by the islands of ocean archipelagos. Darwin's experiences in the Galápagos, Wallace's travels in the East Indies, and Hooker's explorations in the southern oceans had tremendous impact on the thinking of these scientists and consequently on the development of ideas about evolution and related areas of environmental biology that revolutionized scientific thought in the mid-1800s.

Another revolution, a much more modest one but nevertheless a major shift in the direction of scientific thought, occurred in the mid-1900s with the integration of concepts from ecology, evolution, and biogeography. It is difficult to pinpoint the exact beginning of this endeavor, but certainly islands again played a central role. One of the pioneers was David Lack (1947, 1976), who early in his career conducted a classic study of the evolution and ecology of Darwin's finches on the Gálapagos Archipelago and shortly before his death investigated the distribution and ecology of birds in the West Indies. As Lack had followed Darwin to the Galápagos, another ornithologist, Ernst Mayr

(1942, 1963), followed Wallace to the East Indies and returned to make major contributions to the understanding of speciation and other aspects of the evolutionary process. Another pioneer was G.E. Hutchinson (1958, 1959, 1967), who also traveled widely but studied lakes rather than real islands. In his 1959 paper "Homage to Santa Rosalia, Or why are there so many kinds of animals?" Hutchinson called attention to the problem of trying to explain geographic variation in the diversity of species. This has remained the focus of research in ecological biogeography and community ecology right up to the present.

However, if any single contribution can be said to have triggered the recent revolution in ecological biogeography, it was R.H. MacArthur and E.O. Wilson's equilibrium theory of island biogeography (1963, 1967). This seminal work was completed when both men were young, still in their 30s. MacArthur had been a student of Hutchinson at Yale. His doctoral dissertation (1958) was a classic study of competition and coexistence in several closely related, coexisting species of warblers. After completing his degree, he did postdoctoral work in Britain with Lack and then held professorships at the University of Pennsylvania and at Princeton University. Wilson, who has spent his entire career at Harvard, began it as a systematist. Strongly influenced by Mayr, he had worked extensively on the origins and relationships of the ants of the East Indies and South Pacific. He was also a coauthor of the classic paper on character displacement (Brown and Wilson, 1956). Both men had extensive experience with islands: MacArthur in the montane islands of the southwestern United States, in the West Indies, and in the small islands off the coasts of Maine and Panama; Wilson in the East Indies, Polynesia, and the Florida Keys. Both men went on to have illustrious careers. MacArthur died tragically of cancer in 1972 at the age of 42, but he had already produced many theoretical papers on population and community ecology that still motivate much of the research in those fields today. Wilson has continued to work on social insects, especially ants, but his interests have shifted from systematics to biogeography and most recently to animal behavior. Today he is best known as the primary proponent of the controversial new field of sociobiology (Wilson, 1975), which integrates genetics, evolution, and behavior to investigate the organization of animal and human societies.

MacArthur and Wilson's equilibrium theory represented a radical change in biogeographic thought. Prior to their work, investigators had focused on historical problems. The primary questions of biogeography have always been those just addressed in Chapters 8 to 14 of this book: Where did a particular taxonomic group of organisms originate, and how, as a result of subsequent dispersal, speciation, and extinction, did its diversity and distribution change? Obviously these questions, as asked in Chapters 8 and 9, have a historical and phylogenetic focus, and they are ad hoc in the sense that they are normally applied to particular taxa or specific regions.

MacArthur and Wilson deliberately departed from this classical approach and asked radically new kinds of questions to motivate their work. They searched for general patterns in the distributions of diverse kinds of species, independent of their phylogenetic affinities, in the hope that such patterns would have general ecological explanations rather than specific historical ones. Whereas the importance of taxonomy and historical geology were duly recognized, MacArthur and Wilson were primarily interested in patterns that might be explained without invoking unique historical events. Their approach was to focus on variation in plant and animal distributions, variation that appeared to be correlated with the functional attributes of contemporary organisms and with measurable characteristics of their present environments. For example, they were more impressed with the fact that birds and bats are similar both in their ability to fly and in their wide distribution on oceanic islands than with the fact that they differ in many morphological traits that reflect the long period of divergent evolutionary history of the birds and mammals.

Island patterns

It is not surprising that MacArthur and Wilson looked for and found interrelated ecological and biogeographic patterns on islands. Three of these patterns were particularly instrumental in developing their ideas. All of these patterns had already been pointed out separately in the literature, but MacArthur and Wilson's innovation was to consider these observations together and to propose a single, unifying theory to account for all of them.

Species-area relationships. The first pattern is the relationship within archipelagos between the sizes of individual islands and the number of species that comprise their biotas. It had long been recognized that larger islands usually have more species. Darlington (1957) had quantified this relationship for the herpetofauna of the West Indies: "Within the size range of these islands . . . division of area by ten divides the amphibian and reptile fauna by two." The data are plotted in Figure 15.1. Note that as suggested by Darlington's multiplicative rule of thumb, when both axes are logarithmic, a straight line fits the points extremely well. Such a relationship can be expressed by the relationship $S = CA^Z$, where S is the number of species, A is the area of the island, C is a fitted constant that varies with region and taxon, and the exponent z is a fitted parameter that gives the slope of the regression line fitted to log-transformed data. This relationship is called the species-area curve, because the best fit line would be a curve if the data were plotted using linear instead of logarithmic axes.

Preston (1962) had noted that islands were a special case of the general multiplicative in-

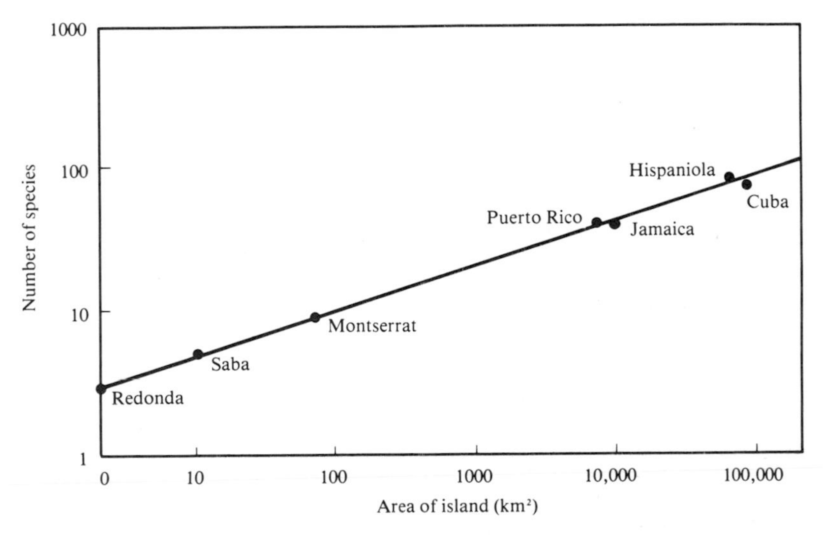

Figure 15.1
Relationship between number of species, *S*, and island area, *A*, for the reptiles and amphibians of the West Indies, plotted from the original data of Darlington (1957). Note that both axes are logarithmic and the points would be well filled by a straight line giving an equation $S = CA^Z$, where C and *Z* are fitted values. (After R.H. MacArthur and E.O. Wilson, *The Theory of Island Biogeography*. Copyright © 1967 by Princeton University Press. Reprinted by permission of Princeton University Press.)

crease in the number of species with an increase in the area sampled. He suggested that this was a consequence of what he termed the canonical lognormal distribution of the number of individuals among species (see also Williams, 1953, 1964). In any region only a few species are extremely common and most are moderately or very rare. The frequency distribution for the number of individuals among species, when plotted on a logarithmic abscissa (x axis), is fairly well fitted by a normal, bell-shaped curve (Figure 15.2). Sometimes this curve is cut off on

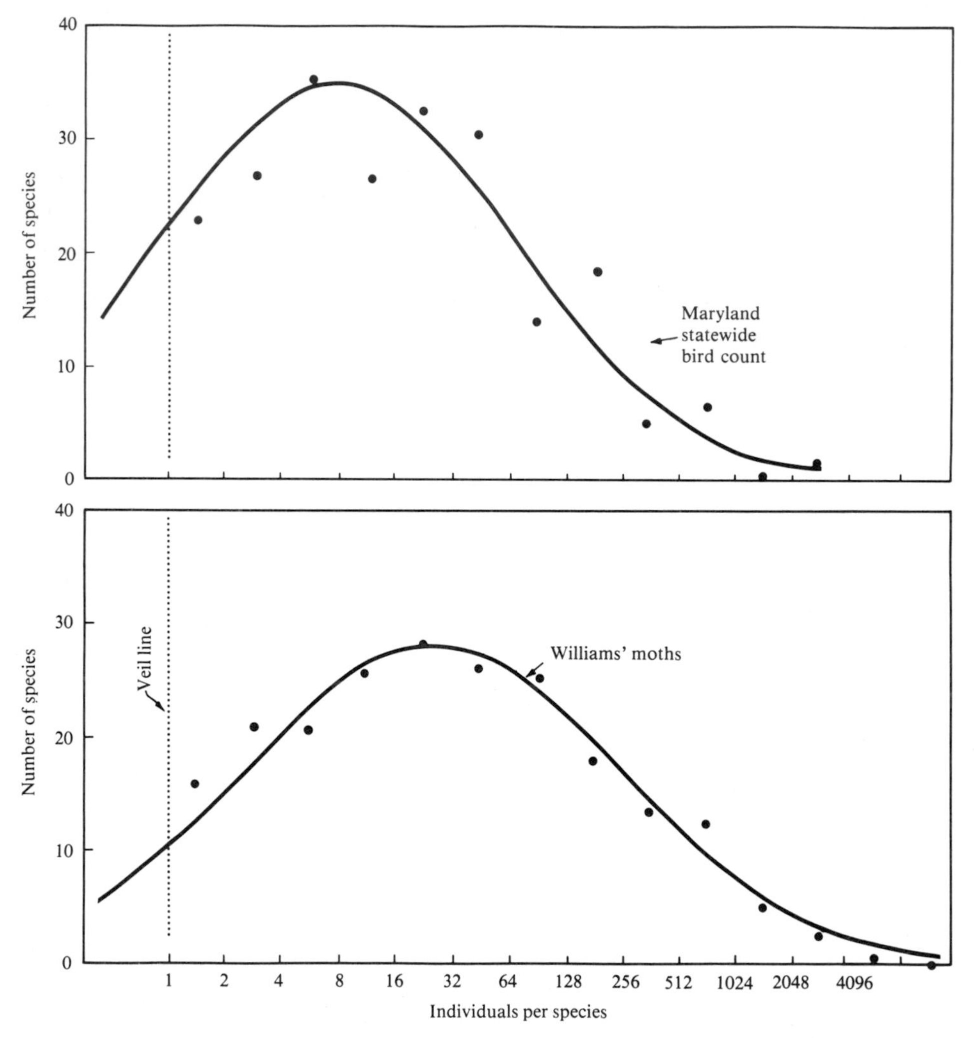

Figure 15.2
Relative abundance of species within a local biota often fits a lognormal distribution; i.e., the frequency distribution approximates a normal curve when abundance is plotted on a logarithmic scale. These data are for a bird census in Maryland (Preston, 1957) and moths coming to a light trap in England (Williams, 1953). Note that often the left-hand tail of the distribution is cut off by what Preston called a veil line. Because the axis is logarithmic this relationship shows that every community contains more rare species than common ones.

the left hand side, and Preston suggested that this happens when the area sampled is so small that some of the rarest species in the community are not observed. It is just this effect that produces the species-area curve. As progressively larger areas of the same kind of habitat are sampled, one obtains not only more individuals but also more species because some of the new individuals will be representatives of rare species that had not yet been seen. Also, as investigators sample increasingly large areas, they will tend to incorporate new kinds of habitats and therefore add those specialized species that are restricted to those environments.

Preston pointed out that small, isolated islands had fewer species per unit area and steeper slopes (higher z values) for the species-area curve than sample areas of comparable size within large regions of continuous habitat on continents (Figure 15.3) (see A. Schoener, 1974a; Sugihara, 1981). The reason for this should be intuitively apparent: small, isolated islands have fewer species than comparable areas on a continent because if a species becomes too rare on an island it is likely to become extinct, whereas on a continent its population can be sustained at low levels by the exchange of individuals between local areas. The effect of such extinctions is much more severe on small islands than on larger ones, resulting in the steep slope of the species-area curve. To make this clear, imagine that mam-

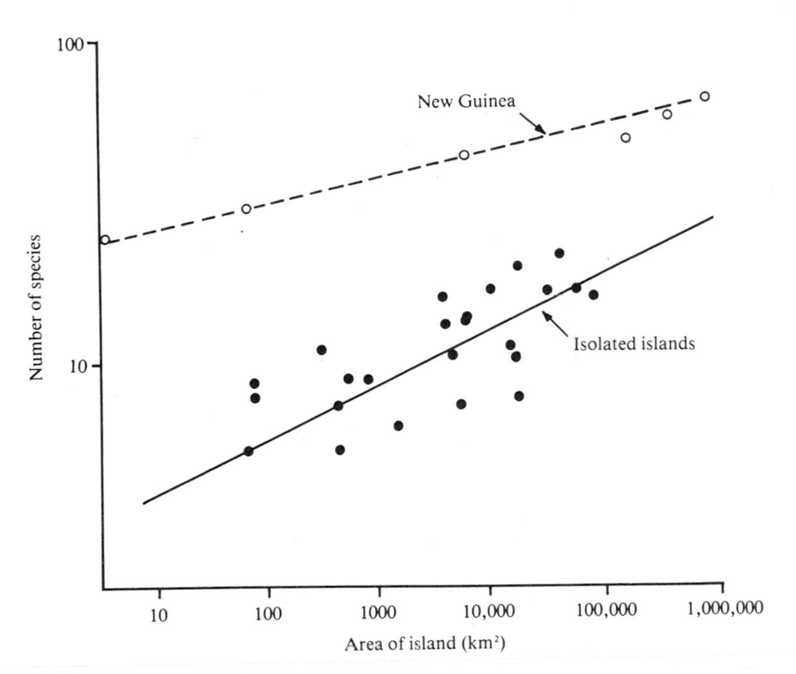

Figure 15.3
Slope of the species-area relationship is much steeper for isolated islands than for sample areas of different size within a single large landmass. These data are for pomerine ants on the Moluccan and Melanesian islands *(below)* and in regions of increasing size on New Guinea *(above)*. The difference between the two curves can be attributed to extinction without replacement by immigration of many of the rare species on the isolated islands. (Reprinted from "The nature of the taxon cycle in the Melanesian ant fauna," *American Naturalist* 95:169-193, by E.O. Wilson, by permission of the University of Chicago Press. Copyright © 1961 by the University of Chicago.)

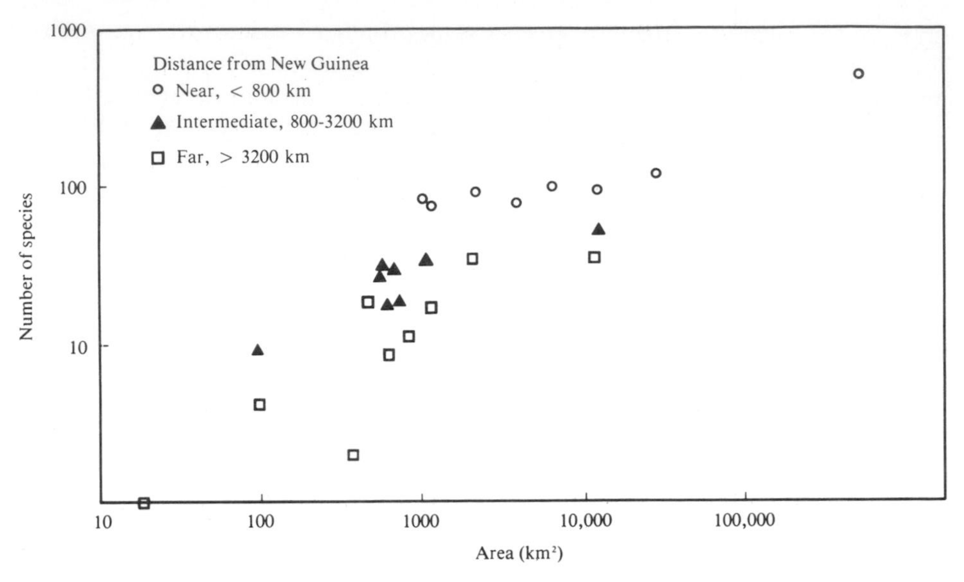

Figure 15.4
Increasingly isolated oceanic islands have fewer species and different species-area relationships. Data for the land and freshwater bird species of the Moluccas, Melanesia, Micronesia, and Polynesia have been divided into three groups: near, less than 800 km from New Guinea; intermediate, 800 to 3200 km from New Guinea; and far, greater than 3200 km from New Guinea. (Modified from MacArthur and Wilson, 1963.)

mal-proof fences were constructed around several patches of identical forest varying in size from hundreds of thousands of hectares to only a few hectares. Some species in all the fenced areas may become extinct, but the smallest areas would lose the most species, dropping and steepening the slope of the species-area curve.

The effect of isolation. The second pattern that impressed MacArthur and Wilson was the effect of isolation by distance in reducing insular species diversity. It was well known that single, isolated islands far out in the ocean support fewer species than islands that are part of major archipelagos or are near to continents. This can be shown more quantitatively by plotting species-area curves for different archipelagos; those islands very distant from continents (or from extremely large islands) have fewer species per island of comparable size, and species-area curves of different slopes, than groups of less isolated islands (Figure 15.4). Systematic studies, such as Wilson's work on the ants of New Guinea, Melanesia, and Polynesia, indicate that insular species have ultimately been derived from continental ancestors. The extreme impoverishment of isolated insular biotas suggests that distance barriers have limited the number of species that could have successfully colonized. This conclusion is supported by the observation that not only do distant islands have few species but also that most of those species present possess special features (such as those described in Chapter 7) that probably made possible their successful dispersal across extensive seawater barriers. It is hardly surprising that birds and bats are much better represented on distant oceanic islands than frogs and terrestrial mammals.

Species turnover. The third pattern that influenced MacArthur and Wilson was the rapid-

Table 15.1
Number of species of land and freshwater birds on Krakatau and Verlaten
The table is based on data gathered by Dammerman (1948) during three census periods after the eruption of Krakatau in 1883 and includes the number of species that became extinct or colonized between censuses. Note that the number of species increased from the census of 1883 to that of 1919-1921 and then remained relatively constant despite extinction of some species and colonization of others. (After R.H. MacArthur and E.O. Wilson, The Theory of Island Biogeography. Copyright © 1967 by Princeton University Press. Adapted by permission of Princeton University Press.)

	Krakatau			Verlaten		
	Nonmigrant	Migrant	Total	Nonmigrant	Migrant	Total
1908	13	0	13	1	0	1
1919-1921	27	4	31	27	2	29
1932-1934	27	3	30	29	5	34

	Krakatau		Verlaten	
	Extinctions	Colonizations	Extinctions	Colonizations
1908 to 1919-1921	2	20	0	28
1919-1921 to 1932-1934	5	4	2	7

ity and nature of the recolonization of Krakatau and Verlaten. Recall from Chapters 2 and 7 that the biotas of these islands were destroyed by a volcanic eruption in 1883. Subsequently, several expeditions visited the islands and inventoried the biota. Recolonization, apparently from the nearby large islands of Java and Sumatra, was rapid; by the 1930s a good tropical rain forest, supporting numerous species of plants (Figure 7.2), birds, and other organisms, was developing. Censuses of the birds, which were particularly complete, are summarized in Table 15.1 (see also Figure 7.2). MacArthur and Wilson noted that the number of species increased rapidly until about 1920, but after that the total number of species remained relatively constant, despite changes in the composition of the avifauna. Species not only colonized rapidly prior to 1920 but also continued to immigrate after 1920. Some of these late arrivals were successful colonists, replacing about an equal number of species that became extinct. These offsetting colonizations might simply have reflected successional changes in the avifauna in response to the development of a tropical forest and the concomitant elimination of open habitats, but they also suggested that continual turnover may be typical of insular biotas. Such turnover might be particularly high when, as is the case for birds on Krakatau and Verlaten, highly dispersive organisms need to cross only modest barriers to reach small islands.

The theory

MacArthur and Wilson produced a single theory to explain what they considered to be the three basic characteristics of insular biotas: (1) the number of species increases with increasing island size, (2) the number of species decreases with increasing distance to the nearest continent or other source of species, and (3) a continual turnover in species composition occurs, owing to recurrent colonizations and extinctions, but the number of species remains approximately the same. MacArthur and Wilson proposed that the number of species inhabiting an island represents an equilibrium between opposing rates of colonization and extinction.

This can be portrayed by plotting these rates as a function of the number of species present on the island in a simple graphic model (Figure 15.5). The number of species could potentially range from zero to a maximum, P, the number in the species pool that are available to colonize the island from the nearest continent or other source area. Now the colonization rate (defined as the rate of immigration of propagules of species not already present on the island) must decline from some maximum value when the island is empty, to zero when the island contains all the species in the pool and there are no more new species to arrive. Conversely, the extinction rate should increase from zero when there are no species present to become extinct, to some maximum value when all species in the mainland pool are hypothetically inhabiting the island. At some number of species, normally intermediate between zero and P, the lines representing the opposing colonization and extinction rates must cross. At this point the two rates are exactly equal, resulting

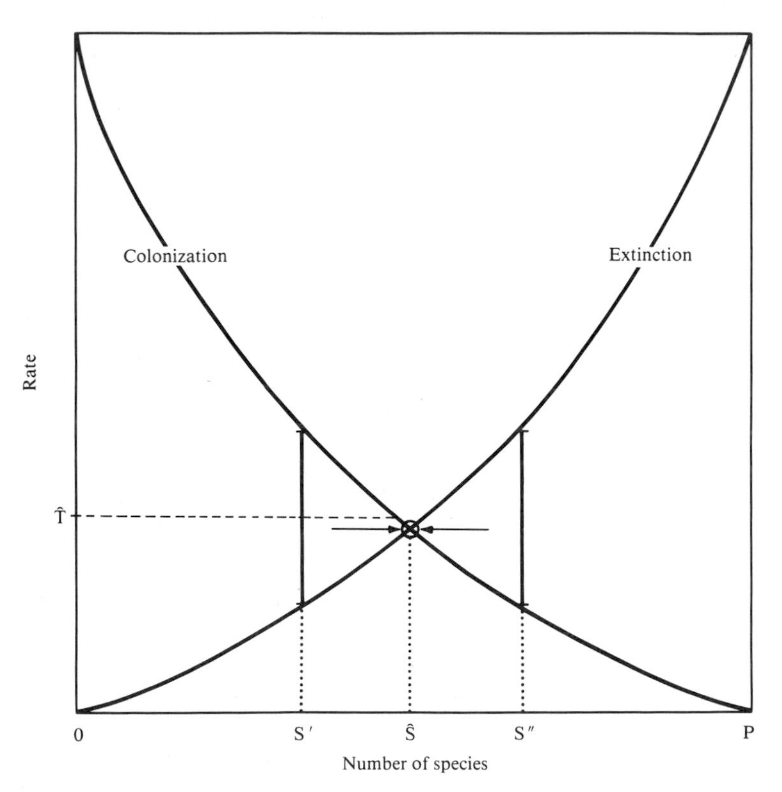

Figure 15.5
Simple model in which the number of species inhabiting an island represents an equilibrium between opposing rates of colonization and extinction. Note that colonization rate declines and extinction rate increases as the number of species increases from zero to P, the number in the mainland species pool. The point of intersection of the two curves represents a stable equilibrium, because if the number of species is displaced from Ŝ to either higher (S″) or lower (S′) numbers, it will return *(arrows)*.

in an equilibrial rate of species turnover, $\hat{T}$, and equilibrial number of species, $\hat{S}$. This is a stable equilibrium because if the number of species is perturbed from this value, it will always return. For example, suppose that a natural disaster, such as a hurricane, caused the extinction of several insular species, temporarily reducing the number of species from $\hat{S}$ to S' (Figure 15.5). Then the colonization rate will exceed the extinction rate and the island will accumulate species until it has again reached $\hat{S}$. Similarly, if S is perturbed from $\hat{S}$ to a larger number, say S'', then the extinction rate will be greater than the colonization rate and species will be lost until $\hat{S}$ is restored.

MacArthur (1972) showed that the lines depicting the colonization and extinction rates would be curves bowed downward, as in Figure 15.5, so long as the species in the pool differ in their probabilities of colonizing and becoming extinct, which is a reasonable assumption. MacArthur's argument goes as follows. Suppose there were only two species, A and B, in the pool. The immigration curve for an island would then actually represent the average of two curves, one for each species (Figure 15.6).

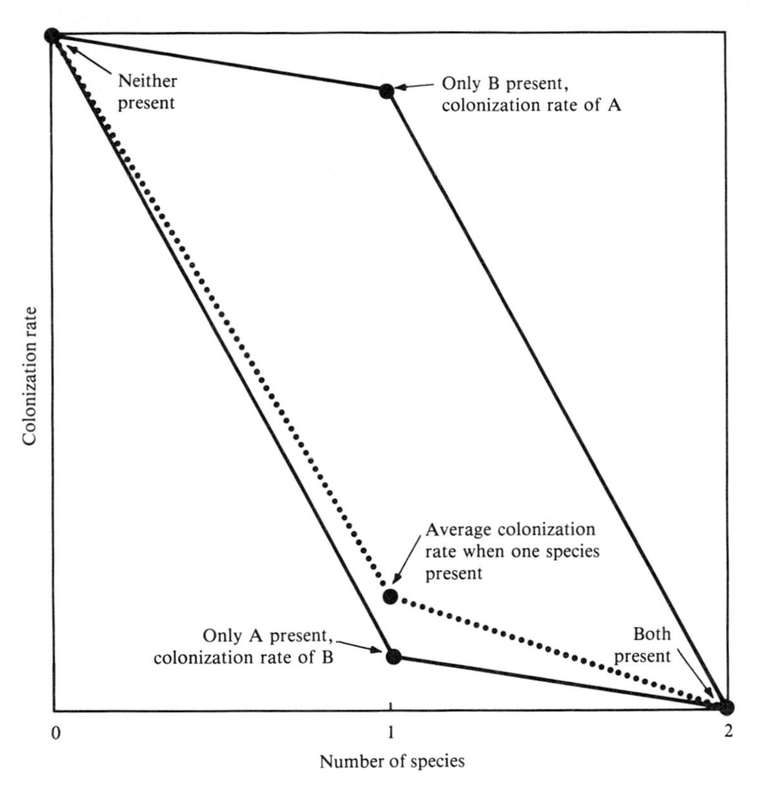

Figure 15.6
Colonization rate curve will be bowed downward if species differ in their probabilities of immigrating. This is the case for two species, a good colonizer *(A)* and a poor one *(B)*. When only one species is present it will usually be *A*, and the colonization rate of *B* will be low. Thus the average colonization rate curve will be bowed as shown. Similar reasoning can be used to show that the extinction rate curve will also be bowed downward if species differ in their probability of becoming extinct. (After Figure 5-12a [p. 98] from *Geographical Ecology: Patterns in the Distribution of Species* by Robert H. MacArthur. Copyright © 1972 by Harper & Row, Publishers, Inc. Reprinted by permission of Harper & Row, Publishers, Inc.)

These must go through the same end points, at $S = 0$ and $S = P = 2$, but unless the species have exactly the same colonization rate, they will be different when only one of the two species is present on the island, so $S = 1$. To visualize this, suppose that the immigration rate of A is much higher than that of B. If B were present on the island, then the probability of arriving would be high; whereas if A were present, the immigration rate of B would be much lower. The average rate of arrival when one species is present would be lower than if the two species had equal colonization rates (Figure 15.6) because A, by virtue of its higher immigration rate would tend to be the species that is present, and the rate of arrival of B would be low. Similar reasoning can be extended to cases in which there are any number of species in the pool and to show that the extinction rate curves also should be bowed downward.

Now let us incorporate the effects first of island size and then of insular isolation into this model. MacArthur and Wilson assumed that the size of an island would affect only the extinction rate. Although they recognized that a large island potentially provides a larger target for dispersing propagules than a small one, they reasoned that such an effect on colonization

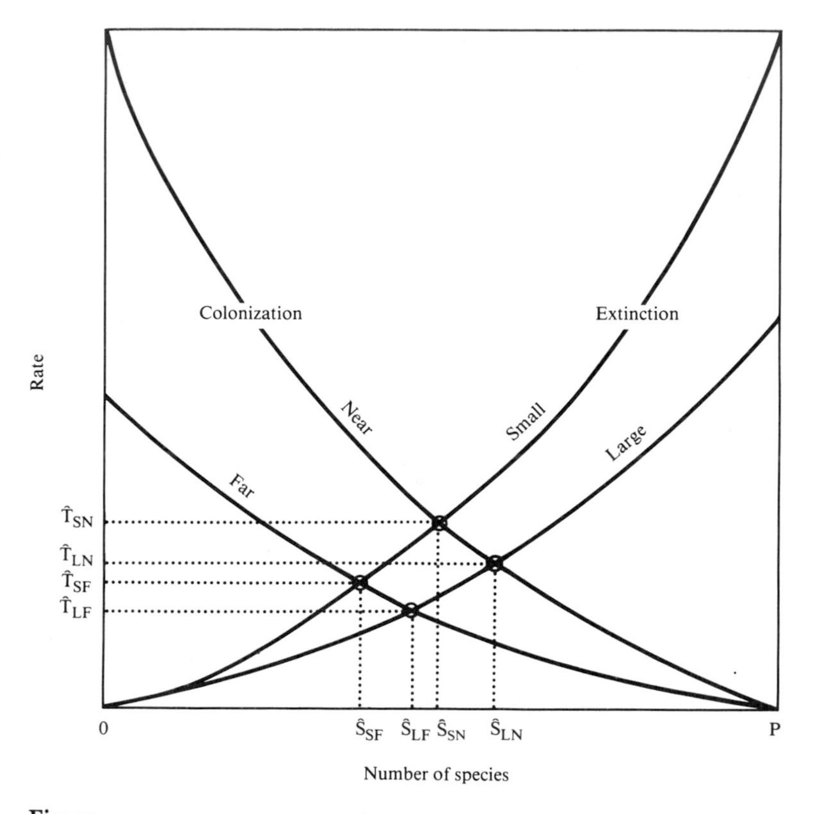

Figure 15.7
MacArthur and Wilson's (1963) equilibrium model of island biogeography, showing the effect of island size (different extinction rate curves) and isolation (different colonization rate curves) on the equilibrium number of species ($\hat{S}$) and rate of species turnover ($\hat{T}$). The intersections of curves for islands of different combinations of size and distance (small far and large near shown here) can be used to predict the relative number of species and turnover rates at equilibrium.

rate would be insignificant compared to the importance of island size in extinction. The population sizes of all species should decrease with decreasing island area, and as discussed in Chapter 6, the probability of extinction increases rapidly as a population gets very small. Because species inevitably differ in their niches and have different carrying capacities, particular species should vary in their probability of extinction on an island of any given size. Consequently, for a pool biota of many species the extinction rate should be substantially greater for a small island than for a larger one. This can be shown in the graphic model by drawing two extinction rate curves, one for a small island that is always higher than the other for a large island (Figure 15.7). Examining the intersections of these curves with the colonization rate curve shows immediately that the small island is predicted to have a smaller equilibrium number of species and a higher equilibrium turnover rate than the large island.

MacArthur and Wilson assumed that the distance of an island from the source pool affects only the colonization rate. No matter what the mechanism of dispersal, if a barrier exerts a filtering effect, then the probability of an organism crossing the barrier decreases as the width of the barrier increases. For example, in any kind of random dispersal, such as dissemination of seeds from a parent plant (recall Figure 7.1), the probability of reaching any given distance decreases with distance from the origin. This effect of isolation by distance can be incorporated into the model by drawing two colonization rate curves, one for an island near a source of species that is always higher than the other for a more isolated island (Figure 15.7). The intersections of these lines with the extinction rate curve predict that at equilibrium near islands should have more species and higher rates of turnover than distant islands.

By combining the effects of both island size and isolation in a single graph (Figure 15.7), one can see the predictions of the model. There are four intersections of colonization and extinction rate curves, one for each combination of island size, large (L) and small (S), and distance, near (N) and far (F). The number of species at equilibrium is predicted to be in the order $\hat{S}_{LN} > \hat{S}_{LF} \cong \hat{S}_{SN} > \hat{S}_{SF}$. (Note that whether a large far or small near island should have more species will depend on the exact shapes of the colonization and extinction curves.) The model obviously predicts the observations that motivated it: namely that the number of species increases with area and decreases with isolation and that there is a continual turnover of species. However, like any good theory, the model goes beyond what is already known to make additional predictions that can only be tested with new observations and experiments. Specifically, it predicts the following order or turnover rates at equilibrium: $\hat{T}_{SN} > \hat{T}_{SF} \cong \hat{T}_{LN} > \hat{T}_{LF}$. The model also predicts the relative rates at which islands of different size and isolation should return to equilibrium if the biota is perturbed by manipulative experiment or some unusual natural event. For example, a near island should return to equilibrium more rapidly than a distant island of the same size because it should have a higher colonization rate but the same extinction rate.

Strengths and weaknesses of the theory

Like most important new ideas, MacArthur and Wilson's theory elicited a mixed response from other scientists. It generated not only great interest and enthusiasm on the part of some investigators but also severe skepticism and criticism from others. Certainly the theory stimulated a new wave of research in ecological biogeography. Most of these studies were on islands or other insular habitats and were designed specifically to evaluate or elaborate on MacArthur and Wilson's model. In a review of the equilibrium theory written only a decade after MacArthur and Wilson's first seminal paper,

Simberloff (1974a) cited 121 references, and the pace of research has continued to accelerate in the last few years.

Several features of the theory contributed to its favorable reception. It not only interjected stimulating new ideas that helped bridge the gap between traditional biogeography and ecology but also it presented them in the form of an elegantly simple model that makes clear, testable predictions. Many mathematical models, especially those used in ecology, are not empirically operational, i.e., they do not explicitly indicate what observations would be necessary to test the theories unequivocally and to reject the models if they are inadequate. Such models can still be valuable, however, if they serve a heuristic function, if they cause one to think about a problem in new and more precise ways. MacArthur and Wilson's model certainly serves such a heuristic role, but it also suggests the kind of data that are necessary for a rigorous test. It predicts qualitative trends (increases or decreases) in numbers of species and turnover rates with island size and isolation. These patterns can be checked using simple lists of species in different taxa inhabiting various archipelagos at different times. Such lists can be compiled by original field work to census the organisms concerned, but the results of previous surveys of the biotas of many islands are also available in the published literature. The only other data required, areas of the islands and distances to probable source areas, can be obtained from standard maps and gazetteers.

The simplicity of the theory, however, has also been the cause for much criticism. Some have argued that it is so simple as to be useless because the approach obscures, rather than clarifies, the patterns and processes that make island biogeography interesting (e.g., Sauer, 1969; Lack, 1970; Carlquist, 1974; Gilbert, 1980). In view of the research the theory has stimulated and the increased understanding of island distributions that has resulted (see Chapter 16), too much criticism on this account is probably unwarranted. All models are intended to help investigators understand nature by presenting a simplified, abstracted concept of a more complex reality. They inevitably sacrifice a certain amount of precision for clarity and generality. Usually new theories are presented initially in a very general, incomplete form and are corrected and refined as a result of subsequent empirical and theoretical research.

If the equilibrium theory of island biogeography survives at all, it will probably experience this fate. Critics have pointed out specific problems with several assumptions of the model, which may limit its applicability to many insular distributions. Much of the recent research has served to identify and evaluate these problems and to modify the model to incorporate different and, it is hoped, more realistic assumptions. We will briefly describe some of the more important problems.

1. Many insular biotas may not be in equilibrium between opposing rates of colonization and extinction. Rather, the number of species may increase or decrease over evolutionary time. This will be particularly likely when colonization and extinction occur on approximately the same time scale as the geologic and climatic events that create, change, and destroy islands. An equilibrium theory may be useful for interpreting these cases, however, because it may be profitable to consider these biotas as approaching a new equilibrium following a historical perturbation.

2. The model assumes that the identities and characteristics of particular species can be ignored. Species are not all assumed to have identical colonization and extinction rates, but nevertheless, the resulting turnover is viewed as a highly stochastic process: new species immigrate and existing ones die out more or less at random, and only the approximate number of species remains the same. To the extent that ecological processes, including interspecific interactions, determine what species can exist on a particular island, this assumption is at least technically incorrect. Eventually it should be possible to construct an alternative theory that

has greater predictive power because it takes the ecological mechanisms of species interactions into account.

3. Colonization and extinction are treated as independent processes. This is probably justified, given that colonization is defined as the arrival of propagules of new species and that secondary succession is not occurring. On the other hand, the immigration rate of additional individuals of species already present will tend to be affected by the same variables that determine colonization rates, and such immigration can potentially reduce extinction rates of insular populations. Consequently, isolation by distance might influence extinction rates as well as colonization rates.

4. Larger islands certainly present larger targets for dispersing propagules, suggesting that the assumption that island size affects only extinction rates may be in error. In many cases, however, target size should be related to the linear dimensions of an island (e.g., the diameter), which will increase only as the square root of the area. Consequently, island size may have a much greater effect on population size, and hence on extinction rate, than on colonization rate.

5. It may be difficult to identify the source of the biota without careful investigation of the systematics and historical distribution of the group. Indeed, the species inhabiting a single island may be derived from several sources, including over-water dispersal from continents and other islands, past connections with other landmasses, and endemic speciation within the island itself. If insular species are acquired by colonization from multiple sources, the theory can potentially be modified to deal with this complication. For example, MacArthur and Wilson considered stepping-stone colonization, in which a species disperses from island to island down a chain of islands. If, on the other hand, insular species are derived by speciation within the island itself, then a basic assumption of the model is clearly violated.

6. Area provides only a very general, often indirect measure of the capacity of islands to support individuals and species. Although the extent of most habitat types increases with increasing island size, so usually does the diversity of different kinds of habitats. Larger islands tend to have higher mountains, more aquatic habitats, and so on, as well as larger areas of most of the vegetation types found on small islands. Consequently, some of the increase in species diversity with island size may be owing to the addition of specialists whose habitat requirements are met only on large islands. In this case a more elaborate model, which also incorporates specific habitat variables, should predict patterns of insular species diversity and distribution better than the MacArthur-Wilson model (e.g., Power, 1972; Johnson, 1975).

These criticisms are serious and should be kept in mind as we proceed to discuss insular distributions in this and the following chapter. Many cases will be seen in which the model, in its original form, is inadequate to account for observed patterns. It will also be found that in following MacArthur and Wilson's general approach, in testing and often rejecting their theory, a great deal has been learned about the ecological and historical processes that determine the distribution of organisms among islands and other insular habitats.

Tests of the model

Beginning in the late 1960s, numerous authors published papers purporting to test MacArthur and Wilson's theory. Often only one or two predictions, especially variation in number of species with area and isolation, were tested. Even these incomplete tests are valuable because when the predictions are not borne out by the data, they indicate that the theory in its present form cannot account for the distributions of these biotas. However, when a few observations concur with the predictions, it is unwarranted to assume, as many authors have done, that this corroborates the theory. It is possible to obtain the right result for the wrong reason because alternative explanations based on different assumptions can generate some of the same predictions. Before accepting a particular model as the explanation for a set of observations, it is important not only to test all possible predictions of the model but also to evaluate the validity of the underlying assumptions and to attempt to imagine and rule out alternative explanations.

Early studies that tested and purported to support the model fell into three categories. Most common were analyses of additional data sets on insular distributions of various taxa at one point in time. Many of these simply confirmed for other groups of organisms and other archipelagos the relationship between number of species, island size, and island isolation pointed out by MacArthur and Wilson (e.g., Hamilton et al., 1964; Hamilton and Armstrong, 1965; Johnson et al., 1968; Johnson and Raven, 1973). Others described similar patterns for insular habitats, such as mountaintops (Vuilleumier, 1970) and caves (Culver, 1970; Vuilleumier, 1973). Thus investigators now have many species-area relationships for various taxa on different kinds of islands. Frequently, ad hoc explanations have been advanced to explain differences among these relationships in such things as z values and goodness of fit of the regression equation to the data points. Although these studies cannot be viewed as rigorous tests of the theory, they indicate the generality of insular diversity patterns.

Estimates of turnover. In 1969 Diamond reported a more direct test of one crucial prediction: continual turnover in species composition resulting from approximately balancing colonizations and extinctions. Diamond recensused the avifauna of the Channel Islands off the coast of southern California almost exactly 50 years after Howell (1917) had published a detailed account of the birds known to breed on each island around the turn of the century. Comparisons of the two censuses revealed striking differences (Table 15.2). Although there had been relatively little change in the number of species breeding on each island, in 1968 Diamond observed a number of species not known to breed prior to 1917 and he failed to find about an equal number of species known to have been present 50 years earlier. From these observations he concluded that at least 20 to 60% of the bird species on each island had turned over since 1917. He pointed out that ac-

Table 15.2
Diamond's (1969) analysis of turnover of breeding land bird species on the California Channel Islands between 1917 and 1968

Island	Area (km^2)	Distance to mainland (km)	Number of species		Extinctions	Introductions (by humans)	Colonizations	Percent turnover*
			1917	1968				
Los Coronados	2.6	13	11	11	4	0	4	36
San Nicholas	57	98	11	11	6	2	4	50
San Clemente	145	79	28	24	9	1	4	25
Santa Catalina	194	32	30	34	6	1	9	24
Santa Barbara	2.6	61	10	6	7	0	3	62
San Miguel	36	42	11	15	4	0	8	46
Santa Rosa	218	44	14	25	1	1	11	32
Santa Cruz	249	31	36	37	6	1	6	17
Anacapa	2.9	21	15	14	5	0	4	31

*Turnover rate, expressed as percent of the resident species per 51 years, is calculated as 100 (extinctions + colonizations)/(1917 species + 1968 species − introductions).

tual turnover rates could well have been even higher; some species might have colonized or become extinct more than once in the 50 years. Although there were not enough islands to test rigorously for the relationship between turnover rate and either island size or isolation, Diamond noted that turnover appeared to be greatest on the islands with fewest species.

Diamond's study has been cited widely as confirming or supporting the equilibrium model (e.g., MacArthur, 1972), but it has been vigorously challenged by Lynch and Johnson (1974). They pointed out that most of the thoroughly documented changes in the avifauna can be attributed directly to human influence, and some of the other changes may have resulted from errors in conducting and interpreting the census. In particular, most of the extinctions involved the disappearance of the large birds of prey, including the osprey *(Pandion haliaetus)*, the bald eagle *(Haliaeetus leucocephalus)*, and the peregrine falcon *(Falco peregrinus)*, which were almost certainly eliminated by pesticide poisoning. Many colonizations were a result of the immigration of house sparrows *(Passer domesticus)* and starlings *(Sturnus vulgaris)* that colonized under their own power as they expanded their ranges after introduction into eastern North America from Europe. Turnovers involving these latter two and probably other species also have been influenced by habitat changes caused by recent human activities. Lynch and Johnson conclude that there is little evidence for natural turnover of breeding birds on the Channel Islands within the last 50 years, as Diamond had claimed.

The debate has continued. Jones and Diamond (1976) surveyed the birds of the California Channel Islands, especially Santa Catalina, for several years in succession. There were year-to-year changes in the breeding status of several species. Diamond and May (1976) analyzed many years of careful records of birds breeding on the small Farnes Islands off Great Britain. Several species were reported to have been present only sporadically during the years for which data were available (Table 15.3). In both of these studies, however, the "turnovers" involved spe-

Table 15.3
Year-by-year turnover of breeding bird species on the Farnes Islands, Great Britain
This table is based on 29 consecutive years of censuses. Note that the 12 species that bred in some years but not in others had very low average populations, whereas those 4 species that were present every year maintained average populations of about 15 pairs per species. (After Diamond and May, 1977.)

Number of breeding bird species	
Present per year	
$\overline{X}$	5.9
Range	4-9
Exhibiting turnover	12
Present every year	4
Percent species turnover per year	
$\overline{X}$	13
Range	3-22
Number of breeding pairs per species	
Of species exhibiting turnover	
$\overline{X}$	0.23
Range	0-6
Of species present every year	
$\overline{X}$	14.87
Range	2-60

cies that were migratory or highly nomadic. Their presence or absence on an island is more a matter of arrival and departure of highly mobile individuals for which a few miles of water represent no significant barrier, rather than the establishment and extinction of real resident populations. Terborgh and Faaborg (1973) reported apparent turnover in the avifauna of Mona, a small island west of Puerto Rico in the West Indies, as a result of comparisons between an early survey and their own recent census. Again, however, the island has been changed substantially by human activity, especially by the impact of introduced goats. Lack (1976) documented changes in the birds on Jamaica in more than 200 years of recorded history. On this large Caribbean island with a resident land avifauna numbering 65 species, there have been two extinctions and one colonization, and all of these can be attributed directly to human influence. All of this is not to say that organisms do not colonize and become extinct naturally on oceanic islands as predicted by equilibrium theory. Rather, the data documenting such turnover of established populations are not convincing, even for birds, which are perhaps the best long-distance colonizers of all organisms.

Experimental defaunation. The most direct and rigorous test of MacArthur and Wilson's theory was conducted by Wilson and his student Daniel Simberloff (Simberloff and Wilson, 1969, 1970; Wilson and Simberloff, 1969). This study has become justly famous as an example of the use of controlled, manipulative experimentation to test theoretical models in biogeography and ecology. The basic design was simple: all species of arthropods were eliminated from tiny islands of red mangrove *(Rhizophora mangle)* in the Florida Keys and subsequent changes were monitored closely. This was a drastic but effective perturbation. Simberloff and Wilson hired an exterminator, who used methyl bromide gas to kill all insects, spiders, mites, and other terrestrial animals while leaving the mangrove vegetation virtually undamaged. Recolonization, monitored by careful surveys, was surprisingly rapid (Figure 15.8). Within less than a year all but the most distant island had recovered their initial number of species. In fact, the number of species increased rapidly and appeared to overshoot the original number before declining and stabilizing close to the initial value. Furthermore, there was a great deal of turnover, even after the number of species had stopped changing significantly. Individual species colonized and disappeared, sometimes repeatedly, during the short-term study (Figure 15.9).

Simberloff and Wilson's results strongly supported several predictions of the model. On these minute islands near a source of colonists (much larger islands of the Florida Keys) the arthropod fauna apparently represented a dynamic equilibrium between recurrent immigration and local extinction. When the biota was perturbed from this equilibrium by elimination of all species, the original number of species was rapidly restored by colonization. Although there were too few islands to test rigorously for the predicted relationships of initial colonization rate, equilibrium turnover rate, and equilibrium number of species to island size and isolation, nevertheless the results were generally consistent with the predictions. For example, the most isolated island (E*1* in Figure 15.8) had the fewest species and the lowest rate of recolonization. The apparent overshoot in the number of species during recolonization suggests that the islands could support more than the equilibrium number of species so long as most of them were rare, but as populations approached their carrying capacities competition and predation eliminated the excess species.

The rescue effect. Another study of arthropods on isolated patches of vegetation points to a potentially important problem with MacArthur and Wilson's equilibrium model. Brown and Kodric-Brown (1977) censused the number of individuals and species of arthropods (mostly insects and spiders) on individual plants of thistle *(Cirsium neomexicanum)* growing in desert shrubland in southeastern Arizona. Although

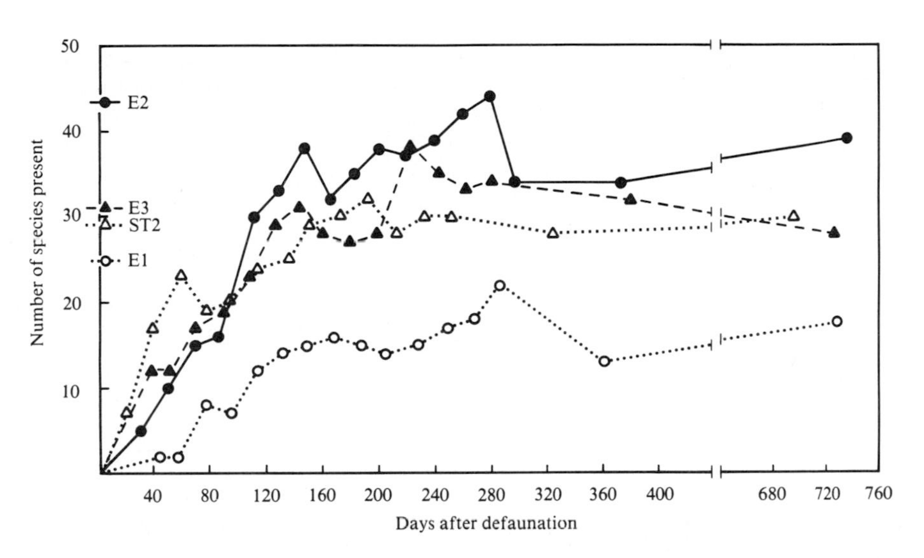

Figure 15.8
Recolonization by terrestrial arthropods of four small mangrove islands as a function of time since the fauna was removed. The initial number of species present is indicated along the vertical axis. Note that after defaunation the number of species increases rapidly, tends to overshoot the initial number, declines, and then increases gradually to approximately the initial number. Island E1 with a lower rate of colonization and smaller number of species was more isolated from a source of colonists than the other islands. (From D.S. Simberloff and E.O. Wilson. Experimental zoogeography of islands. A two-year record of colonization. *Ecology* 51[5]:934-937. Adapted by permission of Duke University Press. Copyright © 1970, Ecological Society of America.)

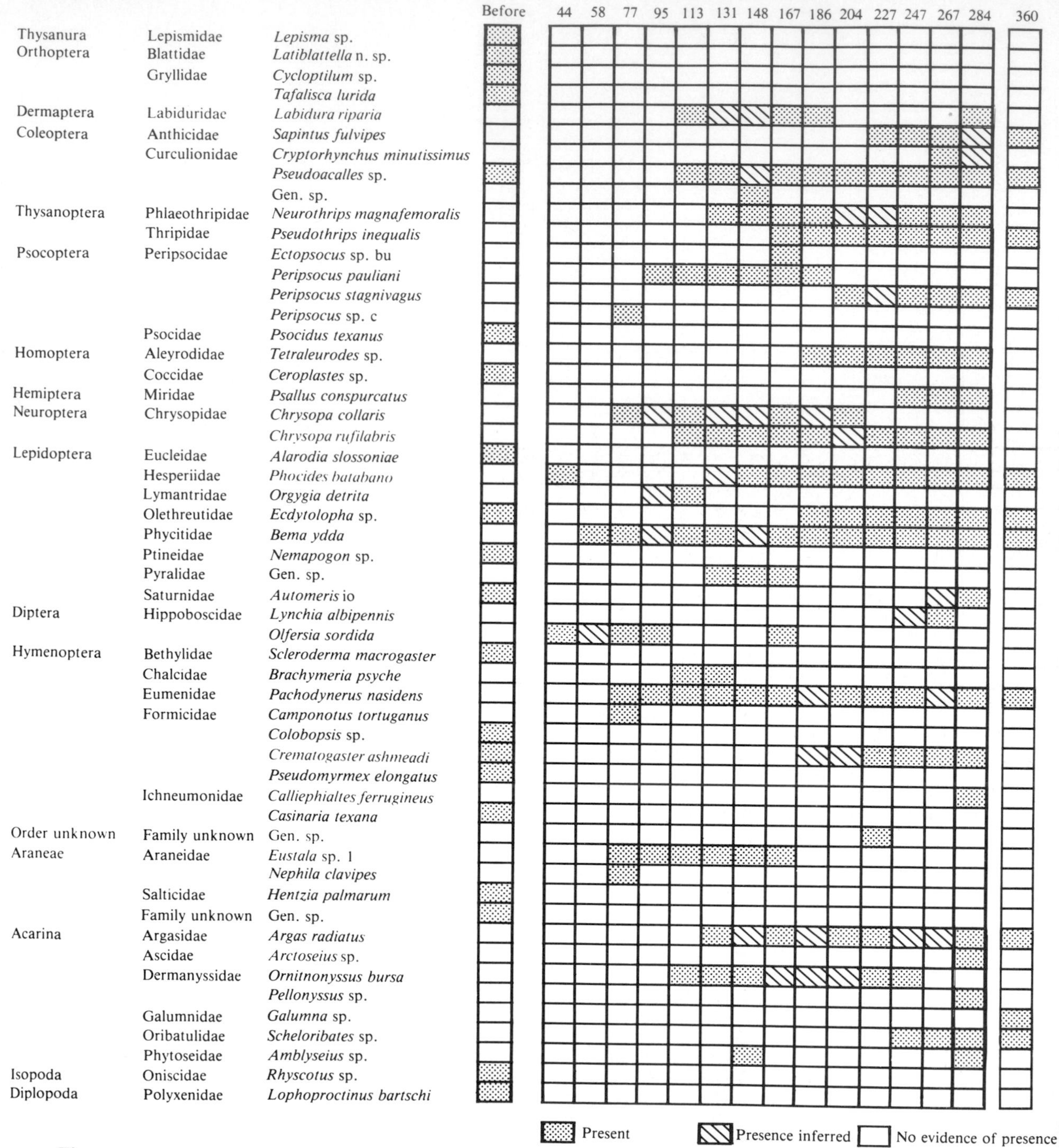

Figure 15.9
Record of the occurrence of the different species of terrestrial arthropods on Simberloff and Wilson's mangrove island E1. Note that not all of the species originally present recolonized and that some of the immigrant species became extinct during the period that the fauna was monitored. (From D.S. Simberloff and E.O. Wilson. Experimental zoogeography of islands. The colonization of empty islands. *Ecology* 50[2]:278-296. Adapted by permission of Duke University Press. Copyright © 1969, Ecological Society of America.)

Table 15.4

Turnover rates of arthropod species on individual thistle plants

Data were gathered over a 5-day period on two sites in southern Arizona. Plants were divided into objective size and isolation categories on the basis of thenumber of flowers and the number of other plants in the immediate vicinity. Note that the numbers of species follow the predictions of the MacArthur-Wilson model, but the turnover rates do not because near plants have lower turnover rates than far ones. (From J.H. Brown and A.K. Brown. Turnover rates in insular biogeography: effect of immigration on extinction. *Ecology* 58[2]:445–449. Adapted by permission of Duke University Press. Copyright © 1977, Ecological Society of America.)

Size-isolation category	Site 1			Site 2		
	Number of plants	$\bar{X}$ number of species*	$\bar{X}$ turnover rate†	Number of plants	$\bar{X}$ number of species*	$\bar{X}$ turnover rate†
Large-near	16	3.82	0.67	9	5.25	0.29
Large-far	7	3.78	0.78	9	4.44	0.42
Small-near	56	1.89	0.78	21	2.21	0.69
Small-far	3	1.33	1.00	11	0.80	0.91

*Number of species is for the second of the two censuses 5 days apart.

†Turnover rate equals the number of species present only in the first census plus the number of species present only in the second census, divided by the total number of species present in both censuses.

the intervening habitat may have been suitable for some of the arthropod species, the thistle plants constituted isolated patches of favorable habitat. Brown and Kodric-Brown censused the plants at 5-day intervals. Results confirmed all but one prediction of the MacArthur-Wilson model (Table 15.4). The number of individuals and species increased with plant size and decreased with increasing distance from the nearest plants. Although the animals did not maintain real populations on the plants, there was a dynamic equilibrium between the rates of arrival and disappearance. Defaunated plants were reinhabited rapidly, those near other thistles more quickly than isolated plants: plants closely surrounded by others regained 94% of their original arthropod biota in 24 hours, whereas isolated plants acquired only 67% of the initial number of species in the same period. The turnover of individuals and species was higher on small plants than on large ones, but contrary to the prediction of the model, turnover rates were higher on plants in close proximity to others than on isolated ones.

This single exceptional result is important, because it suggests a problem with the model that may be as important for organisms on real islands as for arthropods on thistles. The most likely explanation for all of the results taken together is that there is an insular equilibrium maintained by opposing colonization and extinction similar to that envisioned by MacArthur and Wilson, but the factors affecting the arrival of new species are not independent of those influencing the extinction (or departure, in the case of the arthropods on thistles) of species already present. Proximity to a source of species increases the immigration rate of all species, and a continual influx of individuals belonging to species already present tends to prevent the disappearance of these species. In the case of arthropods on thistles, this effect is probably simply statistical: high rates of immigration reduce the probability that a species will temporarily be absent and hence recorded as a turnover. On real islands, however, immigrants may rescue populations from extinction by contributing to the breeding stock and by injecting new genetic variability to counteract the deleterious effects of inbreeding, which can be severe in small, isolated populations. Smith (1980) corroborated the rescue effect of immigrants on extinction, showing that populations of pikas *(Ochotona princeps)* on isolated rockslides had higher turnover rates than those near a source of colonists.

This rescue effect was not anticipated by MacArthur and Wilson, but it is easy to modify their model slightly to incorporate it. This simply involves drawing different extinction rate curves as well as different colonization rate curves for near and far islands (Figure 15.10). Note that when this is done, it can reverse the order of the equilibrium turnover rates ($\hat{T}_F > \hat{T}_N$) from that predicted by the MacArthur-Wilson model, but the predicted relationship for the equilibrium number of species remains unchanged ($\hat{S}_N > \hat{S}_F$).

Brown and Kodric-Brown's study emphasizes the importance of testing all predictions of a model as well as critically evaluating its basic assumptions. Unless this is done, investigators risk misinterpreting data that are merely consistent with the model as corroborating evidence. The most crucial assumptions and predictions of the MacArthur-Wilson model are those re-

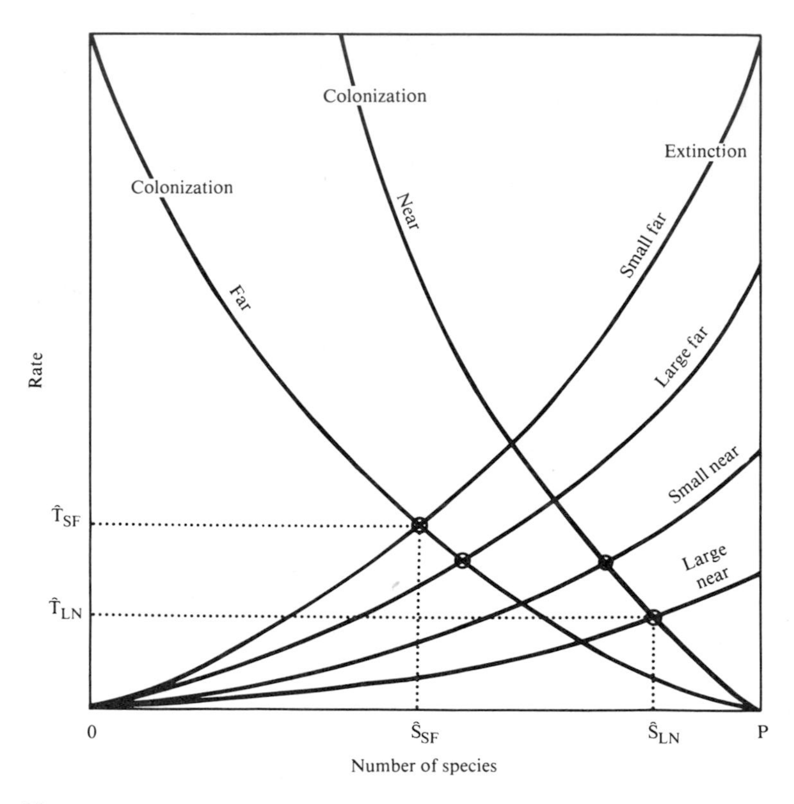

Figure 15.10
Rescue effect of immigration on extinction. Islands with higher colonization rates (nearer to sources of species) will tend also to have lower extinction rates because immigrating individuals of species already present will reduce the probability that those species will become extinct. This can be modeled by drawing additional extinction rate curves for islands of different combinations of size and isolation. This model predicts the same relative order of number of species at equilibrium as the MacArthur-Wilson model, but a different order of turnover rates. Specifically, near islands should have lower rates of species turnover than more distant ones. (From J.H. Brown and A.K. Brown. Turnover rates in insular biogeography: effect of immigration on extinction. *Ecology* 58[2]:445-449. Adapted by permission of Duke University Press. Copyright © 1977, Ecological Society of America.)

garding species turnover. Unfortunately these are also the most difficult to test, because they require repeated observations of undisturbed islands over relatively long periods of time. Although the individuals present on small, isolated habitats, such as patches of mangrove or individual thistle plants, may turn over rapidly, it remains to be shown that breeding populations become established and become extinct with sufficient frequency on real islands to test the critical predictions of the equilibrium model.

Selected references

Historical background

J. Brown (1981); W.L. Brown and Wilson (1956); Cody and Diamond (1975); G. Hutchinson (1958, 1959, 1967, 1978); Lack (1947, 1969, 1976); MacArthur (1958, 1972); MacArthur and Wilson (1963, 1967); Mayr (1942, 1952, 1963, 1965a, 1965b, 1976); Simberloff and Wilson (1969, 1970); E. Wilson (1959, 1961, 1969, 1975).

Island patterns

Carlquist (1965, 1966a, 1966b, 1970, 1974, 1981); Dammerman (1948); Darlington (1938, 1943, 1957, 1965); Docters van Leeuwen (1936); Gressitt (1963, 1970, 1974); Lack (1947, 1976); MacArthur (1972); Maguire (1963); Niering (1963); Preston (1962); T. Schoener (1969b); C. Williams (1953, 1964); M. Williamson (1981); E. Wilson (1959, 1961).

The theory

Gould (1979); Kilburn (1965); MacArthur and Wilson (1963, 1967); MacArthur (1972); Simberloff (1974a); M. Williamson (1981); E. Wilson (1969).

Strengths and weaknesses of the theory

J. Brown (1981); Carlquist (1974); F. Gilbert (1980); N. Johnson (1975); Lack (1970, 1976); Pielou (1979); Power (1972); Sauer (1969); Simberloff (1974a); M. Williamson (1981).

Tests of the model

Baroni Urbani (1971); J. Brown and Kodric-Brown (1977); Connor and McCoy (1979); Culver (1970); Diamond (1969, 1971, 1980); Diamond and May (1976); Gilpin and Diamond (1976); Haas (1975); T. Hamilton and Armstrong (1965); T. Hamilton and Rubinoff (1963, 1964, 1967); T. Hamilton et al. (1964); M. Johnson and Raven (1970, 1973); M. Johnson et al. (1968); H. Jones and Diamond (1972); Lack (1976); Lassen (1975); Lynch and Johnson (1974); MacArthur (1972); Mauriello and Roskoski (1974); McCoy and Connor (1976); Molles (1978); A. Schoener (1974a); Sepkoski and Rex (1974); Simberloff (1969, 1974, 1976a); Simberloff and Wilson (1969, 1970); Solem (1973); Terborgh and Faaborg (1973); Vuilleumier (1970, 1973); Whitcomb (1977); Whitehead and Jones (1969); M. Williamson (1981); E. Wilson (1969).

Chapter 16 Island Patterns and Processes

Many insular distributions cannot be explained solely in terms of a stochastic equilibrium between opposing rates of colonization and extinction, as predicted by the MacArthur-Wilson model. Deterministic patterns, not only in the number of species but also in their ecological characteristics and taxonomic identities, reflect the importance of both contemporary ecological processes and historical events in determining the composition of insular biotas. Geologic history and past climate have influenced the accessibility of islands to colonists. Ecological interactions, both between the organisms and their physical environment and among different species of organisms, have played a major role in determining which of the colonists have been able to persist to comprise the contemporary insular biotas.

Nonequilibrium diversity patterns

Comparing the pattern of insular species diversity with the predictions of the MacArthur-Wilson model reveals many cases in which the diversity clearly does not represent an equilibrium between contemporary rates of colonization and extinction. The number of species on these islands is not remaining approximately constant; instead it is either increasing or decreasing steadily in response to major historical events. Despite, or rather because of, the failure of its predictions, MacArthur and Wilson's approach to island biogeography has proven valuable for detecting and interpreting these nonequilibrium distributions. In fact, many of the patterns can be explained in terms of the biota approaching a new equilibrium number of species following a historical perturbation.

As pointed out in Chapter 14, Pleistocene changes in climate and sea level caused major shifts in the distribution of organisms, especially of terrestrial and freshwater species. The effects on certain insular habitats were profound. On the one hand, once isolated islands were connected by habitat bridges, permitting a free interchange of biotas; on the other hand, new habitat islands were created by the intrusion of formidable barriers to colonization. The last glacial period ended and the distribution of climate, sea level, and major biome types has been approximately stable only within the last 8000 to 10,000 years.

The legacy of these Pleistocene perturbations is still apparent in many insular distributions. Many islands and patches of insular habitat were formed by rising sea levels and climatic changes at the end of the Pleistocene. Sometimes the fragmentation of extensive areas of once continuous habitat left small islands oversaturated with species, and diversity began to decrease as certain species were eliminated by extinction. In other cases completely new insular habitats were created, and these began to acquire species by colonization and speciation. The extent to which the contemporary biota still shows the effects of these historical changes depends largely on the kinds of barriers isolating these post-Pleistocene islands and on the ability of different kinds of organisms to disperse across them.

In one of the first attempts to test the applicability of the MacArthur-Wilson model to the distribution of organisms among habitat islands on continents, Brown (1971b, 1978; see also Johnson, 1975; Behle, 1978) studied the small, nonflying mammals and later the birds inhabiting isolated mountaintops in the western United States. The Great Basin is a vast region of cold desert, lying at about 1500 m elevation between the Rocky Mountains to the east and the Sierra Nevada to the west (Figure 16.1). Isolated mountain ranges, some rising to more than 3000 m, form islands of isolated coniferous forest and other mesic habitats surrounded by a sea of sagebrush desert. These mountaintops are inhabited by a number of boreal mammal and bird species that are restricted to the cool, moist habitats of higher elevations. The patterns of diversity of these mammals and birds exhibit both similarities and differences when compared to each other and to the biotas of oceanic islands. The number of species inhabiting a mountaintop increases with the size (area) of the mountain range, but the slopes of these species-area relationships are very different for the two taxa (Figure 16.2). The z value for mammals (0.33) is higher and that for birds (0.17) is lower than the range usually reported

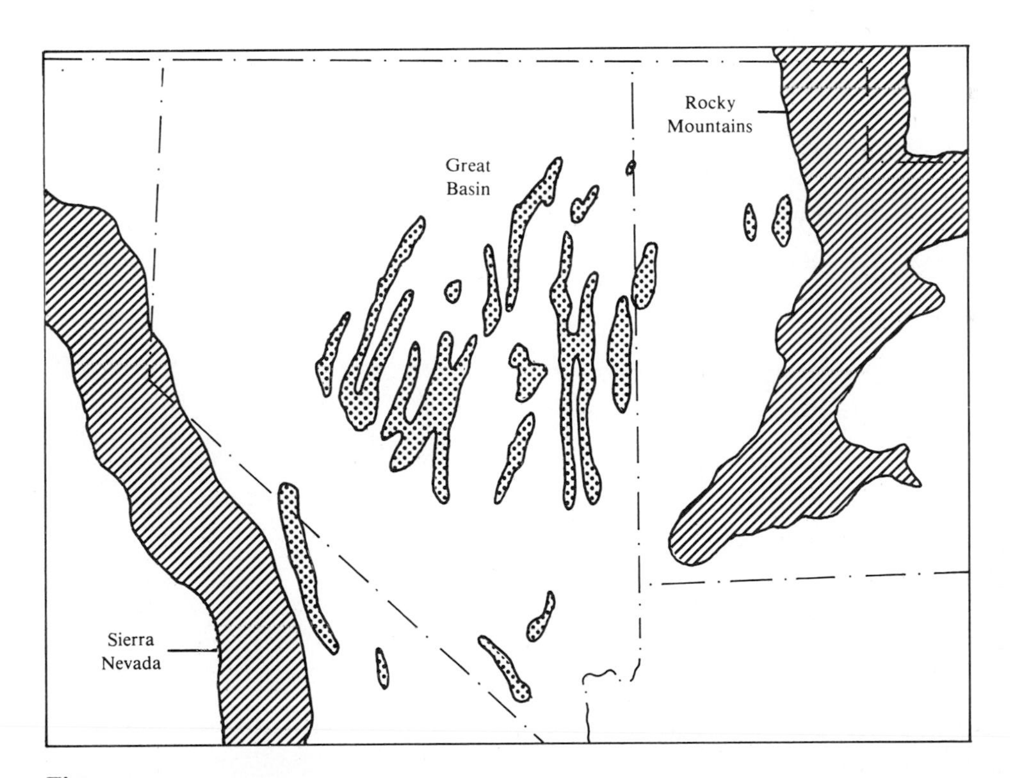

Figure 16.1
Isolated mountain ranges of the Great Basin in western North America are islands of cool, mesic, forest habitat in a sea of sagebrush desert. The ranges shown, with peaks mostly higher than 3000 m, lie between two montane "mainlands": the central mountains of Utah, part of the Rocky Mountains, to the east and the Sierra Nevada to the west. The desert valleys between the mountains are readily crossed by birds, but they are virtually absolute barriers to small mammal dispersal. (Reprinted from "Fish species diversity in lakes," *American Naturalist* 108:473-489, by C.D. Barbour and J.H. Brown, by permission of the University of Chicago Press. Copyright© 1974 by the University of Chicago.)

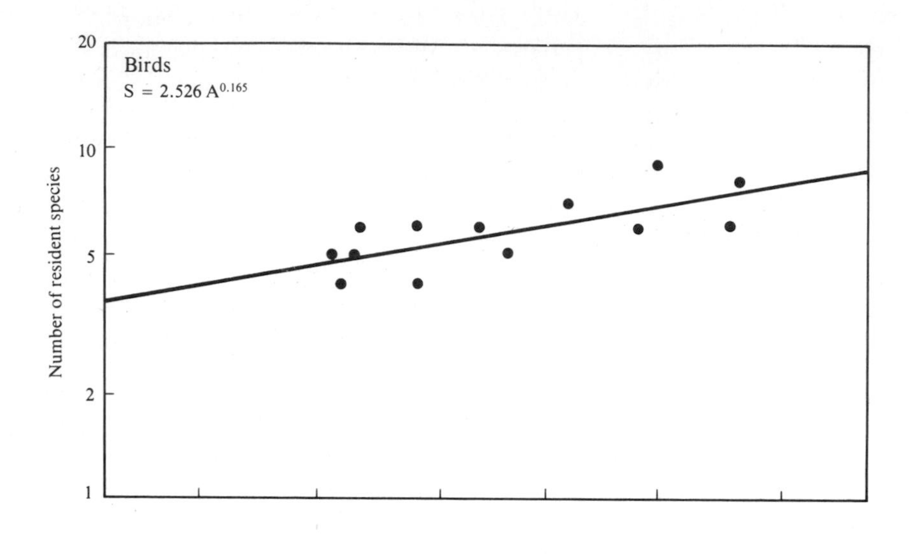

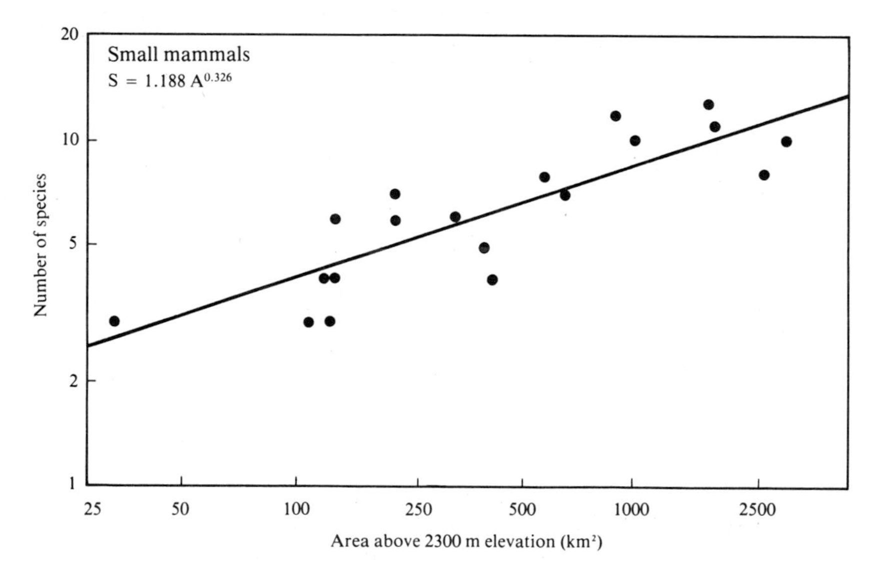

Figure 16.2
Species-area relationships for the boreal resident birds *(above)* and small terrestrial mammals *(below)* inhabiting the isolated mountain ranges of the Great Basin. Note that the slope of the curve for mammals is about twice as steep as that for birds. Birds continually recolonize these mountains, whereas the mammals are relicts of widespread populations during the Pleistocene. (After Brown, 1978.)

for organisms on oceanic islands (0.20 to 0.30). Furthermore, there is no detectable effect of isolation by distance on the diversity of either taxon. The Rocky Mountains and the Sierra Nevada are likely sources of colonists; they support a much higher diversity of boreal forms, including all of the species found on the isolated mountaintops. Nevertheless, except for this overall pattern, no effect of habitat barriers on the number of boreal species is apparent.

Brown proposed that the present boreal mammal populations of these montane islands are relicts, vicariant remnants of once widespread distributions during the Pleistocene. This model is consistent with plant fossils showing that as recently as 10,000 to 12,000 years ago the climate of the Great Basin was cooler and wetter and the vegetation zones were shifted several hundred meters below their present elevations (Wells and Berger, 1967; Wells, 1976, 1979; Thompson and Mead, 1982). This shift would have connected several presently isolated habitats across the entire Great Basin, permitting all islands to be colonized by all species for which appropriate habitat bridges existed. At the end of the Pleistocene, however, the cool, mesic habitats shrank back to higher elevations, completely isolating on the mountaintops those boreal mammal species unable to disperse across the desert valleys. After being isolated, some insular populations became extinct, reducing the diversity of the boreal biota, especially on small mountaintops, but vicariant populations of other species have survived until the present.

Brown's model proposes that in the absence of colonization the mammalian faunas of the isolated mountaintops have been relaxing toward an equilibrium of zero species. Small islands have had high extinction rates and lost most of their fauna in 10,000 years, whereas large islands still retain most of their original species. This effect of extinction, unopposed by colonization, produces the exceptionally steep slopes of the species-area curve in Figure 16.2. Two additional observations support the vicariant model. First, late Pleistocene or more recent fossils of boreal species, including some forms still found on some of the larger isolated ranges, have been found on several mountaintops, indicating that these species were indeed once present and have become extinct within the last 12,000 years (Grayson, 1981; Thompson and Mead, 1982). Second, the virtually absolute effectiveness of barriers of inhospitable habitat only a few kilometers across in preventing dispersal of small terrestrial mammals is supported by the distribution of those species whose habitats were never connected by habitat bridges. Several species are restricted to high alpine habitats that were never connected during the Pleistocene. All of these forms are unknown from any of the isolated mountaintops, even as fossils, although large areas of appropriate habitat are present on some islands.

Differences between the distributions of birds and mammals in this respect suggest that the avian distributions are not limited by dispersal and no longer reflect Pleistocene changes in habitat isolation. In contrast to the mammals, boreal bird species appear to be present on all mountain ranges where there are sufficient areas of suitable habitat (Johnson, 1975; Behle, 1978). Boreal birds have been observed flying across the desert valleys, which appear to pose no significant barriers to their colonization of even the most isolated mountaintops. A high colonization rate apparently continually replenishes bird populations on small islands, resulting in the extremely shallow species-area curve in Figure 16.2. Thus the distribution of boreal birds represents an equilibrium between colonization and extinction. However, even this is not quite the sort of equilibrium predicted by MacArthur and Wilson because the colonization and extinction processes appear to be highly deterministic instead of stochastic. The rate of colonization is so high that habitats are almost completely saturated with those species that can live there. When on rare occasion an insular population does become extinct it probably is replaced rapidly, not by a random species from the mainland pool, but by the reestablishment of a new population of the same species.

The insular distributions of many other terrestrial and freshwater organisms probably can be attributed primarily to the effects of extinction following the fragmentation of once continuous habitats into isolated patches at the end of the Pleistocene. Patterson (1980) showed that the nonflying boreal mammals inhabiting isolated mountain ranges in New Mexico also represent vicariant relicts of a once widespread Pleistocene fauna. The present distribution of freshwater fishes in southwestern North America reflects a history of aquatic habitat connections during the cooler and wetter climate of the Pleistocene, followed by isolation and subsequent extinction (Hubbs and Miller, 1948; Miller, 1948; Smith, 1978). Hard as it may be to believe, only a few thousand years ago Death Valley, in the most arid part of the North American desert, was almost completely filled with a large lake supplied by a major system of permanent rivers and springs (Figure 14.10). At least five genera of fishes inhabited this basin, and they have persisted as relictual populations in isolated springs. In the case of fishes, there can be no doubt about the susceptibility of such small, isolated populations to extinction. Many have disappeared within the last few years as humans have diverted water or introduced exotic species of competing or predatory fishes.

Rising sea levels since the Pleistocene have created many new islands in coastal regions. These are called land bridge or continental islands because, unlike oceanic islands, they were once part of the mainland or at least connected to it by a complete bridge of terrestrial habitats. The 100 to 200 m rise in sea level, which occurred about 10,000 years ago, inundated many land bridges and created numerous continental islands. Diamond (1972, 1975b) found that the influence of past connections to the mainland of New Guinea is apparent in the composition of the avifauna of the satellite islands off the coast. Islands that are separated from the mainland by water less than 200 m deep support a greater number of species than oceanic islands of comparable size and distance from New Guinea but that arose as undersea volcanoes and have never been connected to the mainland. Although the land bridge islands lack many bird species found in comparable habitats on New Guinea, they have several kinds of birds that are found on the mainland but never occur on the oceanic islands. Such continental islands have lost mainland bird species by extinction since the land connections were severed by rising water at the end of the Pleistocene, but for the last 10,000 years they have maintained species that are such poor over-water colonists that they have not become established on the oceanic islands. Of course, there are still other species that occur on both land bridge and oceanic islands. Because these birds obviously have crossed water barriers to colonize the oceanic islands, investigators cannot be sure whether the populations on the continental islands are Pleistocene relicts or whether they have been replenished by subsequent immigration.

In most of the examples discussed so far, the biotas of insular habitats isolated since the Pleistocene are relictual. Once widespread habitats containing diverse biotas have diminished in size and become fragmented. As soon as islands were formed, they are oversaturated with species for their restricted areas and they have been losing species by extinction. In contrast, some isolated habitats were left uninhabited at the end of the Pleistocene, and these undersaturated islands have been acquiring species over the last few thousand years.

An excellent example is provided by the glacier-formed lakes of northern North America and Eurasia. Many lakes, including the Great Lakes and the Finger Lakes in northeastern North America, were gouged out by the advancing continental ice sheets and filled with water as the glaciers retreated (see Figure 1.1). Some groups, such as certain algae, protozoans, and invertebrates that have cysts or other means of effective long-distance dispersal, probably already have approached an equilibrium between rates of colonization and extinction; but other animals, such as fishes (Smith, 1981) and some molluscs, have been limited by their inability to cross land barriers.

The fish faunas of many of these lakes appear to be undersaturated and still gradually increasing in diversity (Barbour and Brown, 1974). Only a handful of fish species can survive for more than a few minutes out of water at any stage of their life cycle, so fishes can colonize new areas only when connections of aquatic habitat are present. They usually colonize lakes from rivers and streams, but most lotic waters contain only a few species that can also be successful in lentic environments. Many glacier-formed alpine lakes lack native fishes entirely because they have never been connected by suitable habitat bridges to waters inhabited by fishes. The fact that several introduced species thrive in some of these lakes is further evidence that the absence of fishes is a consequence of barriers to dispersal. Even many of those glacier-formed lakes that contain fishes appear to be undersaturated with species. This is particularly true of the very large ones, such as the Great Lakes in North America and Lake Baikal in the Soviet Union. There were not enough species in the rivers and streams draining these lakes to fill the niches available to fishes. Some

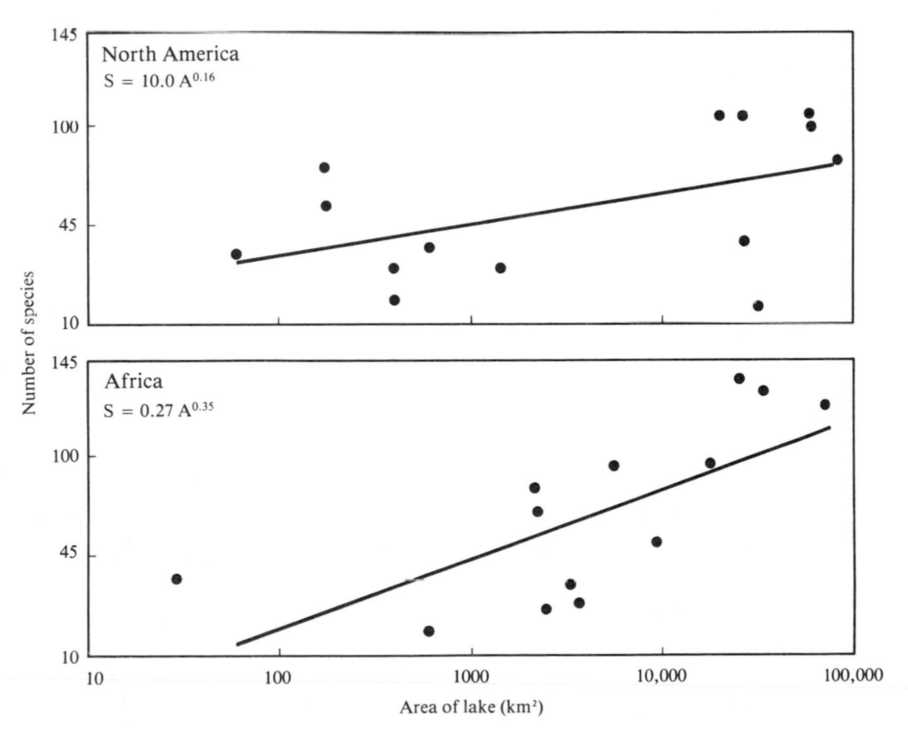

Figure 16.3
Species-area relationships for the fishes inhabiting lakes in northern North America recently formed by glaciers *(above)* and the much older lakes of central and eastern Africa *(below)*. Note that the slope of the curve for the African lakes is about twice as steep as that for the North American lakes. The large African lakes have acquired diverse fish faunas by endemic speciation. Although such speciation is occurring in the North American lakes, it has not yet produced high species richness in the largest lakes. (After Barbour and Brown, 1974.)

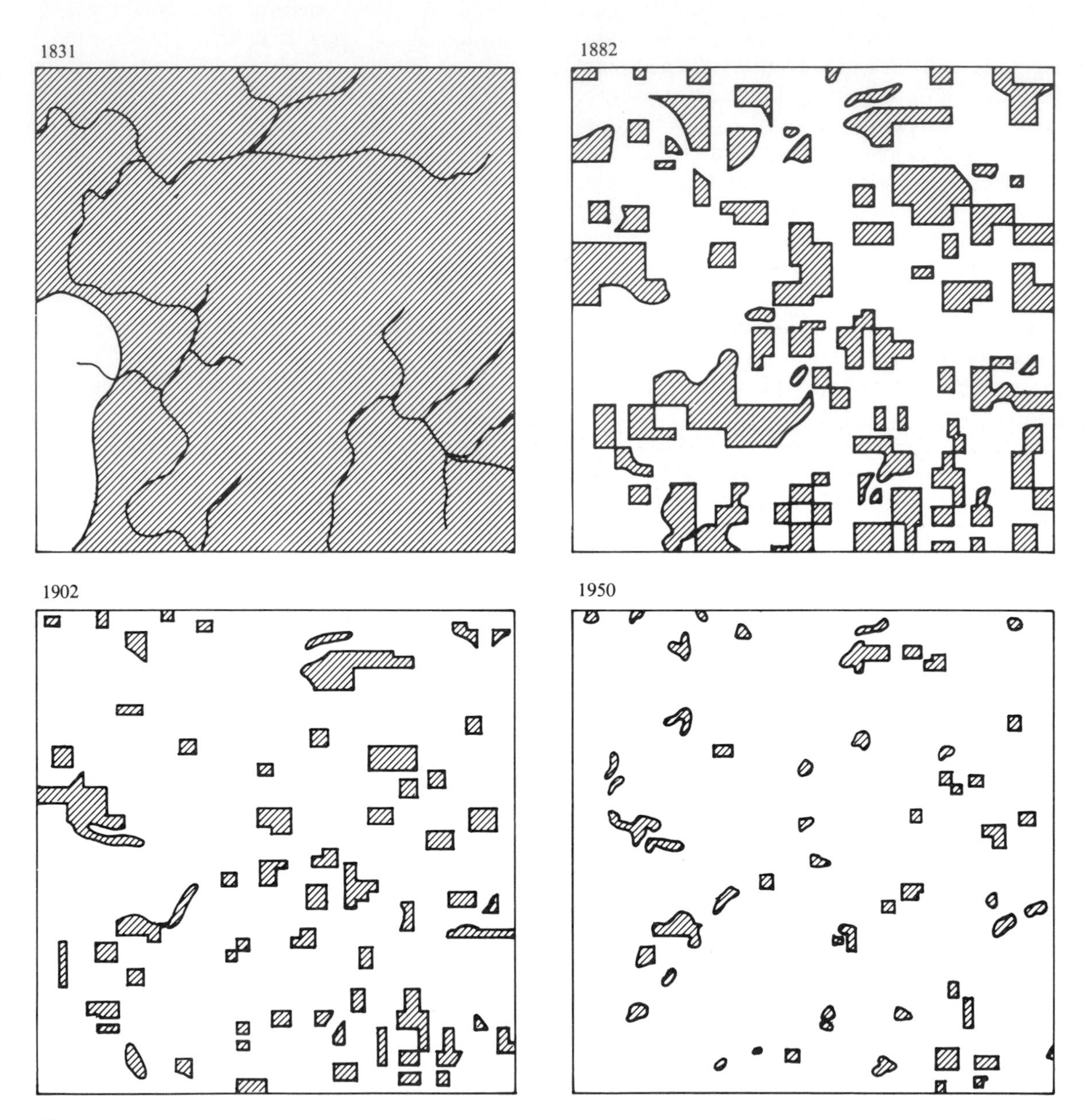

Figure 16.4
Cutting of once extensive forests has produced isolated islands of woodland habitat surrounded by agricultural fields in many parts of the world. This figure shows the reduction and fragmentation of forests in Wisconsin since the early 1800s. (Reprinted from *Man's Role in Changing the Face of the Earth,* W. L. Thomas, Editor, by permission of the Wenner-Gren Foundation for Anthopological Research, Inc., New York, and the University of Chicago Press.)

taxa, such as the ciscoes or whitefish *(Coregonus)* in the Great Lakes, have been diversifying by endemic speciation and adaptive radiation, but there has not been sufficient time since the Pleistocene to achieve an equilibrium between speciation and extinction.

This conclusion is supported by a comparison of species-area relationships for fishes inhabiting the glacier-formed lakes in temperate North America and the much older lakes of tropical Africa (Figure 16.3). The steeper slope for the African lakes is a consequence of the much larger number of species inhabiting the largest lakes. As mentioned in Chapter 6, this great diversity is owing primarily to the spectacular endemic speciation and adaptive radiation in the family Cichlidae (Fryer and Iles, 1972; Greenwood, 1974). Over a period of perhaps 5 million years, the African lakes appear to have attained an approximate equilibrium between speciation and extinction. The North American lakes are probably approaching such an equilibrium, but, considering that they have been in existence for only about 10,000 years, it is not surprising that they are still undersaturated.

Not all nonequilibrial insular distributions are the legacy of Pleistocene events. New insular habitats are continually being created and destroyed, sometimes by natural events such as volcanic eruptions (see discussion of Krakatau in Chapters 7 and 15), but now more often by human activities. A surprisingly large number of species have colonized human-modified habitats and increased in abundance and geographic range during the last few centuries. A far greater number, however, have decreased as their habitats have been reduced and fragmented, and many are now insular species threatened with extinction. This is apparent on many scales. In the eastern United States, once extensive forests have been reduced to isolated woodlots (Figure 16.4), and this has been accompanied by the local extinction of populations that cannot survive on the small islands that remain. Table 16.1 documents the mammal species that remain in woodlots of varying size surrounded by agricultural fields in Iowa.

Table 16.1

Small mammal species inhabiting small woodlots isolated by agricultural fields in Iowa

Although these woodlot islands supported an average of only 1.6 species, nearby tracts of large forest were inhabited by as many as 7 species, with a mean of 5.8 species. *Peromyscus maniculatus, Zapus hudsonius,* and *Cryptotis parva* were present on these mainlands, in addition to the four species listed below. (Data are from Gottfried, 1979.)

Island area (m^2)	Island isolation (km)	Species present				Total number of species
		Peromyscus leucopus	*Microtus pennsylvanicus*	*Microtus ochrogaster*	*Blarina brevicauda*	
639	0.16	X	X	0	X	3
630	1.60	X	X	0	0	2
573	0.32	X	0	0	0	1
510	0.42	X	0	X	X	3
502	0.08	X	X	X	0	3
374	0.64	X	0	0	0	1
350	0.75	X	0	0	0	1
337	2.88	X	0	0	0	1
250	1.83	X	0	0	0	1
93	0.80	X	0	0	0	1

Within the last few decades there has been increased recognition of the threatened loss of much of the earth's biota, and a major effort has been made to establish parks and nature preserves to save unusual natural habitats and endangered species. Recently, several authors (e.g., Terborgh, 1974, 1975; Wilson and Willis, 1975; Diamond, 1975a; Diamond and May, 1976; Forman et al., 1976; Galli et al., 1976; Simberloff and Abele, 1976; Whitcomb et al., 1976; Simberloff, 1978; Higgs, 1981; Hoffman, 1981) have suggested that biogeographic and ecological principles should be applied to design preserves that are maximally effective in preventing the extinction of threatened biota (Figure 16.5). Some of these principles should be intuitively obvious even to those without any scientific training (e.g., a large area is better than a small one of similar habitat), but others are hotly debated even by knowledgeable scientists (e.g., whether a single large preserve should be preferred over several smaller ones of the same total land area). Clearly, any reasonable efforts to preserve endangered habitats and species should be encouraged. Unfortunately, we may already have passed the point of no return for many species, even though we may not witness their extinction during our lifetimes. Many isolated parks undoubtedly are oversaturated with species and can be expected to lose some of their biota in the future. If 1000 km^2 of mountaintop is insufficient to prevent the extinction of several species of small mammals in 10,000 years (Brown, 1971b, 1978), how large an area is necessary to ensure the survival of the grizzly bear, the cheetah, the California condor, or the African elephant?

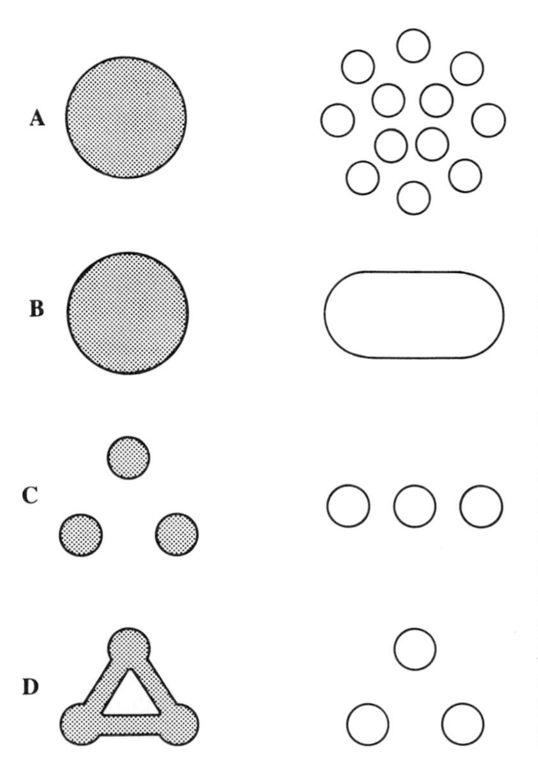

Figure 16.5
Application of biogeographic principles to the optimal design of nature preserves results in several recommendations. In each pair of figures above, Wilson and Willis (1975) suggest that the configuration on the left is to be preferred over that on the right, even though both designs incorporate the same total area. The basic concepts are as follows: **A,** a continuous preserve is better than a fragmented one; **B,** the ratio of area to perimeter should be maximized; **C,** distance between refuges should be minimized; and **D,** dispersal corridors should be provided between isolated fragments. (Reprinted by permission of the publishers from *Ecology and Evolution of Communities,* ed. M.L. Cody and J.M. Diamond, Cambridge, Mass.: The Belknap Press of Harvard University Press. Copyright © 1975 by the President and Fellows of Harvard College.)

Patterns reflecting differential extinction and colonization

As mentioned in the previous chapter, the MacArthur and Wilson approach to insular biogeography has its limitations as well as its advantages. The equilibrium model is a stochastic theory that assumes that all species are essentially equivalent and that colonization and extinction are random processes. The theory does not delve into ecological processes that might result in deterministic patterns of island community organization and that might therefore be used to predict ecological attributes and taxonomic identities as well as the number and turnover rates of insular species. MacArthur and Wilson were well aware that islands do not contain simply random assemblages of species. In fact, MacArthur and his associates (e.g., Crowell, 1962, 1973; MacArthur, 1972; MacArthur et al., 1972; Diamond, 1973, 1975b) played a leading role in searching for patterns in the organization of insular biotas and proposing ecological mechanisms to account for them. Certainly the existence of archipelagos with numerous islands varying in environmental characteristics and species has stimulated much research at the interface of biogeography and ecology. Although Diamond (1975b) claims it is possible to develop assembly rules that not only describe accurately the observed distribution of species among islands but also can be used to make precise predictions about the composition of insular biotas, not all investiga-

Table 16.2
Occurrence of three native fish species in isolated springs in Ash Meadows, Nevada, near Death Valley
Springs are arranged in approximate order of decreasing size. The pattern corresponds to the abundance of the species and the fate of the populations in recent times. Miller (1948) collected 22 specimens of *Empetrichthys merriami*, 515 of *Rhinichthys osculus*, and 3861 of *Cyprinodon nevadensis*. All five populations of *Empetrichthys merriami* and a few populations of the other species have become extinct since 1948.

Spring	*Empetrichthys merriami*	*Rhinichthys osculus*	*Cyprinodon nevadensis*
Big	X	X	X
Jackrabbit	X	X	X
Point of Rocks	X	X	X
Forest	X	X	X
Longstreet	X	X	X
Crystal Pool		X	X
Fairbanks		X	X
Soda		X	X
Tubbs		X	X
Devil's Hole			X*
School			X
North Indian			X
South Indian			X
Scruggs			X
Mexican			X

*This spring is inhabited by an endemic species, *Cyprinodon diabolis*.

tors are so optimistic. The demonstration and explanation of insular patterns has become a highly controversial area of modern biogeography.

Aside from the basic relationships between climate and biome type and the organization of trophic structure that characterize all communities (see Chapter 4) and the general increase in the number of species with increasing island area and habitat diversity (Chapter 15), there are at least two patterns of insular community structure that might be expected to reflect ecological processes. On the one hand, to the extent that general ecological attributes influence the probabilities of extinction and colonization, one can expect to see this reflected in the distribution of species among islands. In such cases the frequency of occurrence of a species on islands is expected to depend primarily on its own ecological characteristics rather than on the attributes of coexisting species. On the other hand, specific biotic interactions among species might also influence insular distributions. In these cases, one would expect the abundance and resource use of particular species on individual islands to depend on the presence or absence of certain other species that are competitors, predators, prey, or mutualists. There is considerable evidence for both kinds of patterns, so we will consider each of them in turn.

The differential survival of certain species in insular habitats isolated since the Pleistocene suggests that extinction has by no means been

Table 16.3

Ecological characteristics and distributions of small boreal mammals on isolated mountain

One can infer that all species were present on all mountain ranges at the end of the Pleistocene, so those not been random but have occurred most frequently in species of large body size, carnivorous diet, and

Species*	Body weight (g)	Diet	Habitat
Eutamias umbrinus	60	Mostly seeds	Generalist
Neotoma cinerea	300	Vegetation	Generalist
Eutamias dorsalis	55	Mostly seeds	Generalist
Spermophilus lateralis	170	Vegetation	Generalist
Microtus longicandus	45	Vegetation	Generalist
Sylvilagus nuttalli	800	Vegetation	Generalist
Marmota flaviventris	3000	Vegetation	Generalist
Sorex vagrans	7	Invertebrates	Generalist
Sorex palustris	14	Invertebrates	Streams
Mustela erminea	50	Small vertebrates	Generalist
Ochotona princeps	120	Vegetation	Talus near meadow
Zapus princeps	25	Mostly seeds	Streams and wet meadows
Spermophilus beldingi	300	Vegetation	Wet meadows
Lepus townsendi	3000	Vegetation	Large open meadows
TOTAL NUMBER OF SPECIES			

*In a few cases includes other species that are sibling species or ecological equivalents (see Brown, 1978).

an entirely random process. Some species have become extinct much more frequently than others during the last 10,000 years, and this differential susceptibility appears to be related to ecological characteristics. Often the pattern of extinction is so regular that ecologically similar islands of varying size within an archipelago form nested sets of species; that is, progressively smaller islands contain an increasingly limited subset of the species found on larger islands (Tables 16.2 and 16.3). Although these patterns do not always exhibit perfect regularity, they are much more ordered than one would expect if the probability of extinction of each species varied randomly from island to island.

Of course, this is just what would be expected if, as suggested in earlier chapters, ecological characteristics of species determine their carrying capacities and these equilibrium population densities in turn largely determine the probability of extinction (e.g., Brown, 1971b, 1978, 1981; Van Valen 1973a, 1973b). In fact, it should be possible to predict the relative abilities of species to persist on isolated land bridge islands based on some easily measured ecological traits. For example, animals of large body size, carnivorous diet, and specialized habitat requirements should have lower carrying capacities and higher extinction rates than species that are smaller, herbivorous, or more generalized in habitat use. This predicted pattern is precisely what is observed for small terrestrial mammals on isolated mountaintops in western

ranges in the Great Basin Desert

species absent today have become extinct within the last 10,000 years. Note that extinctions have specialized habitat requirements. (Data from Brown, 1978; modified to include additional records.)

Mountain ranges																			
Panamint	Sheep	Pilot	Roberts Creek	Diamond	Spruce	Grant	Spring	White Pine	Desafoya	Stansbury	Schell Creek	Deep Creek	Oquirrh	Snake	White	Toquima	Ruby	Toiyabe	Total number of ranges inhabited
	X		X	X	X	X	X	X	X	X	X	X	X	X	X	X	X	X	17
X	X	X		X	X		X	X	X	X	X	X	X	X	X	X	X	X	17
X	X	X		X	X	X	X	X	X	X	X	X	X	X	X	X		X	17
		X		X	X	X	X	X	X		X	X		X	X	X	X	X	14
			X			X		X	X	X	X	X	X	X	X	X	X	X	13
X						X	X	X	X		X	X		X	X	X	X	X	12
								X	X	X	X	X	X	X	X	X	X	X	11
							X			X	X	X	X	X	X	X	X	X	10
			X							X			X	X	X	X	X	X	8
										X		X	X	X	X			X	6
									X						X	X	X	X	5
			X										X			X	X	X	5
																	X	X	2
													X				X		2
3	3	3	4	4	4	5	6	7	8	8	8	9	10	10	11	11	12	13	

North America (Table 16.3). Small herbivores that are habitat generalists form the limited subset of species that are found on virtually all mountain ranges, whereas carnivores or herbivores of larger body size or restricted habitat types have become extinct on many montane islands and persist on only the largest ones. This pattern is particularly impressive because it is repeated on two different geographically isolated sets of mountaintop islands (Brown, 1971b, 1978; Patterson, 1980).

Patterns caused by differential extinction are particularly apparent on islands once connected by habitat bridges. Oceanic islands and other insular habitats that have always been isolated typically contain biotas that are at least as much influenced by dispersal as by extinction. There tends to be a differential representation of taxonomic groups with obvious adaptations for dispersing long distances over inhospitable habitats. For example, birds, flying insects, and plants with bird-transported or airborne diaspores tend to be well represented on distant oceanic islands, whereas few if any native species of terrestrial mammals, reptiles, amphibians, and freshwater fishes are present. Lawlor (in progress) points out that mammals can be divided into two groups, bats and terrestrial forms, on the basis of over-water dispersal ability. Bats show patterns of distribution on oceanic islands very different from those of terrestrial mammals. Bats have naturally colonized such distant outposts as New Zealand, New Caledonia, and the Canary and Hawaiian Islands, and they are represented by 33 genera on the Greater and Lesser Antilles. In contrast, native terrestrial mammals are completely absent from New Zealand, New Caledonia, the Canaries, and the Hawaiian Archipelago, and only 26 genera are known from the Antilles (and all but 5 of the latter have become extinct since the late Pleistocene, probably owing at least in part to human activity).

A similar dichotomy in distribution patterns is apparent in freshwater organisms that have different capabilities for dispersal. Strong flying insects and groups of crustaceans, protozoans, and algae that form resistant eggs, spores, or cysts tend to have wide distributions that include lakes, springs, and streams that have never been connected to other bodies of water. Other forms such as some molluscs (especially bivalves) and weak-flying insects with short-lived terrestrial stages are less widely dispersed. Freshwater fishes have the most limited distributions. It is possible to trace historical connections between freshwater drainages by the taxonomic affinities of the fish species (e.g., Smith, 1978, 1981; Hocutt et al., 1978; Rosen, 1979). Bodies of water that have never had aquatic connections are devoid of fish, although they may support diverse communities of more vagile organisms. Freshwater fishes are also notoriously poor overseas colonists. New Zealand, for example, has no native primary division freshwater fishes (Chapter 10), but introduced trout thrive in its magnificent rivers, producing what is probably the best freshwater sport fishing in the world.

Patterns reflecting interspecific interactions

Several other patterns suggest that the distribution and abundance of particular species on islands are influenced not only by general ecological characteristics that affect overall probabilities of colonization but also by strong interspecific interactions with certain other species. All three kinds of interactions, competition, predation, and mutualism, are probably important, but most studies of insular ecology have emphasized competition. Three patterns that have been hypothesized to be caused by interspecific competition are (1) mutually exclusive distributions of ecologically similar, closely related species among islands, (2) the apparent tendency of species that coexist on the same island to be more dissimilar in ecologically important morphological traits or in taxonomic relatedness than would be expected in a random assemblage of species from the archipelago as a whole or from the mainland species pool, and (3) increases in population density and niche breadth when populations coexist with fewer potentially competing species on small islands than on larger islands or on the mainland. Many of the studies purporting to demonstrate these patterns and to support the competition hypothesis have been criticized severely, but we will first consider the supporting evidence before attempting to evaluate the criticisms.

Distribution patterns. In the absence of direct experiments on or observations of competitive exclusion of one species by another (which are scarce for any kind of natural field situation), one kind of evidence for competition is a strong negative association in the distributions of two or more species with similar resource requirements. Archipelagos consisting of many similar islands appear to provide numerous examples of such patterns. Diamond (1975b) described several mutually exclusive distributions, which he called checkerboard patterns, for pairs of congeneric bird species inhabiting the Bismarck Archipelago. In one example, the flycatcher *Pachycephala melanura dahli* is found on 18 islands and its congener *P. pectoralis* on 11 islands, but the two species never occur together on the same island (Figure 16.6). Other nega-

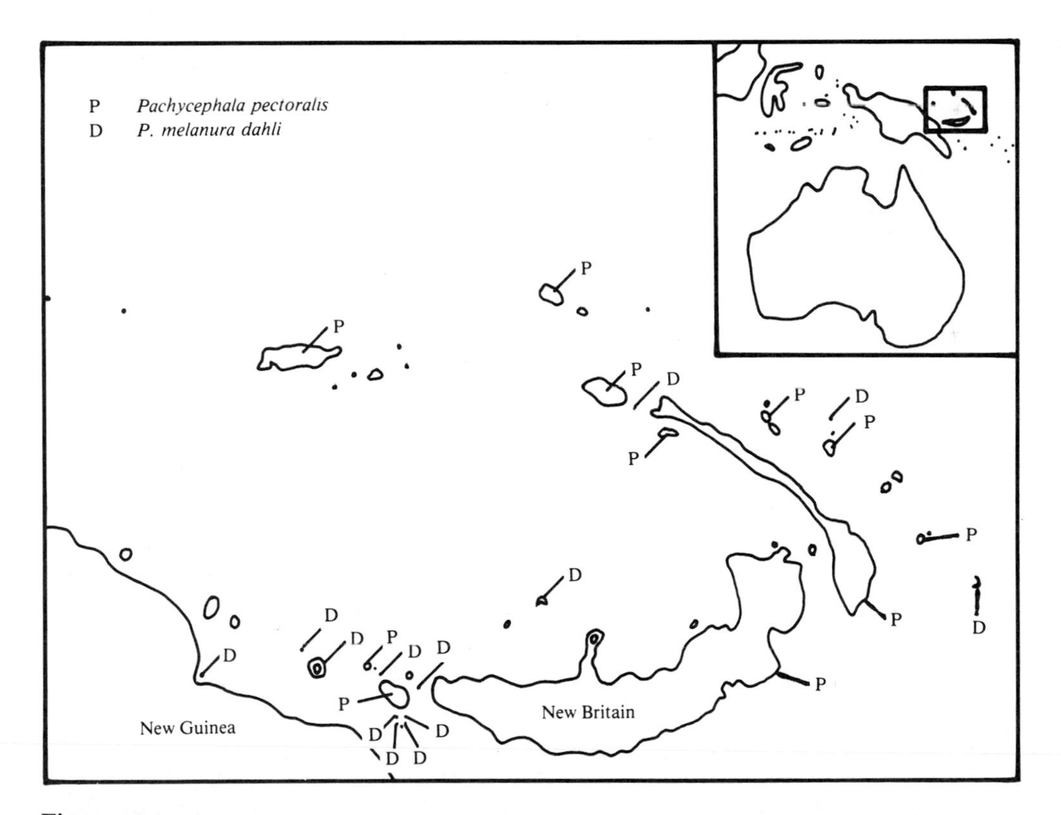

Figure 16.6
Mutually exclusive or checkerboard distributions of *Pachycephala* flycatchers on the Bismarck Islands off New Guinea. There are two species, *P. pectoralis* and *P. melanura dahli.* Most islands have one of these, no islands have both, and a few (especially the smallest islands) have neither. (After Diamond, 1975b. Reprinted by permission of the publishers from *Ecology and Evolution of Communities,* ed. M.L. Cody and J.M. Diamond, Cambridge, Mass.: The Belknap Press of Harvard University Press. Copyright © 1975 by the President and Fellows of Harvard College.)

tive associations resulting from interspecific competition may be more complex than simple checkerboards. Species that do not occur together on small islands may coexist on larger ones, suggesting that they competitively exclude each other when carrying capacities are low but are able to subdivide the more plentiful resources of larger islands. In still other cases, certain combinations of species may be able to coexist whereas others appear to be forbidden by competitive exclusion. Diamond (1975b) cites the example of cuckoo doves in the genera *Macropygia* and *Reinwardtoena*. There are four species in this group, so there are 15 possible combinations of species that would give biotas of from one to four species (Table 16.4). Only six of these combinations are actually observed, however, suggesting that certain sets of species are incompatible. Some combinations would not be expected to coexist simply on the basis of chance, but Diamond's statistical analysis suggests that other combinations would be likely to be observed if they were not forbidden by competitive interactions.

Sometimes diffuse competition from many unidentified species, rather than direct interactions with a few closely related ones, has been implicated in determining insular distributions. Again, perhaps the best examples come from Diamond's (1974, 1975) studies of the birds of the Bismarck Archipelago. He noted that certain species, which he called supertramps, are usually found only on small islands containing few other bird species. Supertramps are also among the few species that have recolonized islands following a recent volcanic eruption. By plotting incidence functions, i.e., the proportion of islands inhabited by a given species as a function of the number of other species present, Diamond was able to quantify some distributional patterns (Figure 16.7). On such graphs, supertramps, such as the flycatcher (*Monarcha cinerascens*) and the honey eater (*Myzomela pammelaena*), are readily distinguished as species that appear to be excellent colonists but poor competitors in diverse bird communities. Other species, including the starling (*Aplonis metallica*) and the incubator bird (*Me-*

Table 16.4
Combinations of four cuckoo dove species present or absent in the Bismarck Islands
Included are the number of islands inhabited by each observed combination. The four species are A, *Macropygia amboinensis;* M, *M. mackinlayi;* N, *M. nigrirostris;* R, *Reinwardtoena* superspecies. (Simplified from Diamond, 1975.)

Number of species	Observed combination: Species	Observed combination: Number of islands inhabited	Missing combination (species)
1	A	3	N
	M	8	R
2	A, M	5	A, N
	A, R	4	M, N
	M, R	2	N, R
3	A, N, R	5	A, M, N
			A, M, R
			M, N, R
4	None		A, M, N, R

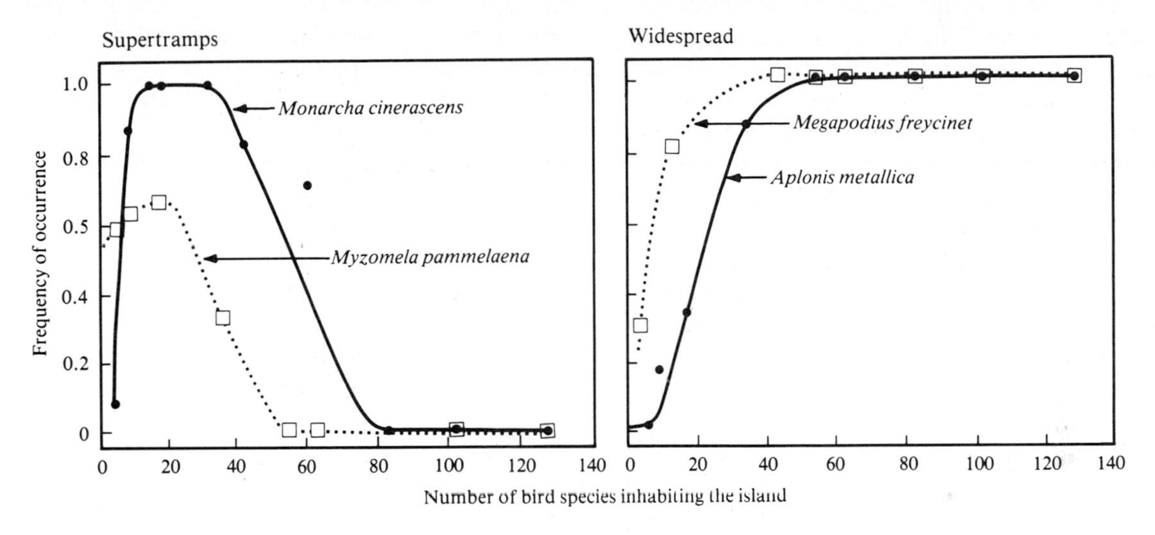

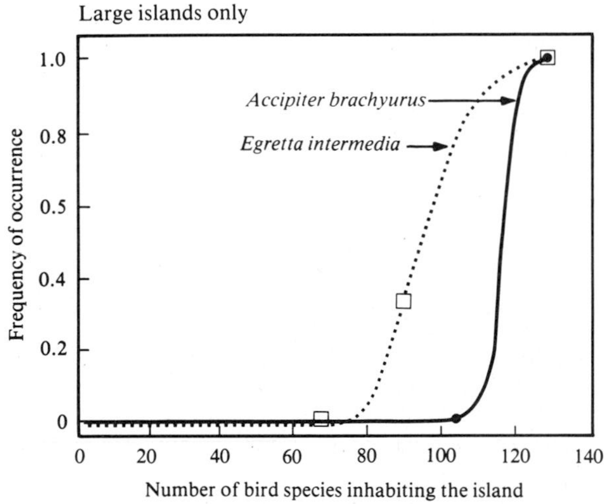

Figure 16.7
Incidence functions for various bird species in the Bismarck Islands. Note that some species, called supertramps, such as *Monarcha cinerascens* and *Myzomela pammelaena,* occur on most islands with few bird species but are absent from islands with diverse avifaunas; other species, such as *Aplonis metallica* and *Megapodius freycinet,* are found on all but the smallest islands; and still other species, including *Accipiter brachyurus* and *Egretta intermedia* occur on only the largest islands with many other bird species. (After Diamond, 1975b. Reprinted by permission of the publishers from *Ecology and Evolution of Communities,* ed. M.L. Cody and J.M. Diamond, Cambridge, Mass.: The Belknap Press of Harvard University Press. Copyright © 1975 by the President and Fellows of Harvard College.)

gapodius freycinet), appear to be both capable colonists and good competitors, as they are found in relatively high frequency on all but the smallest islands. Still other species, especially birds of large body size, specialized diet, or restricted habitat requirements, such as the hawk *(Accipiter brachyurus)* and the heron *(Egretta intermedia),* are restricted to the largest islands, which presumably offer more resources but also contain more bird species.

An alternative approach to searching for apparent effects of interspecific competition by considering only one or a small number of species at a time is to analyze negative associations of species within large biotas. If interspecific competition limits the distribution of species among islands, one would predict that the most ecologically similar species would coexist on islands less frequently than expected by chance. Several investigators have noted that islands contain fewer congeners and species more different in ecologically important morphological traits (such as bill size in birds) than the nearest mainlands from which the insular biotas were derived (e.g., Grant, 1965, 1966b, 1968; MacArthur, 1972; Diamond, 1973, 1975b; Lack, 1973, 1976). Lack (1947, 1969) and Abbott et al. (1977) have described regular patterns of character displacement in Darwin's finches on the Galápagos (Figure 6.10), but the situation is somewhat different there because the group is endemic to the archipelago and comparisons must be made between islands having different numbers of species. For similar patterns in West Indian *Anolis* lizards see Schoener (1967, 1969b, 1970), Williams (1972, 1976, 1983), Roughgarden (1974), Lister (1976b), Lister and McMurtrie (1976), Roughgarden and Fuentes (1977), and Roughgarden et al. (1983). Care must be taken in interpreting these patterns, however, because small islands have fewer species than larger islands and mainlands; even a randomly structured small community should contain species that are more different on the average than those in a more diverse community.

In analyzing the above insular patterns, the critical question is whether differences among the insular species are greater than would be expected if the island biotas were assembled at random from the mainland species pool. Some analyses have attempted to address this problem and have produced results that are consistent with the competition hypothesis. For example, in the case of both Darwin's finches in the Galápagos Islands (Grant and Abbott, 1980; Hendrickson, 1981; Case and Sidell, in press) and hummingbirds in the Antilles (Lack, 1973, 1976; Brown and Bowers, in press; Case et al., in press) it appears that species that coexist on the same island tend to be more different in bill and wing measurements than would be expected if the species associated at random (Table 16.5). In Table 16.5, the bill length of every other species was expressed as a ratio relative to the bill length of the smallest species, and the species with ratios less or greater than 1.8 were designated as small or large, respectively. Note that coexisting pairs of species tend to be of different sizes; the probability of this occurring by chance is less than 0.02. Actually, the pattern is

Table 16.5
Coexistence of pairs of hummingbird species of varying size combinations on the islands of the Greater and Lesser Antilles
Data are from Brown and Bowers (in preparation).

	Both small	One small, one large	Both large
Observed species pairs	6	27	16
Species pairs expected from a random distribution	14.3	18.7	24.3

even more dramatic because when species of similar size occur on the same island they tend to be segregated by elevation and habitat.

Density changes and niche shifts. If competition plays a major role in determining the distribution of species between islands, one might also expect it to affect the abundance and resource utilization of populations within islands. Specifically, one would predict that on small islands, in the absence of many close competitors, populations might show increases in population density and niche breadth compared to those exhibited on larger islands or on continents. Competition is not the only explanation for this pattern, however; the presence of fewer predators on smaller islands could have the same effect as that of fewer competitors. If competition and predation influence the organization of communities by limiting species to only part of their fundamental niches, one would expect insular populations to expand their niches and increase concomitantly in population density when they interact with fewer other species. Comparing the ecological properties of birds on the Atlantic island of Bermuda with that of the same species in similar habitats in eastern North America, Crowell (1962) was the first to focus attention on these patterns of density compensation and niche expansion (see also Diamond, 1970a, 1973, 1975b; MacArthur et al., 1972, 1973).

The patterns and the ecological processes that produce them have probably been most thoroughly studied, however, in the *Anolis* lizards of the West Indies. No field biologist can visit a small Caribbean island without being impressed by the incredible abundance of *Anolis*. These small reptiles seem to be everywhere, from the ground to the tops of the tallest trees, from disturbed habitats along roadsides and in cities to pristine native forests. Indeed, the lizards are much more abundant on most islands than they are anywhere on the tropical American mainland. E.E. Williams of Harvard University and his students T.W. Schoener, G.C. Gorman, J. Roughgarden, B. Lister, and R. Holt have studied the evolution, ecology, and biogeography of the Caribbean *Anolis* in great detail. Many patterns are well documented, but the underlying causal processes have proven more difficult to demonstrate convincingly. Recently several of these investigators have begun to conduct manipulative experiments in the field to clearly distinguish alternative hypotheses.

Anolis is but one of several important lizard genera on the mainland of tropical America, but it is by far the most abundant genus of vertebrates on the Caribbean islands. On the large islands of the Greater Antilles, a few colonizing species have given rise by endemic speciation and adaptive radiation to a diverse *Anolis* fauna. For example, Hispaniola, the second largest and ecologically most diverse island, has at least 35 species probably derived from four separate invasions (Williams, 1976, 1983). These species occupy a variety of ecological niches: there are tiny insectivores and large carnivores with head and body lengths ranging from 40 to more than 200 mm, respectively; there are species morphologically, physiologically, and behaviorally specialized for distinctive habitats and microenvironments from sunny sites to deep shade, from open ground and rocks to grasslands and scrub forests, to different layers in the complex vegetation of mature tropical and montane forests (Figure 16.8). In contrast, the small islands of the Lesser Antilles have only one or two generalized species (Roughgarden and Fuentes, 1977; Roughgarden et al., 1983). When two species coexist on an island they differ in body size, prey size, and habitat; but when only one species is present it is intermediate in size, takes a wide range of prey, and occupies virtually all habitats (Figure 16.9). Clearly, the fundamental niche of an *Anolis* species that has evolved in the absence of congeners is very broad. Some of the observed niche expansion of *Anolis* on small islands may represent immediate behaviorally mediated ecological responses to the absence of competing spe-

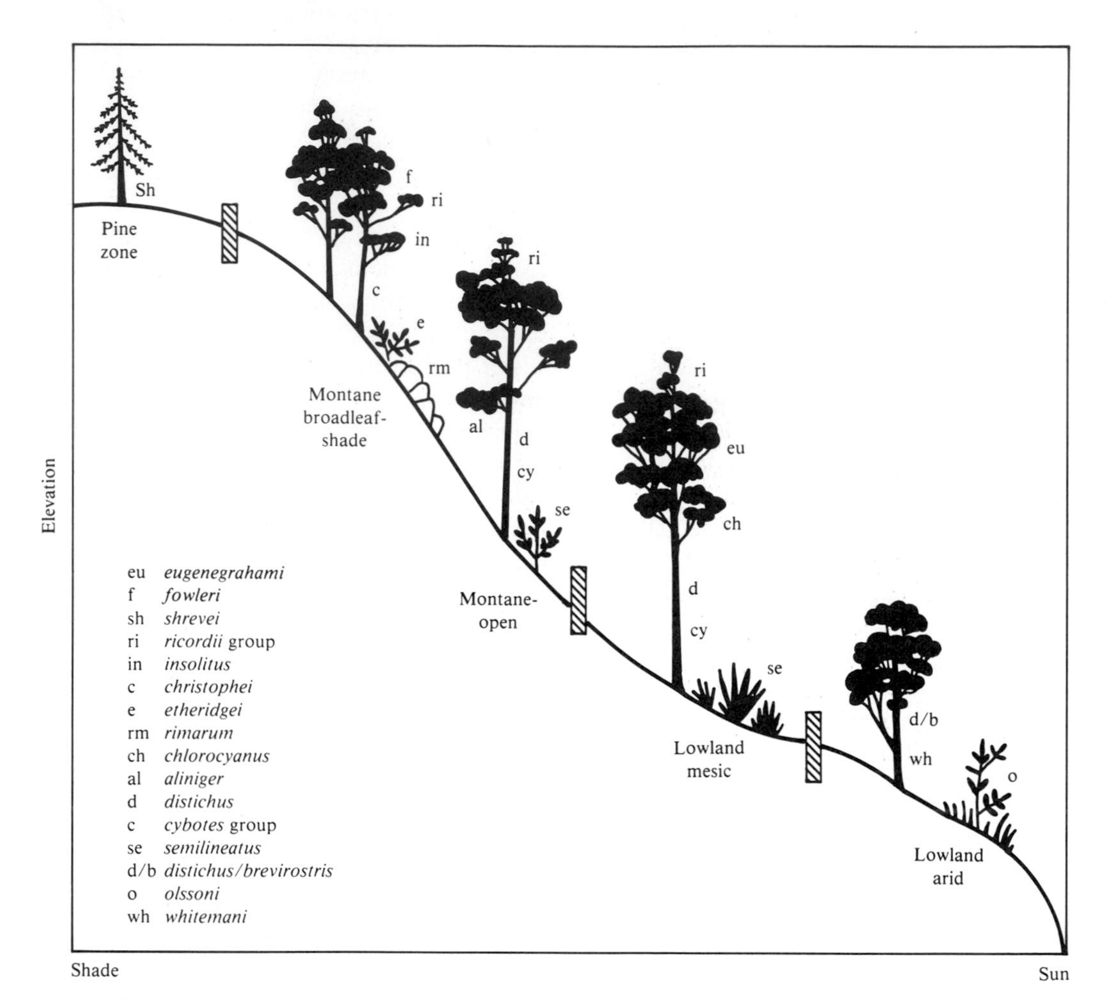

Figure 16.8
Diagrammatic representation of the habitats occupied by different *Anolis* species on the northern part of Hispaniola. This figure shows that the niches differ in elevation, vegetation type, perch height, and position in a gradient of sunlight to shade. The species indicated here, only a fraction of at least 35 species that inhabit the island, have been produced largely by speciation and adaptive radiation within the island. (From E.E. Williams, 1983. "Ecomorphs, faunas, island size and diverse end points in island radiations of *Anolis*," *in* R.B. Huey, E.R. Pianka, and R.W. Schoener [eds.], *Lizard Ecology: Studies of a Model Organism*. Reprinted by permission of Harvard University Press.)

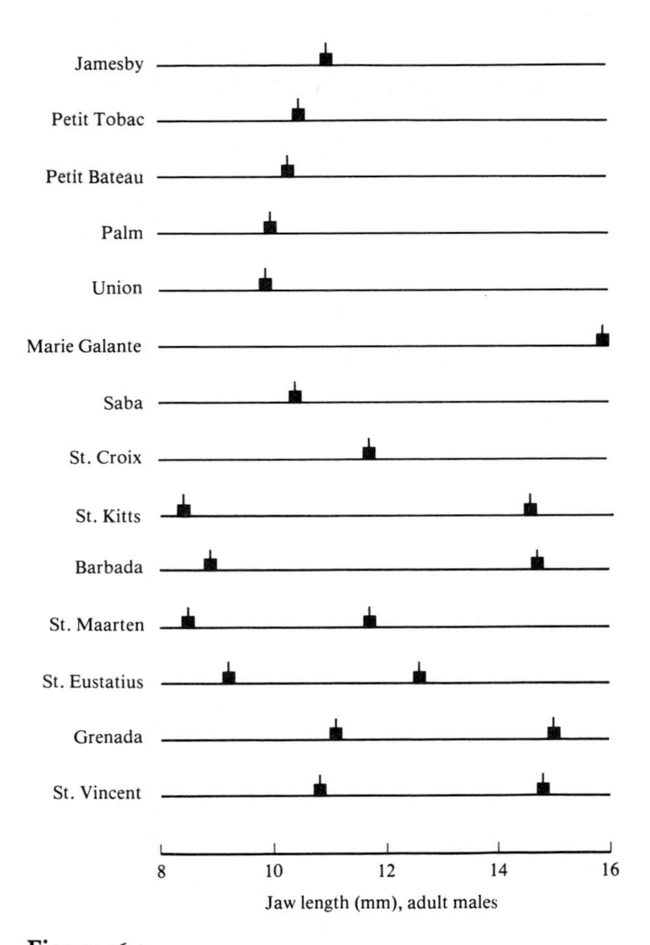

Figure 16.9
Body sizes of *Anolis* lizards on the islands of the Lesser Antilles. Note that all islands have either one or two species. When two species cooccur they tend to be displaced in size, whereas when only one species is present it tends to be of intermediate size. (Plotted from data in Roughgarden, 1974; Roughgarden and Fuentes, 1977; Roughgarden et al., 1983.)

cies. Nevertheless, most of the niche shifts in Caribbean species represent evolutionary adaptations to communities containing different numbers and kinds of species, and these are reflected in morphological, physiological, and behavioral changes.

Variation in *Anolis* densities and niche characteristics between islands and between habitats within islands has been investigated, especially by Schoener (1968a, 1975), Lister (1976a, 1976b), and Holt. They have shown that when a species occurs on an island or in habitats where there are few other lizards, increases in niche breadth (Figure 16.10) are often paralleled by increases in density. It is tempting to attribute these patterns of density compensation and niche expansion to a release from competition by other *Anolis* species, but this may be unwarranted. The small islands where the lizards show the most spectacular increases in density contain not only fewer *Anolis* but also fewer species of all terrestrial animals. Thus the *Anolis* populations are potentially released from competition with congeners, from competition with insectivorous invertebrates (e.g., spiders), birds, frogs, and lizards of other genera, and from predation by birds and other lizards. It is also possible that small islands support higher standing stocks of insect prey. Holt (in progress) studied *Anolis* on islands off Trinidad and concluded that direct competition with other *Anolis* or other lizard species cannot account for observed density changes. He suggested that the absence of predatory birds might be the most important of the various factors contributing to increased density and niche breadth on small islands.

One puzzling aspect of insular density compensation is that the total densities of a few species inhabiting a small island may exceed the combined densities of a much greater number of species of the same taxon occupying similar habitats on a large island or continent. This phenomenon, called density overcompensation or excess density compensation, is well documented, especially for birds on small oceanic islands. In one of the first quantitative studies of insular ecology, Crowell (1962) studied bird populations on Bermuda, 900 km east of North

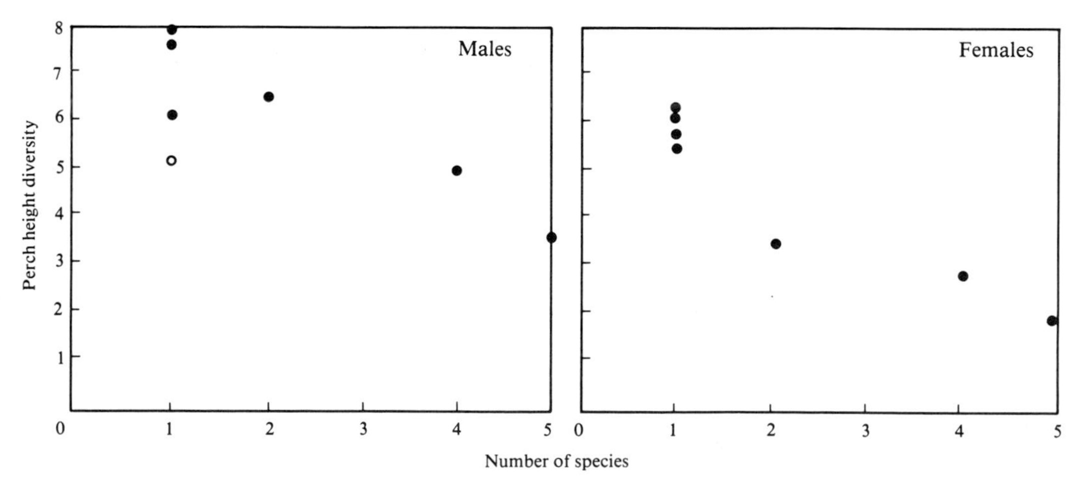

Figure 16.10
Relationship between the diversity of heights of perches used by males and females of *Anolis sangrei* and the number of *Anolis* species occurring in the same habitat on the same island. Note that when *A. sangrei* is the only species present it uses a wide variety of perch heights, but the breadth of this niche dimension contracts and *A. sangrei* is excluded from arboreal habitats as it encounters an increasing number of coexisting species. (After Lister, 1976a.)

Carolina in the Atlantic Ocean, and in similar habitats on the North American mainland. He found that just 10 species of small passerine birds on Bermuda maintained populations 1.5 times greater than the combined densities of 20 to 30 species on the mainland. Subsequent studies by MacArthur et al. (1972) on the Pearl Islands south of Panama, by Diamond (1970b, 1975b) on the Bismarcks and other archipelagos north and east of New Guinea, and by Emlen (1978) on the Bahamas have described similar patterns. Case (1975) documents density overcompensation among the lizards on the islands in the Gulf of California compared to the mainland of Baja California and Sonora.

Several explanations have been proposed for such density overcompensation, especially in birds, although many can also be applied to other taxa. The explanations are as follows:

1. Because bird species of large body size tend to be absent from small islands, the same resources could support substantially larger populations of small birds.
2. The density of birds reflects release from competition not only with missing bird species but also with other taxa, such as mammals and amphibians, which use similar food and other resources but are even less well represented than birds on oceanic islands because they are poorer over-water colonists.
3. The density of birds reflects the absence of avian predators such as hawks, mammals, and snakes, which also tend to be poorly represented on oceanic islands.
4. Oceanic islands are more productive, at least of the foods and other resources required by small birds.
5. On oceanic islands, renewable food resources are harvested at nearer their maximum sustained yields than on mainlands, where intense competition leads to overexploitation of resources.
6. Populations can become more finely adapted to their local environment and hence attain higher densities on isolated islands than on continents, where extensive gene flow between populations occupying different habitats over a wider geographic area tends to prevent specialization for efficient use of local resources.

We know of no study that has compiled sufficient data to evaluate convincingly the relative importance of these processes that might cause overcompensation. Indeed, this will be a difficult task because, although some explanations (such as the first) might be shown to be insufficient by themselves to account for the observed magnitude of overcompensation, all could operate and contribute to the phenomenon.

Testing null hypotheses. Recently a group lead by D. Simberloff and D.R. Strong of Florida State University has criticized many of the studies that have claimed that patterns of insular distribution reflect interspecific competition and other direct biotic interactions. In general this group has taken the approach of reanalyzing the original data statistically to compare observed distributions with those expected on the basis of null hypotheses that do not assume any interspecific interaction. Usually these null hypotheses are generated by using a computer to assign species from an appropriate species pool to islands at random. Various constraints can be placed on the computer simulation to assure that important features of the insular biotas (such as variations in the number of species with island size) are incorporated into the null model to make it biologically realistic, so long as it does not include the influence of competition or other direct interspecific interactions.

Such testing of apparent patterns against null hypotheses is a valuable exercise for two reasons. First, the human mind and senses are so attuned to detecting pattern that they may mistakenly perceive order where none exists. As Cole (1954) elegantly showed by simulating microtine population "cycles" from random numbers, purely stochastic processes can produce amazingly deceptive apparent regularities in data. Second, there is the danger that one may select, from larger data sets, a few subsets that appear nonrandom but that would be expected from chance alone. In any large random data set, such as a million random numbers, it is possible to pick out many subsets, such as strings of five identical numbers or of five consecutive numbers, which would have a very low probability of being observed if only a single subset of the same size were drawn at random. Because biogeographers and ecologists are faced with many combinations of species distributed among many islands or other communities, they should guard against the possibility that they may unwittingly select for their examples the few subsets of a large random data base that appear to exhibit patterns of negative association, regular ratios of body size, or other regularities that suggest interspecific interactions.

Statistical reanalyses of insular distributions and tests of the data against predictions of null hypotheses suggest, at the very least, that some earlier workers should have been much more careful to be sure that they had found real patterns before they invoked deterministic processes such as competition to account for them (e.g., Simberloff, 1974a, 1978; Connor and Simberloff, 1978, 1979; Strong et al., 1979). This reappraisal has raised serious doubts about the conclusions of many earlier studies. One study singled out for especially severe criticism (Connor and Simberloff, 1979) is Diamond's (1975b) derivation of assembly rules describing the role of competition in determining the distribution of birds on the Bismarck Islands and other archipelagos off New Guinea, which we have referred to repeatedly in this chapter. Diamond and Gilpin (1982; Gilpin and Diamond, 1982) have responded with a strong defense of Diamond's conclusions and a severe criticism of the usefulness of null hypotheses in biogeographic analyses.

At present it is uncertain how these issues will be resolved, but certainly the null hypothesis approach has generated a healthy controversy. This debate is likely to continue for some time. One reason is the difficulty in constructing and testing null models that are both biologically reasonable and statistically powerful. In principle null hypotheses are simple, but in practice there is much room for legitimate debate about the biological constraints and statistical tests that should be employed (e.g., see Strong et al., 1979; Grant and Abbott, 1980; Hendrickson, 1981; Strong and Simberloff, 1981; Diamond and Gilpin, 1982; Gilpin and Diamond, 1982; Bowers and Brown, 1982; Wright and Biehl, 1982; Case and Sidell, 1982; Case et al., 1982; Colwell and Winkler, 1982). The ultimate resolution of this controversy will mark a major advance in the understanding of ecological processes in insular biogeography because it will provide a valuable set of analytical tools in addition to a more rigorous evaluation of the role of direct interspecific interactions.

One additional benefit of this controversy is already being reaped. It is apparent to many investigators that they cannot be content simply to infer the operation of dynamic ecological processes from static patterns in the morphology, abundance, and distribution of insular species. Rather, the static patterns should be used to inspire hypotheses about underlying mechanisms, and these hypotheses should be tested by independent observations. In many cases the most convincing tests of hypotheses can be performed by conducting manipulative experiments in the field. It is encouraging to note that several investigators, such as Crowell (1973) investigating mice on islands off the coast of Maine, Schoener and Roughgarden working on *Anolis* in the Caribbean, and Grant studying Darwin's finches in the Galápagos, have begun to perform experiments to test the effects of interspecific interactions on the ecology and biogeography of insular populations.

Taxon cycles and endemism

Susceptibility of insular biotas to invasion. One of the most striking patterns in insular biogeography is that insular populations

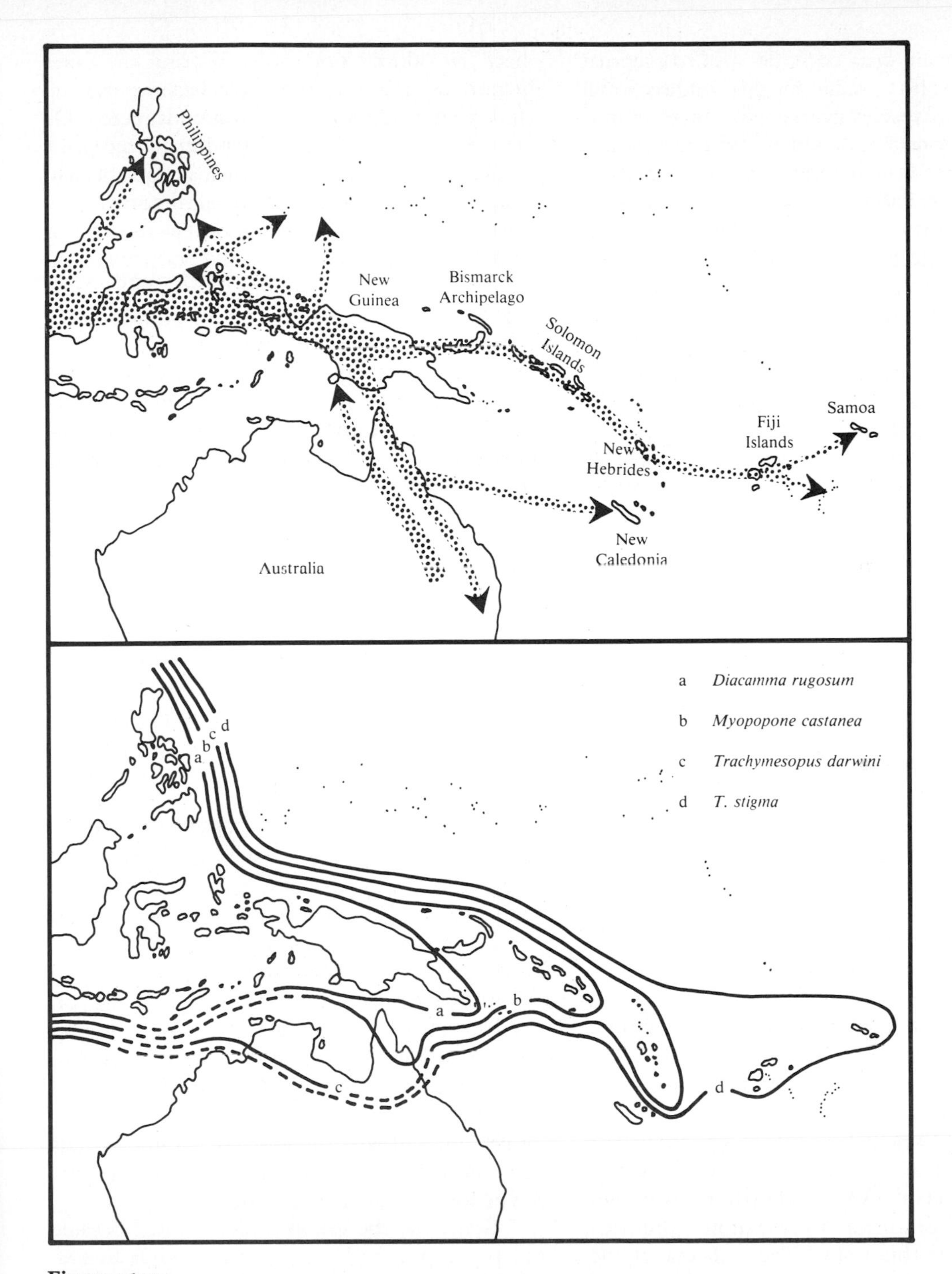

Figure 16.11
Stepping-stone colonization of the islands of the western Pacific by ants originating in southeastern Asia and dispersing eastward through New Guinea and out to the Melanesian islands. *Above,* The general dispersal routes followed by ponerine ants. *Below,* The ranges of four expanding species, showing different degrees of dispersal along the routes outlined above. (After Wilson, 1959.)

are evolutionary dead ends, doomed to eventual extinction. This results in the unidirectional movement of species from continents to islands over evolutionary time scales. Systematic studies show that insular biotas have been derived by the colonization of taxa that originated on larger landmasses. Sometimes there is a stepping-stone effect, so that small or distant islands are colonized by species that have come immediately from other islands that are usually larger or nearer to continents (Figure 16.11). Some island species are formed by endemic speciation within an island or an archipelago, as in the Hawaiian *Drosophila* (Figure 12.16), but continents remain the ultimate source of insular forms. This appears to be a special case of the general tendency of organisms originating in large areas with diverse biotas, whenever the opportunity arises, to invade small areas with lower organic diversity and ultimately to replace the native species (Chapter 17).

Even the exceptions to this pattern support the general rule. The region around Miami in southern Florida has been colonized successfully by several species from the Caribbean islands, including about six *Anolis* species that apparently have been imported either intentionally or accidentally by humans during the last century. However, southernmost Florida is essentially an island, an isolated region of tropical habitat attached to temperate North America by a low-lying peninsula that was frequently inundated by seawater during the Pleistocene. Although the Caribbean species seem to thrive in the lush tropical gardens in the Miami region, they have not extended their ranges northward to invade more temperate habitats.

The extreme susceptibility of islands to invasion by continental species is probably best demonstrated by the successful colonization of islands by species introduced by humans within the last two centuries. Two examples dramatically illustrate this point. The avifauna of the Hawaiian Archipelago is comprised of 42 species of land birds. Of these, 22 are native and 20 (48%) are well-established exotics that have been introduced since 1800. During the same period 14 native species have become extinct, and several others are rare and endangered. Of 94 species known to have been introduced prior to 1940, 53 became established, at least locally and temporarily, and only 41 failed completely. Most of the introduced species were deliberately released to augment the local avifauna with game birds and species of beautiful plumage and song. A strange mixture of species derived from different taxonomic groups and biogeographic provinces now coexists on the archipelago: the mockingbird *(Mimus polyglottos)*, cardinal *(Cardinalis cardinalis)*, and California quail *(Callipepla californica)* from North America; the Indian mynah *(Acridotheres tristis)*, lace-necked dove *(Streptopelia chinensis)*, and Pekin robin (*Leiothrix lutea;* a babbler, not a thrush) from Asia; the house sparrow *(Passer domesticus)*, skylark *(Alauda arvensis)*, and ring-necked pheasant *(Phasianus colchicus)* from Europe; and the Brazilian cardinal *(Paroaria cristata)* from South America (Elton, 1958).

The fate of the Hawaiian Archipelago is by no means unique. Consider the birds and mammals of New Zealand (Table 16.6). The potential impact of these exotic species on the insular ecology can be better appreciated when one realizes the only native New Zealand mammals are two species of bats: all terrestrial mammals and over half of all mammal and bird species have been introduced (Wodzicki, 1950; Elton, 1958; Falla et al., 1966). It is little wonder that these exotics have had tremendous impact on native species of lizards and ground-nesting birds that have evolved in the complete absence of mammalian predators. The same theme is repeated, fortunately often somewhat less dramatically, on virtually all isolated oceanic islands that have been inhabited by Western civilizations for the last two centuries.

Stages of the taxon cycle. Isolated oceanic islands appear to be equally susceptible to natural invasion. Oceanic barriers severely limit the immigration of propagules, but those that manage to gain a foothold have a high probability

Table 16.6
Native and established introduced land birds and mammals of New Zealand
The regions of origin of the species are included. (Data compiled from Wodzicki, 1950; Elton, 1958; Falla et al., 1966.)

	Birds	Mammals	Total
Native			
Families	19	2	21
Species	39	2	41
Introduced			
Families	13	11	24
Species	26	26	52
Source of introduced species			
Europe	16	15	31
Australia	6	2	8
Asia	3	4	7
North America	1	3	4
Polynesia	0	2	2

of success. This infrequent but continued establishment of colonists drives a pattern of evolutionary change in island biotas that Wilson (1961) has termed the taxon cycle. Insular species evolve through a series of stages from newly arrived colonists, indistinguishable from their mainland relatives, to highly differentiated endemics that ultimately become extinct. Although this process can be prevented by sufficient gene flow via immigrants to prevent insular differentiation, and it can be terminated prematurely at any stage by extinction, it is rightly termed a "cycle" because once insular populations begin to differentiate and adapt to island life they appear to be doomed to extinction and to be replaced by new colonists from the mainland.

An excellent example of the taxon cycle is provided by an analysis of the birds of the Lesser Antilles (Ricklefs and Cox, 1972, 1978). The land birds of these Caribbean islands can be divided into four categories that represent successive stages in their evolution from South American ancestors (Figure 16.12). Stage I includes species that occur on most islands, but the insular populations exhibit little differentiation from each other or from mainland populations. The species representing stage II often have spotty distributions among the islands, where they are represented by somewhat distinct populations that are sometimes recognized as endemic subspecies. Stage III species reflect an even longer history of evolution in isolation because they consist of scattered endemic subspecies and species that, nevertheless, still exhibit close affinities to related populations on other islands and on the continent. Stage IV, representing the termination of the taxon cycle, contains highly differentiated endemic species and genera persisting as relicts on single islands.

When Wilson (1961) originally described the taxon cycle, he noted similar stages in the taxonomic differentiation of ants on the islands of Melanesia, north and east of New Guinea in the western Pacific. He also described successive changes in the niches of the ants as they evolved through the taxon cycle. Those species that are good over-water colonists typically occur in coastal or disturbed habitats on New Guinea, and recently arrived, undifferentiated populations are found in similar habitats on the Melanesian islands. As they differentiate in isolation, however, the ants also change their ecological requirements and expand into other

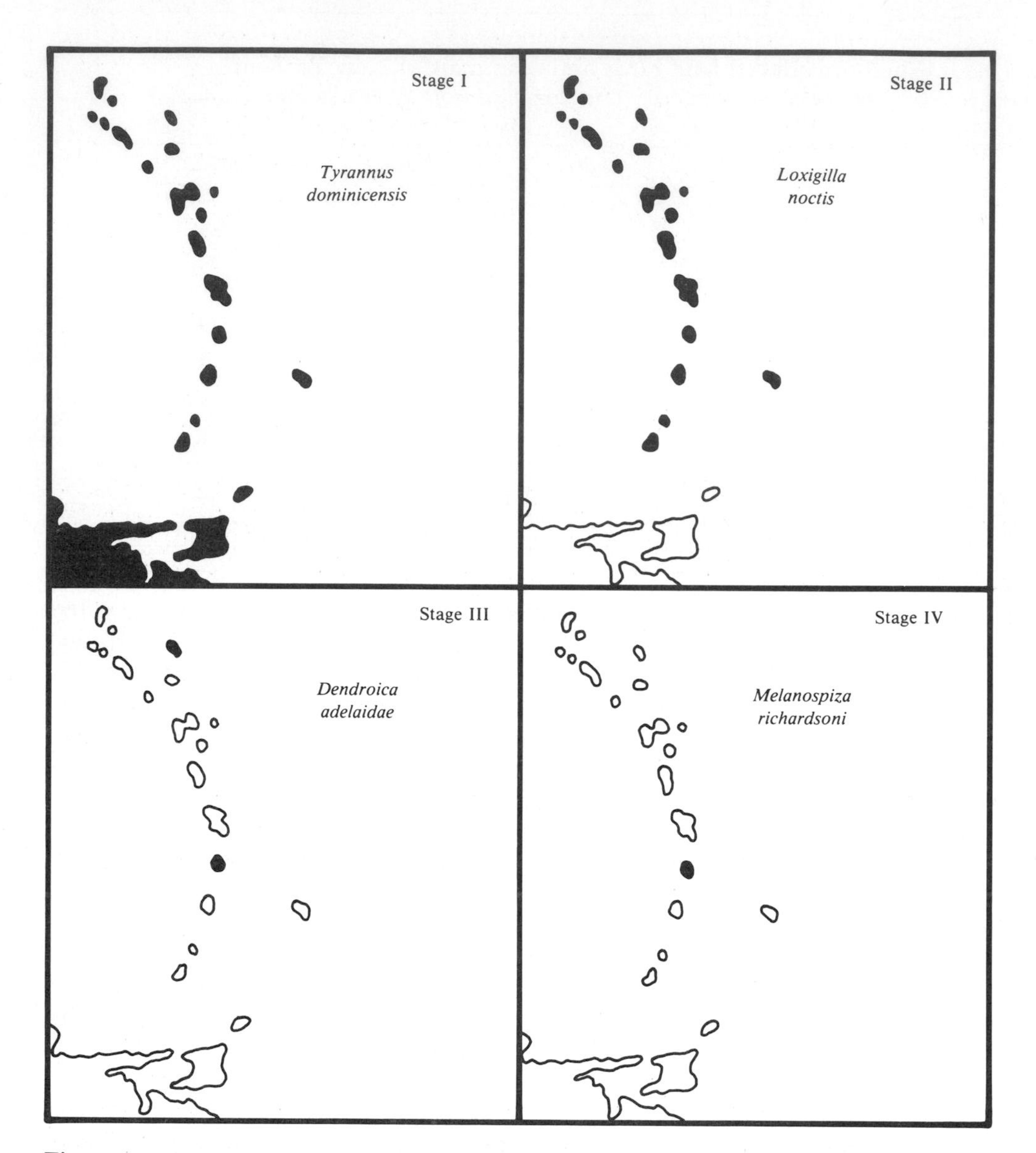

Figure 16.12
Stages of the taxon cycle illustrated by birds of the West Indies. *Stage I,* a widespread undifferentiated species that presumably has recently colonized from South America, represented here by the flycatcher, *Tyrannus dominicensus; stage II,* a widespread form with well-differentiated races (letters) on different islands, exemplified by the finch *Loxigilla noctis; stage III,* with well-differentiated races on only a few islands, such as the warbler *Dendroica adelaidae;* and *stage IV,* endemic species, in this case the finch *Melanospiza richardsoni,* confined to St. Lucia. (Reprinted from "Taxon cycles in the West Indies avifauna," *American Naturalist* 106:195-219, by R.E. Ricklefs and G.W. Cox, by permission of the University of Chicago Press. Copyright © 1972 by the University of Chicago.)

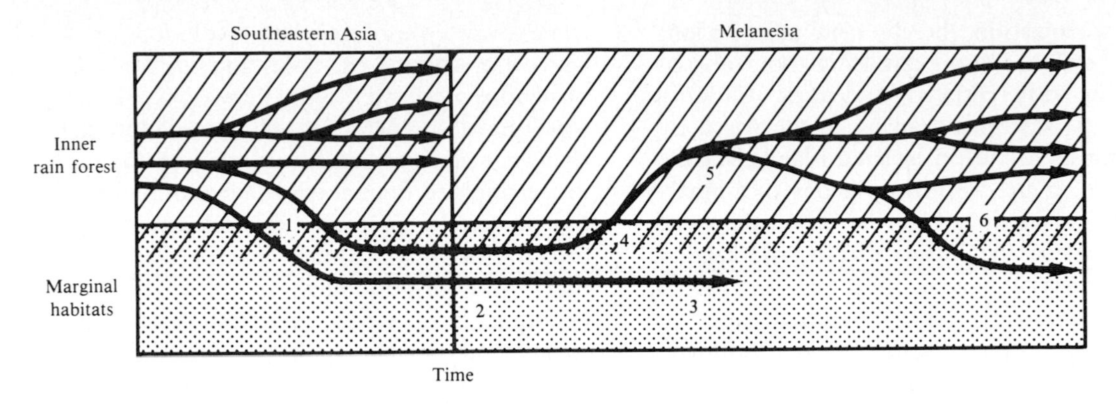

Figure 16.13
Ecological changes accompanying the taxon cycle in the ants of Melanesia. Species of forest ancestry that have secondarily invaded marginal (usually disturbed or coastal) habitats in southeastern Asia (1) tend to be good dispersers and to colonize similar habitats on the islands of Melanesia (2). Then populations become extinct fairly rapidly (3) or else they invade interior rain forest habitats (4), where they may undergo differentiation and adaptive radiation (5). Sometimes they may give rise to forms that secondarily invade marginal habitats and disperse in stepping-stone fashion to more distant islands (6). (After Wilson, 1959.)

habitats, such as native forest (Figure 16.13). Highly differentiated endemic forms, representing the last stage of the taxon cycle, typically are restricted to a narrow range of environments, usually rain forest or montane forest deep in the interior of the islands. Meanwhile, these species have been replaced by a new wave of colonists occupying the beaches and disturbed habitats. One curious result of the taxon cycle is that species inhabiting the interior forests of the oceanic islands are more closely related to species in the source area of New Guinea that occur in disturbed habitats than they are to those species that have similar niches and inhabit mature forest. Ricklefs and Cox (1972, 1978) document similar ecological shifts in West Indian birds with successive stages in the taxon cycle.

Evolutionary trends in insular populations. In order to explain the taxon cycle, we must explain an apparent paradox. As they adapt to the local environment and differentiate in isolation, why do insular populations doom themselves to extinction and to replacement by colonizing species that have never experienced local conditions? This may seem particularly puzzling because species at intermediate stages of the cycle often have very large populations that should have infinitesimally small probabilities of extinction. Even so, these species populations decrease in size and ultimately become extinct as they enter the terminal stages of the taxon cycle. The answer would seem to be that colonizing species drive the cycle. Successful colonization initiates processes that not only result in evolutionary differentiation and habitat shifts of the colonist but also affect the evolution and ecology of established species.

The taxon cycle accounts for the existence on islands of strange endemic forms. These are particularly prevalent on large and extremely isolated islands, which may or may not once have been connected to larger landmasses. Not only do larger islands support a greater diversity of species but also they tend to have larger populations that consequently have a lower probability of extinction and more opportunity to evolve into distinctive endemics. The isolation of islands reduces the probability of successful colonization, thereby tending to prolong taxon cycles and also permitting the evolution of highly differentiated endemics. Carlquist (1965, 1974) documents many of the wierd endemic forms found on isolated islands, such as the dodos, giant flightless pigeons that inhabited three islands (Réunion, Mauritius, and Rodriguez) in the Indian Ocean; the kagu, a single species of cranelike bird that comprises a family endemic to New Caledonia; and the kiwi, a strange New Zealand bird that resembles a mammal in many respects.

Several trends are apparent in the evolution of insular species. These have been noted by many authors and most are discussed by Carlquist (1965, 1974). Many groups of organisms evolve larger body size on islands. Examples of insular gigantism include several kinds of vascular plants, some lizards, several groups of birds, and many small mammals. On the other hand, a few organisms, including large mammalian herbivores and carnivores, have become dwarfed during their evolution on islands. For specific examples and suggested explanations see Foster, 1964; Case, 1978; Heaney, 1978; Lawlor, 1982. Of all the evolutionary changes in insular organisms, perhaps the most interesting and dramatic is the loss of flight that has occurred repeatedly and independently in several groups of birds and insects (Darlington, 1943) and dispersability of plant diaspores (Carlquist, 1965, 1966b, 1974). An equally spectacular example in habitat islands is the loss of eyes and pigments in endemic cave animals. Evolutionists are still debating the mechanisms responsible for the loss of such "unused" structures, but their biogeographic implications are clear. They represent the epitome of the tendency of evolution in insular habitats to produce traits that obligately restrict organisms to specialized environments, prevent dispersal, and ultimately result in extinction.

Mechanism of the taxon cycle. It remains to be explained therefore how colonists are able

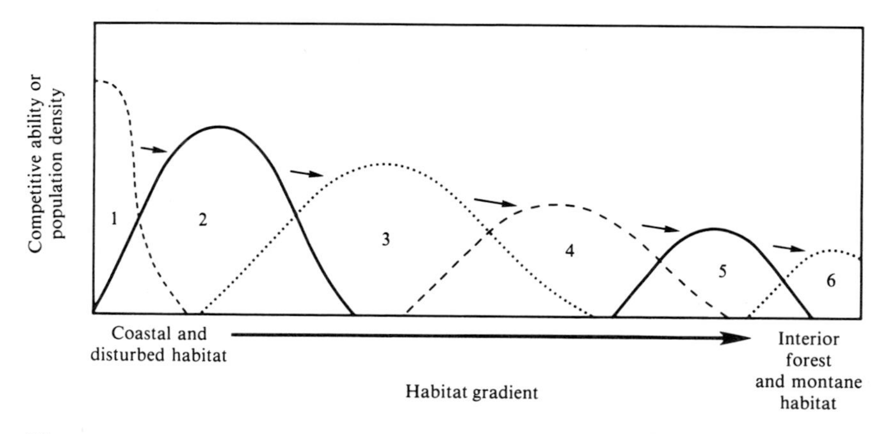

Figure 16.14
Graphic model of the processes suggested to be involved in the insular taxon cycle. The cycle is driven by the colonization of specialized species *(left)*, which evolve to become more generalized, sacrificing competitive ability and experiencing reduced population density within particular habitats *(center)*. Pushed farther along the habitat gradient by superior competitors, species in the terminal stages are forced to specialize again *(right)*, but evolutionary constraints prevent them from becoming well adapted to these new niches and hence from increasing in density and competitive ability. Consequently they evolve into rare endemics and ultimately become extinct.

to gain a foothold, what influences their subsequent differentiation, and why this ultimately results in the extinction of endemic species. In Figure 16.14 we illustrate diagrammatically the three successive processes that appear to cause the taxon cycle. First, colonization is possible because specialized immigrants have a competitive advantage over native species. Although colonists have no previous experience with local conditions on the island, they are able to compete successfully with native species because they are highly adapted to similar kinds of disturbed and coastal habitats on the mainland, where they withstand competition from a much more diverse community. This is an example of the situation described in Chapter 3, in which a highly specialized species (the colonist) coexists with more generalized species (the natives) because it is better at using a narrow range of resources.

The next process involves the specialized colonists expanding their niches to invade other habitats and become more generalized in their use of resources. Because they face fewer and less specialized competitors than they did on the mainland, the colonists gradually evolve to become more generalized and increase their fitness and total population size at the expense of coexisting species. Earlier in this chapter we cited evidence that many insular populations have expanded their niches as compared with their relatives on mainlands. Of course, as recently colonized species become more generalized, they lose their special adaptations for the habitats and resources that they had initially used, thereby paving the way for the successful invasion of the next wave of specialized colonists.

The final process is perhaps the most difficult to understand. At some point in the taxon cycle, insular species attain maximum niche breadth and population size; however, they remain continually in competition with more specialized species that invaded more recently. As they diverge from these competitors, they contract their niches and begin to specialize for new habitats and new resources, driving the species already occupying these niches to even greater specialization. Specialization, however, involves a reduction in population size and an increased probability of extinction. In part this occurs when species are forced to specialize for habitats and resources for which they are not particularly well suited because of evolutionary constraints. Because they began as stage I species specialized for disturbed and shore habitats, insular organisms may possess traits that are not well suited for the mature rain forest and montane habitats into which they are forced as the taxon cycle runs its course.

Roughgarden, Heckel, and Fuentes (1983) describe the evolution of body size in *Anolis* lizards in the Lesser Antilles, which appears to offer a specific example of the processes just described. Of the 24 islands inhabited by *Anolis*, 8 have two species and 16 have only one species. As shown in Figure 16.9, when two species coexist they are displaced in body size, but when only a single species is present it is of intermediate size (on 15 of the 16 islands jaw length is between 10 and 12 mm). The authors suggest that this pattern results from a taxon cycle, which is diagrammed in Figure 16.15. Initially, islands with one species are invaded successfully by large lizards. In response to competition from the large colonists, the original species evolves smaller size and becomes restricted to certain microhabitats; in the process the species suffers reduced population size and an increased probability of extinction. As the small species becomes restricted, and especially once it becomes extinct, the large species expands its niche and evolves toward an intermediate size. This, of course, paves the way for an invasion by another large species and a repeat of the cycle.

Our explanation of taxon cycles suggests that if islands are extremely isolated, the cycle will be slow because the rate of successful colonization will be low. Taxon cycles should also

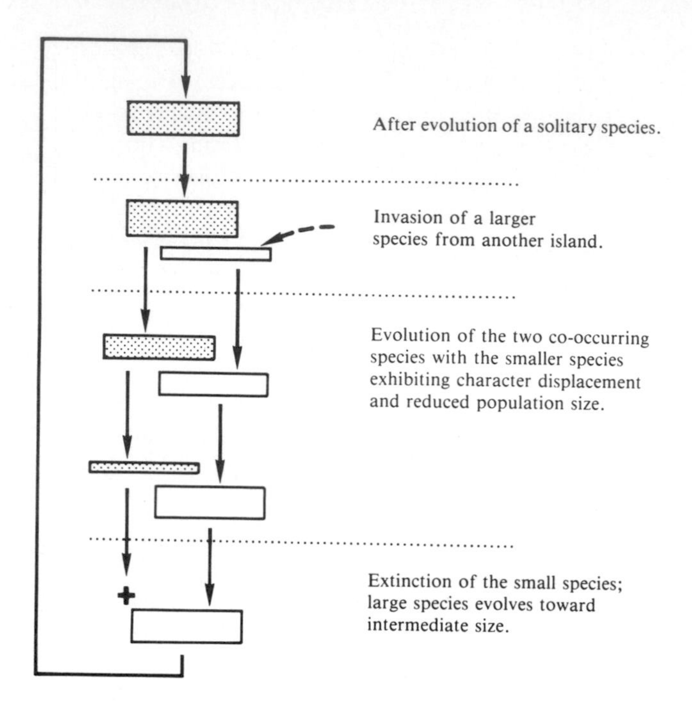

Figure 16.15
Model of the taxon cycle based on patterns of body size and niche characteristics of *Anolis* lizards in the Lesser Antilles. In this case there is a maximum of two species, and the second can invade only when the solitary resident has evolved an intermediate size. Although in theory the taxon cycle could be driven equally well by the colonization of a smaller species, field observations suggest that usually the colonist is a larger species, perhaps because larger species can better cross seawater barriers and can behaviorally dominate the smaller species. Bar height indicates relative population size; the shaded and open bars indicate species of smaller and larger body size, respectively. (Greatly modified from Roughgarden et al., 1983.)

be slower for organisms that are poor dispersers than for those that are good colonists, although high rates of immigration and gene flow could prevent the differentiation of insular populations. These predictions are testable, and existing data seem to support them. Highly differentiated endemic forms are famous for occurring on distant islands such as New Zealand, New Caledonia, and the Hawaiian and Galápagos archipelagos (for examples see Darlington, 1957; Carlquist, 1965, 1970, 1974; Williamson, 1981). Within the same archipelago there tends to be more endemism in taxa that are poorer dispersers. For example, most of the terrestrial mammals inhabiting the Greater and Lesser Antilles are endemic at the generic level, but many of the birds and bats are not even endemic at the species level. If the taxon cycle proceeds slowly, extinctions may be caused as much by changes in the abiotic environment as by biotic interactions with more recent colonists. Pregill and Olson (1981) suggested that the taxon cycle in West Indian birds described by Ricklefs and Cox (1972, 1978) reflects extinctions owing to changing (wetter) climates since the end of the Pleistocene, rather than to interspecific interactions. A thorough analysis of the distribution of the fossil and recent avifauna is necessary to evaluate this contention. However, available information (e.g., Ricklefs, 1970;

Bond, 1971; Ricklefs and Cox, 1972, 1978; Lack, 1976) suggests that in general Antillean endemics are not restricted to arid habitats. On the contrary, many are confined to wet montane forests.

The evolutionary trends in insular populations that result in taxon cycles emphasize the susceptibility of insular biotas to colonization by continental forms and eventually to extinction. The cycles suggest the importance of Van Valen's (1973b) (see Chapter 6) Red Queen hypothesis: that in order to avoid extinction, species must continually evolve in response to a changing environment, and the most important changes are caused by the evolution of coexisting species. Insular species, isolated from the intense biotic interactions of continental communities, lose their ability to withstand such levels of competition and predation and eventually are replaced by mainland taxa.

Selected references

Nonequilibrium diversity patterns

Abbott and Grant (1976); C. Barbour and Brown (1974); J. Brown (1971b, 1978); R. Browne (1981); Culver et al. (1973); Diamond (1972, 1975a, 1975b); Diamond and May (1976); Faeth and Connor (1979); Forman et al. (1976); Galli et al. (1976); Gottfried (1979); Grayson (1981); Higgs (1981); Hubbs and Miller (1948); Juvik and Austring (1979); R.R. Miller (1948, 1958); B. Patterson (1980); Simberloff (1978); Simberloff and Abele (1976); B. Simpson (1974, 1975); G. Smith (1978, 1981); Terborgh (1974, 1975); Thompson and Mead (1982); Vuilleumier and Simberloff (1980); Whitcomb et al. (1976); Wilcox (1978); E. Wilson and Willis (1975).

Patterns reflecting differential extinction and colonization

Abele and Patten (1976); Abbott (1974); J. Brown (1971b, 1978); Cairns and Ruthven (1970); Carlquist (1965, 1966a, 1966b, 1974); Darlington (1938); Diamond (1974, 1975b); Diamond and Marshall (1977); Fosberg (1963); Greenslade (1968a); Gressitt (1963); Gunn et al. (1976); Guppy (1906); Heatwole (1975); Heatwole and Levins (1972, 1973); Holdgate (1960); N. Johnson (1978); Levins and Heatwole (1973); Morse (1971, 1973, 1977); Philbrick (1967); A. Schoener (1974b); Simberloff (1971, 1976b, 1976c, 1980); Vagvolgyi (1975); Van Valen (1973a, 1973b); Wickens (1979).

Patterns reflecting interspecific interactions

Abbott (1981); Abbott et al. (1977); Case (1975, 1978); Case et al. (1979); Connor and Simberloff (1979); Cox and Ricklefs (1977); Crowell (1962, 1973); Diamond (1970a, 1970b, 1974, 1975b); Diamond and Gilpin (1982); J. Emlen (1978); Gilpin and Diamond (1982); P. Grant (1965, 1966a, 1966b, 1968, 1970, 1971); P. Grant and Abbott (1980); Hendrickson (1981); Keast (1971); Lack (1947, 1969, 1973, 1976); Lazell (1972); Lister (1976a, 1976b); MacArthur (1972); MacArthur et al. (1972, 1973); McNab (1971); Pianka (1971); A.S. Rand (1969); Roughgarden (1974); Roughgarden and Fuentes (1977); Roughgarden et al. (1983); T. Schoener (1967, 1968a, 1969b, 1970, 1974b, 1975); T. Schoener and Gorman (1968); Simberloff (1970, 1974a); Van Valen (1965); Van Valen and Grant (1970); E. Williams (1969, 1976, 1983); Willson (1969); M. Williamson (1981); Wright and Biehl (1982); Yeaton (1974); Yeaton and Cody (1974).

Taxon cycles and endemism

R. Berry (1964); Bramwell (1979); Carlquist (1965, 1966b, 1970, 1974); Case (1976, 1979); Corbet (1964); Darlington (1943, 1957, 1970); Elton (1958); Foster (1964); Fryer and Iles (1972); Greenwood (1974); Heaney (1978); M. Johnson and Raven (1973); Keast (1971); Lack (1947, 1976); Lawlor (1982); Lister (1976b), MacArthur and Wilson (1967); D. Porter (1979); Pregill and Olson (1981); Ricklefs (1970); Ricklefs and Cox (1972, 1978); Roughgarden et al. (1983); Sondaar (1971, 1977); Sondaar and Boekschoten (1967); Soulé (1966, 1972); Thaler (1973); Wassersug et al. (1979); Williamson (1981); E. Wilson (1959, 1961) E. Wilson and Taylor (1967).

Chapter 17 Species Diversity in Continental and Marine Habitats

The same kinds of questions that we asked about the distributions of insular biotas can also be asked about the distributions of species on continents and in the oceans. What determines the number and kinds of species that occur together in one place, and why do the number and kinds of species vary from one place to another? The taxonomic identities and affinities of species depend, of course, on the history of phyletic lineages, which in turn have been greatly influenced by the historical geologic and climatic events discussed in Chapters 5 and 10 to 14. In these last two chapters, we will be concerned primarily with describing and attempting to explain patterns in species diversity and in the ecological attributes of species. Because isolated habitats such as lakes, desert oases, caves, and mountaintops can often be viewed from the conceptual perspective developed for islands in the two previous chapters, we will focus here on large-scale continental and marine patterns.

Measurement and terminology

For many biogeographers it is sufficient to be concerned primarily about the number of species that inhabit a local area or geographic region. The rare species may be as interesting and important as the common ones; in fact, many important taxa used in vicariance biogeography are narrow endemics and are not the dominant forms of those biotas. Consequently, one must evaluate all elements of a biota. To compare the similarities of two regions, it is often most useful as well as most convenient simply to count up the total number of species. For well-studied taxa in many geographic regions this measure, often called species richness, can be obtained from the literature of systematic works and of faunal and floral surveys. The use of species richness permits simple, relatively unambiguous comparisons of the organic diversity of different geographic areas.

Many ecologists, on the other hand, are interested primarily in the organization of natural communities. For them, the relative abundance of species and other characteristics indicative of their ecological roles may be of more concern than simply the number of species. Rare species may be important components of ecological systems, but ecologists often deemphasize them and focus on species that dominate the community in terms of abundance, biomass, and cover. Consequently, ecologists have adopted several measures or indexes of species diversity that give greater weight to the dominant species. Many such diversity indexes have been proposed, and several are used frequently in the ecological literature. These are defined and their uses, strengths, and weakness are discussed in several recent reviews (e.g., Pielou, 1975; Whittaker, 1977). Although a thorough understanding of these indexes is important to interpret may ecological studies, a treatment of them is beyond the scope of this book. For our purposes it is sufficient to know that among most natural communities, all of the commonly used diversity indexes are highly correlated with the total number of species. For this reason we will use the term *species diversity* somewhat loosely and synonymously with species richness.

Like most other ecological patterns, species diversity varies with the spatial scale on which it is studied. Three terms are commonly used to divide the continuum of spatial scales into convenient categories (Whittaker, 1975, 1977; Cody, 1975). *Alpha diversity* refers to the diversity within small areas of relatively homogeneous habitat, i.e., the number of species per unit area, such as a hectare of grassland or lowland tropical rain forest. *Beta diversity* refers to the change (or turnover) in species composition over relatively small distances, often between recognizably different but adjacent habitat types. An example of beta diversity would be a comparison of species from two distinct communities on a mountain slope, e.g., between lowland tropical rain forest and montane evergreen forest or in temperate regions between deciduous and coniferous forest. The methods of Whittaker and Cody to obtain values of beta diversity are substantially different, but their goals, to express the overall turnover in species composition, are the same. *Gamma diversity* (sometimes redefined and called *delta diversity*) is used to describe differences in diversity between widely separated geographic regions that have similar habitat types, e.g., lowland tropical rain forests in the Congo and Brazil. The principal goal of gamma diversity analysis is to identify the roles history and ecology have played in the assemblage of communities.

In most of the following discussion we will be concerned primarily with patterns in alpha diversity. The primary data most available for numerous taxa and a variety of geographic regions are species counts for small areas. These usually are published either by systematists describing the results of biotic surveys or by ecologists summarizing censuses or samples of local habitats. The changes in species diversity and composition across the landscape, however, are of great interest to biogeographers. We shall discuss these patterns in beta and gamma diversity in the last part of this chapter and in the following chapter.

Latitudinal diversity gradients

One of the most striking biogeographic and ecological patterns is the gradient of increasing species diversity from the poles to the equator. The earliest explorer-naturalists noticed that the tropics teem with life, the temperate zones have fewer kinds of animals and plants, and the Arctic and Antarctic by comparison are stark and barren (Wallace, 1878). This pattern is a general one in that it holds true not only for organisms as a whole but also for most major taxa (classes, orders, and families) of microbes, plants, and animals. However, there are exceptions; penguins (Spheniscidae) and seals (Phocidae) reach their greatest diversity at high latitudes, and ichneumonid wasps (Ichneumonidae) and coniferous trees (e.g., Pinaceae) are most diverse in the temperate zones. For every such exception, however, there are numerous examples of groups that are confined or nearly so to the tropics; these include the New World fruit bats (Phyllostomidae), the Indo-Pacific giant clams (Tridacnidae) (Figure 7.12), and palms (Arecaceae) (Figure 13.1).

Numerous authors have quantified latitudinal diversity gradients. For groups whose distributions are well known this can be done by counting up the number of species in local areas of approximately equal size and drawing contour maps of species richness. The patterns for North American land birds and mammals are shown in Figure 17.1. Of course, the most striking feature is the rapid increase in diversity from the Arctic to the tropics. Interestingly, the rate of change of species with latitude is not the same in the two taxa. Not only are there more birds than mammals in every region but also bird species richness increases about twelve-fold in the 60° of latitude, whereas mammal diversity increases only eight times. There are also differences among the different families of

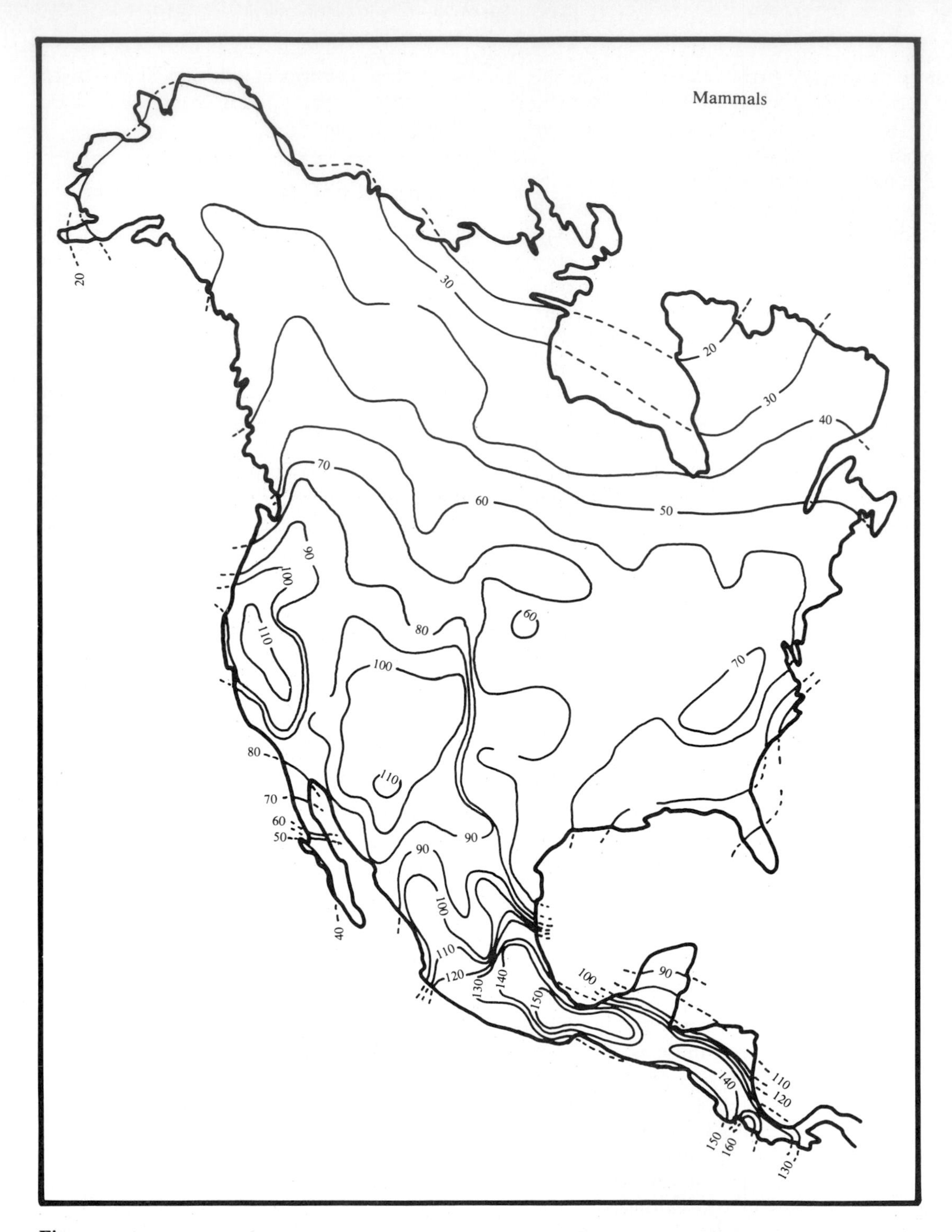

Figure 17.1
Geographic variation in species diversity of North American mammals *(left)* and land birds *(right)*. Note the pronounced latitudinal gradients in both groups and the high diversity in the southwestern United States and northern Mexico, a region of great topographic relief and habitat diversity. (After Simpson, 1964, and Cook, 1969.)

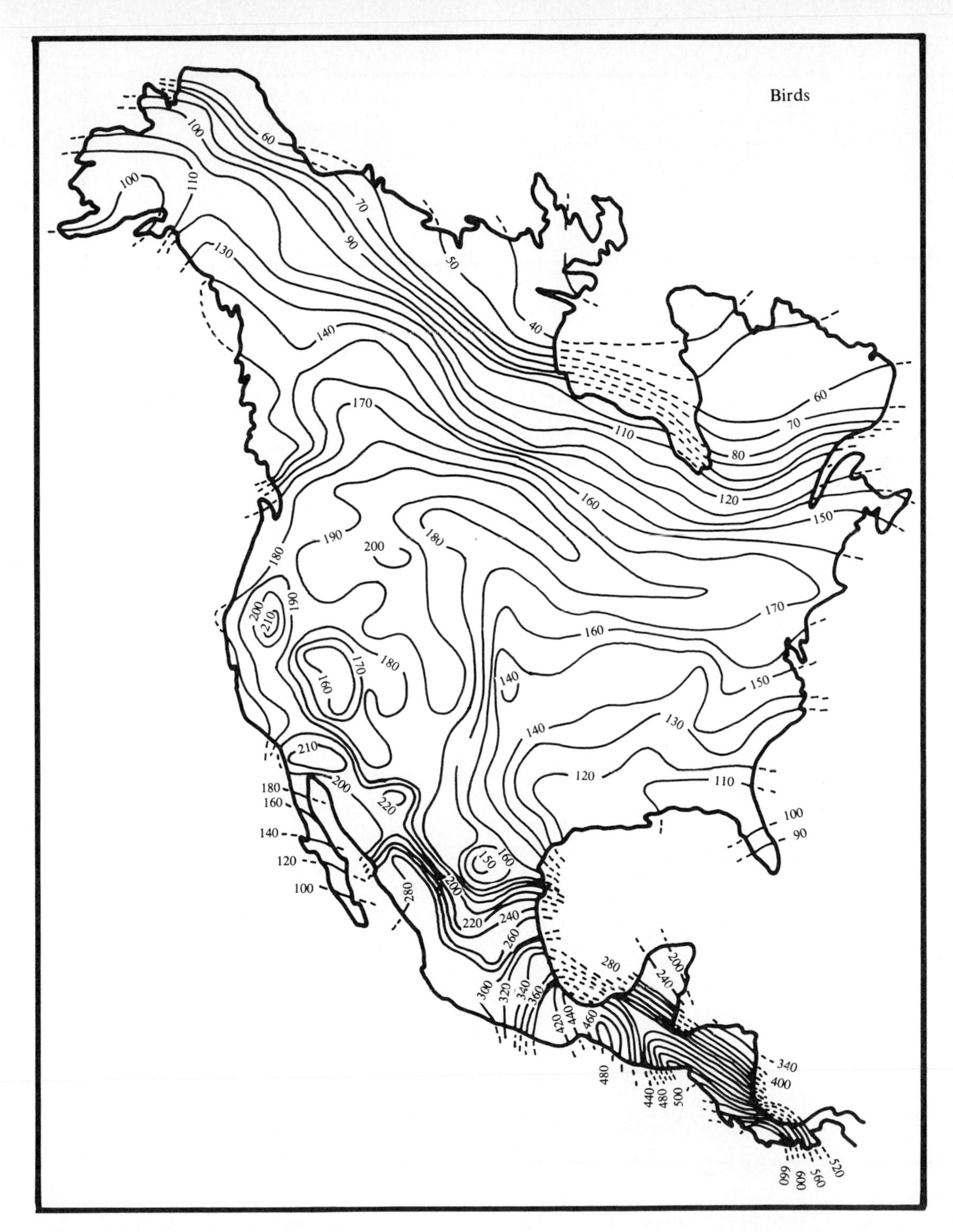

Figure 17.1, cont'd
For legend see opposite page.

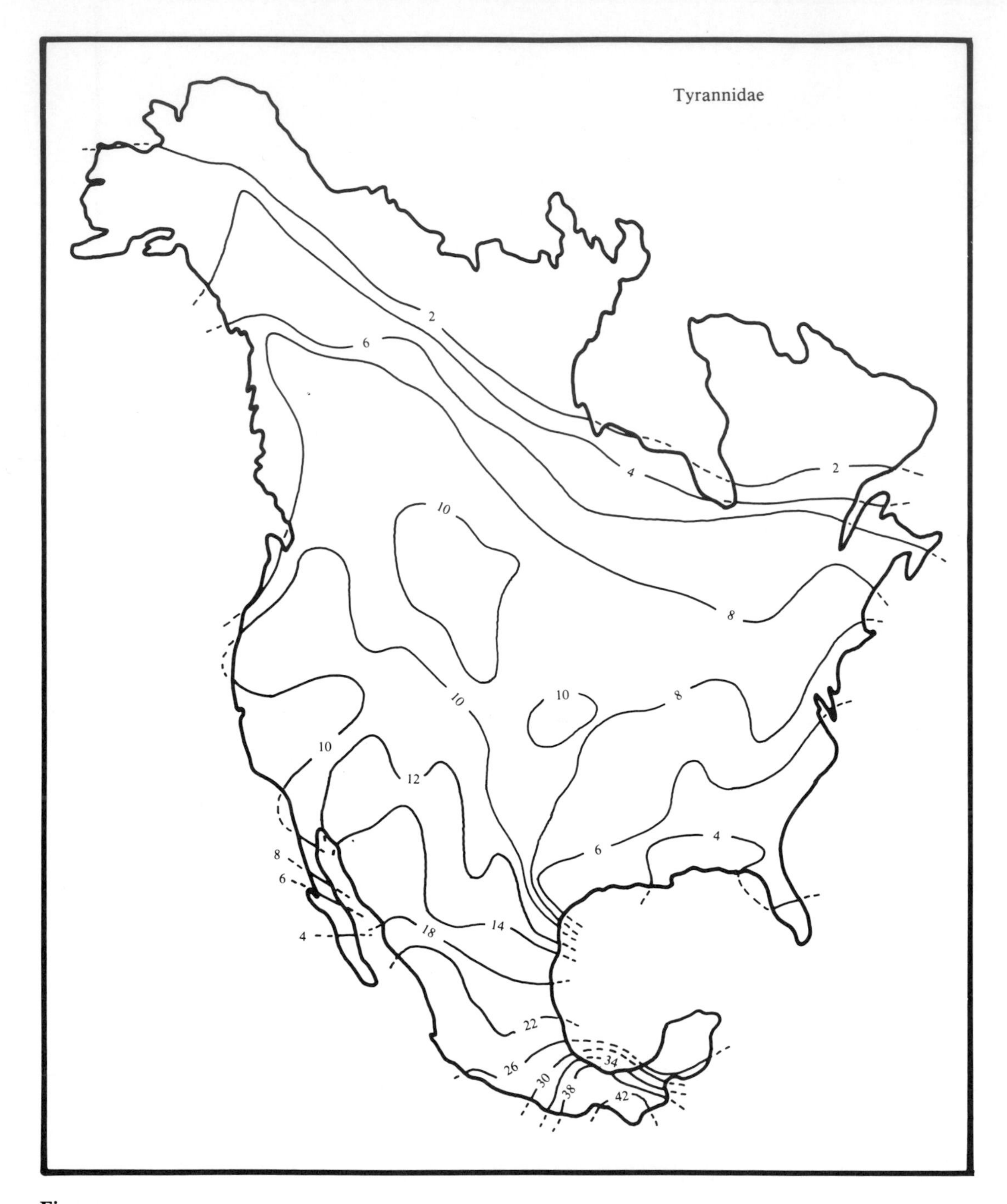

Figure 17.2
Species diversity patterns in two families of North American birds, the flycatchers (Tyrannidae), *left*, which show a typical pattern of increasing diversity toward the tropics; and the sandpipers (Scolopacidae), *right*, one of the exceptional groups that is most diverse at high latitudes. (After Cook, 1969.)

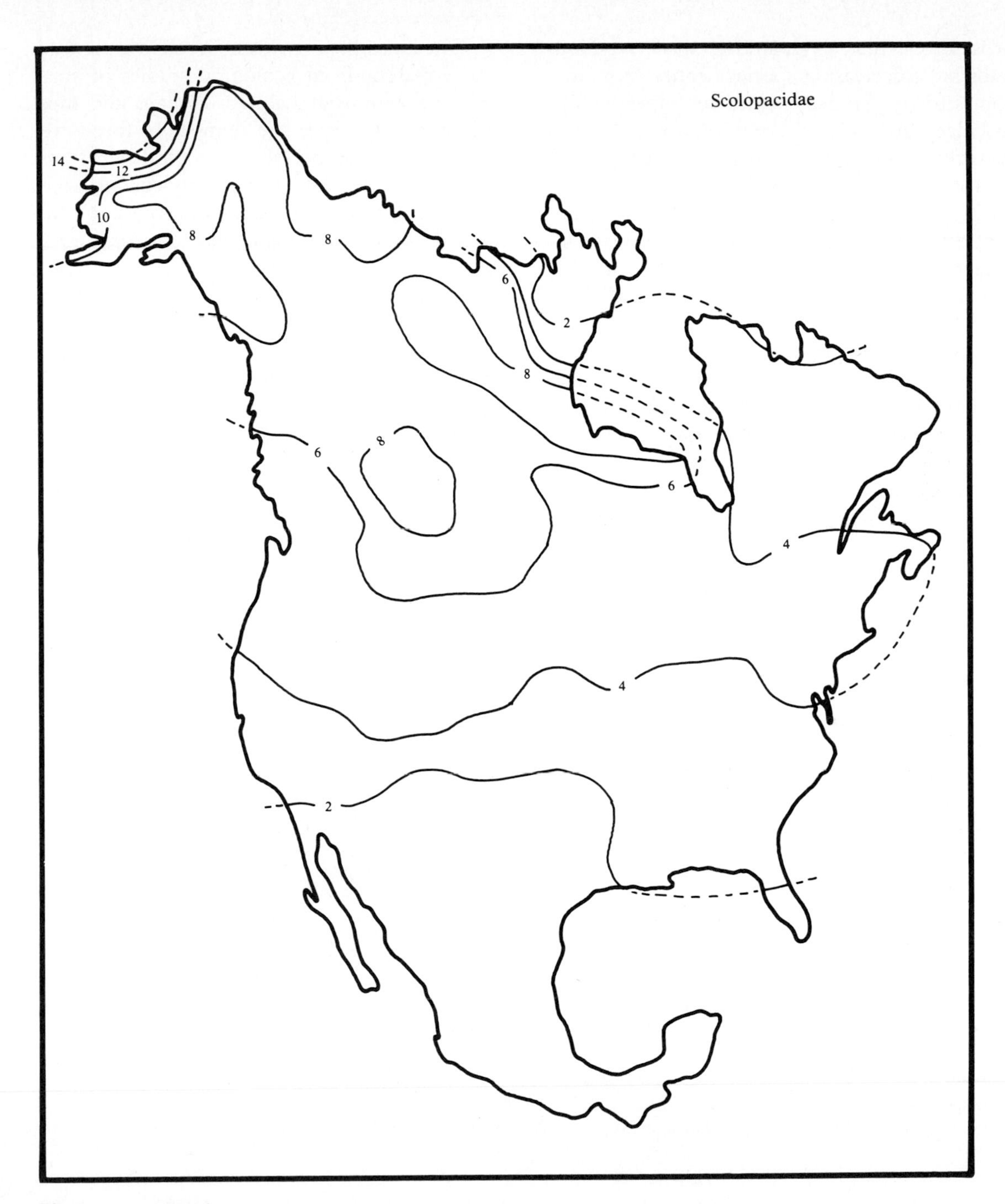

Figure 17.2, cont'd
For legend see opposite page.

birds. Flycatchers (Tyrannidae) show a typical gradient of decreasing species richness from the tropics to the Arctic, but the sandpipers (Scolopacidae) are one of the exceptional groups that exhibit the opposite pattern (Figure 17.2).

Table 17.1 summarizes data on latitudinal diversity patterns in a variety of taxa. Note again that, although all of these groups decrease in species richness with increasing latitude, the rate of change varies from group to group. It is difficult to show the exceptional patterns in such a table. One group that has been particularly well studied is the Ichneumonidae, a large family of parasitic wasps. These apparently are most diverse in temperate regions, with the greatest species richness in North America occurring at about 40° N latitude (Owen and Owen, 1974; Janzen, 1981; see also Hespenheide, 1978).

The latitudinal trends in species richness are also apparent from ecological studies of small areas of terrestrial habitat. Perhaps the most dramatic example is the number of forest tree species. A single hectare of tropical rain forest in South or Central America may contain 40 to 100 different kinds of trees (Richards, 1957; Anderson and Brown, 1980), whereas comparable areas of deciduous forest in eastern North America and of coniferous forest in northern Canada usually have 10 to 30 species and 1 to 5 species, respectively (Braun, 1950; Glenn-Lewin, 1977). Because many insect species are specific herbivores and pollinators of individual plant species, it is expected that many insect taxa will show comparable patterns when the data become available (see Otte, 1976). Fleming (1973) compiled data on the number of mammal species in North American forest habitats. He

Table 17.1
Latitudinal gradients in species diversity in various taxonomic groups

Taxon	Region	Latitudinal range	Range in species richness	Source
Land mammals	North America	8°-66° N	160-20	Simpson, 1964; Wilson, 1974
Bats (Chiroptera)	North America	8°-66° N	80-1	Simpson, 1964; Wilson, 1974
Quadrupedal land mammals (all orders except Chiroptera)	North America	8°-66° N	80-20	Simpson, 1964; Wilson, 1974
Breeding land birds	North America	8°-66° N	600-50	Cook, 1969; MacArthur, 1972
Reptiles	United States	30°-45° N	60-10	Kiester, 1971
Amphibians	United States	30°-45° N	40-10	Kiester, 1971
Marine fishes	California coast	32°-42° N	229-119	Horn and Allen, 1978
Ants	South America	20°-55° S	220-2	Darlington, 1965
Calanid crustacea	North Pacific	0°-80° N	80-10	Fischer, 1960
Gastropod molluscs	Atlantic coast of North America	25°-50° N	300-35	Fischer, 1960
Bivalve molluscs	Atlantic coast of North America	25°-50° N	200-30	Fischer, 1960
Planktonic foraminifera	World oceans	0°-70°	16-2	Stehli, 1968

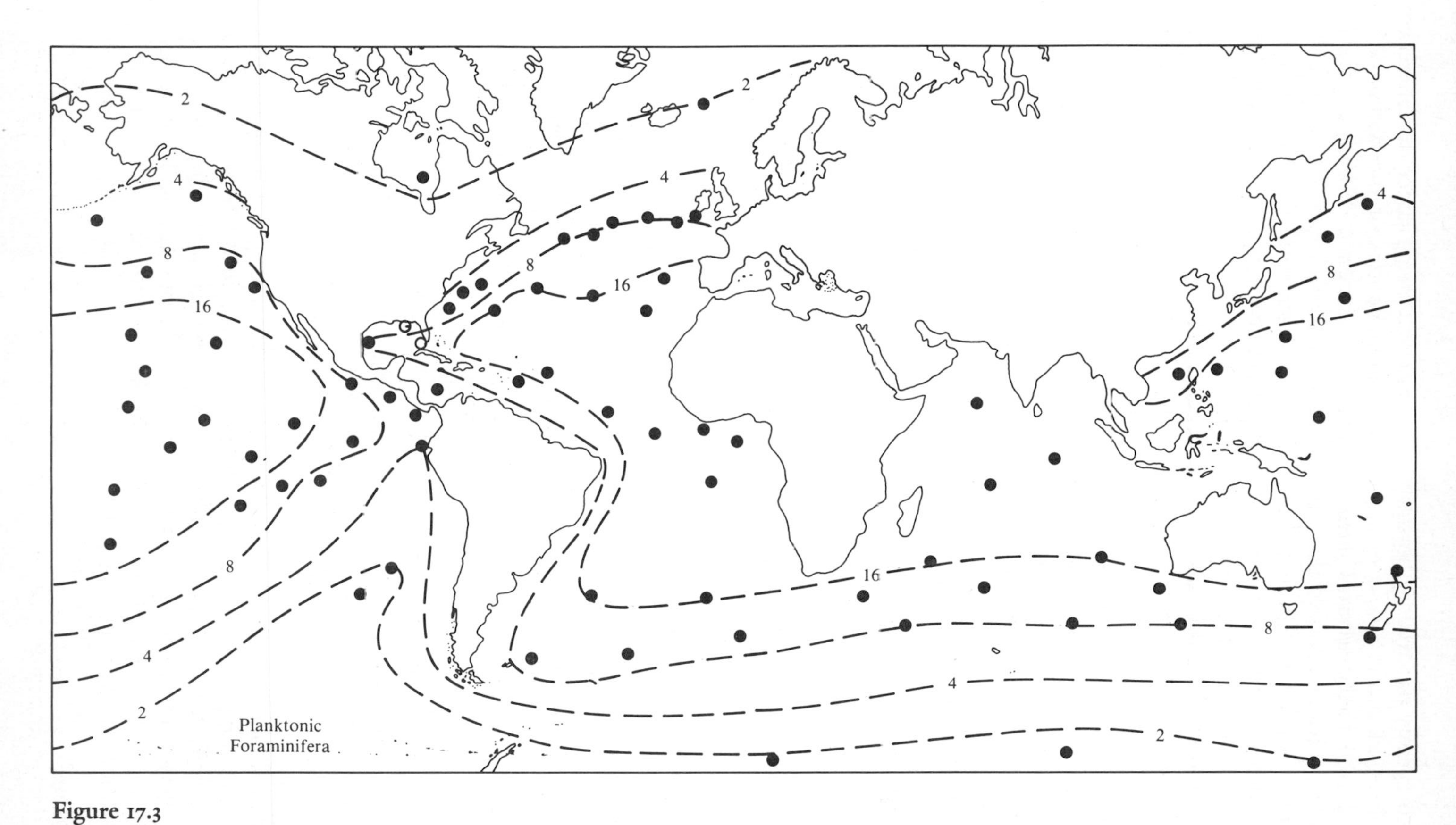

Figure 17.3
Worldwide pattern of species diversity in planktonic foraminifera, showing a positive correlation between species richness and latitude. (After F.G. Stehli, 1968. "Taxonomic diversity gradients in pole locations: the recent model." Pp. 163-227 *in* E.T. Drake [ed.], *Evolution and environment*. Peabody Museum Centennial Symposium. Copyright © 1968 by Yale University.)

noted that 15 to 16 species occurred in boreal coniferous forests in Alaska, 31 to 35 species inhabited deciduous forests in the eastern United States, and 70 species were found in both wet and dry tropical forests in Panama. As implied by Table 17.1, the increase in diversity can be attributed largely to rapidly increasing numbers of bat species at lower latitudes. Mammals with specialized arboreal foraging habits and insect, fruit, and pollen diets also contribute importantly to the composition of the diverse tropical communities. From somewhat more sketchy data in the literature, Fleming concluded that similar latitudinal changes characterize Old World forest mammals.

Qualitatively similar but quantitatively variable relationships also characterize latitudinal species diversity gradients in the oceans. Figure 17.3 and Table 17.1 illustrate the patterns in some marine taxa that have been fairly well studied (see also Stehli, 1968; Stehli et al., 1969; Taylor and Taylor, 1977; Turner, 1981). Sanders (1968) found about five times as many species (100 as opposed to 20) of bottom-dwelling annelids and bivalve molluscs in the Bay of Bengal (20° N latitude) as in comparable areas off Cape Cod, Massachusetts (40° N latitude). These examples reinforce the intuitive impression that tropical coral reefs support the most diverse marine communities, whereas the cold oceans and shores of polar latitudes have fewer species.

Although it was once thought that freshwater organisms were an exception to the gradients of increasing species richness observed in other habitats (Patrick, 1961, 1966), recent evidence suggests this is not the case. In a quantitative study of stream insects, Stout and Vandermeer (1975) usually found 30 to 60 species in tropical American sites compared with 10 to 30 species in the temperate United States. The number of fish species in rivers and lakes around the world also increases with decreasing latitudes (e.g., Barbour and Brown, 1974), although there is a great deal of variation in this pattern, owing largely to differences in the sizes and histories of the bodies of water, as might be expected from our discussion in Chapter 16.

The observation that the qualitative latitudinal gradients in species diversity are general ones, not only to many taxonomic groups of organisms but also to a wide range of environments that include most terrestrial, marine, and freshwater habitats, stimulates the search for a general explanation. In fact, the pattern has intrigued ecological biogeographers for many years, and numerous hypotheses have been advanced to explain it. The most frequently cited ideas are listed in Table 17.2. All explanations can be categorized as being either historical or equilibrial. The historical hypotheses claim that the widespread distribution of contemporary high-latitude habitats is a recent phenomenon in evolutionary time. In contrast, tropical environments were thought to have persisted relatively unchanged for a much longer period. The adaptive radiation of numerous taxa that has filled tropical habitats with diverse species has not had sufficient time to generate comparable diversity in the more recent temperate and arctic environments. Proponents of historical hypotheses usually point to perturbations associated with Pleistocene glaciation or to more ancient climatic and geologic changes to support their case (e.g., Fischer, 1960; Fischer and Arthur, 1977; Stanley, 1979).

In contrast to historical explanations are numerous hypotheses that can be classified as equilibrial because they assume that the latitudinal diversity gradients can be attributed primarily to current conditions that limit the capacities of different environments to support species (e.g., Van Valen, 1973b; Rosenzweig, 1975; Brown, 1981). Proponents argue that although the earth has admittedly experienced a complex climatic and geologic history, cold environments with annual fluctuating photoperiods have long existed at high latitudes. If there has been sufficient time for groups such as pen-

Table 17.2
Processes that have been hypothesized to account for geographic patterns of species diversity

Hypothesis or theory	Mechanism of action
Historical perturbation	Habitats that have experienced historical changes are undersaturated because of inadequate time for species to colonize and adapt.
Productivity	The greater the availability of usable energy, the larger the number of species that can be supported and the greater the specialization of coexisting species.
Harshness	Small, isolated, ephemeral, or physically extreme habitats have lower colonization rates or higher extinction rates than large, continuous, permanent, and physically equable habitats.
Climatic stability	A fluctuating environment may preclude specialization or increase the extinction rate, whereas in a constant environment species can specialize on predictable resources and persist when rare.
Habitat heterogeneity	Diverse physical habitat structure permits finer subdivision of limiting resources and hence greater specialization.
Competition, predation, or mutualism	One or more of these classes of interspecific interactions promote coexistence and specialization.

guins (Spheniscidae), willows *(Salix)*, and many others to invade, adapt to the "rigorous" environment, and radiate, it is hard to claim that there has not been adequate time for these same or other taxa to attain much higher diversity. As an alternative, equilibrium hypotheses suggest that such factors as low productivity or seasonal climatic regimes prevent high-latitude environments from supporting as many species as the tropics.

It is obvious that both historical and equilibrial explanations must rely ultimately on some kind of latitudinal variation in the nature, including the constancy, of the physical environment. Otherwise, given unlimited evolutionary time, one would expect organisms to colonize and radiate until all latitudes supported approximately the same number of species. Many published explanations for latitudinal diversity gradients can be categorized as secondary or tertiary because they focus on characteristics of organisms that affect community organization rather than on the physical factors that must be the primary cause of diversity patterns (Pianka, 1966, 1978). These include suggestions that in the tropics species have narrower niches (i.e., more specialized requirements), that competition and predation are more intense, that speciation rates are higher and extinction rates are lower, and so on. These more proximate, mechanistic hypotheses are important and worth pursuing as part of any historical or equilibrial explanation because they suggest how differences in the physical environment are mediated by ecological and evolutionary processes to produce ecological communities containing different numbers and kinds of species.

We shall return to reconsider the explanations for latitudinal diversity gradients after examining some other patterns of species richness in terrestrial and aquatic habitats.

Diversity in gradients of elevation and aridity

In terrestrial habitats, variation in species diversity along gradients of elevation and available soil moisture are almost as general and striking as latitudinal variation. Just as the number of species decreases in progressively cooler climates as one moves from tropical to polar regions, so it also decreases in the cooler environments as one ascends mountains. Species diversity also declines with decreasing moisture availability, from mesic forests to more arid shrublands and grasslands to xeric deserts. Although these patterns have long been obvious to ecologists and biogeographers, they are less well documented quantitatively than the latitudinal diversity gradients. This is in part because there have been fewer censuses of species on a sufficiently small scale along transects in which these factors vary systematically. However, it is also in part because elevational gradients of temperature are often confounded by variation in moisture, and gradients of aridity are frequently complicated by variations in elevation or latitude. To analyze the influence of one factor it is desirable to find "natural experiments" in which that factor is the only important variable and other conditions remain relatively constant. Unfortunately, for reasons explained in Chapter 2, both temperature and precipitation tend to vary systematically with both elevation and latitude.

Despite the poverty of quantitative studies, the general decline in species diversity with decreasing temperatures at higher elevations is as obvious as the latitudinal pattern. The lowland tropics of the Amazon Basin support a rich rain forest biota, but the peaks of the nearby Andes, covered with bare rock, ice, and snow, are as barren of life as the polar ice caps. The effect of temperature is most unambiguous in elevational gradients in mesic regions, where limiting moisture is not a confounding factor. Yoda (1967) has quantified the dramatic decline in tree species richness with increasing elevation in the Himalayas. Similarly, the number of bird species varies inversely with elevation in New Guinea (Figure 17.4) (Kikkawa and Williams, 1971) and on the eastern slope of the Andes in Peru (Terborgh, 1977). Whittaker (1960, 1977) has shown similar patterns for trees on mesic mountains in North America (Figure 17.5).

As with the latitudinal gradients, certain restricted taxonomic or ecological groups provide exceptions to the general pattern. Herbaceous plant species sometimes reach their greatest richness at intermediate elevations on temperate mountains (Figure 17.5) (Whittaker, 1960, 1977), and orchids (Orchidaceae) attain their greatest diversity on tropical mountainsides in the low-stature forests that occur considerably higher than lowland rain forests (Dressler, 1981). On desert mountains, such as those in the southwestern United States, species diversity in most organisms probably is greatest at intermediate elevations (e.g., Whittaker and Niering, 1965), but this effect can be attributed to the extreme aridity of the lowlands.

Species diversity decreases with diminishing water availability. Deserts are so named because they appear barren and lifeless in comparison with more mesic grasslands, shrublands, and forests. Again, this pattern is obvious but not well documented. Studies of several taxa within desert regions show that species richness of rodents, ants, and perennial plants increases with increasing annual precipitation (Figure 17.6) (Brown, 1973; Davidson, 1977; Brown and Davidson, 1977). Although a few groups, such as the rodent family Heteromyidae (Brown, 1975), reach their greatest diversity within desert regions, in general species richness continues to increase with water availability through grasslands and shrublands, reaching a maximum in mesic forest habitats. The wettest habitats, including swamps and marshes, often have fewer species than drier surrounding habitats. We shall return to this observation shortly when we consider the concepts of environmental favorability and harshness.

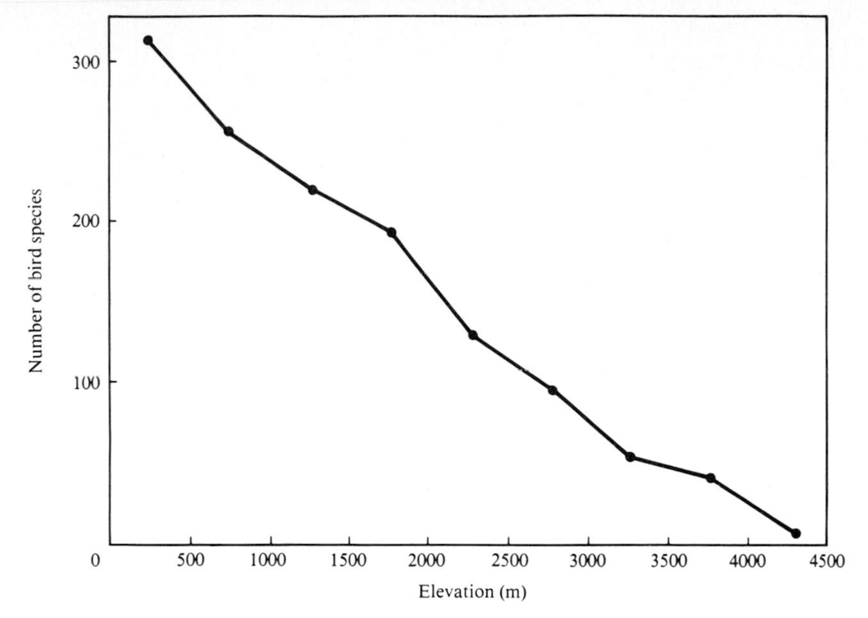

Figure 17.4
Number of bird species in New Guinea as a function of elevation. A similar decrease in bird species richness with increasing elevation was also reported by Terborgh (1977) for the avifauna of the eastern slope of the Andes in Peru. (After Kikkawa and Williams, 1971.)

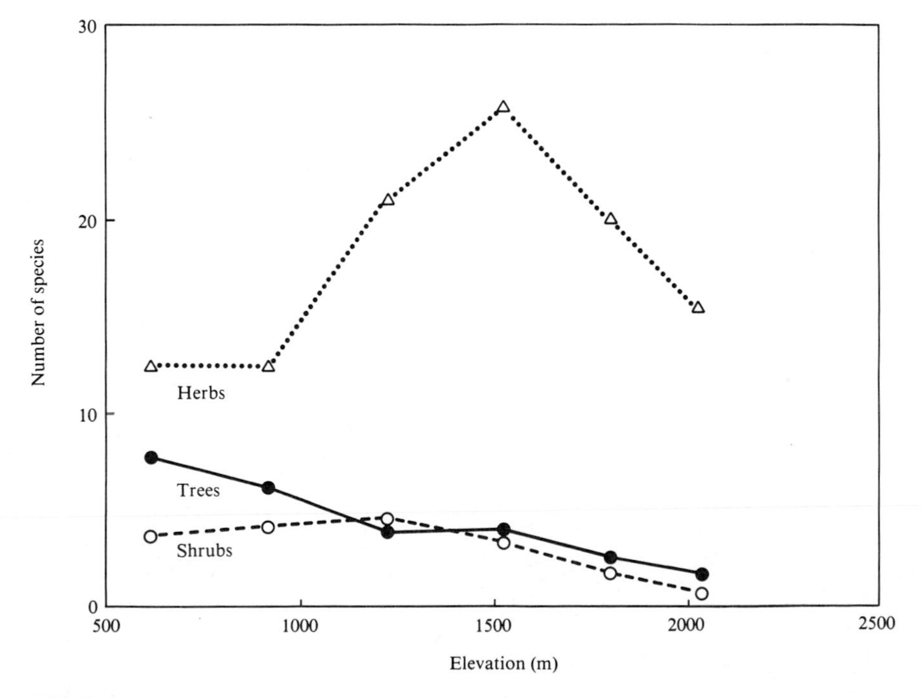

Figure 17.5
Variation in the number of species of trees, shrubs, and herbs with elevation in the Siskijou Mountains of Oregon and California. Note that the number of tree species declines with increasing elevation, but herb and shrub species richness is greatest at intermediate elevations. (Replotted from data in Whittaker, 1960.)

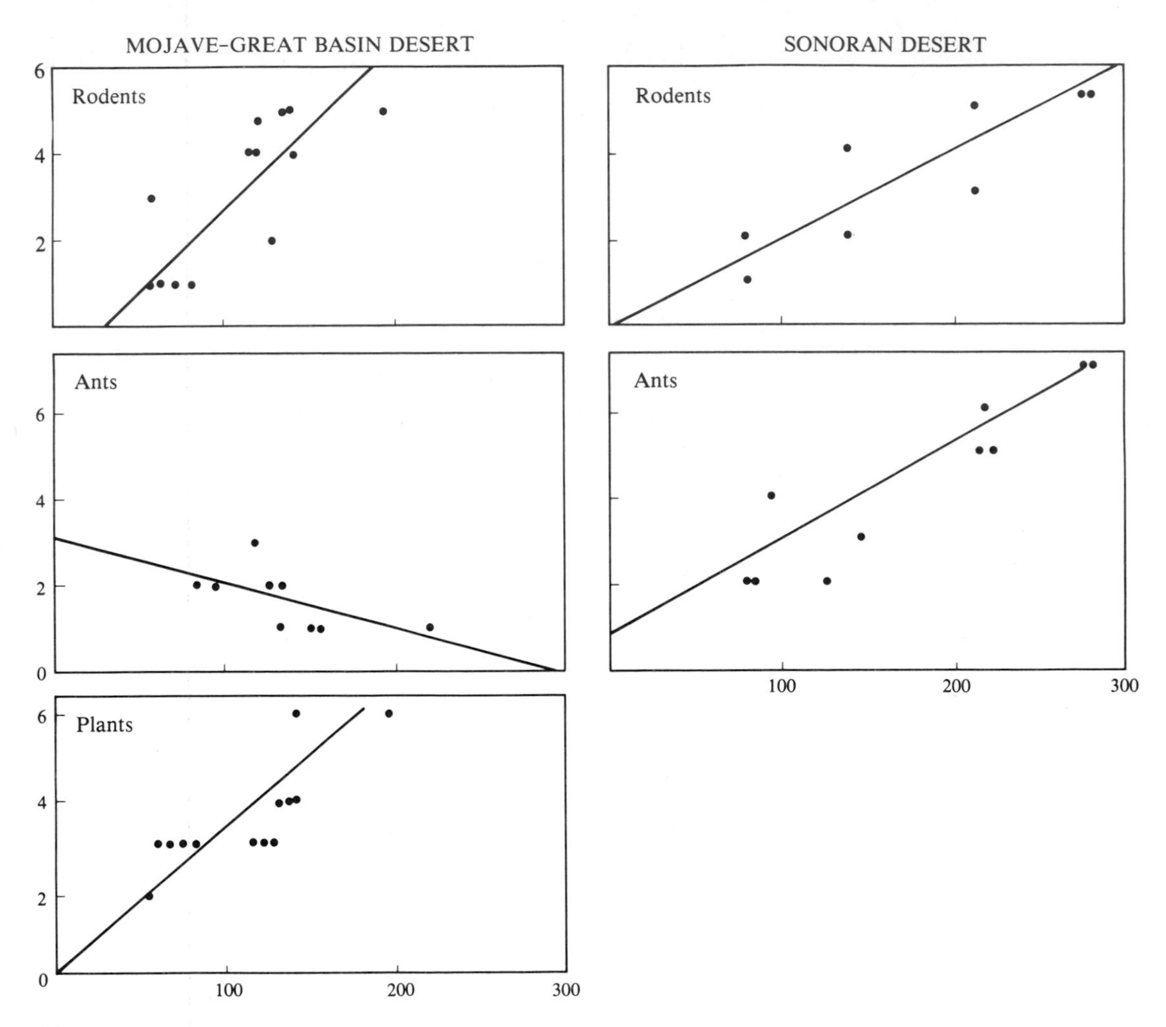

Figure 17.6
Variation in the number of rodent, ant, and perennial plant species with annual precipitation in North American deserts. Note that all groups increase in diversity with increasing precipitation and primary productivity, except for ants in the Great Basin desert, where the wettest sites are also the coldest and low temperatures may limit the activity and distribution of poikilothermic ants. (Replotted from data in Brown, 1973, and Davidson, 1977.)

Diversity in aquatic environments

Some of the patterns of species richness in water are remarkably similar to those on land. We have already mentioned that small lakes, like small islands, have fewer species than large ones, and that the high latitudes support lower diversity than the tropics. In general, species richness varies with water depth in a way comparable to its variation with elevation on land. The cold, dark, oxygen-poor abyssal depths contain fewer species than the shallow surface waters. This is particularly obvious in lakes. Some large lakes, including the Great Lakes and the Finger Lakes in northeastern North America, Lake Baikal in the Soviet Union, and several of the large lakes of central Africa, are more than 100 m deep, but few organisms inhabit the depths. For example, in Chapter 6 we described the great diversity of fishes, especially in the family Cichlidae, which inhabit the rift lakes of central Africa. Although Lake Malawi and Lake Tanganyika are 704 and 1470 m deep, respectively, the approximately 200 species of fishes are most diverse in the surface layers and none are found below about 200 m. The lower depths apparently are inhabited only by a few kinds of anaerobic invertebrates and bacteria.

The same pattern of decreasing diversity with increasing depth probably also holds true for the oceans. Marine plants, including both phytoplankton and larger (usually attached) algae, are confined to the photic zone (see Chapter 2), which rarely extends below 30 m. Pelagic animals, both passively floating plankton and free-swimming nekton, are most diverse in surface waters. Although diversity of these forms decreases rapidly with depth, many kinds of bizarre deep-sea fishes and some unusual invertebrates inhabit the abyssal depths. Bottom-dwelling animals exhibit more complex patterns. Certainly benthic species are not always most diverse in shallow waters, although Vinogradova (1962) showed that the richness of invertebrate species declines rapidly with increasing depth below about 2000 m, the approximate limit of the continental slope (Figure 17.7).

In the shallower waters of the continental slope and shelf the pattern is more complicated. Sanders (1968) performed classic studies on the diversity of benthic bivalve molluscs and polychaete worms collected by dredging in the oceans off both eastern and western North America. Although he used a somewhat unconventional method to express his data (Figure 17.8), the results exhibit two general trends. First, he found consistently higher species richness on the continental slopes than on the shallower continental shelf inshore. Sanders (1968, 1969; Slobodkin and Sanders, 1969) attributed this pattern to the temporal stability of the deep-water sites, which do not experience the pronounced seasonal and long-term fluctuations in temperature and other factors that are characteristic of shallow waters. However, Sanders' explanation and the data on which it is based have recently been criticized (Abele and Walters, 1979). Sanders also found consistently greater species richness in the Pacific Ocean off Washington than at comparable sites in the Atlantic off New England. This difference is characteristic of most habitats along the two coasts and is perhaps even more pronounced in the rocky intertidal zone (Menge and Sutherland, 1976). It is often attributed to greater physical environmental stability or to greater productivity owing to upwelling along the Pacific coast.

It is not true, however, that productive waters always have high species diversity. Some productive marine habitats, such as coral reefs, also are rich in species. On the other hand, small productive areas of ocean water, such as local regions of upwelling and enriched sites near rivers or estuaries, often contain relatively few species, especially of plankton. McGowan and his associates (McGowan and Walker, 1979; Hayward and McGowan, 1979) have studied the plankton of the north central Pacific, the

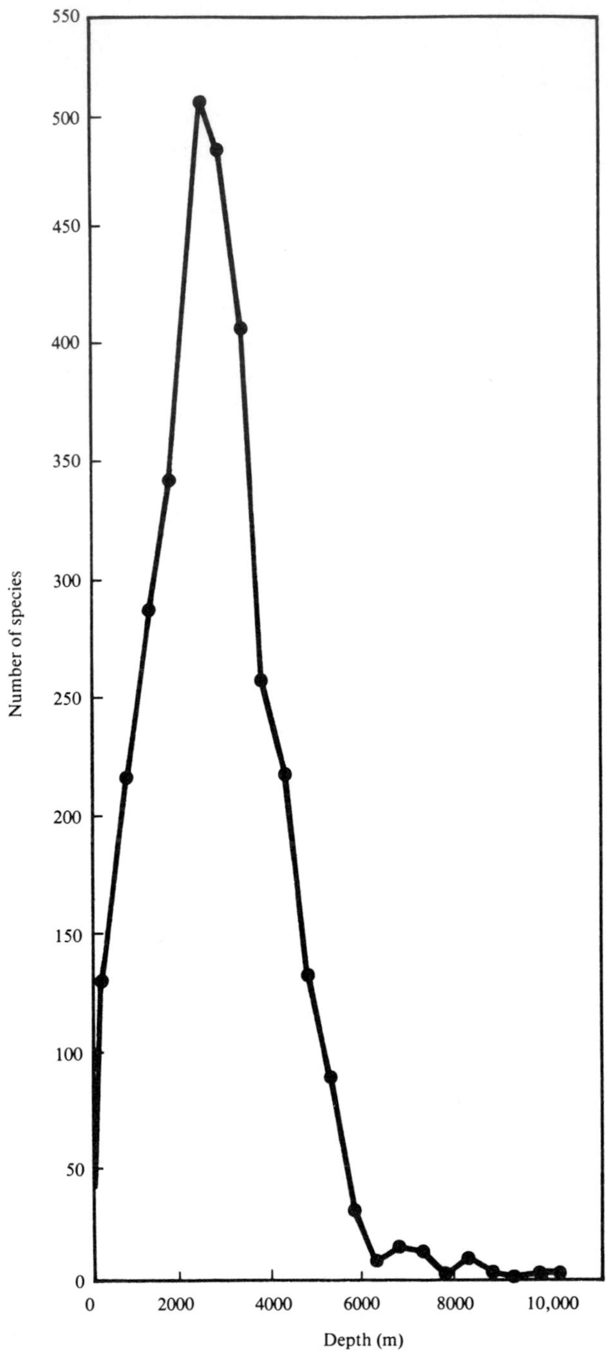

Figure 17.7
Variation in species richness of benthic marine invertebrates with depth. Note that diversity increases from low levels in the shallow waters on the continental shelf, reaches a maximum at a depth of about 2000 m on the continental slope, and then declines to very low levels in the abyssal depths. (After Vinogradova, 1962.)

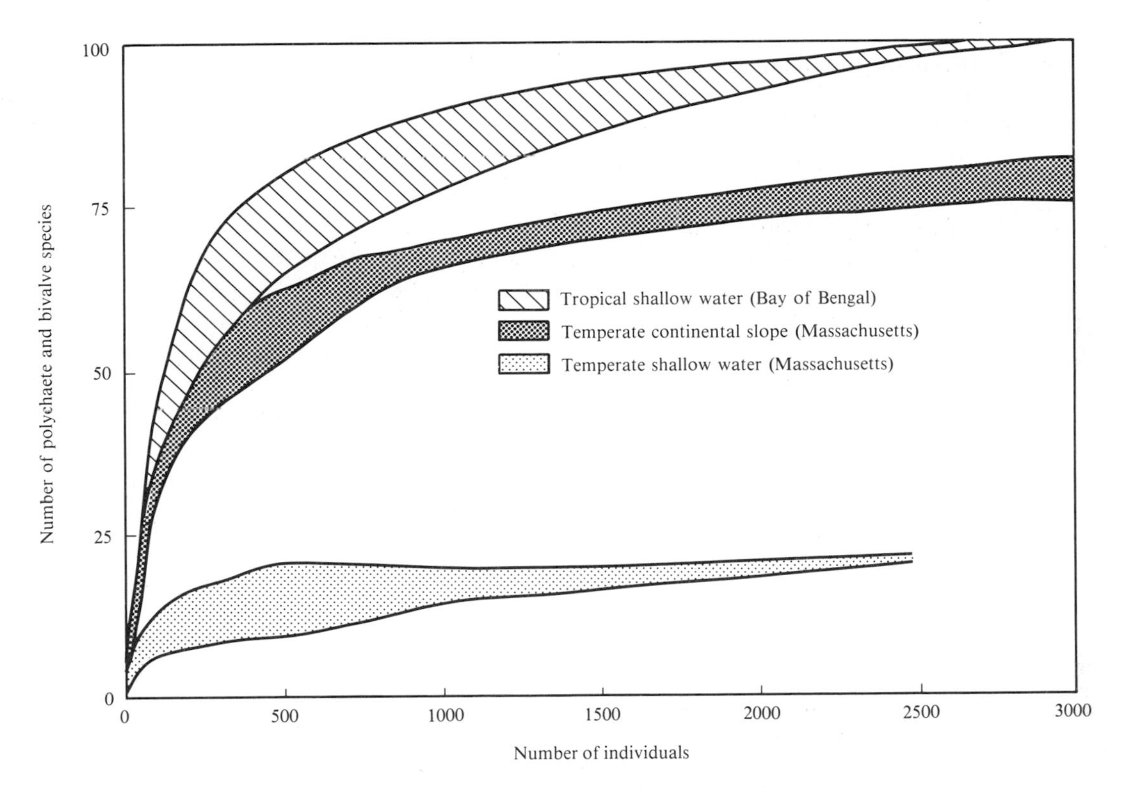

Figure 17.8
Sanders' data on species diversity of bottom-dwelling marine bivalve molluscs and polychaete worms. Cumulative numbers of species are plotted against the number of individuals counted, so diversity is expressed as a curve that approaches an asymptote when enough individuals have been sampled to include the entire biota. The shaded areas show the range of values for replicated samples from the same site. Note that tropical shallow-water areas are more diverse than temperate ones and that at temperate latitudes continental slope sites have more species than shallow-water sites on the continental shelf. (Reprinted from "Marine benthic diversity: a comparative study," *American Naturalist* 102:243-282, by H.L. Sanders, by permission of the University of Chicago Press. © 1968 by the University of Chicago.)

most extensive area of uninterrupted water on the earth. They find that this region is rich in species, even though the water is unproductive and the density of individual planktonic organisms is usually low. Many of the species are extremely widely distributed, ranging thousands of kilometers across the region. Several studies of temperate lakes have shown an inverse relationship between productivity and planktonic diversity (Figure 17.16) (Whiteside and Harmsworth, 1967), although this is confounded by the size of the lakes. Small lakes tend to be shallow and to experience much vertical mixing of water, substantially increasing their productivity. Thus highly productive marine and freshwater environments tend to be small and patchy, whereas unproductive regions are usually large and more extensive. We shall return shortly to consider likely explanations for the associated trends in species diversity.

A final widespread pattern concerns the diversity of benthic species in intertidal and shallow subtidal habitats. A common observation by marine and intertidal ecologists is that soft, unstable substrates such as mud and especially sand support many fewer species than nearby hard, permanent substrates such as rock and coral. Rocky intertidal habitats in the temperate zones and coral reefs in the tropics support particularly diverse communities in which sessile plants and animals extend above the surface and crevice-dwelling and burrowing forms live beneath the surface to provide a complex, three-dimensional structure. In contrast, sandy beaches and mud flats throughout the world usually appear virtually lifeless, although the latter especially may support reasonable numbers of burrowing forms. It has been suggested that these differences can be attributed at least in part to variations in the frequency and extent of perturbations (e.g., Stanley, 1979). Hard substrate habitats and the organisms that are firmly attached to them are relatively resistant to buffeting from severe storms, which can virtually destroy soft substrate habitats and temporarily eliminate most of their inhabitants.

Explanations for the patterns

In discussing the major geographic patterns in species richness we have already mentioned some of the hypotheses advanced to account for them. Now we shall consider these explanations in greater detail. Because many of the proposed mechanisms are not necessarily mutually exclusive, it is probably more realistic to ask how they may contribute to causing the observed patterns rather than to attempt to support or refute each one individually.

Although there is little disagreement about the existence of most of the patterns described above, the same cannot be said for the proposed explanations. One reason for the controversy is that most investigators have favored a single hypothesis to account for just one pattern—most often the latitudinal gradient. Perhaps taking a broader perspective and developing a conceptual framework that includes many patterns and processes (see Pianka, 1966, 1978) will eventually lead to better understanding and more consensus.

Historical perturbations. Several investigators (e.g., Fischer, 1960; Fischer and Arthur, 1977; Stanley, 1979) have stressed the importance of historical geologic and climatic changes in causing latitudinal and other patterns of diversity. As we have shown, especially in Chapters 5 and 14, the earth has provided anything but a constant environment for its inhabitants. Over the last 500 million years continents and oceans have changed position, continental seas and glaciers have covered extensive areas, mountains have risen and rivers have altered course, and climatic regimes have changed drastically. We also know that there have been episodes of widespread, catastrophic extinctions (Chapter 6) that may have been caused by some of the changes listed above but that also could have been caused by some sort of more sudden, cataclysmic event, such as a large meteorite striking the earth. It is certainly reasonable to expect that many extant patterns in diversity might be attributed to such historical perturbations, especially because some of them occurred as recently as 8000 to 12,000 years ago at the end of the Pleistocene.

There can be no doubt that these events changed the geographic distribution of diversity by causing the extinction of many species and the contraction of the ranges of others while enabling still others to invade and speciate in newly available regions. The clearest evidence that legacies of the perturbations affect extant diversity comes from insular habitats that were created or isolated since the end of the Pleistocene, about 10,000 years ago (Chapter 16). It is tempting to attribute latitudinal diversity gradients to a similar cause. There may have been insufficient time for species to adapt to and fill the extensive high-latitude environments that became available with the retreat of the vast continental glaciers and the associated climatic changes. This would be even more likely in cases in which there were no refuges for arctic and temperate species to survive glacial periods. Thus the lower diversity of temperate forest trees in Europe than in North America has been attributed to the fact that in Europe the major mountain ranges, such as the Pyrenees and the Alps, run east to west, whereas in North America the important ranges, the Appalachian and Rocky Mountains and the Sierra Nevada, run north to south. The idea is that in Europe the trees were trapped against the mountains and became extinct, whereas in North America they were able to retreat southward in front of advancing glaciers and deteriorating climates and thus survived.

There are several problems, however, with invoking Pleistocene changes as a general explanation for latitudinal and other (e.g., elevational) patterns of species diversity. First, although it was once generally believed that Pleistocene glaciation and climatic changes were confined to high latitudes, there is now excellent evidence that major shifts in climate and habitats occurred worldwide, even in lowland tropical regions (see Chapter 14). If the rich Amazon rain forest biota survived glacial periods confined to relatively few small refugia in an extensive area of grassland habitat, this greatly weakens the argument that similar habitat restriction can account in large part for the low diversity of species at high latitudes.

Second, it has been seen that the diversity of marine species shows the same qualitative trend from tropics to poles as terrestrial species richness, yet marine organisms have almost certainly been less affected by Pleistocene changes. Water temperatures changed less on the average than the terrestrial climate (CLIMAP, 1976), northern oceans apparently were not covered with much more ice than at present so marine habitats were not much restricted or fragmented, and in any event high-latitude forms could have remained in relatively constant environmental conditions simply by shifting their ranges toward the equator during periods of colder climate.

Third, there is the observation that low diversity is not confined to the cooler regions at high latitudes but also is characteristic of high elevation terrestrial and abyssal aquatic environments. There is every reason to believe deep water habitats have always been cold (approximately 4° C) for basic physical reasons outlined in Chapter 2. Although high mountains in the tropics have been subjected to climatic changes and even glaciation during the Pleistocene (Simpson, 1974, 1975), the shift of climatic and vegetation zones to lower levels would have increased the size and decreased the isolation of these areas, the opposite effect of the same events at high latitudes. Taken together, these patterns suggest that some common influence of low temperatures per se is the simplest explanation for the low species diversity of cold regions.

Other historical explanations for diversity would put the time of important events back beyond the Pleistocene and would invoke one or more of a number of geologic and climatic changes, including shifting poles, long-term climatic shifts, geologic formation of mountains and other habitats, and so on. Although the postulated events may have occurred, there are two major problems with using them to account for contemporary patterns of species diversity. First, the best estimates of the total number of species inhabiting the earth at various times during the past (Sepkoski et al., 1981; see also Raup, 1972, 1976; Boucot, 1975a, 1975b; Sepkoski, 1976; Flessa and Sepkoski, 1978) suggest that there has been only a very gradual increase in total species richness in all of the last 500 million years (Figure 17.9). Even the catastrophic extinctions at the end of the Permian and Cretaceous, the former of which was estimated to have exterminated over 88% of all marine species (Raup, 1979; Raup and Sepkoski, 1982) (see Chapter 6), did not prevent the rapid reattainment of high diversity within a few million years. The history of many taxonomic groups is characterized by rapid early radiation and then a long period of relatively constant diversity. Lillegraven (1972) (Figure 17.10) and Nicklas et al. (1980) (Figure 17.11) give examples of such patterns for fossil mammals and terrestrial plants, respectively.

Second, as evidenced by the numerous species that originated during the Pleistocene, a

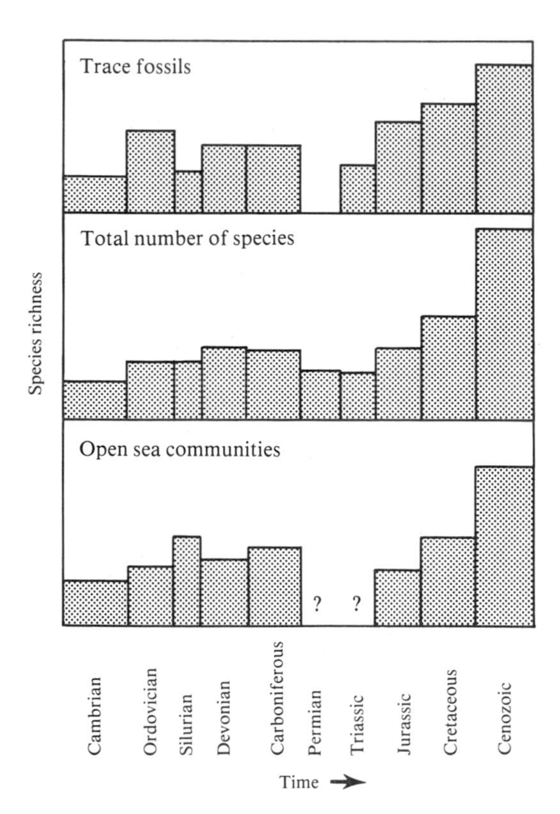

Figure 17.9
General, long-term trends in the diversity of marine organisms during about 500 million years of Phanerozoic history. Three different data sets give similar patterns of changing species richness. Note that the number of species does not appear to increase exponentially with time, but instead remains fairly constant with perhaps a significant increase in the Cenozoic. (After Sepkoski et al., 1981. Reprinted by permission from *Nature* **293**[5832]:435-437. Copyright 1981, Macmillan Journals, Ltd.)

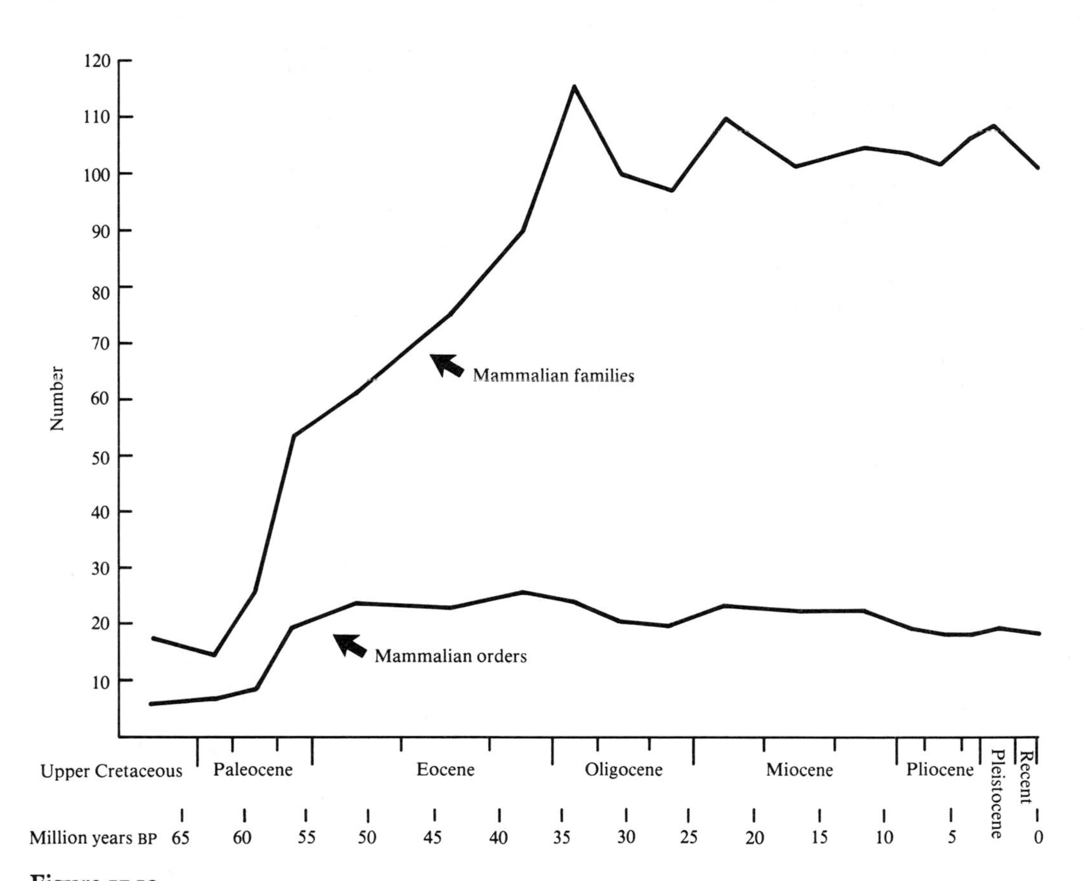

Figure 17.10
Historical trends in fossil mammal diversity in the Cenozoic. Note that the number of both orders and families increased rapidly in the Eocene radiation but then remained remarkably constant. Although accurate data for genera and species have not been collated, it is to be expected that they would follow the pattern for orders and families with a short time lag. (After Lillegraven, 1972.)

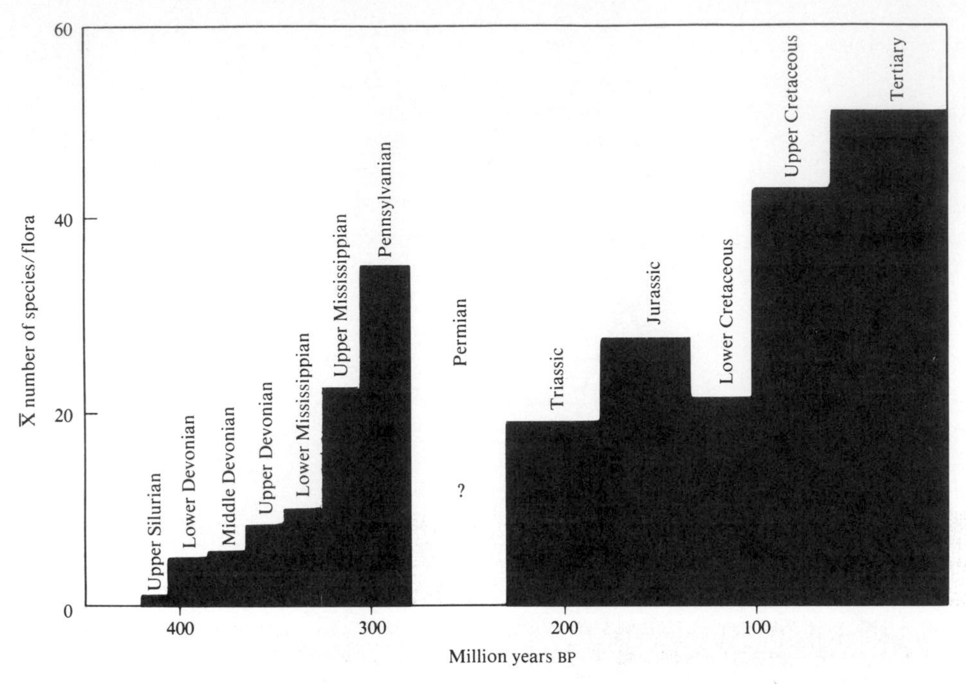

Figure 17.11
Historical pattern in the number of fossil terrestrial plant species over evolutionary time. Although this increase in diversity appears more gradual than that in mammals (Figure 17.10), note that there has not really been much change in species richness of terrestrial floras in the last 300 million years. (Plotted from data in Nicklas et al., 1980.)

million years is sufficient evolutionary time for much adaptation and speciation. If some groups have been able to colonize and adapt to cold environments at high latitudes and elevations and in the abyssal depths, why have not others been able to do the same or why have not the cold-adapted forms speciated and radiated to fill those environments with more species? In the case of some Pleistocene perturbations it may be true that indeed there has not been sufficient time, but even for these cases the historical explanation should be advanced with caution. This is especially true in view of Strong's evidence (1974; Strong et al., 1977) that the number of species of insect pests on introduced crop plants seems to increase rapidly to an equilibrial level determined by the area of the crop under cultivation (Figure 17.12). Similar patterns suggest equilibration on larger spatial and temporal scales. Flessa (1975) found the area of continents to be well correlated with contemporary mammalian diversity. Marshall et al. (1982) have used an equilibrial explanation to account for the changing levels of mammalian diversity in North and South America before, during, and after the faunal interchange across the Panamanian Land Bridge in the Pliocene. Finally, Raup (1972, 1976) has suggested that the changing area of shallow-water marine habitats caused by shifting sea levels and tectonic events might explain, at least in part, the fluctuations in diversity of shallow water benthic marine invertebrates during the Phanerozoic.

To make clear the distinction between the historical hypothesis and the equilibrial explanations that will be discussed below, one final matter warrants discussion. Many investigators have attempted to explain diversity gradients in

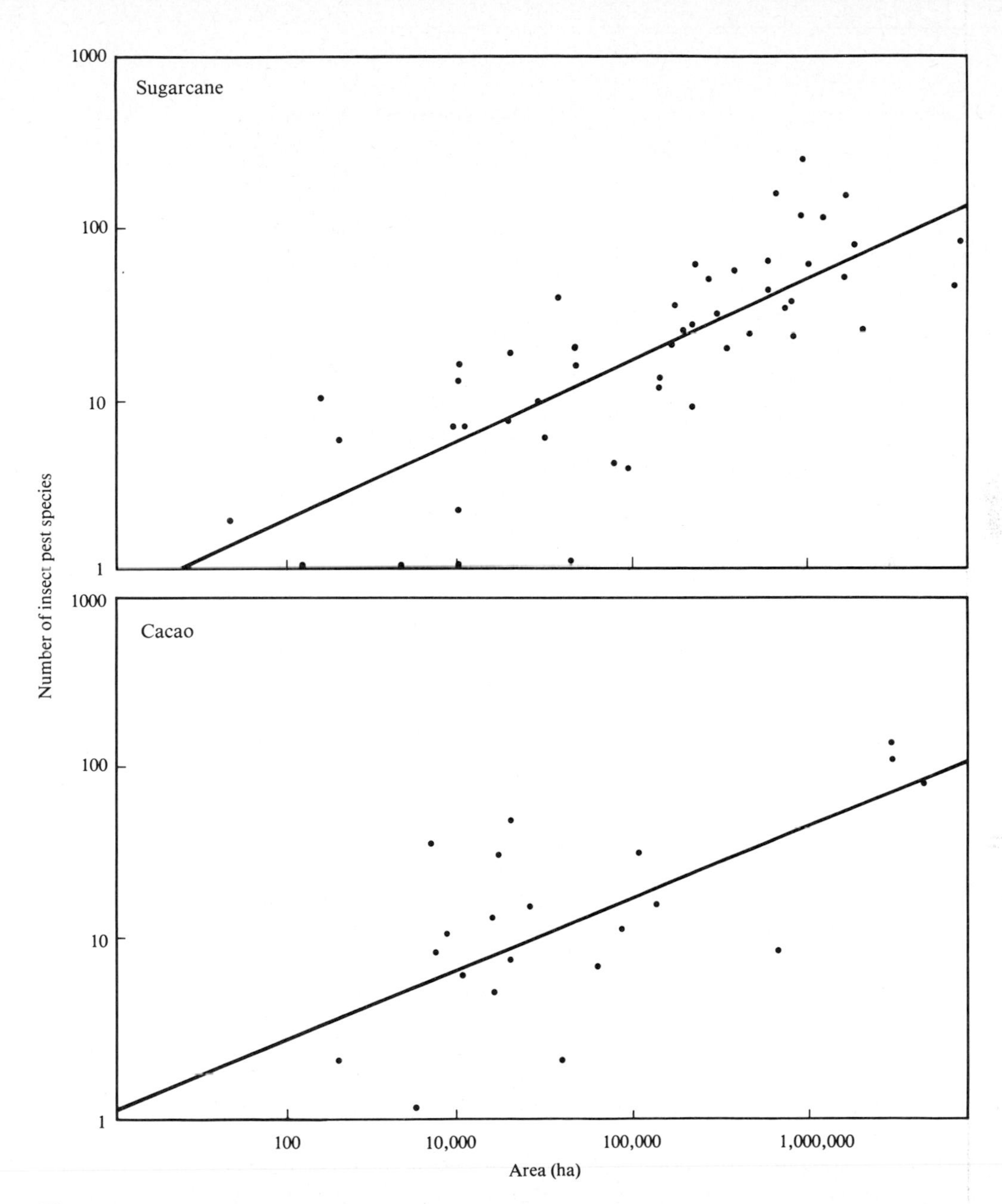

Figure 17.12
Relationship between the number of insect pest species of sugarcane *(above)* and cacao *(below)* and the area under cultivation for these crops. Note the similarity to species-area curves for islands, suggesting that insect species diversity has equilibrated with the total quantity of host plants available within a few centuries. (From D.R. Strong, E.D. McCoy, and J.R. Rey. Time and the number of herbivore species: the pests of sugarcane. Ecology 58[1]:167-175. Adapted by permission of Duke University Press. Copyright © 1977, Ecological Society of America.)

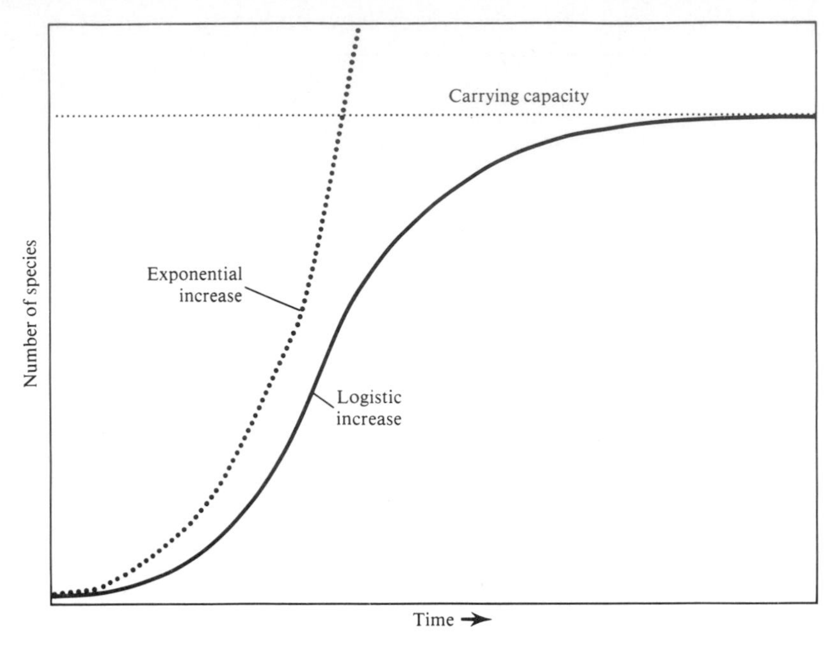

Figure 17.13
Graphic models of increase in species richness with time. Because speciation is a multiplicative branching process (analogous to reproduction of individuals in population growth), in the absence of environmental limiting factors the number of species should increase at an exponential rate *(dotted line)*. A more realistic model (analogous to the logistic model of population growth) suggests that the capacity of the environment to hold living things is limited, so the number of species increases in a sigmoidal fashion toward an equilibrium species diversity *(solid line)*.

terms of differences in speciation and extinction rates, suggesting that speciation rates must be higher or extinction rates lower in regions of high diversity such as the tropics. The rates of these processes are important, but the crucial question is how they change with the number of species already present (e.g., Rosenzweig, 1975). Stanley (1979) perceptively points out the analogy between speciation and extinction and birth and death processes so that the increase in species richness can be analogized to population growth. Because speciation is a branching, or multiplicative, process, in the absence of ecological limits the number of species should increase exponentially just as the number of individuals in a population does (Figure 17.13). However, investigators know that for growing populations some ecological resource eventually becomes scarce and growth levels off. The simplest model to describe this pattern is a sigmoid or logistic one in which population size asymptotically approaches an equilibrium, often referred to as its carrying capacity. This carrying capacity represents an equilibrium between opposing rates of birth and death for individuals or origination (speciation plus colonization) and extinction for species. One might expect that the same kinds of limited resources that ultimately limit the growth of populations would also limit the capacity of environments to hold species (Figure 17.13).

The crucial question for evaluating the relative importance of the historical and equilibrial process concerns the frequency and magnitude of historical perturbations relative to the proliferation of species. If environmental changes are

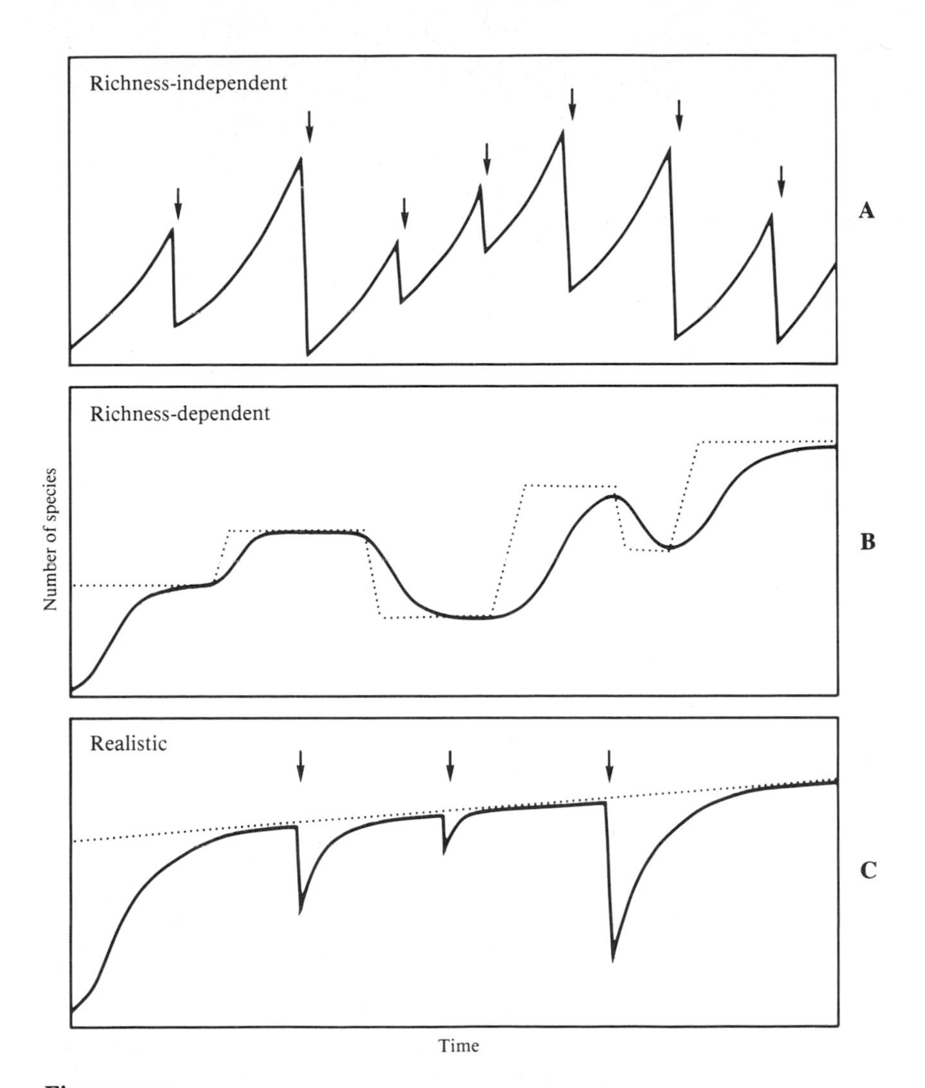

Figure 17.14
Three graphic representations of possible patterns of species richness over evolutionary time. **A,** Richness-independent model in which the number of species increases *(solid line)* exponentially but is periodically drastically reduced by major environmental perturbations *(arrows)*. **B,** Richness-dependent model in which diversity rapidly approaches some fixed carrying capacity *(dashed line)* of the environment for species and is only briefly affected by historical perturbations, here modeled as changes in carrying capacity. **C,** More realistic representation in which richness appears to be limited by a carrying capacity of the environment for species, but the historical perturbations periodically reduce species richness for substantial periods and the carrying capacity itself increases with time as species evolve.

sufficiently frequent and severe so that most of the time species numbers are increasing (or decreasing) almost exponentially, then clearly historical perturbations must be assigned a primary role in the determination of diversity patterns (Figure 17.14, *A*). If, on the other hand, environmental changes are relatively slight or infrequent, then one might expect that most of the time species richness will be approaching the carrying capacity of the environment for the species (Figure 17.14, *B*). Note that during the nearly exponential increase phase the rate of species increase will be a positive function of the number of species already present, and as the equilibrium point is approached it will be a decreasing function.

The actual temporal pattern of species diversity probably includes both exponential and environmentally limited phases (Figure 17.14, *C*). Life on earth has been decimated at times by catastrophic extinctions, but the overall patterns of diversity through time (Figures 17.9 to 17.11) suggest that the exponential increase in species following these catastrophies has been rapid. Most of the time the number of species has remained relatively constant, suggesting it is close to an equilibrium determined by environmental resources. It is unlikely that species richness has ever exactly attained equilibrial levels, however, for two reasons: first, because the environment is always changing and, second, because there has probably been a gradual improvement in the capacity of organisms collectively to use the resources available on earth. This is suggested by the gradual increase in diversity over time (Figures 17.9 to 17.11) and can probably be attributed to both individual and species selection operating on the phyletic lines that have persisted since the origin of life, increasing both the efficiency of individual organisms and the number of ways species can make a living and pack into ecological communities.

Productivity. It is probably not a coincidence that the regions and habitats that contain the greatest numbers of species are also highly productive. Productivity and biomass are much greater in tropical rain forests and coral reefs than in the arctic tundra and abyssal depths (Table 4.1). Although not all productive environments are rich in species, the general correspondence between geographic variation in productivity and the major latitudinal, elevational, aridity, and depth gradients of diversity is striking.

A causal relationship between productivity and diversity has been suggested by Hutchinson (1959), Connell and Orias (1964), MacArthur (1965, 1972), and others. MacArthur (1972) presented a graphic model of resource use that can be used to show the processes likely to be involved (Figure 17.15). Basically, this model suggests that more productive environments support more species because, on average, species can be more specialized and still maintain sufficiently large populations to avoid extinction. It also assumes implicitly that there is a trade-off between being specialized and generalized, so that specialists use their narrow range of resources more efficiently than generalists and thus are superior competitors in the overlapping portions of their niches. However, they are constrained to have smaller populations and hence are more susceptible to extinction than generalists. Because, other things being equal, the areas under the resource use curves in Figure 17.15 are proportional to population size, at an equilibrium between rates of origination and extinction a more productive environment would contain more species of specialists, which have about the same average population sizes as the fewer species of generalists in an environment in which resources are scarce.

Perhaps the best empirical support for this odel comes from geographic studies of rodent species diversity in the deserts of the southwestern United States. The number of species that coexist in local areas of structurally similar, relatively uniform habitat increases with annual precipitation (Figure 17.15, *A*) (Brown, 1973, 1975). In desert regions water is limiting; primary productivity and the production of seeds,

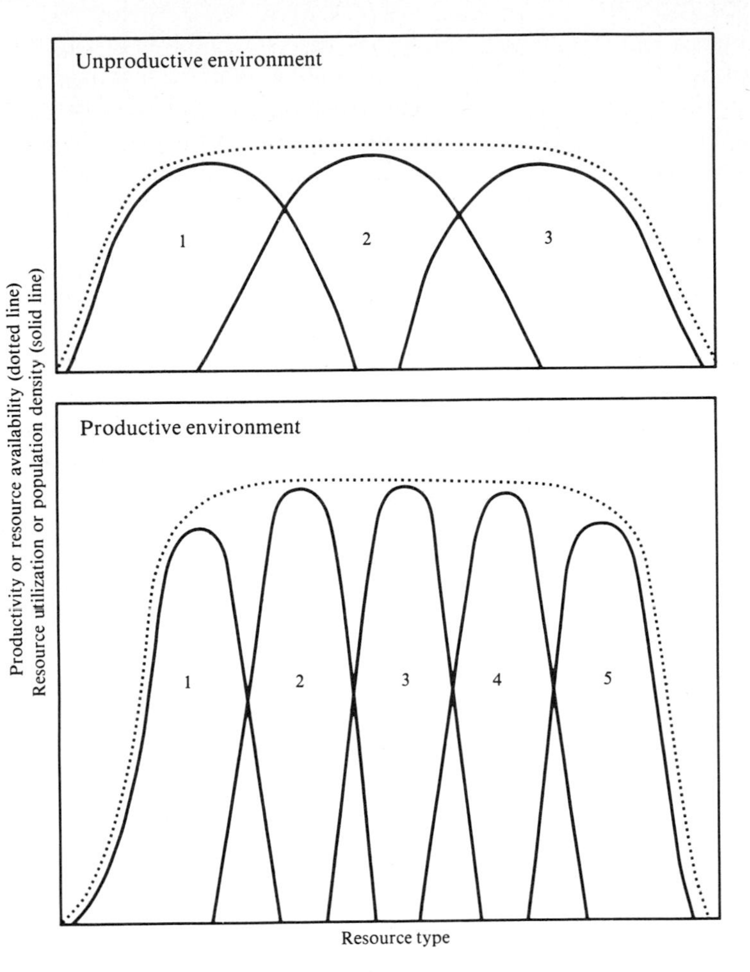

Figure 17.15
Graphic model for the effect of environmental productivity on equilibrial species diversity. This model assumes that species compete for a resource along a single niche dimension. Persistence of species in this community is related to population size (total area under each curve), which in turn depends on niche width or the degree of specialization. A more productive environment can support a greater number of more specialized species at equilibrium. (Greatly modified from MacArthur, 1972.)

the primary food of desert rodents, increase directly with increasing precipitation. A variety of evidence, including morphology, habitat utilization, geographic distribution, and habitat and dietary shifts in wide-ranging species, suggests that specialization of the individual species increases with the number of coexisting species (Brown, 1975; Bowers, in preparation). Furthermore, Abramsky (1978) found that a specialized seed-eating species successfully colonized an area in which he added supplemental seeds. The general, qualitative relationship between specialization, diversity, and productivity is supported by the numerous examples of incredibly specialized rare species from tropical communities. For example, Vermeij (1978) notes that the elaborate morphological adaptations of marine molluscs are much more developed in the rich tropical faunas.

Considerations of limiting factors and community energetics developed in Chapters 3 and 4 bolster the reasoning developed above (see Hutchinson, 1959; Van Valen, 1973b; Brown, 1981). Usable energy is the primary limiting resource required to sustain life. It seems reasonable that the more energy is available, the more species can subdivide the supply, with each still acquiring a sufficient quantity to maintain a population size that will have a low probability of extinction. Thus one might expect close correlations between species diversity and rates of energy supply: rates of solar energy input for plants, rates of primary production for heterotrophic organisms, and rates of food availability for particular animal groups. However, it must be borne in mind that other limiting factors influence the extent to which different kinds of organisms can use various energy sources. Much of the sunlight in desert ecosystems cannot be used by plants because of water shortage, and soil nutrients limit primary production in some other habitats. Similarly, nitrogen and other nutritional requirements limit the capacities of many heterotrophs to use apparently suitable food resources.

Although the logic that predicts a positive relationship at equilibrium between productivity and diversity seems irrefutable, the productivity hypothesis has not been generally accepted as an important cause of geographic patterns of species richness. In part this may be attributed to the practical difficulties in measuring available energy for particular groups of organisms. In part, however, it is because of the obvious exceptions. Small, shallow eutrophic lakes have fewer species of zooplankton and probably of other groups than large, deep oligotrophic lakes. Salt marshes and hot springs are among the most productive ecosystems, but they contain many individuals and a high biomass of a very few species. Clearly, productivity alone is an inadequate explanation for species richness.

Harshness. Many of the habitats that support a small number of species are ones that one would intuitively think of as harsh, that is, having conditions that are inimical for life in general. A list of such environments would include not only unproductive habitats such as deserts, arctic tundra, caves, and abyssal depths but also highly productive areas such as salt marshes, estuaries, hot springs, and temporary ponds. Is it possible to make the concept of harshness sufficiently precise that it becomes a useful tool for explaining patterns of species richness?

Terborgh (1973b; see also Brown, 1981) suggests that harshness, and its opposite, favorableness, be used to reflect relative rates of origination and extinction in different habitats or geographic regions. By this definition harsh environments would be those for which colonization and speciation rates are low or extinction rates are high, or both. Several characteristics of environments determine these rates and hence their relative harshness. Geographic isolation and physical conditions very different from surrounding areas reduce the rate of successful colonization. Small size and ephemeral existence of habitats increase extinction rates and concommitantly, if extinction rates are high enough to prevent evolutionary differentiations, also reduce speciation rates.

To better understand this concept of harshness, consider the example of hypersaline lakes. A few landlocked bodies of water in desert regions, such as the Great Salt Lake and the Dead Sea, are several times more concentrated than seawater. As mentioned in Chapter 3, only a few species occur in these unusual habitats, and these, such as the brine shrimp *(Artemia)* and an alga *(Dunaliella)*, tend to have wide distributions, occurring in many if not all of the widely scattered hypersaline lakes around the world. It is not difficult to imagine why this low diversity persists. The lakes are difficult to colonize, not only because they are small and isolated but also because most potential colonists cannot tolerate the high salt concentrations. Extinction rates must be high, not only because the lakes are small but also because many of them have periodically been much less concentrated, even freshwater lakes, during the pluvial periods of the Pleistocene. Most of the inhabitants, including *Artemia* and *Dunaliella,* are relatives of marine taxa that have specialized to occupy temporarily landlocked hypersaline pools and embayments along seacoasts. The fact that some species can inhabit these environments suggests that the harshness cannot be attributed to high salinity per se. If all the oceans of the world were as saline as the Great Salt Lake, and they had become concentrated gradually over hundreds of millions of years, then it would seem reasonable to assume that organisms would have adapted and radiated in these

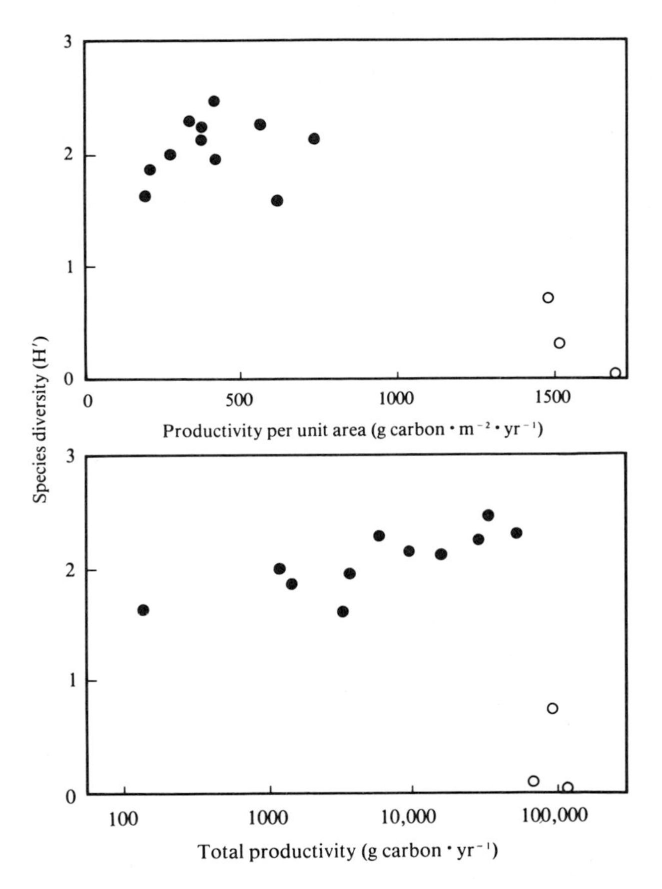

Figure 17.16
Relationship between species diversity, as measured by one of the community-used diversity indexes (H′), of chydroid cladocerans (a kind of zooplankton) and primary productivity in 14 lakes in Indiana. Open circles show three heavily polluted lakes. *Above,* With productivity expressed on a per unit area basis (grams of carbon per square meter per year) there appears to be a negative relationship, especially if the polluted lakes are counted. *Below,* With productivity expressed on a per lake basis (grams of carbon per entire lake per year) there is a significant positive relationship for the unpolluted lakes, although the polluted lakes still have low diversity. Thus these data support, rather than contradict, the productivity hypothesis if one assumes that it is the availability of energy within entire lakes that determines their capacity to support zooplankton species. (Data from Whiteside and Harmsworth, 1967, plotted by D.H. Wright.)

seas pretty much as they have in our present oceans.

Used in this way, the concept of harshness, together with productivity, promises to provide a potentially powerful equilibrial explanation for many contemporary patterns of diversity (Brown, 1981). Most habitats with high species richness are both productive and favorable (the opposite of harsh), whereas habitats with low diversity are either unproductive (e.g., deserts, tundra, the abyssal depths), harsh (e.g., salt marshes, hot springs, and temporary ponds), or both (e.g., caves and alpine lakes). Both the productivity and harshness hypotheses are visualized as exerting their effects primarily by influencing equilibrial population sizes and thus affecting the balance between extinction and origination rates that determine the carrying capacity of environments for species. These mechanisms may even help explain some of the more perplexing patterns, such as the high diversities of plankton in the central north Pacific and of benthic invertebrates at intermediate ocean depths on the continental slopes. Neither of these environments is highly productive, although the latter is more productive than the deeper abyssal habitats. However, both are extremely extensive habitats. Although the local densities of most species populations are low, the species have wide geographic ranges and large total population sizes. Similarly, the inverse relationship between the diversity of zooplankton and the productivity of temperate lakes is resolved if one supposes that the critical factor is the total population size of each species and that this is determined by the total productivity of the lake rather than by the productivity per unit surface area. If total productivity is calculated by multiplying measured productivity per unit of surface area by the surface area of the lakes, the relationship between diversity and productivity becomes positive rather than negative (Figure 17.16).

Climatic stability. Some investigators have suggested that short-term, especially seasonal, climatic fluctuations at high latitudes are a primary cause of low diversity. There are two problems with this as a general explanation. First, there are numerous exceptions. Many regions with low species richness, such as tropical mountaintops and the abyssal depths, do not experience highly seasonal physical environments. Many habitats with high diversity also experience seasonal climates. For example, as pointed out in Chapters 2 and 4, most tropical regions have highly seasonal rainfall regimes. Often the dry season is so severe that the areas support deciduous forest or savanna vegetation rather than evergreen tropical forest. Although these habitats are typically less productive and less rich in species than rain forests, they are more productive and diverse in species than most temperate and arctic regions. Second, there is that old problem: if some species can adapt to the climate by becoming dormant during unfavorable times or by other means, why cannot other species invade and why cannot the present inhabitants speciate to generate greater diversity?

Habitat heterogeneity. In a pioneering study that had great influence on community ecology, MacArthur and MacArthur (1961) showed that the number of bird species in different habitats in the eastern United States was highly correlated with a measure of the number of vertical layers of vegetation present, which they called foliage height diversity (Figure 17.17). Since then many other close correlations between species richness and the variety of vegetation structure have been documented, not only for birds in a wide range of habitats (e.g., MacArthur et al., 1966; Recher, 1969; Cody, 1968, 1974, 1975) but also for a variety of other organisms in different habitats, e.g., desert rodents (Rosenzweig and Winakur, 1969) and desert lizards (Pianka, 1967).

It is important to ask whether these patterns reflect an important causal role of structural habitat heterogeneity in promoting species richness or whether they reflect correlations between habitat structure and some third factor, such as historical perturbation, productivity, or

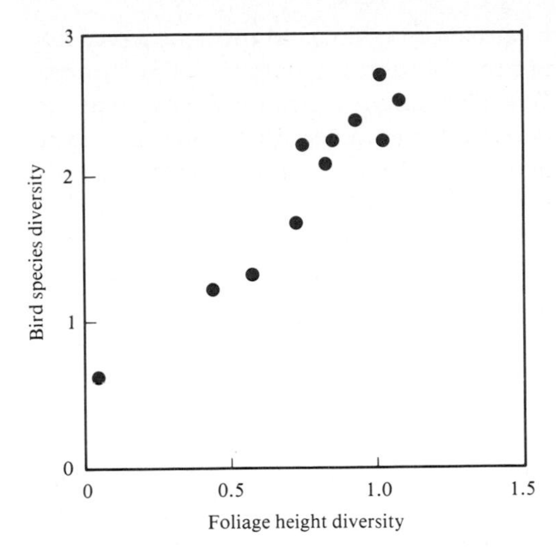

Figure 17.17
Relationship between bird species diversity as measured by the diversity index, H′, and foliage height diversity, the number of equally dense layers of vegetation at different heights above the ground, in deciduous forest areas in eastern North America. This correlation between bird species richness and vegetation structure holds well within regions, but differences in the relationship appear when regions of different habitat and climate are compared. (From R.H. MacArthur and J.W. MacArthur. On bird species diversity. Ecology **42**[3]:594-598, 1961.)

harshness. For many groups it is clear that the relationship holds and that it reflects tendencies of coexisting species to apportion essential resources, such as food, along niche dimensions of habitat structure, such as foliage layers in birds (e.g., MacArthur, 1958; Cody, 1974) and distance from shrub cover in desert rodents (e.g., Rosenzweig and Winakur, 1969; Brown, 1975). This is not good evidence, however, that physical habitat heterogeneity by itself promotes species richness. Troublesome exceptions to the correlations suggest that causal relationships be viewed with caution. For example, MacArthur and his associates have shown that the general relationship between bird species diversity and foliage height diversity holds for a variety of mesic habitats in North America from Panama to the northeastern United States (MacArthur et al., 1966). However, productivity, although it was not measured, also varies in concert with vegetation heterogeneity; tall, multilayered tropical rain forests are more productive than temperate forests, which in turn are more productive than temperate old fields and shrublands. In relatively arid regions of the western United States, Mexico, and Central America the pine forests have tall but widely spaced trees and sparse understory vegetation. These pine forests have comparable foliage height diversity to eastern deciduous forests, but substantially lower bird species diversity and probably also lower productivity (E.W. Stiles and G.H. Orians, personal communication).

Particularly troubling are some aquatic organisms, such as the benthic marine invertebrates of the continental slope and the freshwater and marine plankton. These attain extremely high diversity in habitats that seemingly lack much in the way of structural heterogeneity. Hutchinson (1961) was so impressed by the numbers of plankton species coexisting in seemingly uniform habitats that he termed this "the paradox of the plankton." Recent oceanographic and limnological studies indicate that in fact oceans and even lakes are by no means uniform water masses (e.g., Wiebe, 1976, 1982), but it is not yet clear to what extent the spatial heterogeneity that is being revealed on different scales can resolve Hutchinson's paradox.

Competition, predation, and mutualism. It is often suggested that the high species diversity of the tropics can be attributed to the fact that one or more of the classes of interspecific interactions are more important at low than at high latitudes. One often hears that the intensity of interspecific competition, the relative effect of inter- compared to intraspecific competition, the intensity of predation, the complexity of food webs, or the development of mutualistic relationships is greater in the tropics and perhaps also in other species-rich regions (e.g., Dobzhansky, 1950; Paine, 1966; Janzen, 1970; Connell, 1975; Menge and Sutherland, 1976). Some or all of these assertions may be true. As the number of species increases, from purely statistical considerations one would expect there to be more frequent interspecific interactions and more opportunities to coevolve special relationships with other species. As in the case of habitat heterogeneity, the real challenge is to distinguish the correlates and consequences of diversity from its primary causes.

In one sense this is not difficult. Ultimately, all general patterns of diversity must be attributed to physical causes, either historical perturbations or contemporary variation in the physical environment. To understand this, perform a thought experiment. Freeze the climate and geology of the earth in its present condition for as many millions of years as necessary for species richness in all regions to approach closely to an equilibrium. One of two outcomes is possible: either all areas will have the same diversity and one can unreservedly accept the historical perturbation hypothesis for present patterns, or species richness will vary with variation in the physical environment, perhaps continuing to exhibit patterns similar to those observed today and thus supporting an equilibrial hypothesis. In either case the relationships of species to each other must reflect the more basic constraints set by the relationships between the organisms and their physical environment. In this sense, hypotheses that invoke biotic interactions must always be at least secondary explanations for diversity patterns.

This does not mean, however, that interspecific interactions play no role in producing the geographic patterns in diversity. Indeed, competition, predation, and mutualism determine how the physical resources of a region are allocated among species to result in the observed species diversity and community structure (MacArthur, 1958, 1972; Brooks and Dodson, 1965; Paine, 1966; Harper, 1969; Cody, 1974, 1975; Brown, 1975, 1981; Connell, 1975, 1978; Menge and Sutherland, 1976; Lubchenco, 1978, 1980; Lubchenco and Menge, 1978; Vermeij, 1978). For reasons mentioned above and in Chapter 4, the most important of these resources is solar energy, either in its original form or in its converted form as the organic molecules that constitute the bodies of living and dead organisms. In one sense, all organisms can be viewed as competing for a limited supply of this usable energy; some (green plants) obtain their share directly from sunlight, whereas others (heterotrophs) obtain it from other organisms that ultimately acquire it from the sun.

A detailed consideration of the many interacting effects of competition, predation, and mutualism on the organization of natural communities is beyond the scope of this book. This is an area of extremely active research and frequent controversy in contemporary ecology and we cannot do justice to the complex issues and numerous patterns and processes here. However, the ecological roles of species are reflected not only in overall patterns of species richness but also in the geographic distributions of particular ecological and taxonomic groups of organisms. This is the subject of the next chapter.

Selected references

Measurement and terminology

Cody (1975); Osman and Whitlatch (1978); Pielou (1975); L. Taylor (1978); Whittaker (1977).

Latitudinal diversity gradients

Amerson (1975); Anderson and Brown (1980); Barbour and Brown (1974); Braun (1950); Brunig (1973); Campbell and van der Meulen (1980); Cook (1969); Federov (1966); Fischer (1960); T. Fleming (1973); Glenn-Lewin (1977); Hespenheide (1978); M. Horn and Allen (1978); Hubbell (1979); Janzen (1970, 1981); Jarvinen (1979); Kiester (1971); McCoy and Connor (1980); Monk (1967); Orians (1969); Owen and Owen (1974); Parsons and Cameron (1974); Pianka (1966, 1978); Pianka and Schall (1978); Richards (1957); Sanders (1968, 1969); G. Simpson (1964); Stehli (1968); Stehli et al. (1969); Stout and Vandermeer (1975); J. Taylor and Taylor (1977); R. Taylor and Regal (1978); Tramer (1974); J.T. Turner (1980); Wallace (1878).

Diversity in gradients of elevation and aridity

Beals (1969); J. Brown (1973, 1975); J. Brown and Davidson (1977); Dressler (1981); Huey (1978); Kikkawa and Williams (1971); Terborgh (1977); Whittaker (1960, 1977); Yoda (1967).

Diversity in aquatic environments

Abele (1974, 1976); Abele and Walters (1979); Barbour and Brown (1974); Flessa and Imbrie (1973); Fryer and Iles (1972); Funnell (1971); Hayward and McGowan (1979); McGowan (1974); McGowan and Walker (1979); Menge and Sutherland (1976); Rex (1981); Stanley (1979); Stehli and Wells (1971); Van der Spoel and Pierrot-Bults (1979); Vermeij (1978); Vinogradova (1959, 1962); Whiteside and Harmsworth (1967).

Explanations for the patterns

Abramsky (1978); Boucot (1975a, 1975b); J. Brown (1981); Connell (1975, 1978); Dobzhansky (1950); Fischer (1960); Fischer and Arthur (1977); Flessa (1975); Flessa and Sepkoski (1978); Goodman (1975); J. Harper (1969); G. Hutchinson (1959, 1961); Janzen (1970); Järvinen (1979); M. Johnson et al. (1968); M. Johnson and Raven (1970); Karr and Roth (1971); Krebs (1978); Lillegraven (1972); Lubchenco (1978, 1980); Lubchenco and Menge (1978); MacArthur (1969, 1972); MacArthur and MacArthur (1961); MacArthur et al. (1966); Mark and Flessa (1977); Menge and Sutherland (1976); Nicklas et al. (1980); Osman (1977, 1978); Paine (1966); Pianka (1966, 1967, 1978); Recher (1969); Rosenzweig (1975); Rosenzweig and Winakur (1968, 1969); Sanders (1968, 1969); Schall and Pianka (1978); T. Schopf (1974); T. Schopf et al. (1978); Sepkoski (1976); Sepkoski et al. (1981); Simberloff (1974b); Slobodkin and Sanders (1969); Stanley (1979); Stehli et al. (1969); Strong (1974, 1977); Strong et al. (1977); Van Valen (1973b); Vermeij (1978); Whittaker (1969, 1977).

Chapter 18

Interactions of History and Ecology in the Distribution of Species and Biotas

In the last three chapters we examined patterns in the diversity of species on a variety of scales from small, isolated patches of habitat to vast oceans and continents. We also evaluated the hypotheses proposed to account for these patterns of species richness. None of these explanations can be truly complete, however, unless it can also account for the characteristics of those species that occur together in one place and explain the changes in species composition from place to place and time to time. This is one of the ultimate problems of biogeography, and it should be obvious by now that the problem remains largely unsolved. Even to try to synthesize what is known, however, is a major challenge. It requires us to bring together and interrelate most of the ideas and information presented in this book: the geologic and climatic history of the earth, the mechanisms of ecological interaction and evolutionary change, and the present and past distributions of different taxonomic and ecological groups of organisms. We attempt here to synthesize briefly some of this information in the hope that it will impress the reader that the different approaches to biogeography do complement each other. When properly integrated, the seemingly divergent techniques, data, and concepts suggest interesting interpretations of existing information and focus attention on important questions that investigators hope to answer in the future.

Patterns and randomness

Such an optimistic outlook is, of course, predicated on the assumption that there are general patterns to be discovered and interpreted and that general mechanisms can be found to account for them. The principal goal of science is to uncover patterns. To the extent that history has been a series of unique events and organisms are the products of capricious, unpredictable histories, there may be a relatively limited number of general patterns to discover and explain. Therefore one of the major problems in both biogeography and community ecology is to distinguish patterns from randomness and to understand how stochastic processes and deterministic biological (or physical) mechanisms interact to influence the diversity, abundance, and distribution of organisms.

This problem is complicated by the fact that most biological and physical processes are neither completely stochastic nor entirely deterministic. The degree of apparent randomness is often largely a matter of spatial or temporal scale. Consider, for example, the case of dispersal. There often is a large random component to the distance traveled by a single passively dispersing individual. Thus if one marked a single maple seed before it dehisced, one could not predict with accuracy how far from the parent tree it would travel. On the other hand, if one marked many seeds on trees in a large area for several years and followed their fates, one could quite precisely characterize the distribution of their dispersal distances in terms of such statistics as the mean, variance, skewness, and so on. Furthermore, additional analyses might reveal that this distribution changes in regular ways with the direction of prevailing winds, local topography, density of surrounding vegetation, and other factors. If such distributions were determined for several species, one could make reasonable but falsifiable predictions about the probability of seeds reaching a distant site and hence about the relative rates of colonization of a cleared site or the relative frequencies of the different species on oceanic or habitat islands. Finally, we should note that if the probability of dispersing over a long distance is very low, one might expect on these grounds that the occurrence of the species in widely separated regions would be highly unpredictable. (Of course there might be other historical reasons why the species would or would not occur in these distant places.) We conclude that there is a significant stochastic element to dispersal and that the same process can result in either highly random or quite predictable outcomes depending on the temporal and spatial scales on which it is studied.

The same example of a maple can be used to illustrate how historical events can have mixed random and nonrandom properties. Maples may be living in a forest biome consisting of some species that shared a distributional history and had ranges that were subdivided by identical barriers, producing vicariance. However, in the same forest occur plant and animal taxa that have immigrated in more recent time. Therefore each biota is an assemblage that has been accumulated over geologic time. It is the goal of historical biogeographers to discriminate between the unique and seemingly stochastic histories of individual taxa and the coincident histories of others.

Recently several investigators have emphasized the value of testing biogeographic distributions, species compositions of communities, and phylogenies of taxonomic groups against null models that invoke stochastic processes alone to account for observed distributions (e.g., Raup et al., 1973; Raup and Gould, 1974; Simberloff, 1974a, 1978; Connor and Simberloff, 1978, 1979; Strong et al., 1979). In biogeography this approach has thus far been applied primarily to insular distributions because the small size and the discrete nature of insular biotas makes them convenient for analysis (see above references and Chapter 16). In principle, however, these methods are equally applicable to continental and marine biotas, and they should be particularly valuable for sifting through the diversity and complexity to separate the random features from the deterministic patterns that call for causal explanations (Brooks, 1981).

Ultimately, all biogeographers share a common goal: to describe and explain the distribution of organisms. Despite important differences in philosophy and methodology between historical and ecological biogeographers, they both use many of the same principles and procedures. For example, investigators building and testing vicariance models and their colleagues evaluating the equilibrium theory of island biogeography share the fundamental assumption that similar distribution patterns in different taxonomic groups reflect general causal processes that can eventually be elucidated by the continued interplay of theories and data.

The examples that we discuss in the remainder of this chapter indicate the kinds of problems for which both historical and ecological explanations are possible. Sometimes these explanations can be tested against each other as alternative hypotheses, although there are as yet only a few cases in which such rigorous, quantitative analyses have been attempted. We suspect that combined historical and ecological processes will have to be invoked more frequently to account for the complexities of the real distributions. After all, the distributions of all organisms reflect the influences of both historical events and current environmental conditions, but it is not easy to understand how these processes have combined to determine the range of even a single species or group of closely related forms.

Distributions of closely related species

Overlaps among congeners. Are there general patterns in the distribution of closely related species? Specifically, do close relatives, such as congeners, overlap either more or less in their ranges than one might expect on the basis of chance? There are good a priori reasons for predicting either outcome. On the one hand, closely related forms are likely to occur in the same general region because they share a fairly recent common ancestor, possess similar dispersal capabilities, and have similar requirements for environmental conditions and resources. On the other hand, if species have originated by allopatric speciation, one might expect the vicariant pattern to be preserved, and if they are too similar in their requirements one would expect Gause's principle of competitive exclusion to preclude local coexistence. It is possible to determine which of these patterns predominates and to infer which of the processes has been more important by quantitative analyses of the distributions of well-studied taxa.

It is simplest to consider the overlap of ranges of congeneric species. Pielou (1978) analyzed the distribution of algae along the coasts of North and South America. Within each of three groups, Rhodophyta (red algae), Phaeophyta (brown algae), and Chlorophyta (green algae) on both Atlantic and Pacific coasts, she considered all pairs of species. Each pair could be scored as to whether it was congeneric or more distantly related and whether its distribution was overlapping or disjunct. Results for one group, the Rhodophyta on the Pacific coasts, are as follows.

	Overlapping ranges	Disjunct ranges
Species pairs		
Congeneric	21	16
More distantly related	1175	1489

Statistical analyses of these data provide no grounds for rejecting a null hypothesis of independent (i.e., random) distribution of these species, but there appears to be a qualitative tendency for congeneric pairs to overlap more than distantly related ones. The probability that this could be owing to chance is high ($p \cong 0.2$). This same trend is present in each of the six data sets Pielou examined. Because this in itself is statistically unlikely (< 0.02), Pielou concluded that closely related species tend to overlap in their ranges and that the interspecific competition does not tend to result in mutually exclusive geographic ranges.

Simberloff (1970) used somewhat different methods to analyze the coexistence of congeneric pairs of bird and plant species on islands. He determined the number of species per genus on both islands and on the mainland source area from which the insular biota has been derived. As pointed out in Chapter 16, one expects islands to have fewer species per genus than the mainland because they have fewer species. Simberloff used computer simulations to determine whether the observed number of island congeners was less than the number expected by chance. Like Pielou, he found that the observed distributions were difficult to distinguish from the null hypotheses of random coexistence of congeners, but overall there was a tendency for congeners to occur together more frequently than expected on the basis of chance. Differences in dispersal ability or extinction rates among genera may explain, at least in part, the cooccurrence of species in certain genera on oceanic islands. One also finds highly overlapping distributions of certain congeners on oceanic islands, especially in groups such as the Hawaiian *Drosophila,* Darwin's finches of the Galápagos, and the *Anolis* lizards of the West Indies (see Chapters 6, 12, and 16), where there has been much endemic speciation within archipelagos and even within individual islands.

The fact that these two studies, performed independently using different techniques, give remarkably similar results for widely divergent groups of organisms is significant. They suggest that similar dispersal capabilities and ecological requirements, which are the consequences of shared evolutionary histories, tend to result in overlapping distributions of closely related species. It is important that the influence of neither allopatric speciation nor competitive exclusion is apparent at this level of analysis. We think it is incorrect to conclude that the effects of vicariance and interspecific competition often cannot be detected in the distributions of closely related species. If, however, the influences of such processes on patterns of species distribution are to be demonstrated unequivocally, then more sophisticated analyses will be required. One problem with examining the coexistence of congeners is that the genus is an arbitrary level of classification. Species within a genus are likely to vary considerably in the time since they have split off from a common ancestor. Consequently, they may also be expected to differ in the extent to which they have diverged ecologically and shifted their ranges. This is particularly true of the many species in large genera that are likely to contribute differentially to quantitative statistical analyses.

Avian systematists have long noted that the most closely related species (e.g., sibling species and members of the same subgenus) typically are morphologically very similar and occur in the same geographic area but are separated by habitat and often in mountainous regions by elevation. More distantly related species, even congeners in different species groups or subgenera, are more likely to coexist in the same habitat but to differ in diet and in related traits such as body size and bill size. Similar patterns probably occur in some mammals and lizards. By lumping all congeners together and considering distribution only on a geographic scale, analyses such as those of Pielou and Simberloff are unlikely to detect differences in local and elevational distribution that may reflect the influence of competition on habitat utilization.

Segregation of competitors. Several recent statistical studies suggest that interspecific competition does indeed limit the overlap of both local and geographic ranges, as might be expected from our relatively qualitative treatment in Chapter 3. For example, Bowers and Brown (in preparation) analyzed the distributions of pairs of rodent species in the North American deserts. They used an analysis similar to Pielou's, but instead of scoring species pairs on the basis of whether they were in the same or different genera, they classified them on the basis of similarities in body size and diet. This was based on the assumption that size and diet, rather than taxonomic affinity, provide a better estimate of the likelihood that two species compete. Bowers and Brown found that similar-sized (body weight ratios < 1.5) species of seed-eating rodents occurred together in local habitats and overlapped in their geographic ranges much less frequently than expected on the basis of a null hypothesis of independent distribution (Figure 18.1). In contrast, the cooccurrence of rodent species of different sizes or different diets could not be distinguished from random.

In a somewhat similar study, Brown and Bowers (in preparation) analyzed the co-occurrence of hummingbird species on the islands of the Greater and Lesser Antilles. They found that species of dissimilar morphology are much more likely to be found together on the same island (Table 16.5). Of the traits that they examined, bill length, which of course plays a primary role in the ability of hummingbirds to obtain nectar from different kinds of flowers, is particularly important in the geographic distributions and community composition of hummingbirds. Both the desert rodent and island

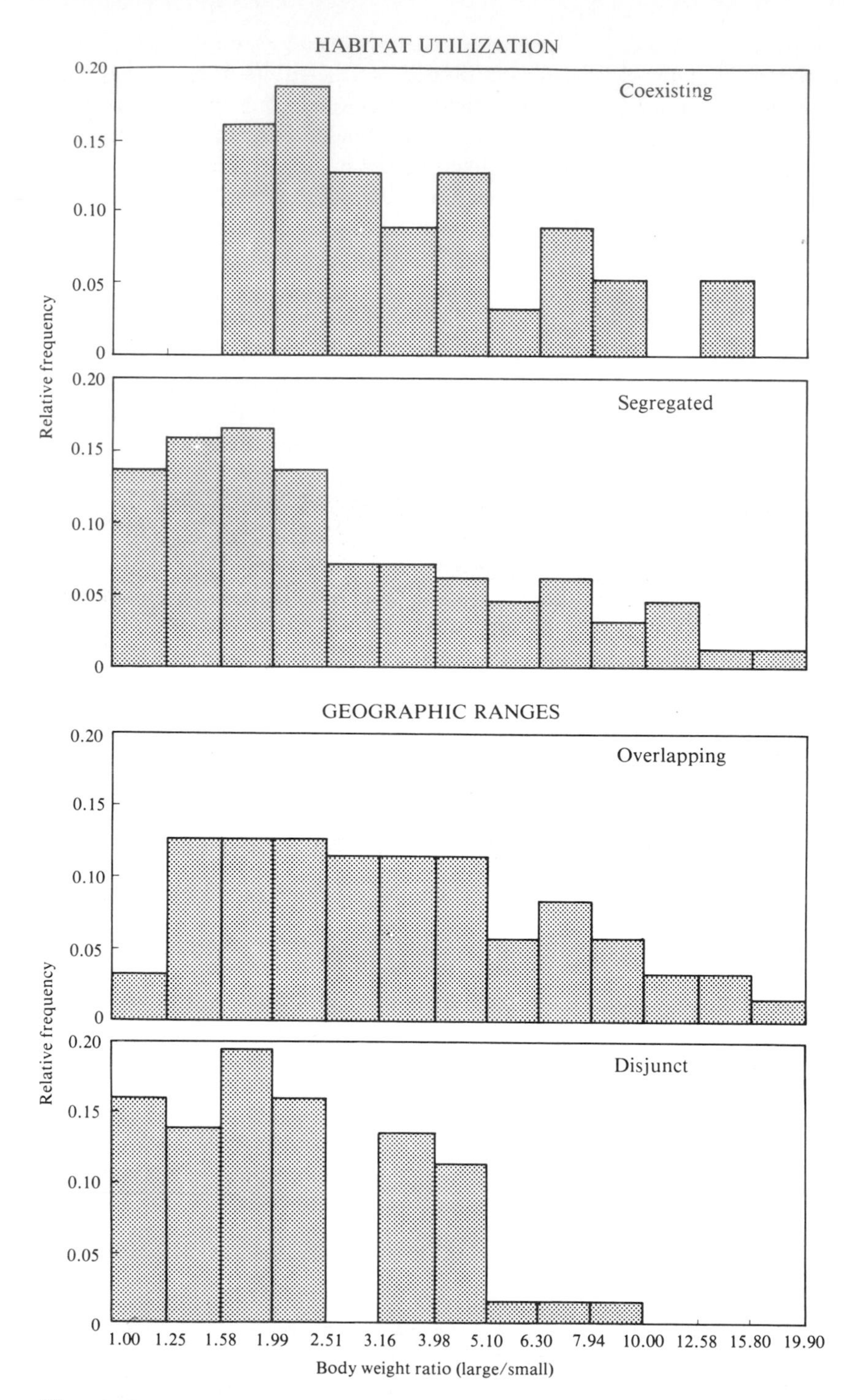

Figure 18.1
For legend see opposite page.

hummingbird studies suggest that interspecific competition can play a significant role in limiting the distributions of ecologically similar, closely related species. The alternative explanation for the observed patterns, namely that the most similar pairs of species are the vicariant products of recent allopatric speciation, appears to be unlikely in these cases because the species most similar in diet and related morphological traits are not necessarily the most closely related (see also Case, Faaborg, and Sidell, 1982).

It should be possible to use somewhat similar analyses to test whether the effects of allopatric speciation can be detected in the distribution of closely related species. One would predict that recently derived sister species would have ranges that overlapped less than those of species that are the products of a more ancient splitting of phyletic lineages. Unfortunately, we know of no direct tests of this prediction. This seems somewhat surprising in view of the fact that it is one of the basic conclusions of vicariance theory, namely that most speciation is allopatric and that the history of speciation events is reflected in the geographic distribution of the resulting species. Closely related (at least congeneric) species often exhibit random or even clumped distributions, and pairs of species with nonoverlapping ranges tend to be ecologically similar but not necessarily particularly closely related. This is, however, not very convincing evidence because the tests were not designed to test directly for vicariant patterns. Such tests should not be difficult to perform on groups whose phylogenies and distribution are well known, such as internal parasites (Brooks, 1981; Brooks et al., 1981). For example, among the Hawaiian *Drosophila* is there any significant tendency for the most closely related species to occur on different islands or in different areas within islands?

Distributions of distantly related species

Boundaries between biogeographic regions. At higher taxonomic levels, such as families and orders, it appears that related species generally are clumped within certain geographic regions and that there is often much overlap in their ranges. As pointed out in Chapter 8, this pattern provides the basis for recognizing biogeographic provinces containing similar biotas. Although the early biogeographers recognized these patterns and defined provinces somewhat subjectively, modern quantitative methods can now be used to assess compositional (taxonomic) similarities among biotas (e.g., Connor and Simberloff, 1978; Flessa and Miyazaki, 1978; Flessa et al., 1979; Flessa, 1981) and to delineate biogeographic provinces (e.g., Kikkawa and Pearse, 1969). Given the magnitude of the task of attempting to perform such quantitative analyses of all major groups, it is encouraging that the more objective methods generally corroborate the findings of their predecessors. The highly clumped distributions of species and genera within many families and orders presumably reflects similar ecological requirements

Figure 18.1
Frequencies of cooccurrence of pairs of seed-eating desert rodent species as a function of the ratios of their body sizes is analyzed on two spatial scales: coexistence within local habitats and overlap of geographic ranges. In each case species pairs are divided into two groups, those that occur together more frequently and less frequently than expected on the basis of chance. Note that similar-sized species (with small body size ratios) rarely coexist frequently in the same habitat or overlap extensively in their geographic ranges, suggesting that interspecific competition prevents cooccurrence. (After Bowers and Brown, in preparation.)

and limits to dispersal, which in turn reflect their common origins.

The long histories of particular taxa within restricted geographic regions have resulted in the somewhat independent evolution of integrated biotas in different parts of the world. It is the distinctiveness of these assemblages that investigators recognize when they delineate biogeographic provinces. The biotas come into contact and interact at the boundaries of the provinces. Ecological processes largely determine the outcome of these interactions: whether the biotas remain distinct or intermingle, or whether one replaces another. One can get some idea of the nature and stability of the present boundaries by examining the characteristics of species in the different regions and by observing the fate of introduced species.

Two characteristics of the boundaries between biogeographic regions are worth noting. First, their locations usually correspond to the sites of natural geographic barriers to dispersal. For example, Figure 1.3 immediately reveals that most of the terrestrial zoogeographic realms are isolated from each other by substantial water gaps and that others are separated by narrow isthmuses, harsh deserts, and high mountains. Flessa (1981) showed that the distinctiveness of the terrestrial mammalian biotas of the classic biogeographic realms, as measured by shared and unshared genera and families, is highly correlated with their present separation by distance (Figure 18.2). Thus the geographic ranges of mammalian taxa, which evolved relatively recently in geologic time, reflect the present distribution of continents; the geographic features that separate the landmasses, especially large expanses of ocean, have effectively reduced interchange among biotas.

The second important feature of the boundaries between biogeographic regions is that their nature and location have changed over

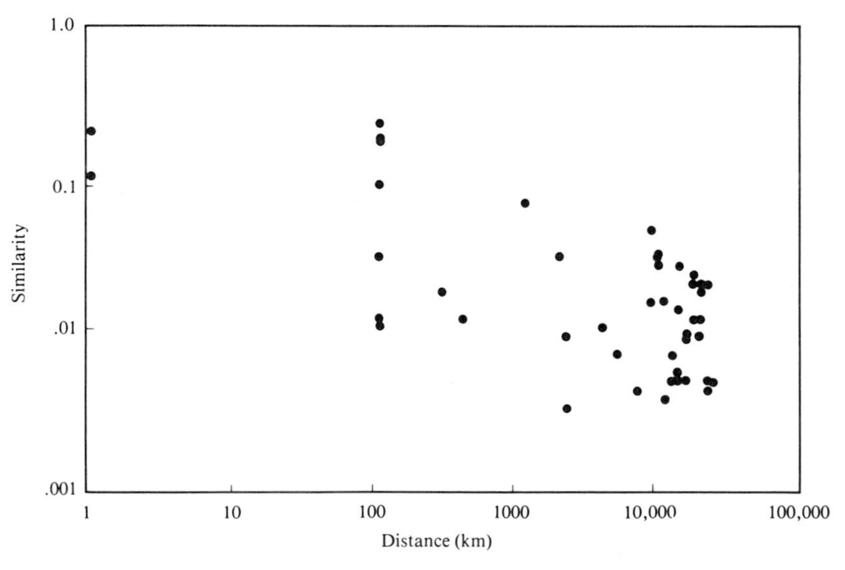

Figure 18.2
Relationship between faunal similarity (measured by the Jaccard index using genera) of terrestrial mammals and the overland distance separating pairs of biogeographic regions. Note that the proportion of the biota shared between regions decreases with increasing distance separating them. Thus the overall distribution of mammalian taxa is closely related to the present geography of the earth and shows little effect of continental drift and prior distribution of landmasses. (After Flessa, 1981.)

time as climates fluctuated, continents drifted, mountains were uplifted, and land bridges were formed and eliminated. Much of the middle section of this book is devoted to the historical dispersal of groups across regions that are now barriers to biotic interchange. We shall return to consider some of the ecological and evolutionary consequences of these historical events after examining the nature of present boundaries between biogeographic provinces.

One intriguing aspect of the current boundaries is that although they almost invariably correspond with physical geographic features that limit dispersal, their effect is almost as great on highly vagile groups as it is on much more sedentary forms. Remember that one of the first to delineate biogeographic provinces was P.L. Sclater (1858), who based his realms on the distribution of birds. Despite the great potential and actual dispersal of birds, the avifaunas of these regions have retained their distinctiveness. For example, there is much dispersal between North America and Eurasia. Every year birdwatchers in the northeastern United States and eastern Canada record several European species that have accidentally crossed the North Atlantic, and there is probably at least as much dispersal in the opposite direction and between Alaska and Siberia across the Bering Sea. Nevertheless, despite this continual interchange of individuals there are several families and numerous genera that range into the northern parts of each continent but are virtually or completely absent from the other landmass. The large passerine families Tyrannidae (tyrant flycatchers; 365 species) and Parulidae (wood warblers; 120 species) are confined entirely to the New World, whereas the Musicapidae (flycatchers; 330 species) are exclusively Old World and Sylviidae (warblers; 400 species) have only 5 genera and 16 species in the New World.

Competition between biotas? It is probably not coincidental that many of these avian groups, such as the warblers and flycatchers just mentioned, are replaced on the other continent by forms that appear to be ecological equivalents. It is tempting to speculate that the distinctiveness of these biotas is maintained by competition between them. Certainly, limited dispersal by itself is not sufficient to account for their failure to cross the boundaries. It is known that three different groups of sylviids have successfully colonized North America, one so recently that the same species, *Phylloscopus borealis* (the arctic warbler), is widely distributed across northern Eurasia and has populations that breed in western Alaska but still migrate across the Bering Sea to winter in tropical Asia (Peterson, 1961). Another species of Asian sylviid and a musicapid occasionally are found in Alaska. Even with these modest footholds, the Old World families have not been very successful in the New World. Yet it is difficult to make more than a weak, circumstantial case for competition limiting these distributions. If it is difficult to demonstrate competition when two closely related, ecologically similar species occupy abutting or overlapping ranges, one can imagine the problems of trying to assess the potentially diffuse interactions among several species in families that occur exclusively on different continents. It is particularly hard to dismiss the obvious alternative, wholly historical explanation: that specialized adaptations for unique features of their own regions, rather than direct competition from species in other provinces, prevents the spread of species and entire biotas across the boundaries.

Even to the extent that competition may be important in maintaining the distinctiveness of biotas, it is usually difficult or impossible to identify the particular species that are responsible. Often there is no single species or small number of species that are sufficiently exact ecological counterparts to be obvious candidates. Similarly, although predation might be invoked as an alternative to competition, it is difficult to attribute the boundaries between entire biotas to the presence or absence of certain kinds of predators. This raises the intriguing possibility that in some way entire biotas have coevolved so that the species are adapted to coexist with each other but are able collectively to resist invasion from exotics that have evolved with other organisms in different regions. In this case the resistance to invasion would be a consequence not simply of direct competitive or predator-prey interactions involving only a few species, but of a wide array of diffuse interactions of all kinds, involving many species. This would be much more difficult to prove or disprove than a simple competition (or predation) hypothesis, but it is worth considering in the light of some supportive evidence.

One boundary between biotas that does not correspond precisely to a present or past natural geographic barrier is that separating the Nearctic from the Neotropical realm (see Figure 1.3). Although the ancient separation and the narrowest point of the present land connection between North and South America is the Isthmus of Panama, the Neotropical biota ranges well north of this and begins to be replaced by Nearctic forms only in southern Mexico. In one sense the reason for this is obvious: the organisms are limited primarily by climate and habi-

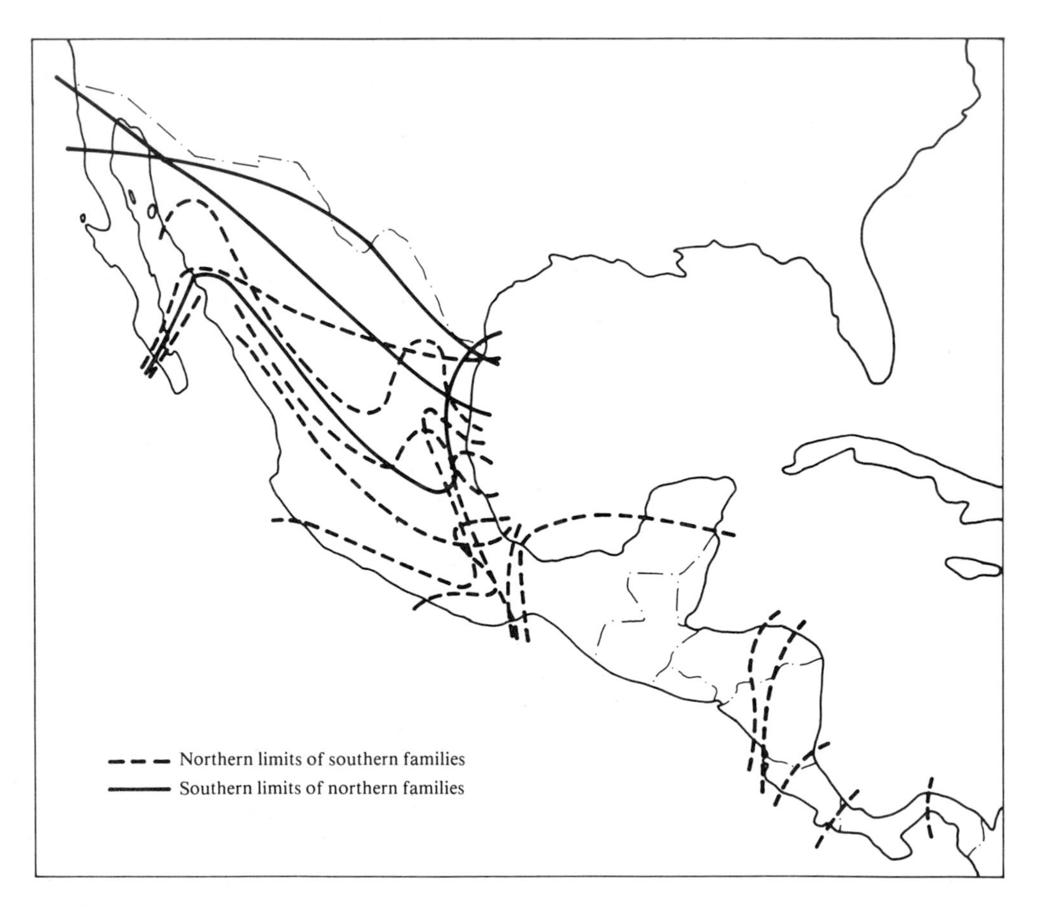

Figure 18.3
Northern limits of Neotropical mammal families *(dashed lines)* and southern limits of Nearctic families *(solid lines)*. Note that the transition between the two mammal faunas occurs over a wide region and that each family has different limits. Similar patterns characterize virtually all taxa (e.g., see Figure 7.15 for fishes).

tat, and the Neotropical biota occurs as far north as a truly tropical environment. In fact there is a gradual replacement of Neotropical species by Nearctic ones throughout those regions of Mexico with a subtropical climate (Figure 18.3). Thus many temperate North American species reach their southern limits somewhere between the United States–Mexican border and the Isthmus of Tehuantepec, 1500 km to the south. Similarly, a few tropical forms, such as a trogon (Trogonidae) and a chachalaca (Cracidae) among the birds and the javelina (Tayassuidae) and coati (Procyonidae) among the mammals, range northward to reach the southwesternmost United States.

Climate is clearly important in determining this boundary and in limiting the distribution of different forms. It is unlikely, however, that climate alone is a sufficient explanation. The tropical climate is equable and well within the range of conditions that most temperate species can tolerate throughout most of the transition region. Other distinctive features of the environment, including the structure of the habitat, are largely consequences of the different kinds of organisms that occur there, so that to invoke these features causes us to advocate some kinds of interspecific interactions.

On the other hand, the gradual transition between the Nearctic and Neotropical realms across a broad region of subtropical Mexico suggests that the biotas are not highly discrete units composed of species all closely interdependent on each other. Furthermore, at the species level some of the forms now restricted to the Nearctic or Neotropical regions originally came from the other continent before completion of the Panamanian Land Bridge in the Pliocene. For example, spotted cats, such as the jaguar *(Felis onca)* and ocelot *(F. pardalis)*, are restricted to tropical regions south of the United States–Mexico border but are originally of northern origin, whereas the porcupine *(Erethizon dorsatum)* is confined to temperate North American habitats but is of South American ancestry. Not only have these species successfully invaded another biota but also they are presently excluded from the region in which they originated and no longer occur together with many forms with which they shared a long evolutionary history.

One way out of this seemingly paradoxical dilemma is to suggest that in any large region most of the niches are filled. Because of historical opportunity, they tend to be filled by species that originated nearby and were adapted to similar environmental conditions. Although all of the species within a local ecological community or larger geographic region may not be coevolved to be closely interdependent on each other, nevertheless if these species have adapted to use most of the available resources efficiently, there may be insufficient resources available to support the invasion of exotic species. Whether or not one wants to use the term *coevolution* to describe such radiation of a biota to fill most niches, it nevertheless provides a possible mechanism by which large continental and marine biotas can resist invasion and maintain their distinctiveness.

Evidence in support of this process comes from the fate of introduced species (for individual case histories see Chapter 4 and Elton, 1958; Udvardy, 1969). Numerous Old World plant and animal species have become established in North America following their intentional or accidental introduction. Although the success of these invaders is impressive, the vast majority of them have established populations exclusively or largely in disturbed habitats. Of the hundreds of exotic plants, most can be classified as weeds, species that occur in successional habitats created primarily by human disturbance. Of the many introduced insects, the majority of successful species are crop pests, associated with introduced plant species, or confined largely to disturbed habitats. This pattern holds even for the vertebrates. Many Eurasian birds have been introduced into North America, but only a few have become established. The two amazingly successful introduced bird species, the house sparrow *(Passer domesticus)* and the starling

(Sternus vulgaris), which have spread to cover most of the North American continent, are largely commensals with humans, using man-made structures for nesting sites and urban and agricultural habitats for food resources. The success of these Eurasian exotics in North America and the much lower success of New World species in the Old World suggests that most of the Eurasian forms are adapted to occupy niches that are dependent on human activity, and they exploit similar niches in North America. As evidence of this we note that not only have the exotics been generally unsuccessful at invading undisturbed native habitats but also it is difficult to point with confidence to extinction of a native species owing to replacement in its niche by an introduced competitor.

One of the exceptional North American species that has successfully colonized the Old World is the muskrat *(Ondatra zibethica)*, a large rodent related to the voles. This valuable furbearer escaped from captivity in Czechoslovakia in 1905 and has since spread to inhabit most of northern Eurasia, an area about as large as its original geographic range in North America (Figure 18.4). The muskrat clearly seems to fill a previously empty niche in the Old World. It is a large, semiaquatic, herbivorous rodent that inhabits marshes and lives in houses that it constructs itself out of reeds and other plant

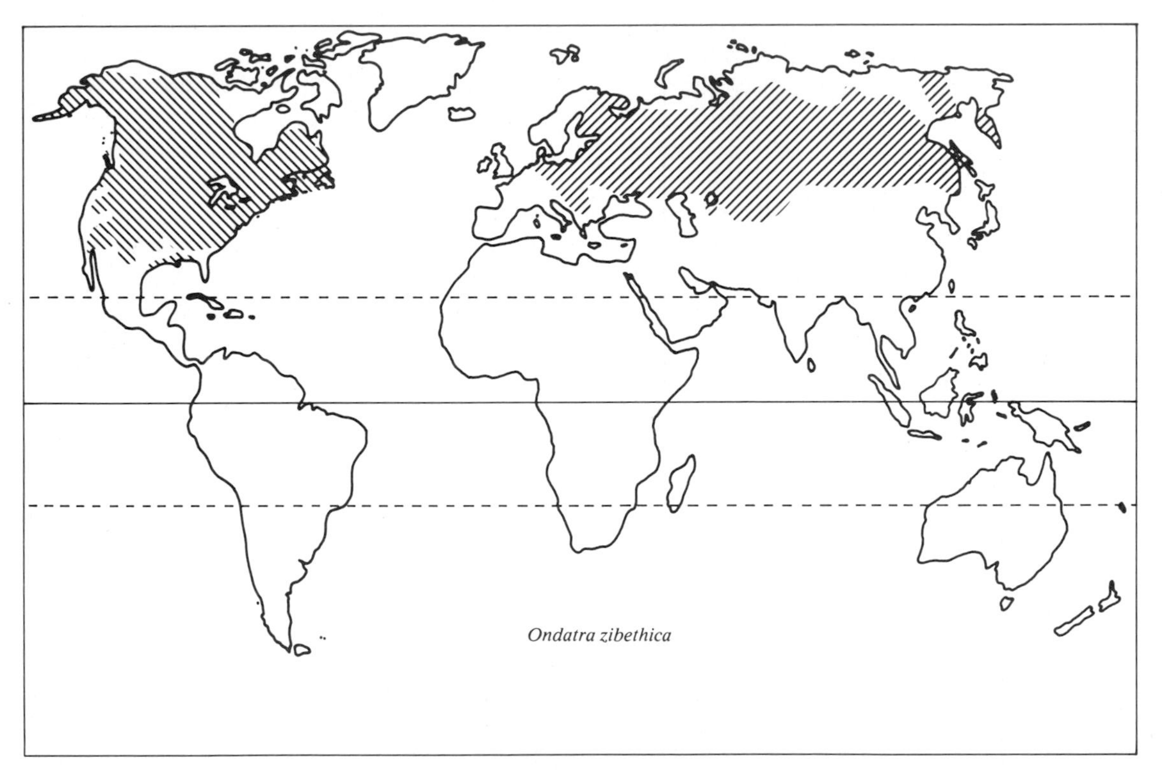

Figure 18.4
Geographic range of the muskrat *(Ondatra zibethica)* in North America where it is native and in Eurasia where it was introduced in 1905. This large, semiaquatic rodent appears to occupy a niche not previously filled by any Old World rodent and it has spread rapidly to colonize an area about as large as its native range in the New World. (Map drawn from several sources.)

material or is found along slow streams, where the animal lives in burrows in the banks. No native species with similar ecological attributes occurred in Eurasia prior to the introduction of the muskrat.

Another exception to the general pattern outlined above is the success of some introduced parasites and pathogens in attacking native species in undisturbed habitats. Conspicuous examples include the chestnut blight (Chapter 6) and Dutch elm disease. Both of these are caused by fungi of European origin that attack native American tree species that are closely related to their original hosts. However, several investigators have noted that hosts can be regarded as evolutionary islands for host-specific parasites. Certain niches for parasites may be unfilled because specialized adaptations are required for parasites to speciate and colonize new hosts, and because extinction rates are potentially high for specialists dependent on limited host populations. As mentioned in Chapter 16, islands are highly susceptible to invasion by introduced species from continental regions. Presumably this is because the few species native to islands tend to have evolved broad niches, and the specialized exotics can exploit unoccupied niches or outcompete the generalized natives for a narrow range of resources.

Historical biotic interchange. The fate of artificially introduced species provides some limited insight into the kinds of interactions that can occur when representatives of one formerly isolated biotic region contact those of another. This has happened naturally in the past, not only when individual species have managed to disperse across barriers but also when the barriers themselves have been abolished, bringing two distinctive, previously isolated biotas into direct contact. This has happened repeatedly as the continents have drifted over the earth and as land and water connections have been formed. Unfortunately, the record of the results of these contacts is often poor because it must be pieced together largely from limited fossil evidence. For example, when the Indian plate slammed into southern Asia it presumably brought a distinctive biota derived from Gondwanaland forms into immediate contact with a large Eurasian biota. It may have been apparent that, in the discussion of groups with Southern Hemisphere affinities that might be traced back to common origins on Gondwanaland, little mention was made of relictual populations in India. Could it be that the Indian biota derived from Gondwanaland was almost completely eliminated by interactions with a more diverse biota from the large Eurasian landmass? This is a plausible explanation, but unfortunately the fossil inhabitants of the Indian subcontinent are so poorly known that at present there is little evidence to support it.

Perhaps the best record of biotic interchange between two previously isolated regions is for the events following the establishment of the inter-American Land Bridge in the Pliocene. Recall from Chapter 5 that for most of its history South America was an island continent. During this time it acquired a distinctive biota derived both from forms that had persisted since its ancient connection to Gondwanaland and from groups that had managed to disperse successfully across the water gaps that isolated it from other landmasses. Its isolation ended abruptly less than 4 million years ago when mountain building in the region that was to become Central America first created a land connection to North America. This brought the previously isolated biotas of the two continents into contact in one of the most impressive natural experiments in biogeography.

The results of this contact have been best documented for mammals because of the extensive work of many paleontologists and the valuable syntheses of Simpson (1950, 1980b; see also Keast et al., 1972; Webb, 1976; Marshall, 1979; Marshall et al., 1982). Simpson's analyses are generally based on the family level of classification because at this level the fossil record is relatively complete. Admittedly, this is a relatively crude basis for trying to understand the historical events and their causes, but fortunately most mammalian families are characterized by distinctive ecological characteristics and morphological traits. Even at the family level the details of the faunal interchange and its effects, especially on the South American biota, are complex and controversial. Simpson himself has changed his interpretation somewhat, in part in response to new evidence that became available between 1950 and 1980. The fossil record is still far from complete and the experts continue to disagree among themselves, so the following simplified treatment should be accepted with caution.

In the early Pliocene, prior to completion of the land bridge to North America, South America had at least 30 families of terrestrial mammals (bats and aquatic forms excluded). These included marsupials, monkeys, armadillos, anteaters, sloths, rodents, and several groups of large grazing mammals. The histories of some of these taxa and the means by which they may have reached South America are discussed in Chapter 11. At the same time the North American mammalian fauna contained at least 32 families. Many of these still inhabit the continent although some, such as the hyenas, horses, camels, and elephants, are now extinct there. The North American fauna reflected a long history of interchange with Eurasia, but not so much as the present mammals because of subsequent exchange across the Bering Land Bridge. Of the 30 South American and 32 North American families, only one, the cricetid rodents, occupied both continents at the beginning of the Pliocene, although at least two more, which apparently crossed the water gaps by island hopping, were shared by the middle Pliocene when the land bridge is thought to have been completed. (For a summary of the exchange of groups see Figure 7.14.)

At present there are 27 mammal families in temperate North America and 32 in South America and tropical North America. (As mentioned earlier, the boundary between the Nearctic and Neotropical biotas is now in central Mexico, not at the Isthmus of Panama.) Of these families, 18 now occur on both continents. Obviously there has been extensive interchange across the Panamanian Land Bridge as well as extinction of a number of families on the two continents. Although the total number of families on each continent has remained relatively constant at about 30, the exchange has been very unequal (Table 18.1). Of the 32 original

Table 18.1

Summary of interchange of land mammal families following completion of inter-American Land bridge in the Pliocene

Note that although the number of families inhabiting each continent remained relatively constant, North American families were much more successful in invading South America than South American forms were in colonizing temperate North America. (Data from Simpson, 1980.)

	Temperate North America	South America
Families that occurred at the beginning of Pliocene	30	32
Survived to present on same continent	20	18
Survived to present but on different continent	6	0
Initially spread to other continent	16	8
Initially spread and survived to present on other continent	13	4
Colonized since the Pliocene from Eurasia	3	1
Occur at present	27	32*

*Includes families inhabiting tropical North America.

North American families, 20 still live there, 16 successfully invaded South America, and 13 of these survive there today. Of the 30 original South American families, 18 still survive, but these mammals were not very successful in invading North America: only eight families reached the temperate part of the continent and only one species in each of three families (opossum, Didelphidae; armadillo, Dasypodidae; and porcupine, Erethizontidae) in addition to the cricetid mice persist in temperate North America today.

How do we interpret this fascinating history? It is tempting to suggest, as Simpson did in 1950, that the two faunas competed with each other and the end result was a differential elimination of many of the original South American families by superior competitors that had invaded from the north. Two aspects of the interchange are particularly consistent with this interpretation. First, despite the extinction of several families, especially in South America, the number of families remained remarkably similar both between the two continents and before and after the contact. This suggests that some equilibrial number of mammalian life forms and corresponding ecological niches was preserved by a process of competitive exclusion (Marshall et al., 1982). Second, several of the South American groups that became extinct were replaced by ecologically similar mammals that had invaded from the north. In particular a diverse group of marsupial carnivores, the Borhyaenidae, disappeared and were replaced by several families of modern placental carnivores including dogs (Canidae), bears (Ursidae), raccoons (Procyonidae), weasels (Mustelidae), and cats (Felidae). Similarly, six families of ancient grazing mammals became extinct and an equal number of families of modern ungulates invaded; the latter include mastodons (Gomphotheridae), horses (Equidae), tapirs (Tapiridae), peccaries (Tayassuidae), camels (Camelidae), and deer (Cervidae).

Unfortunately, on closer examination this story is not quite so neat. The complications include the fact that five families of placental carnivores "replaced" a single family of marsupial predators. Also, in the case of both the carnivores and the grazers, fossil evidence clearly shows that representatives of both the North and South American groups occurred together in South America, in some cases for as long as 2 million years. Finally, some of the North American families that invaded South America subsequently became extinct in the late Pleistocene either in North America (tapirs, peccaries, and camels) or on both continents (mastodons and horses). It has been suggested that humans, colonizing the New World via the Bering Strait, contributed importantly to the demise of these groups (Chapters 6 and 14) (Martin, 1966, 1967, 1973). These facts are certainly not sufficient to rule out competition as an important cause of the elimination of at least some groups. The observation that it took a long time after the invasion for the complete demise of certain groups, for example, does not mean that competition did not occur. It may have taken thousands or even millions of years for some lines of the invaders to adapt sufficiently to the new physical and biotic environments in South America to exclude certain native competitors. On the other hand, the evidence does not show many examples of direct replacement of ecologically equivalent groups that can be attributed unambiguously to competition. It is hoped that the increasing fossil record of South American organisms will shed new light on the results and causes of the great biogeographic experiment that took place in the Pliocene.

There is one important unintentional experiment in which humans have recently brought into contact previously long-isolated marine biotas. This was performed in 1869 by the completion of the Suez Canal between the Mediterranean Sea and the Red Sea (Por, 1971, 1977). Although the hypersaline lakes through which the canal passes constitute an impassible barrier for most marine forms, a few taxa have been able to disperse between the two seas. The biotic exchange has been almost one-sided.

About 30 species of fishes, 20 species of decapod crustaceans, and 40 species of molluscs have colonized the eastern Mediterranean from the Red Sea, but it is hard to document cases of migration in the opposite direction. Two explanations for this unidirectional dispersal have been proposed, and both causes probably contribute to the successful colonization by Red Sea forms. The Gulf of Suez at the southern end of the canal is itself more saline than ocean water, so the species that occurred there may have been preadapted to cross the barriers created by the hypersaline lakes. In addition, the Red Sea is an arm of the Indian Ocean, which contains a far more diverse biota than the Mediterranean. It has been suggested that species of the rich Indo-Pacific biota are more resistant to predation than their Mediterranean counterparts or are competitively superior to them or both. This is supported by observations of declining populations of some endemic Mediterranean species in the eastern part of the sea (e.g., along the coast of Israel) where Red Sea colonists have become well established. For a much more complete discussion of the history and results of exchange through the Suez Canal see Por (1971, 1975, 1977) and Vermeij (1978).

One final point is warranted. The differential extinctions of South American mammals in the face of North American invaders and the unidirectional exchange of marine forms through the Suez Canal are both consistent with a general pattern noted by many biogeographers. As early as 1915, W.D. Matthew (see also Willis, 1922) noted that organisms from diverse biotas on large landmasses were able successfully to invade smaller areas and to replace the native organisms. Darlington (1957, 1959) reemphasized this point, although he argued that the successful forms usually originated in tropical regions, whereas Matthew had thought they came from temperate climates. Because the climates of regions that are now temperate and tropical have changed greatly over their geologic history (Chapters 5 and 14) this point may be difficult to resolve, but the success of organisms from large biotas in colonizing small, previously isolated regions containing fewer native species seems to be an important phenomenon in biogeography. The interactions that have been seen on a continental scale between the North and South American biotas fit the pattern of successful invasion of the island continent Australia by many introduced species from the other larger interconnected continents, as well as natural and human-assisted colonization of islands by continental forms (Chapter 16).

Divergence and convergence of biotas

Historical changes in isolated biotas. We have considered what happens when previously isolated biotas come into contact. Now we will examine the opposite situation, the fate of individual species and entire biotas that were once united but have been separated for varying periods of time in different regions. We shall be particularly concerned about the role of ecological factors in the differentiation of organisms in isolation.

Most of what we have said thus far would cause one to expect that geographically isolated organisms should diverge. For example, much of phylogenetic systematics and vicariance biogeography is based on the assumption that genetically and geographically isolated lineages tend to become more different as they evolve independently from each other. This is certainly true for many attributes of organisms, including most of those that are used in classification. It need not be true of ecological characteristics, however. If groups are isolated in regions of different area, geology, and climate, the differences in the physical environment will tend to promote ecological divergence. If the physical environments are similar, however, distantly related organisms may independently evolve similar adaptations. We call this evolutionary convergence and recognize that it can occur on many levels. Within species it may be restricted to a few traits, or it can involve essentially the entire organism, resulting in convergence in morphology, physiology, and behavior as unrelated forms specialize for similar niches. Convergence can also occur at the level of entire biotas of many species, resulting in geographically isolated ecological communities with similar structures and functions.

One would expect the divergence of forms derived from a common ancestor and a particular geographic region to vary directly with the length of time they have been genetically and geographically isolated. In general this is true, but it is important to reemphasize (see Chapter 6) that the rate of evolutionary change can be extremely variable. The rapidity of possible divergence in a biogeographic setting is illustrated not only by the speciation that has occurred in many groups as a consequence of isolation resulting from climatic and habitat changes during the Pleistocene (Chapter 14) but even more dramatically by the differentiation of exotic species introduced into new continents by modern humans.

Probably the most thorough studies are those of R.F. Johnston (Johnston and Selander, 1964, 1971, 1973; Johnston and Klitz, 1977) on the evolution of the house sparrow *(Passer domesticus)*. This species was brought to North America from Europe and released in New York City in 1852. It became widespread on the continent only within the last century, reaching the Pacific coast in the early 1900s. Presently it is one of the most successful of North American birds, ranging from coast to coast and from Guatemala to central Canada, although it is usually abundant only around areas of human habitation. More interesting from our present viewpoint is the fact that local populations have already evolved distinctive combinations of traits. Birds from the northeastern United States, for example, are large and dark colored, whereas those from the southwestern deserts are small and pale. Because many of the trends parallel those in other birds, it appears that they represent adaptations to local climate and other factors. Differences between some populations are so great that they could easily be used to describe formal subspecies. Interestingly, a closely related species, the European tree sparrow *(Passer montanus)* has also been introduced into North America, but it is established only in a small area in southwestern Illinois and eastern Missouri. Although it is abundant and widespread in Europe, in North America this sparrow has experienced none of the ecological success, geographic expansion, or evolutionary differentiation of its congener.

Given the rapidity with which species in new environments can evolve, it is not surprising that convergence of biotas has occurred during and since the Pleistocene. Brown (1975) has documented the convergent organization of rodent communities in different desert regions of North America (Figure 18.5). Habitats of similar climate, soil, and vegetation, separated by hundreds of kilometers, frequently support comparable numbers of rodent species that have similar morphologies and ecological niches. Two aspects of these similarities are noteworthy. First, species with similar traits in separated regions often are not particularly closely related, indicating that the characteristics have evolved independently, presumably as a result of selection to exploit similar ecological niches. Second, the convergence must be relatively recent. Only a few thousand years ago many of the desert valleys where these communities occur were filled with large pluvial lakes, and the taxonomic affinities of the component species suggest that the communities have been assembled from forms that survived the pluvial periods in different desert refugia.

When the Isthmus of Panama came into being in the Pliocene, it not only connected North and South America by a continuous land bridge but also it completely isolated the tropical Atlantic and Pacific Oceans. This event was perhaps as important for marine organisms as it was for the terrestrial forms we have already discussed. It is especially important in view of the fragile nature of the isthmus as a barrier. This became particularly apparent in the 1960s,

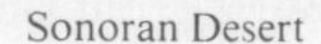

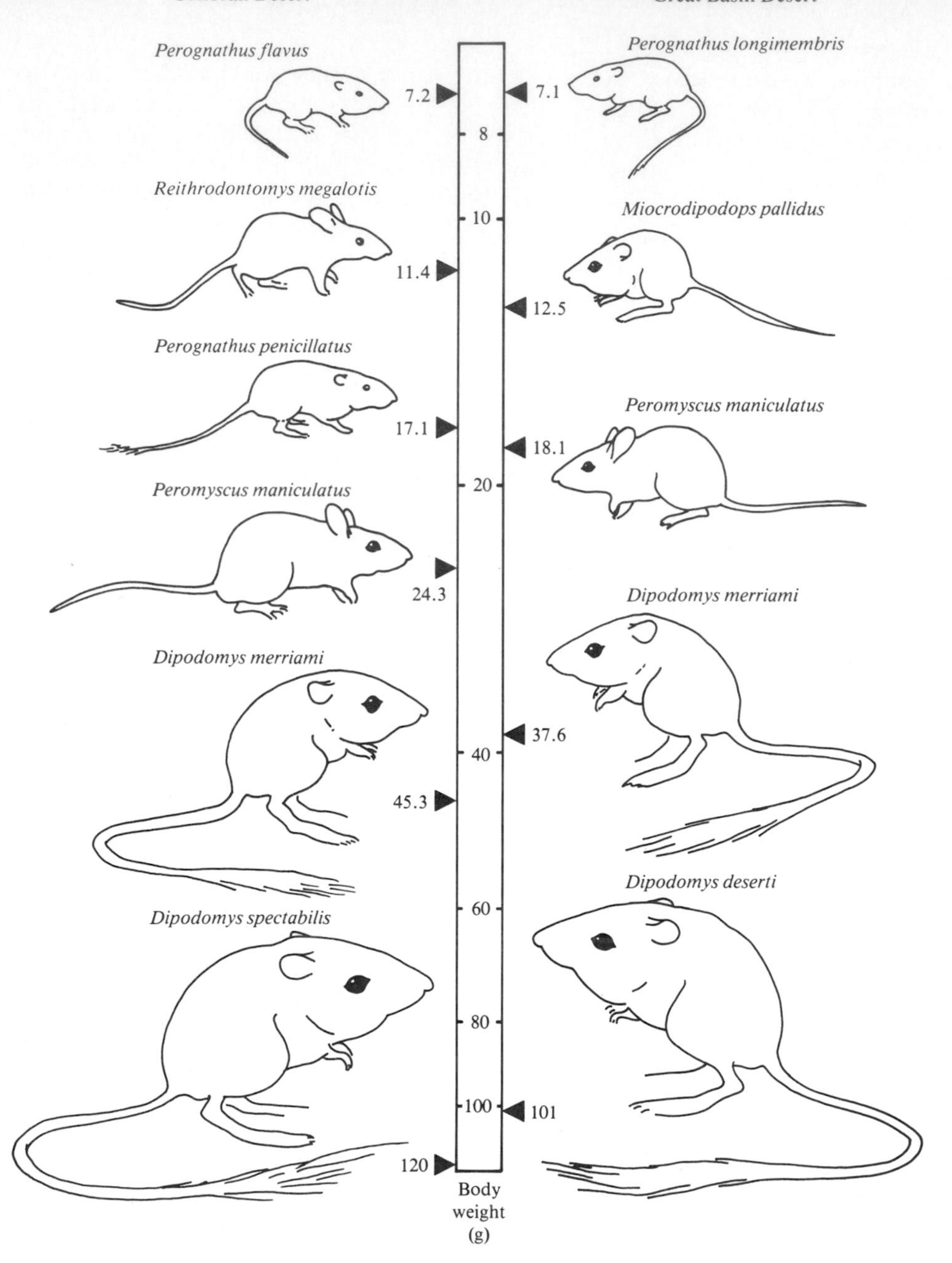

Figure 18.5
Convergent organization of seed-eating desert rodent communities in two widely separated North American deserts. Productive habitats in the Great Basin and Sonoran deserts support similar numbers and kinds of species, even though these forms are often members of different species, genera, and sometimes even families. (After Brown, 1975. Reprinted by permission of the publishers from *Ecology and Evolution of Communities,* ed. M.L. Cody and J.M. Diamond, Cambridge, Mass.: The Belknap Press of Harvard University Press, Copyright © 1975 by the President and Fellows of Harvard College.)

when the possibility of constructing a sea-level canal across the isthmus was being seriously considered. The present Panama Canal, constructed in the early part of the present century, incorporates a large body of fresh water, Lake Gatun, and uses a series of locks to raise and lower ships as they traverse the isthmus. The fresh water effectively prevents interchange between most elements of the Pacific and Caribbean tropical marine biotas, but the construction of a sea-level canal would perform a biogeographic experiment of gigantic proportions. The effect on the marine biotas would be analogous to the influence of the original establishment of the isthmus on the terrestrial forms, with the exception that the Caribbean and Pacific have been isolated for only about 3 million years, whereas North and South America had been separated for at least 135 million years. Controversy surrounding the ecological effects of an interchange of species as a result of a sea level canal stimulated much research on differences and similarities between the Caribbean and eastern tropical Pacific biotas.

There are major differences, especially in species richness, between the marine biotas of the eastern Pacific and the Caribbean (Briggs, 1968, 1974; Rubinoff, 1968; Porter, 1972, 1974; Vermeij, 1978). Most groups are more diverse in the Pacific. Examples include most major taxa of molluscs, crabs, and echinoderms. There are exceptions, however. The Caribbean has about 900 species of shallow-water and coral reef fishes, compared to only about 650 species in the eastern Pacific. Sea grasses and their specialized animal fauna are abundant, widespread, and diverse in the Caribbean but virtually absent from the eastern Pacific, where appropriate, highly productive shallow water habitats are not extensive. In many groups there are closely related sister species on either side of the Isthmus (Jordan, 1908; Vermeij, 1978). The rates of divergence of some of these forms have been of considerable interest to systematists and evolutionists (e.g., Rubinoff and Rubinoff, 1971), because the time of isolation is known quite accurately and is presumably the same for all groups. Although there has been some differentiation, most species pairs remain similar in morphology and presumably in their ecological niches. This suggests that competition between such species pairs might prevent much interchange across a sea-level canal, or, if some forms were superior competitors and able to invade the other ocean, competitive exclusion might result in extinction of some sister species without greatly affecting the diversity of the biotas on each side of the isthmus. The absence of complementary species in a few exceptional groups has caused more concern. Two Pacific taxa in particular that do not have close relatives or obvious ecological counterparts in the Caribbean are the sea snake, *Pelamis,* and the crown of thorns starfish, *Acanthaster planci.* The former, of course, is highly venomous, whereas the latter feeds voraciously on certain corals and occasionally devastates reefs in its native region (Chester, 1969). It is possible that *Acanthaster* might wreak even greater havoc were it able to colonize the rich Caribbean reefs, where the coral species have had no opportunity to evolve resistance (Porter, 1972; Vermeij, 1978).

There are few examples of dramatic convergence of marine taxa between forms in different but separated water masses, such as the eastern tropical Pacific and the Caribbean or the Mediterranean and the Red Sea. In general, the isolation of water masses has been recent in relation to the evolution of most marine groups. The oceans presently are and always have been effectively more interconnected than the continents. Thus although there may be differences between biotas at the species level, most of the genera are widely distributed. There has been a tendency for isolated pairs of closely related species to continue to evolve more or less in parallel but little opportunity for distantly related forms to converge to fill the ecological niches of the missing taxa.

Patterns of convergence. Under favorable circumstances, aquatic organisms have exhibited much convergence, even in the 3 million

Table 18.2

Ecologically equivalent species of cichlid fishes in the three great African lakes

(Modified from Fryer and Iles, 1972.)

Description	Victoria	Malawi	Tanganyika
Algae-eating rock and plant scrapers	*Haplochromis nigricans*	*Pseudotropheus* spp.	*Tropheus* spp.
		Petrotilapia tridentiger	*Petrochromis* spp.
	H. obliquidens	*Cyathochromis obliquidens*	
	Haplochromis sp.	*Hemitilapia oxyrhynchus*	
Invertebrate-pickers	*Paralabidochromis victoriae*	*Labidochromis* spp.	*Tanganicodus irsacae*
	Haplochromis chilotes	*Haplochromis euchilus*	*Lobochilotes labiatus*
		H. lobochilus	
		Melanochromis labrosus	
Invertebrate-eating sand sifters		Some *Lethrinops* spp.	*Callochromis pleurospilus*
			Certain *Xenochromis* spp.
Mollusc eaters	*Haplochromis ishmaeli*	*Haplochromis mola*	*Lamprologus tretocephalus*
	H. pharyngomylus	*H. incola*	*Haplochromis straeleni*
	Astatoreochromis alluaudi	*H. placodon*	
Zooplankton feeders	*Haplochromis* sp.	*Haplochromis* spp. of the Utaka group	*Limnochromis permaxillaris*
Inshore fish-eaters	*H. prognathus*	*Aristochromis christyi*	Certain *Lamprologus* spp.
	H. guiarti	*Haplochromis rostratus*	
	H. mento		
Elongate, slender-bodied fish-eaters	*H. cavifrons*	*Rhamphochromis* spp.	*Bathybates* spp.
	H. longirostris	*Diplotaxodon* spp.	*Boulengerochromis microlepis*
	H. dentex		
	H. macrognathus		
Bilaterally compressed fish-eaters		*Haplochromis compressiceps*	*Lamprologus compressiceps*
Scale scrapers	*H. welcommei*	*Genyochromis mento*	*Plecodus* spp.
		Corematodus spp.	*Perissodus microlepis*
		Haplochromis sp.	

years since the Pliocene vicariance event. For example, consider again the adaptive radiation of cichlid fishes in the great lakes of central Africa (Chapter 6). Recall that these radiations apparently began in the Pliocene when a few stocks of river-dwelling cichlids and other fish groups were isolated when the lakes began to form. The subsequent radiation produced in excess of 200 species, the vast majority of them cichlids (Table 6.1), in the largest lakes. Of particular interest here is the convergence of species from different ancestral stocks to occupy similar niches in the various lakes. Some of these are indicated in Table 18.2. Note that although representatives of different genera often fill the same niches, the complementarity is far from perfect. To some extent this may reflect a certain degree of randomness in the organization of the fish communities and the existence of unfilled niches, but there are also major differences in habitats between the lakes that affect the niches available to fishes. For example, the deep rift lakes, Tanganyika and Malawi, offer much more steep and rocky habitats than Victoria, whose generally shallow, gently sloping bottom is covered with much more sand and mud. If these differences in habitat could be taken into account, the convergence of the fishes might well be even more impressive than indicated in Table 18.2.

Mammals provide some of the best-documented examples of adaptive radiation and convergence on a worldwide scale. Mammals are found on all the continents, but there is much

Table 18.3
Niches of nonmarine mammals of the world
This classification is based on diet and substrate categories defined by Eisenberg (1981). Entries indicate the numbers of genera in the world assigned to each niche. Note that some niches are represented by many genera, others are filled by only a few kinds of mammals, and still others are empty. (Data supplied by J.F. Eisenberg.)

	Substrate category							
Diet category	Fossorial	Semifossorial	Aquatic	Semiaquatic	Flying	Terrestrial	Scansorial	Arboreal
Fish and squid	—	—	48	26	2	—	—	—
Vertebrates	—	1	2	—	5	27	8	3
Nectar	—	—	—	—	20	—	—	2
Sap and gum	—	—	—	—	—	—	—	3
Crustaceans and molluscs	—	—	2	2	—	—	—	—
Ants and termites	—	7	—	—	—	3	2	1
Aerial insects	—	—	—	—	72	—	—	—
Insects from foliage	—	—	—	—	—	—	—	—
Insects and other items	13	9	—	3	19	76	16	20
Fruits and other items	—	—	—	—	56	35	19	22
Seeds and fruits	—	2	—	4	—	115	43	59
Fruits and leaves	11	10	—	1	—	46	12	29
Browser (leaves and stems)	9	2	—	3	—	78	2	11
Grazer (grasses and forbs)	—	1	2	4	—	43	—	—
Plankton	—	—	6	—	—	—	—	—
Blood	—	—	—	—	3	—	—	—

provincialism in the distribution of major groups. Many orders and families are confined to one or a small number of the biogeographic realms (Chapter 11), even though similar environments are found in other regions. J.F. Eisenberg (1981) recently has compiled and synthesized an enormous body of information on the radiation of mammalian groups throughout the world. Eisenberg developed a simple, robust classification of mammalian niches based on diet and substrate use (Table 18.3). One conspicuous feature of this analysis is that the blank spaces show that many niches are unfilled, at least by mammals. There are, for example, no flying crustacean and mollusc eaters among the mammals, although several kinds of birds fill this niche. Even when one examines those niches that are filled by mammals, it is found that they are occupied by different numbers of taxa (genera, in this case). This is not surprising. Some niches simply do not contain sufficient quantities of resources to support more than a few species in any region (e.g., fossorial fruit and seed eaters or arboreal anteaters), whereas others may be rich in resources but have remained relatively unexploited by mammals, perhaps because constraints have prevented mammals from acquiring the necessary adaptations, although these niches are often filled by other taxa (e.g., there are few fish, nectar, and plankton feeders among terrestrial mammals, although many birds use these food resources).

Table 18.4 shows the extent to which the different niches are occupied in the different biogeographic realms and in some smaller regions. Note that those niches that are filled by many taxa on a worldwide basis tend to be filled within each of the biogeographic realms. On the other hand, niches containing only a few mammals often go unfilled, even on large landmasses. Mammals have radiated predictably to fill most of the major niches available to them, but whether they have also occupied the smaller niches appears to be much more a matter of chance.

The first evidence for convergence comes when distantly related taxa (usually species in

Table 18.4
Mammalian genera occupying particular niches in different biogeographic regions
The niche categories are the same as those in Table 18.3 but are abbreviated here. Note that most niches with many genera are filled by mammals in all areas; others, especially those exploited by only a few genera, are filled only in tropical regions or are apparently haphazardly filled. (Data supplied by J.F. Eisenberg, 1981.)

Niche categories		Tropical			Temperate			
Substrate	Diet	Asia	South America	Africa	North America	Europe	Asia	Australia
Fossorial	Insects	1	—	5	4	1	3	1
Fossorial	Fruits and leaves	2	1	6	—	1	2	—
Fossorial	Browser	—	1	—	3	—	—	—
Semifossorial	Vertebrates	—	—	—	1	—	—	—
Semifossorial	Ants	—	4	1	—	—	—	1
Semifossorial	Insects	3	1	—	2	—	4	—
Semifossorial	Seeds and fruits	—	—	—	1	—	—	—
Semifossorial	Fruits and leaves	—	2	—	3	3	5	—
Semifossorial	Browser	—	—	—	1	1	1	—

Table 18.4—cont'd

Mammalian genera occupying particular niches in different biogeographic regions

The niche categories are the same as those in Table 18.3 but are abbreviated here. Note that most niches with many genera are filled by mammals in all areas; others, especially those exploited by only a few genera, are filled only in tropical regions or are apparently haphazardly filled. (Data supplied by J.F. Eisenberg, 1981.)

Niche categories		Tropical			Temperate			
Substrate	Diet	Asia	South America	Africa	North America	Europe	Asia	Australia
Aquatic	Fish	4	3	3	9	3	7	4
Aquatic	Vertebrates	—	—	—	—	—	—	1
Aquatic	Crustaceans	—	—	—	1	—	1	—
Aquatic	Grazer	1	1	—	1	—	—	1
Semiaquatic	Fish	4	8	5	1	4	6	2
Semiaquatic	Crustaceans	—	—	—	—	—	—	1
Semiaquatic	Fruits	—	3	—	—	—	—	—
Semiaquatic	Seeds and fruits	—	1	1	—	—	—	—
Semiaquatic	Fruits	—	1	—	—	—	—	—
Semiaquatic	Browser	—	—	—	3	1	1	—
Semiaquatic	Grazer	—	1	2	—	—	—	—
Flying	Fish	1	1	—	—	—	—	—
Flying	Vertebrates	1	3	1	—	—	—	1
Flying	Nectar	2	9	3	—	—	—	1
Flying	Aerial insects	22	27	25	9	8	8	11
Flying	Insects from foliage	2	9	3	3	2	2	2
Flying	Fruits	14	18	9	—	—	—	4
Flying	Blood	—	3	—	—	—	—	—
Terrestrial	Vertebrates	3	5	16	5	4	6	3
Terrestrial	Ants	—	2	—	—	—	—	1
Terrestrial	Insects	11	8	16	7	3	5	11
Terrestrial	Fruits	5	10	10	8	5	9	2
Terrestrial	Seeds and fruits	5	18	26	14	10	22	5
Terrestrial	Fruits and leaves	5	10	12	4	3	11	2
Terrestrial	Browser	12	9	28	6	9	13	8
Terrestrial	Grazer	3	4	15	7	6	10	7
Scansorial	Vertebrates	4	1	1	2	2	2	2
Scansorial	Ants	1	1	1	—	—	—	—
Scansorial	Insects	5	1	—	—	—	—	1
Scansorial	Fruits	4	5	—	5	2	3	3
Scansorial	Seeds and fruits	10	8	8	5	4	7	2
Scansorial	Fruits and leaves	—	2	1	1	—	—	1
Scansorial	Browser	—	1	—	1	—	—	—
Arboreal	Vertebrates	1	—	1	—	1	—	—
Arboreal	Nectar	—	—	—	—	—	—	2
Arboreal	Sap	—	1	1	—	—	—	—
Arboreal	Ants	—	1	—	—	—	—	—
Arboreal	Insects	2	4	3	—	—	—	4
Arboreal	Fruits	4	12	5	1	1	2	3
Arboreal	Seeds and fruits	16	14	10	2	4	4	2
Arboreal	Fruits and leaves	9	3	5	—	—	4	5
Arboreal	Browser	1	3	1	—	—	—	3
TOTAL		158	220	224	110	76	138	97

Table 18.5
Examples of convergence of mammalian genera to fill similar ecological niches in different regions of the world

Niche	Region	Common name	Genus	Family	Order
Aquatic (stream) invertebrate-eater	North America	Water shrew	*Sorex*	Soricidae	Insectivora
	Central America	Water mouse	*Rheomys*	Cricetidae	Rodentia
	Eurasia	Water shrew	*Neomys*	Soricidae	Insectivora
	Africa	Water shrew	*Micropotamogale*	Potamogalidae	Insectivora
	New Guinea	Water rat	*Crossomys*	Muridae	Rodentia
	Australia	Platypus	*Ornithorhynchus*	Ornithorhynchidae	Monotremata
Terrestrial insectivore	North America	Shrew	*Blarina*	Soricidae	Insectivora
	South America	Rat opossum	*Caenolestes*	Caenolestidae	Marsupialia
	Africa	Elephant shrew	*Elaphantulus*	Macroscelididae	Macroscelida
	Madagascar	Tenrec	*Tenrec*	Tenrecidae	Insectivora
	Asia	Gymnure	*Hylomys*	Erinaceidae	Insectivora
	Australia	Bandicoot	*Perameles*	Peramelidae	Marsupialia
Fossorial insectivore	North America	Mole	*Scalopus*	Talpidae	Insectivora
	Africa	Golden mole	*Amblysomus*	Chrysochloridae	Insectivora
	Australia	Marsupial mole	*Notoryctes*	Notoryctidae	Marsupialia
Fossorial leaf- and tuber-eater	North America	Pocket gopher	*Thomomys*	Geomyidae	Rodentia
	South America	Tucu-tuco	*Ctenomys*	Ctenomyidae	Rodentia
	Eurasia	Mole rat	*Spalax*	Spalacidae	Rodentia
	Southern Africa	Mole rat	*Cryptomys*	Bathyergidae	Rodentia
	Northern Africa	Mole rat	*Tachyoryctes*	Rhizomyidae	Rodentia
Terrestrial anteater	South America	Giant anteater	*Myrmecophaga*	Myrmecophagidae	Edentata
	Africa	Aardvark	*Orycteropus*	Orycteropodidae	Tubilidentata
	Asia (also Africa)	Pangolin	*Manis*	Manidae	Pholidota
	Australia	Spiny anteater	*Tachyglossus*	Tachyglossidae	Monotremata
Arboreal leaf-eater	South America	Howler monkey	*Alouatta*	Cebidae	Primates
	South America	Sloth	*Bradypus*	Bradypodidae	Edentata
	Africa	Colobus monkey	*Colobus*	Cercopithecidae	Primates
	Madagascar	Woolly lemur	*Lichanotus*	Indriidae	Primates
	Asia	Langur monkey	*Presbytis*	Cercopithecidae	Primates
	Australia	Koala	*Phascolarctos*	Phascolarctidae	Marsupialia
Arboreal frugivore and omnivore	South America	Kinkajou	*Potos*	Procyonidae	Carnivora
	Africa	Palm civet	*Nandinia*	Viverridae	Carnivora
	Madagascar	Lemur	*Lemur*	Lemuridae	Primates
	Asia	Binturong	*Arctictis*	Viverridae	Carnivora
	New Guinea	Cuscus	*Phalanger*	Phalangeridae	Marsupialia

different families and orders) fill the same niches on different continents or smaller regions. Interestingly, one finds many examples of convergence among forms that occupy both large and small niches. A few of these are indicated in Table 18.5. However, this table only documents the fact that different taxa have evolved to play similar ecological roles; it cannot indicate the extent to which many of these forms have also evolved strikingly similar morphology, physiology, and behavior. Numerous examples of amazingly complete convergence have been cited in the literature. Similarities between marsupials in Australia and their placental mammalian counterparts on other continents have been illustrated so frequently they will not be repeated here. Figure 18.6 shows some of the convergences between mammalian species inhabiting tropical rain forests in South America and Africa (see Dubost, 1968; Bourliere, 1973).

Some of the less widely heralded but most striking examples of convergence in mammals occur in the rodents. Although the order Rodentia has a virtually cosmopolitan distribution, many families are restricted to particular regions in which, despite these restrictions, they have often evolved amazingly similar forms. For example, each of the biogeographic realms has large areas of desert and semiarid habitat. Only South America, where extensive deserts may have formed only quite recently, lacks a highly specialized desert rodent fauna. On each of the other continents rodents in different families have independently evolved to fill a variety of niches. Perhaps the most striking convergence occurs in the seed eaters. On each landmass a different family has acquired a remarkably similar suite of adaptations for making a living from dry seeds while simultaneously avoiding similar kinds of predators in open desert habitats. These specializations include light, substrate-matching coloration; enlarged ear cavities (auditory bullae) for detecting predators; elongated hind legs and tails for bipedal hopping locomotion; short, long-clawed front legs for digging and collecting seeds; and urine-concentrating kidneys for maintaining water balance on a diet of dry seeds. Although illustrations of these rodents show an impressive degree of morphological convergence (Figure 18.7), they cannot depict the equally striking similarities in physiology and behavior. Thus in the desert rodents one can observe different levels of convergence on different spatial and temporal scales: the convergence of long-isolated families on different continents described here and the convergence of more recently evolved genera and species within different parts of the North American deserts discussed earlier (Figure 18.5).

In cases such as the desert rodents one sees striking similarities between particular distantly related species, presumably because they have adapted to fill extremely similar niches. Such precise morphological, physiological, and behavioral resemblance implies that certain ecological niches are discrete and predictable and that their resources can be most efficiently exploited only by organisms with certain specific characteristics. This need not always be the case, however; there may be alternative ways to solve similar ecological problems. It is possible to have functional ecological convergence at the level of entire diverse biotas that is not necessarily reflected in ecologically equivalent species. This is perhaps best illustrated by recent studies of community organization in regions that experience mediterranean-type climates on several different continents (Cody and Mooney, 1978; Mooney, 1977).

As mentioned earlier (Chapters 1 and 2), several isolated regions around the world (Figure 1.2) on the western sides of continents in warm temperate latitudes have a climate characterized by slight to moderate precipitation (300 to 900 mm per year), of which at least two thirds falls in the winter; hot, dry summers; and cool, foggy, but rarely freezing winters. These areas of mediterranean-type climate support a distinctive evergreen shrub vegetation variously called chaparral, matorral, macchia, maquis, or

Figure 18.6
For legend see opposite page.

Africa

South America

Pigmy hippopotamus (*Choeropsis*)
Artiodactyla, Hippopotamidae

Capybara (*Hydrochoeris*)
Rodentia, Hydrochoeridae

Cheurotain (*Hyemoschus*)
Artiodactyla, Tragulidae

Paca (*Cuniculus*)
Rodentia, Dasyproctidae

Royal antelope (*Neotragus*)
Artiodactyla, Bovidae

Agouti (*Dasyprocta*)
Rodentia, Dasyproctidae

Yellow-backed duiker (*Cephalophus*)
Artiodactyla, Bovidae

Brocket deer (*Mazama*)
Artiodactyla, Cervidae

Terrestrial pangolin (*Manis*)
Pholidota, Manidae

Giant armadillo (*Priodontes*)
Edentata, Dasypodidate

Figure 18.6
Convergence in morphology between pairs of species of African and South American rain forest mammals that appear to occupy similar ecological niches. Note that many of the convergent species are in different families and even different orders. (After F. Bourliere, 1973. By permission of the Smithsonian Institution Press from *Tropical Forest Ecosystems in Africa and South America: A Comparative Review,* B.J. Meggers, E.S. Ayensu, and W.D. Duckworth [eds.], Fig. 1, p. 282, Copyright © Smithsonian Institution, Washington, D.C. 1972.)

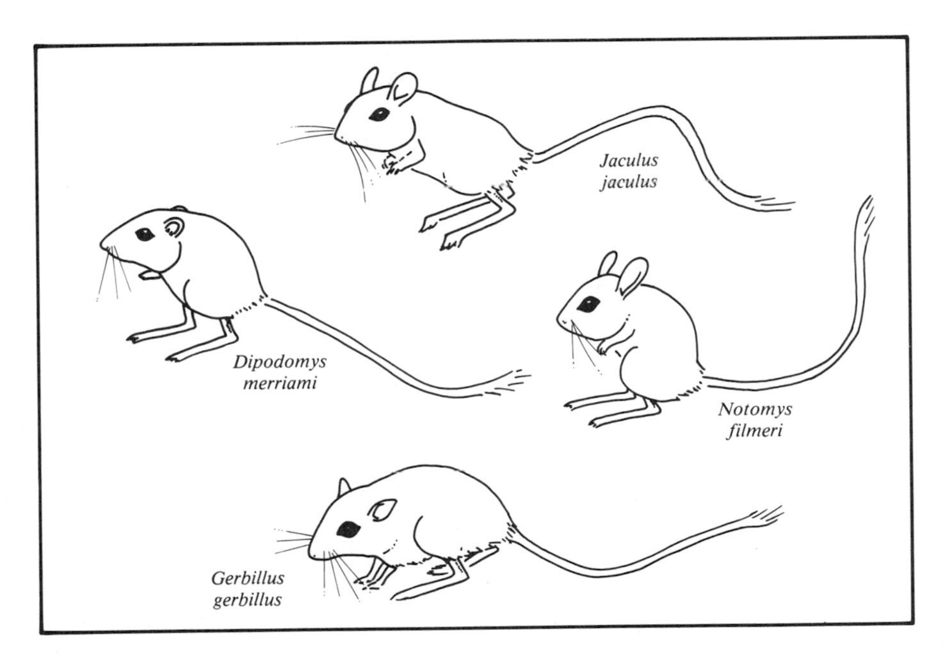

Figure 18.7
Morphological convergence of seed-eating desert rodents on four continents. The representatives shown are *Dipodomys* (Heteromyidae) from North America, *Jaculus* (Dipodidae) from Asia, *Gerbillus* (Cricetidae) from Australia, and *Notomys* (Muridae) from Australia. Each of these groups has independently evolved bipedal hopping rodents similar in morphology, physiology, behavior, and ecology from generalized mouselike ancestors.

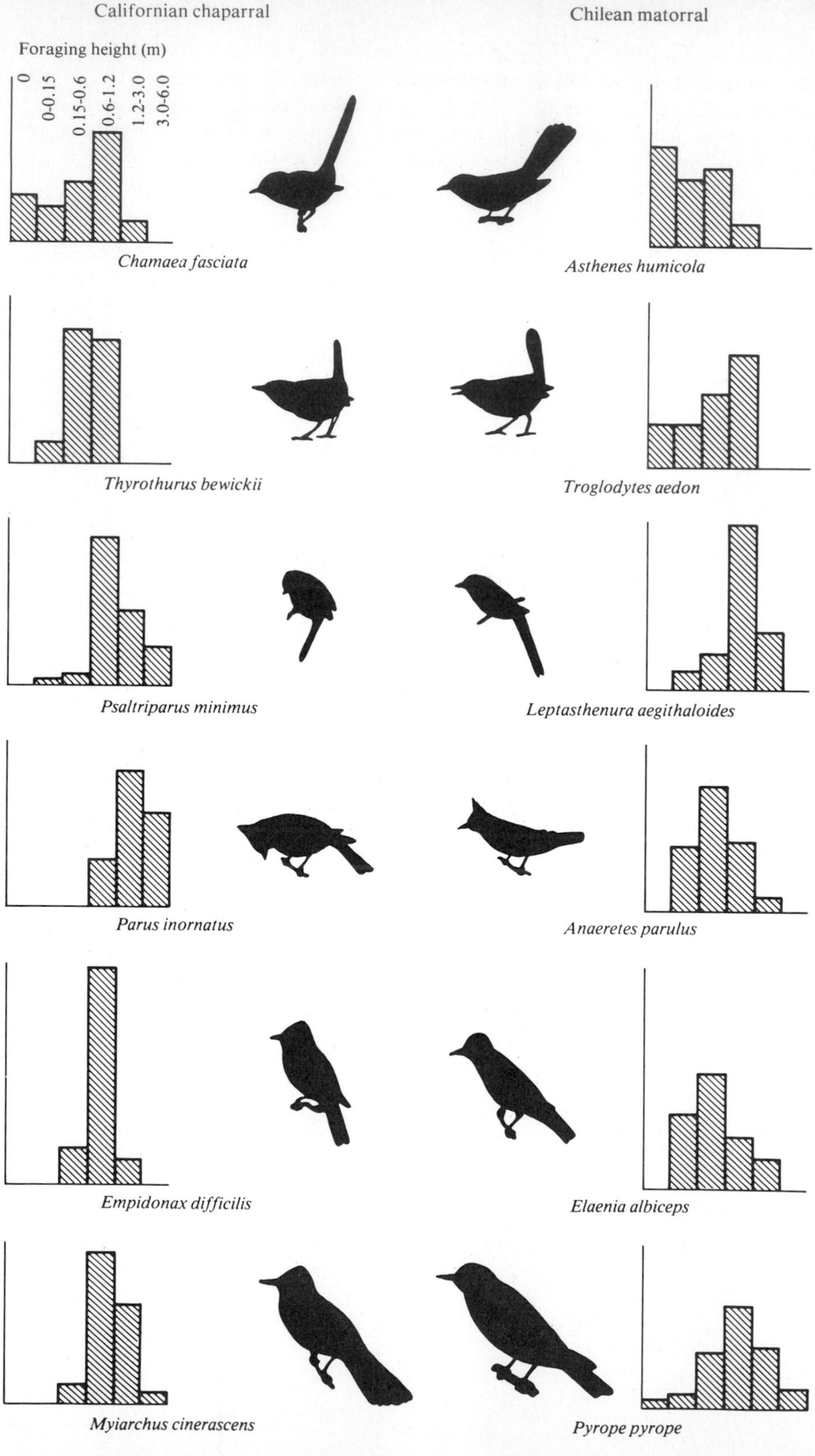

Californian chaparral

Chilean matorral

Foraging height (m)

0
0-0.15
0.15-0.6
0.6-1.2
1.2-3.0
3.0-6.0

Chamaea fasciata

Asthenes humicola

Thyrothurus bewickii

Troglodytes aedon

Psaltriparus minimus

Leptasthenura aegithaloides

Parus inornatus

Anaeretes parulus

Empidonax difficilis

Elaenia albiceps

Myiarchus cinerascens

Pyrope pyrope

Figure 18.8
For legend see opposite page.

fynbos. Although individual species of the dominant shrubs and other plants differ in details of their morphology, physiology, and life history, reflecting in part their diverse historical origins, there are many structural and functional resemblances among the vegetations. The dominant woody shrubs have small to medium-sized, thick, leathery, evergreen leaves (a condition known as sclerophylly). They often grow in dense, mixed species stands forming an almost impenetrable mat of vegetation 1 to 3 m high. Fire plays a major role in these communities, and the shrubs have several adaptations, including fire-enhanced germination and stump sprouting, which result in rapid secondary succession following burning. Recent quantitative studies, especially in California and Chile, have demonstrated other similarities at levels ranging from details of the photosynthetic process to patterns of phenological changes through the seasons to distributions of life forms in climatic gradients of latitude and elevation.

The animals that occur in these distinctive habitats also exhibit many similarities. Perhaps the most thorough studies have been of the birds (Cody, 1974, 1975; Cody and Mooney, 1978; Mooney, 1977) and lizards (Fuentes, 1976; Cody and Mooney, 1977). Although it is sometimes difficult to identify exact ecological replacements among the bird species (but see Figure 18.8), if they are divided into major feeding niches or foraging guilds the four mediterranean-type shrub communities studied by Cody are much more similar in their organization than they are to two other "control" communities from similar or nearby habitats analyzed in the same way (Table 18.6). Nevertheless, note that even this overall convergence is far from perfect, so that some niches remain unfilled, at least by birds, in certain regions (e.g., there are no nectarivores in Sardinian macchia). Lizards have been studied in California, Chile, and Sardinia where there are four, five, and three species, respectively (Cody and Mooney, 1978). Careful comparisons of the Californian and Chilean communities indicate again that it is difficult to document obvious convergence in the characteristics of particular pairs of species, but the communities are organized so that the major substrates and food items available to lizards are used quite similarly by the different combinations of species on each continent (Figure 18.9) (Fuentes, 1976).

Studies of divergence and convergence of biotas in different regions already have contributed substantially to the understanding of biogeography, but they promise to contribute much more as investigators learn more about the relationship between historical geologic and climatic events and ecological and evolutionary processes. Many of the most interesting questions remain unanswered. At best, our present knowledge is tentative and incomplete. Investigators would like to know why, for example, there are no semiaquatic grazing herbivores or arboreal sap feeders among the mammals of southeastern Asia and no nectar feeders among the birds of southern Europe, although these niches are filled in comparable environments on other continents. Are these resources not available in these communities, or are they unexploited or used by other taxa? In the case of nectar-feeding birds, investigators know some

Figure 18.8
Ecological convergence exhibited by some pairs of insectivorous bird species inhabiting evergreen shrub habitats in mediterranean-type climates in California and Chile. Species in different genera and families play similar ecological roles on the two continents, not only by foraging at similar heights in the vegetation (as shown here) but also by using similar behavior to find and consume the same kinds of insect prey. (Simplified from Cody and Mooney, 1978.)

Table 18.6

Bird community organization for four sites with mediterranean-type shrub habitat

Data are for sites on different continents and include two other control sites in different habitat types. Note that although representatives of different families fill the niches on the different continents, the number of species and densities are relatively similar among the mediterranean type habitats. (From Cody and Mooney, 1978.)

Foraging guild	California chaparral		Chilean matorral		Sardinian macchia	
	Number of species by family	Total density pairs/hectare	Number of species by family	Total density pairs/hectare	Number of species by family	Total density pairs/hectare
Foliage insectivores	Chamaeidae (1) Troglodytidae (1) Paridae (2) Vireonidae (1) Sylviidae (1) Parulidae (1)	6.61	Tyrannidae (1) Troglodytidae (1) Furnariidae (2)	5.75	Sylviidae (5) Paridae (3) Troglodytidae (1)	7.74
Sallying flycatchers	Tyrannidae (2)	0.39	Tyrannidae (2)	1.51	Musicapidae (1)	0.30
Nectarivores	Trochilidae (2)	1.18	Trochilidae (2)	1.11		0.00
Ground foragers	Emberizidae (2) Corvidae (1) Phasianidae (1) Columbidae (1) Cuculidae (1) Mimidae (1)	5.31	Emberizidae (2) Mimidae (1) Rhinocryptidae (3) Icteridae (1) Phasianidae (1) Tinamidae (1) Columbidae (1) Tryannidae (1)	7.06	Turdidae (2) Fringillidae (1) Corvidae (1) Laniidae (1) Phasianidae (1)	4.36
Seed and fruit eaters	Fringillidae (2) Emberizidae (1)	0.60	Phytotomidae (1) Fringillidae (1)	1.05	Fringillidae (3)	0.94
Trunk and bark foragers	Picidae (2)	0.35	Picidae (2)	0.08		0.00
Aerial insectivores	Apodidae (1)	common	Hirundinidae (1)	common	Apodidae (1)	rare
Total	17 families 23 species	14.44	15 families 24 species	16.56	10 families 20 species	13.34

Foraging guild	South African fynbos		Californian oak woodland		British successional scrub	
	Number of species by family	Total density pairs/hectare	Number of species by family	Total density pairs/hectare	Number of species by family	Total density pairs/hectare
Foliage insectivores	Zosteropidae (1) Sylviidae (4)	2.18	Paridae (2) Parulidae (2) Emberizidae (1) Troglodytidae (3) Vireonidae (2) Chamaeidae (1) Sylviidae (1) Icteridae (1)	9.43	Sylviidae (5) Paridae (4) Troglodytidae (1)	6.42
Sallying flycatchers	Musicapidae (2)	0.79	Tyrannidae (3)	2.00	Musicapidae (1)	0.10
Nectarivores	Nectariniidae (2) Promeropidae (1)	2.66	Trochilidae (2)	1.25		0.00
Ground foragers	Turdidae (2) Laniidae (1) Phasianidae (1) Ploceidae (1) Columbidae (2) Sturnidae (1)	2.56	Emberizidae (2) Corvidae (1) Columbidae (1) Mimidae (1) Phasianidae (1) Turdidae (1)	4.60	Prunellidae (1) Emberizidae (1) Turdidae (4) Corvidae (2) Phasianidae (1) Motacillidae (1) Columbidae (1) Fringillidae (1)	7.90
Seed- and fruit-eaters	Fringillidae (3) Pycnonotidae (1)	0.93	Emberizidae (1) Fringillidae (4) Ptilogonatidae (1) Columbidae (1)	2.53	Fringillidae (4) Columbidae (1)	4.08
Trunk and bark foragers	Picidae (1)	0.20	Picidae (3)	0.93	Picidae (1)	0.15
Aerial insectivores	Hirundinidae (1)	common	Apodidae (1)	rare	Hirundinidae (1) Apodidae (1)	common
TOTAL	15 families 24 species	9.32	19 families 36 species	20.74	15 families 31 species	18.65

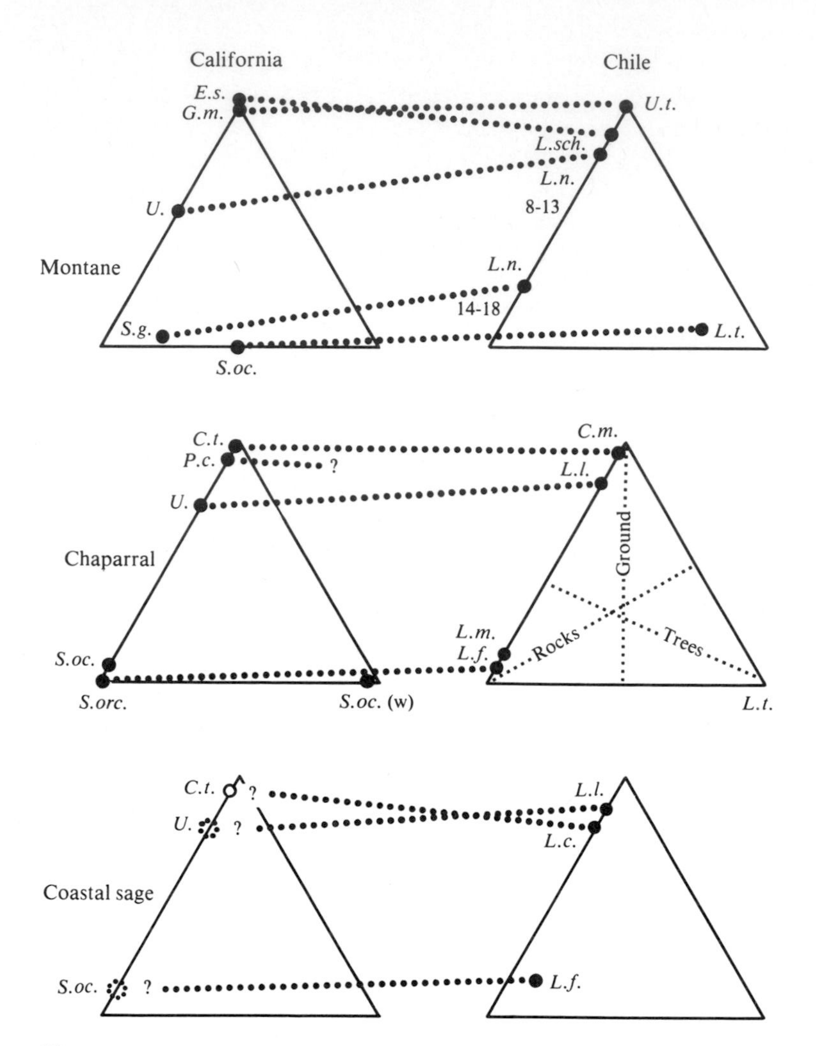

Figure 18.9
Ecological convergence in microhabitat use by lizards in three different habitat types in California and Chile. The triangular graphs show the proportions in which each species *(dots)* use ground, rock, and tree resources. Dotted lines connect closest ecological equivalents, which are always in different genera and sometimes in different families on the two continents. Note the similarities in numbers of species and microhabitat use between the same habitats in the different regions. California taxa are *C.t., Cnemidophorus tiaris; E.s., Eumeces skiltonianus; G.m., Gerrhonotus multicarinatus; S.g., Sceloporus graciosus; S.oc., S. occidentalis; S.oc.* (w), *S. occidentalis* (woodland); *S.orc., S. orcuttii; U., Uta stansburiana; P.c., Phrynosoma coronatum.* Chilean taxa are: *C.m., Callopistes maculatus; L.c., Liolaemus chilensis; L.f., L. fuscus; L.l., L. lemniscatus; L.m., L. monticola; L.n.* 8 to 13, *L. nigroviridis* 8 to 13 mm; *L.n.* 14 to 18, *L. nigroviridis* 14 to 18 mm; *L.sch., L. schroederi; L.t., L. tenuis; U.t., Urostrophus torquatus.* (From E.R. Fuentes. Ecological convergence of lizard communities in Chile and California. Ecology 57[1]:3-17. Adapted by permission of Duke University Press. Copyright © 1976, Ecological Society of America.)

of the answers. There are no native flowers specialized for pollination by nectarivorous birds in Europe; the kinds of plants pollinated by birds on all other continents are serviced by insects in Europe and northern Asia. But this is only a partial answer. Why haven't bird-flower mutualisms coevolved on the largest continental landmass? And, given that they have not yet done so, how resistant are the present biotas to invasion of mutualistic combinations of bird and flower species?

Obviously, the complete answers to these kinds of questions can be neither entirely historical nor solely ecological. Nor can they be obtained by studying the distribution of just a single taxonomic group, even one as large as vascular plants or terrestrial vertebrates. In the past, biogeography was divided into specialized disciplines, such as phytogeography and zoogeography and historical and ecological biogeography, which appear to have different and sometimes even conflicting approaches and goals. Ultimately, however, the goal of all biogeographers must be the same: to explain the distribution of living things on the earth. This goal will be satisfied only when investigators understand how the climatic and geologic history of the earth and the ecological and evolutionary processes of organisms have combined to determine the present and past distributions of individual species and entire biotas. Such a complete understanding is a long way in the future. It will require integration of a tremendous amount of information from both the biological and the physical sciences, and many of the essential data are not yet available. However, we hope that this book has helped to put the established facts and concepts, the still unverified data and interpretations, and the needs for further empirical and theoretical research in a useful perspective. Perhaps some of our readers will make the next important contributions.

Selected references

Patterns and randomness

Bowers and Brown (in preparation); Case and Sidell (1982); Case et al. (1982); K. Cole (1954); Connor and Simberloff (1978, 1979); Diamond and Gilpin (1982); Gilpin and Diamond (1982); Gould et al. (1977); Osman and Whitlatch (1978); Raup and Gould (1974); Raup et al. (1973); Simberloff (1970, 1974a); Stanley et al. (1981); Strong (1980); C. Williams (1964); Wright and Biehl (1982).

Distributions of closely related species

Bowers and Brown (in preparation); Brown and Lieberman (1973); Case et al. (1982); MacArthur and Levins (1967); Pielou (1977b, 1978, 1979); Simberloff (1970); Simberloff and Vuilleumier (1980); Terborgh (1971).

Distributions of distantly related species

Briggs (1968, 1974); J. Brown and Davidson (1977); Boucot (1978); De Vos et al. (1956); Elton (1966); Flessa (1975, 1981); Flessa and Miyazaki (1978); Flessa et al. (1979); Gray et al. (1981; R. Johnston and Klitz (1977); R. Johnston and Selander (1964, 1971, 1973); Marshall and Hecht (1978); Marshall et al. (1982); Por (1971, 1975); J.W. Porter (1972, 1974); Rubinoff (1968); Rubinoff and Rubinoff (1971); Selander and Johnson (1973); G. Simpson (1947b, 1950, 1980b); Topp (1969); Vermeij (1978); S. Webb (1976, 1978).

Divergence and convergence of biotas

Blair et al. (1976); Bourliere (1973); Briggs (1968); Campell and Van der Meulen (1980); J. Brown (1975); Cody (1966, 1973a, 1973b, 1974, 1980); Cody and Mooney (1978); De Vos et al. (1956); Dexter (1972); di Castri and Mooney (1973); Dubost (1968); Eisenberg (1981); Fuentes (1976); Karr (1976); Karr and James (1975); Kornas (1972); Mares (1980); L. Marshall and Hecht (1978); Mooney (1977); Nevo (1979) J. Porter (1972, 1974); Rubinoff (1968); Rubinoff and Rubinoff (1971); Topp (1969); Traband and Lepart (1980); Vauter et al. (1980).

Glossary

abiotic The nonliving components of an ecosystem, such as water, heat, solar radiation, and minerals.

abscissa On a graph, the horizontal (x) axis or the horizontal coordinate of a point on that graph.

abyss In deep bodies of water, the zone between 4000 and 6000 m through which solar radiation does not penetrate (aphotic) and temperature remains at or slightly below 4° C year-round.

abyssal plain The relatively flat floor of a deep ocean, mostly between 4 and 6 km beneath the surface.

acid rain Precipitation with an extremely low pH. The acid condition is caused by the combination of water vapor in the atmosphere with hydrogen sulfide vapor released from the burning of fossil fuels, producing sulfuric acid.

active dispersal The movement of an organism from one point to another by its own motility, such as by active swimming, walking, or flying, rather than by being carried along by some other force; compare with **passive dispersal.**

actualism The philosophical assumption that the physical processes now operating are timeless, and therefore the fundamental laws of nature have remained unchanged; also called **uniformitarianism.**

adaptation Any feature of an organism that substantially improves its ability to survive and leave more offspring over that of other ancestral forms or coexisting phenotypes.

adaptive radiation The evolutionary divergence of a monophyletic taxon (from a single ancestral condition) into a number of very different forms and life-styles (adaptive zones).

adaptive zone A way of life, including such properties as ecological preference and mode of feeding, that has been adopted by a group of organisms.

adiabatic cooling The decrease in air temperature as a result of a decrease in air pressure (not a loss of heat to the outside) as warm air rises and expands. The rate of cooling is about 1° C/100 m for dry air and 0.6° C/100 m for moist air.

aerial Occurring in the air.

aestivation A specialized type of animal behavior and physiology in which the organism lives through the summer in a dormant condition.

age and area hypothesis According to Willis, an explanation that states the greater the age of a taxon, the larger its distributional range.

allele One of two or more alternative forms of a gene located at a single point (locus) on a chromosome.

Allen's rule Among homeotherms, the ecogeographic trend for limbs and extremities to become shorter and more compact, respectively, in colder climates than in warmer ones.

allochthonous Having originated outside the area in which it now occurs.

allometry The manner in which the relative size of one part of an organism increases in relation to the size increase of the entire organism; also known as **scaling.**

allopatric Occurring in geographically different places, i.e., ranges that are mutually exclusive.

allopatric speciation The formation of a new species that occurs when the populations are geographically separated.

allopolyploid A hybrid polyploid formed following the union of two gametes, usually from distantly related species, with nonhomologous chromosomes.

alpha diversity The number of species within a small area of a relatively homogeneous habitat.

amphitropical Occurring in subtropical or temperate areas on opposite sides of the tropics; extratropical.

anadromous Living in salt water but breeding in fresh water.

anagenesis The process of evolution that produces entirely new levels of structural organization (grades).

ancestor The individual or population that gave rise to subsequent individuals or populations with different features.

aneuploidy The formation of a new chromosomal arrangement resulting in the increase or decrease of the chromosome number by one pair; often caused by an uneven meiotic division.

anthropochory The unintentional transport of disseminules by humans.

aphotic zone The lower zone in a water column, usually below 50 to 100 m, in which the intensity of solar radiation is too low to permit photosynthesis by plants.

apomixis Reproduction without the union of sexual cells (gametes).

apomorphy In cladistics or phylogenetic systematics, a derived character state.

apterous Wingless; often used to contrast these forms with their primitive ancestors that had wings and the ability to fly.

aquatic Living exclusively or for most of the time in water.

arboreal Living predominantly or entirely in the canopies of trees.

arborescent Treelike.

arctic Pertaining to all nonforested areas north of the coniferous forests in the Northern Hemisphere, especially everything north of the Arctic Circle.

arid Exceedingly dry; strictly defined as any region receiving less than 10 cm of annual precipitation.

asthenosphere A fluid, viscous zone of the upper mantle on which the continental and oceanic plates float (ride) and over which they move.

austral Pertaining to the temperate and subtemperate zone of the Southern Hemisphere.

Australasia The continental fragments of the original Australian Plate, including Australia, New Zealand, New Guinea, Tasmania, Timor, New Caledonia, and several smaller islands.

autochthonous Having originated in the area in which it presently occurs.

autogamous Able to produce offspring sexually by the fusion of gametes from the same individual, such as the pollen and ovules from the same plant.

autopolyploid A polyploid possessing more than two sets of homologous chromosomes.

autosome A chromosome that is not a sex chromosome; a somatic chromosome.

autotroph An organism that uses carbon dioxide occurring in the environment as the primary source of cellular carbon.

avifauna All the species of birds inhabiting a specified region.

baccate Berrylike.

barrier Any abiotic or biotic feature that totally or partially restricts the movement (flow) of genes or individuals from one population or locality to another.

basin-and-range topography In geology, a geographic region characterized by wide, parallel valleys periodically interrupted by small but high mountain ranges.

bathyal Pertaining to anything in the deep sea, in particular those occurring within the aphotic zone but above 4000 m.

bathymetry The depth and configuration of the bottom of a body of water.

Benioff zones Zones of high earthquake activity, located on the back sides of trenches, which are caused by the subduction of a plate. The earthquakes of a Benioff zone are shallow near the trench and progressively deeper at greater distances.

benthic Living at, in, or associated with structures on the bottom of a body of water.

Bergmann's rule An ecogeographic rule stating that in homeotherms populations from cooler climates tend to have larger body size and hence have smaller surface-to-volume ratios than related populations living in warmer climates.

Beringia The geographic area of western Alaska, the Aleutians, and eastern Siberia that was connected in the Cenozoic by a land bridge when the Bering Sea and adjacent shallow waters became emergent.

beta diversity The rate of change (turnover) in species composition with relatively small changes in habitat.

biogeography The study of the geographic distributions of organisms, both past and present.

biological species A group of potentially interbreeding populations that are reproductively isolated from all other populations.

biomass The amount of living material of an organism, population, or community.

biome A major type of natural vegetation that occurs wherever a particular set of climatic and edaphic conditions prevail but that may have different taxa in the different regions, e.g., temperate grassland.

biosphere Collectively, all the living things of the earth and the areas they may inhabit.

biota All species of plants, animals, and microbes inhabiting a specified region.

biotemperature The mean annual temperature of a locality, determined by averaging over all hours and treating all subzero periods as zero.

biotic Pertaining to the components of an ecosystem that are living or came from a once-living form.

bipedal Using two hind limbs for locomotion, usually by hopping or jumping, such as a kangaroo or a kangaroo rat.

bipolar Occurring at both poles, in the cold or subtemperate zones.

bivoltine Breeding twice per year.

boreal Occurring in the temperate and subtemperate zone of the Northern Hemisphere that characteristically has coniferous (evergreen) forests and some types of deciduous forest.

bottleneck In evolutionary biology, the general term for any stressful situation that greatly reduces the size of a population.

brachypterous Short-winged, as in certain species of insects that have long- (macropterous) and short-winged (brachypterous) forms.

brackish Water that has salt concentrations greater than fresh water (> 0.5 ‰) and less than seawater (35 ‰).

breeding area In migratory land animals, the area in which populations mate and produce offspring.

browser An animal that feeds on plant materials, especially on woody parts of trees and shrubs.

calcareous In soil biology, pertaining to a soil whose horizons are rich in calcium carbonate and have a basic reaction.

calcification The formation of a soil under conditions of a continental climate with relatively low moisture and hot to cool temperature, resulting in a soil rich in calcium carbonate ($CaCO_3$) because rainfall is not sufficient to leach calcium from the upper soil horizons.

calciphile A plant that grows exclusively or best on a soil rich in calcium carbonate.

caliche A hard, often rocklike layer of calcium carbonate that forms in soils of arid regions at the level in which the leached calcium salts from the upper soil horizon are precipitated.

canonical distribution A log-normal distribution of the number of individuals or species in an insular habitat according to the mathematical formulation of Preston (1962).

carnivore An animal that feeds mostly or entirely on animal prey.

carrying capacity The total number of individuals of a species that the resources of a habitat can support.

catadromous Living in fresh water but breeding in seawater.

catastrophic extinction A major episode of extinction for many taxa, occurring fairly suddenly in the fossil record.

catastrophic speciation A term used to describe a sudden speciation event in which a new species arises from a population that was recently reduced to a few individuals by a natural catastrophe.

chaparral The type of sclerophyllous scrub occurring in the southwestern region of North America with a mediterranean-type climate.

character Any feature or attribute of an organism.

character displacement Divergence of a feature of two markedly similar species wherever their ranges overlap in order to avoid the effects of competition, so that where the two species cooccur each uses some different resources.

character state One of several alternative forms of a character, e.g., the ancestral form or one of several derived forms.

chromosome An organelle consisting of genetic material on long-stranded deoxyribonucleic acid (DNA) wrapped with proteins.

circumboreal Occurring in the temperate or subtemperate zones of the New and Old World portions of the Northern Hemisphere.

clade Any evolutionary branch in a phylogeny, especially one that is based on genealogical relationships.

cladistics The method of reconstructing the evolutionary history (phylogeny) of a taxon by identifying the branching sequence of differentiation through analysis of shared (nested) derived character states. See **phylogenetic systematics.**

cladogenesis The process of evolution that produces a series of branching events.

cladogram A line diagram derived from a cladistic analysis showing the hypothesized branching sequence (genealogy) of a monophyletic taxon and using shared derived character states (synapomorphies) to determine when each branch diverged.

cleistogamy Self-pollination within a flower that never opens; a form of autogamy.

climax Any community that perpetuates itself under the prevailing climatic and edaphic conditions, therefore the last stage (sere) in secondary succession.

cline The change in one or several heritable characteristics of populations along a geographic transect, attributable to changes in the frequencies of certain alleles and often correlated with a gradual change in the environment.

coevolution The simultaneous, interdependent evolution of two unrelated species that have strong ecological interactions, such as a flower and its pollinator or a predator and its prey.

coexistence Living together in the same local community.

colonization The immigration of a species into a new habitat and the founding of a population.

commensalism An interspecific relationship in which one species draws benefits from the association and the other one is unaffected.

community An assemblage of organisms that live in a particular habitat and interact with each other.

community ecology The study of interactions among cooccurring organisms living in a particular habitat.

competition Any interaction that is mutually detrimental to both participants. Interspecific competition occurs between species that share requirements for limited resources.

competitive exclusion The principle that when two species that have similar requirements for resources cooccur, one species eventually outcompetes and causes the extinction of the other.

congeners Species belonging to the same genus.

continental drift The model first proposed by Alfred Wegener that the continents were once united and have since become independent structures that have been displaced over the surface of the globe.

continental island Islands that were formed as part of a continent and that have a nucleus of continental (sialic) rocks.

convergent evolution The development of two or more species with strong superficial resemblances from totally unlike and unrelated ancestors.

Cope's rule A special type of directional evolution in a taxon (orthogenesis) in which there occurs a trend toward increased body size.

coprolite Fossil excrement.

Coriolis effect The physical consequence of the law of conservation of angular momentum whereby, as a result of the earth's rotation, a moving object veers to the right in the Northern Hemisphere and to the left in the Southern Hemisphere.

corridor A route that permits the direct spread of many or most taxa from one region to another.

corroborate In scientific inquiries, to support an hypothesis with evidence.

cosmopolitan Occurring essentially worldwide, as on all habitable landmasses or in all major oceanic regions.

craton The stable crustal nucleus of a continent or continental island, which is older than 600 million years; also called **Precambrian shield.**

crust The outermost rock layer of the earth, covering the mantle.

cryptic species Species of a genus that are morphologically so similar that they cannot be visually distinguished by superficial features.

Curie point The temperature at which remanent magnetism develops in cooling minerals, e.g., 680° C for hemitite and 580° C for magnetite.

deciduous In plants, having leaves that are shed for at least one season, usually in response to the onset of cold or drought.

decomposer An organism (usually a bacterium or fungus) capable of metabolically breaking down organic materials into simple organic and inorganic compounds and releasing them into the ecosystem.

deductive reasoning A method of analysis in which one reasons from general constructs to specific cases.

defaunation The elimination of animal life from a particular area.

dehiscence The act of splitting along a natural line to discharge the contents, such as the dehiscence of an anther to release pollen grains or of a capsule to release seeds.

demersal Referring to aquatic eggs that are deposited at the bottom, e.g., attached to a benthic substrate.

density compensation In island biogeography, the case in which the density of a species inhabiting an island habitat increases when one or more taxonomically similar competitors are absent.

density overcompensation (excess density compensation) In island biogeography, the case in which the total densities of a few species inhabiting a small island exceed the combined densities of a much greater number of species of the same taxon occupying similar habitats on a large island or continent.

desert A general term for an extremely dry habitat, especially where water is unavailable for plant growth most of the year; in particular, a habitat with long periods of water stress and sparse coverage by plants, often with perennials covering less than 10% of the total area.

desilification The removal of silica from the soil by leaching, especially in wet tropical environments.

deterministic Having been determined or controlled by some regulatory force, such as natural selection; compare with **stochastic.**

detritivore An organism that feeds solely on detritus.

detritus Freshly dead or partially decomposed organic matter.

diadromous Referring to an aquatic animal that must migrate between fresh water and seawater to complete its life cycle, such as certain lampreys and eels.

diapause An arrested state of development in the life cycle, especially of many insects, during which the organism has reduced metabolism and is more resistant to stressful environmental conditions, such as cold, heat, or drought.

diaspore Any part of a plant or stage in the life cycle that is adapted for dispersal.

diffuse competition A type of competition in which one species is interfered with by numerous other species that deplete resources, although no single competitor can be identified.

diffusion The passive movement of a solution, gas, or heat from a condition of high concentration (high free energy) to one of lower concentration (low free energy).

dimorphic Having two distinct forms in a population.

dioecious Having individuals with only male or only female reproductive systems.

diploid Having two sets of chromosomes (2N).

disharmonic Having a biota that is biased toward those groups with good dispersal capabilities and at the same time lacking those with poor dispersal properties; therefore a disharmonic biota is not a random subset of the source area from a biological viewpoint.

disjunct Any taxon, e.g., a species or some higher category, whose range is geographically isolated from its closest relatives.

disjunction A discontinuous range of a monophyletic taxon in which at least two closely related populations are separated by a wide geographic distance.

dispersal The movement of organisms away from their point of origin; this may result in extending the range on the margin of an existing population by the colonization of a new habitat within the range of the population or by the colonization of a distant location across a major physical barrier or unfavorable habitat.

dispersal biogeography A loosely defined term for the study of the distribution patterns of organisms that places strong emphasis on the dispersal capabilities and ecological properties of each species when evaluating the origin of elements in a particular biota; compare with **vicariance biogeography.**

dispersion The spatial distribution of individual organisms within a local population.

disseminule Any part of a plant that is used for dispersal; occasionally restricted to include only seeds and seed-bearing structures. See also **diaspore.**

doldrums A narrow equatorial zone characterized by long periods of calm or light shifting winds, caused by the upward movement of air masses from this region of high atmospheric pressure to higher latitudes with relatively lower pressure.

dominant Any species having great influence on the composition and structure of a community by virtue of its abundance, size, or aggressive behavior.

double fertilization In plants, the synchronous process involving the fusion of one sperm nucleus with an egg nucleus to form a zygote and the fusion of a second sperm nucleus with another embryo sac nucleus (diploid or polyploid) to form endosperm.

downwelling The vertical downward movement of surface water to abyssal depths.

dynamism The sum total of major and minor and

regional and local fluctuations in the range of each taxon.

ecogeographic rule Any generalized statement that accounts for a regular change in the design of organisms along a geographic (and climatic) transect, e.g., Allen's rule or Bergmann's rule.

ecological biogeography The study of the ecological factors influencing the distributions of organisms.

ecological niche See **niche.**

ecological time The period in which populations may interact with their environment and respond to environmental fluctuations without undergoing substantial evolutionary modification.

ecology The study of the abundance and distribution of organisms. The relationships between organisms and their biotic and abiotic environments.

ecosystem The set of biotic and abiotic components in a given environment.

ecotone A zone of transition between two habitats or communities.

ectoparasite A parasite that lives on the exterior of its host, such as a louse.

ectotherm An animal whose body temperature is determined largely by the temperature of the environment, not by its own oxidative metabolism; compare with **endotherm.**

edaphic Pertaining to soil.

electrolyte A substance that in solution disassociates as ions and that can act as an electric conductor.

emigration Dispersal from the place of origin to a new residence.

endemic A taxon restricted to the geographic area specified, such as a continent, lake, biome, or island.

endosperm The nutritive tissue in the embryo or seed of an angiosperm.

endotherm An animal whose body temperature is maintained largely by its own metabolic heat production; compare with **ectotherm.**

entropy A measure of the unavailable energy in a closed thermodynamic system.

epeiric sea A large body of salt water that lies over a part of a continent; an epicontinental sea.

epibiotic A taxon that was at one time very wide ranging but that is now narrowly restricted following the extinction of most populations.

epicenter The point on the earth's surface directly above the origin (focus) of an earthquake.

epicontinental sea See **epeiric sea.**

epifaunal Living on a substrate, such as a barnacle or coral.

epipelagic Living in open water, mostly within the upper 100 m.

epiphyllous Living on a leaf.

epiphyte A plant that usually lives on another plant (not rooted in soil) and derives its moisture and nutrients from atmospheric precipitation and whatever materials are released by the organisms in the immediate vicinity.

equatorial submergence The occurrence in the tropics, in deep, cold, aphotic zones, of species that occur in shallow water near the poles; compare with **polar emergence.**

equilibrium A condition of balance between opposing forces, such as birth and death rates or immigration and extinction rates.

equilibrium theory of island biogeography The theory of MacArthur and Wilson that explains the species number on any island as an equilibrium between the opposing rates of colonization and extinction.

equilibrium turnover rate The change in species composition per unit time when immigration equals extinction.

equinox Either of two times (March 21 and September 22) in a year when the sun passes the equator so that day and night are the same length everywhere on earth.

establishment The successful start of a population.

Ethiopian Region The region of Africa south of the Sahara Desert plus Madagascar and other nearby islands.

eukaryotic Referring to any organism in which the genetic material of a cell is organized into chromosomes and included within a nucleus.

euryhaline Having a tolerance to an extremely wide range of salt concentrations.

eurytopic Having a tolerance to an extremely wide range of habitats and environmental conditions.

eutherian A placental mammal.

eutrophic lake A lake that is rich in dissolved nutrients but that is usually shallow and seasonally deficient in oxygen.

evapotranspiration The sum total of water lost by evaporation from land and water and by transpiration of plants.

evergreen A plant having leaves at all times of the year.

evolution In the strictest sense, any irreversible change in the genetic composition of a population.

evolutionary species A discrete cluster of individuals and populations, evolving separately from other clusters, that exhibits a clear pattern of parental ancestry and descent; compare with **biological species.**

evolutionary systematics The method of reconstructing the evolutionary history (phylogeny) of a taxon by analyzing the evolution of major features along with the distribution of both shared primitive and shared derived characteristics; compare with **cladistics**.

evolutionary time The period during which a population can evolve and become adapted to an environment by means of genetic changes.

exoskeleton An external skeleton on an animal, as of a clam or insect.

extant Living at this time.

extinct No longer living.

extratropical Occurring outside the tropical zone.

facultative response A response to changing or varying environmental conditions that is mediated by behavioral or physiological processes.

family A taxonomic category above the rank of genus and below the rank of order.

fattening area In migratory aquatic animals, the region in which individuals consume most of the food needed for the subsequent reproductive investment; compare with **spawning area**.

fault In geology, a weakness in the earth's crust along which there can be crustal motion and displacement.

filter A geographic or ecological barrier that restricts some dispersal between regions and blocks passage of certain forms but not others.

filter feeder An aquatic animal that feeds on plankton or other minute organic particules by using one of a variety of filtering mechanisms.

fitness (or Darwinian fitness) The relative ability of a genotype to leave offspring in the next generation or succeeding generations as compared with other genotypes.

flyway An established air route used by vast numbers of migratory birds.

forest Any of a variety of vegetation types dominated by trees and usually having a fairly well developed or closed canopy when trees have leaves.

fossil A remnant, impression, or other trace of a living organism from the past.

fossorial Referring to an animal that lives in and forages on plants from a burrow.

founders principle Greatly simplified, the principle that when a new isolated population is started, such as on an island, by one or a few colonists (founders), the features of the new population may be markedly different than the ancestral population because the gene pool of the founders may be a biased and small sample of the source.

fragment In plate tectonics, a portion of a former landmass; also called a **microplate.**

fresh water In the strictest sense, water that has less than 0.5 ‰ of salt concentration.

frugivore An animal that feeds mainly on juicy fruits.

fundamental niche The total range of environmental conditions in which a species can survive and reproduce.

fynbos The type of sclerophyllous scrub occurring in the region of South Africa with a mediterranean-type climate.

gamete One of two cells, usually from different parents, that fuse to form a zygote.

gametophyte In plants, the structure in a life cycle that produces gametes.

gamma diversity The diversity differences between similar habitats in widely separated geographic regions.

Gause's principle See **competitive exclusion.**

geminate species Pairs of species that are morphologically similar and that formed when a geographic barrier split an initial population; also called **twin species** and **vicariants.**

gene The small unit of a DNA molecule that is coded for a specific protein to produce one of the chemical, physiological,and structural attributes of an organism.

genealogy The study of the exact sequence of descent from an ancestor to all derived forms.

gene flow The movement of alleles within a population or between populations caused by the dispersal of gametes or offspring.

gene frequency (allelic frequency) The proportions of gene forms (alleles) in a population.

generalized track In vicariance biogeography, a line drawn on a map representing the coincident distributions of numerous disjunct taxa.

genetic drift Change in gene frequency of a population caused solely by chance, i.e., which individuals happened to mate and leave offspring in the next generation without an influence by natural selection. This event is important in small populations.

genome A full set of chromosomes.

genotype The total genetic message found in a cell or an individual.

genus A taxonomic category for classifying species derived from a common ancestor; a rank below that of family and tribe.

geographic isolation Spatial separation of two potentially interbreeding populations; **allopatry.**

geographic speciation See **allopatric speciation.**

gleization The formation of a soil under moist and cool or cold conditions, resulting in an acid soil with a great amount of organic matter and iron present in a reduced state (FeO).

Gondwanaland The supercontinent, consisting of all southern continental landmasses and India, which was united for at least 1 billion years but broke up during the Upper Mesozoic and Lower Cenozoic.

grade A level of organization, such as the structure of a jaw or the condition of homeothermy, which may or may not have evolved only once from the same ancestry.

granivore An animal, such as many rodents, that feeds mainly on dry seeds and fruits; a special type of herbivore.

grassland Any of a variety of vegetation types composed mostly of grasses and other herbaceous plants (forbs) but few if any trees and shrubs.

gravid Referring to a female bearing fertilized ova or unborn young.

greenhouse effect The retention of heat in the atmosphere when clouds (water vapor) and carbon dioxide absorb the infrared (heat) radiation reradiated from the earth rather than permitting the heat to escape.

guide fossil A fossil used to date the age of a sedimentary stratum. Also called **index fossil.**

guyot A flat-topped submarine volcano.

habitat island Any geographically isolated or patchy habitat, such as a pond, mountaintop, or cave, that can be studied in the same ways as oceanic islands for patterns of colonization and extinction.

habitat selection The preference for particular habitat types, e.g., grassland or coniferous forest.

hadal In deep oceans, the deepest zones below 6000 meters, e.g., in the oceanic trenches, which have constant environments with no light and year-round temperature near 4° C.

half-life The amount of time needed for half of the radioactive material in a rock to decay to a stable element.

haploid Having one set of chromosomes (N).

harmonic biota (balanced biota) A balanced assemblage of organisms inhabiting a specified region, being similar to the biota of the source area or areas.

herbivore An animal that feeds mostly or entirely on plants.

hermaphroditic Having both male and female reproductive structures in the same individual.

heterochromatin Dark-staining regions on the chromosomes that tend to replicate later than other regions of the DNA strand.

heterogeneity The state of being mixed in composition, as in genetic or environmental heterogeneity.

heterotroph An organism that uses organic carbon (compounds made by living organisms) as its source of cellular carbon.

historical biogeography The study that attempts to determine the relationship of present and past distributions of organisms to the physical history of the earth.

Holarctic Region The extratropical zone of the Northern Hemisphere.

homeostasis The maintenance of a constant internal state despite fluctuations in the external environment.

homeotherm An animal with a fairly constant body temperature.

homologous chromosomes Two chromosomes (in a diploid) that have essentially the same gene sequence and that are similar enough to pair during meiosis.

horizon In soils, the major stratifications or zones, each of which has particular structural and chemical characteristics.

horse latitudes The zones of dry descending air between 30° and 40° N and S latitude, where many deserts of the world are located.

hot spot In plate tectonics, a stationary weak point in the upper mantle that discharges magma (molten rock) as a plate passes over it, producing a narrow chain of islands (archipelago).

hybridization The production of offspring by parents of two different species or dissimilar populations or genotypes.

hyperosmotic Referring to an environment in which water will diffuse from the organism because the external solution has a higher salt concentration.

hypersaline Water that has a higher salt concentration than normal seawater.

hypoosmotic Referring to an environment in which water will diffuse into the organism because the internal solution has a higher salt concentration.

hypothetico-deductive reasoning A method of analysis in which one starts with a new, tentative idea and then tests the predictions and assumptions one by one in a quest to falsify the original idea.

ichthyofauna All the species of fishes inhabiting a specified region.

immigration The arrival of new individuals to a population or habitat.

included niche A niche of a specialized species characterized by a narrow range of conditions and lying entirely within the larger fundamental niche of a more generalized species.

index fossil See **guide fossil.**

inductive reasoning A method of analysis in which one uses specific observations to derive a general principle.

infaunal Living within a substrate, such as clams that bury themselves in sand or mud and that feed by means of a long siphon.

insectivore An animal that feeds mainly on insects.

interglacial During the Quaternary, a phase when glacial ice sheets retreated and climate was more equable.

intertidal zone The zone above the low-tide mark and below the high-tide mark of a body of water; the **littoral zone.**

inversion A complete reversal in orientation of a portion of a chromosome from an ancestral condition; a type of chromosomal mutation.

isocryme A line drawn on a map to show all locations with the same mean minimum monthly sea-surface temperature.

isolating mechanism Any structural, physiological, ecological, or behavioral mechanism that blocks or strongly interferes with hybridization or gene exchange between two populations.

isostasy In geology, the term that describes how continental blocks float on a viscous layer of the mantle. While doing this the block may rise or fall to achieve an equilibrium.

isotherm A line on a map connecting all locations with the same mean temperature.

karyotype The morphological appearance of a set of chromosomes in the first metaphase of meiosis.

laterite In tropical soils, a hard, rocklike layer composed principally of ferric oxide, produced when this compound accumulates in high concentration in the soil.

laterization The formation of a soil under conditions of abundant moisture and high decomposer activity (from high temperature), resulting in a soil from which bases and silica have been removed (leached) and leaving behind a clay rich in ferric and aluminum oxides.

Laurasia The northern half of the supercontinent Pangea, including North America, Europe, and parts of Asia.

Laurasian distribution Any distribution in the Old and New World portion of Laurasia; but having this distribution does not indicate that the distribution was produced in ancient rather than fairly recent times.

leach In soil science, to remove soluble substances with water.

lecithotrophic Referring to larvae that have large yolk reserves so they do not have to spend time in the plankton during development.

lentic Referring to standing freshwater habitats, such as ponds and lakes.

leptokurtic A mathematical distribution characterized by a sharply peaked curve with a long tail.

liana A climbing woody or herbaceous vine that is especially common in wet tropical forests.

Liebig's law of the minimum An early ecological generalization, now discredited or greatly modified, that abundance and distribution are limited by a single factor in the shortest supply. See also **limiting factor.**

life zones The characteristic changes in vegetation composition and form that occur in an elevational or latitudinal gradient.

limiting factor The resource or environmental parameter that most limits the abundance and distribution of a population.

lithosphere The earth's crust exclusive of water (hydrosphere) and living organisms (biosphere).

littoral Referring to the marginal zone of the sea; intertidal. In fresh water, the shallow zone that may contain rooted plants.

log-transformed Referrng to the change of values of a data set by taking the log of each.
long-distance dispersal A general term for the ability of certain organisms to "jump" inhospitable environments to colonize a favorable distant habitat.
lotic Referring to running freshwater habitats, such as brooks and rapids.
macchia In the Old World, a name applied to sclerophyllous scrub vegetation in a region with a mediterranean-type climate.
macroevolution A general term for any evolution above the population level.
macropterous Longwinged. See **brachypterous.**
magnetic reversals In geology, the episodes during which the direction of magnetism has been reversed, approximately twice every million years.
magnetic stripes In geology, the long alternating stripe of normally and reversely magnetized basaltic rock on the ocean floor.
malacofauna All the species of molluscs inhabiting specified region.
Malesia The area including Malaysia and Melanesia, hence the tropical region from southeast Asia to New Guinea.
mantle The second and thickest layer of the earth.
maquis A name applied to sclerophyllous scrub vegetation in the Mediterranean region with its characteristic climate.
marine Living in salt water.
maritime climate In general, a coastal or island environment with little or no freezing, much cloud cover and fog, and less variance in temperature, resulting in a mild climate year-round in comparison to nearby inland or mainland localities.
mattoral In Chile, a name applied to sclerophyllous scrub vegetation in a region with a mediterranean-type climate.
mediterranean-type climate A semiarid climatic regime that typically has mild, rainy winters and hot, dry summers.
megafauna A general term for the large terrestrial vertebrates inhabiting a specified region.
mesic Relatively moist and benign.
Messinian crisis The sudden drainage of the Mediterranean Sea in the Cenozoic.
metabolism The sum total of the positive and negative chemical reactions in an organism or a cell that provide the energy and chemical substances necessary for existence.
metamorphosis A major change in the form of an individual animal during development, e.g., a caterpillar becomes a moth or a butterfly and a tadpole becomes a frog or a toad.
metatherian A marsupial mammal.
microclimate The fine-scale environmental regime.
microcosm A small community that represents in miniature the components and processes of a larger ecosystem.
microhabitat The fine-scale environment that often determines the presence or absence of each kind of organism.
microphyllous In ecology and biogeography, having a small, narrow leaf; characteristic of many plants in very dry habitats.
midden A solid mass of collected organic debris left by an animal.
midoceanic ridge In plate tectonics, a submarine mountain chain, within which sea-floor spreading of the oceanic plates occurs.
migration The process of spatial displacement from one habitat to another by individuals that are morphologically, behaviorally, and physiologically adapted to make the move in a particular direction and by using specific environmental cues; hence an intentional and directional form of animal transport.
mimicry The marked resemblance of an organism to another organism or a background (e.g., a leaf, tree bark, or sand) to deceive predators or prey by "disappearing" (crypsis) or by causing the predator or prey to confuse the mimic with something it is not.
mobility The ability to move or be moved.
monoecious In botany, having separate male and female flowers on the same plant.
monomorphic Having only one form in a population.
monophyletic Having arisen from one ancestral form, in the strictest sense from an initial population.
monospecific Having only one species in the genus.
monotypic Having only one species in the taxon.
motility The ability to move under one's own power, as by wings or a flagellum.
multiarea cladogram A hierarchical statement with three or more branches that hypothesizes the historical relationships of the areas concerned based on the hypothesized relationships of the disjunct taxa living in those areas.

multivoltine Breeding three or more times per year.

mutation Any change in the genetic information that results either in an alteration in a gene (point mutation) or in a major modification in the karyotype (chromosomal mutation), neither of which is usually reversible in the strictest sense.

mutualism An interspecific relationship in which both species receive positive benefits from the interaction.

mycorrhiza A symbiotic relationship between a fungus and a plant root that benefits the plant by providing a source of useful nitrogen, which is manufactured by the fungus.

natural selection The process of eliminating from a population through differential survival and reproduction those individuals with inferior phenotypes.

Nearctic Region The extratropical region of North America.

nectarivore An animal that feeds on plant nectar.

nektonic Free-swimming in the upper zone of open water and strong enough to move against ocean currents.

neoendemic An endemic that evolved in fairly recent times.

Neotropical Region The region from southern Mexico and the West Indies to southern South America.

neritic Referring to the shallow water adjoining a seacoast, especially the zone over a continental shelf.

niche The total requirements of a population or species for all resources and physical conditions.

niche breadth Given that different populations can live and reproduce under different conditions, the range of resources and physical conditions used by a particular population relative to that of other populations.

niche expansion An increase in the range of habitats or resources used by a population, which may occur when a potential competitor is absent.

nomadic Having no fixed pattern of migration.

null hypothesis A statistical hypothesis to be tested in order to determine whether an observation can be a result of chance alone or instead must be the result of some directing force; the null hypothesis states what one expects by chance alone.

nunatak An area in a glaciated region that was not covered by an ice sheet; hence, a refugium.

obligate relationship An interaction between species in which one organism cannot survive or reproduce without the other, e.g., many host-parasite relationships, in which a single species of host is required.

oceanic In marine ecology, referring to open ocean with very deep water.

oceanic island An island that was formed de novo from the floor of the ocean through volcanic activity and that has never been attached or adjacent to a mainland; hence its biota is biased in favor of those groups that have superior dispersal properties.

oligotrophic lake A deep lake with low primary productivity that is not subject to frequent oxygen depletion.

omnivore An animal that feeds on either plants or animals.

order A taxonomic category above the rank of family and below the rank of class.

ordinant On a graph, the vertical (y) axis or the vertical coordinate of a point.

Oriental Region The tropical zone of southeast Asia eastward to the margin of the continental shelf (Wallace's line).

orthogenesis The supposed intrinsic tendency to evolve steadily in a particular direction, e.g., to become larger or smaller, in spite of the influence of natural selection.

osmoconformer An organism that does not osmoregulate but instead has internal salt concentrations in osmotic balance with seawater.

osmoregulation The process of maintaining homeostasis by keeping the same internal concentration of body fluids in changing external solutions.

outbreak area In organisms that plague or irrupt, the area into which populations expand during peak population densities.

outcrossing Having gametes that are exchanged between different genotypes.

oversaturated island An island that has more taxa of a particular kind than expected on the basis of an equilibrium between origination and extinction.

overturn Vertical mixing of a water column in lakes caused by temperature changes over the seasons.

ovoviviparous Pertaining to animals that produce eggs internally, which then hatch within the female or immediately after they are extruded.

Pacifica A hypothesized ancient continent that may have existed somewhere in the Pacific Basin.

Palearctic Region The region of extratropical climates in Eurasia and the coastal area of northernmost Africa with a mediterranean-type climate.

paleocirculation The ocean currents of the past.

paleoclimatology The study of past climates, as elucidated mainly through the analysis of fossils.

paleoecology The branch of paleontology that attempts to reconstruct the structure of and processes affecting ancient populations and communities.

paleoendemic An endemic that evolved in the distant past, e.g., in the Lower Tertiary; compare with **neoendemic.**

paleoflora All the species of plants inhabiting a specified region in the past.

paleomagnetism The magnetism or magnetically induced orientation of microstructures that has existed in a rock since its origin.

paleontology The study devoted to describing, analyzing, and explaining the fossil record.

paleotropical Occurring in the Old World tropics and subtropics, i.e., in Africa, Madagascar, India, and southeast Asia.

Pangea In plate tectonics, the supercontinent of the Permian that was composed of essentially all continents and major continental islands.

panmixis The condition whereby interbreeding within a large population is totally random.

pantropical Occurring in all major tropical areas around the world.

parallel evolution The development of species with strong resemblances from fairly closely related ancestors; hence two or more taxa that evolved in the same direction from related ancestors.

páramo Tropical alpine vegetation, characteristic of high mountains at the equator, which has a low and compact perennial cover as a response to the perpetually wet, cold, and cloudy environment.

parapatric Having contiguous but nonoverlapping distributions.

parapatric speciation A mode of speciation in which differentiation occurs when two populations have contiguous but nonoverlapping ranges, often representing two distinct habitat types.

paraphyletic In cladistics, referring to taxa that are classified chiefly on the basis of shared primitive character states.

parasitism An interspecific relationship in which one species (the parasite) draws nutrition from or is somehow dependent for survival on the other species (the host), and the host is negatively affected by the interaction.

paratemperate (sensu Wolfe) An area in which the mean annual temperature ranges from 3° to 10° C.

paratropical (sensu Wolfe) An area in which the mean annual temperature ranges from 20° to 25° C.

parsimony The logical principle or rule of thumb that the simplest solution (i.e., involving the fewest logical steps or conditions) should be chosen when trying to pick from two or more conflicting explanations; also called Ockham's razor.

parthenogenesis The development of eggs without fertilization by a male gamete.

passive dispersal The movement of an organism from one location to another by means of a stronger force, such as wind or water or via a larger animal.

pattern Nonrandom, repetitive organization.

pedogenic regimes The soil-forming processes, e.g., laterization, podzolization, calcification, and gleization.

pelagic Occurring in open water and away from the bottom.

peripheral fish Any fish that can osmoregulate and can therefore live in either fresh or salt water.

peripheral population Any population of a species that occurs on the edges of a range, around either the perimeter or the elevational limit. Similar to but not necessarily identical with marginal population, which refers to a population that has difficulty in surviving as a result of limiting abiotic or biotic factors.

permanence theory The widely held view before continental drift was accepted that the distribution pattern of ocean basins and continents has remained relatively constant over the history of the earth.

perturbation Any event that greatly upsets the equilibrium or changes the state or direction of change of a system.

phenetics The study of the overall similarities of organisms; compare with **phylogeny.**

phenotype The way in which the genetic message of an individual is expressed in its morphology, physiology, and behavior.

phoresy The transportation of organisms by other organisms.

photic (euphotic) The uppermost zone in a water column that receives adequate solar radiation to permit photosynthesis of plants.

photoautotroph An organism that uses light as the energy source and carbon dioxide as the carbon source for basic metabolism; hence an organism that is photosynthetic.

photosynthesis The chemical process of using pigments to capture sunlight and then using the energy, water, and carbon dioxide to make organic compounds (sugars), releasing oxygen in the process.

phylad A general term for a monophyletic group.

phyletic gradualism An evolutionary process whereby a species is gradually transformed over time into a different organism; compare with **punctuated equilibrium.**

phyletic sequencing The technique of ordering taxa in a phylogenetic classification in such a way that a cladogram can be constructed if there are more than two taxa of the same rank, e.g., six families in a superfamily, by having each taxon be the sister taxon of the one above it.

phylogenetic systematics See **cladistics;** compare with **evolutionary systematics.**

phylogeny The evolutionary relationships between an ancestor and all known descendents.

physiognomy The external aspect of a landscape, e.g., the topography and other physical characteristics of a land form and its vegetation.

physiological ecology The study of how the physiological characteristics of organisms relate to the abundance and distribution of these organisms in natural habitats.

phytogeography (plant geography) The study of the distribution of plants.

phytosociology (plant sociology) The quantitative study of the composition of plant communities and how these relate to environmental factors.

placental In mammals, any organism that has a placenta connecting the mother to the fetus.

plankton Small organisms (especially tiny plants, small invertebrates, and juvenile stages of larger animals) that inhabit water and are transported mainly by water currents and wave action rather than by individual locomotion.

planktotrophic Referring to aquatic larvae that have no long-term storage of nutrients and so must feed on small organisms in the plankton during their development.

plate In plate tectonics, a portion of the upper earth's surface, about 100 km thick, that moves over the asthenosphere during sea-floor spreading.

plate tectonics The study of the origin, movement, and destruction of plates and how these events have been involved in the evolution of the earth's crust.

plesiomorphy In cladistics, a primitive or ancestral character state.

pluvial Referring to periods with fairly high rainfall and water runoff.

podzolization The formation of a soil under conditions of adequate moisture and low decomposer activity, resulting in a soil from which the bases, humic acids, colloids, and ferric and aluminum oxides have been removed (leached) from the upper horizon.

poikilotherm An animal with a relatively variable body temperature, often ectothermic, i.e., relying on an external heat source to produce and control body temperature.

polar emergence The occurrence of cold-water species that occur in deep tropical water near the surface in polar latitudes. See also **equatorial submergence.**

pollen rain A massive shower of pollen, primarily from wind-pollinated species.

pollination The transfer of pollen grains to a receptive stigma, usually by wind or flower-visiting animals.

polygamous In plants, having unisexual and hermaphroditic (bisexual) flowers on the same individual.

polymorphic Having several distinct forms in a population.

polyphagous Feeding on a variety of different kinds of food.

polyploid Any organism or cell that has three or more sets of chromosomes.

Popperian philosophy The use of strict hypothetico-deductive reasoning in scientific research.

Precambrian shield See **craton.**

predation The act of feeding on other organisms; an interspecific interaction that has negative effects on both species.

predator In any antagonistic relationship, the species that consumes other organisms.

prey In any mutually detrimental relationship, the species that is consumed or used by the other organism.

primary division freshwater fish Any fish that is totally intolerant of salt water.

primary production The production of biomass by green plants.

primary succession The gradual transformation of bare rock or another sterile substrate into a soil that supports a living ecological community.

progression rule The idea that a primitive form remains in the center of origin whereas derived forms are found at greater distances from the center.

prokaryotic Referring to any organism whose cell lacks a true membrane-bounded nucleus.

propagule Any part of an organism or stage in the life cycle that can reproduce the species and thus establish a new population.

provincialism The coincident occurrence of large numbers of well-differentiated endemic forms in an area; regional or provincial distinctiveness.

proximate ecological factor Any characteristic of an environment, such as photoperiod or temperature, used by individual organisms as a cue to shift into a different physiological or behavior mode, e.g., a cue for flowering in plants or migration of birds; compare with **ultimate ecological factor.**

punctuated equilibrium The model that describes macroevolution as a rapid period of differentiation (often accompanying speciation) followed by a long period in which few if any characters evolve.

quadrupedal Having four limbs for locomotion.

radiation In evolutionary biology, a general term for the expansion of a group, implying at the least that many new species have been produced.

rafting The transport over water of living organisms on large floating mats of debris.

Rassenkreis A species composed of a ring of subspecies or races in a cline in which the ones at both ends, which may be sympatric, are very different and may not be interfertile.

realized niche The actual environmental conditions in which a species survives and reproduces in nature: a subset of the fundamental niche.

Red Queen hypothesis The hypothesis that states that a species must continually evolve in order to keep pace with an environment that is perpetually changing, because all other species are evolving, altering the availability of resources and the nature of biotic interactions.

refugium An area that has remained unchanged, e.g., with the same climate and vegetation type, while areas surrounding it have changed markedly; hence the area has served as a refuge for species requiring specific habitats.

relict A species of a group that was once widespread (epibiotic) or diverse (taxonomic relict).

reproductive isolation The separation of one population from others by the inability to produce viable offspring when the individuals from two populations are mated.

rescue effect On near islands, the continual influx of individuals belonging to species already present that tends to prevent the disappearance of those species in part by contributing to the genetic variability of the population.

resident A species that lives year-round in the habitat or location specified.

reticulate evolution The formation of new species by the hybridization of dissimilar populations, e.g., interspecific hybridization.

saline Having high concentrations of salts and especially ions of chloride and sulfate.

scaling See **allometry.**

scansorial Capable of climbing and active on both the ground surface and in above-ground vegetation.

scavenger An animal that feeds on carrion.

sclerophyllous Having tough, thick, evergreen leaves.

scrub Any of a wide variety of vegetation types dominated by low shrubs; in exceedingly dry locations scrub vegetation has few or no trees and widely spaced low shrubs, but in areas of fairly high rainfall, scrub has trees and grades into either woodland or forest.

sea-floor spreading In plate tectonics, the process of adding crustal material at a midoceanic ridge and thus displacing older rocks, usually on both sides, away from their point of origin.

seamount A peaked submarine volcano.

secondary division freshwater fish A fish that prefers fresh water but can live for short periods in salt water.

secondary succession A series of radical changes in the vegetational composition of an environment in response to disturbance, involving the gradual and regular replacement of species and ending with the return to a stable state (climax).

self-compatible Having the ability to produce offspring without requiring gametes from another individual.

self-incompatible Requiring two individuals to exchange gametes in order to produce offspring.

semiaquatic Living partly in or adjacent to water.

semiarid Having a fairly dry climate with low precipitation, usually 25 to 60 cm per year, and a high evapotranspiration rate so that potential loss of water to the environment exceeds the input.

semidesert A semiarid habitat characterized by low vegetation, e.g., small, widely spaced shrubs.

sere A stage in secondary succession.

serpentine A rock or soil type rich in magnesium but deficient in calcium.

sessile Organisms that remain fixed in the same spot as adults (e.g., most plants and certain benthic aquatic invertebrates).

shrubland Any of a wide variety of vegetation types dominated by shrubs that may form a fairly solid cover, e.g., sclerophyllous scrub (chaparral), or sparse cover, e.g., desert scrub.

sial Rock rich in silica-aluminum; the principal component of continental rocks.

siliceous Containing silica.

sima Rock rich in silica-magnesium; the principal component of basalt and the rock type of oceanic plates.

similarity index An estimate of the similarity or relatedness of two communities, biota, or taxa.

sister taxa In cladistics, the two taxa that are most closely (and therefore most recently) related.

sociobiology The study of the biological bases of animal behavior and sociology.

solstice Either of two times in a year (June 22, December 22) when the sun reaches the highest latitude (23.5° N and S, respectively).

spawning area In migrating aquatic animals, the region in which egg laying or birth takes place.

speciation The process in which two or more contemporaneous species evolve from a single ancestral population.

species The fundamental taxonomic category for organisms; a group of organisms that are morphologically and reproductively more similar to each other than to other populations and that share a singular ancestor-descendent heritage.

species-area curve A log to log plot of the numbers of species of a particular taxon, e.g., plants or ants, against area, e.g., of islands or other biogeographic regions.

species pool All the organisms present on neighboring source areas that are theoretically available to colonize a particular habitat or island.

species selection An analog of natural selection at the species level, in which some species with certain characteristics increase while others decrease or become extinct.

stasipatric speciation See **parapatric speciation.**

stenohaline Having the ability to tolerate only a narrow range of salt concentrations.

stenothermal Having a tolerance for only a narrow range of temperatures.

step cline A cline characterized by abrupt, discontinuous variation.

stochastic Random, expected (statistically) by chance alone; compare with **deterministic.**

stratigraphy The branch of geology dealing with the sequence of deposition of rocks and fossils as well as their composition, origin, and distribution.

stream capture The interception and incorporation of a stream or part of a stream by another stream or watershed.

subduction In plate tectonics, the movement of one plate beneath another, leading to the heating and subsequent remelting of the lower plate.

subfamily A taxonomic category below the family, used whenever the genera can be grouped into two or more distinct lineages.

sublittoral Referring to the coastal marine zone below the intertidal zone and therefore below the point at which sea bottom is periodically exposed to the atmosphere.

suborder A taxonomic category used for grouping families within an order.

subspecies A taxonomic category used by some systematists to designate a genetically distinct set of populations of a species that have a discrete range.

subtemperature (sensu Wolfe) An area in which the mean annual temperature is less than 3° C.

subterranean Living underground.

subtropical (sensu Wolfe) An area in which the mean annual temperature ranges from 13° to 20° C.

superfamily A taxonomic category for grouping families below the rank of suborder.

superpáramo A vegetation of the high elevation in the equatorial Andes mountains.

superspecies A group of closely related species.

supertramp A species that has excellent colonizing abilities but that is a poor competitor in a diverse community.

supragenus A monophyletic group of closely related genera.

sweepstakes dispersal The partly stochastic colonization of some elements of a biota across a severe barrier, resulting in establishment of a disharmonic biota.

symbiosis A long-term interspecific relationship in which two unrelated and unlike organisms live together in a close association so that each receives some adaptive benefit.

sympatric In the strictest sense, two taxa living in the same local community, close enough that the organisms of each population could interact; in the more general sense of many authors, having broadly overlapping geographic ranges.

sympatric speciation The differentiation of two reproductively isolated species from one initial population within the same local area; hence speciation occurs under conditions in which much gene flow potentially could or does actually occur. Compare with **allopatric** and **parapatric speciation.**

symplesiomorphy In cladistics, the term used to identify a shared primitive character state between taxa.

synapomorphy In cladistics, the term used to identify a shared derived character state between taxa.

systematics The study of the evolutionary relationships between organisms.

taxon (pl. taxa) A convenient and general term for any taxonomic category, e.g., a species, genus, family, or order.

taxon cycle A proposed series of stages for the evolution of a newly arrived colonist on an isolated oceanic island, from the state of being indistinguishable from mainland relatives to that of highly differentiated endemics to extinction.

taxonomy In the strictest sense, the study of the names of organisms, but often used to include the entire process of classification; see **systematics.**

tectonic Referring to any process involved in the production or deformation of the earth's crust.

temperate (sensu Wolfe) An area in which the mean annual temperature ranges from 10° to 13° C.

terrestrial Living on land.

tetrapod Any four-legged vertebrate, including amphibians, reptiles, and mammals.

thermocline In a water column, the subsurface zone in which the temperature drops sharply.

tillites Glacial rock deposits.

timberline The uppermost limit of forest vegetation at high elevation.

track In historical biogeography, a line drawn on a map connecting the geographically isolated ranges of species in a taxon that are closest relatives (vicariants).

trade winds (trades) Winds blowing toward the equator between the horse latitudes and the doldrums in the Northern and Southern hemispheres.

transform fault A fault in an oceanic plate, perpendicular to the midoceanic ridge, which divides the plate into smaller units.

transformation series A linear statement representing the evolutionary history of a feature (character) from a primitive condition to the most highly derived state.

translocation A kind of chromosomal mutation in which a segment of one chromosome becomes attached to a different chromosome.

transpiration The loss of water vapor from plants through pores called stomates.

trench In plate tectonics, an exceedingly deep cut in the ocean floor where an oceanic plate is descending to be consumed in the mantle.

tribe A taxonomic category below the rank of subfamily used for grouping genera.

triple junction In plate tectonics, a point at which three oceanic plates meet; the position of that junction shifts because each of the plates grows at a different rate.

trophic status The position or role of an organism in the nutritional structure of a community, such as a primary producer, herbivore, or top carnivore.

tropical (sensu Wolfe) An area in which the mean annual temperature is greater than 25° C and no freezing occurs.

turnover The rate of replacement of species in a particular area when some taxa become extinct but others immigrate from outside.

ultimate ecological factor The actual resource, physical condition, or biotic interaction that affects the fitness of an organism; compare with **proximate ecological factors.**

undersaturated island An island that has fewer taxa of a particular kind than would be predicted from an equilibrium between colonization and extinction.

uniformitarianism See **actualism.**

unisexual Referring to an individual that has only one sex.

univoltine Breeding only once per year.

upwelling The vertical movement of deep water, including dissolved nutrients from the bottom, to the surface.

vagility The ability to be transported or to move actively from one place to another.

vicariance biogeography The study of distribution patterns of organisms that attempts to reconstruct historical events through cladistic methods, often with little or no attention to dispersal capabilities or ecological properties.

vicariants Two disjunct species that are most closely related to each other and that are assumed to have been created when an initial range of the ancestor was split by some historical event.

viking funeral ship The term coined by McKenna for a landmass, such as a fragment, that has fossils that were laid down when the land was in one location but that were transported to a completely different locality via continental drift.

waif In dispersal biology, a diaspore or any type of individual that is carried passively by waves or air currents to a distant place, e.g., most colonizers of oceanic island beaches.

Wallace's line The most famous biogeographic line, between the islands of Borneo and Celebes, which demarks the eastward boundary of many landlocked Eurasian organisms and the boundary of the Oriental Region.

Weberian apparatus A modification of the anterior vertebrae of certain fishes (Ostariophysi) that connects the air bladder with the inner ear.

Wegenerism The general idea of continental drift according to Alfred Wegener.

West Gondwanaland The portion of the ancient supercontinent Gondwanaland that includes Africa and South America.

wintering area In migratory land animals, the area in which populations spend the cold season and feed but do not breed; similar to fattening area in migratory aquatic animals.

woodland Any of a variety of vegetation types consisting of small, widely spaced trees with or without substantial undergrowth.

xerophyte A plant with particular physiological and structural features adapting it for a habitat that receives strong diurnal or seasonal water stress.

zoogeography The study of the distributions of animals.

Bibliography

Abele, L.G. 1974. Species diversity of decapod crustaceans in marine habitats. Ecology **55**:156-161.

Abele, L.G. 1976. Comparative species richness in fluctuating and constant environments: coral-associated decapod crustaceans. Science **192**:461-463.

Abele, L.G., and W.K. Patten. 1976. The size of coral heads and the community biology of associated decapod crustaceans. J. Biogeogr. **3**:35-47.

Abele, L.G., and K. Walters. 1979. Marine benthic diversity: a critique and alternative explanation. J. Biogeogr. **6**:115-126.

Abbott, I. 1974. Numbers of plant, insect and land bird species on nineteen remote islands in the Southern Hemisphere. Biol. J. Linn. Soc. **6**:143-152.

Abbott, I. 1981. The composition of landbird faunas of islands round south-western Australia: is there evidence for competitive exclusion? J. Biogeogr. **8**:135-144.

Abbott, I., L.K. Abbott, and P.R. Grant. 1977. Comparative ecology of Galapagos ground finches (*Geospiza* Gould): evaluation of the importance of floristic diversity and interspecific competition. Ecol. Monogr. **47**:151-184.

Abbott, I., and P.R. Grant. 1976. Nonequilibrial bird faunas on islands. Am. Nat. **110**:507-528.

Abramsky, Z. 1978. Small mammal community ecology. Changes in species diversity in response to manipulated productivity. Oecologia **34**:113-123.

Ackermann, R. 1976. The philosophy of Karl Popper. Amherst, Mass., University of Massachusetts Press.

Adams, C.C. 1902. Southeastern United States as a center of geographical distribution of flora and fauna. Biol. Bull. **3**:115-131.

Adams, C.C. 1909. An ecological survey of Isle Royale, Lake Superior. Lansing, Mich., Michigan Biological Survey.

Adams, C.G., and D.V. Ager (eds.). 1967. Aspects of Tethyan biogeography. The Systematics Association Pub. no. 7. Wetteren, Universa.

Ager, D.V. 1963. Principles of paleoecology. New York, McGraw-Hill Book Co.

Albrecht, F.O. 1967. Polymorphisme phasaire et biologie des acridiens migrateurs. Paris, Masson Editeur.

Albrecht, P.W. 1976. The cranial arteries of turtles and their evolutionary significance. J. Morphol. **149**:159-182.

Allard, G.O., and V.J. Hurst. 1969. Brazil-Gabon geologic link supports continental drift. Science **163**:528-532.

Allee, W.C., O. Park, A.E. Emerson, T. Park, and K.P. Schmidt. 1949. Principles of animal ecology. Philadelphia, W.B. Saunders Co.

Allen, J., J. Golson, and R. Jones (eds.). 1977. Sunda and Sahul. Prehistoric studies in southeast Asia, Melanesia, and Australia. Melbourne, Academic Press, Ltd.

Allen, J.A. 1878. The geographic distribution of mammals. Bull. U.S. Geol. Geogr. Surv. **4**:339-343.

Alvarez, L.W., W. Alvarez, F. Asaro, and H.V. Michel. 1980. Extraterrestrial cause for the Cretaceous-Tertiary extinction. Science **208**:1095-1108.

Amadon, D. 1950. The Hawaiian honeycreepers (Aves, Drepaniidae). Bull. Am. Mus. Nat. Hist. **95**(4):153-262.

Amadon, D. 1973. Birds of the Congo and Amazon forests: a comparison, p. 267-277. *In* B.J. Meggars, E.S. Ayensu, and W.D. Duckworth (eds.), Tropical forest ecosystems in Africa and South America: a comparative review. Washington, D.C., Smithsonian Institution Press.

Amerson, A.B., Jr. 1975. Species richness on the nondisturbed northwestern Hawaiian Islands. Ecology **56**:435-444.

Amiran, D.H.K., and A.W. Wilson (eds.). 1973. Coastal deserts. Their natural and human environments. Tucson, Ariz., University of Arizona Pres.

Andersen, N.M., and J.T. Polhemus. 1976. Water-striders (Hemiptera: Gerridae, Veliidae, etc.), p. 187-224. *In* L. Cheng (ed.), Marine insects. Amsterdam, North-Holland Publishing Co.

Anderson, A.B., and W.W. Brown. 1980. On the number of tree species in Amazonian forests. Biotropica **12**:235-237.

Anderton, R., P.H. Bridges, M.R. Leeder, and B.W. Sellwood. 1979. A dynamic stratigraphy of the British Isles. A study in crustal evolution. London, George Allen & Unwin (Publishers), Ltd.

Andrewartha, H.G., and L.C. Birch. 1954. The distribution and abundance of animals. Chicago, University of Chicago Press.

Antonovics, J. 1971. The effects of a heterogeneous environment on the genetics of natural populations. Am. Sci. **59**:593-599.

Antonovics, J., and A.D. Bradshaw. 1970. Evolution in closely adjacent plant populations. VIII. Clinal patterns at a mine boundary. Hereditas (London) **25**:349-362.

Antonovics, J., A.D. Bradshaw, and R.G. Turner. 1971. Heavy metal tolerance in plants. Adv. Ecol. Res. **7**:1-85.

ARCYANA. 1975. Transform faults and rift valley from bathyscaph and diving saucer. Science **190**:108-116.

Ashton, P.S. 1969. Speciation among tropical trees: some deductions in the light of recent evidence. Biol. J. Linn. Soc. **1**:155-196.

Atchley, W.R., and D. Woodruff (eds.). 1981. Evolution and speciation. Essays in honor of M.J.D. White. Cambridge, Cambridge University Press.

Atlas of American Agriculture. 1936. Washington, D.C., U.S. Government Printing Office.

Atwater, T. 1970. Implications of plate tectonics for the Cenozoic tectonic evolution of western North America. Geol. Soc. Am. Bull. **81**:3513-3536.

Auffenberg, W. 1971. A new fossil tortoise, with remarks on the origin of South American testudines. Copeia (1):106-117.

Auffenberg, W. 1974. Checklist of fossil land tortoises (Testudinidae). Bull. Florida State Mus. **18**:121-251.

Axelrod, D.I. 1950. Studies in late Tertiary paleobotany. VI. Evolution of desert vegetation in western North America. Washington, D.C., Carnegie Institute, Pub. no. **590**:215-306.

Axelrod, D.I. 1952. A theory of angiosperm evolution. Evolution **6**:29-60.

Axelrod, D.I. 1958. Evolution of the Madro-Tertiary geoflora. Bot. Rev. **24**:433-509.

Axelrod, D.I. 1960. The evolution of flowering plants, p. 227-305. *In* S. Tax (ed.), The evolution of life. Chicago, University of Chicago Press.

Axelrod, D.I. 1963. Fossil floras suggest stable not drifting continents. J. Geophys. Res. **68**:3257-3263.

Axelrod, D.I. 1966. Origin of deciduous and evergreen habits in temperate forests. Evolution **20**:1-15.

Axelrod, D.I. 1967. Quaternary extinctions of large mammals. Univ. California Pub. Geol. **74**:1-42.

Axelrod, D.I. 1970. Mesozoic paleogeography and early angiosperm history. Bot. Rev. **36**:277-319.

Axelrod, D.I. 1972. Ocean-floor spreading in relation to ecosystematic problems, p. 15-68. *In* R.T. Allen and F.C. James (eds.), A symposium on ecosystematics. University of Arkansas Museum Occas. Pap. no. 4.

Axelrod, D.I. 1975. Evolution and biogeography of madrean-Tethyan sclerophyllous vegetation. Ann. Missouri Bot. Gard. **62**:280-334.

Axelrod, D.I. 1976. History of the coniferous forests, California and Nevada. Univ. Calif. Pub. Bot. **70**:1-62.

Axelrod, D.I. 1977. Outline history of California vegetation, p. 139-193. *In* M.G. Barbour and J. Major (eds.), Terrestrial vegetation of California. New York, John Wiley & Sons, Inc.

Axelrod, D.I. 1979. Age and origin of the Sonoran Desert vegetation. California Acad. Sci. Occas. Pap. no. **132**:1-74.

Axelrod, D.I., and H.P. Bailey. 1969. Palaeotemperature analysis of Tertiary floras. Palaeogeogr. Palaeoclimatol. Palaeoecol. **6**:163-195.

Axelrod, D.I., and P.H. Raven. 1982. Paleobiogeography and origin of the New Guinea flora, p. 919-941. *In* J.L. Gressitt (ed.), Biogeography and ecology of New Guinea, vol. 2. Monographiae Biologicae 42, The Hague, Dr. W. Junk BV, Publishers.

Backhuys, W. 1975. Zoogeography and taxonomy of the land and freshwater molluscs of the Azores. Amsterdam, Backhuys and Meesters.

Bacon, F. 1620. Instauratio Magna (Novum Organum). London, Billium.

Bagnara, J.T., and W. Ferris. 1975. The presence of Phyllomedusinae melanosomes and pigments in Australian hylids. Copeia (3):592-595.

Baker, H.A., and E.G.H. Oliver. 1967. Ericas in southern Africa. Purnell & Sons.

Baker, H.G., and G.L. Stebbins. (eds.). 1965. The genetics of colonizing species. New York, Academic Press, Inc.

Baker, R.H. 1968. Habitats and distribution, p. 98-126. *In* J.A. King (ed.), Biology of *Peromyscus*. Am. Soc. Mammal. Spec. Pub. no. 2.

Baker, R.R. 1978. The evolutionary ecology of animal migration. London, Hodder & Stoughton, Ltd.

Baker, R.R. (ed.). 1981. The mystery of migration. New York, The Viking Press.

Bakker, R.T. 1978. Dinosaur feeding behavior and the origin of flowering plants. Nature **274**:661-663.

Baldwin, B., P.J. Coney, and W.R. Dickinson. 1974. Dilemma of a Cretaceous time scale and rates of sea-floor spreading. Geology **2**:267-270.

Balgooy, M.M.J. van. 1966. Distribution maps of Pacific plants, p. 1-312. *In* C.G. G.J. van Steenis and M.M.J. van Balgooy (eds.), Pacific plant areas, vol. 2. Blumea **14**, Suppl. 5.

Balgooy, M.M.J. van. 1969. A study on the diversity of island floras. Blumea **17**:139-178.

Balgooy, M.M.J. van. 1971. Plant-geography of the Pacific. Blumea **19**, Suppl. **6**:1-222.

Ball, I.R. 1975. Nature and formulation of biogeographical hypotheses. Syst. Zool. **24**:407-430.

Ball, I. 1980. The status of historical biogeography, p. 1283-1288. *In* R. Nöhring (ed.), Acta XVII Congressus Internationalis Ornithologici, vol. 2. Berlin, Verlag der Deutschen Ornithologen-Gesellschaft.

Ballard, R.D., W.B. Bryan, J.R. Heirtzler, G. Keller, J.G. Moore, and Tj. van Andel. 1975. Manned submersible observations in the FAMOUS area: Mid-Atlantic Ridge. Science **190**:103-108.

Banfield, A.W.F. 1974. The mammals of Canada. Toronto, University of Toronto Press.

Banks, H.P. 1975. Early vascular land plants: proof and conjecture. Bioscience **25**:730-737.

Bannister, P. 1976. Introduction to physiological plant ecology. New York, Halsted Press.

Barbour, C.D., and J.H. Brown. 1974. Fish species diversity in lakes. Am. Nat. **108**:473-489.

Barbour, M.G., J.H. Burk, and W.D. Pitts. 1980. Terrestrial plant ecology. Menlo Park, Calif., The Benjamin/Cummings Publishing Co., Inc.

Barigozzi, C. (ed.) 1982. Mechanisms of speciation. Progr. Clin. Biol. Res. **96**. New York, Alan R. Liss, Inc.

Barker, J.S.F., and W.T. Starmer (eds.). 1982. Ecological genetics and evolution: the cactus-yeast-*Drosophila* model system. Sydney, Academic Press, Ltd.

Barnett, S.A. (ed.). 1958. A century of Darwin. Cambridge, Mass., Harvard University Press.

Baroni Urbani, C. 1971. Studien zur Ameisenfauna Italiens. XI. Die Ameisen des Toskanischen Archipels. Betrachtungen zur Herkunft der Inselfaunen. Rev. Súisse Zool. **78**:1037-1067.

Barry, R.G., and R.J. Chorley. 1970. Atmosphere, weather, and climate. New York, Holt, Rinehart & Winston, Inc.

Bartholomew, B. 1970. Bare zone between California shrub and grassland communities: the role of animals. Science **170**:1210-1212.

Bartholomew, G.A. 1958. The role of physiology in the distribution of terrestrial vertebrates, p. 81-95. *In* C.I. Hubbs (ed.), Zoogeography. Washington, D.C., Am. Assoc. Advancement Sci. Pub. 51.

Battistini, R., and G. Richard-Vindard (eds.). 1972. Biogeography and ecology in Madagascar. Monographiae Biologicae 21. The Hague, Dr. W. Junk BV, Publishers.

Beadle, N.C.W. 1966. Soil phosphate and its role in molding segments of the Australian flora and vegetation, with special reference to xeromorphy and sclerophylly. Ecology **47**:991-1007.

Beadle, N.C.W. 1981. The vegetation of Australia. Stuttgart, Gustav Fischer Verlag.

Beals, E.W. 1969. Vegetational change along altitudinal gradients. Science **165**:981-985.

Beard, J.S. 1977. Tertiary evolution of the Australian flora in the light of latitudinal movements of the continent. J. Biogeogr. **4**:111-118.

Beck, C.B. (ed.). 1976. Origin and early evolution of angiosperms. New York, Columbia University Press.

Beckwith, S.L. 1954. Ecological succession on abandoned farm lands and its relationship to wildlife management. Ecol. Monogr. **24**:349-376.

Beddall, B.G. 1968. Wallace, Darwin, and the theory of natural selection. A study in the development of ideas and attitudes. J. Hist. Biol. **1**:261-323.

Behle, W.H. 1978. Avian biogeography of the Great Basin and Intermontane Region. Great Basin Nat. Mem. **2**:55-80.

Beilmann, A.P., and L.G. Brenner. 1951. The recent intrusion of forests in the Ozarks. Ann. Missouri Bot. Gard. **38**:261-282.

Benioff, H. 1954. Orogenesis and deep crustal structure: additional evidence from seismology. Geol. Soc. Am. Bull. **65**:385-400.

Benson, W.W., K.S. Brown, and L.E. Gilbert. 1975. Coevolution of plants and herbivores: passion flower butterflies. Evolution **19**:659-680.

Bent, A.C. 1965. Life histories of North American blackbirds, orioles, tanagers and allies. New York, Dover Publications, Inc.

Bentley, P.J. 1971. Endocrines and osmoregulation. A comparative account of the regulation of water and salt in vertebrates. New York, Springer-Verlag New York, Inc.

Bequaert, J.C., and W.B. Miller. 1973. The mollusks of the arid Southwest. Tucson, Ariz., University of Arizona Press.

Berg, R.Y. 1975. Myrmecochorous plants in Australia and their dispersal by ants. Australian J. Bot. **23**:475-508.

Berggren, W.A., and C.D. Hollister. 1977. Plate tectonics and paleocirculation–commotion in the ocean. Tectonophysics **38**:11-48.

Bergh, H.W., and I.O. Norton. 1976. Prince Edward fracture zone and the evolution of the Mozambique Basin. J. Geophys. Res. **81**:5221-5239.

Bergmann, C. 1847. Über die Verhältnisse der Wärmeökonomie der Thiere zu ihren Grösse. Göttinger Studien **1**:595-708.

Bernabo, J.C., and T. Webb. 1977. Changing patterns in the Holocene pollen record from northeastern North America: a mapped summary. Quat. Res. **8**:64-96.

Berra, T.M. 1981. An atlas of distribution of the freshwater fish families of the world. Lincoln, Nebr. University of Nebraska Press.

Berry, R.J. 1964. The evolution of an island population of the house mouse. Evolution **18**:468-483.

Berry, W.B.N. 1973. Silurian-Early Devonian graptolites, p. 81-87. *In* A. Hallam (ed.), Atlas of palaeobiogeography. Amsterdam, Elsevier Scientific Publishing Co.

Bickman, J.W. 1976. A meiotic analysis of four species of turtles. Genetica **46**:193-198.

Bierhorst, D.W. 1971. Morphology of vascular plants. New York, Macmillan Publishing Co., Inc.

Bigalke, R. C. 1972. The contemporary mammal fauna of Africa, p. 141-194. *In* A. Keast, F.C. Erk, and B. Glass (eds.), Evolution, mammals, and southern continents. Albany, N.Y., State University of New York Press.

Biju-Duval, B., and L. Montadert (eds.). 1977. Structural history of the Mediterranean basins. International Symposium of the 25th Plenary Congress Assembly of the International Commission for the Scientific Exploration of the Mediterranean. Paris, Société des Éditions Technip.

Billings, W.D. 1952. The environmental complex in relation to plant growth and distribution. Quart. Rev. Biol. **27**:251-265.

Biological Sciences Curriculum Study. 1963. Biological science: molecules to man. Boston, Houghton Mifflin Co.

Bird, J.W. (ed.). 1980. Plate tectonics. 2nd ed. Selected papers from publications of the American Geophysical Union. Washington, D.C., American Geophysical Union.

Bird, J.M., and B. Isacks (eds.). 1972. Plate tectonics. Washington, D.C., American Geophysical Union.

Birge, E.A., and C. Juday. 1911. The inland lakes of Wisconsin. Bull. Wisc. Geol. Nat. Hist. Surv. **22**:1-259.

Bisby, G.R. 1943. Geographical distribution of fungi. Bot. Rev. **9**:466-482.

Blackett, P.M.S., E. Bullard, and S.K. Runcorn (eds.). 1965. A symposium on continental drift. Philos. Trans. Roy. Soc. London, ser. A 258.

Blackwelder, R.E. 1977. Twenty-five years of taxonomy. Syst. Zool. **26**:107-137.

Blair, W.F., A.C. Hulse, and M.A. Mares. 1976. Origins and affinities of vertebrates of the North American Sonoran Desert and Monte Desert of northwestern Argentina. J. Biogeogr. **3**:1-18.

Blanc, C.P. 1972. Les reptiles de Madagascar et des iles voisines, p. 501-611. *In* R. Battistini and G. Richard-Vindard (eds.), Biogeography and ecology in Madagascar. Monographiae Biologicae 21. The Hague, Dr. W. Junk BV, Publishers.

Bock, C.E., and L.W. Lepthian. 1976. Synchronous eruptions of boreal seed-eating birds. Am. Nat. **110**:559-571.

Bock, W.J. 1973. Philosophical foundations of classical evolutionary classification. Syst. Zool. **22**:375-392.

Böcher, T.W., and O.B. Lyshede. 1968. Anatomical studies in xerophytic apophyllous plants. I. *Monttea aphylla, Bulnesia retama* and *Bredemeyera colletioides*. Biol. Skr. Dan. Videnskabernes Selskab **16**(3):1-44.

Boer, P.J. den. 1970. On the significance of dispersal power for populations of carabid-beetles (Coleoptera, Carabidae). Oecologia **4**:1-28.

Bond, C.E. 1979. Biology of fishes. Philadelphia, W.B. Saunders Co.

Bond, J. 1971. Birds of the West Indies. 2nd ed. London, William Collins Sons & Co., Ltd.

Bonde, N. 1977. Cladistic classification as applied to vertebrates, p. 741-804. *In* M.K. Hecht, P.C. Goody, and B.M. Hecht (eds.), Major patterns in vertebrate evolution. New York, Plenum Press.

Bormann, F.H., G.E. Likens, T.G. Siccama, R.S. Pierce, and J.S. Eaton. 1974. The export of nutrients and recovery of stable conditions following deforestation at Hubbard Brook. Ecol. Monogr. **44**:255-277.

Boucot, A.J. 1975a. Standing diversity of fossil groups in successive intervals of geologic time in the light of changing levels of provincialism. J. Paleontol. **49**:1105-1111.

Boucot, A.J. 1975b. Evolution and extinction rate controls. New York, Elsevier North-Holland, Inc.

Boucot, A.J. 1978. Community evolution and rates of cladogenesis, p. 545-655. *In* M.K. Hecht, W.C. Steere, and B. Wallace (eds.), Evolutionary biology 11. New York, Plenum Press.

Boucot, A.J., and J. Gray. 1979. Epilogue: a Paleozoic Pangaea? p. 465-484. *In* J. Gray and A.J. Boucot (eds.), Historical biogeography, plate tectonics, and the changing environment. Corvallis, Ore., Oregon State University Press.

Boucot, A.J., and J.G. Johnson. 1973. Silurian brachiopods, p. 59-65. *In* A. Hallam (ed.), Atlas of palaeobiogeography. Amsterdam, Elsevier Scientific Publishing Co.

Boulin, J. 1981. Afghanistan structure, greater India concept and eastern Tethyan evolution. Tectonophysics **72**: 261-287.

Bourliere, F. 1973. The comparative ecology of rain forest mammals in Africa and tropical America: some introductory remarks, p. 279-292. *In* B.J. Meggers, E.S. Ayensu, and W.D. Duckworth (eds.), Tropical forest ecosystems in Africa and South America: a comparative review. Washington, D.C., Smithsonian Institution Press.

Bousfield, E.L., and M.L.H. Thomas. 1975. Postglacial changes in the distributions of littoral marine invertebrates in the Canadian Atlantic region, p. 47-60. *In* J.G. Ogden and M.J. Harvey (eds.), Environmental changes in the Maritimes. Halifax, Nova Scotia Institute of Science.

Bowers, M.A., and J.H. Brown. 1982. Body size and coexistence in desert rodents: chance or community structure? Ecology **63**:391-400.

Boyd, E.M., and S.A. Nunneley. 1964. Banding records substantiating the changed status of 10 species of birds since 1900 in the Connecticut Valley. Bird-Banding **35**:1-8.

Bradshaw, A.D. 1971. Plant evolution in extreme environments, p. 20-50. *In* R. Geed (ed.), Ecological genetics and evolution. London, Blackwell Scientific Publications, Ltd.

Brady, N.C. 1974. Nature and property of soils. 8th ed. New York, Macmillan Publishing Co., Inc.

Brakenridge, G.R. 1978. Evidence for a cold, dry full-glacial climate in the American Southwest. Quat. Res. **9**: 22-40.

Bramwell, D. (ed.). 1979. Plants and islands. London, Academic Press, Inc.

Brattstrom, B.H. 1961. Some new fossil tortoises from western North America with remarks on the zoogeography and paleontology of tortoises. J. Paleontol. **35**:543-560.

Brattstrom, B.H. 1964. Evolution of the pit vipers. Trans. San Diego Soc. Nat. Hist. **13**:185-268.

Braun, E.L. 1950. Deciduous forests of eastern North America. New York, Hafner Press.

Braun-Blanquet, J. 1965. Plant sociology. The study of plant communities. New York, Hafner Publishing Co.

Breder, C.M., Jr., and D.E. Rosen. 1966. Modes of reproduction in fishes. Garden City, N.Y., Natural History Press.

Brenan, J.P.M. 1978. Some aspects of the phytogeography of tropical Africa. Ann. Missouri Bot. Gard. **65**: 437-478.

Brice, W.C. (ed.). 1978. The environmental history of the New and Middle East since the last Ice Age. London, Academic Press, Inc.

Briggs, J.C. 1968. Panama sea-level canal. Science **162**:511-513.

Briggs, J.C. 1974. Marine zoogeography. New York, McGraw-Hill Book Co.

Briggs, J.C. 1979. Ostariophysan zoogeography: an alternative hypothesis. Copeia (1):111-118.

Brinton, E. 1962. The distribution of Pacific euphausids. Bull. Scripps Inst. Oceanogr. Univ. Calif. **8**:51-270.

Brodkorb, P. 1971a. Catalogue of fossil birds. Bull. Florida State Museum Biol. Sci. **15**(4):163-226.

Brodkorb, P. 1971b. Origin and evolution of birds, p. 19-55. *In* D.S. Farner and J.R. King (eds.), Avian biology, vol. 1. New York, Academic Press, Inc.

Brooks, D.R. 1981. Hennig's parasitological method: a proposed solution. Syst. Zool. **30**:229-249.

Brooks, D.R., T.B. Thorson, and M.A. Mayes. 1981. Freshwater stingrays (Potamotrygonidae) and their helminth parasites: testing hypotheses of evolution and coevolution, p. 147-175. *In* V.A. Funk and D.R. Brooks (eds.), Advances in cladistics. Proceedings of the first meeting of the Willi Hennig Society. New York, New York Botanical Society.

Brooks, J.L., and S.I. Dodson. 1965. Predation, body size, and composition of plankton. Science **150**:28-35.

Brower, L.P. 1977. Monarch migration. Nat. Hist. **86**(6):40-53.

Brown, A.L. 1977. Ecology of fresh water. Cambridge, Mass., Harvard University Press.

Brown, D.E., N.B. Carmony, C.H. Lowe, and R.M. Turner. 1976. A second locality for native California fan palms *(Washingtonia filifera)* in Arizona. J. Arizona Acad. Sci. **11**:37-41.

Brown, D.S. 1978. Freshwater molluscs, p. 1155-1180. *In* M.J.A. Werger and A.C. van Bruggen (eds.), Biogeography and ecology of southern Africa, vol. 2. Monographiae Biologicae 31. The Hague, Dr. W. Junk BV, Publishers.

Brown, D.S. 1979. Biogeographical aspects of African freshwater gastropods. Malacologia **18**:79-102.

Brown, D.S. 1980. Freshwater snails of Africa and their medical importance. London, Taylor & Francis, Ltd.

Brown, J.H. 1968. Adaptation to environmental temperature in two species of woodrats, *Neotoma cinerea* and *N. albingula*. University of Michigan Museum Zool. Misc. Pub., no. **135**:1-48.

Brown, J.H. 1971a. Mechanisms of competitive exclusion between two species of chipmunks *(Eutamias)*. Ecology **52**:306-311.

Brown, J.H. 1971b. Mammals on mountaintops: nonequilibrium insular biogeography. Am. Nat. **105**:467-478.

Brown, J.H. 1971c. The desert pupfish. Sci. Am. **225** (5):104-110.

Brown, J.H. 1973. Species diversity of seed-eating desert rodents in sand dune habitats. Ecology **54**:775-787.

Brown, J.H. 1975. Geographical ecology of desert rodents, p. 315-341. *In* M.L. Cody and J.M. Diamond (eds.), Ecology and evolution of communities. Cambridge, Mass., Belknap Press.

Brown, J.H. 1978. The theory of insular biogeography and the distribution of boreal birds and mammals. Great Basin Nat. Mem. **2**:209-227.

Brown, J.H. 1981. Two decades of homage to Santa Rosalia: toward a general theory of diversity. Am. Zool. **21**:877-888.

Brown, J.H., and M.A. Bowers. On the relationship between morphology and ecology: community organization in hummingbirds. Ecology **63** (in preparation).

Brown, J.H., W.A. Calder, and A. Kodric-Brown. 1978. Correlates and consequences of body size in nectar-feeding birds. Am. Zool. **18**:687-700.

Brown, J.H., and D.W. Davidson. 1977. Competition between seed-eating rodents and ants in desert ecosystems. Science **196**:880-882.

Brown, J.H., and C.R. Feldmeth. 1971. Evolution in constant and fluctuating environments: thermal tolerances of desert pupfish *(Cyprinodon)*. Evolution **25**:390-398.

Brown, J.H., and A. Kodric-Brown. 1977. Turnover rates in insular biogeography: effect of immigration on extinction. Ecology **58**:445-449.

Brown, J.H., and G.A. Lieberman. 1973. Resource utilization and coexistence of seed-eating desert rodents in sand dune habitats. Ecology **54**:788-797.

Brown, J.H., O.J. Reichman, and D.W. Davidson. 1979. Granivory in desert ecosystems. Ann. Rev. Ecol. Syst. **10**:201-227.

Brown, K.S., Jr. 1976. Geographical patterns of evolution in neotropical Lepidoptera: systematics and derivation of known and new Heliconiini (Nymphalidae: Nymphalinae). J. Entomol., ser. B **44**:201-242.

Brown, K.S., Jr. 1982. Paleoecology and regional patterns of evolution in neotropical forest butterflies, p. 255-308. *In* G.T. Prance (ed.), Biological diversification in the tropics. New York, Columbia University Press.

Brown, W.H., E.D. Merrill, and H.S. Yates. 1919. The revegetation of Volcano Island, Luzon, Philippine Islands, since the eruption of Taal Volcano in 1911. Philippine J. Sci., sect. C, vol. 12.

Brown, W.L., and E.O. Wilson. 1956. Character displacement. Syst. Zool. **5**:49-64.

Browne, R.A. 1981. Lakes as islands: biogeographic distribution, turnover rates, and species composition in the lakes of central New York. J. Biogeogr. **8**:75-83.

Bruggen, A.C. van. 1977. A preliminary analysis of African non-marine Gastropoda Euthyneura families. Malacologia **16**:75-80.

Bruggen, A.C. van. 1980. Gondwanaland connections in the terrestrial molluscs of Africa and Australia. J. Malacol. Soc. Australia **4**:215-222.

Bruhnes, B. 1906. Recherches sur la direction d'aimentation des roches volcaniques (1). J. Physique, 4e sér., **5**:705-724.

Brundin, L. 1965. On the real nature of transantarctic relationships. Evolution **19**:496-505.

Brundin, L. 1966. Transantarctic relationships and their significance, as evidenced by chironomid midges. Kungliga Svenska Vetenskapsakademiens Handlingar, 4th ser. 11, no. **1**:1-472.

Brundin, L. 1967. Insects and the problem of austral disjunctive distribution. Ann. Rev. Entomol. **12**:149-168.

Brundin, L. 1972. Phylogenetics and biogeography. Syst. Zool. **21**:69-79.

Brunig, E.F. 1973. Species richness and stand diversity in relation to site and succession of forests in Sarawak and Brunei (Borneo). Amazonia **4**:293-320.

Bullard, E. 1975. The emergence of plate tectonics: a personal view. Ann. Rev. Earth Planet. Sci. **3**:1-26.

Bullard, E.C., J.E. Everett, and A.G. Smith. 1965. Fit of continents around Atlantic, p. 41-75. *In* P.M.S. Blackett, E.C. Bullard, and S.K. Runcorn (eds.), A symposium on continental drift. Philos. Trans. Roy. Soc. London, ser. A 258.

Burbridge, N.T. 1960. The phytogeography of the Australian region. Australian J. Bot. **8**:75-212.

Bureau de Recherches Géologiques et Minières. 1980a. Colloque C5. Géologie des chaînes alpines issues de la Téthys. Mém. Bur. Rech. Géol. Min. no. 115.

Bureau de Recherches Géologies et Minières. 1980b. Colloque C6. Géologie de l'Europe du Précambrien aux bassins sedimentaires post-hercyniens. Mém. Bur. Rech. Géol. Min. no. 108.

Burke, K. 1976. The Chad Basin: an active intracontinental basin. Tectonophysics **36**:197-206.

Burke, K., J.F. Dewey, and W.S.F. Kidd. 1977. World distribution of sutures–the sites of former oceans. Tectonophysics **40**:69-99.

Burke, K.C., and J.T. Wilson. 1976. Hot spots on the earth's surface. Sci. Am. **235**(2):46-57.

Burke, M.J., L.V. Gusta, H.A. Quamme, C.J. Weiser, and P.H. Li. 1976. Freezing and injury in plants. Ann. Rev. Plant Physiol. **27**:507-528.

Burtt, R.B. 1929. A record of fruits and seeds dispersed by mammals and birds from the Singida district of Tanganyika territory. J. Ecol. **17**:351-355.

Bush, G.L. 1975. Modes of animal speciation. Ann. Rev. Ecol. Syst. **6**:339-364.

Bush, G.L., S.M. Case, A.C. Wilson, and J.L. Patten. 1977. Rapid speciation and chromosomal evolution in mammals. Proc. Natl. Acad. Sci. U.S.A. **74**:3942-3946.

Bussing, W.A. 1976. Geographic distribution of the San Juan ichthyofauna of Central America with remarks on its origin and ecology, p. 157-175. *In* T.B. Thorson (ed.), Investigations of the ichthyofauna of Nicaraguan lakes. Lincoln, Nebr., University of Nebraska School of Life Sciences.

Butzer, K.W. 1971. Environment and archeology. Chicago, Aldine-Atherton.

Butzer, K.W. 1976. Pleistocene climates. Geoscience and Man (Baton Rouge) **13**:27-44.

Butzer, K.W. 1977. Environment, culture and human evolution. Am. Sci. **65**:572-584.

Butzer, K.W. 1978. Climatic patterns in an unglaciated continent. Geogr. Mag. **51**(3):201-208.

Butzer, K.W., R. Stuckenrath, A.J. Bruzewicz, and D.M. Helgren. 1978. Late Cenozoic paleoclimates of the Gaap Escarpment, Kalahari Margin, South Africa. Quat. Res. **10**:310-339.

Bystrak, D. 1979. The breeding bird survey. Sialia **1**:74-79.

Cabrera, A.L., and A. Willink. 1973. Biogeografia de America Latina. Washington, D.C., Programa Regional de Desarrollo Cientifico y Tecnologico, Departamento Asuntos Cientificos, Secretario General de la Organizacion de los Estados Americanos.

Cain, S.A. 1944. Foundations of plant geography. New York, Harper.

Cairns, J., and J.A. Ruthven. 1970. Artificial microhabitat size and the number of colonizing protozoan species. Trans. Am. Micros. Soc. **89**:100-109.

Cameron, R.A.D., and M. Redfern. 1976. A synopsis of the land snails (Mollusca: Gastropoda). Synopses of the British fauna, n.s. no. 6. New York, Academic Press, Inc.

Campbell, B.M., and F. van der Meulen. 1980. Patterns of plant species diversity in fynbos vegetation, South Africa. Vegetatio **43**:43-47.

Candolle, A. de. 1855. Géographie botanique raisonnée. 2 vol. Paris, Masson Editeur.

Carlquist, S. 1965. Island life. Garden City, N.Y., Natural History Press.

Carlquist, S. 1966a. The biota of long-distance dispersal. I. Principles of dispersal and evolution. Quart. Rev. Biol. **41**:247-270.

Carlquist, S. 1966b. The biota of long-distance dispersal. II. Loss of dispersibility in the Hawaiian flora. Brittonia **18**:310-335.

Carlquist, S. 1966c. The biota of long-distance dispersal. IV. Genetic systems in floras of oceanic islands. Evolution **20**:433-455.

Carlquist, S. 1970. Hawaii: a natural history. Garden City, N.Y., Natural History Press.

Carlquist, S. 1974. Island biology. New York, Columbia University Press.

Carlquist, S. 1975. Ecological strategies of xylem evolution. Berkeley, Calif., University of California Press.

Carlquist, S. 1981. Chance dispersal. Am. Sci. **69**:509-515.

Carey, S.W. 1955. The orocline concept of geotectonics, part 1. Proc. Roy. Soc. Tasmania, Papers **89**:255-288.

Carey, S.W. (ed.). 1958a. Continental drift; a symposium. Hobart, University of Tasmania, Geology Department.

Carey, S.W. 1958b. The tectonic approach to continental drift, p. 177-358. *In* S.W. Carey (ed.), Continental drift: a symposium. Hobart, University of Tasmania, Geology Department.

Carpenter, F.M. 1977. Geological history and evolution of the insects, p. 63-70. *In* D. White (ed.), Proceedings of the 15th International Congress of Entomology, Washington. College Park, Md., Entomological Society of America.

Carothers, S.W., and R.R. Johnson. 1974. Population structure and social organization of southwestern riparian birds. Am. Zool. **14**:97-108.

Carr, A. 1952. Handbook of turtles. Ithaca, N.Y., Cornell University Press.

Carson, H.L. 1970. Chromosome tracers of the origin of species. Science **168**:1414-1418.

Carson, H.L. 1971. Speciation and the founder principle. Univ. Missouri Stadler Symp. **3**:51-70.

Carson, H.L. 1975. The genetics of speciation at the diploid level. Am. Nat. **109**:83-92.

Carson, H.L. 1981. Microevolution in insular ecosystems, p. 471-482. *In* D. Mueller-Dombois, K.W. Bridges, and H.L. Carson (eds.), Island ecosystems. Biological organization in selected Hawaiian communities. Stroudsburg, Penn., Hutchinson Ross Publishing Co.

Carson, H.L., D.E. Hardy, H.T. Spieth, and W.S. Stone. 1970. The evolutionary biology of the Hawaiian Drosophilidae, p. 437-543. *In* M.K. Hecht and W.C. Steere (eds.), Essays in evolution and genetics in honor of Theodosius Dobzhansky. New York, Appleton-Century-Crofts.

Carson, H.L., and K.Y. Kaneshiro. 1976. *Drosophila* of Hawaii: systematics and ecological genetics. Ann. Rev. Ecol. Syst. **7**:311-346.

Case, T.J. 1975. Species numbers, density compensation and the colonizing ability of lizards on islands in the Gulf of California. Ecology **56**:3-18.

Case, T.J. 1976. Body size differences between populations of the chuckwalla *Sauromalus obesus*. Ecology **57**:313-323.

Case, T.J. 1978. A general explanation for insular body size trends in terrestrial vertebrates. Ecology **59**:1-18.

Case, T.J. 1979. Optimal body size and an animal's diet. Acta Biotheoretica **28**(1):54-69.

Case, T.J., J. Faaborg, and R. Sidell. 1982. The role of body size in the assembly of West Indian bird communities. Evolution (in press).

Case, T.J., M.E. Gilpin, and J.M. Diamond. 1979. Overexploitation, interference competition, and excess density compensation in insular faunas. Am. Nat. **113**:843-854.

Case, T.J., and R. Sidell. 1982. Pattern and chance in the structure of model and natural communities. Evolution (in press).

Chambers, K.L. 1963. Amphitropical species pairs in *Microseris* and *Agoseris* (Compositae: Cichorieae). Quart. Rev. Biol. **38**:124-140.

Chapman, V.J. 1976. Mangrove vegetation. Vaduz, Liechtenstein, Cramer.

Chappell, M.A. 1978. Behavioral factors in the altitudinal zonation of chipmunks *(Eutamias)*. Ecology **59**:565-579.

Charig, A.J. 1973a. Jurassic and Cretaceous dinosaurs, p. 339-352. *In* A. Hallam (ed.), Atlas of palaeobiogeography. Amsterdam, Elsevier Scientific Publishing Co.

Charig, A.J. 1973b. Kurtén's theory of ordinal variety and the number of the continents, p. 231-245. *In* D.H. Tarling and S.K. Runcorn (eds.), Implications of continental drift to the earth sciences, vol. 1. London, Academic Press, Inc.

Cheatham, A.H., and J.E. Hazel. 1969. Binary (presence-absence) similarity coefficients. J. Paleontol. **43**:1130-1136.

Cheng, L. 1974. Notes on the ecology of the oceanic insect *Halobates*. Mar. Fish. Rev. **36**:1-7.

Cheng, L. (ed.). 1976. Marine insects. Amsterdam, North Holland Publishing Co.

Chester, R.H. 1969. Destruction of Pacific corals by the sea star *Acanthaster planci*. Science **165**:280-283.

Ciochon, R.L., and A.B. Chiarelli (eds.). 1980. Evolutionary biology of the New World monkeys and continental drift. New York, Plenum Press.

Clagg, H.B. 1966. Trapping of air-borne insects in the Atlantic-Antarctic area. Pacific Insects **8**:455-466.

Clague, D.A., and R.D. Jarrard. 1973. Tertiary plate motion deduced from the Hawaiian-Emperor chain. Geol. Soc. Am. Bull. **84**:1135-1154.

Clausen, J., D.D. Keck, and W.M. Hiesey. 1948. Experimental studies on the nature of species. III. Environmental responses of climatic races of *Achillea*. Washington, D.C., Carnegie Institute, Pub. no. 581.

Clegg, J.A., M. Almond, and P.H.S. Stubbs. 1954. The remanent magnetism of some sedimentary rocks in Britain. Philos. Mag. **45**:583-598.

Clemens, W.A., Jr. 1977. Phylogeny of the marsupials, p. 51-68. *In* B. Stonehouse and D. Gilmore (eds.), The biology of the marsupials. Baltimore, Md., University Park Press.

Clemens, W.A., Jr. 1982. Interrelationships of South and North American, Cretaceous and early Paleogene mammalian faunas. Ann. Missouri Bot. Gard. (in press).

Clements, F.E. 1916. Plant succession: an analysis of the development of vegetation. Washington, D.C., Carnegie Institute Pub. no. 242.

Clements, F.E. 1949. Dynamics of vegetation. New York, Hafner Press.

CLIMAP project members. 1976. The surface of the ice-age earth. Science **191**:1131-1137.

Cluver, M.A., and N. Hotton III. 1979. The dicynodont genus *Diictodon* (Reptilia, Therapsida) and its significance, p. 176-198. *In* B. Laskar and C.S.R. Rao (eds.), Papers of the Fourth International Gondwana Symposium, vol. 1. Delhi, Hindustan Publishing Corp.

Cockrum, E.L. 1962. Introduction to mammalogy. New York, Ronald Press.

Cockrum, E.L. 1969. Migration of the guano bat, *Tadarida brasiliensis*. University of Kansas Museum Nat. Hist. Misc. Pub. **51**:303-336.

Cody, M.L. 1966. The consistency of intra- and inter-continental grassland bird species counts. Am. Nat. **100**:371-376.

Cody, M.L. 1968. On the methods of resource division in grassland bird communities. Am. Nat. **102**:107-137.

Cody, M.L. 1973a. Parallel evolution and bird niches, p. 307-338. *In* F. di Castri and H.A. Mooney (eds.), Mediterranean type ecosystems. Origin and structure. Ecological studies 7. New York, Springer-Verlag. New York, Inc.

Cody, M.L. 1973b. Coexistence, coevolution, and convergent evolution in seabird communities. Ecology **54**: 31-44.

Cody, M.L. 1974. Competition and the structure of bird communities. Monographs in population biology no. 7. Princeton, N.J., Princeton University Press.

Cody, M.L. 1975. Towards a theory of continental species diversity, p. 214-257. *In* M.L. Cody and J.M. Diamond (eds.), Ecology and evolution of communities. Cambridge, Mass., Belknap Press.

Cody, M.L. 1980. Evolution of habitat use: geographic perspectives, p. 1013-1018. *In* R. Nöhring (ed.), Acta XVII Congressus Internationalis Ornithologici, vol. 2. Berlin, Verlag der Deutschen Ornithologen-Gesellschaft.

Cody, M.L., and J.M. Diamond (eds.). 1975. Ecology and evolution of communities. Cambridge, Mass., Belknap Press.

Cody, M.L., and H.A. Mooney. 1978. Convergence versus nonconvergence in mediterranean-climate ecosystems. Ann. Rev. Ecol. Syst. **9**:265-321.

Coffey, D.J. 1977. Dolphins, whales, and porpoises: an encyclopedia of sea mammals. New York, Macmillan Publishing Co., Inc.

Cogger, H.G. 1975. Sea snakes of Australia and New Guinea, p. 59-139. *In* W.A. Dunson (ed.), The biology of sea snakes. Baltimore, Md., University Park Press.

Cogger, H.G. 1979. Reptiles and amphibians of Australia. Rev. 1st ed. Hollywood, Fla., Ralph Curtis Books.

Colbert, E.H. 1971. Tetrapods and continents. Quart. Rev. Biol. **46**:250-269.

Colbert, E.H. 1972. *Lystrosaurus* and Gondwanaland, p. 157-177. *In* Th. Dobzhansky, M.K. Hecht, and W.C. Steere (eds.), Evolutionary biology 6. New York, Appleton-Century-Crofts.

Colbert, E.H. 1973a. Wandering lands and animals. New York, E.P. Dutton & Co., Inc.

Colbert, E.H. 1979. Gondwana vertebrates, p. 135-143. *In* B. Laskar and C.S. Rao (eds.), Papers of the Fourth International Gondwana Symposium, vol. 1. Delhi, Hindustan Publishing Corp.

Colbert, E.H. 1980. Evolution of the vertebrates. A history of the backboned animals through time. New York, John Wiley & Sons, Inc.

Cole, J.W., and K.B. Lewis. 1981. Evolution of the Taupo-Hikurangi subduction system. Tectonophysics **72**:1-21.

Cole, K.L. 1982. Late Quaternary zonation of vegetation in the eastern Grand Canyon. Science **217**:1142-1145.

Cole, L. 1954. Some features of random cycles. J. Wildl. Managem. **18**:107-109.

Coleman, P.J. (ed.). 1973. The Western Pacific. Island arcs, marginal seas, geochemistry. New York, Crane, Russak & Co., Inc., and Nedlands, University of West Australia Press.

Colinvaux, P.A. 1964. The environment of the Bering Land Bridge. Ecol. Monogr. **34**:297-325.

Colinvaux, P.A. 1981. Historical ecology in Beringia: the southland bridge coast at St. Paul Island. Quat. Res. **16**:18-36.

Colwell, R.K. 1973. Competition and coexistence in a simple tropical community. Am. Nat. **107**:737-760.

Colwell, R.K. 1979. The geographical ecology of hummingbird flower mites in relation to their host plants and carriers, p. 461-468. *In* J.S. Rodriguez (ed.), Recent advances in acarology, vol. 2. New York, Academic Press, Inc.

Colwell, R.K., and D.W. Winkler. 1982. A null model for null models in biogeography, *In* D. Simberloff and D.R. Strong, Jr. (eds.) (in press).

Coney, P.J. 1979. Mesozoic-Cenozoic cordilleran plate tectonics, p. 33-50. *In* R.B. Smith and G.P. Eaton (eds.), Cenozoic tectonics and regional geophysics of the Western Cordillera. Geol. Soc. Am. Mem. no. 152.

Coney, P.J. 1980. Cordilleran metamorphic core complexes: an overview, p. 7-31. *In* M.D. Crittender, Jr., P.J. Coney, and G.H. Davis (eds.), Cordilleran metamorphic core complexes. Geol. Soc. Am. Mem. no. 153.

Coney, P.J. 1982. Plate tectonic constraints on biogeographic connections between North and South America. Ann. Missouri Bot. Gard. (in press).

Coney, P.J., D.L. Jones, and J.W.H. Monger. 1980. Cordilleran suspect terrains. Nature **288**:329-333.

Connell, J.H. 1961. The influence of interspecific competition and other factors on the distribution of the barnacle *Chthamalus stellatus*. Ecology **42**:710-723.

Connell, J.H. 1975. Some mechanisms producing structure in natural communities: a model and evidence from field experiments, p. 460-490. *In* M.L. Cody and J.M.

Diamond (eds.), Ecology and evolution of communities. Cambridge, Mass., Belknap Press.

Connell, J.H. 1978. Diversity in tropical rain forests and coral reefs. Science **199**:1302-1310.

Connell, J.H., and E. Orias. 1964. The ecological regulation of species diversity. Am. Nat. **98**:399-414.

Connell, J.H., and R.O. Slatyer. 1977. Mechanisms of succession in natural communities and their role in community stability and organization. Am. Nat. **111**:1119-1144.

Connor, E.F., and E.D. McCoy. 1979. The statistics and biology of the species-area relationship. Am. Nat. **113**:791-833.

Connor, E.F., and D. Simberloff. 1978. Species number and compositional similarity of the Galápagos flora and avifauna. Ecol. Monogr. **48**:219-248.

Connor, E.F., and D. Simberloff. 1979. The assembly of species communities: chance or competition? Ecology **60**:1132-1140.

Constance, L. 1963. Introduction and historical review. Quart. Rev. Biol. **38**:109-116.

Cook, R.E. 1969. Variation in species density of North American birds. Syst. Zool. **18**:63-84.

Cooke, H.B.S. 1972. The fossil mammal fauna of Africa, p. 89-139. *In* A. Keast, F.C. Erk, and B. Glass (eds.), Evolution, mammals, and southern continents. Albany, N.Y., State University of New York Press.

Cooper, A.K., D.W. School, and M.S. Marlow. 1976. Plate tectonic model for the evolution of the eastern Bering Sea Basin. Geol. Soc. Am. Bull. **87**:1119-1126.

Corbet, G.B. 1961. The origin of the British insular races of small mammals and of the "Lusitanian" fauna. Nature **191**:1037-1040.

Corbet, G.B. 1978. The mammals of the Palaearctic Region: A taxonomic review. Ithaca, N.Y., Cornell University Press.

Corbet, G.B., and J.E. Hill. 1980. A world list of mammalian species. Ithaca, N.Y., Comstock Publishing Associates.

Cox, A. (ed.). 1973. Plate tectonics and geomagnetic reversals. San Francisco, W.H. Freeman & Co., Publishers.

Cox C.B. 1973a. The distribution of Triassic terrestrial tetrapod families, p. 369-371. *In* D.H. Tarling and S.K. Runcorn (eds.), Implications of continental drift to the earth sciences, vol. 1. New York, Academic Press, Inc.

Cox, C.B. 1973b. Triassic tetrapods, p. 213-223. *In* A. Hallam (ed.), Atlas of palaeobiogeography. Amsterdam, Elsevier Scientific Publishing Co.

Cox, C.B. 1974. Vertebrate palaeodistributional patterns and continental drift. J. Biogeogr. **1**:75-94.

Cox, C.B., I.N. Healey, and P.D. Moore. 1980. Biogeography: an evolutionary and ecological approach. 3rd ed. Oxford, Blackwell Scientific Publications, Ltd.

Cox, G.W., and R.E. Ricklefs. 1977. Species diversity, ecological release and community structuring in Caribbean land bird faunas. Oikos **29**:60-66.

Cracraft, J. 1973. Continental drift, paleoclimatology, and the evolution and biogeography of birds. J. Zool. (London) **169**:455-545.

Cracraft, J. 1974a. Phylogeny and evolution of the ratite birds. Ibis **116**:494-521.

Cracraft, J. 1974b. Phylogenetic models and classification. Syst. Zool. **23**:71-90.

Cracraft, J. 1974c. Continental drift and vertebrate distribution. Ann. Rev. Ecol. Syst. **5**:215-262.

Cracraft, J. 1975a. Mesozoic dispersal of terrestrial faunas around the southern end of the world. Mém. Mus. Nat. Hist. Zool. (Paris) **88**:29-52.

Cracraft, J. 1975b. Historical biogeography and earth history: perspectives for a future synthesis. Ann. Missouri Bot. Gard. **62**:227-250.

Cracraft, J. 1977. Avian evolution on southern continents: influences of palaeogeography and palaeoclimatology, p. 40-52. *In* H.J. Frith and J.H. Calaby (eds.), Proceedings of the 16th International Ornithological Congress. Canberra, Australian Academy of Sciences.

Cracraft, J. 1980. Avian phylogeny and intercontinental biogeographic patterns, p. 1302-1308. *In* R. Nöhring (ed.), Acta XVII Congressus Internationalis Ornithologici, vol. 2. Berlin, Verlag der Deutschen Ornithologen-Gesellschaft.

Cracraft, J. 1981. Toward a phylogenetic classification of the Recent birds of the world (class Aves). Auk **98**:681-714.

Cracraft, J., and N. Eldredge (eds.). 1979. Phylogenetic analysis and paleontology. New York, Columbia University Press.

Craddock, C., T.W. Bastien, R.H. Rutford, and J.J. Anderson. 1965. *Glossopteris* discovered in West Antarctica. Science **148**:634-637.

Craw, R.C. 1979. Generalized tracks and dispersal in biogeography: a response to R.M. McDowall. Syst. Zool. **28**:99-107.

Crawford, A.R. 1974. A greater Gondwanaland. Science **184**:1179-1181.

Creer, K.M., E. Irving, and S.K. Runcorn. 1954. The direction of the geomagnetic field in remote epochs in Great Britain. J. Geomag. Geoelect. **6**:163-168.

Creer, K.M., E. Irving, and S.K. Runcorn. 1957. Geophysical interpretation of palaeomagnetic directions from Great Britain. Philos. Trans. Roy. Soc. London, ser. A **250**:144-156.

Crisp, D.J. (ed.). 1964. The effects of the severe winter of 1962-63on marine life in Britain. J. Anim. Ecol. **33**:165-210.

Croizat, L. 1952. Manual of phytogeography. The Hague, Dr. W. Junk BV, Publishers.

Croizat, L. 1958. Panbiogeography. 2 vol. Caracas, published by the author.

Croizat, L. 1960. Principia botanica. Caracas, published by the author.

Croizat, L. 1964. Space, time, form: the biological synthesis. Caracas, published by the author.

Croizat, L., G.J. Nelson, and D.E. Rosen. 1974. Centers of origin and related concepts. Syst. Zool. **23**:265-287.

Cronquist, A. 1981. An integrated system of classification of flowering plants. New York, Columbia University Press.

Crook, K.A.W. 1981. The break-up of the Australian-Antarctic segment of Gondwanaland, p. 1-14. *In* A. Keast (ed.), Ecological biogeography of Australia, vol. 1. Monographiae Biologicae 41. The Hague, Dr. W. Junk BV, Publishers.

Crook, K.A.W., and L. Belbin. 1978. The southwest Pacific area during the last 90 million years. J. Geol. Soc. Australia **25**(1):23-40.

Crosby, G.T. 1972. Spread of the cattle egret in the Western Hemisphere. Bird-Banding **43**:205-212.

Crowell, K.L. 1962. Reduced interspecific competition among the birds of Bermuda. Ecology **43**:75-88.

Crowell, K.L. 1973. Experimental zoogeography: introductions of mice to small islands. Am. Nat. **107**:535-558.

Cruden, R.W. 1966. Birds as agents of long-distance dispersal for disjunct plant groups of the temperate Western Hemisphere. Evolution **20**:517-532.

C.S.I.R.O. 1970. The insects of Australia. Melbourne, Melbourne University Press.

Cuellar, O. 1977. Animal parthenogenesis. Science **197**: 837-843.

Cuellar, O., and A.G. Kluge. 1972. Natural parthenogenesis in the gekkonid lizard *Lepidodactylus lugubris*. J. Genet. **61**:14-26.

Culver, D.C. 1970. Analysis of simple cave communities. I. Caves as islands. Evolution **29**:463-474.

Culver, D.C., J.R. Holsinger, and R. Baroody. 1973. Toward a predictive cave biogeography: the Greenbrier Valley as a case study. Evolution **27**:689-695.

Cumber, R.A. 1953. Some aspects of the biology and ecology of bumblebees bearing upon the yields of red clover seed in New Zealand. New Zealand J. Sci. Techn. Bull. **34**:227-240.

Curtis, J.T. 1956. The modification of mid-latitude grasslands and forests by man, p. 721-736. *In* W.L. Thomas, Jr. (ed.), Man's role in changing the face of the earth. Chicago, University of Chicago Press.

Curtis, J.T. 1959. The vegetation of Wisconsin. Madison, Wisc., University of Wisconsin Press.

Curtis, J.W. 1973. Plate tectonics and the Papua–New Guinea–Solomon Islands region. J. Geol. Soc. Australia **20**:21-36.

Cushing, E.J., and H.E. Wright, Jr. (eds.). 1967. Quaternary paleoecology. New Haven, Conn., Yale University Press.

Dahlgren, R.M.T. 1980. A revised system of classification of the angiosperms. Bot. J. Linn. Soc. **80**:91-124.

Dahlgren, R.M.T., S. Rosendal-Jensen, and B.J. Nielsen. 1981. A revised classification of the angiosperms with comments on correlation between chemical and other characters, p. 149-199. *In* D.A. Young and D.S. Seigler (eds.), Phytochemistry and angiosperm phylogeny. New York, Praeger Publishers.

Dalrymple, G.B., E.A. Silver, and E.D. Jackson. 1973. Origin of the Hawaiian Islands. Am. Sci. **61**:294-308.

Daly, H.V., J.T. Doyen, and P.R. Ehrlich. 1978. Introduction to insect biology and diversity. New York, McGraw-Hill Book Co.

Dalziel, I.W.D., R.H. Dott, Jr., R.D. Winn, Jr., and R.L. Bruhn. 1975. Tectonic relations of South Georgia Island to the southernmost Andes. Geol. Soc. Am. Bull. **86**:1034-1040.

Damon, P.E., J.C. Lerman, and A. Long. 1978. Temporal fluctuations of atmospheric ^{14}C: causal factors and implications. Ann. Rev. Earth Planet. Sci. **6**:457-494.

Dammermann, K.W. 1948. The fauna of Krakatau, 1883-1933. Koninklijke Nederlandsche Akad. Wetenschappen Verhandelingen **44**:1-594.

Dana, J.D. 1853. The question whether temperature determines the distribution of marine species of animals in depth. Am. J. Sci. Arts, ser. 2, **15**:204-207.

Dansereau, P.M. 1957. Biogeography; an ecological perspective. New York, Ronald Press.

Darlington, P.J., Jr. 1938. The origin of the fauna of the Greater Antilles, with discussion of dispersal of animals over water and through the air. Quart. Rev. Biol. **13**:274-300.

Darlington, P.J., Jr. 1943. Carabidae of mountains and islands: data on the evolution of isolated faunas and on atrophy of wings. Ecol. Monogr. **13**:37-61.

Darlington, P.J., Jr. 1957. Zoogeography: the geographical distribution of animals. New York, John Wiley & Sons, Inc.

Darlington, P.J., Jr. 1959a. Darwin and zoogeography. Proc. Am. Philos. Soc. **103**:307-319.

Darlington, P.J., Jr. 1959b. Area, climate and evolution. Evolution **13**:488-510.

Darlington, P.J., Jr. 1965. Biogeography of the southern end of the world: distribution and history of far southern life and land, with an assessment of continental drift. Cambridge, Mass., Harvard University Press.

Darlington, P.J., Jr. 1970. Carabidae on tropical islands, especially in the West Indies. Biotropica **2**:7-15.

Darwin, C. 1859. On the origin of species by means of natural selection, or the preservation of favoured races in the struggle for life. London, John Murray. (Facsimile edition, E. Mayr [ed.], 1964. Cambridge, Mass., Harvard University Press.)

Darwin, C. 1962. The voyage of the Beagle. Introduced by L. Engel. Garden City, N.J., Doubleday Publishing Co.
Darwin, F. 1887. The life and letters of Charles Darwin, including an autobiographical chapter. 3 vol. London, John Murray.
Dary, D.A. 1974. The buffalo book: the full saga of the American animal. Chicago, The Swallow Press, Inc.
Daubenmire, R.F. 1943. Vegetational zonation in the Rocky Mountains. Bot. Rev. **9**:325-393.
Daubenmire, R.F. 1954. Alpine timberlines in the Americas and their interpretation. Butler Univ. Bot. Stud. **11**:119-136.
Daubenmire, R. 1968. Plant communities; a textbook of plant synecology. New York, Harper & Row Publishers, Inc.
Daubenmire, R. 1978. Plant geography with special reference to North America. New York, Academic Press, Inc.
Davidson, D.W. 1977. Species diversity and community organization in desert seed-eating ants. Ecology **58**:525-537.
Davis, M.B. 1969. Palynology and environmental history during the Quaternary Period. Am. Sci. **57**:317-332.
Davis, M.B. 1976. Pleistocene geography of temperate deciduous forests. Geosci. Man **13**:13-26.
Dawson, W.R., and C. Carey. 1976. Seasonal acclimation to temperature in cardueline finches. J. Comp. Physiol. **112**:317-333.
Dayton, P.K. 1971. Competition, disturbance, and community organization: the provision and subsequent utilization of space in a rocky intertidal community. Ecol. Monogr. **41**:351-389.
Dayton, P.K. 1979. Ecology: a science and a religion, p. 3-18. *In* R.J. Livingston (ed.), Ecological processes in coastal and marine systems. New York, Plenum Press.
De Laubenfels, D.J. 1975. Mapping the world's vegetation. Syracuse, N.Y., Syracuse University Press.
Delgadillo, M.C. 1979. Mosses and phytogeography of the *Liquidambar* forest of Mexico. The Bryologist **82**:432-449.
de Vos, A., R.H. Manville, and R.G. Van Gelder. 1956. Introduced mammals and their influence on native biota. Zoologica **41**:163-194.
de Vries, A.L. 1971. Freezing resistance in fishes, p. 157-190. *In* W.S. Hoar and D.J. Randall (eds.), Fish physiology, vol. 6. New York, Academic Press, Inc.
de Wet, J.M.J. 1979. Origins of polyploids, p. 3-15. *In* W.H. Lewis (ed.), Polyploidy. Biological relevance. New York, Plenum Press.
De Wit, M.J. 1977. The evolution of the Scotia Arc as a key to the reconstruction of southwestern Gondwanaland. Tectonophysics **37**:53-81.
Dewey, J.F. 1977. Suture zone complexities: a review. Tectonophysics **40**:53-67.
Dexter, D. 1972. Comparison of community structure in a Pacific and Atlantic Panamanian sandy beach. Bull. Mar. Sci. **22**:449-462.
Diamond, J.M. 1969. Avifaunal equilibria and species turnover rates on the Channel Islands of California. Proc. Natl. Acad. Sci. U.S.A. **64**:57-63.
Diamond, J.M. 1970a. Ecological consequences of island colonization by Southwest Pacific birds. I. Types of niche shifts. Proc. Natl. Acad. Sci. U.S.A. **67**:529-536.
Diamond, J.M. 1970b. Ecological consequences of island colonization by Southwest Pacific birds. II. The effect of species diversity on total population density. Proc. Natl. Acad. Sci. U.S.A. **67**:1715-1721.
Diamond, J.M. 1971. Comparison of faunal equilibrium turnover rates on a tropical island and a temperate island. Proc. Natl. Acad. Sci. U.S.A. **68**:2742-2745.
Diamond, J.M. 1972. Biogeographic kinetics: estimation of relaxation times for avifaunas of southwest Pacific islands. Proc. Natl. Acad. Sci. U.S.A. **69**:3199-3203.
Diamond, J.M. 1973. Distributional ecology of New Guinea birds. Science **179**:759-769.
Diamond, J.M. 1974. Colonization of exploded volcanic islands by birds: the supertramp strategy. Science **184**:803-806.
Diamond, J.M. 1975a. The island dilemma: lessons of modern biogeographic studies for the design of natural reserves. Biol. Conserv. **7**:129-146.
Diamond, J.M. 1975b. Assembly of species communities, p. 342-444. *In* M.L. Cody and J.M. Diamond (eds.), Ecology and evolution of communities. Cambridge, Mass., Belknap Press.
Diamond, J.M. 1980. Species turnover in island bird communities, p. 777-782. *In* R. Nöhring (ed), Acta XVII Congressus Internationalis Ornithologici, vol. 2. Berlin, Verlag der Deutschen Ornithologen-Gesellschaft.
Diamond, J.M., and M.E. Gilpin. 1982. Examination of the "null" model of Connor and Simberloff for species co-occurrence on islands. Oecologia **52**:64-74.
Diamond, J.M., and A.G. Marshall. 1977. Distributional ecology of New Hebridean birds: a species kaleidoscope. J. Anim. Ecol. **46**:703-727.
Diamond, J.M., and R.M. May. 1976. Island biogeography and the design of natural preserves, p. 163-186. *In* R.M. May (ed.), Theoretical ecology: principles and applications. Philadelphia, W.B. Saunders Co.
Diamond, J.M., and R.M. May. 1977. Species turnover rates on islands: dependence on census interval. Science **197**:266-270.
di Castri, F., and H.A. Mooney (eds.). 1973. Mediterranean type ecosystems. Origin and structure. Ecological studies 7. New York, Springer-Verlag New York, Inc.
Dice, L.R. 1947. Effectiveness of selection by owls of deer mice *(Peromyscus maniculatus)* which contrast in color with their background. Contr. Lab. Vert. Biol. Univ. Mich. **34**:1-20.
Dice, L.R. 1952. Natural communities. Ann Arbor, Mich., University of Michigan Press.

Dickinson, W.R. 1973. Reconstruction of past arc-trench systems from petrotectonic assemblages in the land arcs of the western Pacific, p. 569-601. *In* P.J. Coleman (ed.), The western Pacific. Island arcs, marginal seas, geochemistry. New York, Crane, Russak & Co.

Dietz, R.S. 1961. Continent and ocean basin evolution by spreading of the sea floor. Nature **190**:854-857.

Dietz, R.S., and J.C. Holden. 1970a. Reconstruction of Pangaea: breakup and dispersion of continents, Permian to present. J. Geophys. Res. **75**:4939-4955.

Dietz, R.S., and J.C. Holden. 1970b. The break-up of Pangaea. Sci. Am. **223**(4):30-41.

Dingle, H. 1980. Ecology and evolution of migration, p. 1-101. *In* S.A. Gauthreaux, Jr. (ed.), Animal migration, orientation, and navigation. New York, Academic Press, Inc.

Dixon, J.R. 1979. Origin and distribution of reptiles of lowland tropical rain forest of South America, p. 217-240. *In* W.E. Duellman (ed.), The South American herpetofauna: its origin, evolution, and dispersal. Lawrence, Kans., University of Kansas Museum Nat. Hist. Monogr. no. 7.

Dobzhansky, Th. 1950. Evolution in the tropics. Am. Sci. **38**:209-221.

Dobzhansky, Th. 1951. Genetics and the origin of species. Rev. 3rd ed. New York, Columbia University Press.

Docters van Leeuwen, W.M. 1936. Krakatau, 1883-1933. Ann. Jard. Bot. Buitenzorg **46-47**:1-506.

Dodd, A.P. 1959. The biological control of prickly pear in Australia, p. 565-577. *In* A. Keast (ed.), Biogeography and ecology in Australia. Monographiae Biologicae 8. The Hague, Dr. W. Junk BV, Publisher.

Dorf, E. 1964. The use of fossil plants in palaeoclimatic interpretations, p. 13-31. *In* A.E.M. Nairn (ed.), Problems in palaeoclimatology. London, Interscience Publishers.

Dorst, J. 1962. The migrations of birds. Boston, Houghton Mifflin Co.

Dorst, J. 1972. The evolution and affinities of the birds of Madagascar, p. 615-660. *In* R. Battistini and G. Richard-Vindard (ed.), Biogeography and ecology in Madagascar. Monographiae Biologicae 21. The Hague, Dr. W. Junk BV, Publishers.

Dorst, J. 1974. The life of birds. London, Weidenfeld and Nicolson.

Downhower, J.F. 1976. Darwin's finches and the evolution of sexual dimorphism in body size. Nature **263**:558-563.

Doyle, J.A. 1969. Cretaceous angiosperm pollen of the Atlantic Coastal Plain and its evolutionary significance. J. Arnold Arbor. **50**:1-35.

Doyle, J.A. 1977. Patterns of evolution in early angiosperms, p. 501-546. *In* A. Hallam (ed.), Patterns of evolution, as illustrated by the fossil record. Amsterdam, Elsevier Scientific Publishing Co.

Doyle, J.A. 1978. Origin of angiosperms. Ann. Rev. Ecol. Syst. **9**:365-392.

Dressler, R.L. 1981. The orchids: natural history and classification. Cambridge, Mass., Harvard University Press.

Drude, O. 1887. Atlas der Pflanzenverbreitung. 5 Abt. Gotha, Berghaus Physikalischer Atlas.

Drury, W.H., and I.C.T. Nisbet. 1973. Succession. J. Arnold Arbor. **54**:331-368.

Dubost, G. 1968. Les niches écologiques des forêts tropicales sud-américaines et africaines, sources des convergences remarquables entre Rongeurs et Artiodactyles. Terre Vie **22**:3-28.

Duellman, W.E. 1966. The Central American herpetofauna: an ecological perspective. Copeia (4):700-719.

Duellman, W.E. 1975. On the classification of frogs. University of Kansas Museum Nat. Hist. Occas. Pap. **42**:1-14.

Duellman, W.E. (ed.). 1979. The South American herpetofauna: its origin, evolution and dispersal. Lawrence, Kans., University of Kansas Museum Nat. Hist. Monogr. no. 7.

Dunham, A.E., D.W. Tinkle, and J.W. Gibbons. 1978. Body size in island lizards: a cautionary tale. Ecology **59**:1230-1239.

Dunmire, W.W. 1960. An altitudinal survey of reproduction in *Peromyscus maniculatus*. Ecology **41**:174-182.

Dunson, W.A. (ed.). 1975. The biology of sea snakes. Baltimore, Md., University Park Press.

du Toit, A.L. 1927. A geological comparison of South America with South Africa. Washington, D.C., Carnegie Institute Pub. no. **381**:1-157.

du Toit, A.L. 1937. Our wandering continents. Edinburgh, Oliver & Boyd.

du Toit, A.L. 1944. Tertiary mammals and continental drift. Am. J. Sci. **242**:145-163.

Dyer, R.A. 1975. The genera of southern African flowering plants, vol. 1. Pretoria, Department of Agricultural Technical Services, Botanical Research Institute.

Dykyjova, D., and J. Kvet (eds.). 1978. Pond littoral systems, structure and function. Berlin, Springer-Verlag.

Dzitzer, C. 1975. Biogeography of the Laminariales. University of California, Los Angeles, Department of Geography, M.A. thesis.

Edmunds, G.F., Jr. 1972. Biogeography and evolution of Ephemeroptera. Ann. Rev. Entomol. **17**:21-42.

Edmunds, G.F., Jr. 1975. Phylogenetic biogeography of mayflies. Ann. Missouri Bot. Gard. **62**:251-263.

Edmunds, G.F., Jr. 1981. Discussion, p. 287-297. *In* G. Nelson and D.E. Rosen (eds.), Vicariance biogeography: a critique. New York, Columbia University Press.

Edwards, C.A., and J.R. Lofty. 1977. Biology of earthworms. London, Chapman & Hall, Ltd.

Edwards, J.L. 1976. Spinal nerves and their bearing on salamander phylogeny. J. Morphol. **148**:305-327.

Egerton, F.N. (ed.). 1978a. History of American ecology: an original anthology. New York, Arno Press, Inc.

Egerton, F.N. (ed.). 1978b. Ecological phytogeography in the nineteenth century: an original anthology. New York, Arno Press, Inc.

Ehrlich, P.R. 1961. Intrinsic barriers to dispersal in the checkerspot butterfly. Science **134**:108-109.

Ehrlich, P.R. 1965. The population biology of the butterfly *Euphydryas editha*. II. The structure of the Jaspar Ridge colony. Evolution **19**:327-336.

Erlich, P.R., and R.W. Holm. 1963. The process of evolution. New York, McGraw-Hill, Inc.

Eiseley, L. 1958. Darwin's century. Evolution and the men who discovered it. New York, Doubleday Publishing Co.

Eisenberg, J.F. 1981. The mammalian radiations. An analysis of trends in evolution, adaptation, and behavior. Chicago, University of Chicago Press.

Ekman, S. 1953. Zoogeography of the sea. London, Sidgwick & Jackson, Ltd.

Eldredge, N. 1979. Cladism and common sense, p. 165-197. *In* J. Cracraft and N. Eldredge (eds.), Phylogenetic analysis and paleontology. New York, Columbia University Press.

Eldredge, N., and J. Cracraft. 1980. Phylogenetic patterns and the evolutionary process. New York, Columbia University Press.

Eldredge, N., and S.J. Gould. 1972. Punctuated equilibria: an alternative to phyletic gradualism, p. 82-115. *In* T.J.M. Schopf (ed.), Models in paleobiology. San Francisco, Freeman, Cooper & Co.

Eldredge, N., and S.J. Gould. 1976. Rates of evolution revisited. Paleobiology **2**:174-177.

Elliot, D.H., E.H. Colbert, W.J. Breed, J.A. Jensen, and J.S. Powell. 1970. Triassic tetrapods from Antarctica: evidence for continental drift. Science **169**:1197-1201.

Elliot, G.F. 1951. On the geographical distribution of terebratelloid brachiopods. Ann. Mag. Nat. Hist., ser. 12, **4**:305-334.

Elton, C.S. 1958. The ecology of invasions by animals and plants. London, Methuen & Co., Ltd.

Elton, C.S. 1966. The patterns of animal communities. London, Methuen & Co., Ltd.

Embleton, B.J.J. 1973. The palaeolatitude of Australia through Phanerozoic time. J. Geol. Soc. Australia **19**:475-482.

Embleton, B.J.J., M.W. McElhinny, A.R. Crawford, and G.R. Luck. 1974. Palaeomagnetism and the tectonic evolution of the Tasman orogenic zone. J. Geol. Soc. Australia **21**:187-193.

Embleton, B.J.J., and P.W. Schmidt. 1977. Revised palaeomagnetic data for the Australian Mesozoic and a synthesis of Late Palaeozoic-Mesozoic results for Gondwanaland. Tectonophysics **38**:355-364.

Embleton, B.J.J., and D.A. Valencio. 1977. Paleomagnetism and the reconstruction of Gondwanaland, p. 1-12. *In* M.W. McElhinny (ed.), The past distribution of continents. Tectonophysics **40**.

Emiliani, C. 1966. Paleotemperature analysis of Caribbean cores P6304-8 and P6304-9 and a generalized temperature curve for the past 425,000 years. J. Geol. **74**:109-126.

Emiliani, C. 1971. The amplitude of Pleistocene climatic cycles at low latitudes and the isotopic composition of glacial ice, p. 183-197. *In* K. Turikian (ed.), The late Cenozoic glacial ages. New Haven, Conn., Yale University Press.

Emlen, J.T. 1978. Density anomalies and regulation mechanisms in land bird populations on the Florida peninsula. Am. Nat. **112**:265-286.

Emlen, S.T. 1969. Bird migration: influence of physiological state upon celestial orientation. Science **165**:716-718.

Emsley, M.G. 1965. Speciation in *Heliconius* (Lep., Nymphalidae). Morphology and geographic distribution. Zoologica (New York) **50**(4):191-254.

Endler, J.A. 1977. Geographic variation, speciation, and clines. Monographs in population biology, no. 10. Princeton, N.J., Princeton University Press.

Ernst, W.G. (ed.). 1975. Subduction zone metamorphism. Benchmark papers in geology, no. 19. Stroudsburg, Penn., Dowden, Hutchinson & Ross, Inc.

Erwin, T.L., G.E. Ball, D.R. Whitehead, and A.L. Halpern (eds.). 1979. Carabid beetles: their evolution, natural history, and classification. The Hague, Dr. W. Junk BV, Publishers.

Estes, R., and O.A. Reig. 1973. The early fossil record of frogs. A review of the evidence, p. 11-63. *In* J.L. Vial (ed.), Evolutionary biology of the anurans. Columbia, Mo., University of Missouri Press.

Eyre, S.R. 1968. Vegetation and soils: a world picture. 2nd ed. London, Edward Arnold (Publishers), Ltd.

Eyre, S.R. (ed.). 1971. World vegetation types. New York, Columbia University Press.

Faeth, S.H., and E.F. Connor. 1979. Supersaturated and relaxing island faunas: a critique of the species-age relationship. J. Biogeogr. **6**:311-316.

Falla, R.A. 1960. Oceanic birds as dispersal agents. Proc. Roy. Soc. London, ser. B **152**:655-659.

Falla, R.A., R.B. Sibson, and E.G. Turbott. 1966. A field guide to the birds of New Zealand and outlying islands. London, William Collins Sons & Co., Ltd.

Fallaw, W.C. 1978. Reply. Geol. Soc. Am. Bull. **89**:478-480.

Farris, J.S. 1976. Phylogenetic classification of fossils with recent species. Syst. Zool. **25**:271-282.

Farris, J.S., A.G. Kluge, and M.J. Eckardt. 1970. A numerical approach to phylogenetic systematics. Syst. Zool. **19**:172-191.

Federov, A.A. 1966. The structure of the tropical rain forest and speciation in the humid tropics. J. Ecol. **54**:1-11.

Feduccia, A. 1980. The age of birds. Cambridge, Mass., Harvard University Press.

Fell, H.B. 1967. Cretaceous and Tertiary surface currents of the oceans. Ann. Rev. Oceanogr. Mar. Biol. **5**:317-341.

Fellows, D.P., and W.B. Heed. 1972. Factors affecting host plant selection in desert-adapted cactiphilic *Drosophila*. Ecology **53**:850-858.

Felsenstein, J. 1978. Cases in which parsimony or compatibility methods will be positively misleading. Syst. Zool. **27**:401-410.

Fernald, M.L. 1925. Persistence of plants in unglaciated areas of boreal America. Mem. Am. Acad. Arts Sci. **15**:237-342.

Ferris, V.R., C.G. Goseco, and J.M. Ferris. 1976. Biogeography of free-living soil nematodes from the perspective of plate tectonics. Science **193**:508-510.

Findley, J.S., and C. Jones. 1964. Seasonal distribution of the hoary bat. J. Mammal. **45**:461-470.

Finerty, J.P. 1980. The population ecology of cycles in small mammals. Mathematical theory and biological fact. New Haven, Conn., Yale University Press.

Fink, S.V., and W.L. Fink. 1981. Interrelationships of the ostariophysan fishes (Teleostei). Zool. J. Linn. Soc. **72**:297-353.

Fischer, A.G. 1960. Latitudinal variation in organic diversity. Evolution **14**:64-81.

Fischer, A.G., and M.A. Arthur. 1977. Secular variations in the pelagic realm. Soc. Econ. Paleontol. Minerol. Spec. Pub. **25**:19-50.

Fisher, D.R. 1968. A study of faunal resemblances using numerical taxonomy and factor analysis. Syst. Zool. **17**:48-63.

Fisher, R.L., and R. Revelle. 1955. The trenches of the Pacific. Sci. Am. **153**(5):36-41.

Fittkau, E.J., J. Illies, H. Klinge, G.H. Schwabe, and H. Sioli (eds.). 1968-1969. Biogeography and ecology in South America. 2 vol. Mongraphiae Biologicae 20. The Hague, Dr. W. Junk BV, Publishers.

Fittkau, E.J. 1969. The fauna of South America, p. 624-658. *In* E.J. Fittkau, J. Illies, H. Klinge, G.H. Schwabe, and H. Sioli (eds.), Biogeography and ecology in South America, vol. 2. Monographiae Biologicae 20. The Hague, Dr. W. Junk BV, Publishers.

Fleming, C.A. 1975. The geological history of New Zealand and its biota, p. 1-86. *In* G. Kuschel (ed.), Biogeography and ecology in New Zealand. Monographiae Biologicae 27. The Hague, Dr. W. Junk BV, Publishers.

Fleming, T.H. 1973. Numbers of mammal species in North and Central American forest communities. Ecology **54**:555-563.

Flenley, J.R. 1979a. The equatorial rain forest: a geological history. London, Butterworth & Co. (Publishers), Ltd.

Flenley, J.R. 1979b. The Late Quaternary vegetational history of the equatorial mountains. Progr. Phys. Geogr. **3**:488-509.

Flessa, K.W. 1975. Area, continental drift and mammalian diversity. Paleobiology **1**:189-194.

Flessa, K.W. 1980. Biological effects of plate tectonics and continental drift. Bioscience **30**:518-523.

Flessa, K.W. 1981. The regulation of mammalian faunal similarity among the continents. J. Biogeogr. **8**:427-438.

Flessa, K.W., S.G. Barnett, D.B. Cornue, M.A. Lomaga, N. Lombardi, J.M. Miyazaki, and A.S. Murer. 1979. Geologic implications of the relationship between mammalian faunal similarity and geographic distance. Geology **7**:15-18.

Flessa, K.W., and J. Imbrie. 1973. Evolutionary pulsations: evidence from Phanerozoic diversity patterns, p. 247-285. *In* D.H. Tarling and S.K. Runcorn (eds.), Implications of continental drift to the earth sciences, vol. 2. New York, Academic Press, Inc.

Flessa, K.W., and J.M. Miyazaki. 1978. Trends in trans-North Atlantic commonality among Phanerozoic invertebrates, and plate tectonic events: discussion and reply. Geol. Soc. Am. Bull. **89**:476-477.

Flessa, K.W., and J.J. Sepkoski, Jr. 1978. On the relationship between phanerozoic diversity and changes in habitable area. Paleobiology **4**:359-366.

Flint, R.F. 1971. Glacial and Quaternary geology. New York, John Wiley & Sons, Inc.

Flohn, H. 1969. Climate and weather. New York, World University Library, McGraw-Hill Book Co.

Florin, R. 1963. The distribution of conifer and taxad genera in time and space. Acta Hort. Bergiani (Uppsala) **20**:121-312.

Foin, T.C., J.W. Valentine, and F.J. Ayala. 1975. Extinction of taxa and Van Valen's law. Nature **257**:514-515.

Fooden, J. 1972. Breakup of Pangaea and isolation of relict mammals in Australia, South America, and Madagascar. Science **175**:894-898.

Ford, E.B. 1975. Ecological genetics. 4th ed. London, Chapman & Hall, Ltd.

Forman, R.T.T. (ed.). 1979. Pine barrens: ecosystem and landscape. New York, Academic Press, Inc.

Forman, R.T.T., A.E. Galli, and C.F. Leck. 1976. Forest size and avian diversity in New Jersey woodlots with some land-use implications. Oecologia **26**:1-8.

Fosberg, F.R. 1963. Plant dispersal in the Pacific, p. 273-281. *In* J.L. Gressitt (ed.), Pacific Basin biogeography. Honolulu, Bishop Museum Press.

Foster, J.B. 1964. Evolution of mammals on islands. Nature **202**:234-235.

Fowler, H.W. 1959. Fishes of Fiji. Fiji, Suva.

Free, J.B. 1970. Insect pollination of crops. New York, Academic Press, Inc.

Frenzel, B. 1968. The Pleistocene vegetation of northern Eurasia. Science **161**:637-649.

Fretwell, S. 1980. Evolution of migration in relation to factors regulating bird numbers, p. 517-528. *In* A. Keast and E.S. Morton (eds.), Migrant birds in the Neotropics: ecology, behavior, distribution, and conservation. Washington, D.C., Smithsonian Institute Press.

Frey, D.G. (ed.). 1963. Limnology in North America. Madison, Wisc., University of Wisconsin Press.

Fritts, H.C. 1976. Tree rings and climate. New York, Academic Press, Inc.

Fritts, H.C., G.R. Lofgren, and G.A. Gordon. 1979. Variations in climate since 1602 as reconstructed from tree rings. Quat. Res. **12**:18-46.

Frost, S.H. 1977. Miocene to Holocene evolution of Caribbean Province reef-building corals, p. 353-359. *In* Proceedings of the Third International Coral Reef Symposium, Miami.

Fryer, G., and T.D. Iles. 1972. The cichlid fishes of the Great Lakes of Africa: their biology and evolution. Edinburgh, Oliver & Boyd.

Fryxell, P.A. 1967. The interpretation of disjunct distributions. Taxon **16**:316-324.

Fuentes, E.R. 1976. Ecological convergence of lizard communities in Chile and California. Ecology **57**:3-17.

Funk, V.A. 1981. Species concerns in estimating plant phylogenies, p. 73-86. *In* Funk, V.A., and D.R. Brooks (eds.), 1981. Advances in cladistics. Proceedings of the first meeting of the Willi Hennig Society. New York, New York Botanical Garden.

Funk, V.A., and T.F. Stuessy. 1978. Cladistics for the practicing plant taxonomist. Syst. Bot. **3**:159-178.

Funnell, B.M. 1971. Post-Cretaceous biogeography of oceans: with especial reference to plankton, p. 191-198. *In* F.A. Middlemiss and P.F. Rawson (eds.), Faunal provinces in space and time. Liverpool, Steel House Press.

Gaffney, E.S. 1975. A phylogeny and classification of the higher categories of turtles. Bull. Am. Mus. Nat. Hist. **155**:387-436.

Gaffney, E.S. 1977. The side-necked turtle family Chelidae: a theory of relationships using shared derived characters. Am. Mus. Novitates no. **2620**:1-28.

Gaines, M.S., and L.R. McClenaghan, Jr. 1980. Dispersal in small mammals. Ann. Rev. Ecol. Syst. **11**:163-196.

Galler, S.R., K. Schmidt-Koenig, G.J. Jacobs, and R.E. Belleville. (eds.). 1972. Animal orientation and navigation. Washington, D.C., U.S. Government Printing Office (N.A.S.A.).

Galli, A.E., C.F. Leck, and R.T.T. Forman. 1976. Avian distribution patterns in New Jersey woodlots with some land-use implications. Auk **93**:356-364.

Ganapathy, R. 1980. A major meteorite impact on the earth 65 million years ago: evidence from the Cretaceous-Tertiary boundary clay. Science **209**:921-923.

Gans, C., and T.S. Parsons (eds.). 1969-1981. Biology of the Reptilia. 11 vol. London, Academic Press, Inc.

Garrido, O.H., and F.G. Montaña. 1975. Catálogo de las Aves de Cuba. La Habana, Academia de Ciencias de Cuba.

Gastil, G., G. Morgan, and D. Krummenacher. 1981. The tectonic history of peninsular California and adjacent Mexico, p. 284-306. *In* W.G. Ernst (ed.), The geotectonic development of California, Rubey vol. 1. Englewood Cliffs, N.J., Prentice-Hall, Inc.

Gates, D.M. 1962. Energy exchange in the biosphere. New York, Harper & Row, Publishers, Inc.

Gates, D.M. 1970. Animal climates: where animals must live. Environ. Res. **3**:132-144.

Gates, D.M. 1980. Biophysical ecology. New York, Springer-Verlag New York, Inc.

Gause, G.F. 1934. The struggle for existence. Baltimore, Md., The Williams & Wilkins Co.

Gauthreaux, S.A., Jr. 1980a. Animal migration, orientation, and navigation. New York, Academic Press, Inc.

Gauthreaux, S.A., Jr. 1980b. The influences of long-term and short-term climatic changes on the dispersal and migration of organisms, p. 103-174. *In* S.A. Gauthreaux (ed.), Animal migration, orientation, and navigation. New York, Academic Press, Inc.

Geiger, R. 1957. The climate near the ground. Cambridge, Mass., Harvard University Press.

George, W. 1962. Animal geography. London, William Heinemann, Ltd.

Georgi, G. 1962. Memoirs of Alfred Wegener, p. 309-324. *In* S.K. Runcorn (ed.), Continental drift. New York, Academic Press, Inc.

Gervasio, F.C. 1973. Geotectonic development of the Philippines, p. 307-324. *In* P.J. Coleman (ed.), The western Pacific. Island arcs, marginal seas, geochemistry. New York, Crane, Russak & Co.

Géry, J. 1969. The fresh-water fishes of South America, p. 828-848. *In* E.J. Fittkau, J. Illies, H. Klinge, G.H. Schwabe, and H. Sioli (eds.), Biogeography and ecology in South America, vol. 2. Monographiae Biologicae 20. The Hague, Dr. W. Junk BV, Publishers.

Géry, J. 1977. Characoids of the world. Hong Kong, T.F.H. Publications, Inc., Ltd.

Gibson, A.C., and K.E. Horak. 1978. Systematic anatomy and phylogeny of Mexican columnar cacti. Ann. Missouri Bot. Gard. **65**:999-1057.

Gilbert, F.S. 1980. The equilibrium theory of island biogeography: fact or fiction? J. Biogeogr. **7**:209-235.

Gilbert, L.E. 1975. Ecological consequences of a coevolved mutualism between butterflies and plants, p. 210-240. *In* L.E. Gilbert and P.H. Raven (eds.), Coevolution of animals and plants. Austin, Tex., University of Texas Press.

Gilbert, L.E., and M.C. Singer. 1975. Butterfly ecology. Ann. Rev. Ecol. Syst. **6**:365-398.

Gilpin, M.E., and J.M. Diamond. 1976. Calculations of immigration and extinction curves from the species-area-distance relation. Proc. Natl. Acad. Sci. U.S.A. **73**:4130-4134.

Gilpin, M.E., and J.M. Diamond. 1982. Factors contributing to non-randomness in species co-occurrences on islands. Oecologia **52**:75-84.

Gingerich, P.D. 1976a. Paleontology and phylogeny: patterns of evolution at the species level in Early Tertiary mammals. Am. J. Sci. **276**:1-28.

Gingerich, P.D. 1976b. Cranial anatomy and evolution of early Tertiary Plesiadapidae. Univ. Michigan Pap. Paleontol. no. **15**:1-140.

Gingerich, P.D. 1979. The stratophenetic approach to phylogeny reconstruction in vertebrate paleontology, p. 41-77. *In* J. Cracraft and N. Eldredge (eds.), Phylogenetic analysis and paleontology. New York, Columbia University Press.

Girdler, R.W., and P. Styles. 1978. Seafloor spreading in the western Gulf of Aden. Nature **271**:615-617.

Gjaeveroll, O. 1963. Survival of plants on nunataks in Norway during the Pleistocene glaciation, p. 261-283. *In* A. Löve and D. Löve (eds.), North Atlantic biota and their history. Oxford, Pergamon Press, Ltd.

Glaessner, M.F. 1969. Decapoda, p. 400-532. *In* R.C. Moore (ed.), Treatise on invertebrate paleontology, Part R, Arthropoda, vol. 2. Boulder, Colo., Geological Society of America.

Glantz, M.H. (ed.). 1976. The politics of natural disaster. The case of the Sahel drought. New York, Praeger Publishers.

Gleason, H.A. 1917. The structure and development of the plant association. Bull. Torrey Bot. Club **53**:7-26.

Gleason, H.A. 1926. The individualistic concept of plant associations. Bull. Torrey Bot. Club **53**:7-26.

Glenn-Lewin, D.C. 1977. Species diversity in North American temperate forests. Vegetatio **33**:153-162.

Godwin, Sir H. 1975. The history of the British flora. 2nd ed. Cambridge, Cambridge University Press.

Goin, C.J., O.B. Goin, and G.R. Zug. 1978. Introduction to herpetology. 3rd ed. San Francisco, W. H. Freeman & Co., Publishers.

Goldblatt, P. 1978. An analysis of the flora of southern Africa: its characteristics, relationships, and origins. Ann. Missouri Bot. Gard. **65**:369-436.

Golley, F.B. 1960. Energy dynamics of a food chain of an old-field community. Ecol. Monogr. **30**:187-206.

Gómez, L.D. 1982. The origin of Central American pteridophytes. Ann. Missouri Bot. Gard. (in press).

Good, R. 1974. The geography of flowering plants. 4th ed. New York, John Wiley & Sons, Inc. (1st ed. 1947.)

Goodman, D. 1975. The theory of diversity-stability relationships in ecology. Quart. Rev. Biol. **50**:237-266.

Gordon, M.S., G.A. Bartholomew, A.D. Grinnell, C.B. Jørgensen, and F.N. White. 1981. Animal physiology: principles and adaptations. 4th ed. New York, Macmillan, Inc.

Gottfried, B.M. 1979. Small mammal populations in woodlot islands. Am. Midl. Nat. **102**:105-112.

Gould, S.J. 1965. Is uniformitarianism necessary? Am. J. Sci. **263**:223-228.

Gould, S.J. 1977. Ever since Darwin. Reflections on natural history. New York, W.W. Norton & Co., Inc.

Gould, S.J. 1979. An allometric interpretation of the species-area curves: the meaning of the coefficient. Am. Nat. **114**:335-343.

Gould, S.J., and C.B. Calloway. 1980. Clams and brachiopods–ships that pass in the night. Paleobiology **6**:383-396.

Gould, S.J., and N. Eldredge. 1977. Punctuated equilibria: the tempo and mode of evolution reconsidered. Paleobiology **3**:115-151.

Gould, S.J., D.M. Raup, J.J. Sepkoski, Jr., T.J.M. Schopf, and D.S. Simberloff. 1977. The shape of evolution: a comparison of real and random clades. Paleobiology **3**:23-40.

Graham, A. (ed.). 1972. Floristics and paleofloristics of Asia and eastern North America. New York, Elsevier North-Holland, Inc.

Graham, A. (ed.). 1973. Vegetation and vegetational history of northern Latin America. Amsterdam, Elsevier Publishing Co., Inc.

Grant, P.R. 1965. The adaptive significance of some size trends in island birds. Evolution **19**:355-367.

Grant, P.R. 1966a. The density of land birds on Tres Marías Islands in Mexico. I. Numbers and biomass. Can. J. Zool. **44**:391-400.

Grant, P.R. 1966b. Ecological compatibility of bird species on islands. Am. Nat. **100**:451-462.

Grant, P.R. 1967. Bill length variability in birds of the Tres Marías Islands, Mexico. Can. J. Zool. **45**:805-815.

Grant, P.R. 1968. Bill size, body size and the ecological adaptations of bird species to competitive situations on islands. Syst. Zool. **17**:319-333.

Grant, P.R. 1970. Colonization of islands by ecologically dissimilar species of mammals. Can. J. Zool. **48**:545-553.

Grant, P.R. 1971. Experimental studies of competitive interactions in a two-species system. III. *Microtus* and *Peromyscus* species in enclosures. J. Anim. Ecol. **40**:323-350.

Grant, P.R. 1972a. Interspecific competition among rodents. Ann. Rev. Ecol. Syst. **3**:79-106.

Grant, P.R. 1972b. Convergent and divergent character displacement. Biol. J. Linn. Soc. **4**:39-68.

Grant, P.R., and I. Abbott. 1980. Interspecific competition, island biogeography and null hypotheses. Evolution **34**:332-341.

Grant, V. 1963. The origin of adaptations. New York, Columbia University Press.

Grant, V. 1971. Plant speciation. 2nd ed. New York, Columbia University Press.

Grant, V., and K.A. Grant. 1965. Pollination in the phlox family. New York, Columbia University Press.

Gray, J., and A.J. Boucot (eds.). 1979. Historical biogeography, plate tectonics, and the changing environment. Corvallis, Ore., Oregon State University Press.

Gray, J., A.J. Boucot, and W.B.N. Berry (eds.). 1981. Communities of the past. Stroudsburg, Penn., Hutchinson Ross Publishing Co.

Grayson, D.K. 1977. Pleistocene avifaunas and the overkill hypothesis. Science **195**:691-692.

Grayson, D.K. 1979. Vicissitudes and overkill: the development of Pleistocene extinctions, p. 199-237. *In* M.B. Schiffer (ed.), Advances in archaeological method and theory, vol. 3. New York, Academic Press, Inc.

Grayson, D.K. 1981. A mid-Holocene record for the heather vole, *Phenacomys* cf. *intermedius,* in the central Great Basin and its biogeographic significance. J. Mammal. **62**:115-121.

Greenslade, P.J.M. 1968a. The distribution of some insects of the Solomon Islands. Proc. Linn. Soc. London **179**:189-196.

Greenslade, P.J.M. 1968b. Island patterns in the Solomon Islands bird fauna. Evolution **22**:751-761.

Greenslade, P.J.M. 1969. Insect distribution patterns in the Solomon Islands. Philos. Trans. Roy. Soc. London, ser. B **225**:271-285.

Greenwood, P.H. 1974. The cichlid fishes of Lake Victoria, East Africa: the biology and evolution of a species flock. Bull. Brit. Mus. Nat. Hist. Suppl. **6**:1-134.

Greenwood, P.H., R.S. Miles, and C. Patterson (eds.). 1973. Interrelationships of fishes. London, Academic Press for Linnean Society of London, Zool. J. Linn. Soc. **53**, Suppl. no. 1.

Gressitt, J.L. (ed.). 1963. Pacific Basin biogeography. Honolulu, Bishop Museum Press.

Gressitt, J.L. 1970. Subantarctic entomology and biogeography. Pacific Insects Monogr. **23**:295-374.

Gressitt, J.L. 1974. Insect biogeography. Ann Rev. Entomol. **19**:293-321.

Gressitt, J.L. (ed.). 1982. Biogeography and ecology of New Guinea. 2 vol. Monographiae Biologicae 42. The Hague, Dr. W. Junk BV, Publishers.

Gressitt, J.L., and C.M. Yoshimoto. 1963. Dispersal of animals in the Pacific, p. 283-292. *In* J.L. Gressitt (ed.), Pacific Basin biogeography. Honolulu, Bishop Museum Press.

Griffiths, G.C.D. 1973. Some fundamental problems in biological classifications. Syst. Zool. **22**:338-343.

Griffiths, J.R., and P.M. Austin. 1977. Discussion and reply. Paleogeographic and paleotectonic models for the New Zealand geosyncline in eastern Gondwanaland. Geol. Soc. Am. Bull. **88**:1203-1210.

Griffiths, J., and C. Burnett. 1973. Were southeast Asia and Indonesia parts of Gondwanaland? Nature **245**:92-93.

Griffiths, M. 1978. The biology of the monotremes. New York, Academic Press, Inc.

Griggs, R.F. 1946. The timberlines of northern Alaska America and their interpretation. Ecology **27**:275-289.

Grime, J.P. 1979. Plant strategies and vegetation processes. Chichester, England, John Wiley & Sons, Ltd.

Grime, J.P., and J.G. Hodgson. 1969. An investigation of the ecological significance of lime-chlorosis by means of large-scale comparative experiments, p. 67-99. *In* I.H. Rorison (ed.), Ecological aspects of the mineral nutrition of plants. Oxford, Blackwell Scientific Publications, Ltd.

Grinnell, J. 1914. The Colorado River as a hindrance to the dispersal of species. Univ. California Pub. Zool. **12**:100-107.

Grinnell, J. 1917. The niche-relationship of the California thrasher. Auk **34**:427-433.

Grinnell, J. 1922. The role of the "accidental." Auk **39**:373-380.

Guilday, J.E. 1967. Differential extinction during Late Pleistocene and Recent times, p. 121-140. *In* P.S. Martin and H.E. Wright, Jr. (eds.), Pleistocene extinctions: the search for a cause. New Haven, Conn., Yale University Press.

Gruson, E.S. 1976. Checklist of the world's birds: a complete list of the species, with names, authorities, and areas of distribution. New York, Quadrangle.

Gunn, C., J.V. Dennis, and P.J. Paradine. 1976. World guide to tropical drift seeds and fruits. New York, New York Times Book Co.

Gunn, P.J. 1975. Mesozoic-Cainozoic tectonics and igneous activity: southeastern Australia. J. Geol. Soc. Australia **22**:215-221.

Guppy, H.B. 1906. Observations of a naturalist in the Pacific between 1891 and 1899. 2 vol. London, The Macmillan Co.

Gwinner, E. 1977. Circannual rhythms in bird migrations. Ann. Rev. Ecol. Syst. **8**:381-406.

Haas, P.H. 1975. Some comments on use of the species-area curve. Am. Nat. **109**:371-372.

Hackman, B.D. 1980. The geology of Guadalcanal, Solomon Islands. Overseas Mem. 6, Inst. Geol. Sci. London, Her Majesty's Stationary Office.

Haffer, J. 1969. Speciation in Amazonian forest birds. Science **165**:131-137.

Haffer, J. 1974. Avian speciation in tropical South America, with a systematic survey of the toucans (Ramphastidae) and jacamars (Galbulidae). Cambridge, Mass., Nuttall Ornithological Club.

Haffer, J. 1978. Distribution of Amazon forest birds. Bonn. Zool. Beitr. **29**:38-78.

Haffer, J. 1981. Aspects of neotropical bird speciation during the Cenozoic, p. 371-394. *In* G. Nelson and D.E. Rosen (eds.), Vicariance biogeography: a critique. New York, Columbia University Press.

Hagmeir, E.M. 1966. A numerical analysis of the distributional patterns of North American mammals. II. Re-evaluation of the provinces. Syst. Zool. **15**:279-299.

Hagmeir, E.M., and C.D. Stults. 1964. A numerical analysis of the distributional patterns of North American mammals. Syst. Zool. **13**:125-155.

Haile, N.S., M.W. McElhinny, and I. McDougall. 1977. Palaeomagnetic data and radiometric ages from the Cretaceous of West Kalimantan (Borneo), and their significance in interpreting regional structure. J. Geol. Soc. London **133**:133-144.

Hall, B.P., and R.E. Moreau. 1970. Atlas of speciation of African birds. London, British Museum of Natural History.

Hall, E.R. 1981. The mammals of North America. 2 vol. 2nd ed. New York, John Wiley & Sons, Inc.

Hallam A. 1972. Continental drift and the fossil record. Sci. Am. **227**(5):56-66.

Hallam, A. 1973a. Provinciality, diversity and extinction of Mesozoic marine invertebrates in relation to plate movements, p. 287-294. *In* D.H. Tarling and S.K. Runcorn (eds.), Implications of continental drift to the earth sciences, vol. 1. New York, Academic Press, Inc.

Hallam, A. (ed.). 1973b. Atlas of palaeobiogeography. Amsterdam, Elsevier Scientific Publishing Co.

Hallam, A. 1973c. A revolution in the earth sciences: from continental drift to plate tectonics. Oxford, Clarendon Press.

Hallam, A. 1975a. Alfred Wegener and the hypothesis of continenttal drift. Sci. Am. **232**(2):88-97.

Hallam, A. (ed.). 1975b. Patterns of evolution, as illustrated by the fossil record. Amsterdam, Elsevier Scientific Publishing Co.

Hallam, A. 1981. Relative importance of plate movements, eustasy, and climate in controlling major biogeographic changes since the early Mesozoic, p. 303-330. *In* G. Nelson and D.E. Rosen (eds.), Vicariance biogeography: a critique. New York, Columbia University Press.

Hallam, A., and B.W. Sellwood. 1976. Middle Mesozoic sedimentation in relation to tectonics in the British area. J. Geol. **84**:301-321.

Hambrey, M.J., and W.B. Harland. 1981. Earth's pre-Pleistocene glacial record. Cambridge, Cambridge University Press.

Hamilton, A. 1976. The significance of patterns of distribution shown by forest plants and animals in tropical Africa for the reconstruction of Upper Pleistocene paleoenvironments: a review. Palaeoecol. Africa **9**:63-97.

Hamilton, T.H., and N.E. Armstrong. 1965. Environmental determination of insular variation in bird species abundance in the Gulf of Guinea. Nature **207**:148-151.

Hamilton, T.H., R.H. Barth, Jr., and I. Rubinoff. 1964. The environmental control of insular variation in bird species abundance. Proc. Natl. Acad. Sci. U.S.A. **52**:132-140.

Hamilton, T.H., and I. Rubinoff. 1963. Isolation, endemism, and multiplication of species in the Darwin's finches. Evolution **17**:388-403.

Hamilton, T.H., and I. Rubinoff. 1964. On models predicting abundance of species and endemics for the Darwin finches in the Galápagos Archipelago. Evolution **18**:339-342.

Hamilton, T.H., and I. Rubinoff. 1967. On predicting insular variation in endemism and sympatry for the Darwin finches in the Galápagos Archipelago. Am. Nat. **101**:161-171.

Hamilton, T.H., I. Rubinoff, R. Barth, and G.L. Bush. 1963. Species abundance: natural regulation of insular variation. Science **142**:1575-1577.

Hansen, T.A. 1980. Influence of larval dispersal and geographic distribution on species longevity in neogastropods. Paleobiology **6**:193-207.

Haq, B.U., W.A. Berggren, and J.A. van Couvering. 1977. Corrected age of the Pliocene/Pleistocene boundary. Nature **269**:483-488.

Hardin, G. 1960. The competitive exclusion principle. Science **131**:1292-1297.

Hardisty, M.V. 1979. Biology of the cyclostomes. London, Chapman & Hall, Ltd.

Hardy, A. 1971. The open sea: its natural history. Boston, Houghton Mifflin Co.

Hare, F.V. 1954. The boreal conifer zone. Geogr. Stud. **1**:4-18.

Harlan, J.R., and J.M.J. deWet. 1975. On Ö Winge and a prayer: the origins of polyploidy. Biol. Rev. **41**:361-390.

Harper, J.L. 1969. The role of predation in vegetational diversity. Brookhaven Symp. Biol. **22**:48-62.

Harper, J.L. 1977. Population biology of plants. New York, Academic Press, Inc.

Harris, M.P. 1973. The Galápagos avifauna. Condor **75**: 265-278.

Harris, P., D. Peschken, and J. Milroy. 1969. The status of biological control of the weed *Hypericum perforatum* in British Columbia. Can. Entomol. **101**:1-15.

Harris, V.T. 1952. An experimental study of habitat selection by prairie and forest races of the deer mouse, *Peromyscus maniculatus*. Contrib. Lab. Vert. Biol. Univ. Michigan **56**:1-53.

Harrison, R.G. 1980. Dispersal polymorphisms in insects. Ann. Rev. Ecol. Syst. **11**:95-118.

Hastings, J.R., and R.M. Turner. 1965. The changing mile. Tucson, Ariz., University of Arizona Press.

Hastings, J.R., R.M. Turner, and D.K. Warren. 1972. An atlas of some plant distributions in the Sonoran Desert. Tucson, Ariz., University of Arizona Institute of Atmospheric Physics, Technical Reports on the Meteorology of Arid Regions no. 20.

Hawksworth, F.G., and D. Wiens. 1972. Biology and classification of dwarf mistletoes *(Arceuthobium)*. Washington, D.C., U.S. Dept. Agr. Agricultural Handbook no. 401.

Hayward, T.L., and J.A. McGowan. 1979. Pattern and structure in an oceanic zooplankton community. Am. Zool. **19**:1045-1055.

Heaney, L.R. 1978. Island area and body size of insular mammals; evidence from the tri-colored squirrel *(Callosciurus prevosti)* of Southwest Africa. Evolution **32:** 29-44.

Heatwole, H. 1975. Biogeography of reptiles on some of the islands and cays of eastern Papua–New Guinea. Atoll Res. Bull. no. 180.

Heatwole, H., and R. Levins. 1972. Trophic structure stability and faunal change during recolonization. Ecology **53**:531-534.

Heatwole, H., and R. Levins. 1973. Biogeography of the Puerto Rican bank: species turnover on a small cay, Cayo Ahogado. Ecology **54**:1042-1055.

Hecht, A.D. (ed.), R. Barry, H. Fritts, J. Imbrie, J. Kutzbach, J.M. Mitchell, and S.M. Savin. 1979. Paleoclimatic research: status and opportunities. Quat. Res. **12**:6-17.

Hecht, A.D., and J. Imbrie. 1979. Editorial introduction: toward a comprehensive theory of climatic change. Quat. Res. **12**:2-5.

Hecht, M.K., and J.L. Edwards. 1977. The methodology of phylogenetic inference above the species level, p. 3-51. *In* M.K. Hecht, P.C. Goody, and B.M. Hecht (eds.), Major patterns of vertebrate evolution. New York, Plenum Press.

Hecht, M.K., P.C. Goody, and B.M. Hecht (eds.). 1977. Major patterns of vertebrate evolution. New York, Plenum Press.

Heckard, L.R. 1963. The Hydrophyllaceae. Quart. Rev. Biol. **38**:117-123.

Hedgpeth, J.W. 1957. Classification of marine environments, p. 17-28. *In* J.W. Hedgpeth (ed.), Treatise on marine ecology and paleoecology, vol. 1, ecology. Mem. Geol. Soc. Amer. **67.**

Hedgpeth, J.W. 1979. Prologue: at sea with provinces and plates, p. 1-8. *In* J. Gray and A.J. Boucot (eds.), Historical biogeography, plate tectonics, and the changing environment. Corvallis, Ore., Oregon State University Press.

Heed, W.B. 1982. Origins of the Sonoran Desert *Drosophila*. *In* J.S.F. Barker and W.T. Starmer (eds.), Ecological genetics and evolution: the cactus-yeast-*Drosophila* model system. Sydney, Academic Press, Ltd.

Heezen, B.C., M. Tharp, and M. Ewing. 1959. The floors of the oceans. I. The North Atlantic. Geol. Soc. Am. Spec. Pap. 65.

Heirtzler, J.R., and W.B. Bryan. 1975. The floor of the Mid-Atlantic Rift. Sci. Am. **233**(2):79-90.

Heirtzler, J.R., G.O. Dickson, E.M. Herron, W.C. Pitman, III, and X. Le Pichon. 1968. Marine magnetic anomalies, geomagnetic field reversals, and motions of the sea floor and continents. J. Geophys. Res. **73**:2119-2136.

Hemmingsen, A.M. 1960. Energy metabolism as related to body size and respiratory surfaces, and its evolution. Copenhagen, Reports of the Steno Memorial Hospital and the Nordisk Insulin laboratorium. Vol. 9(2).

Hendrickson, J.A., Jr. 1981. Community-wide character displacement reexamined. Evolution **35**:794-809.

Hennig, W. 1950. Grundzüge einer theorie der phylogenetischen Systematik. Berlin, Deutscher Zentralverlag.

Hennig, W. 1966. Phylogenetic systematics. 2nd ed. Tr. by D.D. Davis and R. Zanderl. Urbana, Ill., University of Illinois Press.

Hennig, W. 1979. Phylogenetic systematics. 3rd ed. Tr. by D.D. Davis and R. Zanderl. Urbana, Ill., University of Illinois Press.

Hershkovitz, P. 1972. The recent mammals of the Neotropical Realm: a zoogeographic and ecological review, p. 311-431. *In* A. Keast, F.C. Erk, and B. Glass (eds.), Evolution, mammals, and southern continents. Albany, N.Y., State University of New York Press.

Hershkovitz, P. 1977. Living New World monkeys. (Platyrrhini), vol. 1. Chicago, University of Chicago Press.

Hespenheide, H.A. 1978. Are there fewer parasitoids in the tropics? Am. Nat. **112**:766-769.

Hess, H.H. 1946. Drowned ancient islands of the Pacific basin. Am. J. Sci. **244**:772-791.

Hess, H.H. 1955. The oceanic crust. J. Mar. Res. **14**:423-439.

Hess, H.H. 1962. History of ocean basins, p. 599-620. *In* A.E.J. Engel, H.L. James, and B.F. Leonard (eds.), Petrological studies: a volume in honor of A.F. Buddington. New York, Geological Society of America.

Hesse, R., W.C. Allee, and K.P. Schmidt. 1951. Ecological animal geography. 2nd ed. New York, John Wiley & Sons, Inc.

Hester, J. 1967. The agency of man in animal extinctions, p. 169-192. *In* P.S. Martin and H.E. Wright (eds.), Pleistocene extinctions, the search for a cause. New Haven, Conn., Yale University Press.

Hey, R. 1977. Tectonic evolution of the Cocos-Nazca spreading center. Geol. Soc. Am. Bull. **88**:1404-1420.

Higgs, A.J. 1981. Island biogeography and nature reserve design. J. Biogeogr. **8**:117-124.

Hilde, T.W.C., S. Uyeda, and L. Kroenke. 1977. Evolution of the western Pacific and its margin. Tectonophysics **38**:145-165.

Hilden, O. 1965. Habitat selection in birds: a review. Ann. Zool. Fennici **2**:53-75.

Hnatiuk, S.H. 1979. A survey of germination of seeds of some vascular plants found on Aldabra Atoll. J. Biogeogr. **6**:105-114.

Hoar, W.S., and D.J. Randall (ed.). 1969. Fish physiology, vol. 1. New York, Academic Press, Inc.

Hochachka, P.W., and G.N. Somero. 1973. Strategies of biochemical adaptation. Philadelphia, W.B. Saunders Co.

Hocker, H.W., Jr. 1956. Certain aspects of climate as related to the distribution of loblolly pine. Ecology **37**:824-834.

Hocutt, C.H., R.F. Denoncourt, and J.R. Stauffer, Jr. 1978. Fishes of the Greenbrier river, West Virginia, with drainage history of the Central Appalachians. J. Biogeogr. **5**:59-80.

Hoffmann, R.S. 1981. Different voles for different holes: environmental restrictions on refugial survival of mammals, p. 25-45. *In* G.G.E. Scudder and J.L. Reveal (eds.), Evolution today. Proceedings of the Second International Congress of Systematics and Evolutionary Biology. Pittsburgh, Hunt Institute of Botanical Documentation.

Hoffstetter, R. 1972. Relationships, origins, and history of the ceboid monkeys and caviomorph rodents: a modern reinterpretation, p. 323-347. *In* Th. Dobzhansky, M.C. Hecht, and W.C. Steere (eds.), Evolutionary biology 6. New York, Appleton-Century-Crofts.

Holdgate, M.W. 1960. The fauna of the mid-Atlantic islands. Proc. Roy. Soc. London ser. B **152**:550-567.

Holdridge, L.R. 1947. Determination of world plant formations from simple climatic data. Science **105**:367-368.

Holdridge, L.R., W.C. Grenke, W.H. Hatheway, P. Liang, and J.A. Tosi, Jr. 1971. Forest environments in tropical life zones, a pilot study. New York, Pergamon Press.

Holland, P.G. 1981. Pleistocene refuge areas, and the revegetation of Nova Scotia, Canada. Progr. Phys. Geogr. **5**:535-562.

Holloway, J.D. 1979. A survey of the Lepidoptera, biogeography and ecology of New Caledonia. The Hague, Dr. W. Junk BV, Publishers.

Holloway, J.D., and N. Jardine. 1968. Two approaches to zoogeography: a study based on the distributions of butterflies, birds and bats in the Indo-Australian area. Proc. Linn. Soc. London **179**:153-188.

Holmes, A. 1931. The thermal history of the earth. J. Washington Acad. Sci. **23**:169-195.

Holmes. A. 1944. The machinery of continental drift: the search for a mechanism, p. 505-509. *In* A. Holmes (ed.), Principles of physical geology. New York, Ronald Press.

Honacki, J.H., K.E. Kinman, and J.W. Koeppl (eds.). 1982. Mammal species of the world. Lawrence, Kans., Allen Press, Inc., and the Association of Systematics Collections.

Hooker, J.D. 1860. On the origin and distribution of species: introductory essay to the flora of Tasmania. Am. J. Sci. Arts ser. 2 **29**:1-25, 305-326.

Hooper, E.T. 1942. An effect on the *Peromyscus maniculatus* Rassenkreis of land utilization in Michigan. J. Mammal. **23**:193-196.

Hooykaas, R. 1956. The principle of uniformity in geology, biology, and theology. J. Trans. Victoria Inst. **88**:101-116.

Hopkins, D.M. 1959. Cenozoic history of the Bering Land Bridge. Science **129**:1519-1528.

Hopkins, D.M. (ed.). 1967. The Bering Land Bridge. Stanford, Calif., Stanford University Press.

Hopkins, D.M. 1979. Landscape and climate of Beringia during Late Pleistocene and Holocene times, p. 15-41. *In* W.S. Laughlin and A.B. Harper (eds.), The first Americans: origins, affinities, and adaptations. New York, Gustav Fischer Verlag.

Hopkins, D.M. 1980. Vegetation of the Bering Land Bridge revisited. Rev. Archaeol. **1**:2-15.

Hopkins, D.M., and P.A. Smith. 1981. Dated wood from Alaska and the Yukon: implications for forest refugia in Beringia. Quat. Res. **15**:217-249.

Hopper, S.D. 1979. Biogeographical aspects of speciation in the southwest Australian flora. Ann. Rev. Ecol. Syst. **10**:399-422.

Horn, H.S. 1974. The ecology of secondary succession. Ann. Rev. Ecol. Syst. **5**:25-37.

Horn, M.H., and L.G. Allen. 1978. A distributional analysis of California coastal marine fishes. J. Biogeogr. **5**:23-42.

Horton, D.R. 1974a. Species movement in zoogeography. J. Biogeogr. **1**:155-158.

Horton, D.R. 1974b. Dominance and zoogeography of the southern continents. Syst. Zool. **23**:440-445.

Howard, R., and A. Moore. 1980. A complete checklist of the birds of the world. Oxford, Oxford University Press.

Howarth, F.G. 1980. The zoogeography of specialized cave animals: a bioclimatic model. Evolution **34**:394-406.

Howden, H.F. 1974. Problems in interpreting dispersal of terrestrial organisms as related to continental drift. Biotropica **6**:1-6.

Howden, H.F. 1981. Random and infrequent dispersals, p. 188-194. *In* G. Nelson and D.E. Rosen (eds.), Vicariance biogeography: a critique. New York, Columbia University Press.

Howell, A.B. 1917. Birds of the islands off the coast of southern California. Pacific Coast Avifauna no. 12. Hollywood, Calif., Cooper Ornithogical Club.

Howell, T.R. 1969. Avian distribution in Central America. Auk **86**:293-326.

Hsü, K.J. 1972. When the Mediterranean dried up. Sci. Am. **227**(6):26-36.

Hubbell, S.P. 1979. Tree dispersion, abundance, and diversity in a dry tropical forest. Science **203**:1299-1309.

Hubbs, C.L. (ed.). 1974. Zoogeography. New York, Arno Press, Inc.

Hubbs, C.L., and R.R. Miller. 1948. The zoological evidence: correlation between fish distribution and hydrographic history in the desert basins of western United States, p. 17-166. *In* The Great Basin, with emphasis on glacial and postglacial times. Bull. Univ. Utah Biol. ser. no. 107.

Huey, R.B. 1978. Latitudinal pattern of between altitude faunal similarity: mountains might be "higher" in the tropics. Am. Nat. **112**:225-229.

Huffaker, C.B., and C.E. Kennett. 1959. A ten-year study of vegetational changes associated with biological control of Klamath weed. J. Range Managem. **12**:69-82.

Hughes, N.F. 1973. Organisms and continents through time. Paleontol. Soc. London Spec. Pap. no. 12.

Hughes, N.F. 1976. Palaeobiology of angiosperm origins. Cambridge, Cambridge University Press.

Hull, D. 1970. Contemporary systematic philosophies. Ann. Rev. Ecol. Syst. **1**:19-54.

Hull, D.L. 1973. Darwin and his critics: the reception of Darwin's theory of evolution by the scientific community. Cambridge, Mass., Harvard University Press.

Hull, D.L. 1974. Philosophy of biological sciences. Englewood Cliffs, N.J., Prentice-Hall, Inc.

Hull, D.L. 1980. Individuality and selection. Ann. Rev. Ecol. Syst. **11**:311-322.

Hultén, E. 1937. Outline of the history of arctic and boreal biota during the Quaternary Period. Stockholm, Bokforlags Aktiebolaget Thule.

Hultén, E. 1958. The amphi-Atlantic plants and their phytogeographical connections. Stockholm, Almqvist & Wiksell Förlag.

Hultén, E. 1963a. Phytogeographical connections of the North Atlantic, p. 45-72. *In* A. Löve and D. Löve (eds.), North Atlantic biota and their history. Oxford, Pergamon Press, Ltd.

Hultén, E. 1963b. The distributional conditions of the flora of Beringia, p. 7-22. *In* J.L. Gressitt (ed.), Pacific Basin biogeography. Honolulu, Bishop Museum Press.

Hultén, E. 1964. The circumpolar plants. II. Dicotyledons. Stockholm, Almqvist & Wiksell Förlag.

Hultén, E. 1968. Flora of Alaska and neighboring territories. Stanford, Calif., Stanford University Press.

Humphries, C.J. 1981. Biogeographical methods and the southern beeches (Fagaceae: *Nothofagus*), p. 177-207. *In* V.A. Funk and D.R. Brooks (eds.), Advances in cladistics. Proceedings of the first meeting of the Willi Hennig Society. New York, New York Botanical Society.

Hunziker, J.H., R.A. Palacios, A.G. de Valesi, and L. Poggio. 1972. Species disjunctions in *Larrea:* evidence from morphology, cytogenetics, phenolic compounds, and seed albumins. Ann. Missouri Bot. Gard. **59**:224-233.

Hurlbert, S.H. 1981. A gentle depilation of the niche: Dicean resource sets in resource hyperspace. Evol. Theory **5**:177-184.

Hurley, P.M. 1968. The confirmation of continental drift. Sci. Am. **218**(4):52-62.

Hurley, P.M., and J.R. Rand. 1969. Pre-drift continental nuclei. Science **164**:1229-1242.

Hutchins, L.W. 1947. The bases for temperature zonation in geographical distribution. Ecol. Monogr. **17**:325-335.

Hutchinson, G.E. 1957. A treatise on limnology, vol. 1. New York, John Wiley & Sons, Inc.

Hutchinson, G.E. 1958. Concluding remarks. Cold Spring Harbor Symp. Quant. Biol. **22**:415-427.

Hutchinson, G.E. 1959. Homage to Santa Rosalia, or Why are there so many kinds of animals? Am. Nat. **93**:145-159.

Hutchinson, G.E. 1961. The paradox of the plankton. Am. Nat. **95**:137-145.

Hutchinson, G.E. 1967. A treatise on limnology, vol. 2. New York, John Wiley & Sons, Inc.

Hutchinson, G.E. 1978. An introduction to population ecology. New Haven, Conn., Yale University Press.

Hutchinson, T.C., and M. Havas (eds.). 1978. Effects of acid precipitation on terrestrial ecosystems. New York, Plenum Press.

Hutton, J. 1795. Theory of the earth with proofs and illustrations. Edinburgh.

Hynes, H.B.N. 1970. The ecology of running water. Toronto, University of Toronto Press.

Illies, J. 1974. Introduction to zoogeography. Translated by W.D. Williams. London, Macmillan Publishers, Ltd.

Imbrie, J., and N.G. Kipp. 1971. A new micropaleontological method for quantitative paleoclimatology: applications to a Late Pleistocene Caribbean core, p. 71-81. *In* K. Turikian (ed.), The late Cenozoic glacial ages. New Haven, Conn., Yale University Press.

Irving, E. 1956. Paleomagnetic and paleoclimatological aspects of polar wandering. Pure Appl. Geophys. **33**:23-41.

Irving, E. 1959. Paleomagnetic pole positions. J. Roy. Astron. Soc. Geophys. **2**:51-77.

Isacks, B., J. Oliver, and L.R. Sykes. 1968. Seismology and the new global tectonics. J. Geophys. Res. **73**:5855-5899.

Ives, J.D. 1974. Biological refugia and the nunatak hypothesis, p. 605-636. *In* J.D. Ives and R.G. Barry (eds.), Arctic and alpine environments. London, Methuen & Co., Ltd.

Jablonski, D. 1982. Evolutionary rates and modes in Late Cretaceous gastropods: role of larval ecology. *In* B. Mamet and M.J. Copeland (eds.), Proceedings of the Third North American Paleontological Convention. Toronto, The Convention.

Jablonski, D., and R.A. Lutz. 1980. Molluscan shell morphology: ecological and paleontological applications, p. 323-377. *In* D.C. Rhoads and R.A. Lutz (eds.), Skeletal growth of aquatic organisms. New York, Plenum Press.

Jackson, E.D., E.A. Silver, and G.B. Dalrymple. 1972. Hawaiian-Emperor chain and its relation to Cenozoic circumpacific tectonics. Geol. Soc. Am. Bull. **83**:601-618.

Jackson, J.B.C. 1974. Biogeographic consequences of eurytopy and stenotopy among marine bivalves and their evolutionary significance. Am. Nat. **108**:541-560.

Jaeckel, S.G.A., Jr. 1969. Die Mollusken SüdAmerikas, p. 794-827. *In* E.J. Fittkau, J. Illies, H. Klinge, G.H. Schwabe, and H. Sioli (eds.), Biogeography and ecology in South America, vol. 2. Monographiae Biologicae 20. The Hague, Dr. W. Junk BV, Publishers.

Jamieson, B.N. 1974. The Oligochaeta of Tasmania, p. 195-228. *In* W.D. Willimas (ed.), Biogeography and ecology in Tasmania. Monographiae Biologicae 25. The Hague, Dr. W. Junk BV, Publishers.

Jamieson, B.N. 1981. Historical biogeography of Australsian Oligochaeta, p. 885-921. *In* A. Keast (ed.), Ecological biogeography of Australia, vol. 2. Monographiae Biologicae 41. The Hague, Dr. W. Junk BV, Publishers.

Jannasch, H.W., and C.O. Wirsen. 1980. Chemosynthetic primary production at East Pacific sea floor spreading center. Bioscience **29**:592-598.

Janzen, D.H. 1966. Coevolution of mutualism between ants and acacias in Central America. Evolution **20**:249-275.

Janzen, D.H. 1967. Why mountain passes are higher in the tropics. Am. Nat. **101**:233-249.

Janzen, D.H. 1970. Herbivores and the number of tree species in tropical forests. Am. Nat. **104**:501-528.

Janzen, D.H. 1981. The peak in North American ichneumonid species richness lies between 38° and 42° N. Ecology **62**:532-537.

Jarrard, R.D., and D.A. Clague. 1977. Implications of Pacific island and seamount ages for the origin of volcanic chains. Rev. Geophys. Space Phys. **15**:57-76.

Järvinen, O. 1979. Geographical gradients of stability in European land bird communities. Oecologia **38**:51-69.

Jeffrey, R.C. 1975. Probability and falsification: critique of the Popperian program. Methodologies: Bayesian and Popperian. Synthese **30**:95-117.

Jeffrey, S.W. 1981. Phytoplankton ecology–with particular reference to the Australasian region, p. 241-291. *In* M.N. Clayton and R.J. King (eds.), Marine botany: an Australasian perspective. Melbourne, Longman Cheshire Pty., Ltd.

Jenny, H. 1979. The soil resource. Origin and behavior. Ecological studies 37. New York, Springer-Verlag New York, Inc.

Johnsgard, P.A. 1981. The plovers, sandpipers, and snipes of the world. Lincoln, Nebr., University of Nebraska Press.

Johnson, A.W., and J.G. Packer. 1965. Polyploidy and environment in arctic Alaska. Science **148**:237-239.

Johnson, A.W., J.G. Packer, and G. Reese. 1965. Polyploidy, distribution, and environment, p. 497-507. *In* H.E. Wright, Jr., and D.G. Frey (eds.), The Quaternary of the United States. Princeton, N.J., Princeton University Press.

Johnson, C.G. 1963. The aerial migration of insects. Sci. Am. **209**(6):132-138.

Johnson, C.G. 1969. Migration and dispersal of insects by flight. London, Methuen & Co., Ltd.

Johnson, C.G., and J. Bowden. 1973. Problems related to the transoceanic transport of insects, especially between the Amazon and Congo areas, p. 207-222. *In* B.J. Meggers, E.S. Ayensu, and W.D. Duckworth (eds.), Tropical forest ecosystems in Africa and South America: a comparative review. Washington, D.C., Smithsonian Institute Press.

Johnson, D.L. 1978. The origin of island mammoths and the Quaternary land bridge history of the Northern Channel Islands, California. Quat. Res. **10**:204-225.

Johnson, D.L. 1980. Problems in the land vertebrate zoogeography of certain islands and the swimming powers of elephants. J. Biogeogr. **7**:383-398.

Johnson, D.L. 1981. More comments on the Northern Channel Island mammoths. Quat. Res. **15**:105-106.

Johnson, G.A.L. 1973. Closing of the Carboniferous sea in western Europe, p. 843-850. *In* D.H. Tarling and S.K. Runcorn (eds.), Implications of continental drift to the earth sciences, vol. 2. New York, Academic Press, Inc.

Johnson, L.A.S., and B.G. Briggs. 1975. On the Proteaceae: the evolution and classification of a southern family. Bot. J. Linn. Soc. **70**(2):83-182.

Johnson, M.P., L.G. Mason, and P.H. Raven. 1968. Ecological parameters and species diversity. Am. Nat. **102**:297-306.

Johnson, M.P., and P.H. Raven. 1970. Natural regulation of plant species diversity, p. 127-162. *In* Th. Dobzhansky, M.K. Hecht, and W.C. Steere (eds.), Evolutionary biology 4. New York, Appleton-Century-Crofts.

Johnson, M.P., and P.H. Raven. 1973. Species number and endemism: the Galápagos Archipelago revisited. Science **179**:893-895.

Johnson, N.K. 1972. Origin and differentiation of the avifauna of the Channel Islands, California. Condon **74**:295-315.

Johnson, N.K. 1975. Controls of the number of bird species on montane islands in the Great Basin. Evolution **29**:545-567.

Johnson, N.K. 1978. Patterns of avian biogeography and speciation in the intermountain region. Great Basin Nat. Mem. **2**:137-160.

Johnston, M.C. 1963. Past and present grasslands of southern Texas and northeastern Mexico. Ecology **44**: 456-466.

Johnston, R.F., and W.J. Klitz. 1977. Variation and evolution in a granivorous bird: the house sparrow, p. 15-51. *In* J. Pinowski and S.C. Kendeigh (eds.), Granivorous birds in ecosystems. Internatl. Biol. Prog. no. 12. Cambridge, Cambridge University Press.

Johnston, R.F., and R.K. Selander. 1964. House sparrows: rapid evolution of races in North America. Science **144**:548-550.

Johnston, R.F., and R.K. Selander. 1971. Evolution in the house sparrow. II. Adaptive differentiation in North American populations. Evolution **25**:1-28.

Johnston, R.F., and R.K. Selander. 1973. Variation, adaptation, and evolution in the North American house sparrows, p. 301-326. *In* S.C. Kendeigh and J. Pinowski (eds.), Productivity, population dynamics and systematics of granivorous birds. Warsaw, Polish Scientific Publishers.

Jones, F.R.H. 1968. Fish migration. London, Edward Arnold (Publishers), Ltd.

Jones, H.L., and J.M. Diamond. 1976. Short-time-base studies of turnover in breeding bird populations on the California Channel Islands. Condor **78**:526-549.

Jones, J.K., Jr., and H.H. Genoways. 1970. Chiropteran systematics, p. 3-21. *In* B.H. Slaughter and D.W. Walton (eds.), About bats. Dallas, Southern Methodist University Press.

Jones, J.R.E. 1949. A further ecological study of calcareous streams in the "Black Mountain" district of South Wales. J. Anim. Ecol. **18**:142-159.

Jordan, C.F. 1971. A world pattern in plant energetics. Am. Sci. **59**:425-433.

Jordan, D.S. 1908. The law of geminate species. Am. Nat. **42**:73-80.

Jubb, R.A. 1967. Freshwater fishes of southern Africa. Cape Town, A.A. Balkema, Publishers.

Juvik, J.O., and A.P. Austring. 1979. The Hawaiian avifauna: biogeographic theory in evolutionary time. J. Biogeogr. **6**:205-223.

Kalandadze, N.N. 1974. Intercontinental connections of tetrapod faunas of the Triassic period. Paleontol. J. **74**:352-362.

Kalela, O. 1949. Changes in geographic ranges in the avifauna of northern and central Europe in relation to recent changes in climate. Bird-Banding **20**:77-103.

Karl, D.M., C.O. Wirsen, and H.W. Jannasch. 1980. Deep-sea primary production at the Galápagos hydrothermal vents. Science **207**:1345-1347.

Karr, J.R. 1976. Within- and between-habitat avian diversity in African and neotropical lowland habitats. Ecol. Monogr. **46**:457-481.

Karr, J.R. 1980. Patterns in the migration systems between north temperate zone and the tropics, p. 529-544. *In* A. Keast and E.S. Morton (eds.), Migrant birds in the Neotropics: ecology, behavior, distribution, and conservation. Washington, D.C., Smithsonian Institution Press.

Karr, J.R. 1982. Avian extinction on Barro Colorado Island, Panama: a reassessment. Am. Nat. **119**:220-239.

Karr, J.R., and F.C. James. 1975. Ecomorphological configurations and convergent evolution, p. 258-291. *In* M.L. Cody and J.M. Diamond (eds.), Ecology and evolution of communities. Cambridge, Mass., Belknap Press.

Karr, J.R., and R.R. Roth. 1971. Vegetation structure and avian diversity in several New World areas. Am. Nat. **105**:423-435.

Keast, A. 1961. Bird speciation on the Australian continent. Bull. Mus. Compar. Zool. (Harvard) **123**(8):305-495.

Keast, A. 1968. Australian mammals: zoogeography and evolution. Quart. Rev. Biol. **43**:373-408.

Keast, A. 1971. Adaptive evolution and shifts in niche occupation in island birds, p. 39-53. *In* W.L. Stern (ed.), Adaptive aspects of insular evolution. Pullman, Wash., Washington State University Press.

Keast, A. 1972a. Continental drift and the biota of the mammals on southern continents, p. 23-87. *In* A. Keast, F.C. Erk, and B. Glass (eds.), Evolution, mammals, and southern continents. Albany, N.Y., State University of New York Press.

Keast, A. 1972b. Australian mammals: zoogeography and evolution, p. 195-246. *In* A. Keast, F.C. Erk, and B. Glass (eds.), Evolution, mammals, and southern continents. Albany, N.Y., State University of New York Press.

Keast, A. 1972c. Comparisons of contemporary mammal faunas of southern continents, p. 433-501. *In* A. Keast, F.C. Erk, and B. Glass (eds.), Evolution, mammals, and southern continents. Albany, N.Y., State University of New York Press.

Keast, A. 1976. The origins of adaptive zone utilizations and adaptive radiations, as illustrated by the Australian Meliphagidae, p. 71-82. *In* H.J. Frith and J.H. Calaby (eds.), Proceedings of the 16th International Ornithological Congress. Canberra, Australian Academy of Science.

Keast, A. 1977a. Zoogeography and phylogeny: the theoretical background and methodology to the analysis of mammal and bird fauna, p. 249-312. *In* M.K. Hecht, P.C. Goody, and B.M. Hecht (eds.), Major patterns in vertebrate evolution. New York, Plenum Press.

Keast, A. 1977b. Historical biogeography of the marsupials, p. 69-95. *In* B. Stonehouse and D. Gilmore (eds.), The biology of marsupials. Baltimore, Md., University Park Press.

Keast, A. 1981. Ecological biogeography of Australia. 3 vol. Monographiae Biologicae 41. The Hague, Dr. W. Junk BV, Publishers.

Keast, A., F.C. Erk, and B. Glass (eds.). 1972. Evolution, mammals, and southern continents. Albany, N.Y., State University of New York Press.

Keast, A., and E.S. Morton (eds.). 1980. Migrant birds in the Neotropics: ecology, behavior, distribution, and conservation. Washington, D.C., Smithsonian Institution Press.

Kellman, M.C. 1975. Plant geography. New York, St. Martin's Press, Inc.

Kellogg, C.E. 1957. We seek: we learn, p. 1-11. *In* U.S. Dept. Agr., Soil, 1957 Yearbook of Agriculture. Washington, D.C., U.S. Department of Agriculture.

Kendeigh, S.C. 1974. Ecology with special reference animals and man. Englewood Cliffs, N.J., Prentice-Hall, Inc.

Kessel, B. 1953. Distribution and migration of the European starling in North America. Condor **55**:49-67.

Kiester, A.R. 1971. Species density of North American amphibians and reptiles. Syst. Zool. **20**:127-137.

Kikkawa, J., and K. Pearse. 1969. Geographical distribution of land birds in Australia: a numerical analysis. Australian J. Zool. **17**:821-840.

Kikkawa, J., and E.E. Williams. 1971. Altitudinal distribution of land birds in New Guinea. Search **2**:64-69.

Kilburn, P.D. 1966. Analysis of the species-area relation. Ecology **47**:831-843.

Kinne, O. (ed.). 1970-1972. Marine ecology: a comprehensive integrated treatise on life in oceans and coastal water, vol. 1, environmental factors. London, Wiley-Interscience.

Kinne, O. (ed.). 1975. Marine ecology: a comprehensive integrated treatise on life in oceans and coastal water, vol. 2, physiological mechanisms. London, Wiley-Interscience.

Kircher, H.W., W.B. Heed, J.S. Russell, and J. Grove. 1967. Senita cactus alkaloids: their significance to Sonoran Desert *Drosophila* ecology. J. Insect Physiol. **13**:1869-1874.

Kirsch, J.A.W. 1977. The classification of marsupials, p. 1-51. *In* D. Hunsaker II (ed.), The biology of marsupials. New York, Academic Press, Inc.

Kirsch, J.A.W., and J.H. Calaby. 1977. The species of living marsupials: an annotated list, p. 9-26. *In* B. Stonehouse and D. Gilmore (eds.), The biology of marsupials. Baltimore, Md., University Park Press.

Kitching, J.A., and F.J. Ebling. 1967. Ecological studies at Lough Ine. Adv. Ecol. Res. **4**:197-291.

Klein, H.A. 1972. Oceans and continents in motion. Philadelphia, J.B. Lippincott Co.

Klopfer, P.H., and J.P. Hailman. 1965. Habitat selection in birds. Adv. Study. Behav. **1**:279-303.

Koopman, K.F. 1968. Taxonomic and distributional notes on Lesser Antillean bats. Am. Mus. Novitates no. 2333.

Koopman, K.F. 1970. Zoogeography of bats, p. 29-50. *In* B.H. Slaughter and D.W. Walton (eds.), About bats. Dallas, Southern Methodist University Press.

Koopman, K.F. 1975. Bats of the Virgin Islands in relation to those of the Greater and Lesser Antilles. Am. Mus. Novitates no. 2581.

Koopman, K.F. 1976. Zoogeography, p. 39-47. *In* R.S. Baker, J.K. Jones, and D.C. Carter (eds.), Biology of bats of the New World family Phyllostomatidae. Part I. Texas Tech Univ. Spec. Pub. no. 10.

Koopman, K.F. 1978. Zoogeography of Peruvian bats with special emphasis on the role of the Andes. Am. Mus. Novitates no. 2651.

Koopman, K.F., and J.K. Jones. 1970. Classification of bats, p. 22-28. *In* B.H. Slaughter and D.W. Walton (eds.), About bats. Dallas, Southern Methodist University Press.

Köppen, W. 1900. Versucheiner Klassifikation der Klimate, vorzugsweise nach ihren Beziehungen zur Pflanzenwelt. Geogr. Z. **6**:593-611, 657-679.

Köppen, W., and R. Geiger. 1936. Handbuch der Klimatologie. 5 vol. Berlin, Borntraeger.

Kornas, J. 1972. Corresponding taxa and their ecological background in the forests of temperate Eurasia and North America, p. 37-59. *In* D.H. Valentine (ed.), Taxonomy, phytogeography and evolution. New York, Academic Press, Inc.

Kozlowski, T.T., and C.E. Ahlgren (eds.). 1974. Fire and ecosystems. New York, Academic Press, Inc.

Kramer, G. 1951. Body proportions of mainland and island lizards. Evolution **5**:193-206.

Krebs, C.J. 1978. Ecology: the experimental analysis of distribution and abundance. New York, Harper & Row, Publishers, Inc.

Kremp, G.O.W. 1974. A re-evaluation of global plant geographic provinces of the Late Paleozoic. Rev. Palaeobot. Palynol. **17**:113-132.

Kristoffersen, Y., and M. Talwani. 1977. Extinct triple junction south of Greenland and the Tertiary motion of Greenland relative to North America. Geol. Soc. Am. Bull. **88**:1037-1049.

Kruckeberg, A.R. 1954. The ecology of serpentine soils: a symposium. III. Plant species in relation to serpentine soils. Ecology **35**:267-274.

Kruckeberg, A.R. 1969. Soil diversity and the distribution of plants, with examples from western North America. Madrōno **20**:129-154.

Küchler, A.W. 1964. Potential natural vegetation of the conterminous United States. Am. Geogr. Soc. Spec. Pub. no. 36.

Kunkel, G.G. (ed.). 1976. Biogeography and ecology in the Canary Islands. Monographiae Biologicae 30. The Hague, Dr. W. Junk BV, Publishers.

Kurtén, B. 1969. Continental drift and evolution. Sci. Am. **220**(3):54-64.

Kurtén, B. 1971. The age of mammals. London, Weidenfeld & Nicolson, Ltd.

Kurtén, B. 1972. The Ice Age. London, Rupert Hart-Davis, Ltd.

Kurtén, B., and E. Anderson. 1980. Pleistocene mammals of North America. New York, Columbia University Press.

Kuschel, G. (ed.). 1975. Biogeography and ecology in New Zealand. Monographiae Biologicae 27. The Hague, Dr. W. Junk BV, Publishers.

Kvasov, D.D., and M.Ya. Verbitsky. 1981. Causes of Antarctic glaciation in the Cenozoic. Quat. Res. **15**:101-104.

Lack, D. 1947. Darwin's finches. Cambridge, Cambridge University Press.

Lack, D. 1969. Subspecies and sympatry in Darwin's finches. Evolution **23**:252-263.

Lack, D. 1970. Island birds. Biotropica **2**:29-31.

Lack, D. 1973. The numbers of species of hummingbirds in the West Indies. Evolution **27**:326-337.

Lack, D. 1976. Island biology illustrated by the land birds of Jamaica. Studies in ecology, vol. 3. Berkeley, University of California Press.

Ladd, J.W. 1976. Relative motion of South America with respect to North America and Caribbean tectonics. Geol. Soc. Am. Bull. **87**:969-976.

Lai, D.Y., and P.L. Richardson. 1977. Distribution and movement of Gulf Stream rings. J. Phys. Oceanogr. **7**:670-683.

La Marche, V.C. 1973. Holocene climatic variations inferred from treeline fluctuations in the White Mountains, California. Quat. Res. **3**:632-660.

La Marche, V.C. 1978. Tree-ring evidence of past climatic variability. Nature **276**:334-338.

Lande, R. 1982. Rapid origin of sexual isolation and character divergence in a cline. Evolution **36**:213-223.

Landry, S. 1965. The status of the theory of the replacement of the Multituberculata by the Rodentia. J. Mammal. **46**:280-286.

Laporte, L.F. 1968. Ancient environments. Englewood Cliffs, N.J., Prentice-Hall, Inc.

Larcher, W. 1980. Physiological plant ecology. 2nd ed. Berlin, Springer-Verlag.

Laseron, C. 1969. Ancient Australia. Revised by R.O. Brunnschewiler. New York, Taplinger Publishing Co., Inc.

Lasiewski, R.C. 1962. The energetics of migrating hummingbirds. Condor **64**:324.

Lassen, H.H. 1975. The diversity of freshwater snails in view of the equilibrium theory of island biogeography. Oecologia **19**:1-18.

Lauff, G. (ed.). 1967. Estuaries. Washington, D.C., American Association for the Advancement of Science.

Lawlor, T.E. 1982. The evolution of body size in mammals: evidence from insular populations in Mexico. Am. Nat. **119**:54-72.

Lazell, J.D. 1972. The anoles (Sauria, Iguanidae) of the Lesser Antilles. Bull. Mus. Comp. Zool. (Harvard) no. **143**:1-108.

Leggett, W.C. 1977. The ecology of fish migrations. Ann. Rev. Ecol. Syst. **8**:285-308.

Lehner, C.E. 1979. A latitudinal gradient analysis of rocky shore fishes of the eastern Pacific. University of Arizona, Department of Ecology and Evolutionary Biology, Ph.D. dissertation.

Leigh, E.G. 1981. The average lifetime of a population in a varying environment. J. Theor. Biol. **90**:213-239.

Leis, J.M., and J.M. Miller. 1976. Offshore distributional patterns of Hawaiian fish larvae. Mar. Biol. **36**:359-367.

Le Pichon, X. 1968. Sea-floor spreading and continental drift. J. Geophys. Res. **73**:3661-3697.

Le Pichon, X., J.-C. Sibuet, and J. Francheteau. 1977. The fit of the continents around the North Atlantic Ocean. Tectonophysics **38**:169-209.

Leroy, J.-F. 1978. Composition, origin, and affinities of the Madagascan vascular flora. Ann. Missouri Bot. Gard. **65**:535-589.

Levin, D.A., and H.W. Kerster. 1974. Gene flow in seed plants, p. 139-220. *In* Th. Dobzhansky, M.C. Hecht, and B. Wallace (eds.), Evolutionary biology 7. New York, Plenum Press.

Levins, R. 1969. Some demographic and genetic consequences of environmental heterogenity for biological control. Entomol. Soc. Am. Bull. **15**:237-240.

Levins, R., and H. Heatwole. 1973. Biogeography of the Puerto Rican bank: introduction of species onto Palominitos Island. Ecology **54**:1056-1064.

Levitt, J. 1980. Responses of plants to environmental stresses, vol. 1. Chilling, freezing, and high temperature. 2nd ed. New York, Academic Press, Inc.

Lewis, H. 1953. The mechanism of evolution in the genus *Clarkia*. Evolution **7**:1-20.

Lewis, H. 1962. Catastrophic selection as a factor in speciation. Evolution **16**:257-271.

Lewis, H. 1966. Speciation in flowering plants. Science **152**:167-172.

Lewis, W.H. (ed.). 1979. Polyploidy. Biological relevance. New York, Plenum Press, Inc.

Lewontin, R.C., and L.C. Birch. 1966. Hybridization as a source of variation for adaptation to new environments. Evolution **20**:315-336.

Li, H.L. 1952. Floristic relationships between eastern Asia and eastern North America. Trans. Am. Philos. Soc. n.s. **42**:371-429.

Liebig, J. 1840. Chemistry in its application to agriculture and physiology. London, Taylor & Walton.

Lieth, H. 1972. Modeling the primary productivity of the world. Trop. Ecol. **13**:125-130.

Lieth, H. 1973. Primary production: terrestrial ecosystems. Human Ecol. **1**:303-332.

Likens, G.E., and F.H. Bormann. 1972. Nutrient cycling in ecosystems, p. 25-67. *In* J.A. Wiens (ed.), Ecosystem structure and function. Corvallis, Ore., Oregon State University Press.

Likens, G.E., and F.H. Bormann. 1974. Acid rain: a serious regional environmental problem. Science **184**:1176-1179.

Lillegraven, J.A. 1972. Ordinal and familial diversity in Cenozoic mammals. Taxon **21**:261-274.

Lillegraven, J.A., Z. Kielan-Jaworowska, and W.A. Clemens. 1979. Mesozoic mammals. The first two-thirds of mammalian history. Berkeley, University of California Press.

Lindroth, C.H. 1957. The faunal connections between Europe and North America. Stockholm, Almqvist & Wiksell Förlag.

Lindsey, C.C. 1964. Problems in zoogeography of the lake trout, *Salvelinus namaycush*. J. Fish. Res. Bd. Can. **21**:977-994.

Lindsey, C.C. 1981. Arctic refugia and the evolution of arctic biota, p. 7-10. *In* G.G.E. Scudder and J.L. Reveal (eds.), Evolution today. Proceedings of the Second International Congress of Systematics and Evolutionary Biology. Pittsburgh, Hunt Institute of Botanical Documentation.

Lintz, J., and D.S. Simonett (eds.). 1976. Remote sensing of environment. Reading, Mass., Addison-Wesley Publishing Co., Inc.

Lister, B.C. 1976a. The nature of niche expansion in West Indian *Anolis* lizards. I. Ecological consequences of reduced competition. Evolution **30**:659-676.

Lister, B.C. 1976b. The nature of niche expansion in West Indian *Anolis* lizards. II. Evolutionary consequences. Evolution **30**:677-692.

Lister, B.C., and R.E. McMurtrie. 1976. On size variation in anoline lizards. Am. Nat. **110**:311-314.

Livingstone, D.A. 1975. Late Quaternary climate change in Africa. Ann. Rev. Ecol. Syst. **6**:249-280.

Longwell, C.R. 1944. Some thoughts on the evidence for continental drift. Am. J. Sci. **242**:218-231.

Lord, R.D. 1960. Litter size and latitude in North Ameri.an mammals. Am. Midl. Nat. **64**:488-499.

Löve, A., and D. Löve. 1963. North Atlantic biota and their history. New York, Pergamon Press, Inc.

Lowe, C.H., J.W. Wright, C.J. Cole, and R.L. Bezy. 1970. Chromosomes and evolution of the species groups of *Cnemidophorus* (Reptilia: Teiidae). Syst. Zool. **19**:128-141.

Lubchenco, J. 1978. Plant species diversity in a marine intertidal community: importance of herbivore food preferences and algal competitive abilities. Am. Nat. **112**:23-39.

Lubchenco, J. 1980. Algal zonation in the New England rocky intertidal community: an experimental analysis. Ecology **61**:333-344.

Lubchenco, J., and S.D. Gaines. 1981. A unified approach to marine plant-herbivore interactions. I. Populations and communities. Ann. Rev. Ecol. Syst. **12**:405-438.

Lubchenco, J., and B.A. Menge. 1978. Community development and persistence in a low rocky intertidal zone. Ecol. Monogr. **48**:67-94.

Lyell, C. 1969. Principles of geology. 3 vol. Reprint of 1830-1833 ed. Introd. by M.J.S. Rudwick. New York, Johnson Reprint Corporation.

Lynch, J.D. 1979. The amphibians of the lowland tropical forests, p. 189-215. *In* W.E. Duellman (ed.), The South American herpetofauna: its origin, evolution and dispersal. Lawrence, Kans., Museum of Natural History, University of Kansas Monograph no. 7.

Lynch, J.D., and N.V. Johnson. 1974. Turnover and equilibria in insular avifaunas, with special reference to the California Channel Islands. Condor **76**:370-384.

Maarel, E. van der, and M.J.A. Werger (eds.). 1978. Plant species and plant communities. Proceedings of an international symposium in honor of Prof. Dr. V. Westhoff on the occasion of his 60th birthday. The Hague, Dr. W. Junk BV, Publishers.

Mabry, T.J., J.H. Hunziker, and D.R.D. Feo, Jr. (eds.). 1977. Creosote bush: biology and chemistry of *Larrea* in New World deserts. Stroudsburg, Penn., Dowden, Hutchinson & Ross, Inc.

Macan, T.T. 1970. Biological studies of English lakes. New York, Elsevier Scientific Publishing Co., Inc.

Macan, T.T. 1973. Ponds and lakes. New York, Crane, Russak & Co., Inc.

Macan, T.T., and E.B. Worthington. 1952. Life in lakes and rivers. London, William Collins Sons & Co., Ltd.

MacArthur, R.H. 1958. Population ecology of some warblers of northeastern coniferous forests. Ecology **39**:599-619.

MacArthur, R.H. 1960. On the relative abundance of species. Am. Nat. **94**:25-36.

MacArthur, R.H. 1965. Patterns of species diversity. Biol. Rev. **40**:510-533.

MacArthur, R.H. 1969. Patterns of communities in the tropics. Biol. J. Linn. Soc. **1**:19-30.

MacArthur, R.H. 1972. Geographical ecology; patterns in the distributions of species. New York, Harper & Row, Publishers, Inc.

MacArthur, R.H., and J.H. Connell. 1966. The biology of populations. New York, John Wiley & Sons, Inc.

MacArthur, R.H., J.M. Diamond, and J. Karr. 1972. Density compensation in island faunas. Ecology **53**:330-342.

MacArthur, R., and R. Levins. 1967. The limiting similarity, convergence, and divergence of coexisting species. Am. Nat. **101**:377-385.

MacArthur, R.H., and J.W. MacArthur. 1961. On bird species diversity. Ecology **42**:594-598.

MacArthur, R.H., J. MacArthur, D. MacArthur, and A. MacArthur. 1973. The effect of island area on population densities. Ecology **54**:657-658.

MacArthur, R., H. Recher, and M. Cody. 1966. On the relation between habitat selection and species diversity. Am. Nat. **100**:319-332.

MacArthur, R.H., and E.O. Wilson. 1963. An equilibrium theory of insular zoogeography. Evolution **17**:373-387.

MacArthur, R.H., and E.O. Wilson. 1967. The theory of island biogeography. Monographs in population biology no. 1. Princeton, N.J., Princeton University Press.

MacFadden, B.J. 1980. Rafting mammals or drifting islands?: Antillean insectivores *Nesophontes* and *Solenodon*. J. Biogeogr. **7**:11-22.

Mackie, G.O. 1974. Locomotion, floatation, and dispersal, p. 313-357. *In* L. Muscatine and H.M. Lenhoff (eds.), Coelenterate biology. Reviews and new perspectives. New York, Academic Press, Inc.

Magee, B. 1974. Karl Popper: the useful philosopher. New York, McGraw-Hill Book Co.

Maglio, V.J. 1973. Origin and evolution of the Elephantidae. Trans. Philos. Soc. Philadelphia, n.s. **63**(3):1-149.

Maglio, V.J., and H.B.S. Cooke (eds.). 1978. Evolution of African mammals. Cambridge, Mass., Belknap Press.

Maguire, B. 1963. The passive dispersal of small aquatic organisms and their colonization of isolated bodies of water. Ecol. Monogr. **33**:161-185.

Mahe, J. 1972. The Malagasy subfossils, p. 339-365. *In* R. Battistini and G. Richard-Vindard (eds.), Biogeography and ecology in Madagascar. Monographiae Biologicae 21. The Hague, Dr. W. Junk BV, Publishers.

Major, J. 1963. A climatic index to vascular plant activity. Ecology **44**:485-498.

Major, J. 1977. California climate in relation to vegetation, p. 11-74. *In* M.G. Barbour and J. Major (eds.), Terrestrial vegetation of California. New York, John Wiley & Sons, Inc.

Malfait, B.T., and M.G. Dinkelman. 1972. Circum-Caribbean tectonic and igneous activity and the evolution of the Caribbean plate. Geol. Soc. Am. Bull. **83**:251-272.

Maloiy, G.M.O. (eds.). 1979. Comparative physiology of osmoregulation in animals. 2 vol. London, Academic Press, Ltd.

Malthus, T.R. 1798. An essay on the principle of population. London, J. Johnson.

Mani, M.S. 1968. Ecology and biogeography of high altitude insects. The Hague, Dr. W. Junk Publisher, Inc.

Mani, M.S. (eds.). 1974. Ecology and biogeography in India. Monographiae Biologicae 23. The Hague, Dr. W. Junk BV, Publishers.

Mares, M.A. 1980. Convergent evolution among desert rodents: a global perspective. Bull. Carnegie Mus. Nat. Hist. **16**:1-51.

Mark, G.A., and K.W. Flessa. 1977. A test for evolutionary equilibria: Phanerozoic brachiopods and Cenozoic mammals. Paleobiology **3**:17-22.

Marks, G., and W.K. Beatty. 1976. Epidemics. New York, Charles Scribner's Sons.

Marshall, L.G. 1979. Evolution of metatherian and eutherian (mammalian) characters: a review based on cladistic methodology. Zool. J. Linn. Soc. **66**:369-410.

Marshall, L.G., and R.S. Corruccini. 1978. Variability, evolutionary rates, and allometry in dwarfing lineages. Paleobiology **4**:101-119.

Marshall, L.G., and M.K. Hecht. 1978. Faunal equilibrium? Mammalian faunal dynamics of the great American interchanges: an alternative explanation (with a reply to an alternative interpretation by S.D. Webb). Paleobiology **4**:203-209.

Marshall, L.G., S.D. Webb, J.J. Sepkoski, Jr., and D.M. Raup. 1982. Mammalian evolution and the great American interchange. Science **215**:1351-1357.

Martin, P.G. 1977. Marsupial biogeography and plate tectonics, p. 97-115. *In* B. Stonehouse and D. Gilmore (eds.), The biology of marsupials. Baltimore, Md., University Park Press.

Martin, P.S. 1964. Problems in paleoclimatolgy. London, Wiley-Interscience.

Martin, P.S. 1966. Africa and Pleistocene overkill. Nature **212**:339-342.

Martin, P.S. 1967. Prehistoric overkill, p. 75-120. *In* P.S. Martin and H.E. Wright, Jr. (eds.), Pleistocene extinctions: the search for a cause. New Haven, Conn., Yale University Press.

Martin, P.S. 1973. The discovery of America. Science **180**:969-974.

Martin, P.S., and H.E. Wright, Jr. (eds.). 1967. Pleistocene extinctions: the search for a cause. New Haven, Conn. Yale University Press.

Martin, R.A., and S.D. Webb. 1974. Late Pleistocene mammals from the Devil's Den fauna, Levy County, p. 114-145. *In* S.D. Webb (eds.), Pleistocene mammals of Florida. Gainesville, Fla., University of Florida Press.

Marvin, U.B. 1973. Continental drift. The evolution of a concept. Washington, D.C., Smithsonian Institution Press.

Marx, H., and G.B. Rabbil. 1970. Phyletic analysis of fifty characters of advanced snakes. Fieldiana, Zool. **63**:1-321.

Mather, J.R. 1974. Climatology: fundamentals and applications. New York, McGraw-Hill Book Co.

Mather, J.R. 1978. The climatic water budget in environmental analysis. Farnborough, England, Hauts, Teakfield, Ltd.

Matthew, W.D. 1915. Climate and evolution. Ann. New York Acad. Sci. **24**:171-318.

Matthews, L.H. 1980. The natural history of the whale. New Columbia University Press.

Mauriello, D., and J.P. Roskoski. 1974. A reanalysis of Vuilleumier's data. Am. Nat. **108**:711-714.

May, R.M. 1978. The dynamics and diversity of insect faunas, p. 188-204. *In* L.A. Mound and N. Waloff (eds.), Diversity of insect faunas. New York, Blackwell Scientific Publications, Ltd.

Mayr, E. 1942. Systematics and the origin of species. New York, Columbia University Press.

Mayr, E. 1944a. Wallace's Line in the light of recent zoogeographic studies. Quart. Rev. Biol. **19**:1-14.

Mayr, E. 1944b. The birds of Timor and Sumba. Bull. Am. Mus. Nat. Hist. **83**:127-194.

Mayr, E. (ed.). 1952. The problem of land connections across the South Atlantic, with special reference to the Mesozoic. Bull. Am. Mus. Nat. Hist. **99**(3):79-258.

Mayr, E. 1957. Evolutionary aspects of host specificity among parasites of vertebrates, p. 7-14. *In* U.N.E.S.C.O., Premier symposium sur la specificite parasitaire des parasites de vertebres. Union Internationale des sciences biologiques, ser. B. no. 32.

Mayr, E. 1963. Animal species and evolution. Cambridge, Mass., Harvard University Press.

Mayr, E. 1965a. The nature of colonization in birds, p. 30-47. *In* H.G. Baker and G.L. Stebbins (eds.), The genetics of colonizing species. New York, Academic Press, Inc.

Mayr, E. 1965b. Avifauna: turnover on islands. Science **150**:1587-1588.

Mayr, E. 1969. Principles of systematic zoology. New York, McGraw-Hill Book Co.

Mayr, E. 1970. Populations, species, and evolution. London, Oxford University Press.

Mayr, E. 1974. Cladistic analysis or cladistic classification? Z. Zool. Syst. Evolut.-forsch. **12**:94-128.

Mayr, E. 1976. Evolution and the diversity of life. Selected essays. Cambridge, Mass., Belknap Press.

McAtee, W.L. 1947. Distribution of seeds by birds. Am. Midl. Nat. **38**:214-223.

McClure, H.E. 1974. Migration and survival of the birds of Asia. Bangkok, Applied Scientific Research Corporation of Thailand.

McCoy, E.D., and E.F. Connor. 1976. Experimental determinants of island species number in the British Isles: a reconsideration. J. Biogeogr. **3**:381-382.

McCoy, E.D., and E.F. Connor. 1980. Latitudinal gradients in the species diversity of North American mammals. Evolution **34**:193-203.

McDowall, R.M. 1978a. New Zealand fishes: a guide and natural history. Auckland, Heinemann Educational Books.

McDowall, R.M. 1978b. Generalized tracks and dispersal biogeography. Syst. Zool. **27**:88-104.

McDowall, R.M., and A.M. Whitaker. 1975. The freshwater fishes, p. 277-299. *In* G. Kuschel (ed.), Biogeography and ecology in New Zealand. Monographiae Biologicae 27. The Hague, Dr. W. Junk BV, Publishers.

McDowell, S.B. 1969. Notes on the Australian sea snake *Ephalophis greyi* M. Smith (Serpentes: Elapidae, Hydrophiinae) and the origin and classification of sea snakes. Zool. J. Linn. Soc. **48**:333-349.

McDowell, S.B. 1972. The genera of sea snakes of the *Hydrophis* group (Serpentes: Elapidae). Trans. Zool. Soc. London **32**:189-247.

McDowell, S.B. 1974. Additional notes on the rare and primitive sea snake, *Ephalophis greyi*. J. Herpetol. **8**:123-128.

McElhinny, M.W. 1973a. Palaeomagnetism and plate tectonics. Cambridge, Cambridge University Press.

McElhinny, M.W. 1973b. Palaeomagnetism and plate tectonics of eastern Asia, p. 407-414. *In* P.J. Coleman (ed.), The Western Pacific. Island arcs, marginal seas, geochemistry. New York, Crane, Russak & Co., Inc.

McElhinny, M.W. 1976. The palaeoposition of Madagascar: remanence and magnetic properties of Late Paleozoic sediments. Earth Planet. Sci. Letters **31**:101-112.

McElhinny, M.W., N.S. Haile, and A.R. Crawford. 1974. Palaeomagnetic evidence shows Malay Peninsula was not a part of Gondwanaland. Nature **252**:641-645.

McGowan, J.A. 1974. The nature of oceanic ecosystems, p. 9-28. *In* C.B. Miller (ed.), The biology of the oceanic Pacific. Corvallis, Ore., Oregon State Press.

McGowan, J.A., and P.W. Walker. 1979. Structure in the copepod community of the North Pacific central gyre. Ecol. Monogr. **49**:195-226.

McGowran, B. 1973. Rifting and drift of Australia and the migration of mammals. Science **180**:759-761.

McIntosh, R.P. 1967. The continuum concept of vegetation. Bot. Rev. **33**:130-187.

McKenna, M.C. 1972a. Eocene final separation of the Eurasian and Greenland–North American landmasses. 24th Internatl. Geol. Congr., Sect. 7:275-281.

McKenna, M.C. 1972b. Was Europe conncected directly to North America prior to the Middle Eocene? p. 179-188. *In* Th. Dobzhansky, M.K. Hecht, and W.C. Steere (eds.), Evolutionary biology 6. New York, Appleton-Century-Crofts.

McKenna, M.C. 1972c. Possible biological consequences of plate tectonics. Bioscience **22**:519-525.

McKenna, M.C. 1973. Sweepstakes, filters, corridors, Noah's Arks, and beached viking funeral ships in palaeogeography, p. 295-308. *In* D.H. Tarling and S.K. Runcorn (eds.), Implications of continental drift to the earth sciences, vol. 1. New York, Academic Press, Inc.

McKenna, M.C. 1975a. Fossil mammals and Early Eocene North Atlantic land continuity. Ann. Missouri Bot. Gard. **62**:335-353.

McKenna, M.C. 1975b. Toward a phylogenetic classification of the Mammalia, p. 21-46. *In* W.P. Luckett and F.S. Szalay (eds.), Phylogeny of the primates. New York, Plenum Press.

McKenzie, D.P. 1972. Plate tectonics and sea-floor spreading. Am. Sci. **60**:425-435.

McKenzie, D.P., and W.J. Morgan. 1969. The evolution of triple junctions. Nature **226**:239-243.

McKenzie, D.P., and F. Richter. 1976. Convection currents in the earth's mantle. Sci. Am. **235**(5):72-84, 89.

McNab, B.K. 1963. Bioenergetics and the determination of home range size. Am. Nat. **97**:133-140.

McNab, B. 1971. The structure of tropical bat faunas. Ecology **52**:352-358.

McNeill, W.H. 1976. Plagues and peoples. Garden City, N.Y., Anchor Press.

Means. R.B. 1975. Competitive exclusion along a habitat gradient between two species of salamanders *(Desmognathus)* in western Florida. J. Biogeogr. **2**:253-263.

Medawar, P.B. 1969. Induction and intuition in scientific thought. Mem. Am. Philos. Soc. 75.

Meggers, B.J., E.S. Ayensu, and W.D. Duckworth (eds.). 1973. Tropical forest ecosystems in Africa and South America: a comparative review. Washington, D.C., Smithsonian Institution Press.

Melville, R. 1981. Vicariance plant distributions and paleogeography of the Pacific region, p. 238-274. *In* G. Nelson and D.E. Rosen (eds.), Vicariance biogeography: a critique. New York, Columia University Press.

Menard, H.W. 1960. The East Pacific Rise. Science **132**:1737-1746.

Menge, B.A., and J.P. Sutherland. 1976. Species diversity gradients: synthesis of the roles of predation, competition, and temporal heterogeneity. Am. Nat. **110**:351-369.

Mensching, H.G. 1980. The Sahelian zone and the problem of desertification, p. 257-266. *In* E.M. van Zinderen Bakker, Sr. and J.A. Coetzee (eds.), Palaeoecology of Africa and the surrounding islands, vol. 12. Rotterdam, A.A. Balkema.

Menzies, R.J., R.Y. George, and G.T. Rowe, 1973. Abyssal environment and ecology of the world oceans. New York, John Wiley & Sons, Inc.

Menzies, R.J., O.H. Pilkey, B.W. Blackwelder, D. Dexter, P. Huling, and L. McCloskey. 1966. A submerged reef off North Carolina. Internatl. Rev. Ges. Hydrobiol. **51**:393-431.

Merriam, C.H. 1890. Results of a biological survey of the San Francisco Mountian region and the desert of the Little Colorado, Arizona. North American Fauna no. **3**:1-136.

Merriam, C.H. 1894. Laws of temperature control of the geographic distribution of terrestrial animals and plants. Natl. Geogr. Mag. **6**:229-238.

Mertz, D.B. 1971. The mathematical demography of the California Condor population. Am. Nat. **105**:437-453.

Metcalf, H., and J.F. Collins. 1911. The control of the chestnut bark disease. Farmer's Bull. U.S. Dept. Agr. no. **467**:1-24.

Meyer, F..G 1966. Chile–United States Botanical Expedition to Juan Fernández Islands, 1965. Antarctic J. U.S. **1**:232-240.

Meyer de Schauensee, R. 1966. The species of birds of South America and their distribution. Philadelphia, Academy of Natural Sciences.

Meyerhoff, A.A. 1968. Arthur Holmes: originator of spreading ocean floor hypothesis. J. Geophys. Res. **73**:6563-6565.

Michaelsen, W. 1922. Die Verbreitung der Oligochäten im Lichte der Wegenerschen Theorie der Kontinentenverschiebungun und andere Fragen zur Stammesgeschichte und Verbreitung dieser Tiergruppe. Verh. nat. Ver. Hamburg **3**(29):45-79.

Michener, C.D. 1975. The Brazilian bee problem. Ann. Rev. Entomol. **20**:399-416.

Michener, C.D. 1979. Biogeography of the bees. Ann. Missouri Bot. Gard. **66**:277-347.

Middlemiss, F.A., P.F. Rawson and G. Newall (eds.). 1971. Faunal provinces in space and time. Liverpool, Seel House Press.

Mileikovsky, S.A. 1971. Types of larval development in marine botton invertebrates, their distribution and ecological signifiance: a re-evaluation. Mar. Biol. **10**:193-213.

Miles, R.S. 1977. Dipnoan (lungfish) skulls and the relationships of the group: a study based on new species from the Devonian of Australia. Zool. J. Linn. Soc. **61**:1-328.

Miller, R.R. 1948. The cyprinodont fishes of the Death Valley system of eastern California and southwestern Nevada. University of Michigan Museum Zool. Misc. Pub. **42**:1-80.

Miller, R.R. 1958. Origin and affinities of the freshwater fish fauna of western North America, p. 187-222. *In* C.L. Hubbs (ed.), Zoogeography. Washington, D.C., Am. Assoc. Advancement Sci. Pub. 51.

Miller, R.R. 1961a. Man and the changing fish fauna of the American Southwest. Pap. Michigan Acad. Sci. Arts Letters **46**:365-404.

Miller, R.R. 1961b. Speciation rates in some freshwater fishes of western North America, p. 537-560. *In* W.F. Blair (ed.), Vertebrate speciation. Austin, Tex., University of Texas Press.

Miller, R.R. 1966. Geographical distribution of Central American freshwater fishes. Copeia (4):773-802.

Miller, R.S. 1964. Ecology and distribution of pocket gopher (Geomyidae) in Colorado. Ecology **45**:256-272.

Miller, R.S. 1967. Pattern and process in competition. Adv. Ecol. Res. **4**:1-74.

Minton, S.A. 1975. Geographic distribution of sea snakes, p. 21-31. *In* W.A. Dunson (ed.), The biology of sea snakes. Baltimore, Md., University Park Press.

Miranda, F., and A.J. Sharp. 1950. Characteristics of the vegetation in certain temperate regions of eastern Mexico. Ecology **31**:313-333.

Mitchell-Thomé, R.C. 1976. Geology of the Middle Atlantic Islands. Berlin, Gebrüder Borntraeger.

Mohr, P.A. 1970. The Afar triple junction and sea-floor spreading. J. Geophys. Res. **75**:7340-7352.

Molles, M.C., Jr. 1978. Fish species diversity on model and natural reef patches: experimental insular biogeography. Ecol. Monogr. **48**:289-305.

Molnar, R.E., T.F. Flannery, and T.H.V. Rich. 1981. An allosaurid theropod dinosaur from the Early Cretaceous of Victoria, Australia. Alcheringa **5**:141-146.

Monk, C.D. 1967. Tree species diversity in the eastern deciduous forest with particular reference to north central Florida. Am. Nat. **101**:173-187.

Monk, C.D. 1968. Successional and environmental relationships of the forest vegetation of north central Florida. Am. Midl. Nat. **79**:441-457.

Monteith, J.L. 1973. Principles of environmental physics. New York, American Elsevier Publishing Co., Inc.

Mooney, H.A. (ed.). 1977. Convergent evolution in Chile and California. Mediterranean climate ecosystems. US/IBP synthesis ser. 5. Stroudsburg, Penn., Dowden, Hutchinson, & Ross, Inc.

Moore, D.M., J.B. Harborne, and C.A. Williams. 1970. Chemotaxonomy, variation and geographical distribution of the Empetraceae. Bot. J. Linn. Soc. **63**: 277-293.

Moore, H.E., Jr. 1973. Palms in the tropical forest ecosystems of Africa and South America, p. 63-88. *In* B.J. Meggers, E.S. Ayensu, and W.D. Duckworth (eds.), Tropical forest ecosystems in Africa and South America: a comparative review. Washington, D.C., Smithsonian Institution Press.

Moorehead, A. 1969. Darwin and the Beagle. London, Hamish Hamilton, Ltd.

Morafka, D.J. 1977. Is there a Chihuahuan desert? A quantitative evaluation through a herpetofaunal perspective, p. 437-454. *In* R.H. Wauer and D.H. Riskind (eds.), Transactions of the symposium on the biological resources of the Chihuahuan Desert region United States and Mexico. U.S. Dept. Interior National Park Service Transactions and Proceedings, ser. no. 3.

Moreau, R.E. 1966. The bird faunas of Africa and its islands. New York, Academic Press, Inc.

Moreau, R.E., 1972. The Palaearctic-African bird migration systems. New York, Academic Press, Inc.

Morgan, W.J. 1968. Rises, trenches, great faults and crustal blocks. J. Geophys. Res. **73**:1959-1982.

Morgan, W.J. 1971. Convection plumes in the lower mantle. Nature **230**:42-43.

Morgan, W.J. 1972a. Deep mantle convection plumes and plate motion. Bull. Am. Assoc. Petrol. Geol. **56**:203-213.

Morgan, W.J. 1972b. Plate motions and deep mantle convection. Mem. Geol. Soc. Am. **132**:7-22.

Morris, I. (ed.). 1980. The physiological ecology of phytoplankton. Studies in ecology 7. Oxford, Blackwell Scientific Publications, Ltd.

Morse, D.H. 1971. The foraging of warblers isolated on small islands. Ecology **52**:216-228.

Morse, D.H. 1973. The foraging of small populations of Yellow Warblers and American Redstarts. Ecology **54**:346-355.

Morse, D.H. 1977. The occupation of small islands by passerine birds. Condor **79**:399-412.

Mosimann, J.E., and P.S. Martin. 1975. Simulating overkill by paleoindians. Am. Sci. **63**:304-313.

Mueller-Dombois, D. 1981. Island ecosystems: what is unique about their ecology, p. 485-501. *In* D. Mueller-Dombois, K.W. Bridges, and H.L. Carson (eds.), Island ecosystems. Biological organization in selected Hawaiian communities. Stroudsburg, Penn., Hutchinson Ross Publishing Co.

Muller, J. 1970. Palynological evidence on early differentiation of angiosperms. Biol. Rev. Cambridge Philos. Soc. **45**:417-450.

Müller, P. 1973. The dispersal centres of terrestrial vertebrates in the Neotropical Realm. The Hague, Dr. W. Junk BV, Publishers.

Müller, P. 1974. Aspects of zoogeography. The Hague, Dr. W. Junk BV, Publishers.

Muller, S.W., and A. Campbell. 1954. The relative number of living and fossil species of animals. Syst. Zool. **3**: 168-170.

Munro, I.S. 1967. The fishes of New Guinea. Port Moresby, Papua–New Guinea, Department of Agriculture, Stock and Fisheries.

Murray, D.F. 1981. The role of arctic refugia in the evolution of the arctic vascular flora: a Beringian perspective, p. 11-20. *In* G.G.E. Scudder and J.L. Reveal (eds.), Evolution today. Proceedings of the Second International Congress of Systematics and Evolutionary Biology. Pittsburgh, Hunt Institute of Botanical Documentation.

Murray, G.F. 1961. Geology of the Atlantic and Gulf Coastal Province of North America. New York, Harper & Row, Publishers, Inc.

Musick, H.B. 1976. Phosphorus toxicity in seedlings of *Larrea divaricata* grown in solution culture. Bot. Gaz. **139**:108-111.

Myers, G.S. 1938. Fresh-water fishes and West Indian zoogeography. Smithsonian Report for **1937**:339-364.

Naiman, R.J., and D.L. Soltz. 1981. Fishes in North American deserts. New York, John Wiley & Sons, Inc.

Nairn, A.E.M. (ed.). 1961. Descriptive palaeoclimatology. London, Interscience Publishers.

Nairn, A.E.M. 1965. Uniformitarianism and environment. Palaeogeogr., Palaeoclimatol. Palaeoecol. **1**:5-11.

Navarra, J.G. 1979. Atmosphere, weather and climate: an introduction to meteorology. Philadelphia, W.B. Saunders Co.

Neill, W.T. 1969. The geography of life. New York, Columbia University Press.

Nelson, G.J. 1969a. The problem of historical biogeography. Syst. Zool. **18**:243-246.

Nelson, G.J. 1969b. Gill arches and the phylogeny of fishes, with notes on the classification of vertebrates. Bull. Am. Mus. Nat. Hist. **141**:475-552.

Nelson, G. 1973. Comments on Leon Croizat's biogeography. Syst. Zool. **22**:312-320.

Nelson, G. 1974. Historical biogeography: an alternative formalization. Syst. Zool. **23**:555-558.

Nelson, G. 1978. From Candolle to Croizat: comments on the history of biogeography. J. Hist. Biol. **11**:269-305.

Nelson, G., and N.I. Platnick. 1978. The perils of plesiomorphy: widespread taxa, dispersal, and phenetic biogeography. Syst. Zool. **27**:474-477.

Nelson, G., and N. Platnick. 1980. A vicariance approach to historical biogeography. Bioscience **30**:339-343.

Nelson, G., and N. Platnick. 1981. Systematics and biogeography. Cladistics and vicariance. New York, Columbia University Press.

Nelson, G., and D.E. Rosen (eds.). 1981. Vicariance biogeography: a critique. New York, Columbia University Press.

Nelson, J.B. 1978. The Sulidae. Gannets and boobies. Oxford, Oxford University Press for University of Aberdeen.

Nelson, J.S. 1976. Fishes of the world. New York, John Wiley & Sons, Inc.

Nevo, E. 1979. Adaptive convergence and divergence of subterranean mammals. Ann. Rev. Ecol. Syst. **10**: 269-308.

Nevo, E., and H. Bar-El. 1976. Hybridization and speciation in fossorial mole rats. Evolution **30**:831-840.

Newbigin, M.T. 1968. Plant and animal geography. 3rd ed. London, Methuen & Co., Ltd.

Newell, N.D. 1967. Revolutions in the history of life. Geol. Soc. Am. Spec. Pap. **89**:63-91.

New Zealand Department of Scientific and Industrial Research. 1979. Proceedings of the International Symposium on Marine Biogeography and Evolution in the Southern Hemisphere, Auckland, New Zealand, 17-20 July, 1978. 2 vol. New Zealand OSIR Information Ser. 137.

Niethammer, G. 1958. Tiergeographie (Bericht uber die Jahren 1950-56). Fortschr. Zool. **11**:35-141.

Niklas, K.J., B.H. Tiffney, and A.H. Knoll. 1980. Apparent changes in the diversity of fossil plants, p. 1-89. *In* M.C. Hecht, W.C. Steere, and B. Wallace (eds.), Evolutionary biology 12. New York, Plenum Press.

Nix, H.A. 1976. Environmental control of breeding, post-breeding dispersal and migration of birds in the Australian Region, p. 272-305. *In* H.J. Frith and J.H. Calaby (eds.), Proceedings of the 16th International Ornithological Congress. Canberra, Australian Academy of Science.

Nizamuddin, M. 1970. Phytogeography of the Fucales and their seasonal growth. Bot. Mar. **13**:131-139.

Nobel, P.S. 1974. Introduction to biophysical plant physiology San Francisco, W.H. Freeman & Co., Publishers.

Nobel, P.S. 1978. Surface temperatures of cacti–influences of environmental and morphological factors. Ecology **59**:986-996.

Nobel, P.S. 1980a. Morphology, nurse plants, and minimum apical temperatures for young *Carnegiea gigantea*. Bot. Gaz. **141**:188-191.

Nobel, P.S. 1980b. Morphology, surface temperatures, and northern limits of columnar cacti in the Sonoran Desert. Ecology **61**:1-7.

Norris, K.S. (ed.). 1966. Whales, dolphins, and porpoises. Berkeley, University of California Press.

Norton, I.O., and J.G. Sclater. 1979. A model for the evolution of the Indian Ocean and the breakup of Gondwanaland. J. Geophys. Res. **84**:6803-6830.

Numata, M. 1974. The flora and vegetation of Japan. Amsterdam, Elsevier Scientific Publishing Co.

Nur, A., and Z. Ben-Avraham. 1977. Lost Pacifica continent. Nature **270**:41-43.

Nur, A., and Z. Ben-Avraham. 1981. Lost Pacifica continent: a mobilistic speculation, p. 341-358. *In* G. Nelson and D.E. Rosen (eds.), Vicariance biogeography: a critique. New York, Columbia University Press.

Odening, W.R., B.R. Strain, and W.C. Oechel. 1974. The effects of decreasing water potential on net CO_2 exchange of intact desert shrubs. Ecology **55**:1086-1095.

Odum, E.P. 1969. The strategy of ecosystem development. Science **164**:262-270.

Odum, E.P. 1971. Fundamentals of ecology. 3rd ed. Philadelphia, W.B. Saunders Co.

Odum, H.T. 1957. Trophic structure and productivity of Silver Springs, Florida. Ecol. Monogr. **27**:55-112.

Olson, E.C. 1971. Vertebrate paleozoology. New York, Wiley-Interscience.

Olson, S.L., and A. Feduccia. 1980. Relationships and evolution of flamingos (Aves: Phoenicopteridae). Smithsonian Contr. Zool. no. 316.

Orians, G.H. 1969. The number of bird species in some tropical forests. Ecology **50**:783-801.

Orians, G.H., and M.F. Willson. 1964. Interspecific territories of birds. Ecology **45**:736-745.

Orshan, G. 1953. Note on the application of Raunkiaer's system of life forms in arid regions. Palest. J. Bot. (Jerusalem) ser. **6**:120-122.

Osman, R.W. 1977. The establishment and development of a marine epifaunal community. Ecol. Monogr. **47**:37-63.

Osman, R.W. 1978. The influence of seasonality and stability on the species equilibrium. Ecology **59**:383-399.

Osman, R.W., and R.B. Whitlatch. 1978. Patterns of species diversity: fact or artifact? Paleobiology **4**:41-54.

Otte, D. 1976. Species richness patterns of New World desert grasshoppers in relation to plant diversity. J. Biogeogr. **3**:197-209.

Owen, D.F., and J. Owen. 1974. Species diversity in temperate tropical Ichneumonidae. Nature **249**:583-584.

Paine, R.T. 1966. Food web complexity and species diversity. Am. Nat. **100**:65-76.

Paijmans, K. 1976. New Guinea vegetation. Amsterdam, Elsevier Scientific Publishing Co.

Palmer, R.S. 1962. Handbook of North American birds, vol. 1. Loons through flamingos. New Haven, Conn., Yale University Press.

Palmer R.S. 1976. Handbook of North American birds. vol. 2 and 3. Waterfowl. New Haven, Conn., Yale University Press.

Panchen, A.L. 1980. The terrestrial environment and the land vertebrates. London, Academic Press, Ltd.

Papavero, N. 1977. The world Oestridae (Diptera), mammals and continental drift. The Hague, Dr. W. Junk BV, Publishers.

Parin, N.V. 1970. Ichthyofauna of the epipelagic zone. Translated from Russian. Jerusalem, Israel Program for Scientific Translations.

Park, T. 1948. Experimental studies of interspecific competition. I. Competition between populations of the flour beetles *Tribolium confusum* Duval and *Tribolium castaneum* Herbst. Ecol. Monogr. **18**:265-307.

Park, T. 1954. Experimental studies of interspecies competition. II. Temperature, humidity, and competition in two species of *Tribolium*. Physiol. Zool. **27**:177-238.

Park, T. 1962. Beetles, competition, and populations. Science **138**:1369-1375.

Parker, J. 1963. Cold resistance in woody plants. Bot. Rev. **29**:123-201.

Parker, J. 1969. Further studies of drought resistance in woody plants. Bot. Rev. **35**:317-371.

Parsons, R.F., and D.G. Cameron. 1974. Maximum plant species diversity in terrestrial communities. Biotropica **6**:202-203.

Patrick, R. 1961. A study of the numbers and kinds of species found in rivers in eastern United States. Proc. Acad. Nat. Sci. Philadelphia **113**:215-258.

Patrick, R. 1966. The Catherwood Foundation Peruvian Amazon Expedition: limnological and systematic studies. Monogr. Acad. Nat. Sci. Philadelphia **14**:1-495.

Patrick, R. 1967. The effect of invasion rate, species pool and size of area on the structure of the diatom community. Proc. Natl. Acad. Sci. U.S.A **58**:1335-1342.

Patterson, B.D. 1980. Montane mammalian biogeography in New Mexico. Southwest. Nat. **25**:33-40.

Patterson, B., and R. Pascual. 1972. The fossil mammal fauna of South America, p. 247-309. *In* A. Keast, F.C. Erk, and B. Glass (eds.), Evolution, mammals, and southern continents. Albany, N.Y., State University of New York Press.

Patterson, C. 1981a. Methods of paleobiogeography, p. 446-489. *In* G. Nelson and D.E. Rosen (eds.), Vicariance biogeography: a critique. New York, Columbia University Press.

Patterson, C. 1981b. Significance of fossils in determining evolutionary relationships. Ann. Rev. Ecol. Syst. **12**:195-223.

Patterson, C. (ed.). 1982. Methods of phylogenetic reconstruction. Zool. J. Linn. Soc. **74**(3).

Patton, J.L., J.C. Hafner, M.S. Hafner, and M.F. Smith. 1979. Hybrid zones in *Thomomys bottae* pocket gophers: genetics, phenetics, and ecologic concordance. Evolution **33**:860-876.

Patton, W.W., Jr., and I.L. Tailleur. 1977. Evidence in the Bering Strait region for differential movement between North America and Eurasia. Geol. Soc. Am. Bull. **88**:1298-1304.

Paylore, P. 1976. Desertification: a world bibliography. Tucson, Ariz., University of Arizona Office of Arid Lands.

Peabody, F.E., and J.M. Savage. 1958. Evolution of a coast range corridor in California and its effect on the origin and dispersal of living amphibians and reptiles, p. 159-186. *In* C.L. Hubbs (ed.), Zoogeography. Washington, D.C, Am. Assoc. Advancement Sci. Pub. 51.

Pears, N. 1978. Basic biogeography. London, Longman Group, Ltd.

Pearson, T.G. (ed.). 1936. Birds of America. Garden City, N.Y., Garden City Publishing Co.

Penman, H.L. 1956. Evaporation: an introductory survey. Neth. J. Agr. Sci. **4**:9-29.

Perfit, M.R., and B.C. Heezen. 1978. The geology and evolution of the Cayman Trench. Geol. Soc. Am. Bull. **89**:1155-1174.

Perkins, E.J. 1974. The biology of estuaries and coastal waters. London, Academic Press, Ltd.

Peterson, R.L., and S.D. Downing. 1956. Distributional records of the opossum in Ontario. J. Mammal. **37**: 431-435.

Peterson, R.T. 1961. A field guide to Western birds. Houghton Mifflin, Boston.

Philbrick, R.N. (ed.). 1967. Proceedings of the symposium on the biology of the California Islands. Santa Barbara, Santa Barbara Botanical Garden.

Phillips, J.D., and D.W. Forsyth. 1972. Plate tectonics, palaeomagnetism and the opening of the Atlantic. Geol. Soc. Am. Bull. **83**:1579-1600.

Phillips, W.S. 1963. Photographic documentation. Vegetational changes in northern Great Plains. Tucson, Ariz., University of Arizona Agricultural Experiment Station Report no. 214.

Pianka, E.R. 1966. Latitudinal gradients in species diversity: a review of concepts. Am. Nat. **100**:33-46.

Pianka, E.R. 1967. On lizard species diversity: North American flatland deserts. Ecology **48**:331-351.

Pianka, E.R. 1975. Niche relations of desert lizards, p. 292-314. *In* M.L. Cody and J.M. Diamond (eds.), Ecology and evolution of communities. Cambridge, Mass., Belknap Press.

Pianka, E.R. 1978. Evolutionary ecology. 2nd ed. New York, Harper & Row, Publishers, Inc.

Pichard, G.L. 1975. Descriptive physical oceanography. An introduction. 2nd ed. Oxford, Pergamon Press, Ltd.

Pielou, E.C. 1975. Ecological diversity. New York, John Wiley & Sons, Inc.

Pielou, E.C. 1977a. Mathematical ecology. New York, John Wiley & Sons, Inc.

Pielou, E.C. 1977b. The latitudinal spans of seaweed species and their patterns of overlap. J. Biogeogr. **4**:299-311.

Pielou, E.C. 1978. Latitudinal overlap of seaweed species: evidence for quasi-sympatric speciation. J. Biogeogr. **5**:227-238.

Pielou, E.C. 1979. Biogeography. New York, John Wiley & Sons, Inc.

Pijl, L. van der. 1972. Principles of dispersal in higher plants. 2nd ed. New York, Springer-Verlag New York, Inc.

Pitman, W.C., and M. Talwani. 1972. Sea-floor spreading in the North Atlantic. Geol. Soc. Am. Bull. **83**:619-646.

Platnick, N.I. 1976. Concepts of dispersal in historical biogeography. Syst. Zool. **25**:294-295.

Platnick, N.I., and E.S. Gaffney. 1977. Systematics: a Popperian perspective. Syst. Zool. **26**:360-365.

Platnick, N.I., and E.S. Gaffney. 1978a. Evolutionary biology: a Popperian perspective. Syst. Zool. **27**:137-141.

Platnick, N.I., and E.S. Gaffney. 1978b. Systematics and the Popperian paradigm. Syst. Zool. **27**:381-388.

Platnick, N.I., and G. Nelson. 1978. A method of analysis for historical biogeography. Syst. Zool. **27**:1-16.

Platt, J. 1969. What we must do. Science **166**:1115-1121.

Plumstead, E.P. 1973. The enigmatic *Glossopteris* flora and uniformitarianism, p. 411-424. *In* D.H. Tarling and S.K. Runcorn (eds.), Implications of continental drift to the earth sciences, vol. 1. New York, Academic Press, Inc.

Polunin, N. 1960. Introduction to plant geography and some related sciences. London, Longmans Group, Ltd.

Pomeroy, L.R. 1970. The strategy of mineral cycling. Ann. Rev. Ecol. Syst. **1**:171-190.

Popham, E.J. 1961. Life in fresh water. Cambridge, Mass., Harvard University Press.

Popper, K.R. 1968a. The logic of scientific discovery. 2nd ed. New York, Harper & Row, Publishers, Inc.

Popper, K.R. 1968b. Conjectures and refutations. New York, Harper & Row, Publishers, Inc.

Popper, K.R. 1972. Objective knowledge, an evolutionary approach. Oxford, Clarendon Press.

Por, F.D. 1971. One hundred years of Suez Canal: a century of Lessepsian migration: retrospect and viewpoints. Syst. Zool. **20**:138-159.

Por, F.D. 1975. Pleistocene pulsation and preadaptation of biotas in mediterranean seas: consequences for Lessepsian migration. Syst. Zool. **24**:72-78.

Por, F.D. 1977. Lessepsian migration. The influx of Red Sea biota into the Mediterranean by way of the Suez Canal. Ecological studies 23. Berlin, Springer-Verlag.

Port, G.R., and J.R. Thompson. 1980. Outbreaks of insect herbivores on plants along motorways in the United Kingdom. J. Appl. Ecol. **17**:649-656.

Porter, D.M. 1979. Endemism and evolution in Galapagos Islands vascular plants, p. 225-256. *In* D. Bramwell (ed.), Plants and islands. London, Academic Press, Ltd.

Porter, J.W. 1972. Ecology and species diversity of coral reefs on opposite sides of the Isthmus of Panama. Bull. Biol. Soc. Washington **2**:89-116.

Porter, J.W. 1974. Community structure and coral reefs on opposite sides of the Isthmus of Panama. Science **186**:543-545.

Porter S.C. 1979. Hawaiian glacial ages. Quat. Res. **12**: 161-187.

Poulson, T.L., and W.B. White. 1969. The cave environment. Science **165**:971-981.

Powell, A.M., and B.L. Turner. 1977. Aspects of the plant biology of the gypsum outcrops of the Chihuahuan Desert, p. 315-325. *In* R.H. Wauer and E.H. Riskind (eds.), Transactions of the symposium on the biological resources of the Chihuahuan Desert region United States and Mexico. U.S. Dept. Interior National Park Service Transactions and Proceedings, ser. no. 3.

Power, D.M. 1972. Numbers of bird species on the California Islands. Evolution **26**:451-463.

Prance, G.T. 1973. Phytogeographic support for the theory of Pleistocene forest refugia in the Amazon basin, based on evidence from distribution patterns in Caryocaraceae, Chrysobalanaceae, Dichapetalaceae, and Lecythidaceae. Acta Amazonica 3(3):5-28.

Prance, G.T. 1978. The origin and evolution of the Amazon flora. Interciencia **3**:207-222.

Prance, G.T. 1981. Discussion, p. 395-405. *In* G. Nelson and D.E. Rosen (eds.), Vicariance biogeography: a critique. New York, Columbia University Press.

Prance, G.T. (ed.). 1982. The biological model of diversification in the tropics. New York, Columbia University Press.

Pregill, G.K., and S.L. Olson. 1981. Zoogeography of West Indian vertebrates in relation to Pleistocene climatic cycles. Ann. Rev. Ecol. Syst. **12**:75-98.

Preston, F.W. 1957. Analysis of Maryland statewide bird counts. Maryland Birdlife (Bull. Maryland Ornithological Soc.) **13**:63-65.

Preston, F.W. 1960. Time and space and the variation of species. Ecology **41**:611-627.

Preston, F.W. 1962. The canonical distribution of commonness and rarity: Part I. Ecology **43**:185-215. Part II. **43**:410-432.

Price, P.W. 1980. Evolutionary biology of parasites. Monographs in population biology no. 15. Princeton, N.J., Princeton University Press.

Pritchard, P.C.H. 1979. Taxonomy, evolution, and zoogeography, p. 1-42. *In* M. Harless and H. Morlock (eds.), Turtles: perspectives and research. New York, John Wiley & Sons, Inc.

Proctor, J., and S.R.J. Woodell. 1975. The ecology of serpentine soils. Adv. Ecol. Res. **9**:255-366.

Proctor, V.W. 1968. Long-distance dispersal of seeds by retention in digestive tract of birds. Science **160**:321-322.

Quézel, P. 1978. Analysis of the flora of Mediterranean and Saharan Africa. Ann. Missouri Bot. Gard. **65**:479-534.

Rabinowitz, D., and J.K. Rapp. 1980. Seed rain in a North American tall grass prairie. J. Appl. Ecol. **17**:793-802.

Rabinowitz, D., and J.K. Rapp. 1981. Dispersal abilities of seven sparse and common grasses from a Missouri prairie. Am. J. Bot. **68**:616-624.

Raikow, R.J. 1976. The origin and evolution of the Hawaiian honeycreepers (Drepaniidae). Living Bird **15**:95-117.

Rand, A.L., and E.T. Gilliard. 1967. The handbook of New Guinea birds. London, Weidenfeld & Nicolson, Ltd.

Rand, A.S. 1969. Competitive exclusion among anoles (Sauria: Iguanidae) on small islands in the West Indies. Breviora no. **319**:1-16.

Raunkiaer, C. 1934. The life forms of plans and statistical plant geography. Oxford, Clarendon Press.

Raup, D.M. 1972. Taxonomic diversity during the Phanerozoic. Science **177**:1065-1071.

Raup, D.M. 1976. Species diversity in the Phanerozoic: an interpretation. Paleobiology **2**:289-297.

Raup, D.M. 1979. Size of the Permo-Triassic bottleneck and its evolutionary implications. Science **206**:217-218.

Raup, D.M., and S.J. Gould. 1974. Stochastic simulation and evolution of morphology: towards a nomothetic paleontology. Syst. Zool. **23**:305-322.

Raup, D.M., S.J. Gould, T.J.M. Schopf, and D. Simberloff. 1973. Stochastic models of phylogeny and the evolution of diversity. J. Geol. **81**:525-542.

Raup, D.M., and J.J. Sepkoski, Jr. 1982. Mass extinctions in the marine fossil record. Science **215**:1501-1503.

Raup, D.M., and S.M. Stanley, 1978. Principles of paleontology. 2nd ed. San Francisco, W.H. Freeman & Co., Publishers.

Raven, H.C. 1935. Wallace's Line and the distribution of Indo-Australian mammals. Bull. Am. Mus. Nat. Hist. **68**:179-293.

Raven, P.H. 1963. Amphitropical relationships inthe floras of North and South America. Quart. Rev. Biol. **38**:151-177.

Raven, P.H. 1964. Catastrophic speciation and edaphic endemism. Evolution **18**:336-338.

Raven, P.H. 1972. Plant species disjunctions: a summary. Ann. Missouri Bot. Gard. **59**:234-246.

Raven, P.H. 1976. Systematics and plant population biology. Syst. Bot. **1**:284-316.

Raven, P.H. 1977. A suggestion concerning the Cretaceous rise to dominance of the angiosperms. Evolution **31**:451-452.

Raven, P.H., and D.I. Axelrod. 1972. Plate tectonics and Australasian paleobiogeography. Science **176**:1379-1386.

Raven, P.H., and D.I. Axelrod. 1974. Angiosperm biogeography and past continental movements. Ann. Missouri Bot. Gard. **61**:539-673.

Raven, P.H., and D.I. Axelrod. 1975. History of the flora and fauna of Latin America. Am. Sci. **63**:420-429.

Rawlinson, P.A. 1974. Biogeography and ecology of the reptiles of Tasmania and the Bass Straits area, p. 291-338. *In* W.D. Williams (ed.), Biogeography and ecology in Tasmania. Mongraphiae Biologicae 25. The Hague, Dr. W. Junk BV, Publishers.

Recher, H., 1969. Bird species diversity and habitat diversity in Australia and North America. Am. Nat. **103**:75-80.

Reed, T.M. 1980. Turnover frequency in island birds. J. Biogeogr. **7**:329-335.

Rensch, B. 1960. Evolution above the species level. New York, Columbia University Press.

Repenning, C.A. 1976. Adaptive evolution of sea lions and walruses. Syst. Zool. **25**:375-390.

Repenning, C.A., C.E. Ray, and D. Grigorescu. 1979. Pinniped biogeography, p. 357-370. *In* J. Gray and A.J. Boucot (eds.), Historical biogeography, plate tectonics, and changing environments. Corvallis, Ore., Oregon State University Press.

Rex, M.A. 1981. Community structure in the deep-sea benthos. Ann. Rev. Ecol. Syst. **12**:331-354.

Rice, D.W., and A.A. Wolman. 1971. The life history and ecology of the gray whale *(Eschrichtius robustus)*. Spec. Pub. Am. Soc. Mammal.

Richards, H.G. 1971. Sea level during the past 11,000 years as indicated by data from North and South America. Quaternaria **14**:7-16.

Richards, P.W. 1957. The tropical rainforest. Cambridge, Cambridge University Press.

Richards, P.W. 1969. Speciation in the tropical forest and the concept of the niche. Biol. J. Linn. Soc. **1**:149-153.

Richter-Dyn, N., and N.S. Goel. 1972. On the extinction of a colonizing species. Theor. Pop. Biol. **3**:406-433.

Ricklefs, R.E. 1970. Stage of taxon cycle and distribution of birds on Jamaica, Greater Antilles. Evolution **24**: 475-477.

Ricklefs, R.E. 1979. Ecology. 2nd ed. Newton, Mass., Chiron Press.

Ricklefs, R.E., and G.W. Cox. 1972. Taxon cycles in the West Indian avifauna. Am. Nat. **106**:195-219.

Ricklefs, R.E., and G.W. Cox. 1978. Stage of taxon cycle, habitat distribution, and population density in the avifauna of the West Indies. Am. Nat. **112**:875-895.

Ridd, M.F. 1971. South-east Asia as a part of Gondwanaland. Nature **234**:531-533.

Ridley, H.N. 1930. The dispersal of plants throughout the world. Ashford, England, L. Reeve & Co.

Roberts, T.R. 1973. Interrelationships of ostariophysans, p. 373-395. *In* P.H. Greenwood, R.S. Miles, and C. Patterson (eds.), Interrelationships of fishes. London, Academic Press, Ltd. for Linnean Society of London, Zool. J. Linn. Soc. 53, Suppl. no. 1.

Robinson, E., and J.F. Lewis. 1971. Field guide to aspects of the geology of Jamaica, p. 2-29. *In* International Field Institute guidebook to the Caribbean island-arc system. Am. Geol. Inst. (Jamaica section).

Roe, F.G. 1970. The North American buffalo. 2nd ed. Toronto, University of Toronto Press.

Romer, A.S. 1968. Vertebrate paleontology. 3rd ed. Chicago, Chicago University Press.

Romer, A.S. 1975. Intercontinental correlations of Triassic Gondwanaland vertebrate faunas, p. 469-474. *In* K.S.W. Campbell (ed.), Gondwana geology. Canberra, Australian National University Press.

Rorabacker, J.A. 1970. The American buffalo in transition: a historical and economic survey of the bison in America. Saint Cloud, Minn., North Star Press.

Rosen, D.E. 1974. Review of Croizat, L. 1964. Space, time, form: the biological synthesis. Syst. Zool. **23**:288-290.

Rosen, D.E. 1975. A vicariance model of Caribbean biogeography. Syst. Zool. **24**:431-464.

Rosen, D. 1978. Vicariant patterns and historical explanations in biogeography. Syst. Zool. **27**:159-188.

Rosen, D.E. 1979. Fishes from the uplands and intermontane basins of Guatemala: revisionary studies and comparative geography. Bull. Am. Mus. Nat. Hist. **162**: 267-376.

Rosen, D.E., and P.H. Greenwood. 1976. A fourth neotropical species of synbranchid eel and the phylogeny and systematics of synbranchiform fishes. Bull. Am. Mus. Nat. Hist. **157**:1-69.

Rosen, D.E., G. Nelson, and C. Patterson. 1979. Foreword, p. vii-xiii. *In* W. Hennig, Phylogenetic systematics, 3rd ed. Tr. by D.D. Davis and R. Zangerl. Urbana, Ill., University of Illinois Press.

Rosenberg, N.J. 1974. Microclimate: the biological environment. New York, Wiley-Interscience.

Rosenkrantz, R.D. 1976. Inference, method, and decision. Towards a Bayesian philosophy of science. Dordrecht, Netherlands, D. Reidel Publishing Co.

Rosenzweig, M.L. 1968. Net primary productivity of terrestrial communities: predictions from climatological data. Am. Nat. **102**:67-74.

Rosenzweig, M.L. 1973. Habitat selection experiments with a pair of coexisting heteromyid rodent species. Ecology **54**:111-117.

Rosenzweig, M.L. 1975. On continental steady states of species diversity, p. 121-141. *In* M.L. Cody and J.M. Diamond (eds.), Ecology and evolution in communities. Cambridge, Mass., Belknap Press.

Rosenzweig, M.L. 1978. Competitive speciation. Biol. J. Linn. Soc. **10**:275-289.

Rosenzweig, M.L., and R.H. MacArthur. 1963. Graphical representation and stability conditions of predator-prey interactions. Am. Nat. **97**:209-223.

Rosenzweig, M.L., and J. Winakur. 1969. Population ecology of desert rodent communities: habitats and environmental complexity. Ecology **50**:558-572.

Rosewater, J. 1965. The family Tridacnidae in the Indo-Pacific. Indo-Pacific Mollusca **1**(6):347-396.

Ross, H.H. 1958. Affinities and origins of the northern and montane insects of western North America, p.231-252. *In* C.L.. Hubbs (ed.), Zoogeography. Am. Assoc. Advancement Sci. Pub. 51.

Roughgarden, J. 1974. Niche width: biogeographic patterns among *Anolis* lizard populations. Am. Nat. **108**: 429-442.

Roughgarden, J., and E.R. Fuentes. 1977. The environmental determinants of size in solitary populations of West Indian *Anolis* lizards. Oikos **29**:44-51.

Roughgarden, J., D. Heckel, and E.R. Fuentes. 1983. How coevolutionary theory explains the biogeography and community structure of the *Anolis* lizard communities in the Lesser Antilles. *In* R.B. Huey, E.R. Pianka, and T.W. Schoener (eds.), Lizard ecology: studies of a model organism. Cambridge, Mass., Belknap Press (in press).

Rubinoff, I. 1968. Central American sea-level canal: possible biological effects. Science **161**:857-861.

Rubinoff, R.W., and I. Rubinoff. 1971. Geographic and reproductive isolation in Atlantic and Pacific populations of Panamanian *Bathygobius*. Evolution **25**:88-97.

Runcorn, S.K. 1956. Paleomagnetic comparisons between Europe and North America. Proc. Geol. Assoc. Canada **8**:77-85.

Runcorn, S.K. 1962. Paleomagnetic evidence for continental drift and its geophysical cause, p. 1-40. *In* S.K. Runcorn (ed.), Continental drift. Internatl. Geophys. Ser. 3. New York, Academic Press, Inc.

Russell, S.M. 1964. A distributional study of the birds of British Honduras. Ornithol. Monogr., Am. Orithol. Union **1**:1-195.

Rusterholz, K.A., and R.W. Howe. 1979. Species-area relations of birds on small islands in a Minnesota lake. Evolution **33**:468-477.

Ryther, J.H. 1969. Photosythesis and fish production in the sea. Science **166**:72-76.

Rzedowski, J. 1975. An ecological and phytogeographical analysis of the grasslands of Mexico. Taxon **24**(1): 67-80.

Sabins, F.F., Jr. 1978. Remote sensing: principles and interpretation. San Francisco, W.H. Freeman & Co., Inc.

Sanders, H.L. 1968. Marine benthic diversity: a comparative study. Am. Nat. **102**:243-282.

Sanders, H.L. 1969. Benthic marine diversity and the stability-time hypothesis. Brookhaven Symp. Biol. **22**:71-81.

Sauer, J. 1969. Oceanic islands and biogeographic theory: a review. Geogr. Rev. **59**:582-593.

Saunders, P.M. 1971. Anticyclonic eddies formed from shoreward meander of the Gulf Stream. Deep-Sea Res. **18**:1207-1220.

Savage, D.E. 1958. Evidence from fossil land mammals on the origin and affinities of the western Nearctic fauna, p. 97-129. *In* C.L. Hubbs (ed.), Zoogeography. Washington, D.C., Am. Assoc. Advancement Sci. Pub. 51.

Savage, J.M. 1966. The origin and history of the Central American herpetofauna. Copeia (4):719-766.

Savage, J.M. 1973. The geographic distribution of frogs: patterns and predictions, p. 351-445. *In* J.L. Vial (ed.), Evolutionary biology of the anurans. Columbia, Mo., University of Missouri Press.

Savile, D.B.O. 1956. Known dispersal rates and migratory potentials as clues to the origin of North American biota. Am. Midl. Nat. **56**:434-453.

Savilov, A.I. 1961. The distribution of the ecological forms of the by-the-wind sailor, *Velella lata,* Ch. and Eys., and the Portuguese man-of-war *Physalia utriculus* (La Martiniere) Esch., in the North Pacific. Tr. Inst. Okeanol., Akad. Nauk SSSR **45**:223-239. (in Russian).

Schaal, B.A., and D.A. Levin. 1978. Morphological differentiation and neighborhood size in *Liatris cylindracea.* Am. J. Bot. **65**:923-928.

Schall, J.J., and E.R. Pianka. 1978. Geographical trends in numbers of species. Science **201**:679-686.

Schanda, E. (ed.). 1976. Remote sensing for environmental sciences. Berlin, Springer-Verlag.

Scheltema, R.S. 1971. Larval dispersal as a means of genetic exchange between geographically separated populations of shallow-water benthic marine gastropods. Biol. Bull. Woods Hole no. **140**:284-322.

Scheltema, R.S. 1977. On the relationship between dispersal of pelagic veliger larvae and the evolution of marine prosobranch gastropods, p. 303-322. *In* B. Battaglia and J.A. Beardmore (eds.), Marine organisms. Genetics, ecology, and evolution. New York, Plenum Press.

Schimper, A.F.W. 1898. Pflanzengeographie auf physiologischer Grundlage. 1st ed. Jena.

Schimper, A.F.W. 1903. Plant-geography upon a physiological basis. Tr. by W.R. Fisher. Oxford, Clarendon Press.

Schlinger, E.I. 1974. Continental drift, *Nothofagus,* and some ecologically associated insects. Ann. Rev. Entomol. **19**:323-343.

Schmidt, K.P. 1954. Faunal realms, regions, and provinces. Quart. Rev. Biol. **29**:322-331.

Schmidt-Koenig, K. 1975. Migration and homing in animals. Berlin, Springer-Verlag.

Schmidt-Nielsen, K. 1975. Animal physiology. Adaptation and environment. London, Cambridge University Press.

Schmincke, H-U. 1976. The geology of the Canary Islands, p. 67-184. *In* G. Kunkel (ed.), Biogeography and ecology in the Canary Islands. Monographiae Biologicae 30. The Hague, Dr. W. Junk BV, Publishers.

Schnell, R. 1970-1971. Introduction a la phytogeographie des pays tropicaux. 2 vol. Paris, Société Gauthier-Villars.

Schoener, A. 1974a. Experimental zoogeography: colonization of marine mini-islands, Am. Nat. **108**:715-738.

Schoener, A. 1974b. Colonization curves for planar marine islands. Ecology **55**:818-827.

Schoener, T.W. 1967. The ecological significance of sexual dimorphism in size in the lizard *Anolis conspersus*. Science **155**:474-477.

Schoener, T.W. 1968a. The *Anolis* lizards of Bimini: resource partitioning in a complex fauna. Ecology **49**:704-726.

Schoener, T.W. 1968b. Sizes of feeding territories among birds. Ecology **49**:123-141.

Schoener, T.W. 1969a. Models of optimal size for solitary predators. Am. Nat. **103**:277-313.

Schoener, T.W. 1969b. Size patterns in West Indian *Anolis* lizards. I. Size and species diversity. Syst. Zool. **18**: 386-401.

Schoener, T.W. 1970. Size patterns in West Indian *Anolis* lizards. II. Correlations with the sizes of particular sympatric species-displacement and divergence. Am. Nat. **104**:155-174.

Schoener, T.W. 1974a. The species-area relationship within archipelagos: models and evidence from island land birds, p. 629-642. *In* H.J. Firth and J.H. Calaby (eds.), Proceedings of the 16th International Ornithological Congress. Canberra, Australian Academy of Science.

Schoener, T.W. 1974b. Resource partitioning in ecological communities. Science **185**:27-39.

Schoener, T.W. 1975. Presence and absence of habitat shift in some widespread lizard species. Ecol. Monogr. **45**:233-258.

Schoener, T.W., and G.C. Gorman. 1968. Some niche differences in three Lesser Antillean lizards of the genus *Anolis*. Ecology **49**:819-830.

Schofield, W.B., and H.A. Crum. 1972. Disjunctions in bryophytes. Ann. Missouri Bot. Gard. **59**:174-202.

Schopf, J.M. 1970a. Gondwana paleobotany. Antarctic J. U.S. **5**:62-66.

Schopf, J.M. 1970b. Relation of floras of the Southern Hemisphere to continental drift. Taxon **19**:657-674.

Schopf, J.M. 1975. Precambrian paleobiology: problems and perspectives. Ann. Rev. Earth Planet. Sci. **3**:213-249.

Schopf, J.M. 1976. Morphologic interpretation of fertile structures in glossopterid gymnosperms. Rev. Palaeobot. Palynol. **21**:25-64.

Schopf, T.J.M. (ed.). 1972. Models in paleobiology. San Francisco, Freeman, Cooper & Co.

Schopf, T.J.M. 1974. Permo-Triassic extinctions: relation to sea-floor spreading. J. Geol. **82**:129-144.

Schouw, F. 1823. Grunzüge einer allgemeinen Pflanzengeographie. Berlin.

Schulze, R.E., and O.S. McGee. 1978. Climatic indices and classifications in relation to the biogeography of southern Africa, p. 19-52. *In* M.J.A. Werger and A.C. van Bruggen (eds.), Biogeography and ecology of southern Africa, vol. 1. Monographiae Biologicae 31. The Hague, Dr. W. Junk BV, Publishers.

Schuster, R.M. 1969. Problems of antipodal distribution in lower land plants. Taxon **18**:46-91.

Schwarzbach, M. 1980. Alfred Wegener und die Drift der Kontinente. Grosse Naturstorscher, bd. 42. Stuttgart, Wissenschaftliche Verlagsgesellschaft.

Sclater, P.L. 1858. On the general geographical distribution of the members of the class Aves. Zool. J. Linn. Soc. **2**:130-145.

Scott, R.A., P.L. Williams, L.C. Craig, E.S. Barghoorn, L.J. Hickey, and H.D. MacGinitie. 1972. "Pre-Cretaceous" angiosperms from Utah: evidence for Tertiary age of the palm woods. Am. J. Bot. **59**:886-896.

Scrutton, R.A., and R.V. Dingle. 1976. Observations on the processes of sedimentary basin formation at the margins of southern Africa, p. 143-156. *In* M.H.P. Bott (ed.), Sedimentary basins of continental margins and cratons. Tectonophysics 36.

Seddon, B. 1971. Introduction to biogeography. London, Gerald Duckworth & Co., Ltd.

Selander, R.K., and W.E. Johnson. 1973. Genetic variation among vertebrate species. Ann. Rev. Ecol. Syst. **4**: 75-91.

Sellers, W.D. 1965. Physical climatology. Chicago, University of Chicago Press.

Sepkoski, J.J., Jr. 1976. Species diversity in the Phanerozoic: species-area effects. Paleobiology **2**:298-303.

Sepkoski, J.J., Jr., R.K. Bambach, D.M. Raup, and J.W. Valentine. 1981. Phanerozoic marine diversity and the fossil record. Nature **293**:435-437.

Sepkoski, J.J., Jr., and M.A. Rex. 1974. Distribution of freshwater mussels: coastal rivers as biogeographic islands. Syst. Zool. **23**:165-188.

Serventy, D.L. 1960. Geographical distribution of living birds. p. 95-126. *In* A. J. Marshall (ed.), Biology and comparative physiology of birds, vol. 1. New York, Academic Press, Inc.

Shantz, H.L., and B.L. Turner. 1958. Photographic documentation of vegetational changes in Africa over a third of a century. Tucson, Ariz., University of Arizona College of Agriculture Report no. 169.

Sheard, J.W. 1977. Paleogeography, chemistry, and taxonomy of the lichenized ascomycetes *Dimelaena* and *Thamnolia*. The Bryologist **80**:100-118.

Short, L.L. 1970. The affinity of African with neotropical woodpeckers. Am. Mus. Novitates no. 2467.

Shreve, F. 1915. The vegetation of a desert mountain range as conditioned by climatic factors. Washington, D.C., Carnegie Institute Pub. no. 217.

Shreve, F. 1922. Conditions indirectly affecting vertical distribution on desert mountains. Ecology **3**:269-274.

Shreve, F. 1951. Vegetation of the Sonoran Desert. Washington, D.C., Carnegie Institute Pub. no. 591.

Shrode, J.B. 1975. Developmental temperature tolerance of a Death Valley pupfish *(Cyprinodon nevadensis)*. Physiol. Zool. **48**:378-389.

Shuntov, V.P. 1974. Sea birds and the biological structure of the ocean. Tr. by National Technical Information Service. Washington, D.C., U.S. Department of Commerce.

Simberloff, D.S. 1969. Experimental zoogeography of islands. A model for insular colonization. Ecology **50**:296-314.

Simberloff, D.S. 1970. Taxonomic diversity of island biotas. Evolution **24**:23-47.

Simberloff, D.S. 1971. Population sizes of congeneric birds on islands. Am. Nat. **105**:190-193.

Simberloff, D. 1972. Models in biogeography, p. 160-191. *In* T.J.M. Schopf (ed.), Models in paleobiology. San Francisco, Freeman, Cooper & Co.

Simberloff, D.S. 1974a. Equilibrium theory of island biogeography and ecology. Ann. Rev. Ecol. Syst. **5**:161-182.

Simberloff, D.S. 1974b. Permo-Triassic extinctions: effects of area on biotic equilibrium. J. Geol. **82**:267-274.

Simberloff, D.S. 1976a. Experimental zoogeography of islands: effects of island size. Ecology **57**:629-648.

Simberloff, D.S. 1976b. Species turnover and equilibrium island biogeography. Science **194**:572-578.

Simberloff, D.S. 1976c. Trophic structure determination and equilibrium in an arthropod community. Ecology **57**:395-398.

Simberloff, D.S. 1978. Using island biogeographic distributions to determine if colonization is stochastic. Am. Nat. **112**:713-726.

Simberloff, D. 1980. Dynamic equilibrium island biogeography: the second stage, p. 1289-1295. *In* R. Nöhring (ed.), Acta XVII Congressus Internationalis Ornithologici, vol. 2. Berlin, Verlag der Deutschen Ornithologen-Gesellschaft.

Simberloff, D.S., and L.G. Abele. 1976. Island biogeography theory and conservation practice. Science **191**: 285-286.

Simberloff, D., K.L. Heck, E.D. McCoy, and E.F. Conner. 1981a. There have been no statistical tests of cladistic biogeographical hypotheses, p. 40-63. *In* G. Nelson and D.E. Rosen (eds.), Vicariance biogeography: a critique. New York, Columbia University Press.

Simberloff, D., K.L. Heck, E.D. McCoy, and E.F. Conner. 1981b. Response, p. 85-93. *In* G. Nelson and D.E. Rosen (eds.), Vicariance biogeography: a critique. New York, Columbia University Press.

Simberloff, D.S., and E.O. Wilson. 1969. Experimental zoogeography of islands: the colonization of empty islands. Ecology **50**:278-296.

Simberloff, D.S., and E.O. Wilson. 1970. Experimental zoogeography of islands. A two-year record of colonization. Ecology **51**:934-937.

Simmons, I.G. 1979. Biogeography: natural and cultural. London, Edward Arnold (Publishers), Ltd.

Simpson, B.B. 1974. Glacial migrations of plants: island biogeographical evidence. Science **185**:697-700.

Simpson, B.B. 1975. Pleistocene changes in the flora of the high tropical Andes. Paleobiology **1**:273-294.

Simpson, B.B., and J. Haffer. 1978. Speciation patterns in the Amazonian forest biota. Ann. Rev. Ecol. Syst. **9**:497-518.

Simpson, G.G. 1936. Data on the relationships of local and continental mammalian faunas. J. Paleontol. **10**: 410-414.

Simpson, G.G. 1937. Superspecific variation in nature and in classification from the viewpoint of paleontology. Am. Nat. **71**:236-267.

Simpson, G.G. 1940. Mammals and land bridges. J. Washington Acad. Sci. **30**:137-163.

Simpson, G.G. 1943a. Mammals and the nature of continents. Am. J. Sci. **241**:1-31.

Simpson, G.G. 1943b. Turtles and the origin of the fauna of Latin America. Am. J. Sci. **241**:413-429.

Simpson, G.G. 1945. The principles of classification and a classification of mammals. Bull. Am. Mus. Nat. Hist. **85**:i-xvi, 1-350.

Simpson, G.G. 1946. Tertiary land bridges. Trans. New York Acad. Sci., ser. 2. **8**:255-258.

Simpson, G.G. 1947a. Holarctic mammalian faunas and continental relationships during the Cenozoic. Geol. Soc. Am. Bull. **58**:613-688.

Simpson, G.G. 1947b. Evolution, interchange and resemblance of the North American and Eurasian Cenozoic mammalian faunas. Evolution **1**:218-220.

Simpson, G.G. 1950. History of the fauna of Latin America. Am. Sci. **38**:361-389.

Simpson, G.G. 1951. The species concept. Evolution **5**: 285-298.

Simpson, G.G. 1952a. Periodicity in vertebrate evolution. J. Paleontol. **26**:359-370.

Simpson, G.G. 1952b. Probabilities of dispersal in geologic time. Bull. Am. Mus. Nat. Hist. **99**:163-176.

Simpson, G.G. 1953. The major features of evolution. New York, Columbia University Press.

Simpson, G.G. 1956. Zoogeography of West Indian land mammals. Am. Mus. Novitates no. 1759.

Simpson, G.G. 1960. Notes on the measurement of faunal resemblance. Am. J. Sci., Bradley Volume, no. **258-A**: 300-311.

Simpson, G.G. 1961a. Principles of animal taxonomy. New York, Columbia University Press.

Simpson, G.G. 1961b. Historical zoogeography of Australian mammals. Evolution **15**:413-446.

Simpson, G.G. 1964. Species density of North American Recent mammals. Syst. Zool. **13**:57-73.

Simpson, G.G. 1965. The geography of evolution. Philadelphia, Chilton Book Co.

Simpson, G.G. 1969. South American mammals, p. 879-909. *In* E.J. Fittkau, J. Illies, H. Klinge, G.H. Schwabe, and H. Sioli (eds.), Biogeography and ecology in South America, vol. 2. Monographiae Biologicae 20. The Hague, Dr. W. Junk BV, Publishers.

Simpson, G.G. 1969b. The first three billion years of community evolution, p. 162-176. *In* G.M. Woodwell and H.H. Smith (eds.), Diversity and stability in ecological systems. Brookhaven Symp. Biol. 22.

Simpson, G.G. 1970a. Uniformitarianism. An inquiry into principle, theory, and method in geohistory and biohistory, p. 43-96. *In* M.K. Hecht and W.C. Steere (eds.), Essays in evolution and genetics in honor of Theodosius Dobzhansky. New York, Appleton-Century-Crofts.

Simpson, G.G. 1970b. Drift theory: Antarctica and central Asia. Science **170**:678.

Simpson, G.G. 1976a. The compleat paleontologist? Ann. Rev. Earth Planet. Sci. **4**:1-13.

Simpson, G.G. 1976b. Penguins, past and present, here and there. New Haven, Conn., Yale University Press.

Simpson, G.G. 1977. Too many lines; the limits of the Oriental and Australian zoogeographic regions. Proc. Am. Philos. Soc. **121**:107-120.

Simpson, G.G. 1978. Early mammals in South America: fact, controversy, and mystery. Proc. Am. Philos. Soc. **122**:318-328.

Simpson, G.G. 1980a. Why and how. Some problems and methods in historical biology. Oxford, Pergamon Press, Ltd.

Simpson, G.G. 1980b. Splendid isolation; the curious history of mammals in South America. New Haven, Conn., Yale University Press.

Simpson, S., and J. Cracraft. 1981. The phylogenetic relationships of the Piciformes (Class Aves). Auk **98**: 481-494.

Sims, R.W. 1978. Megadrilacea (Oligochaeta), p. 663-676. *In* M.J.A. Werger and A.C. van Bruggen (eds.), Biogeography and ecology of southern Africa, vol. 2. Monographiae Biologicae 31. The Hague, Dr. W. Junk BV, Publishers.

Sinclair, W.A. 1964. Comparisons of recent declines of white ash, oaks and sugar maple in northeastern woodlands. Cornell Plant. **20**:62-67.

Slatkin, M. 1973. Gene flow and selection in a cline. Genetics **75**:733-756.

Slaughter, B.H. 1967. Animal ranges as a cue to Late-Pleistocene extinctions, p. 155-167. *In* P.S. Martin and H.E. Wright, Jr. (eds.), Pleistocene extinctions: the search for a cause. New Haven, Conn., Yale University Press.

Slijper, E.J. 1962. Whales. Tr. A.J. Pomerans. London, Hutchinson Educational.

Slobodkin, L.B., and H.L. Sanders. 1969. On the contribution of environmental predictability to species diversity, p. 82-93. *In* G.M. Woodwell and H.H. Smith (eds.), Diversity and stability in ecological systems. Brookhaven Symp. Biol. 22.

Slud, P. 1976. Geographic and climatic relationships of avifaunas with special reference to comparative distribution in the Neotropics. Smithsonian Contr. Zool. no. 212.

Smith, A.C. 1967. The presence of primitive angiosperms in the Amazon Basin and its significance in indicating migrational routes. Atas Simpos. Biota Amaz. **4**:37-59.

Smith, A.C. 1970. The Pacific as a key to flowering plant history. Honolulu, University of Hawaii Harold L. Lyon Arbor. Lecture no. 1.

Smith, A.C. 1973. Angiosperm evolution and the relationship of the floras of Africa and America, p. 49-61. *In* B.J. Meggers, E.S. Ayensu, and W.D. Duckworth (eds.), Tropical forest ecosystems in Africa and South America: a comparative review. Washington, D.C., Smithsonian Institution Press.

Smith, A.G., and J.C. Briden. 1977. Mesozoic and Cenozoic paleocontinental maps. Cambridge, Cambridge University Press.

Smith, A.G., and A. Hallam. 1970. The fit of the southern continents. Nature **225**:139-144.

Smith, A.T. 1974. The distribution and dispersal of pikas: consequences of insular population structure. Ecology **55**:1112-1119.

Smith, A.T. 1980. Temporal changes in insular populations of the pika *(Ochotona princeps)*. Ecology **61**:8-13.

Smith, B.J. 1979. Survey of non-marine molluscs of south-eastern Australia. Malacologia **18**:103-105.

Smith, G.R. 1978. Biogeography of intermountain fishes. Great Basin Nat. Mem. **2**:17-42.

Smith, G.R. 1981. Late Cenozoic freshwater fishes of North America. Ann. Rev. Ecol. Syst. **12**:163-193.

Smith, M.H., and J.T. McGinnis. 1968. Relationships of latitude, altitude, and body size to litter size and mean annual production of offspring in *Peromyscus*. Res. Pop. Ecol. **10**:115-126.

Smith, R.L. 1974. Ecology and field biology. 2nd ed. New York, Harper & Row, Publishers, Inc.

Sneath, P.H.A., and R.R. Sokal. 1973. Numerical taxonomy. San Francisco, W.H. Freeman & Co., Publishers.

Snider-Pelligrini, A. 1859. La Création et ses mystères dévoilées. Paris, A. Franck.

Soil Conservation Service. 1975. Soil taxonomy. Agricultural Handbook no. 436. Washington, D.C., U.S. Department of Agriculture, Soil Conservation Service.

Sokal, R.R. 1975. Mayr on cladism: and his critics. Syst. Zool. **24**:257-262.

Solbrig, O.T. 1972. The floristic disjunctions between the "Monte" in Argentina and the "Sonoran Desert" in Mexico and the United States. Ann. Missouri Bot. Gard. **59**:218-223.

Solbrig, O.T. (ed.). 1980. Demography and evolution in plant populations. Botanical Monographs, vol. 15. Berkeley, Calif., University of California Press.

Solbrig, O.T., S. Jain, G.B. Johnson, and P.H. Raven. (eds.). 1979. Topics in plant population biology. New York, Columbia University Press.

Solem, A. 1959. Systematics of the land and fresh-water Mollusca of the New Hebrides. Fieldiana, Zool. **43**: 1-359.

Solem, A. 1973. Island size and species diversity in Pacific land snails. Malacologia **14**:397-400.

Solem, A. 1979a. A theory of land snail biogeographic patterns through time, p. 225-249. *In* S. van der Spoel, A.C. van Bruggen, and J. Lever (eds.), Pathways in malacology. Amsterdam, Sixth European Malacological Congress.

Solem, A. 1979b. Biogeographic significance of land snails, Paleozoic to Recent, p. 277-287. *In* Gray and A.J. Boucot (eds.), Historical biogeography, plate tectonics, and the changing environment. Corvallis, Ore., Oregon State University Press.

Solem, A. 1981. Land-snail biogeography: a true snail's pace of change, p. 197-221. *In* G. Nelson and D.E. Rosen (eds.), Vicariance biogeography: a critique. New York, Columbia University Press.

Sondaar, P.Y. 1971. Paleozoogeography of the Pleistocene mammals from the Aegean, p. 65-69. *In* A. Strid (ed.), Evolution in the Aegean. Opera Botanica 30. Lund, Sweden, C.W.K. Gleerup Bokförlag.

Sondaar, P.Y. 1977. Insularity and its effect on mammal evolution, p. 671-707. *In* M.K. Hecht, P.C. Goody, and B.M. Hecht (eds.), Major patterns of vertebrate evolution. New York, Plenum Press.

Sondaar, P.Y., and G.J. Boekschoten. 1967. Quaternary mammals in the South Aegean Island Arc; with notes on other fossil mammals from the coastal regions of the Mediterranean: I. Proc. Kon. Nede. Akad. Wetensch., ser. B, Phys. Sci. 70(5):556-564.

Soulé, M. 1966. Trends in the insular radiation of a lizard. Am. Nat. **100**:47-64.

Soulé, M. 1972. Phenetics of natural populations. III. Variation in insular populations of a lizard. Am. Nat. **106**:429-446.

Soumalainen, E. 1962. Significance of parthenogenesis in the evolution of insects. Ann. Rev. Entomol. **7**:349-366.

Southern California Coastal Water Research Project. 1973. The ecology of the Southern California bight: implications for water quality management. Tech. Report **104**:1-531. El Segundo, Calif.

Southwood, T.R.E. 1962. Migration of terrestrial arthropods in relation to habitat. Biol. Rev. **37**:171-214.

Souza, W.P., S.C. Schroeter, and S.D. Gaines. 1981. Latitudinal variation in intertidal algal community structure. Oecologia **48**(3):297-303.

Sowunmi, M.A. 1981. Aspects of Late Quaternary vegetational changes in West Africa. J. Biogeogr. **8**:457-474.

Spencer, A.W., and H.W. Steinhoff. 1968. An explanation of geographical variation in litter size. J. Mammal. **49**:281-286.

Sprugel, D.G., and F.H. Borman. 1981. Natural disturbance and the steady state in high-altitude fir forests. Science **211**:390-393.

Stanley, S.M. 1975. A theory of evolution above the species level. Proc. Natl. Acad. Sci. U.S.A. **72**:646-650.

Stanley, S.M. 1979. Macroevolution. Pattern and process. San Francisco, W.H. Freeman & Co., Publishers.

Stanley, S.M., R.W. Signor, III, S. Lidgard, and A.F. Karr. 1981. Natural clades differ from "random" clades: simulations and analyses. Paleobiology **7**:115-127.

Starmer, W.T., W.B. Heed, and E.S. Rockwood-Sluss. 1977. Extension of longevity in *Drosophila mojavensis* by environmental ethanol: differences between subraces. Proc. Natl. Acad. Sci. U.S.A. **74**:387-391.

Starmer, W.T., H.W. Kircher, and H.J. Phaff. 1980. Evolution and speciation of host-plant specific yeasts. Evolution **34**:137-146.

Starmühlner, F. 1979. Distribution of freshwater molluscs in mountain streams of tropical Indo-Pacific islands (Madagascar, Ceylon, New Caledonia). Malacologia **18**:245-255.

Stearn, C.W., R.L. Carroll, and T.H. Clark. 1979. Geological evolution of North America. 3rd ed. New York, John Wiley & Sons, Inc.

Stebbins, G.L. 1950. Variation and evolution in plants. New York, Columbia University Press.

Stebbins, G.L. 1971a. Adaptive radiation of reproductive characteristics in angiosperms. 2. Seeds and seedlings. Ann. Rev. Ecol. Syst. **2**:237-260.

Stebbins, G.L. 1971b. Chromosomal evolution in higher plants. London, Edward Arnold.

Stebbins, G.L. 1974. Flowering plants. Evolution above the species level. Cambridge, Mass., Belknap Press.

Stebbins, R.C. 1949. Speciation in salamanders of the plethodontid genus *Ensatina*. Berkeley, Calif., University of California Press.

Steele, J.H. 1974. The structure of marine ecosystems. Cambridge, Mass., Harvard University Press.

Steenbergh, W.F., and C.H. Lowe. 1976. Ecology of the saguaro. I. The role of freezing weather in a warm-desert plant population, p. 49-92. *In* Research in the parks. Natl. Park Serv. Symp., ser. no. 1. Washington, D.C., U.S. Government Printing Office.

Steenbergh, W.F., and C.H. Lowe. 1977. Ecology of the saguaro. II. Reproduction, germination, establishment, growth, and survival of the young plant. Natl. Park Serv. Sci. Monogr., ser. no. 8. Washington, D.C., U.S. Government Printing Office.

Steenis, G.G.G.J. van. 1962. The land-bridge theory in botany. Blumea **11**:235-372.

Stehli, F.G. 1968. Taxonomic diversity gradients in pole locations: the recent model, p. 163-227. *In* E.T. Drake (ed.), Evolution and environment. Peabody Museum Centennial Symposium. New Haven, Conn., Yale University Press.

Stehli, F.G., R.G. Douglas, and N.D. Newell. 1969. Generation and maintenance of gradients in taxonomic diversity. Science **164**:947-949.

Stehli, F.G., and J.W. Wells. 1971. Diversity and age patterns in hermatypic corals. Syst. Zool. **20**:115-126.

Sterba, G. 1966. Freshwater fishes of the world. London, Studio Vista.

Stoddart, D.R. 1981. Biogeography: dispersal and drift. Prog. Phys. Geogr. **5**:575-590.

Stone, D.B. 1977. Plate tectonics, paleomagnetism and the tectonic history of the N.E. Pacific. Geophys. Surveys **3**:3-37.

Stone, D.B., and D.R. Packer. 1977. Tectonic implications of Alaska Peninsula paleomagnetic data, p. 183-201. *In* S. Uyeda (ed.), Subduction zones, midocean ridges, ocean trenches and geodynamics. Tectonophysics 37.

Stout, J., and J. Vandermeer. 1975. Comparison of species richness for stream-inhabiting insects in tropical and mid-latitude streams. Am. Nat. **109**:263-280.

Strahler, A.N. 1975. Physical geography. 4th ed. New York, John Wiley & Sons, Inc.

Strahler, A.N., and A.H. Strahler. 1978. Modern physical geography. New York, John Wiley & Sons, Inc.

Strain, B.R., and W.D. Billings (eds.). 1974. Vegetation and environment, part 6. The Hague, Dr. W. Junk BV, Publishers.

Street, F.A., 1981. Tropical palaeoenvironments. Prog. Phys. Geogr. **5**:157-185.

Strong, D.R., Jr. 1974. Rapid asymptotic species accumulation in phytophagous insect communities: the pests of cacao. Science **185**:1064-1066.

Strong, D.R., Jr. 1977. Epiphyte loads, treefalls, and perennial disruption: a mechanism for maintaining higher tree species richness in the tropics without animals. J. Biogeogr. **4**:215-218.

Strong, D.R., Jr. 1980. Null hypotheses in ecology. Synthese **43**:271-285.

Strong, D.R., Jr., E.D. McCoy, and J.R. Rey. 1977. Time and the number of herbivore species: the pests of sugarcane. Ecology **58**:167-175.

Strong, D.R., Jr., and D. Simberloff, 1981. Straining at gnats and swallowing ratios: character displacement. Evolution **35**:810-812.

Strong, D.R., Jr., L.A. Szyska, and D. Simberloff. 1979. Tests of community-wide character displacement against null hypotheses. Evolution **33**:897-913.

Stuessy, T.F. 1977. Heliantheae–systematic review, p. 621-671. *In* V.H. Heywood, J.B. Harborne, and B.L. Turner (ed.), The biology and chemistry of the Compositae, vol. 2. London, Academic Press, Ltd.

Sugihara, G. 1981. $S = CA^z$, $z = 1/4$: a reply to Connor and McCoy. Am. Nat. **117**:790-793.

Sullivan, W. 1974. Continents in motion. The new earth debate. New York, McGraw-Hill Book Co.

Sutton, G.H., M.H. Manghnani, R. Moberly, and E.U. McAfee. 1976. The geophysics of the Pacific Ocean basin and its margin. Washington, D.C., Am. Geophys. Union Monogr. 19.

Swain, P.H., and S.M. Davis. 1978. Remote sensing: the quantitative approach. New York, McGraw-Hill Book Co.

Swainson, W. 1835. A treatise on the geography and classification of animals. London, Longman, Green.

Symposium. 1963. The seventh approximation. A symposium. Soil Sci. Soc. Am. Proc. **27**:212-228.

Szafer, W. 1975. General plant geography. Warsaw, Pánstwowe Wydawnictwo Naukowe.

Szalay, F.S. (ed.). 1975. Approaches to primate paleobiology. Basel, S. Karger AG, Medical and Scientific Publishers.

Szalay, F.S. 1977. Ancestors, descendents, sister groups and testing of phylogenetic hypotheses. Syst. Zool. **26**:1-11.

Tait, R.V. 1981. Elements of marine ecology. London, Butterworth & Co. Publishers, Ltd.

Takhtajan, A. 1969. Flowering plants: origin and dispersal. Tr. from Russian by C. Jeffrey. Edinburgh, Oliver & Boyd.

Takhtajan, A. 1978. The floristic regions of the world. (In Russian.) Leningrad, Nauka.

Takhtajan, A.L. 1980. Outline of the classification of flowering plants (Magnoliophyta). Bot. Rev. **46**:225-359.

Tansley, A.G. 1920. The classification of vegetation and the concept of development. Ecology **8**:118-149.

Tarling, D.H. 1962. Tentative correlation of Samoan and Hawaiian Islands using "reversals" of magnetism. Nature **196**:882-883.

Tarling, D.H. 1972. Another Gondwanaland. Nature **238**:92-93.

Tarling, D.H., and S.K. Runcorn (eds.). 1973. Implications of continental drift to the earth sciences. 2 vol. New York, Academic Press, Inc.

Tarling, D.H., and M.P. Tarling. 1975. Continental drift. 2nd ed. New York, Doubleday Publishing Co.

Taylor, B.W. 1954. An example of long-distance dispersal. Ecology **35**:369-372.

Taylor, F.B. 1910. Bearing of the Tertiary mountain belt on the origin of the earth's plan. Geol. Soc. Am. Bull. **21**:179-226.

Taylor, J.D., and C.N. Taylor. 1977. Latitudinal distribution of predatory gastropods on the eastern Atlantic shelf. J. Biogeogr. **4**:73-81.

Taylor, L.R. 1978. Bates, Williams, Hutchinson–a variety of diversities, p. 1-18. *In* L.A. Mound and M. Waloff (eds.), Diversity of insect faunas. Symp. Roy. Entomol. Soc. London **9**. Oxford, Blackwell Scientific Publishers, Ltd.

Taylor, R.J., and P.J. Regal. 1978. The peninsular effect on species diversity and the biogegraphy of Baja California. Am. Nat. **112**:583-593.

Teal, J.M. 1957. Community metabolism in a temperate cold spring. Ecol. Monogr. **27**:283-302.

Teal, J.M. 1962. Energy flow in the salt marsh ecosystem of Georgia. Ecology **43**:614-624.

Tedford, R.H. 1974. Marsupials and the new paleogeography, p. 109-126. *In* C.A. Ross (ed.), Paleogeographic provinces and provinciality. Soc. Econ. Paleontol. Mineral. Spec. Pub. no. 21.

Templeton, A.R. 1980a. The theory of speciation *via* the founder principle. Genetics **94**:1011-1038.

Templeton, A.R. 1980b. Modes of speciation and inferences based on genetic distances. Evolution **34**:719-729.

Templeton, A.R. 1981. Mechanisms of speciation: a population genetic approach. Ann. Rev. Ecol. Syst. **12**:23-48.

Terborgh, J. 1971. Distribution on environmental gradients: theory and a preliminary interpretation of distributional patterns in the avifauna of the Cordillera Vilcabamba, Peru. Ecology **52**:23-40.

Terborgh, J. 1973a. Chance, habitat, and dispersal in the distribution of birds in the West Indies. Evolution **27**: 338-349.

Terborgh, J. 1973b. On the notion of favorableness in plant ecology. Am. Nat. **107**:481-501.

Terborgh, J. 1974. Preservation of natural diversity: the problem of extinction prone species. Bioscience **24**:715-722.

Terborgh, J. 1975. Faunal equilibria and the design of wildlife preserves, p. 369-380. *In* F.B. Golley and E. Medina (eds.), Tropical ecological systems, trends in terrestrial and aquatic research. New York, Springer-Verlag, New York, Inc.

Terborgh, J. 1977. Bird species diversity on an Andean elevational gradient. Ecology **58**:1007-1019.

Terborgh, J.W. 1980. The conservation status of neotropical migrants: present and future, p. 21-30. *In* A. Keast and E.S. Morton (eds.), Migrant birds in the Neotropics: ecology, behavior, distribution, and conservation. Washington, D.C., Smithsonian Institution Press.

Terborgh, J. 1981. Discussion, p. 64-68. *In* G. Nelson and D.E. Rosen (eds.), Vicariance biogeography: a critique. New York, Columbia University Press.

Terborgh, J., and J. Faaborg. 1973. Turnover and ecological release in the avifauna of Mona Island, Puerto Rico. Auk **90**:759-779.

Terborgh, J., J. Faaborg, and H.J. Brockmann. 1978. Island colonization by Lesser Antillean birds. Auk **95**:59-72.

Thaler, L. 1973. Nanisme et gigantisme insulaires. Recherche **4**:741-750.

Thompson, R.S., and J.I. Mead. 1982. Late Quaternary environments and biogeography in the Great Basin. Quat. Res. **17**:39-55.

Thomson, D.A., and M.R. Gilligan. 1983. The rocky-shore fishes: marine insular biogeography. *In* T.J. Case and M.L. Cody (eds.), Biogeography of the islands in the Sea of Cortez. Berkeley, University of California Press (in press).

Thomson, D.A., and C.E. Lehner. 1976. Resilence of a rocky intertidal fish community in a physically unstable environment. J. Exp. Mar. Biol. Ecol. **22**:1-29.

Thorne, R.F. 1963. Biotic distribution patterns in the tropical Pacific, p. 311-350. *In* J.L. Gressitt (ed.), Pacific Basin biogeography. Honolulu, Bishop Museum Press.

Thorne, R.F. 1965. Floristic relationships of New Caledonia. Univ. Iowa Stud. Nat. Hist. **20**(7):1-14.

Thorne, R.F. 1968. Synopsis of a putatively phylogenetic classification of the flowering plants. Aliso **6**(4):57-66.

Thorne, R.F. 1969. Floristic relationships between New Caledonia and the Solomon Islands. Philos. Trans. Roy. Soc. (London) ser. B **255**:595-602.

Thorne, R.F. 1972. Major disjunctions in the geographic ranges of seed plants. Quart. Rev. Biol. **47**:365-411.

Thorne, R.F. 1973. Floristic relationships between tropical Africa and tropical America, p. 27-47. *In* B.J. Meggers, E.S. Ayensu, and W.D. Duckworth (eds.), Tropical forest ecosystems in Africa and South America: a comparative review. Washington, D.C., Smithsonian Institute Press.

Thorne, R.F. 1976a. A phylogenetic classification of the Angiospermae, p. 55-106. *In* M.K. Hecht, W.C. Steere, and B. Wallace (eds.), Evolutionary biology 9. New York, Plenum Press.

Thorne, R.F. 1976b. Where and when might the tropical angiospermous flora have originated? Bull. Singapore Bot. Gard. **29**:183-189.

Thorne, R.F. 1978. Plate tectonics and angiosperm distribution. Notes Roy. Bot. Gard. (Edinburgh) **36:** 297-315.

Thorne, R.F. 1981. Phytochemistry and angiosperm phylogeny. A summary statement, p. 233-295. *In* D.A. Young and D.S. Seigler (eds.), Phytochemistry and angiosperm phylogeny. New York, Praeger Publishers.

Thornthwaite, C.W. 1931. The climates of North America, according to a new classification. Geogr. Rev. **21:** 633-655.

Thornthwaite, C.W. 1948. An approach towards a rational classification of climate. Geogr. Rev. **38:**55-94.

Thornthwaite, C.W., and F.K. Hare. 1955. Climatic classification in forestry. Unasylva **9:**50-59.

Thornthwaite, C.W., and J.R. Mather. 1955. The water balance. Drexel Inst. Technol. Pub. Climatol. **8:**1-86.

Thornthwaite, C.W., and J.R. Mather. 1957. Instructions and tables for computing potential evapotranspiration and the water balance. Drexel Inst. Technol. Pub. Climatol. **10:**183-311.

Thorson, G. 1950. Reproductive and larval ecology of marine bottom invertebrates. Biol. Rev. **25:**1-45.

Thorson, T.B. (ed.). 1976. Investigations of the ichthyofauna of Nicaraguan lakes. Lincoln, Nebr., University of Nebraska School of Life Sciences.

Throckmorton, L.H. 1965. Similarity versus relationship in *Drosophila*. Syst. Zool. **14:**221-236.

Throckmorton, L.H. 1968. Concordance and discordance of taxonomic characters in *Drosophila* classification. Syst. Zool. **17:**355-387.

Throckmorton, L.H. 1974. The phylogeny, ecology, and geography of *Drosophila*, p. 421-469. *In* R.C. King (ed.), Handbook of genetics, vol. 3. Invertebrates of genetic interest. New York, Plenum Press.

Thrower, N.J.W., and D.E. Bradbury (eds.). 1977. Chile-California Mediterranean scrub atlas: a comparative analysis. Stroudsburg, Penn., Dowden, Hutchinson & Ross, Inc.

Thunell, R.C. 1979. Eastern Mediterranean Sea during the last glacial maximum: an 18,000-year B.P. reconstruction. Quat. Res. **11:**353-372.

Tidwell, W.D., P.A. Medlyn, and G.F. Thayn. 1972. Fossil palm from the Tertiary Dipping Vat Formation of central Utah. Great Basin Nat. **32:**1-15.

Tidwell, W.D., S.R. Rushforth, J.L. Reveal, and H. Behunin. 1969. Petrified palm wood from the Arapien Shale (Jurassic) of Utah. Geol. Soc. Am. Abstr. part **5:**82-83.

Tivy, J. 1971. Biogeography: a study of plants in the ecosystem. Edinburgh, Oliver & Boyd.

Topp, R.W. 1969. Interoceanic sea-level canal: effects on the fish faunas. Science **165:**1324-1327.

Traband, L., and J. Lepart. 1980. Diversity and stability in garrique ecosystems after fire. Vegetatio **43:**49-57.

Tramer, E.J. 1974. On latitudinal gradients in avian diversity. Condor **76:**123-130.

Traub, R. 1980. The zoogeography and evolution of some fleas, lice and mammals, p. 94-172. *In* R. Traub and H. Starche (eds.), Fleas. Proceedings of the 1977 International Conference on Fleas, Ashton Wold. Rotterdam, A.A. Balkema.

Trewartha, G.T., A.H. Robinson, and E.H. Hammond. 1976. Fundamentals of physical geography. 3rd ed. New York, McGraw-Hill Book Co.

Turner, J.R.G. 1977. Butterfly mimicry: the genetical evolution of an adaptation, p. 163-206. *In* M.C. Hecht, W.C. Steere, and B. Wallace (eds.), Evolutionary biology 10. New York, Plenum Press.

Turner, J.R.G. 1981. Adaptation and evolution in *Heliconius:* a defense of neoDarwinism. Ann. Rev. Ecol. Syst. **12:**99-121.

Turner, J.T. 1980. Latitudinal patterns of a calanoid and cyclopoid copepod diversity in estuarine waters of eastern North America. J. Biogeogr. **8:**369-382.

Udvardy, M.D.F. 1969. Dynamic zoogeography. New York, Van Nostrand Reinhold Co.

Underwood, G. 1976. A systematic analysis of boid snakes, p. 151-175. *In* A. d'A. Bellairs and C.B. Cox (eds.), Morphology and biology of reptiles. Linn. Soc. Symp. no. 3. London, Academic Press, Ltd.

Urquhart, F.A. 1960. The monarch butterfly. Toronto, University of Toronto Press.

U.S. Department of Agriculture. 1941. Climate and man. Washington, D.C., U.S. Government Printing Office.

Uyeda, S. 1978. The new view of the earth. Moving continents and moving oceans. San Francisco, W.H. Freeman & Co., Publishers.

Uyeda, S., R.W. Murphy, and K. Kobayashi (eds.). 1979. Geodynamics of the western Pacific. J. Phys. Earth Suppl. Tokyo, Center for Academic Publications, Japan Scientific Societies Press.

Vagvolgyi, J. 1975. Body size, aerial dispersal, and origin of the Pacific land snail fauna. Syst. Zool. **24:**465-488.

Valentine, D.H. (ed.). 1972. Taxonomy, phytogeography and evolution. New York, Academic Press, Inc.

Valentine, J.W. 1966. Numerical analysis of marine molluscan ranges on the extratropical northeastern Pacific shelf. Limnol. Oceanogr. **11:**198-211.

Valentine, J.W. 1969. Niche diversity, niche size patterns in marine fossils. J. Paleont. **43:**905-915.

Valentine, J.W. 1973. Evolutionary paleoecology of the marine biosphere. Englewood Cliffs, N.J., Prentice-Hall, Inc.

Valentine, J.W., and D. Jablonski. 1982. Larval strategies and patterns of brachiopod diversity in space and time. Geol. Soc. Am. Abstr. **14:**241.

Vance, R.R. 1973a. On reproductive strategies in marine benthic invertebrates. Am. Nat. **107:**339-352.

Vance, R.R. 1973b. More on reproductive strategies in marine benthic invertebrates. Am. Nat. **107**:353-361.

Vandermeer, J.H. 1972. Niche theory. Ann. Rev. Ecol. Syst. **3**:107-132.

Van der Spoel, S., and A.C. Pierrot-Bults. 1979. Zoogeography and diversity of plankton. New York, Halsted Press.

Van Devender, T.R. 1977. Holocene woodlands in the southwestern deserts. Science **198**:189-192.

Van Devender, T.R., and W.G. Spaulding. 1979. Development of vegetation and climate in the southwestern United States. Science **204**:701-710.

Van Valen, L. 1965. Morphological variation and width of the ecological niche. Am. Nat. **99**:377-390.

Van Valen, L. 1970. Late Pleistocene extinctions, p. 469-485. *In* Proceedings of the North American Pelontologist Conference. Lawrence, Kans., Allen Press, Inc.

Van Valen, L. 1973a. Body size and the number of plants and animals. Evolution **27**:27-35.

Van Valen, L. 1973b. A new evolutionary law. Evol. Theor. **1**:1-33.

Van Valen, L., and R.R. Grant. 1970. Variation and niche width reexamined. Am. Nat. **104**:589-590.

Van Valen, L., and R.E. Sloan. 1966. The extinction of the multituberculates. Syst. Zool. **15**:261-278.

van Waterschoot van der Gracht, W.A.J.M. (ed.). 1928. Theory of continental drift: a symposium on the origin and movement of land masses both inter-continental and intra-continental, as proposed by Alfred Wegener. New York, American Association of Petroleum Geologists.

Vanzolini, P.E., and E.E. Williams. 1970. South American anoles: the geographic differentiation and evolution of the *Anolis chrysolepis* species group (Sauria, Iguanidae). Arq. Zool. (São Paulo) **19**:1-298.

Varley, G.C. 1970. The concept of energy flow applied to a woodland community, p. 389-405. *In* A. Watson (ed.), Animal populations in relation to their food resources. London, Blackwell Scientific Publications, Ltd.

Vaughan, T.A. 1978. Mammalogy. 2nd ed. Philadelphia, W.B. Saunders Co.

Vaughan, T.A., and R.M. Hansen. 1964. Experiments on interspecific competition between two species of pocket gophers. Am. Midl. Nat. **72**:444-452.

Vaurie, C. 1965. The birds of the Palearctic fauna. 2 vol. London, H.F. & G. Witherby, Ltd.

Vaurie, C. 1980. Taxonomy and geographical distribution of the Furnariidae (Aves, Passeriformes). Bull. Am. Mus. Nat. Hist. **166**:5-357.

Vawter, A.T., E. Rosenblatt, and G.C. Gorman. 1980. Genetic divergence among fishes of the eastern Pacific and Caribbean: support for the molecular clock. Evolution **34**:705-711.

Vereshchagin, N.K. 1967. The mammals of the Caucasus: a history of the evolution of the fauna. Jerusalem, Israel Program for Scientific Translations.

Vermeij. G.J. 1974. Marine faunal dominance and molluscan shell form. Evolution **28**:656-664.

Vermeij, G.J. 1978. Biogeography and adaptation. Patterns of marine life. Cambridge, Mass., Harvard University Press.

Vester, H. 1940. Die Areale und Arealtypen der Angiospermen-Familien. Bot. Arch. **41**:203-275, 295-356, 520-577.

Vial, J.L. (ed.). 1973. Evolutionary biology of the anurans: comtemporary research on major problems. Columbia, Mo., University of Missouri Press.

Vine, F.J. 1966. Spreading of ocean floor: new evidence. Science **154**:1405-1415.

Vine, F.J., and D.H. Matthews. 1963. Magnetic anomalies over a young oceanic ridge. Nature **199**:947-949.

Vine, F.J., and J.T. Wilson. 1965. Magnetic anomalies over a young oceanic ridge off Vancouver Island. Science **150**:485-489.

Vinogradova, N.G. 1959. The zoogeographical distribution of the deep-water bottom fauna in the abyssal zone of the ocean. Deep-Sea Res. **5**:205-208.

Vinogradova, N.G. 1962. Vertical zonation in the distribution of deep-sea benthic fauna in the ocean. Deep-Sea Res. **8**:245-250.

Voris, H.K. 1977. A phylogeny of the sea snakes (Hydrophiidae). Fieldiana, Zool. 70(4):79-166.

Vuilleumier, B.S. 1971. Pleistocene changes in the fauna and flora of South America. Science **173**:771-780.

Vuilleumier, F. 1970. Insular biogeography in continental species. I. The Northern Andes of South America. Am. Nat. **104**:373-388.

Vuilleumier, F. 1973. Insular biogeography in continental regions. II. Cave faunas from Tesin, southern Switzerland. Syst. Zool. **22**:64-76.

Vuilleumier, F. 1975. Zoogeography, p. 421-496. *In* D. Farner and J.R. King (eds.), Avian biology, vol. 5. New York, Academic Press, Inc.

Vuilleumier, F., and D. Simberloff. 1980. Ecology versus history as determinants of patchy and insular distributions in high Andean birds, p. 235-379. *In* M.K. Hecht, W.C. Steere, and B. Wallace (eds.), Evolutionary biology 12. New York, Plenum Press.

Wagner, W.H., Jr. 1972. Disjunctions in homosporous vascular plants. Ann. Missouri Bot. Gard. **59**:203-217.

Wahrman, R., R. Goitein, and E. Nevo. 1969. Mole rat *Spalax:* evolutionary significance of chromosome variation. Science **164**:82-84.

Wake, D.B. 1966. Comparative osteology and evolution of the lungless salamanders, family Plethodontidae. Mem. Southern California Acad. Sci. **4**:1-111.

Wake, D.B. 1970. The abundance and diversity of tropical salamanders. Am. Nat. **104**:211-213.

Walker, D. (ed.). 1973. Bridge and barrier: the natural and cultural history of Torres Strait. Canberra, Research School of Pacific Sciences, Australian National University.

Walker, D., and R.G. West (eds.). 1970. Studies in the vegetational history of the British Isles. Cambridge, Cambridge University Press.

Walker, E.P. 1975. Mammals of the world. 3rd ed. 2 vol. Baltimore, Md., Johns Hopkins University Press.

Walker, T.J. 1962. Whale primer, with special attention to the California gray whale. Cabrillo Historical Association.

Wallace, A.R. 1860. On the zoological geography of the Malay Archipelago. J. Linn. Soc. London **4**:172-184.

Wallace, A.R. 1869. The Malay Archipelago: the land of the orangutan, and the bird of paradise. New York, Harper.

Wallace, A.R. 1876. The geographical distribution of animals. 2 vol. London, Macmillan.

Wallace, A.R. 1878. Tropical nature and other essays. New York, Macmillan.

Wallace, A.R. 1880. Island life, or the phenomena and causes of insular faunas and floras. London, Macmillan.

Wallace, B. 1960. The influence of genetic systems on geographical distribution. Cold Spring Harbor Symp. Quant. Biol. **24**:193-204.

Wallwork, J.A. 1976. The distribution and diversity of soil fauna. London, Academic Press, Ltd.

Waloff, Z. 1966. The upsurges and recessions of the desert locust plague: an historical survey. Anti-Locust Mem. **8**:1-111.

Walter, H. 1971. Ecology of tropical and subtropical vegetation. Tr. by D. Mueller-Dombois. Edinburgh, Oliver & Boyd.

Walter, H. 1973. Vegetation of the earth in relation to climate and the eco-physiological conditions. New York, Springer-Verlag New York, Inc.

Walter, H. 1979. Vegetation of the earth and ecological systems of the geo-biosphere. 2nd ed. Berlin, Springer-Verlag.

Walter, H., E. Harnickell, and D. Mueller-Dombois. 1975. Climate-diagram maps of the individual continents and the ecological climate regions of the earth. English ed. New York, Springer-Verlag New York, Inc.

Walter, H., and H. Lieth. 1967. Klimmadiagramm-Weltatlas. Jena, East Germany, VEB Gustav Fischer-Verlag.

Warburton, F.E. 1967. The purposes of classification. Syst. Zool. **16**:241-245.

Waring, R.H., and J. Major. 1964. Some vegetation of the California coastal redwood region in relation to gradients of moisture, nutrients, light and temperature. Ecol. Monogr. **34**:167-215.

Warming, E. 1895. Plantesamfund. Copenhagen.

Warming, E. 1909. Oecology of plants: an introduction to the study of plant communities. Oxford, Oxford University Press.

Wassersug, H.Y., J.J. Sepkoski, Jr., and D.M. Raup. 1979. The evolution of body size on islands: a computer simulation. Am. Nat. **114**:287-295.

Watts, A.B., J.K. Weissel, and R.L. Larson. 1977. Sea-floor spreading in marginal basins of the western Pacific. Tectonophysics **37**:167-181.

Watts, D. 1971. Principles of biogeography. New York, McGraw-Hill Book Co.

Watts, W.A. 1980. The Late Quaternary vegetational history of the southeastern United States. Ann. Rev. Ecol. Syst. **11**:387-410.

Weatherley, A.H. (ed.). 1967. Australian inland waters and their fauna. Canberra, Australian National University Press.

Webb, S.D. 1969. Extinction-origination equilibria in Cenozoic land mammals of North America. Evolution **23**:688-702.

Webb, S.D. (ed.). 1974. Pleistocene mammals of Florida. Gainesville, Fla., The University Presses of Florida.

Webb, S.D. 1976. Mammalian faunal dynamics of the great American interchange. Paleobiology **2**:220-234.

Webb, S.D. 1977. A history of savanna vertebrates in the New World, part 1. Ann. Rev. Ecol. Syst. **8**:355-380.

Webb, S.D. 1978. A history of savanna vertebrates in the New World, part 2. South America and the great interchange. Ann. Rev. Ecol. Syst. **9**:393-426.

Weber, F.R., T.D. Hamilton, D.M. Hopkins, C.A. Repenning, and H. Haas. 1981. Canyon Creek: a Late Pleistocene vertebrate locality in interior Alaska. Quat. Res. **16**:167-180.

Webster, D., and M. Webster. 1974. Comparative vertebrate morphology. New York, Academic Press, Inc.

Webster, P.J., and N.A. Streten. 1978. Late Quaternary Ice Age climates of tropical Australasia: interpretations and reconstructions. Quat. Res. **10**:229-309.

Webster, R. 1977. Quantitative and numerical methods in soil classification and survey. Oxford, Clarendon Press.

Wecker, S.C. 1963. The role of early experience in habitat selection by the prairie deer-mouse, *Peromyscus maniculatus bairdi*. Ecol. Monogr. **33**:307-325.

Wecker, S.C. 1964. Habitat selection. Sci. Am. **211**(4): 109-116.

Wegener, A. 1912a. Die Entstehung der Kontinente. Petermanns Geogr. Mitt. **58**:185-195, 253-256, 305-308.

Wegener, A.L. 1912b. Die Entstehung der Kontinente. Geol. Rundsch. **3**:276-292.

Wegener, A. 1915. Die Entstehung der Kontinente und Ozeane. Braunschweig, Vieweg. (Other editions 1920, 1922, 1924, 1929, 1936.)

Wegener, A. 1966. The origin of continents and oceans. Tr. of 1929 edition by J. Biram. New York, Dover Publications, Inc.

Wells, P.V. 1976. Macrofossil analysis of wood rat *(Neotoma)* middens as a key to the Quaternary vegetational history of arid America. Quat. Res. **6**:223-248.

Wells, P.V. 1978. Postglacial origin of the present Chihuahuan Desert less than 11,500 years ago, p. 67-83. *In* R.H. Wauer and D.H. Riskind (eds.), Transactions of the symposium on the biological resources of the Chihuahuan Desert region United States and Mexico, U.S. Department of Interior National Park Service Transactions and Proceedings, ser. no. 3.

Wells, P.V. 1979. An equable glaciopluvial in the West: pleniglacial evidence of increased precipitation on a gradient from the Great Basin to the Sonoran and Chihuahuan Deserts. Quat. Res. **12**:311-325.

Wells, P.V., and R. Berger. 1967. Late Pleistocene history of coniferous woodland in the Mohave Desert. Science **155**:1640-1647.

Wenner, A.D., and D.L. Johnson. 1980. Land vertebrates on the California Channel Islands: sweepstakes or bridges? p. 497-530. *In* D.M. Power (ed.), The California islands: proceedings of a multidisciplinary symposium. Santa Barbara, Calif., Santa Barbara Museum of Natural History.

Werger, M.J.A. 1978. Biogeographical division of southern Africa, p. 145-170. *In* M.J.A. Werger and A.C. van Bruggen (eds.), Biogeography and ecology of southern Africa, vol. 1. Monographiae Biologicae 31. The Hague, Dr. W. Junk BV, Publishers.

Werger, M.J.A., and A.C. van Bruggen (eds.). 1978. Biogeography and ecology of southern Africa. 2 vol. Monographiae Biologicae 31. The Hague, Dr. W. Junk BV, Publishers.

Werner, P.A. 1975. A seed trap for determining patterns of seed deposition in terrestrial plants. Can. J. Bot. **58**: 810-813.

West, R.G. 1977. Pleistocene geology and biology. 2nd ed. London, Longman Group, Ltd.

Westing, A.H. 1966. Sugar maple decline: an evaluation. Econ. Bot. **20**:196-212.

Wetzel, R.G. 1975. Limnology. Philadelphia, W.B. Saunders Co.

Weyl, P.K. 1970. Oceanography. An introduction to the marine environment. New York, John Wiley & Sons, Inc.

Wheeler, A. 1977. The origin and distribution of the freshwater fishes of the British Isles. J. Biogeogr. **4**:1-24.

Whitcomb, R.F. 1977. Island biogeography and "habitat islands" of eastern forests. I. Introduction. Am. Birds **31**:3-5.

Whitcomb, R.F., J.F. Lynch, P.A. Opler, and C.S. Robbins. 1976. Island biogeography and conservation: strategy and limitations. Science **193**:1030-1032.

White, M.J.D. 1968. Models of speciation. Science **159**:1065-1070.

White, M.J.D. 1973. Animal cytology and evolution. 3rd ed. Cambridge, Cambridge University Press.

White, M.J.D. 1978. Modes of speciation. San Francisco, W.H. Freeman & Co., Publishers.

White, T.C.R. 1976. Weather, food, and plagues of locusts. Oecologia **22**:119-134.

Whitehead, D.R., and C.E. Jones. 1969. Small islands and the equilibrium theory of insular biogeography. Evolution **23**:171-179.

Whiteside, M.C., and R.V. Harmsworth. 1967. Species diversity in chydorid (Cladocera) communities. Ecology **48**:664-667.

Whitmore, F.C., Jr., and R.H. Stewart. 1965. Miocene mammals and Central American seaways. Science **148**:180-185.

Whitmore, T.C. 1975. Tropical rain forests of the Far East. Oxford, Clarendon Press.

Whitney, G.G., and S.D. Adams. 1980. Man as a maker of new plant communities. J. Appl. Ecol. **17**:431-448.

Whittaker, R.H. 1956. Vegetation of the Great Smoky Mountains. Ecol. Monogr. **22**:1-44.

Whittaker, R.H. 1960. Vegetation of the Siskiyou Mountains, Oregon and California. Ecol. Monogr. **30**: 279-338.

Whittaker, R.H. 1967. Gradient analysis of vegetation. Biol. Rev. **42**:207-264. Cambridge University Press.

Whittaker, R.H. 1969. Evolution of diversity in plant communities, p. 178-195. *In* G.M. Woodwell and H.H. Smith (eds.), Diversity and stability in ecological systems. Brookhaven Symp. Biol. 22.

Whittaker, R.H. 1975. Communities and ecosystems. 2nd ed. New York, Macmillan Publishing Co., Inc.

Whittaker, R.H. 1977. Evolution of species diversity in land communities, p. 1-67. *In* M.K. Hecht, W.C. Steere, and B. Wallace (eds.), Evolutionary biology 10. New York, Plenum Press.

Whittaker, R.H. (ed.). 1978a. Classification of plant communities. The Hague, Dr. W. Junk BV, Publishers.

Whittaker, R.H. (ed.). 1978b. Ordination of plant communities. The Hague, Dr. W. Junk BV, Publishers.

Whittaker, R.H., and S.A. Levin (eds.). 1975. Niche. Theory and application. Benchmark papers in ecology, vol. 3. Stroudsburg, Penn., Dowden, Hutchinson & Ross, Inc.

Whittaker, R.H., and G.E. Likens. 1973. Carbon in the biota, p. 281-300. *In* G.M. Woodwell and E.V. Pecan (eds.), Carbon and the biosphere, Conf. 72501. Springfield, Va., National Technical Information Service.

Whittaker, R.H., and W.A. Niering. 1965. Vegetation of the Santa Catalina Mountains, Arizona: a gradient analysis of the south slope. Ecology **46**:429-452.

Whittaker, R.H., and W.A. Niering. 1968. Vegetation of the Santa Catalina Mountains, Arizona. IV. Limestone and acid soils. J. Ecol. **56**:523-544.

Wichler, G. 1961. Charles Darwin, the founder of the theory of evolution and natural selection. New York, Pergamon Press, Inc.

Wickens, G.E. 1979. Speculations on seed dispersal and the flora of the Aldabran archipelago. Philos. Trans. Roy. Soc. London, ser. B **286**:85-97.

Wiebe, P.H. 1976. The biology of cold-core rings. Oceanus **19**:69-76.

Wiebe, P.H. 1982. Rings of the Gulf Stream. Sci. Am. **246**(3):60-70.

Wielgolaski, F.E., and Th. Rosswall. 1972. Tundra biome. Proceedings of the fourth international meeting on biological productivity of tundra (Leningrad), I.B.P. Stockholm, Swedish I.B.P. Committee.

Wilcox, B.A. 1978. Supersaturated island faunas: a species-age relationship for lizards on post-Pleistocene land-bridge islands. Science **199**:996-998.

Wiley, E.O. 1976. The phylogeny and biogeography of fossil and recent gars (Actinopterygii: Lepisosteidae). University of Kansas Museum Nat. Hist. Misc. Pub. **64**:1-111.

Wiley, E.O. 1981. Phylogenetics, the theory and practice of phylogenetic systematics. New York, Wiley-Interscience.

Williams, C.B. 1953. The relative abundance of different species in a wild animal population. J. Anim. Ecol. **22**:14-31.

Williams, C.B. 1964. Patterns in the balance of nature and related problems in quantitative ecology. New York, Academic Press, Inc.

Williams, E.E. 1969. The ecology of colonization as seen in the zoogeography of anoline lizards on small islands. Quart. Rev. Biol. **44**:345-389.

Williams, E.E. 1972. The origin of faunas. Evolution of lizard congeners in a complex island fauna: a trial analysis, p. 47-90. *In* Th. Dobzhansky, M.K. Hecht, and W.C. Steere (eds.), Evolutionary biology, vol. 6. New York, Appleton-Century-Crofts.

Williams, E.E. 1976. West Indian anoles: a taxonomic and evolutionary summary. I. Introduction and a species list. Breviora no. 440.

Williams E.E. 1983. Ecomorphs, faunas, island size and diverse end points in island radiations of *Anolis*. *In* R.B. Huey, E.R. Pianka, and T.W. Schoener (eds.), Lizard ecology: studies on a model organism. Cambridge, Mass., Belknap Press (in press).

Williams, W.D. (ed.). 1974. Biogeography and ecology in Tasmania. Monographiae Biologicae 25. The Hague, Dr. W. Junk BV, Publishers.

Williamson, M. 1981. Island populations. Oxford, Oxford University Press.

Williamson, P.G. 1981. Paleontological documentation of speciation in Cenozoic molluscs from Turkana Basin. Nature **293**:437-443.

Willis, E.O. 1974. Populations and local extinctions of birds on Barro Colorado Island, Panama. Ecol. Monogr. **44**:153-169.

Willis, J.C. 1922. Age and area. Cambridge, Cambridge University Press.

Willson, M.F. 1969. Avian niche size and morphological variation. Am. Nat. **103**:531-542.

Wilson, E.O. 1959. Adaptive shift and dispersal in a tropical ant fauna. Evolution **13**:122-144.

Wilson, E.O. 1961. The nature of the taxon cycle in the Melanesian ant fauna. Am. Nat. **95**:169-193.

Wilson, E.O. 1969. The species equilibrium, p. 38-47. *In* G.M. Woodwell and H.H. Smith (eds.), Diversity and stability in ecological systems. Brookhaven Symp. Biol. 22.

Wilson, E.O. 1975. Sociobiology. Cambridge, Mass., Belknap Press.

Wilson, E.O., and D.S. Simberloff. 1969. Experimental zoogeography of islands: defaunation and monitoring techniques. Ecology **50**:267-278.

Wilson, E.O., and R.W. Taylor. 1967. An estimate of the potential evolutionary increase in species diversity in the Polynesian ant fauna. Evolution **21**:1-10.

Wilson, E.O., and E.O. Willis. 1975. Applied biogeography, p. 522-534. *In* M.L. Cody and J.M. Diamond (eds.), Ecology and evolution of communities. Cambridge, Mass., Belknap Press.

Wilson, J.T. 1963a. A possible origin of the Hawaiian Islands. Can J. Phys. **41**:863-870.

Wilson, J.T. 1963b. Evidence from islands on the spreading of the ocean floors. Nature **197**:536-538.

Wilson, J.T. 1963c. Continental drift. Sci. Am. **209**(4): 86-100.

Wilson, J.T. 1965. A new class of faults and their bearing on continental drift. Nature **207**:343-347.

Wilson, J.T. 1966. Did the Atlantic close and then re-open? Nature **211**:676-681.

Wilson, J.W., III. 1974. Analytical zoogeography of North American mammals. Evolution **28**:124-140.

Wilson, L.G. (ed.), 1970. Sir Charles Lyell's scientific journals on the species question. New Haven, Conn. Yale University Press.

Windley, B.F. 1977. The evolving continents. London, John Wiley & Sons, Ltd.

Winterer, E.L. 1976. Anomalies in the tectonic evolution of the Pacific, p. 269-278. *In* G.H. Sutton, M.H. Manghnani, and R.M. Moberly (eds.), The geophysics of the Pacific basin and its margin. Am. Geophys. Union Monogr. no. 19.

Wodzicki, K.A. 1950. Introduced mammals of New Zealand: an ecological and economic survey. Bull. Dir. Sci. Indust. Res., New Zealand **98**:1-255.

Woillard, G.M. 1978. Grande Pile peat bog: a continuous pollen record for the last 140,000 years. Quat. Res. **9**:1-21.

Wolfe, J.A. 1969a. Paleogene floras from the Gulf of Alaska region. U.S. Geol. Survey open-file report.

Wolfe. J.A. 1969b. Neogene floristic and vegetational history of the Pacific Northwest. Madroño **20**:83-110.

Wolfe, J.A. 1971. Tertiary climatic fluctuations and methods of analysis of Tertiary floras. Palaeogeogr. Palaeoclimat. Palaeoecol. **9**:27-57.

Wolfe, J.A. 1972. An interpretation of Alaskan Tertiary floras, p. 201-233. *In* A. Graham (ed.), Floristic and paleofloristics of Asia and eastern North America. Amsterdam, Elsevier Scientific Publishing Co.

Wolfe, J.A. 1975. Some aspects of plant geography of the Northern Hemisphere during the Late Cretaceous and Tertiary. Ann. Missouri Bot. Gard. **62**:264-279.

Wolfe, J.A. 1978. A paleobotanical interpretation of Tertiary climates in the Northern Hemisphere. Am. Sci. **66**:694-703.

Wolfe, J.A. 1979. Temperature parameters of humid to mesic forests of eastern Asia and relation to forests of other regions of the Northern Hemisphere and Australasia. U.S. Geol. Survey Prof. Pap. no. 1106. Washington, D.C., U.S. Government Printing Office.

Wolfe, J.A. 1981. Vicariance biogeography of angiosperms in relation to paleobotanical data, p. 413-427. *In* G. Nelson and D.E. Rosen (eds.), Vicariance biogeography: a critique. New York, Columbia University Press.

Wood, C.E., Jr. 1971. Some floristic relationships between the southern Appalachians and western North America, p. 331-404. *In* P.C. Holdt (ed.), The distributional history of the biota of the southern Appalachians. Part II. Flora. Res. Div. Monogr. 2, Virginia Polytechnical and State University.

Wood, C.E., Jr. 1972. Morphology and phytogeography: the classical approach to the study of disjunctions. Ann. Missouri Bot. Gard. **59**:107-124.

Wright, S.J., and C.C. Biehl. 1982. Island biogeographic distributions: testing for random, regular, and aggregated patterns of species occurrence. Am. Nat. **119**: 345-357.

Wulff, E.V. 1943. An introduction to historical plant geography. Tr. E. Brissenden. Waltham, Mass., Chronica Botanica.

Yeaton, R.I. 1974. An ecological analysis of chaparral and pine forest bird communities on Santa Cruz Island and mainland California. Ecology **55**:959-973.

Yeaton, R.I. 1981. Seedling morphology and the altitudinal distribution of pines in the Sierra Nevada of central California: a hypothesis. Madroño **28**:67-77.

Yeaton, R.I., and M.L. Cody. 1974. Competitive release in island song sparrow populations. Theor. Pop. Biol. **5**:42-58.

Yeaton, R.I., R.W. Yeaton, and J.E. Horenstein. 1981. The altitudinal replacement of digger pine by ponderosa pine on the western slopes of the Sierra Nevada. Bull. Torrey Bot. Club **107**:487-495.

Yoda, K. 1967. A preliminary survey of the forest vegetation of eastern Nepal. II. General description, structure and floristic composition of sample plots chosen from different vegetation zones. J. Coll. Arts Sci. Chi'ba Univ. Nat. Sci. ser. **5**:99-140.

Zaret, T.M., and R.T. Paine. 1973. Species introduction in a tropical lake. Science **182**:449-455.

Zeller, E.J. 1966. Age of frigid condition in Antarctica. Antarctic J. U.S. **1**:201-202.

Zimmerman. E.C. 1948. The insects of Hawaii, vol. 1. Honolulu, University of Hawaii Press.

Zimmerman, E.C. 1958. Three hundred species of *Drosophila* in Hawaii? A challenge to geneticists and evolutionists. Evolution **12**:557-558.

Index

A

C

E

F

G

H

I

M

Q

R

S

T

U

V